Dopamine

Dopamine

Endocrine and Oncogenic Functions

Nira Ben-Jonathan

CRC Press
Taylor & Francis Group
Boca Raton London New York

CRC Press is an imprint of the
Taylor & Francis Group, an **informa** business

First edition published 2020
by CRC Press
6000 Broken Sound Parkway NW, Suite 300, Boca Raton, FL 33487-2742

and by CRC Press
2 Park Square, Milton Park, Abingdon, Oxon, OX14 4RN

First issued in paperback 2022

ISBN 13: 978-1-03-240016-7 (pbk)
ISBN 13: 978-1-138-39223-6 (hbk)
ISBN 13: 978-0-429-40227-2 (ebk)

DOI: 10.1201/9780429402272

Typeset in Leawood
by Lumina Datamatics Limited

Visit the book's webpage at https://www.crcpress.com/9781138392236

Contents

Preface

Dopamine is the first molecule in the biosynthetic pathway of catecholamines, which also include norepinephrine (noradrenaline) and epinephrine (adrenaline). Dopamine is a major neurotransmitter in the brain, where it is involved in multiple neurological functions. Its dysregulation is associated with diseases such as schizophrenia, bipolar disorder, depression, various addictions, substance abuse, and Parkinson's disease. Consequently, the largest class of pharmaceuticals comprises drugs that alter dopamine functions.

Given its prominence as a critical neuromodulator, there is a vast literature, encompassing basic research and clinical investigation, on dopamine actions within the brain. In fact, several excellent and comprehensive books, dedicated to brain dopamine, were published in the 1980s, 1990s, and early 2000s. Yet, little attention has been given to the accumulating evidence, especially in the last 20 years, on the hormonal and oncogenic functions of both central and peripheral dopamine.

I have been studying many aspects of dopamine homeostasis since the 1970s. Initially, my focus was on the hypothalamo-pituitary system, but my interests later expanded to the reproductive organs, pregnancy, fetal development, obesity/adipose tissue, and cancer. My research has used a variety of animal species, including sheep, chickens, rabbits, rats, and mice. Whenever feasible, I also employed human subjects and human tissues and cell lines. The extensive and long-term engagement with multiple systems and pathophysiological conditions and my familiarity with the related literature served as the impetus for undertaking the very challenging endeavor of summarizing this information in the form of a book.

The book provides a systematic account of dopamine as an endocrine and autocrine/paracrine hormone. To this end, the first two chapters summarize the current knowledge on dopamine homeostasis, including its biosynthesis, transport, release, and receptors, as well as the canonical and noncanonical mechanism of action. The next two chapters cover the anatomical and endocrine characteristics of the central dopaminergic systems. From there on, the remainder of the book is dedicated to peripheral dopamine in terms of its tissues of origin, unique features, and its involvement in endocrinology, pathophysiology, and tumorigenesis in multiple systems. Throughout the book, the therapeutic potential, as well as adverse effects, of dopamine-altering drugs is critically evaluated.

Given its integrative nature, this book should appeal to basic research scientists from multiple disciplines, including endocrinology, physiology, pharmacology, neuroscience, cell biology, and biochemistry. In addition, clinicians specializing in endocrinology, neurology, psychiatry, oncology, and pathology should find many sections of this book of great interest to their discipline. The book is structured to provide basic information to graduate and medical students who are entering these fields as well as to deliver advanced information that is targeted to more seasoned clinicians and basic researchers who wish to expand their knowledge on novel aspects of this multifaceted molecule.

Author

Nira Ben-Jonathan is Emeritus Professor of Cancer Biology at the University of Cincinnati, Ohio, USA. She earned her BSc in Biology from Tel Aviv University, Israel, and PhD in Physiology and Biochemistry at the University of Illinois, Champaign-Urbana, USA. She previously served on the faculties of the University of Texas in Dallas and Indiana University in Indianapolis. The main focus of her research has been on hormones and neurotransmitters as regulators of multiple systems in both the brain and the periphery. She has published 171 manuscripts, edited one book, and contributed 12 chapters to textbooks and encyclopedias. She has mentored 65 students, fellows, and research scientists. She was awarded the National Institutes of Health (NIH) Research Career Development Award, was elected Fellow of the AAAS, and Chairman of the Gordon Research Conference on Prolactin and was elected member of the Royal Society of Medicine, UK. Her research has been supported by multiple U.S. federal and private foundations. She has received the Rieveschl Award for Outstanding Scientific Research, and the Edward Merker Lectureship in Translational Endocrinology. Over the years, she has served as a member on numerous study sections of the NIH, the Department of Defense, and the Komen Foundation and as chairman on five NIH study sections. She presently has appointments as a visiting professor at both the Weizmann Institute of Science, Rehovot Israel, and the Hebrew University of Jerusalem, Israel.

ILLUSTRATOR

Uri Inks earned his Bachelor of Design degree in visual communication with expertise in animation from the Bezalel Academy of Art and Design in Jerusalem, Israel. After graduation, he worked for several years as a free-lance director, animator, illustrator, graphic designer, and storyboard artist directing clips for TV shows and commercials. He then served as the professional director of the creative department in the "Time to Know" organization, and subsequently became the supervisor creative director. Presently, he has an appointment as a lecturer in the 3D section of the department of screen-based arts at the Bezalel Academy.

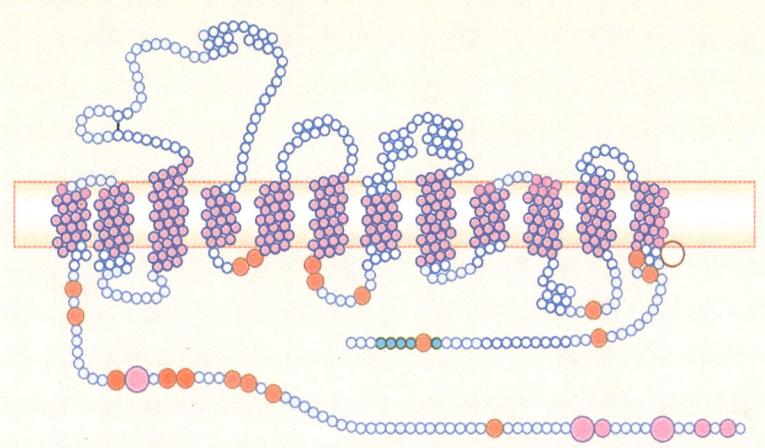

Homeostasis of Dopamine

1

1.1 INTRODUCTION

Dopamine (DA) is a member of the catecholamine family, which is composed of biogenic amines with a catechol ring structure. The family includes three members: DA, norepinephrine (NE), also known as noradrenaline, and epinephrine (Epi), also known as adrenaline. The term catecholamines is derived from their basic structure, which couples an amine side chain with a dihydroxyphenyl (catechol) ring. Historically, the catecholamines were discovered in the reversed order of their position in the biosynthetic sequence. This was due to the early recognition of their tissue location and ease of experimental manipulation (i.e., Epi in the adrenal gland, NE in sympathetic neurons, and DA as the dominant catecholamine in the brain).

Adrenaline, the prototypical catecholamine, was isolated from the bovine adrenal gland in 1901, while Dopa decarboxylase (DDC), the second enzyme in the biosynthesis of catecholamines, was first described in 1936 [1]. In subsequent years, all aspects of catecholamine homeostasis, including biosynthesis, metabolism, storage, release, and receptors, have been extensively investigated and become well documented. Notably, DA itself was recognized as an independent neurotransmitter (not only as a precursor of NE) only in the late 1950s. The concepts of the storage and reuptake of catecholamines, which were initially deduced from their pharmacological behavior, were firmly established following the identification and characterization of the large family of monoamine membrane transporters [2].

Homeostasis refers to the processes by which a living organism, tissues and/or cells keep their internal environment stable in spite of continuous changes in the conditions around them. The synthesis, metabolism, storage, release, and reuptake of DA are interrelated dynamic processes which differ in several respects between the "closed" system of the brain dopaminergic neurons, and the "open-ended" dopaminergic system in peripheral organs (**Figure 1.1**). In the closed configuration of the neuron/synapse/neuron, the concentration of released DA inside the very small space of the synaptic cleft can be as high as 5–10 μM. On the other hand, peripheral DA-producing cells can be quite remote from their target cells. Thus, the concentration of circulating DA, once it reaches the target cells, does not exceed 20–30 nM.

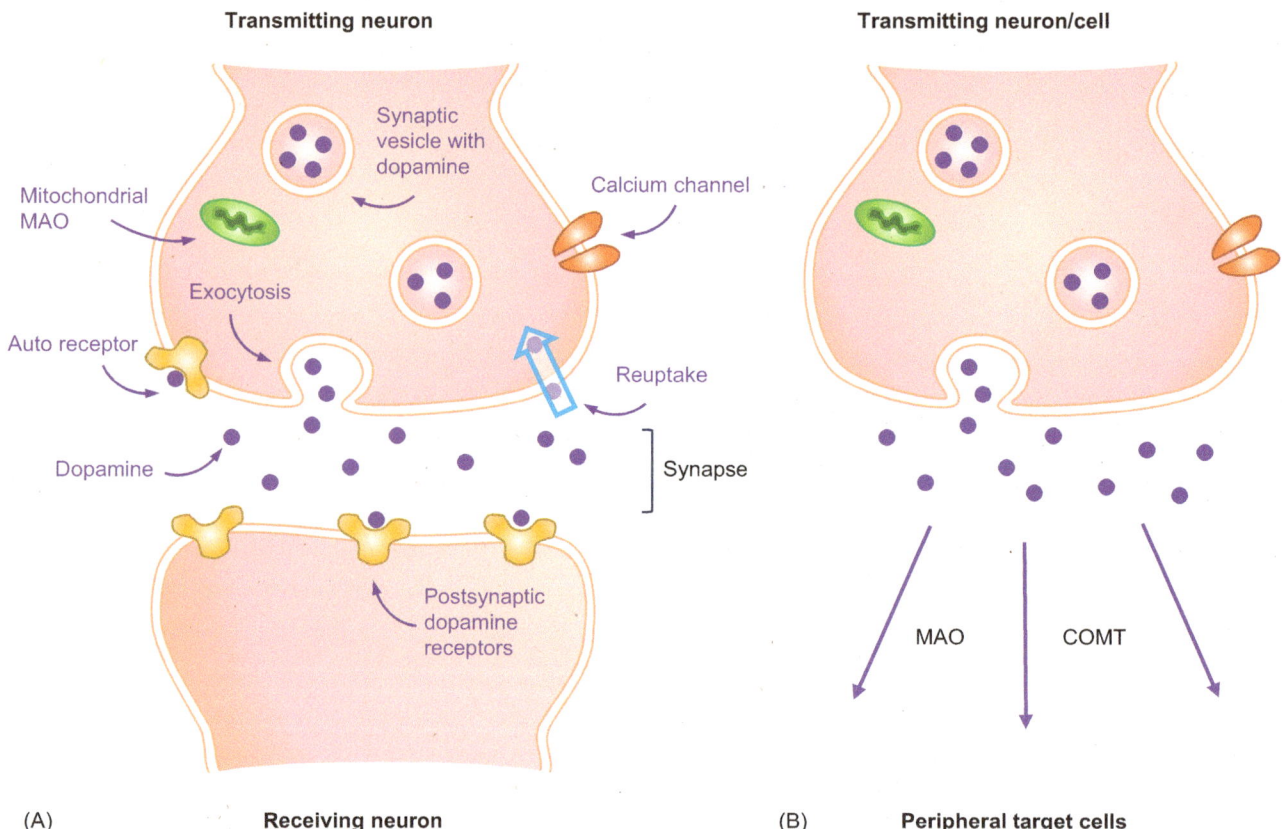

Figure 1.1 Model of a close (Panel A) and an open (Panel B) dopaminergic systems. Most dopaminergic neurons in the brain operate as a "closed system" whereby a transmitting neuron is associated with a receiving neuron through a small gap: the synapse. DA, released by exocytosis, can bind to presynaptic or postsynaptic receptors, and activate various signaling cascades. The action of DA is terminated by its reuptake into the transmitting neuron. Inside the neuron, DA is either stored in synaptic vesicles or is degraded by mitochondrial monoamine oxidase (MAO). In the open-ended system, the released DA is distributed to remote targets by the circulation. In the open system, there is no reuptake mechanism and no autoreceptors on the transmitting neuron/cell.

The wide disparity in the actual concentrations of DA at the target sites between the brain and peripheral sites should be taken into consideration when evaluating results of *in vitro* studies, many of which have used DA at high micromolar levels. This discrepancy may also explain some of the differences in the ability of certain dopaminergic drugs to alter the functions of brain vs. peripheral dopaminergic systems.

Most of the experimental data on DA homeostasis were obtained from research that was focused on brain DA. Nonetheless, PC12 cells, a cell line that was derived from a rat adrenal medullary pheochromocytoma and represents non-differentiated neuroblastic cells, have also been heavily used to study all aspects of catecholamine homeostasis. Whenever appropriate, similarities and differences in the features and regulation of central vs. peripheral DA are emphasized in this and the following chapters.

1.2 BIOSYNTHETIC ENZYMES

The three catecholamines, DA, NE, and Epi, are synthesized by four enzymes that act in sequence, as presented in **Figure 1.2**. The first enzyme, tyrosine hydroxylase (TH), converts tyrosine to L-dihydroxyphenylalanine (L-Dopa). TH serves as the rate-limiting step in the biosynthetic pathway of the catecholamines and has a rather restricted tissue expression. The more widely expressed second enzyme, DDC, also known as aromatic L-amino acid decarboxylase, generates DA from Dopa. Cells that express the third enzyme, dopamine β-hydroxylase

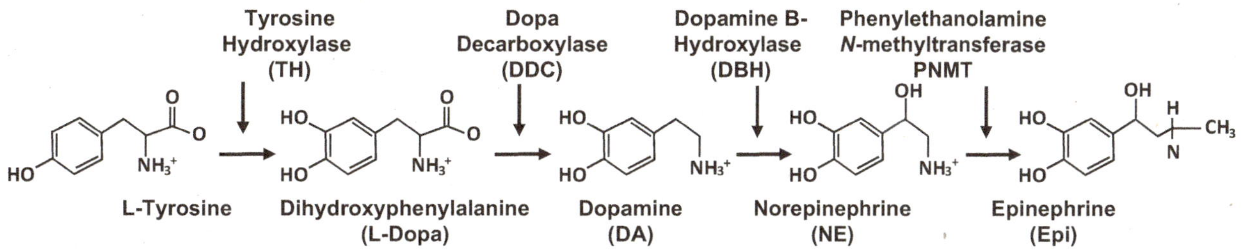

Figure 1.2 Sequential enzymatic steps in the biosynthesis of catecholamines. Catecholamine biosynthesis starts with L-tyrosine, which is converted to L-Dopa by tyrosine hydroxylase (TH), the rate-limiting enzyme. TH introduces a second hydroxyl group into the phenol ring, converting it to a catechol ring. Dopa decarboxylase (DDC) removes the carboxyl group from the side chain of L-Dopa and converts it to dopamine. DBH introduces a hydroxyl group on the side chain of dopamine and converts it to norepinephrine. The final enzyme, phenylethanolamine N-methyl transferase (PNMT), introduces a methyl group to the side chain of norepinephrine, converting it to epinephrine.

Table 1.1 Characteristics of catecholamine biosynthetic enzymes

Enzyme	Gene location	Protein structure	Substrates	Inhibitors
Tyrosine hydroxylase (TH)	11p15.5	Tetramer	L-tyrosine; L-phenylalanine	α-methyl-p-tyrosine; 3-iodotyrosine
Dopa decarboxylase (DDC)	7p12.1	Homodimer	L-Dopa; 5-HTP; L-histidine; trace amines	Carbidopa; benserazide
Dopamine beta-hydroxylase (DBH)	9q34	Tetramer	Dopamine; tyramine	Disulfiram; tropolone: nepicastat
Phenylethanolamine-N-methyltransferase (PNMT)	17q12	Homodimer	Norepinephrine; octopamine; phenylethanolamine	SK&F 64139; tetra-hydroisoquinolines

(DBH), can synthesize NE as their major product, while those cells that also express the fourth enzyme, phenylethanolamine N-methyltransferase (PNMT), can produce Epi. **Table 1.1** depicts the gene location, protein structure, major substrates, and selected inhibitors of TH, DDC, DBH, and PNMT.

1.2.1 Tyrosine hydroxylase

TH is a mixed function oxidase that uses L-tyrosine and molecular oxygen as substrates, and L-tetrahydrobiopterin (BH4) and ferrous iron (Fe^{2+}) as cofactors [3]. Tyrosine is one of the 20 standard amino acids used by cells to synthesize proteins. It is a nonessential amino acid with a polar side chain group. Given its natural abundance, catecholamine levels are not influenced either by changing the dietary levels of tyrosine or by its parenteral administration, even at large amounts. Because of its essential role as a cofactor in TH enzymatic activity, a deficiency in BH4 can cause systemic deficiencies of catecholamines. One example of BH4 deficiency is the development of dopamine-responsive dystonia, characterized by increased muscle tone and Parkinsonian features. This condition can be treated with carbidopa/levodopa which directly restores dopamine levels within the brain.

TH is a 240-kDa tetrameric cytosolic enzyme which acts by introducing a second hydroxyl group on the phenol ring, thereby converting it into a catechol ring. TH has a lesser affinity for L-phenylalanine and no activity toward D-tyrosine, tyramine, and L-tryptophan. Effective inhibitors of TH include amino acid analogs such as α-methyl-p-tyrosine (α-MPT), α-methyl-3-iodotyrosine, and 3-iodotyrosine, all of which act by competing with the tyrosine substrate. TH is expressed in specific regions of the brain where catecholaminergic neurons are located, including the striatum, substantia nigra, locus ceruleus, olfactory bulb, and medulla oblongata [3]. As discussed in more detail in subsequent chapters, TH is also expressed in the heart, adrenal gland, gastrointestinal (GI) tract, and in several other normal and malignant peripheral tissues.

Figure 1.3 Human tyrosine hydroxylase (TH) isoforms and structure/phosphorylation sites of a mature TH. Four isoforms of tyrosine hydroxylase (TH) are generated by alternative splicing of exon 2 (**A**). Numbers below TH4 designate the location of amino acids after splicing. The mature TH protein is divided into regulatory and catalytic domains. The regulatory domain shows positions of four serines that are targeted by phosphorylation (**B**). See text for other explanations. Panel A. (Redrawn and modified from Meiser, J. et al., *Cell Commun. Signal.*, 11, 34, 2013; panel B. (Redrawn and modified from Kumer, S.C. and Vrana, K.E., *J. Neurochem.*, 67, 443–462, 1996.)

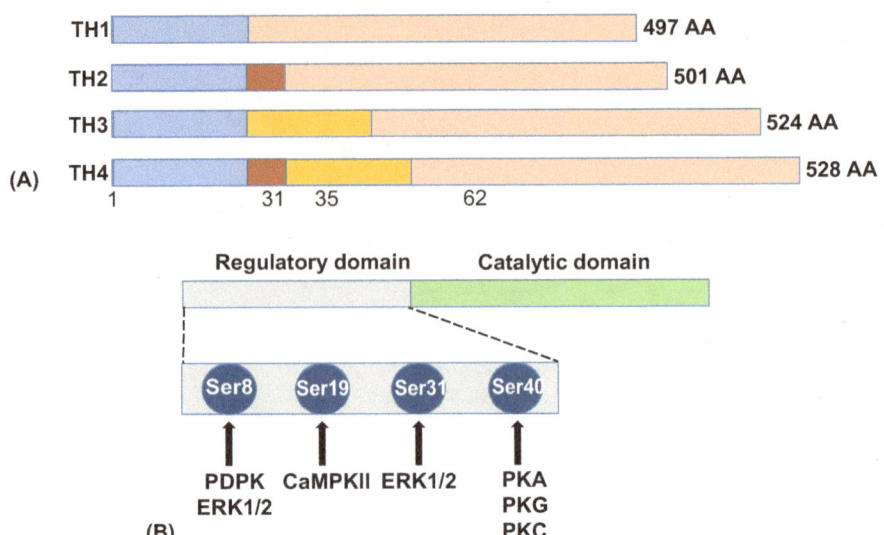

The human *TH* gene is located in chromosome 11p15.5 and is composed of 14 exons, spanning 8.5 kb. Targeted *TH* gene deletion in mice results in an early embryonic lethality, presumably because of cardiac failure. This may explain the absence of records in the clinical literature of a complete TH deficiency. Alternative splicing of the human *TH* gene in exon 2 generates four different mRNAs, which are translated into four TH subunits composed of 497–528 residues (**Figure 1.3**A). Each subunit contains an inhibitory regulatory domain at the *N*-terminus, and a catalytic domain at the C-terminus. Serines 8, 19, 31 and 40 in the *N*-terminal regulatory domain serve as phosphorylation sites that are involved in acute enzyme activation. Two histidine residues (His^{331} and His^{336}), located within the pterin binding site at the catalytic domain, function as iron-binding sites.

TH is subject to both short- and long-term regulation [4]. Short-term regulation occurs rapidly at posttranslational levels and involves feedback inhibition by catecholamines, allosteric regulation, and enzyme phosphorylation. Each of the catecholamines, the end-product of the TH reaction, can inhibit enzyme activity by competing with the pterin cofactor. This results in a reversible enzyme inhibition by converting its active/labile form to an inactive/stable form. In dopaminergic neurons within the brain, end-product inhibition is often associated with the binding of DA to autoreceptors which are localized to various regions of the presynaptic neurons. Such a situation does not generally occur in peripheral DA-producing cells, most of which do not express DA autoreceptors. Allosteric effectors such as heparin, phospholipids, and polyanions do not directly alter the hydroxylation of tyrosine but, rather, increase the affinity of the enzyme for the BH4 cofactor.

TH is phosphorylated in response to nerve stimulation, as well as upon exposure to cAMP analogs, sodium nitroprusside, nerve growth factor, and several other effectors. The ensuing signaling cascade activates a number of protein kinases that act upon specific serine residues in the TH regulatory domain as follows: (1) the proline-directed protein kinase (PDPK), and extracellular regulated kinase 1/2 (ERK 1/2) phosphorylate Ser-8, (2) the Ca^{2+}/calmodulin dependent protein kinase II (CaM-PKII) phosphorylates Ser-19, (3) ERK1/2 also phosphorylates Ser-31, and (4) protein kinase A (PKA), protein kinase C (PKC), and cGMP-dependent protein kinase (PKG), all phosphorylate Ser-40 (**Figure 1.3**B). The enhanced enzyme activity following phosphorylation reflects a combination of increased enzyme V$_{max}$, decreased K$_m$ for BH4, and decreased affinity for end-product inhibition. In addition, serine phosphorylation at one site can influence the phosphorylation at another site through hierarchal phosphorylation [3]. By all criteria, Ser-40 is the most promiscuous phosphorylation site, with the residues around it displaying consensus sequences for recognition by PKA and PKC. In addition, protein phosphatase 2A (PP2A) reverses the TH activation by removing specific phosphates from each of the phosphorylation sites.

Long-term regulation of TH operates both at the transcriptional and translational levels and includes transcriptional regulation, alternative RNA splicing, changes in RNA stability, translational regulation, and altered enzyme stability. TH expression increases in response to environmental challenges such as stress (which is often mediated by increased glucocorticoid release), hypoxia, and certain drugs (e.g., reserpine, nicotine, and cocaine). Chronic or repeated exposures to stressors, including cold exposure or immobilization, result in increased TH mRNA and protein levels in the brain [7] as well as in the sympathoadrenal system [8]. Several putative transcriptional regulatory elements have been identified within the promoter region of the TH gene. Among these are a glucocorticoid response element (GRE), a cAMP response element (CRE), and an activator protein-1 (AP-1) element. In PC12 cells, both cAMP and phorbol ester increase the transcription rate of the *TH* gene via CRE through distinct signaling pathways.

1.2.2 Dopa decarboxylase

DDC is a pyridoxal phosphate-dependent enzyme that catalyzes the decarboxylation of L-Dopa to DA by removing the carboxyl moiety from the catechol side chain (**Figure 1.1**). Under normal conditions, the decarboxylation of L-Dopa is done very rapidly and therefore, L-Dopa is not detectable in most tissues. The DDC enzyme is a cytosolic homodimer, composed of two identical 50-kDa subunits [9]. DDC plays an important role in dopaminergic neurons in the brain and participates in the uptake and decarboxylation of amine precursors in many peripheral tissues. DDC is also named aromatic L-amino acid decarboxylase because, in addition to catecholamines, the enzyme catalyzes the biosynthesis of serotonin (Ser) from 5-hydroxytryptophan (5-HTP), histamine from L-histidine, as well as the synthesis of some trace amines (**Table 1.1**).

The amino acid sequence of DDC is evolutionarily conserved across many species, where it fulfills different functions. For example, DDC is a key enzyme during molting of several insect species, and is also associated with epidermal tanning and melanization [10]. In mammals, DDC has been detected in CNS neurons and glia, as well as in neuronal and nonneuronal cells in many peripheral organs, as discussed in later chapters.

The human *DDC* gene exists as a single copy, located on chromosome 7p12.1 (**Table 1.1**). The gene extends over 107.6 kb, and consists of 15 exons. Alternative splicing generates several mRNA isoforms. Tissue-specific expression of the *DDC* gene is controlled by two spatially distinct promoters—neuronal and nonneuronal—with the mRNA isoforms differing in the composition of the 5′ UTR and the presence of alternative exons. In humans, a deficiency in DDC synthesis, due to frameshift mutations, or alterations in its activity due to single amino acid substitutions, result in impaired cognitive and physiological homeostasis, and/or in some neuropsychiatric disorders [11].

Although DDC is largely considered a nonregulated enzyme, its expression and protein levels are altered by agonists and antagonists of dopaminergic and adrenergic receptors, as well as by some neuroleptics [12]. Unlike TH, DDC is not a rate-limiting step in DA synthesis under normal conditions, but it becomes so in patients with Parkinson's disease who are treated with L-Dopa because DA itself does not cross the blood–brain barrier (BBB). When given as a drug to these patients, L-Dopa is rapidly converted to DA in the blood and only a small portion of a given dose reaches the brain. By adding DDC inhibitors such as carbidopa or benserazide, more L-Dopa can reach the striatum, where it is converted to DA by the local DDC. DDC expression has also been studied in several peripheral cancers. Elevated DDC mRNA levels, along with high enzyme activity, have been detected in small cell carcinoma of the lung, pheochromocytoma, neuroblastoma, ganglio-neuroblastoma and gastric cancer [13], leading to the suggestion that DDC expression could serve as a biomarker for some neuroendocrine-related cancers.

1.2.3 Dopamine beta-hydroxylase and phenylethanolamine-N-methyltransferase

Both DBH and PNMT are responsible for the two sequential biosynthetic steps beyond DA production (**Figure 1.2**). A brief description of their properties, tissue distribution, and regulation is provided here for a better understanding of the presence (or absence) of DA as an end-product in some catecholamine producing cells (**Table 1.1**).

DBH is a 290-kDa copper-containing oxygenase which requires ascorbate as a cofactor and consists of four identical subunits. DBH is expressed in NE nerve terminals of the central and peripheral nervous systems, as well as in chromaffin cells of the adrenal medulla and in few other peripheral cells. In addition to DA, DBH hydroxylates trace amines (e.g., *p*-tyramine to *p*-octopamine) and also participates in the metabolism of xenobiotics such as amphetamine. In humans, aberrations in the expression or activity of DBH have been associated with several disorders, including hypertension, hypotension, congestive heart failure, depression, and idiopathic Parkinson's disease as well as disorders related to copper deficiency [14].

The *DBH* gene is mapped to chromosome 9q34 and consists of a 1.854-Kb transcript, encoding a glycoprotein enzyme composed of 617 amino acids. The enzyme exists in soluble and membrane-bound forms, depending on the absence or presence, respectively, of a signal peptide. In humans, mutations in the *DBH* gene result in deficits in autonomic and cardiovascular functions such as hypotension and ptosis (drooping of the upper eyelid). Polymorphisms in the *DBH* gene are associated with a variety of psychiatric disorders. Expression of DBH is elevated in response to a subset of conditions that increases TH. Targeted disruption of the *DBH* gene in mice results in prenatal mortality at embryonic age 11–12 days [15]. About 10% of DBH-deficient mice are born alive, presumably because of the transfer of sufficient maternal catecholamines across the placenta to rescue these fetuses.

DBH is the only catecholamine biosynthetic enzyme which is localized within synaptic and secretory vesicles, where it is present as both soluble and membrane-bound forms. During transmitter release, soluble DBH is secreted together with NE, Epi, and other vesicular contents, although it may also be secreted via a constitutive exocytotic pathway. By virtue of its secretion into the extracellular space, DBH is present in the cerebrospinal fluid (CSF), and plasma. The relative ease by which DBH levels and/or activity can be measured in plasma has raised the expectations that circulating DBH could serve as a biomarker for sympathetic nervous system activity and for the diagnosis of NE-associated disorders. However, no clear consensus with regard to the diagnostic value of these measurements has yet emerged [16].

PNMT is the final enzyme in the catecholamine biosynthetic pathway (**Figure 1.1**). It converts NE to Epi by introducing a methyl group onto the side chain, utilizing *S*-adenosylmethionine as a methyl donor and coenzyme. PNMT is primarily expressed in the adrenal medulla but is also found in a small group of NE neurons in the brain as well as in limited amounts in the heart, lung, kidney, liver, and pancreas [17]. The *PNMT* gene, located in chromosome 17q12, consists of four exons and spans about 2.8 kb in length. PNMT is a 30-kDa dimeric protein that shares many properties with other methyltransferases, including catechol-O-methyltransferase (COMT), one of the major metabolic enzymes of catecholamines (see below).

Expression of PNMT is under both neural and hormonal regulation. In the adrenal medulla, the major source of circulating NE and the sole source of circulating Epi, PNMT is controlled by the splanchnic nerve which provides cholinergic innervation to the medulla. Acetylcholine, acting via nicotinic and muscarinic receptors, can activate two separate intracellular signaling pathways which stimulate transcription of the *PNMT* gene. Glucocorticoids, acting via the GRE located in the *PNMT* promoter, are the major hormones that regulate enzyme expression [18]. Within the adrenal medulla, cortisol is delivered to the chromaffin cells from glucocorticoid producing cells that are located in the zona fasciculate.

1.3 METABOLIC ENZYMES

Multiple enzymes and converging pathways are involved in the metabolic inactivation of catecholamines. The two major enzymes are monoamine oxidase (MAO), which carries out oxidative deamination, and COMT which does *O*-methylation (**Figure 1.4**). In addition, some of the metabolites undergo conjugation by glucuronidases and sulfation by sulfotransferases (see below). Two dehydrogenases, alcohol dehydrogenase and aldehyde dehydrogenase, also participate in the formation of the final metabolic products.

Of particular relevance for understanding the metabolic fate of the various catecholamines are the intracellular localization of specific metabolic enzymes, as well as their overall tissue distribution. For example, MAO is located in the outer membrane of the mitochondria, is found within the brain in presynaptic dopaminergic neurons, and is also expressed in many peripheral sites. On the other hand, COMT is located at both the endoplasmic reticulum (ER) and the cytoplasm; does not reside in presynaptic dopaminergic neurons; is present at very low levels in postsynaptic neurons; is abundant in ependymal cells of the cerebral ventricles, choroid plexus, and glia; and is particularly high in peripheral organs such as liver, gut, and kidney. Sulfotransferases, which are especially important for peripheral DA homeostasis, are found only at very low levels in the brain, but are very abundant in the GI tract. The levels of the various catecholamine metabolites in different body fluids (CSF, serum, urine) serve as diagnostic tools for a number of central and peripheral disorders.

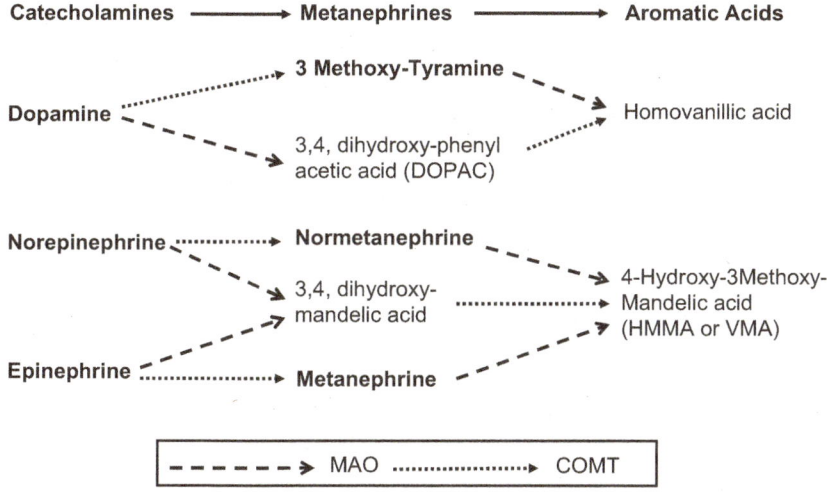

Figure 1.4 Overview of the metabolic inactivation of catecholamines. The main metabolites of dopamine, norepinephrine, and epinephrine are shown in bold. The enzymes which carry out these reactions are identified by two types of dashed arrows.

1.3.1 Monoamine oxidases (MAOs)

MAOs catalyze the oxidative deamination of biogenic amines that include catecholamines, indoleamines, trace amines, and xenobiotics (Table 1.2). MAO action alone produces 3,4, dihydroxyphenylacetic acid (DOPAC) by oxidation from DA and 3,4, dihydroxymandelic acid by oxidation from NE and Epi. Humans have two types of MAO: MAO-A and MAO-B, which are the products of separate genes [19,20]. The MAO isoenzymes share high homology (~70%) and similar intracellular location and structural characteristics but differ in their pharmacological profiles, ontogeny, tissue distribution, and functional roles. MAO-A preferentially oxidizes serotonin (Ser; 5-HT) and NE, while MAO-B has the highest affinity for the trace amine β-phenylethylamine (PEA). DA is catabolized by both isoenzymes, with a species-specific different level of affinity for the two isoenzymes: DA is primarily a substrate for MAO-A in rodents but is a substrate for MAO-B in humans and other primates. Despite their physiological divergence, each isoenzyme contributes to the metabolism of non-preferred substrates in the absence of the other enzyme.

The genes for MAO A and B are mapped to adjacent sites on the X chromosome (Xp11.23). Each gene has 15 exons and an identical exon–intron organization, likely resulting from a tandem duplication of a common ancestral gene. Both genes are deleted in patients with Norrie disease, a rare X-linked recessive neurological disorder which is characterized by blindness, hearing loss, mental retardation, and autistic-like behavior [21]. Another rare genetic disease is a selective MAO-A congenital deficiency, named Brunner syndrome. Affected subjects have an unusual aggressive and violent behavior, mild cognitive impairments, and some stereotyped hand movement and sleep disturbances [19].

The *MAO-A* and *MAO-B* genes encode two proteins with molecular weights of 59,700 and 58,000, respectively. The strong conservation of amino acid sequence of each isoenzyme across mammalian species indicates an evolutionary pressure to maintain the physiological functions of these important enzymes. However, interpretation of the phenotypes of targeted deletion of each MAO gene in mice has been complicated by a significant overlap of the distribution and activity of each enzyme. There is no information whether mice lacking both enzymes are viable.

The ontogeny of MAO-A and MAO-B follows different time courses. In most species, MAO-A activity is predominant during early development, while MAO-B is not detectable during the perinatal stages but, rather, increases with aging. MAO-B is one of the few enzymes whose expression and activity are enhanced by aging. This raises the prospect that increased enzyme activity contributes to some aging-related degenerative processes, possibly via the production of H_2O_2. Indeed, higher levels of brain MAO-B have been detected in patients with Parkinson's and Alzheimer's diseases, and MAO-B inhibitors appear to improve the quality of life in the elderly.

Table 1.2 Characteristics of catecholamine metabolic enzymes

Enzyme	Gene location	Protein structure	Substrates	Inhibitors
Monoamine oxidase (MAO)	Xp11.23	MAO-A MAO-B	Catecholamines; indoleamines; phenylethylamine	Tranylcypromine; phenelzine; rasagiline
Catechol-O-methyltransferase (COMT)	22q11.2	Membrane-bound and soluble forms	Catecholamines; catecholestrogens; flavonoids	L-Dopa; entacapone; tolcapone
UDP-glucuronosyl-transferases (UGTs)	Multiple	Monomers	Multiple	Glucosides; canagliflozin
Phenolsulfo-transferases (PSTs)	16p11.2 (*SALT1A3*)	Homodimer	Catecholamines; catecholestrogens	p-Nitrophenol

The regional distribution of the monoamine oxidases in the human brain reveals very high levels of both isoenzymes in the hypothalamus and hippocampus, and a large amount of MAO-B, but very little MAO-A, in the striatum and globus pallidus [22]. The cortex has high levels of MAO-A only, except for the cingulate cortex, which contains similar amounts of both isoenzymes. Increased concentrations of MAO-A are seen in brain regions that are dense in serotonergic neurons, whereas MAO-B correlates mostly with NE innervation. Outside the brain, MAO-A is expressed in the liver, vascular endothelium, GI tract, and placenta, while MAO-B is found primarily in platelets.

Monoamine oxidase inhibitors (MAOIs) such as tranylcypromine (Parnate) and phenelzine were introduced in the 1950s as the first drugs which were approved for treating clinical depression [23]. The rationale for their use was that depression and anxiety are associated with low levels of DA, NE, and Ser. By blocking MAO and preventing their degradation, the concentration of the three neurotransmitters should increase and, consequently, the symptoms associated with depression such as sadness or anxiety should be relieved. However, the MAOIs later fell out of favor because of concerns about their adverse interactions with certain foods (e.g., cheese) that are high in tyramine, numerous interactions with other drugs, and the eventual development of more effective antidepressants. Today, MAOIs are still used as effective agents for panic disorder and social phobia and for providing an alternative for migraine prophylaxis as well as in the treatment of Parkinson's disease (e.g., deprenyl) by targeting MAO-B to specifically affect dopaminergic neurons. Currently used oral MAOIs include rasagiline (Azilect), selegiline (Eldepryl), isocarboxazid (Marplan), phenelzine (Nardil) and tranylcypromine (Parnate).

1.3.2 Catechol-*o*-methyltransferase

COMT, in the presence of magnesium, catalyzes the transfer of a methyl group from S-adenosyl-L-methionine to one of the phenolic groups of a catechol structure, primarily at the 3-hydroxy position [24,25]. COMT action alone produces 3-methoxytyramine from DA, normetanephrine from NE, and metanephrine from Epi (**Figure 1.4**). The *COMT* gene encodes two distinct enzyme isoforms: membrane-bound (MB-COMT) and soluble (S-COMT). COMT is ubiquitously expressed throughout the body, with particularly high levels in liver, kidneys, brain, adrenals, and lungs. There are substantial variations, however, in the relative expression of the two isoenzymes in different tissues, with S-COMT being the dominant enzyme in most peripheral tissues, while MB-COMT predominates in the brain. Within the brain, MB-COMT expression and enzyme activity are high in cerebral cortical areas and the hypothalamus. COMT is found in some neuronal cells (e.g., pyramidal neurons, cerebellar Purkinje cells, granular cells, and striatal spiny neurons), but not in major long-projection neurons. Both COMT isoforms are abundant in microglia and astroglia and in intestinal macrophages.

COMT also methylates catecholestrogens, which possess weak estrogenic and/or anti-estrogenic activity and are involved in estrogen-induced carcinogenesis. In addition to being the substrates for inactivation by COMT, catecholestrogens are potent competitive inhibitors of both COMT and TH and, therefore, can affect catecholamine biosynthesis and metabolism. There is sexual dimorphism in both expression and functions of COMT, with men having a greater enzyme activity than women. The latter is attributed to the down-regulation of COMT expression by estrogens [25].

The human *COMT* gene is mapped to chromosome 22q11.2 (**Table 1.2**). Its transcription is regulated by two promoters that generate 1.5 and 1.3-kb mRNA species, encoding the MB-COMT and S-COMT enzymes, respectively. The *COMT* gene is associated with several allelic variants, the best studied is val158met (a valine to methionine mutation at position 158). Overexpression of this variant is associated with a 2- to 4-fold decrease in COMT enzymatic activity, resulting in higher synaptic DA levels and ultimately increased dopaminergic stimulation of postsynaptic neuron. COMT-dependent DA degradation is of particular importance in brain regions that have low expression of the presynaptic DA transporter (DAT), such as the prefrontal cortex. Hence, Val158Met polymorphism appears to exert its effects on cognitive functions, schizophrenia, and obsessive-compulsive disorder.

Expression of COMT is affected by hypoxia, blood vessel occlusion, and traumatic brain injury. The increase in enzyme activity has been attributed to its up-regulation in microglia, representing a compensatory mechanism for terminating excessive catecholamine signaling in injured brain regions [26]. Given that the regulation of catecholamines is impaired in a number of medical conditions, several drugs have been developed which target COMT in order to alter its activity and catecholamine availability. For example, COMT inhibitors like entacapone have been introduced as adjunct drugs of L-Dopa therapy. When given together with DDC inhibitors (carbidopa or benserazide), L-Dopa is optimally protected from peripheral degradation. This "triple therapy" has become a standard treatment of Parkinson's disease.

1.3.3 Glucuronosyltransferases and sulfotransferases

The UDP-glucuronosyltransferases (UGTs) represent a multigenic family of membrane-bound monooxygenase enzymes (**Table 1.2**). The enzymes act by transferring the glucuronic acid moiety from a high energy donor, UDP-α-D-glucuronic acid, to xenobiotics, drugs, and endogenous compounds such as bilirubin, catecholamines, and steroid hormones [27,28]. One of the major functions of UTGs is in the detoxication of xenobiotics, thereby contributing to the protection of the organism against hazardous chemicals. Detoxication is a two-phase process (phase I and phase II) of metabolism. Overall, detoxification involves the biotransformation of a toxic lipophilic compound to a more hydrophilic form that can be excreted by the kidneys.

As shown in **Figure 1.5**, the bulk of phase I enzymatic activity takes place in the liver by cytochrome P450-dependent monooxygenases, which

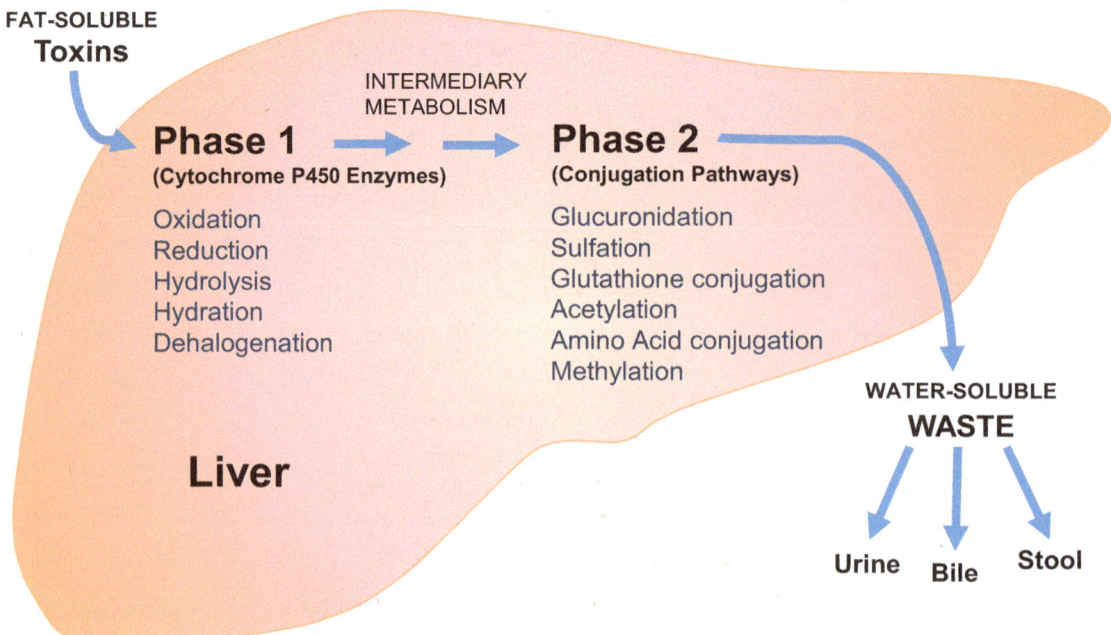

Figure 1.5 The process of detoxification. Phase I of detoxification primarily occurs in the liver by the cytochrome p450 enzyme complex. In phase II, which occurs in the liver and in other tissues, the intermediary metabolites undergo conjugations and other reactions, resulting in formation of water-soluble molecules that can be excreted by the kidney and the gastrointestinal tract.

carry out a variety of oxidation, reduction, hydrolysis, hydration, and dehalogenation reactions. Products of phase I reactions are the major substrates of the UGTs, with glucuronidation constituting the main process of phase II xenobiotic biotransformation. The UGTs are predominantly associated with the luminal side of the ER, while the cytochromes P450 monooxygenases are expressed on the cytosolic surface of the ER membranes. The intramembrane co-localization and interactions of cytochromes P450 and UGTs ensure a topological and functional coupling of the two different enzymes for an efficient stepwise drug biotransformation.

Humans have 19 UGT isoforms which are divided into three subfamilies: UGT1A, UGT2A, and UGT2B. In the brain, UGTs are mainly expressed in endothelial cells and astrocytes of the BBB, but are also associated with brain regions devoid of the BBB, such as the circumventricular organ, pineal gland, pituitary gland, and neuro-olfactory tissues [28]. In addition to their key role as a detoxication barrier, UGTs participate in the maintenance of steady-state of endogenous compounds such as steroids or DA. The UGT isoforms 1A, 2A, 2B, and 3A are expressed at various levels in brain tissue and have distinct, but overlapping, substrate specificity. Of all the human UGTs, only one enzyme, UGT1A10, is known to catalyze the glucuronidation of DA at substantial rates, yielding both dopamine-4-O-glucuronide and dopamine-3-O-glucuronide. In humans, UGT1A10 is primarily expressed in the small intestine, colon, and adipose tissue, whereas only low levels of UGT1A10 are found in the trachea, stomach, liver, testis, and prostate. Unexpectedly, UGT1A10 was not detected in the brain.

The sulfation reaction is catalyzed by members of the sulfotransferase (SULT) superfamily (**Table 1.2**). Sulfation is involved in the conjugation of numerous endogenous and xenobiotic chemicals, thus exerting considerable influence over their biological activity [29]. The sulfation reaction entails the enzymatic transfer of a sulfonate group (SO_3^{-1}) from a universal donor, 3'-phosphoadenosine, 5'-phosphosulfate, to recipient molecules. Sulfation is a major contributor to the homeostasis and regulation of catecholamines, steroids, and iodothyronines, as well as the detoxication of xenobiotics. SULT enzymes are widely expressed in a variety of human tissues, including liver, intestine, and brain. They are classified on the basis of their substrate specificity and amino acid sequence into two subfamilies: SULT1 (phenol sulfotransferases or PST) and SULT2 (steroid sulfotransferases).

Within the brain, the highest PST activities are found in the temporal and frontal cortex [30]. Activity levels about one-tenth those in the brain are seen in the parietal and occipital lobes, amygdala, hypothalamus, and hippocampus, whereas the nucleus accumbens, caudate nucleus, and substantia nigra have the lowest activities. In Parkinson's patients treated with L-Dopa, a significant reduction of PST activities has been observed in the hypothalamus, frontal and temporal cortex, amygdaloid nucleus, and occipital and parietal cortex. The depletion of PST activity is less severe in the hippocampus, nucleus accumbens, putamen, and substantia nigra, while PST activity was increased in the caudate nucleus. These data suggest that selective changes in PST activity in the Parkinson's brain underscore the importance of this enzyme as a regulator of DA storage and metabolism under pathological conditions.

The human SULT1 family consists of six homo-dimeric enzymes, having amino acid sequence identities that range from 47% to 96%. Unlike rodents and several other animal species, humans rely more heavily on sulfation as a means of modulating the activity catecholamines and other biogenic amines and facilitating their transport. This is particularly true for DA because up to 98% of circulating DA in humans is sulfoconjugated [31]. Under basal conditions, serum DA-S at ~5 ng/mL exceeds by 10- to 15-fold the levels of free DA (0.3 ng/mL), NE (0.2 ng/mL) or Epi (0.05 ng/mL) combined. Yet, the presence of DA-S in human serum has been overlooked by most investigators, likely because DA-S is undetectable by routine analytical methods and requires special extraction methods.

Sulfoconjugation is the major form of DA inactivation in human serum, while glucuronidation predominates in rats [32]. No ortholog of SULT1A3 is known in rodents, underlying the greater importance of DA sulfoconjugation in humans than in laboratory animals [33]. SULT1A3 has high specificity for both catecholamines and catecholestrogens [29]. The highest activity of SULT1A3 is found in the GI tract, with moderate activities present in liver, lung, pancreas, and platelets. A single amino acid substitution (Glu 146) confers the enzyme with higher affinity for DA than for NE or Epi [34]. Notably, ingestion of a meal after fasting in humans induced a 50-fold rise in serum DA sulfate (DA-S) levels [32], indicating that food consumption stimulates DA production and/or its sulfoconjugation in the GI tract. Most importantly, unlike the irreversible glucuronidation reaction, sulfoconjugation is reversible by sulfatases [35], among which is arylsulfatase A (ARSA), which converts the biologically inactive DA-S to a nonconjugated, active DA. This issue is discussed in great detail in **Chapter 6**.

1.4 MEMBRANE AND VESICULAR TRANSPORTERS

Small molecules can penetrate the cell membrane through two major processes: passive diffusion and carrier-mediated transport [36]. In general, lipophilic substances have high membrane permeability and move across the cell membrane down their concentration gradients without a need for energy input. Hydrophilic substances, on the other hand, have low membrane permeability, and their efficient uptake into the cell requires a carrier-mediated transport. The latter must be coupled to an energy source to power the uphill transport of a given compound against its concentration gradient. In addition, carrier-mediated transport is saturable, inhibitable, and depends on the specific properties of the transporters expressed in a given tissue or cell type. Clinically, transporters are of great attraction for the development of specific drugs and also for understanding interindividual variability in drug responsiveness.

Common to all monoaminergic neurotransmitters are reuptake mechanisms in which the released neurotransmitter is taken back into the secreting cell by membrane-embedded transporters. Reuptake fulfills two important functions: (1) guarding against neuronal overstimulation through the removal of the released messenger from the synaptic cleft, resulting in the rapid termination of its actions, and (2) enabling energy conservation by recycling and reutilizing existing molecules rather than producing energy-costly new transmitters [37]. Once the released molecule is brought back into the secreting cell, the next step is its repackaging into storage/secretory vesicles (**Figure 1.6**). Repackaging is accomplished by two vesicular monoamine transporters, VMAT1 and VMAT2, whose main roles are to (1) protect the transmitter from degradation by intracellular MAO, (2) maintain an adequate intraneuronal storage/secretory capacity to ensure prompt responses to subsequent stimuli, and (3) enable a regulated release of the neurotransmitter from storage vesicles rather than its unregulated release by diffusion [38].

1.4.1 Transporters of monoamine neurotransmitters

Recent years have witnessed a surge of interest in the distribution, regulation, and functions of transporters for monoamine neurotransmitters. Historically, two distinct transport systems, named uptake$_1$ and uptake$_2$, were recognized [39]. Uptake$_1$ was viewed as Na$^+$ and Cl$^-$-dependent,

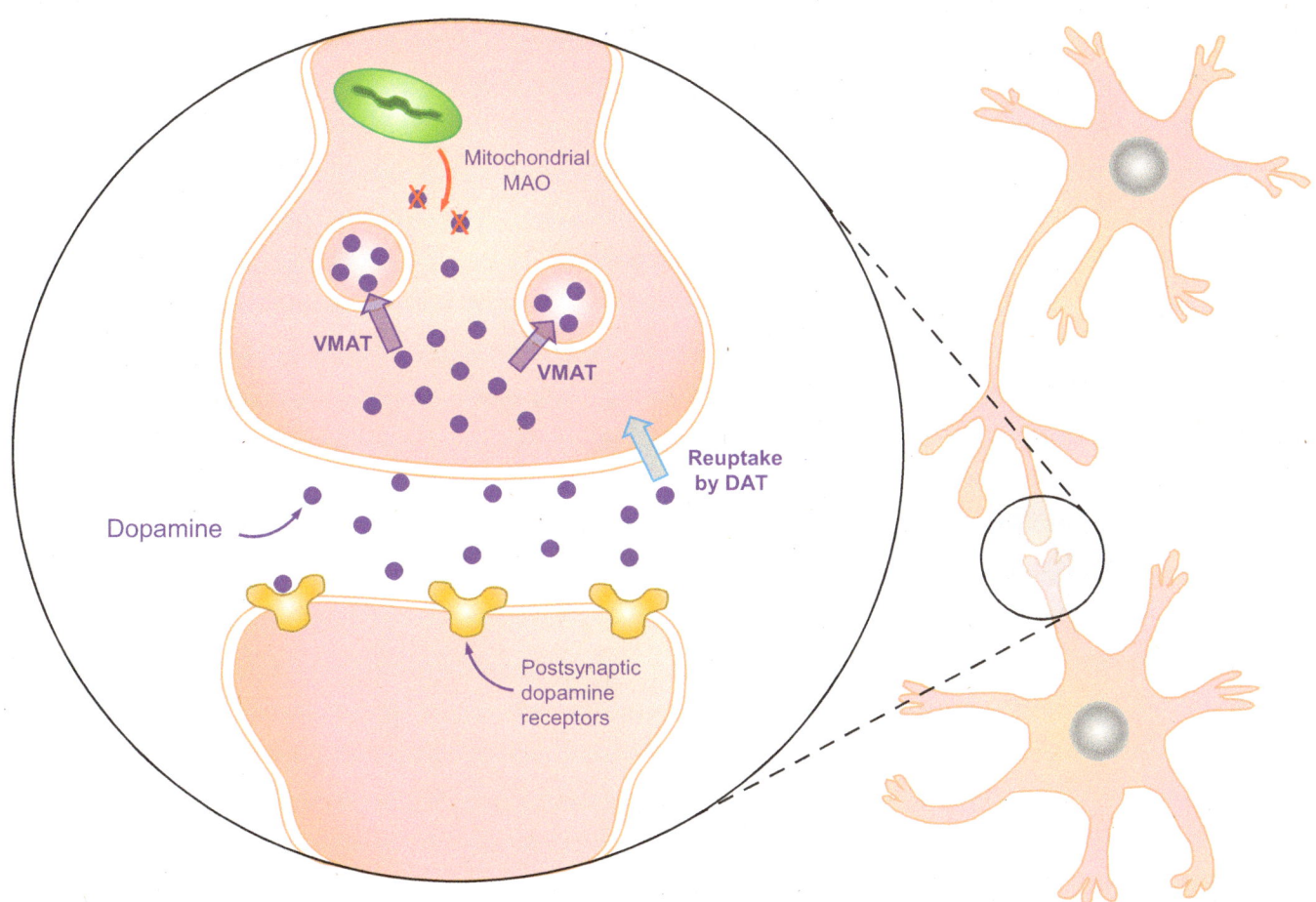

Figure 1.6 The roles played by transporters in the reuptake and repackaging of dopamine. After dopamine (DA) is released into the synaptic cleft, it is taken up into the secreting neuron by the dopamine transporter (DAT). Repackaging into the storage/secretory vesicles is done by the vesicular monoamine transporters (VMAT1 or VMAT2). Cytosolic DA can be degraded by mitochondrial monoamine oxidase (MAO).

high-affinity transporters that are predominantly expressed in nerve endings of monoaminergic neurons and include the Ser (SERT), DA (DAT), and NE (NET) transporters. Uptake$_2$ was defined as a Na^+ and Cl^--independent, low-affinity, high-capacity transport system with a lower substrate specificity, which is found predominantly in peripheral tissues [40]. Molecular cloning, and biochemical and pharmacological analyses have subsequently identified three large families of plasma membrane-embedded transporters: solute carrier transporters (SLC), organic cation transporters (OCT), and plasma membrane monoamine transporters (PMAT).

The SLC superfamily consists of evolutionary and structurally related transporters that actively translocate amino acids and transmitters into cells against their concentration gradient. The family is comprised of ~350 members that are organized into as many as 55 families [41,42]. Among these, the SLC6 subfamily is of particular interest to this discussion. As illustrated in **Figure 1.7**, SLC6 is composed of about 20 structurally related symporters (cotransporters) which are subdivided into four major groups of transporters for (1) Monoamines: Ser (SERT or SLC6A4), DA (DAT or SL6A3), and NE (NET or SLC6A); (2) GABA, four different transporters (GAT1 to GAT4); (3) Amino acids: glycine, two different transporters (GLYT1 and GLYT2), proline, and taurine; and (4) Orphan transporters. All the transporters belonging to the SLC6 subfamily move the solute together with extracellular Na^+ (symport), which provides the driving force for substrate translocation against a chemical gradient.

Figure 1.7 Classification of transporters that belong to the SLC6 subfamily. The monoamine transporters in the SLC6 family include those for dopamine (DAT), norepinephrine (NET), and serotonin (SERT). The GABA transporters include four different entities (GAT1 to GAT4). Within the amino acid transporters are two glycine (Glyt1 and Glyt2), proline and taurine transporters. The orphan subfamily includes an assortment of transporters for amino acid and other constituents.

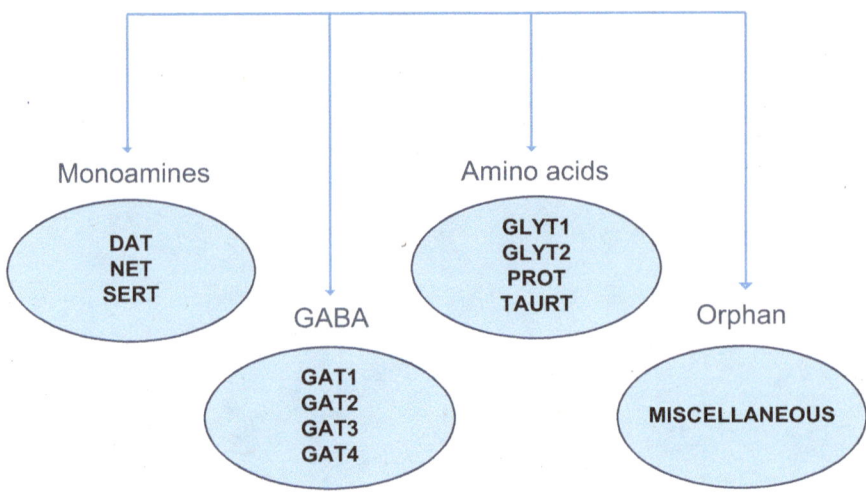

1.4.2 The dopamine transporter

The DAT plays critical roles in many physiological functions, including locomotion, behavior, cognition, and motivation. Disturbances in DAT are associated with many neurological and psychiatric disorders, including attention deficit hyperactivity disorder (ADHD), bipolar disorder, depression, and alcoholism [43,44]. Consequently, DAT has been a major target of multiple therapeutic drugs aimed at alleviating these disorders, and it is also a target for psychostimulants such as methamphetamine and cocaine.

The *DAT* gene is mapped to the human chromosome 5p15.3, is about 64-kb long, and consists of 15 coding exons. The regulatory regions of the *DAT* gene include an 18-kb promoter region and two functional tandem repeats of variable length that are located in intron 8 and the 3'-UTR. *DAT* gene activity is regulated by various internal as well as external factors. Studies with isolated cells and laboratory animals have shown that DAT mRNA levels are altered by several drugs that include the DAT inhibitors cocaine and bupropion, the NET inhibitor desipramine, and also by insulin, estrogen, β-cytotoxin, and some diabetes-inducing drugs. Pathway analysis revealed transcriptional regulation involves multiple signaling pathways among which are histone deacetylation, AKT, and PKC [45].

The *DAT* gene encodes an 80-kDa glycoprotein, made of 620 amino acids, with no known isoforms. The proposed membrane topology of DAT was based on hydrophobic sequence analysis, sequence similarities with the leucine transporter, X-ray crystallography of the *Drosophila melanogaster* DAT (albeit it has only 50% homology with hDAT), and *in silico* modeling. Collectively, these approaches predicted that the DAT is configured in 12 transmembrane domains (TMDs), with a large extracellular loop between the third and fourth TMDs, and cytoplasmic N- and C-termini (**Figure 1.8**).

DAT is expressed in all areas of the brain with well-documented dopaminergic innervation, including the nigrostriatal, mesolimbic, and mesocortical pathways and their associated nuclei. Unlike amino acid transporters, which are found on both neurons and glia, the expression of DAT is restricted to dopaminergic neurons, making DAT a well-accepted experimental and pathological marker for these neurons. At the cellular level, DAT is found in both dendrites and cell bodies of neurons in the substantia nigra and ventral tegmental areas. Against expectations, however, immunocytochemical studies using electron microscopy showed that DAT is present in perisynaptic areas of the dopaminergic neurons rather than within the synapse itself [47]. This and other findings supported the view that following its release, DA diffuses out of the synapse before it is taken up into the neuron by DAT.

DAT is synthesized in the cell body of the dopaminergic neuron, packaged in the Golgi, and is transported to dendrites, axons, and nerve

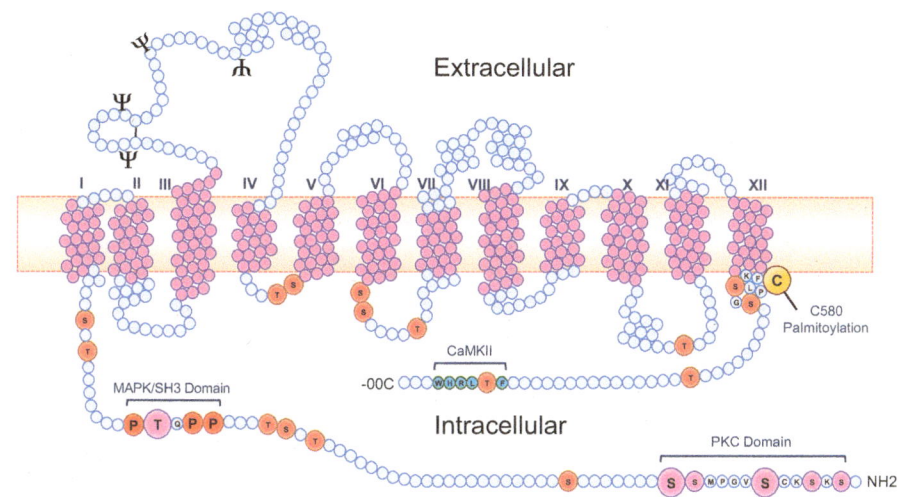

Figure 1.8 Diagram of the dopamine transporter (DAT). DAT is predicted to configure in 12 transmembrane domains (Roman numerals above the diagram) and have cytoplasmic C- and N-termini. Binding domains for calcium-calmodulin dependent kinase (CaMKII) and palmitoylation site (via cysteine) are shown in the C-terminus, while the MAP kinase (MAPK)/SH3 and protein kinase C (PKC) domains are shown in the N-terminus. Several critical serines (S) and tyrosine (T) residues are shown, as are the four glycosylation sites (ψ) on the third extracellular loop. (Redrawn and modified from Foster, J.D. and Vaughan, R.A., *J. Chem. Neuroanat.*, 83–84, 10–18, 2017.)

terminals. The DAT is also expressed in peripheral tissues that produce DA. Nonetheless, not all DA-producing cells in the periphery are neuron-like and most do not have axons or adjacent synapses. Details on the distribution and functions of peripheral DAT are presented in many of the following chapters.

Like other transporters, DAT undergoes dynamic recycling between intracellular compartments and the plasma membrane. Within the plasma membrane, DAT resides in microdomains which facilitate its motility and substrate recognition via interactions with cytoskeletal proteins such as synuclein, a presynaptic protein which has been implicated in Parkinson's disease [44]. Overall, DAT is regulated by three overlapping mechanisms: posttranslational modifications, protein–protein interactions, and intracellular localization [48].

DAT is classified as a symporter that moves DA across the cell membrane through its coupling to sodium ions that move from high to low concentrations into the cells. DAT operates by a sequential binding and cotransport of DA with two Na^+ ions and one Cl^- ion [48]. The driving force for DA reuptake is the ion concentration gradient generated by Na^+/K^+ ATPase located in the plasma membrane. DA binds to DAT in a pocket formed by TMs 1,3,6 and 8. Bindings of two Na^+ and one Cl^- to this pocket cause a conformational change of the transporter from outward facing to inward facing, enabling the release of the cargo to the cytoplasm. Upon DA release, the DAT switches back to an unoccupied, outward-facing conformation. A model named "alternating access mechanism" [49] proposes that the transition between the outward-facing conformation for extracellular (EC) binding of DA and Na^+/Cl^- ions and the inward-facing conformation for intracellular (IC) DA release is intercepted by intermediate configurations that are occluded to both EC and IC regions.

1.4.3 Pharmacology of DAT

Two drugs of abuse, cocaine and amphetamines/methamphetamines, elicit their pleasurable sensation by interacting with DAT. Both drugs act by increasing the availability of DA to the postsynaptic neurons, albeit via different mechanisms. Cocaine and cocaine analogs act as competitive inhibitors of DA binding to the DAT, by having a binding site which overlaps with that of DA [50]. Amphetamines, on the other hand, act by a more complex mechanism. This involves (1) induction of rapid DAT internalization, (2) entrance into the synaptic vesicles through VAMT2 and stimulating DA released into the cytosol, and (3) increased intracellular calcium which activates PKC, causing DAT phosphorylation and a further induction of DA efflux [51].

Several toxic substances which structurally resemble DA are also substrates for DAT and, therefore, can accumulate in DA neurons and cause specific local damage. A prime example is 1-methyl-4-phenyl-1,2,3,6-tetrahydropyridine (MPTP), whose neurotoxic properties were serendipitously discovered in the early 1980s when several young drug addicts who were inadvertently exposed to MPTP as a contaminant of heroin, developed an acute form of Parkinsonism [52]. MPTP is first converted by MAO to MPP$^+$ before it is recognized by DAT and is taken up exclusively by dopaminergic neurons. Once inside the neuron, MPP$^+$ accumulates in the mitochondria, where it interferes with the mitochondrial electron transport chain complex I. MPP$^+$ also alters catecholamine metabolism, leading to high oxidative stress and cell death. The selective dopaminergic toxicity of MPP$^+$ has been widely exploited in the efforts to develop cellular and animal models of Parkinson's disease.

Another highly toxic compound which is specific to catecholaminergic neurons is 6-hydroxydopamine (6-OHDA), also known as oxidopamine [53]. It is a synthetic compound used by researchers to selectively destroy dopaminergic and noradrenergic neurons in the brain. 6-OHDA is structurally similar to DA and NE and has high affinity for their transporters, DAT and NET, respectively. Administration of selective noradrenaline reuptake inhibitors, such as dismethylimipramine or imipramine, before 6-OHDA protects the noradrenergic neurons from damage in these animals and selectively destroys dopaminergic neurons. Because 6-OHDA does not cross the BBB, it is injected directly into the substantia nigra pars compacta, medial forebrain bundle (MFB), which consists of efferent fibers from nigral neuronal cell bodies to the striatum, or into the striatum, where it specifically kills DA and NE neurons.

1.4.4 DAT-deficient mice

Important information on DAT functions comes from studies with DAT-deficient (DAT-knockout, or DAT-KO) mice [54]. These mice are hyperactive, have cognitive deficits, and also some sleep dysregulation. The clearance rates of DA from brain synapses in the DAT-KO mice are 100-fold higher than those in the controls. Such mice fail to respond to cocaine or amphetamines, reinforcing the notion that DAT is their receptor. There was also a profound decrease in TH expression in these mice, explaining the relatively mild increase in extracellular DA levels because of decreased biosynthesis. In addition, expression of both D1R and D2R was decreased. Although the mutant mice have normal social interactions, females have impaired capacity to care for the offspring, likely due to anterior pituitary hypoplasia-related hormonal dysregulation. Some abnormalities in skeletal structure and GI tract motility were also observed.

1.4.5 Vesicular monoamine transporters

The VMAT prototype is a transport protein that is integrated into the membrane of storage/synaptic vesicles of monoamine-producing cells. Unlike the high selectivity of membrane transporters such as DAT for DA, and SET for Ser, the VMATs are not very discriminatory. In fact, the VMATs can transport from the cytosol into the vesicles a large variety of monoamines: DA, NE, Epi, Ser, and histamine. Concomitant with their transport functions, VMATs play important roles in the sorting, storing, and release of the neurotransmitters and contribute to their protection from autoxidation. Because dysregulation of DA and Ser neurotransmission is involved in the etiology of major depressive disorder, bipolar disorder, and schizophrenia, VMATs have been considered suitable targets for the development of neuropsychiatric drugs for treating these disorders.

Two types of VAMTs are recognized in humans: VMAT1 and VMAT2. Originally, VMAT1 was found in large dense core vesicles in peripheral neuroendocrine cells, whereas VAMT2 was found to be expressed in synaptic-like vesicles in the brain, the sympathetic nervous system, as well as in mast cells, pancreatic beta cells, and blood platelets [48,55]. However, later studies showed that VMAT1 is also expressed in the rat and human brains [56].

VMAT1 and VMAT2 are products of separate genes, each encoding glycoproteins of ~70 kDa.

Both transporters are transmembrane proteins with a predicted 12 TMDs, similar to, but not identical with, the structural configuration of DAT or SET (**Figure 1.9**). Among the two transporters, VMAT1 shows higher affinity for Ser, whereas VMAT2 is the only one which transports histamine. Several studies have implicated the VMAT2 transporter in Parkinson's disease, addiction, and other psychiatric disorders, while genetic polymorphisms in *VMAT1* have been associated with bipolar disorders, schizophrenia, and anxiety-related personality traits [56].

A strong energy source is required for packaging a large number of neurotransmitter molecules at high concentrations (up to 0.5 M) into the small space of the vesicles. For that, the VMATs utilize the pH and electrochemical gradient that are generated by vesicular H^+-ATPase. The proposed model of VMAT actions implies that it operates as an antiporter (i.e., an efflux of two protons against the H^+ gradient is coupled with an influx of one monoamine molecule) [57]. The first proton causes a conformational change in the transporter, exposing the high affinity binding site for the monoamines. The second proton causes a second conformational change that results in the release of the monoamine into the vesicle, and concomitantly reduces the affinity of the binding site for the amines. Mutational analyses showed that His419, located between TMDs 10 and 11, plays a key role in the first conformational changes, while Asp431, located on TMD 11, is important for the second change.

Research done in the 1960s with isolated bovine chromaffin granules have identified a vesicular transporter, originally named chromaffin granule amine transporter (CGAT). This transporter was later identified as VMAT1, and is now classified as solute carrier family 18 member 1 (SLC18A1). The *VMAT1* gene (*SLC18A1*) is located on human chromosome 8p21.3, while the *VMAT2* gene (*SLC18A2*) is located on chromosome 10q25. Studies using cultures of chromaffin cells and sympathetic ganglia indicated that increased stimulation and calcium influx up-regulate VMAT transcription, while only minimal changes in gene expression were seen *in vivo* in response to VMAT inhibition by various drugs. As is the case for other integral membrane proteins, the vesicular transporters are synthesized in the ER and are posttranslationally modified in the ER and Golgi, where they also undergo N-linked glycosylation.

Homozygosity is lethal in VMAT2 knockout mice. Heterozygotes show an increase in locomotor behavior in response to apomorphine, cocaine, and

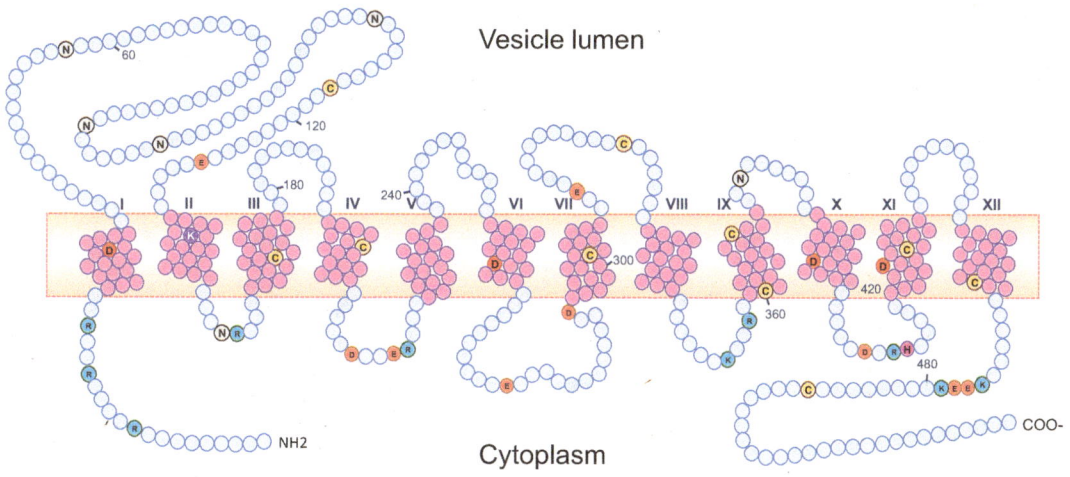

Figure 1.9 Diagram of the vesicular monoamine transporter 2 (VMAT2). VMAT2 is predicted to form 12 transmembrane domains (Roman numerals above the diagram) and have cytoplasmic N and C termini. Several residues (e.g., cysteines, arginines, and asparagines), which are important for the structural integrity and functions of the transporter, are indicated. (Redrawn and modified from Wimalasena, K., *Med. Res. Rev.*, 31, 483–519, 2011.)

amphetamine, suggesting an elevated synaptic response to amine release. Yet, these mice do not appear to undergo sensitization in response to chronic cocaine administration. VMAT2 heterozygotes also show reduced preference for amphetamine and increased locomotor response to ethanol.

Overall, VMATs are regulated by changes in transcription, mRNA splicing of exons, and posttranscriptional modifications such as phosphorylation. The inactivation of transport by VMATs is facilitated by heterotrimeric G proteins. These membrane-associated proteins, which are activated by GPCR, are made up of *alpha*, *beta*, and *gamma* subunits. Activated Gαq down-regulates VMAT2-mediated Ser transport in blood platelets and inhibits VMAT2 activity in the brain. The exact signaling pathway for G protein-mediated regulation of VMATs is not completely clear.

The most common VMAT inhibitors are reserpine and tetrabenazine. Reserpine is an indole alkaloid derived from the plant *Rauwolfia serpentine*. It was introduced in 1954 as an antipsychotic and an antihypertensive medication, and it was instrumental in the development of the "monoamine hypothesis" of affective disorders [58]. Reserpine continues to be used as a research tool in rodent models of depression and anxiety, but its clinical use has declined because of excessive drug interactions and side effects such as lethargy and clinical depression. Yet, reserpine has gained some credence as a potential treatment for cocaine addiction, and recent clinical trials documented that it is well tolerated and is safe. Reserpine binds with high affinity to both VMAT1 and VMAT2 near the substrate-binding site on the cytoplasmic side of the transporter.

Tetrabenazine has been approved for use in the United States for the treatment of choreiform movements in Huntington disease. Tetrabenazine is structurally related to several antipsychotics and has similar side effects, including akathisia, Parkinsonism, dizziness, and depression. Tetrabenazine interacts with VMAT at a different site than does reserpine and differentially inhibits VMAT1 and VMAT2. Despite differences in their binding sites, tetrabenazine inhibits reserpine binding, presumably via allosteric interactions. Two DAT inhibitors, GBR 12909 and 12935, also inhibit VMAT2 at low nanomolar concentrations [38]. Additional VMAT inhibitors include ketanserin, amiodarone, and some derivatives of 3-amino-2-phenylpropene (APP).

1.4.6 Organic cation transporters and plasma membrane monoamine transporters

The OCT transporters belong to the SLC22 family and include three subtypes: OCT1, OCT2 and OCT3. In addition to transporting monoamines, they mediate cellular uptake and efflux of many drugs, thus influencing their disposition as well as their pharmacological and toxicological activities [36]. OCT3 (SLC22A3) was originally identified as a corticosterone-sensitive extra-neuronal monoamine transporter [59]. The abundance of the OCTs within the brain is rated as OCT3>OCT1>>OCT2. The OCTs are expressed in the cerebellum, subfornical organ, dorsal raphe, hippocampus and hypothalamic nuclei and are especially enriched in brain microvessels. Using brain-derived endothelial cells, a proton-coupled OCT antiporter was found to play a role in the transport of apomorphine, a DA agonist, across the BBB [60]. OCT3 is also expressed in skeletal muscle, placenta, salivary glands, heart, adrenal gland, small intestine, kidney, and uterus [36], and it has variable affinity for DA, NE, Ser, and histamine [61]. In the kidney, the OCTs mediate tubular DA uptake [62], but only scant information is available on their role in DA homeostasis in other peripheral sites where they are expressed. OCT knockout mice are viable and fertile, have reduced uptake of MPP+ into the heart, but otherwise they show no significant phenotypic differences from wild-type mice [63].

Another transporter, PMAT (SLC29A4) is not homologous with SERT, NET or DAT and has partial homology with several SCL29 family members [64]. Immunostaining showed expression of the PMAT protein in the human choroid plexus, cerebellum, intestine, kidney, and heart. PMAT is a low-affinity, high-capacity transporter for both Ser and DA, with lesser affinity to NE, Epi or histamine [65]. By virtue of having low affinity and high capacity, as well as its Na+ and Cl−-independent actions, PMAT fits the older definition of uptake2. The *PMAT* gene is localized to human chromosome 7p22.1 and encodes a membrane protein of 530 amino acid residues with a molecular mass of 58 kDa.

The human PMAT transporter is predicted to configure in 11 TMDs, with a long intracellular N-terminus and a short extracellular C-terminus. PNMT has one N-linked glycosylation site and several putative PKC and cAMP-dependent kinase phosphorylation sites. The best-established function of PMAT is in the clearing of neurotoxins and drugs from the CSF via the choroid plexus. Theoretically, PMAT plays a role in clearing released neurotransmitters that have escaped reuptake by the uptake1 transporter, but definitive experimental evidence for this action is lacking. Evidently, more should be investigated with respect to the role of PMAT in the management of either central or peripheral DA.

1.5 STORAGE, RELEASE, AND REUPTAKE

Except for the hypothalamic tuberoinfundibular dopaminergic neurons which release DA into the hypophysial portal vasculature connecting the median eminence of the hypothalamus with the pituitary gland [66], all other brain dopaminergic neurons are linked via synapses to recipient neurons in a "closed" configuration (**Figure 1.2**). Several reviews cover the major processes that govern DA storage, release, and reuptake in the brain [5,67,68]. This information is briefly summarized here as a background for assessing the distinct properties of peripheral DA, which by and large utilizes the "open" configuration system.

The classical paradigm on brain DA posits that soon after its synthesis in the neuronal cytoplasm, DA is transferred by the ATP-dependent VMAT into synaptic vesicles. Within the vesicles, DA is concentrated up to 100–1,000 times above its cytosolic levels and is protected from degradation by mitochondrial MAO. Active transport of DA into the vesicles is controlled by VMATs, which maintain an H+-electrochemical gradient between cytoplasm and vesicle. In response to incoming action potentials, DA is released by a calcium-mediated exocytosis from the producing neurons into the synaptic cleft. DA then binds to DA receptors (DARs) localized either postsynaptically (recipient receptors) or presynaptically (autoreceptors). In postsynaptic neurons, DA triggers action potentials and activates a variety of ion channels and signaling pathways (covered in **Chapter 2**). DA then rapidly disassociates from the receptors and undergoes one or more of the following events: It is (1) taken back ("reuptake") into the presynaptic terminal by DAT, (2) repackaged into the synaptic vesicles by VMATs, (3) becomes deaminated by intracellular MAO to dihydroxyphenylacetic acid (DOPAC), and/or (4) becomes O-methylated by membranous or soluble COMT to 3-methoxytyramine, as illustrated in **Figures 1.2** and **1.4** and **Table 1.2**.

1.5.1 Functional coupling between DA synthesis and storage

The processes of DA synthesis by TH and DDC and its transport into storage vesicles by VMATs have long been viewed as two separate and independent events. However, several lines of evidence, employing co-immunoprecipitation, overexpression, mutagenesis, and *in vitro* binding assays, revealed that VMAT2 and the DA biosynthetic enzymes are physically and functionally coupled at the synaptic vesicle membrane [69]. The coupling complex also includes scaffolding proteins such as 14-3-3 protein and synuclein. Although TH is commonly considered a cytosolic enzyme, it exists in both cytosolic and membrane-bound forms. Cytosolic TH is enriched in neuronal somato–dendritic compartments of the substantia nigra and ventral tegmental area, whereas membrane-bound TH is more common in brain areas enriched in axon terminals (e.g., striatum and nucleus accumbens).

The coupling between DA synthesis and vesicle loading has several important physiological implications. First, it provides a newly synthesized DA to the vicinity of the synaptic vesicle membrane, allowing more efficient and faster transport into the vesicle. Second, it provides better control over vesicular refilling and the regulation of its quantal size during high frequency stimulation. Third, the proximity of DA to TH provides a balanced regulation of enzyme activity via end-product inhibition. Fourth, by minimizing the amount of free cytoplasmic DA, there is less potential for its oxidation and toxicity. Increased DA metabolism leads to toxicity because of the generation of reactive oxygen species (ROS) and neurotoxic quinines. Indeed, the progressive degeneration of dopaminergic neurons in the substantia nigra pars compacta in Parkinson's disease has been attributed, in part, to a chronic exposure to cytosolic DA.

1.5.2 Characteristics of synaptic vs. secretory vesicles

All cells possess a constitutive secretory pathway whereby vesicles that originate in the Golgi complex contain newly synthesized proteins (i.e., enzymes, growth factors, receptors, and extracellular matrix components) and carry them to the cell surface. Once there, the vesicles contact the plasma membrane and either release their content to the cell exterior (e.g., hormones or neurotransmitters), or their enclosed proteins become embedded within the plasma membrane (e.g., receptors). Neurons and endocrine/neuroendocrine cells are highly specialized cells that are dedicated to intercellular communication and store their chemical signals in committed secretory vesicles. Upon receiving appropriate stimuli, these cells release their content to the cell exterior by a calcium-regulated exocytosis.

Secretory vesicles of neuroendocrine cells (exemplified by adrenal medullary chromaffin granules) differ from synaptic vesicles of the CNS in size, morphology, content, biogenesis, and speed of release [57]. Secretory vesicles are larger; have a diameter of 270 nm; contain biogenic amines, ions, peptides, and proteins; are derived from the Golgi apparatus; and are retrieved after exocytosis, but they do not directly recycle. The condensed proteins within the secretory vesicles are visualized in electron micrographs as a solid core and have, therefore, been identified as dense core secretory granules or vesicles. On the other hand, synaptic vesicles are much smaller, have a diameter of 40–60 nm, contain only neurotransmitters, and appear as clear vesicles in electron micrographs. Synaptic vesicles are derived from the endosome, the major sorting compartment within the cell. Synaptic vesicles also release their content very rapidly, and they completely recycle by endocytosis after emptying. Given that synaptic vesicles are not formed at the Golgi complex, they do not contain newly synthesized proteins. Instead, they take up the neurotransmitters from the cytoplasm, using the VMAT intracellular transporters.

The exocytotic pathway of synaptic vesicles in neuronal terminals [70], as compared with the pathway that governs peripheral catecholamine-containing secretory vesicles [71], are briefly summarized here. As illustrated in **Figure 1.10**, synaptic vesicles within presynaptic dopaminergic neurons undergo several progressive steps that include formation, loading, translocation, fusion, pore expansion, content release, retrieval, and refilling. The complete cycle of a synaptic vesicle has been estimated to take no longer

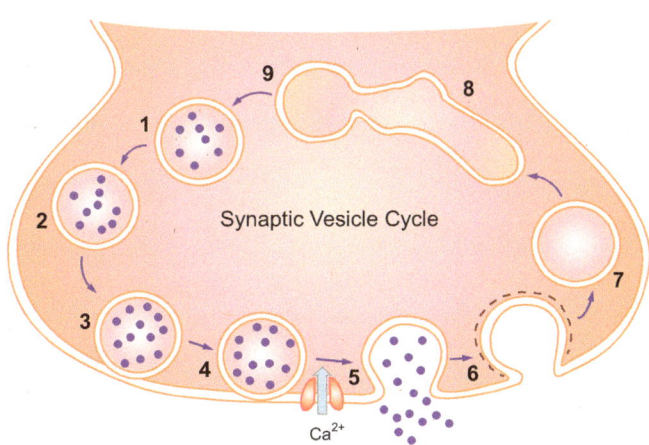

Figure 1.10 A model of the synaptic vesicle cycle. The synaptic vesicle cycle can be divided into nine stages: (1) Empty vesicles take up the neurotransmitters by active transport, using energy provided by a proton pump; (2) Filled vesicles are translocated to the active zone; (3) Docking of vesicles in the active zone; (4) Priming prepares the vesicles for fusion; (5) Calcium influx through voltage-gated channels completes the fusion and triggers exocytosis; (6) Empty vesicles are coated with clathrin and associated proteins; (7) Vesicles are translocated inwards after shedding the clathrin coat and acidification; (8) Vesicles fuse with an early endosome as intermediates; and (9) budding of newly made and recycled vesicles that bypass the endosome. (Redrawn and modified from Sudhof, T.C. and Rizo, J., *Cold Spring Harb. Perspect. Biol.*, 3, 1–16, 2011.)

than 1.2 min and is well coordinated to ensure appropriate release of its content in response to frequent oncoming stimuli.

Synaptic vesicles are formed by budding from the endosome or from the presynaptic membrane after their disengagement. Empty vesicles take up DA by a VMAT-mediated active transport into their lumen, utilizing the electrochemical gradient established by the proton pump. Loaded vesicles are translocated by a calcium-mediated process to specialized sites in the presynaptic terminal, termed the "active zones." Within the active zone, the loaded vesicles are docked and partially fuse with the presynaptic membrane by a complex of SNARE proteins, as described in detail in the subsequent section. In response to action potentials, a calcium influx through voltage-gated channels completes the fusion process, initiates pore opening, and enables the release of the vesicular content into the synaptic cleft.

It has been estimated that at any given time no more than 10–30 vesicles are attached to the active zone of a synapse, while hundreds of vesicles are free in the cytosol [70]. Soon after releasing their content, empty synaptic vesicles become coated by clathrin in preparation for endocytosis. Clathrin is a scaffold protein composed of three heavy and three light chains. It polymerizes around the cytoplasmic face of the invaginated membrane and acts as a reinforced mold around the vesicle. Once separated from the active zone, the empty vesicles shed their clathrin coat, become acidified, and can either become incorporated into an endosome, or skip the endosome and become immediately available for reloading with the neurotransmitter.

Both adrenal chromaffin cells and PC12 cells, their counterpart pheochromocytoma-derived cell line, have been extensively used as models for studying exocytosis of catecholamine-containing secretory vesicles outside the CNS [71,72]. Chromaffin cells originate from the embryonic neural crest and represent a hybrid of the nervous and the endocrine systems. As such, they share some fundamental properties with both cell types. Different populations of adrenal chromaffin cells release Epi (80%) and NE (20%). Small amounts of DA are also released from chromaffin cells, but there are no exclusive DA containing secretory vesicles. Adrenal chromaffin cells also produce enkephalins, chromogranins, and tissue plasminogen activator, all of which are packed within the secretory vesicles together with catecholamines, ATP, ascorbate, and Ca^{2+}. Notably, both membrane-bound and soluble DBH are also constitutive constituents of chromaffin granules, and soluble DBH is secreted from the granules during the process of exocytosis [73].

The biogenesis, targeting, docking, and fusion of secretory vesicles in chromaffin cells have a similar general pattern to that of the exocytotic machinery of neurons but differ in the following respects. First, secretory granules originate from the trans-Golgi network, where they are loaded with catecholamines, peptide hormones, and chromogranins, with the latter serving as the driving force in their biogenesis [74]. Second, newly formed secretory granules are transported along microtubules to the F-actin-rich cell cortex, where they undergo a maturation process. A Ca^{2+}-induced F-actin rearrangement allows the formation of channel-like structures that are perpendicular to the plasma membrane and serve as conduits for directing the granules to the exocytotic sites. Third, chromaffin cells lack morphologically distinct sites for vesicle fusion such as the active zones in presynaptic neurons and, instead, have hot-spots for Ca^{2+} entry. Fourth, acetylcholine, released from the splanchnic nerve, is the main physiological stimulus for exocytosis in adrenal chromaffin cells. Following its binding to cholinergic receptors, acetylcholine depolarizes the cells and elicits action potentials that stimulate Ca^{2+} influx via voltage-dependent Ca^{2+} channels, thereby triggering exocytosis of the secretory granules. Finally, chromaffin granules are functionally heterogeneous and constitute two discrete pools: a large slowly releasable pool, and a smaller, fully mature, readily releasable pool.

1.5.3 Mechanism of exocytosis

Despite the abovementioned differences in the origin and trafficking between synaptic and secretory vesicles, three over-encompassing models of exocytosis and its coupling to endocytosis have emerged [75]. The first model has been coined "full-collapse fusion." According to this model, in response to a stimulus, vesicles are collapsed into the plasma membrane, followed by a clathrin-dependent endocytosis that involves membrane invagination and vesicular reformation. The full-collapse fusion is associated with a rapidly expanding vesicular pore and a complete release of the vesicular content. The second model is named "kiss-and-run." According to this model, the fusion pore opens and closes rapidly, and it is associated with a fast disengagement of the vesicle from the active zone. The third model is known as the "compound exocytosis." According to this model, exocytosis includes very large vesicles that are formed by vesicle–vesicle fusion, followed by bulk endocytosis that retrieves giant vesicles. A schematic illustration of the various models of exocytosis is shown in **Figure 1.11**.

Vesicular fusion with the plasma membrane is an energetically demanding process which entails the movement of different proteins within the membrane and the subsequent disruption of the lipid bilayer. This is followed by the reformation of a highly curved membrane structure. To bring together two membranes and overcome the repulsive electrostatic forces between them requires energy. A complex of proteins named SNARE (standing for Soluble N-ethylmalemide-sensitive factor Attachment protein Receptor) and

Figure 1.11 A model of the interrelations between exocytosis and endocytosis. Three modes of exocytosis are shown: (1) kiss and run and (2) full collapse, both of which are coupled to classical endocytosis, and (3) compound exocytosis, which is coupled to bulk endocytosis. (Redrawn and modified from Wu, L.G. et al., *Annu. Rev. Physiol.*, 76, 301–331, 2014.)

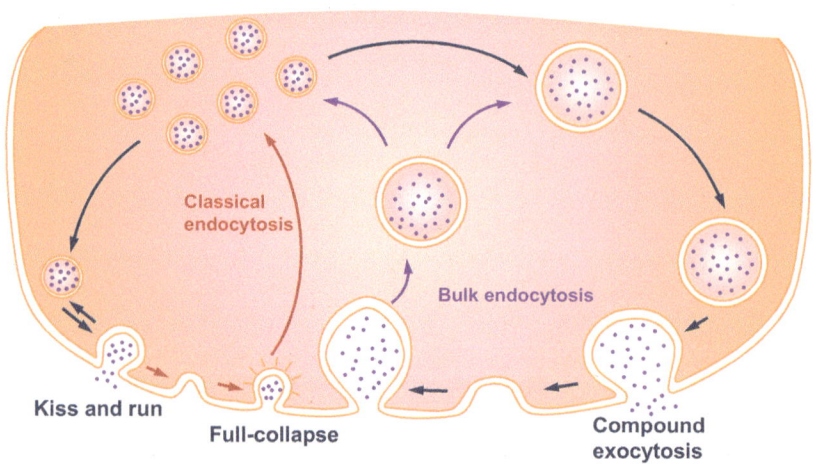

SM (standing for Sec1/Munc18-like) play key roles in the processes of vesicular fusion, pore opening, and retrieval [70,76]. The SNARE proteins were originally identified in the 1970s as targets of clostridium botulinum and tetanus toxins, which enter presynaptic terminals and act as proteases that block membrane fusion.

The SNARE complex is formed by members of the synaptosomal-associated protein 25 (SNAP-25), vesicle-associated membrane protein (VAMP) and members of the syntaxins family. Interactions between these proteins create a four-helix bundle, formed by two helices of SNAP-25, one vesicular-transmembrane VAMP and one presynaptic plasma membrane syntaxin that brings together the vesicular and plasmatic membranes. Other proteins that interact with the SNARE complex include Munc-18, complexin, synaptophysin, and synaptotagmin [77]. In addition, synaptotagmin serves as a calcium sensor and regulates the SNARE zipping. The SM proteins are evolutionary conserved cytosolic proteins that serve as essential partners for SNARE proteins in fusion. Among these is Munc 18, which primarily interacts with syntaxin-1 and whose function is tightly regulated by calcium.

The SNARE proteins are broadly divided into two categories: vesicular or v-SNAREs, which are incorporated into the membranes of the vesicles during budding, and target or t-SNAREs, which are associated with membranes of the nerve terminal. The SNAREs are small, abundant proteins which are often posttranslationally inserted into membranes via a C-terminal transmembrane domain. Seven of the 38 known SNAREs, including SNAP-25, do not have a transmembrane domain and are attached to the membrane through lipid modifications such as palmitoylation.

The first step in fusion of the synaptic vesicles is tethering, where the vesicles are translocated from a reserve pool and make a physical contact with the cell membrane (**Figure 1.12**). At the membrane, Munc-18 is initially bound to syntaxin 1A in a closed structure. The dissociation of Munc-18 from the complex frees syntaxin 1A to bind with the v-SNARE proteins. The next step is docking of the vesicles, whereby the v- and t-SNARE proteins transiently associate in a calcium-independent manner. During vesicular priming, the SNARE motifs form a stable interaction between the vesicle and membrane, while complexin proteins stabilize the primed SNARE-complex and render the vesicles ready for exocytosis.

The span of presynaptic membrane that contains the primed vesicles and the dense collection of SNARE proteins is referred to as the active zone. Voltage-gated calcium channels are highly concentrated around the active

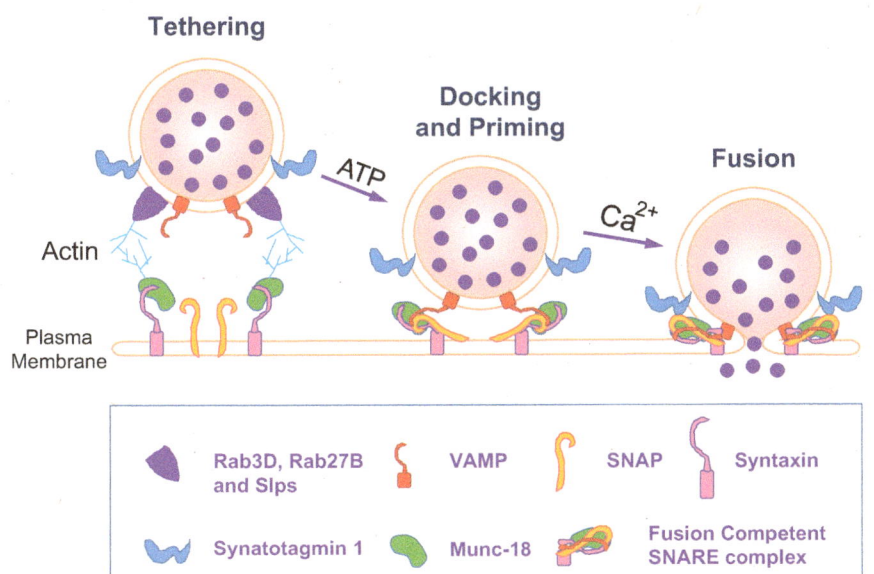

Figure 1.12 Role of SNARE complexes in exocytosis of a secretory granule. The process exocytosis (as seen in pancreatic and parotid acinar cells) proceeds through tethering, docking, priming, and fusion steps. Putative acinar secretory granule tethering factors include Rab3D, Rab27B and Slps. Docking and priming involve SNARE formation, which is facilitated by interactions between Munc and syntaxins. Synaptotagmin 1 and complexin 2 (not shown) trigger VAMP2-mediated granule fusion in response to elevated Ca^{2+}. See text for other details. (Redrawn and modified from Messenger, S.W. et al., *Cell Calcium.*, 55, 369–375, 2014.)

zones and open in response to membrane depolarization at the synapse. The influx of calcium is sensed by synaptotagmin 1, which in turn dislodges the complexin protein and allows the vesicle to fuse with the presynaptic membrane and release the neurotransmitter. It has also been shown that the voltage-gated calcium channels directly interact with the t-SNAREs syntaxin 1A and SNAP-25, as well as with synaptotagmin 1. During exocytosis, v-SNAREs (e, g., synaptobrevin) and t-SNAREs (e.g., syntaxin and SNAP-25) assemble into a core trans-SNARE complex. This complex plays multiple roles during the various stages of exocytosis, including priming, fusion, pore formation, and expansion, eventually resulting in the release of the vesicle content.

Beyond neurotransmission, the SNARE proteins are involved in many biological functions during embryonic development, including fertilization, neural development, and synaptic plasticity. Homozygous deletion of the *snap25* gene in mice results in embryonic lethality, while heterozygous deletion results in mice with small eyes due to failure to separate the cornea from lens epithelium during eye development. Such mice also exhibit head bobbing and locomotor hyperactivity [77]. Given their central role in neurotransmitter release, several genes related to the SNARE complex such as *SNAP25, VAMP1, VAMP2, STX1A, SYT1 and SYT2* have been examined in psychiatric disorders. Although genome-wide association studies have not yet implicated the above genes in specific disorders, several SNARE polymorphisms have been associated with attention deficit hyperactivity disorder, autism spectrum disorders, major depressive disorder, bipolar disorder, and schizophrenia susceptibilities [77].

REFERENCES

1. Starke K. History of catecholamine research. *Chem Immunol Allergy.* 2014;100:288–301.

2. Iversen L. Neurotransmitter transporters and their impact on the development of psychopharmacology. *Br J Pharmacol.* 2006;147(Suppl 1):S82–S88.

3. Johnson ME, Salvatore MF, Maiolo SA, Bobrovskaya L. Tyrosine hydroxylase as a sentinel for central and peripheral tissue responses in Parkinson's progression: Evidence from clinical studies and neurotoxin models. *Prog Neurobiol.* 2018;165:1–25.

4. Lenartowski R, Goc A. Epigenetic, transcriptional and posttranscriptional regulation of the tyrosine hydroxylase gene. *Int J Dev Neurosci.* 2011;29(8):873–883.

5. Meiser J, Weindl D, Hiller K. Complexity of dopamine metabolism. *Cell Commun Signal.* 2013;11(1):34.

6. Kumer SC, Vrana KE. Intricate regulation of tyrosine hydroxylase activity and gene expression. *J Neurochem.* 1996;67(2):443–462.

7. Kiss A, Mravec B, Palkovits M, Kvetnansky R. Stress-induced changes in tyrosine hydroxylase gene expression in rat hypothalamic paraventricular, periventricular, and dorsomedial nuclei. *Ann N Y Acad Sci.* 2008;1148:74–85.

8. Xu L, Chen X, Sun B, Sterling C, Tank AW. Evidence for regulation of tyrosine hydroxylase mRNA translation by stress in rat adrenal medulla. *Brain Res.* 2007;1158:1–10.

9. Pons R, et al. Aromatic L-amino acid decarboxylase deficiency: Clinical features, treatment, and prognosis. *Neurology.* 2004;62(7):1058–1065.

10. Wang MX, et al. Expression and functions of dopa decarboxylase in the silkworm, *Bombyx mori* was regulated by molting hormone. *Mol Biol Rep.* 2013;40(6):4115–4122.

11. Bertoldi M. Mammalian dopa decarboxylase: Structure, catalytic activity and inhibition. *Arch Biochem Biophys.* 2014;546:1–7.

12. Burkhard P, Dominici P, Borri-Voltattorni C, Jansonius JN, Malashkevich VN. Structural insight into Parkinson's disease treatment from drug-inhibited DOPA decarboxylase. *Nat Struct Biol.* 2001;8(11):963–967.

13. Chalatsa I, Nikolouzou E, Fragoulis EG, Vassilacopoulou D. L-Dopa decarboxylase expression profile in human cancer cells. *Mol Biol Rep.* 2011;38(2):1005–1011.

14. Sabban EL, Nankova BB. Multiple pathways in regulation of dopamine beta-hydroxylase. *Adv Pharmacol.* 1998;42:53–56.

15. Thomas SA, Palmiter RD. Examining adrenergic roles in development, physiology, and behavior through targeted disruption of the mouse dopamine beta-hydroxylase gene. *Adv Pharmacol.* 1998;42:57–60.

16. Cubells JF, Zabetian CP. Human genetics of plasma dopamine beta-hydroxylase activity: Applications to research in psychiatry and neurology. *Psychopharmacology.* 2004;174(4):463–476.

17. Wong DL, Anderson LJ, Tai TC. Cholinergic and peptidergic regulation of phenylethanolamine N-methyltransferase gene expression. *Ann N Y Acad Sci.* 2002;971:19–26.

18. Mannelli M, et al. Glucocorticoid-phenylethanolamine-*N*-methyltransferase interactions in humans. *Adv Pharmacol*. 1998;42:69–72.

19. Bortolato M, Floris G, Shih JC. From aggression to autism: New perspectives on the behavioral sequelae of monoamine oxidase deficiency. *J Neural Transm (Vienna)*. 2018;125(11):1589–1599.

20. Boulton AA, Eisenhofer G. Catecholamine metabolism: From molecular understanding to clinical diagnosis and treatment—Overview. *Adv Pharmacol*. 1998;42:273–292.

21. Vossler DG, Wyler AR, Wilkus RJ, Gardner-Walker G, Vlcek BW. Cataplexy and monoamine oxidase deficiency in Norrie disease. *Neurology*. 1996;46(5):1258–1261.

22. Shih JC, Chen K, Ridd MJ. Monoamine oxidase: From genes to behavior. *Annu Rev Neurosci*. 1999;22:197–217.

23. Fiedorowicz JG, Swartz KL. The role of monoamine oxidase inhibitors in current psychiatric practice. *J Psychiatr Pract*. 2004;10(4):239–248.

24. Mannisto PT, Kaakkola S. Catechol-*O*-methyltransferase (COMT): Biochemistry, molecular biology, pharmacology, and clinical efficacy of the new selective COMT inhibitors. *Pharmacol Rev*. 1999;51(4):593–628.

25. Tunbridge EM. The catechol-*O*-methyltransferase gene: Its regulation and polymorphisms. *Int Rev Neurobiol*. 2010;95:7–27.

26. Redell JB, Dash PK. Traumatic brain injury stimulates hippocampal catechol-*O*-methyl transferase expression in microglia. *Neurosci Lett*. 2007;413(1):36–41.

27. Burchell B, Brierley CH, Monaghan G, Clarke DJ. The structure and function of the UDP-glucuronosyltransferase gene family. *Adv Pharmacol*. 1998;42:335–338.

28. Ouzzine M, Gulberti S, Ramalanjaona N, Magdalou J, Fournel-Gigleux S. The UDP-glucuronosyltransferases of the blood-brain barrier: Their role in drug metabolism and detoxication. *Front Cell Neurosci*. 2014;8:349.

29. Strott CA. Sulfonation and molecular action. *Endocr Rev*. 2002;23(5):703–732.

30. Baran H, Jellinger K. Human brain phenolsulfotransferase: Regional distribution in Parkinson's disease. *J Neural Transm Park Dis Dement Sect*. 1992;4:267–276.

31. Eisenhofer G, Coughtrie MW, Goldstein DS. Dopamine sulphate: An enigma resolved. *Clin Exp Pharmacol Physiol Suppl*. 1999;26:S41–S53.

32. Goldstein DS, et al. Sources and physiological significance of plasma dopamine sulfate. *J Clin Endocrinol Metab*. 1999;84(7):2523–2531.

33. Dajani R, et al. X-ray crystal structure of human dopamine sulfotransferase, SULT1A3: Molecular modeling and quantitative structure-activity relationship analysis demonstrate a molecular basis for sulfotransferase substrate specificity. *J Biol Chem*. 1999;274(53):37862–37868.

34. Dajani R, Hood AM, Coughtrie MW. A single amino acid, glu146, governs the substrate specificity of a human dopamine sulfotransferase, SULT1A3. *Mol Pharmacol*. 1998;54(6):942–948.

35. Ghosh D. Human sulfatases: A structural perspective to catalysis. *Cell Mol Life Sci*. 2007;64(15):2013–2022.

36. Wagner DJ, Hu T, Wang J. Polyspecific organic cation transporters and their impact on drug intracellular levels and pharmacodynamics. *Pharmacol Res*. 2016;111:237–246.

37. Rudnick G, Clark J. From synapse to vesicle: The reuptake and storage of biogenic amine neurotransmitters. *Biochim Biophys Acta*. 1993;1144(3):249–263.

38. Wimalasena K. Vesicular monoamine transporters: Structure-function, pharmacology, and medicinal chemistry. *Med Res Rev*. 2011;31(4):483–519.

39. Gainetdinov RR, Caron MG. Monoamine transporters: From genes to behavior. *Annu Rev Pharmacol Toxicol*. 2003;43:261–284.

40. Duan H, Wang J. Selective transport of monoamine neurotransmitters by human plasma membrane monoamine transporter and organic cation transporter 3. *J Pharmacol Exp Ther*. 2010;335(3):743–753.

41. Rudnick G, Kramer R, Blakely RD, Murphy DL, Verrey F. The SLC6 transporters: Perspectives on structure, functions, regulation, and models for transporter dysfunction. *Pflugers Arch*. 2014;466(1):25–42.

42. Kristensen AS, et al. SLC6 neurotransmitter transporters: Structure, function, and regulation. *Pharmacol Rev*. 2011;63(3):585–640.

43. Uhl GR. Dopamine transporter: Basic science and human variation of a key molecule for dopaminergic function, locomotion, and parkinsonism. *Mov Disord*. 2003;18(Suppl 7):S71–S80.

44. Williams JM, Galli A. The dopamine transporter: A vigilant border control for psychostimulant action. *Handb Exp Pharmacol*. 2006;175:215–232.

45. Zhao Y, et al. Human dopamine transporter gene: Differential regulation of 18-kb haplotypes. *Pharmacogenomics*. 2013;14(12):1481–1494.

46. Foster JD, Vaughan RA. Phosphorylation mechanisms in dopamine transporter regulation. *J Chem Neuroanat*. 2017;83–84:10–18.

47. Kuhar MJ. Recent biochemical studies of the dopamine transporter: A CNS drug target. *Life Sci*. 1998;62(17–18):1573–1575.

48. German CL, Baladi MG, McFadden LM, Hanson GR, Fleckenstein AE. Regulation of the dopamine and vesicular monoamine transporters: Pharmacological targets and implications for disease. *Pharmacol Rev*. 2015;67(4):1005–1024.

49. Forrest LR, et al. Mechanism for alternating access in neurotransmitter transporters. *Proc Natl Acad Sci USA*. 2008;105(30):10338–10343.

50. Beuming T, et al. The binding sites for cocaine and dopamine in the dopamine transporter overlap. *Nat Neurosci.* 2008;11(7):780–789.

51. Vaughan RA, Foster JD. Mechanisms of dopamine transporter regulation in normal and disease states. *Trends Pharmacol Sci.* 2013;34(9):489–496.

52. Ramsay RR, Singer TP. Energy-dependent uptake of *N*-methyl-4-phenylpyridinium, the neurotoxic metabolite of 1-methyl-4-phenyl-1,2,3,6-tetrahydropyridine, by mitochondria. *J Biol Chem.* 1986;261(17):7585–7587.

53. Le W, Sayana P, Jankovic J. Animal models of Parkinson's disease: A gateway to therapeutics? *Neurotherapeutics.* 2014;11(1):92–110.

54. Giros B, Jaber M, Jones SR, Wightman RM, Caron MG. Hyperlocomotion and indifference to cocaine and amphetamine in mice lacking the dopamine transporter. *Nature.* 1996;379(65):606–612.

55. Yaffe D, Forrest LR, Schuldiner S. The ins and outs of vesicular monoamine transporters. *J Gen Physiol.* 2018;150(5):671–682.

56. Lohoff FW, et al. Association between polymorphisms in the vesicular monoamine transporter 1 gene (VMAT1/SLC18A1) on chromosome 8p and schizophrenia. *Neuropsychobiology.* 2008;57(1–2):55–60.

57. Henry JP, Sagne C, Bedet C, Gasnier B. The vesicular monoamine transporter: From chromaffin granule to brain. *Neurochem Int.* 1998;32(3):227–246.

58. Baumeister AA, Francis JL. Historical development of the dopamine hypothesis of schizophrenia. *J Hist Neurosci.* 2002;11(3):265–277.

59. Grundemann D, Schechinger B, Rappold GA, Schomig E. Molecular identification of the corticosterone-sensitive extraneuronal catecholamine transporter. *Nat Neurosci.* 1998;1(5):349–351.

60. Okura T, Higuchi K, Kitamura A, Deguchi Y. Proton-coupled organic cation antiporter-mediated uptake of apomorphine enantiomers in human brain capillary endothelial cell line hCMEC/D3. *Biol Pharm Bull.* 2014;37(2):286–291.

61. Miura Y, et al. Characterization of murine polyspecific monoamine transporters. *FEBS Open Bio.* 2017;7(2):237–248.

62. Kouyoumdzian NM, et al. Atrial natriuretic peptide stimulates dopamine tubular transport by organic cation transporters: A novel mechanism to enhance renal sodium excretion. *PLoS One.* 2016;11(7):e0157487.

63. Zhu HJ, Appel DI, Grundemann D, Markowitz JS. Interaction of organic cation transporter 3 (SLC22A3) and amphetamine. *J Neurochem.* 2010;114(1):142–149.

64. Wang J. The plasma membrane monoamine transporter (PMAT): Structure, function, and role in organic cation disposition. *Clin Pharmacol Ther.* 2016;100(5):489–499.

65. Daws LC. Unfaithful neurotransmitter transporters: Focus on serotonin uptake and implications for antidepressant efficacy. *Pharmacol Ther.* 2009;121(1):89–99.

66. Ben-Jonathan N. Dopamine: A prolactin-inhibiting hormone. *Endocr Rev.* 1985;6(4):564–589.

67. Elsworth JD, Roth RH. Dopamine synthesis, uptake, metabolism, and receptors: Relevance to gene therapy of Parkinson's disease. *Exp Neurol.* 1997;144(1):4–9.

68. Eisenhofer G, Kopin IJ, Goldstein DS. Catecholamine metabolism: A contemporary view with implications for physiology and medicine. *Pharmacol Rev.* 2004;56(3):331–349.

69. Cartier EA, et al. A biochemical and functional protein complex involving dopamine synthesis and transport into synaptic vesicles. *J Biol Chem.* 2010;285(3):1957–1966.

70. Sudhof TC, Rizo J. Synaptic vesicle exocytosis. *Cold Spring Harb Perspect Biol.* 2011;3(12):1–16.

71. Marengo FD, Cardenas AM. How does the stimulus define exocytosis in adrenal chromaffin cells? *Pflugers Arch.* 2018;470(1):155–167.

72. Burgoyne RD, Morgan A. Secretory granule exocytosis. *Physiol Rev.* 2003;83(2):581–632.

73. Rush RA, Geffen LB. Dopamine beta-hydroxylase in health and disease. *Crit Rev Clin Lab Sci.* 1980;12(3):241–277.

74. Elias S, et al. Chromogranin A induces the biogenesis of granules with calcium- and actin-dependent dynamics and exocytosis in constitutively secreting cells. *Endocrinology.* 2012;153(9):4444–4456.

75. Wu LG, Hamid E, Shin W, Chiang HC. Exocytosis and endocytosis: Modes, functions, and coupling mechanisms. *Annu Rev Physiol.* 2014;76:301–331.

76. Dhara M, Mohrmann R, Bruns D. v-SNARE function in chromaffin cells. *Pflugers Arch.* 2018;470(1):169–180.

77. Cupertino RB, et al. SNARE complex in developmental psychiatry: Neurotransmitter exocytosis and beyond. *J Neural Transm.* 2016;123(8):867–883.

78. Messenger SW, Falkowski MA, Groblewski GE. Ca^{2+}-regulated secretory granule exocytosis in pancreatic and parotid acinar cells. *Cell Calcium.* 2014;55(6):369–375.

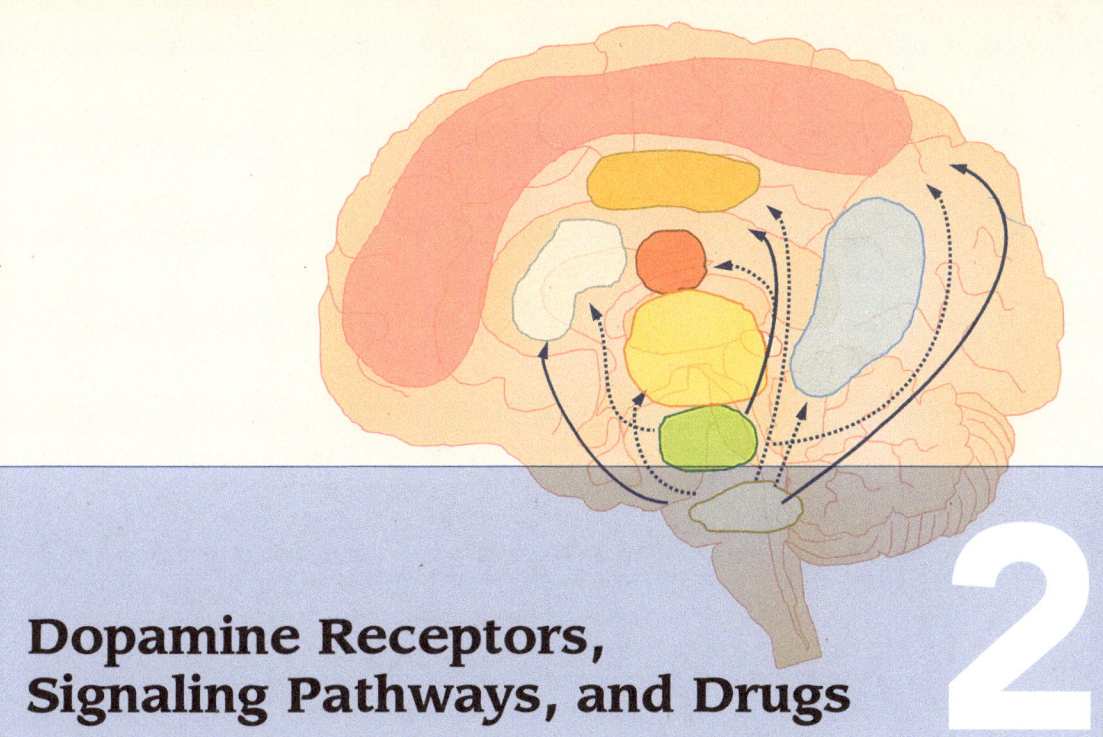

Dopamine Receptors, Signaling Pathways, and Drugs

2

2.1 INTRODUCTION

The receptors for dopamine (DA) were first recognized by their biological activities, radioligand-binding analyses, and the empirical rank order of actions of agonists and antagonists [1]. In the early 1970s, two types of DA receptors (DARs) were identified: those that were positively coupled to **adenylate cyclase (AC)** and increased cAMP production, and those that were negatively coupled to AC and reduced intracellular cAMP levels [2]. Progress with molecular cloning techniques in the 1980s led to the cloning of the rhodopsin and β2 adrenergic receptors, both of which were characterized as seven transmembrane domain (7TM) receptors. The rhodopsin receptor transmits light signals from the retina to the brain through interaction with a G protein known as transducin, while the β2 adrenergic receptors bind norepinephrine (NE) and epinephrine (Epi) and interact with another G protein, known as G stimulatory or Gs. Elucidation of the 7TM architecture, as well as the establishment of the functional linkage of these receptors to the family of G proteins, prompted an extensive search for similar receptors. Many receptors were eventually identified, and subsequently became known as G protein-coupled receptors or GPCRs. In 1988, a new era in the DAR field began with the cloning of the first receptor, the D2 receptor (D2R). Within the next few years, another four DARs were cloned and their structures elucidated.

Based on structural, pharmacological, and biochemical properties, the five known DARs were subdivided into D1-like (D1R and D5R), which increase cAMP, and D2-like (D2R, D3R and D4R), which inhibit cAMP. This early classification, which was solely based on changes in cAMP levels, is now considered oversimplified and not sufficiently encompassing. Although the same DAR classification is still in use today, the signaling cascades which are activated by the DARs are much more variable and complex. The newer concepts are based on advanced knowledge of the noncanonical signaling pathways induced by the DARs, the combinatorial signaling outcome that results from receptor oligomerization, and the roles that are played by proximal effectors such as G protein-independent **β-arrestins**, and G protein-coupled receptor kinases **(GRKs)**.

Information on the properties of the DARs and their signaling pathways has been mostly obtained using the following experimental approaches: (1) analysis of specific targets within the brain; (2) examination of DAR-deficient transgenic mice, albeit often without the ability to distinguish between the brain and peripheral sites of DA actions; (3) the use of agonists and antagonists that potentially have dissimilar effects on the central and peripheral dopaminergic systems; and (4) the analysis of receptor overexpression in null cells that are often deprived of potential cross-talks with other receptors. Consequently, the information that has been gathered by some of these approaches may not fully represent the nature and/or the mode of action of the peripheral DA/DAR systems.

The first section of this chapter presents an overview of the general characteristics of GPCRs and serves as a foundation for a more detailed description of the structure, regulation, distribution, and major functions of each of the five DARs. This is followed by a discussion on receptor oligomerization, desensitization and constitutive activity. Next, canonical vs. noncanonical signal transduction pathways that are activated by the various DARs are evaluated. A special emphasis is placed on the role of the **guanylate cyclase (GC)/cGMP** signaling pathway, which governs many peripheral dopaminergic systems. Appraisal of the various dopaminergic agonists and antagonists in clinical practice then follows, with a focus on those drugs that act selectively on the peripheral dopaminergic systems. The last section compares the phenotypes of the various DAR-mutated mice with respect to their impact on both brain and peripheral organs.

2.2 DOPAMINE RECEPTORS: STRUCTURE–FUNCTION RELATIONSHIP

2.2.1 Overview of G protein-coupled receptors (GPCR)

The DARs belong to the superfamily of GPCRs, all of which lack an intrinsic kinase activity and mediate their signal transduction through the heterotrimeric G proteins [3]. The GPCRs are encoded by a conserved gene family and represent the largest class of cell surface receptors. More than 800 GPCRs have been recognized in the human genome and have overlapping expression in various tissues throughout the body. The GPCRs are activated by a wide variety of molecules that include neurotransmitters, amino acids, peptide and protein hormones, ions, visual signals, taste effectors, and odorants. The responsiveness to such a diverse repertoire of extracellular inputs has made these receptors highly desirable targets for therapeutic intervention. As many as 30% of the currently marketed drugs function as modulators of GPCR activities. These drugs, acting either as agonists or antagonists of GPCRs, are prescribed for the treatment of various ailments, including asthma, hypertension, and peptic ulcers and a multitude of neurological and psychiatric disorders.

The GPCRs are characterized by a seven-transmembrane α-helical configuration, predicted to form three extracellular loops (ECLs) and three intracellular loops (ICLs). The transmembrane loops are flanked by an extracellular N-terminus and an intracellular C-terminus (**Figure 2.1**). Most GPCRs have two cysteines that form an intramolecular disulfide bridge between the first and second ECLs. This bridge enhances the stability of the receptor and is involved in facilitating ligand binding and receptor activation. The helices are arranged in a circular orientation within the membrane plane. In many receptors, a crevice in the middle of the helices contains the ligand-binding site, which is accessible to ligands from the extracellular side. Binding of an agonist induces conformational changes in the cytoplasmic face of the receptor, enabling the linkage of the receptor to G proteins and

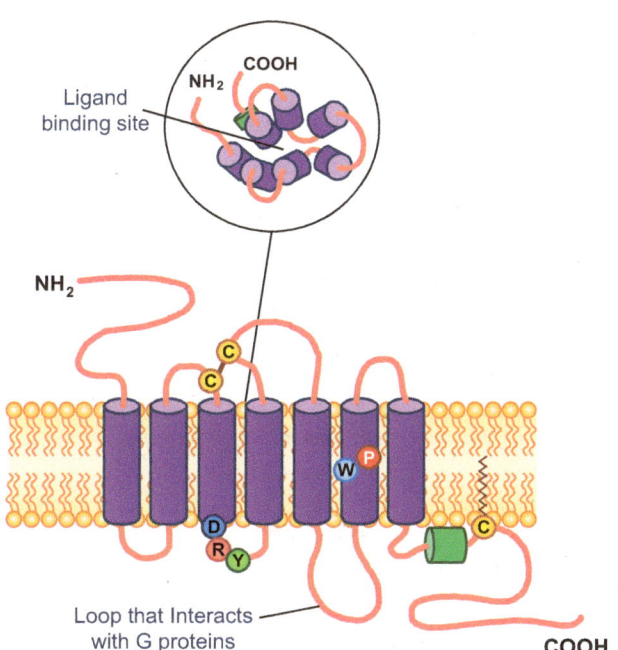

Figure 2.1 Schematic presentation of the structure of GPCRs. The G protein-coupled receptors (GPCRs) are configured in seven transmembrane-spanning domains with an extracellular N-terminus and an intracellular C-terminus. Two cysteines (C) form an intramolecular disulfide bridge between the first and second extracellular loops that stabilizes the receptor. Several residue motifs typical of most GPCRs are also shown. A site of palmitoylation, a covalent bridge between a Cys residue and the fatty acid palmitate, is shown near the cytoplasmic tail. The third intracellular loop is the region which interacts with G proteins. The inset shows the circular arrangement of the transmembrane helices which form the ligand-binding site.

their activation. In addition, activated receptors can become associated with GRKs and interact with β-arrestins. All the above changes are followed by activation of several downstream signaling cascades and are often accompanied by receptor desensitization, internalization and degradation [4,5], as is discussed in more detail in a later section.

2.2.2 Structural and functional classification of the GPCRs

Over the years, numerous schemes have been proposed for the classification of the GPCRs. One of the most widely accepted paradigms is based on the phylogenetic relationships of these receptors [6]. According to this classification, the GPCR superfamily is grouped into five main subfamilies: (1) Rhodopsin, (2) Adhesion, (3) Glutamate, (4) Secretin, and (5) Frizzled/Smoothend (**Figure 2.2**).

The **Rhodopsin subfamily** is the largest and most diverse among the GPCRs. It comprises about 720 receptors that respond to light, odorants, catecholamines, small peptides, and glycoprotein hormones. There are approximately 240 non-odorant receptors and the remainder are odorant

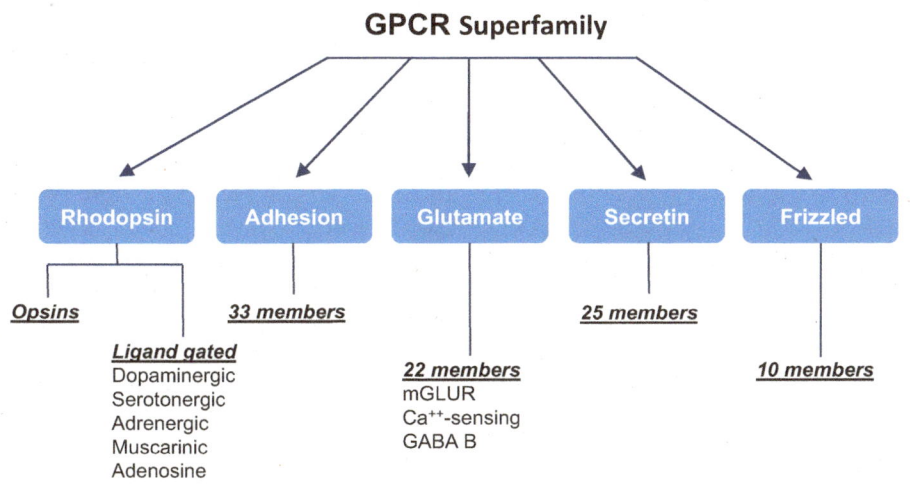

Figure 2.2 Classification of the GPCR superfamily based on their phylogenetic relationship. Among the G protein-coupled receptors (GPCR) subfamilies, the Rhodopsin subgroup is the largest, including more than 600 receptors, among which are the dopaminergic receptors. The number of receptors in the other subfamilies are shown in the diagram.

receptors, accounting for the highly diverse responsiveness of these receptors to olfactory cues. Several consensus motifs are unique to this class of GPCRs, including an E/DRY sequence in TM3 and an NPXXY motif in TM7, which together contribute to a common activation mechanism. There are also distinct structural characteristics that are involved in receptor stabilization and ligand-induced conformational changes. For instance, studies using electron microscopy and confirmed by X-ray crystallography have shown that the TM helices are bent at points positioned near proline residues or Gly-Gly sequences. These regions appear to act as hinges for the relative movement of transmembrane domains within the helical bundle.

The **Adhesion group**, with 33 members, comprises the second largest subfamily of GPCRs. Receptors in this subgroup are characterized by long N-termini that are implicated in cell–cell and cell–matrix interactions. The Adhesion receptors form a separate phylogenetic cluster among the GPCRs, and their functional characteristics differ from the Secretin GPCRs. It is the only GPCR subfamily with a long N-terminus that contains multiple functional domains and the only subfamily that shows some similarities with other receptors and membrane-embedded proteins such as tyrosine kinase receptors, cadherins and integrins. A prominent member of this family is the epidermal growth factor receptor (EGFR).

Members of the **Glutamate subgroup** include the metabotropic glutamate, gamma-aminobutyric acid (GABA), calcium sensing, mammalian pheromone, and taste receptors. They differ from the other subgroups by having an exceptionally large N-terminal domain with a bi-lobed configuration. Although these receptors do not contain any of the conserved residues found in the other family members, they do possess the conserved cysteines in ECLs 1 and 2.

The **Secretin subgroup** includes approximately 25 receptors for peptide hormones such as glucagon, secretin, calcitonin, and vasoactive intestinal peptide (VIP). All of these receptors are coupled to a stimulatory G protein (Gs) and are characterized by a relatively long N-terminus that is critical for targeting the receptor to the plasma membrane. The N-terminus contains a putative hormone-binding site with a conserved series of cysteines and tryptophans, which are required for the interaction of the peptide ligand with the receptor.

Frizzled receptors are composed of 10 known genes in the human genome. These receptors range in length from 500 to 700 amino acids. Their extracellular N-terminus contains a cysteine-rich domain, followed by a hydrophilic linker region of 40–100 amino acids. The intracellular C-terminal domain has a variable length and is not well conserved among the different members of the subfamily. Members of this subgroup serve primarily as receptors for the Wnt signaling pathway. Under some classification schemes, the 13 taste receptors are also included within this subgroup.

2.2.3 Coupling of the GPCRs to G proteins

The G proteins belong to a large group of enzymes that function as GTPases. When bound to GTP, G proteins are in an "on," active conformation, whereas when bound to GDP, they are at an "off," inactive state [7,8]. The extent of G protein activation by the GPCRs is usually proportional to the concentration of the bound ligand. Within the brain, the GPCRs act primarily, but not exclusively, as mediators of slow neuromodulators rather than fast neurotransmitters [7]. The GPCRs mediate a large variety of neurological events, among which are the chemosensory recognition systems (vision, olfaction, taste), movement regulation, and complex behavioral events. In the periphery, GPCRs mediate the actions of many paracrine peptides and circulating hormones, serving as an integral component of endocrine regulation. In the immune system, GPCRs participate in the control of lymphocyte trafficking and motility and are involved in T-cell activation. There are over 160 orphan GPCRs whose natural ligands are still unknown.

G proteins consist of three functional subunits: α, β, and γ. By some estimates, there are 15 α-subunits, 5 β-subunits, and 14 γ-subunits. The α-subunits are grouped by their sequence and functional similarities into the following categories: (1) Gαs (activation of AC), (2) Gαi/olf (inhibition of certain AC isoforms), (3) Gqα11 (activation of phospholipase β), (4) Gα12 (activation of membrane Rho kinase), and (5) Gα13 (activation of GC). Each α-subunit contains a guanine nucleotide-binding site. When inactive, the α-subunit is bound to GDP and to the $\beta\gamma$-complex, together forming a trimeric protein complex. **Table 2.1** provides a list of the general features of the α-subunits with representative effectors, second messengers, and common effects.

The G$_{\beta\gamma}$ subunit is a dimer, composed of two polypeptides that functionally behave as a monomer. The G$_\beta$ and Gγ subunits do not separate and do not appear to act independently. The G$_\gamma$ subunit is considerably smaller than G$_\beta$ and is unstable on its own. It requires interactions with G$_\beta$ for folding, underlying the obligatory association of the dimer. In the G$_{\beta\gamma}$ dimer, the G$_\gamma$ subunit wraps around the outside of G$_\beta$, interacting through hydrophobic association and forming a tightly associated $\beta\gamma$-complex.

Binding of an agonist to a GPCR activates the receptor, enabling it to act as a guanine nucleotide exchanged factor. This involves the release of GDP from the α-subunit, followed by GTP binding to the α-subunit at nanomolar affinity in the presence of magnesium. The exchange of GDP for GTP induces the rapid dissociation of the α-subunit from the $\beta\gamma$-complex and results in its activation (**Figure 2.3**). Both the activated α-GTP and the freed $\beta\gamma$-subunit can then interact with intracellular effectors such as various enzymes, transporters, or ion channels. Upon GTP hydrolysis, the GDP-bound α-subunit and the $\beta\gamma$-subunit re-associate into an inactive trimeric G protein, and the transmission activity of the receptor subsides.

Hydrolysis of GTP occurs through three main mechanisms: (1) intrinsic GTPase activity of the α-subunit, (2) specific GTPase-activating proteins, named regulators of G protein signaling (RGSs), and (3) some effectors. The α-subunits have two domains, a *ras*-like GTPase domain which includes the sites for guanine nucleotide binding and effector interactions, and a helical domain. The guanine nucleotide resides in a cleft between the two domains, and the helical domain is engaged in slowing down GDP release in the inactive state. Mutational analyses [8] have identified two amino acids in the α-subunit as being essential for its intrinsic GTPase activity: a glutamine residue in the N-terminus (at position 204 in Gαi1 and 227 in Gαs), and a conserved arginine (at position 201 in Gαs, and 178 in Gαi1).

The **RGSs** belong to a family of about 40 enzymes with GTPase activity [9]. They are characterized by a 125-amino acid sequence, named the RGS box, or RH homology domain, which binds the GTP-bound G protein subunits and accelerates the rate of GTP hydrolysis. By shortening the lifetimes of the Gα–GTP and the $\beta\gamma$-subunit complexes, the RGS proteins act as negative modulators of G protein signaling, affecting both the potency and efficacy of agonist actions and downstream signaling.

Table 2.1 Typical relationship of GPCR-associated G proteins with effectors and signaling pathways							
G protein Type	αS	αolf	αi	αo	αq/11	α12	α13
Effector	Adenylate cyclase	Adenylate cyclase	Adenylate cyclase	Calcium channels	Phospholipase C	Rho kinase	Guanylate cyclase
Signaling	cAMP	cAMP	cAMP	Ca^{2+}	IP3	RhoA	cGMP
Changes	↑	↑	↓	↓	↑	↑	↑

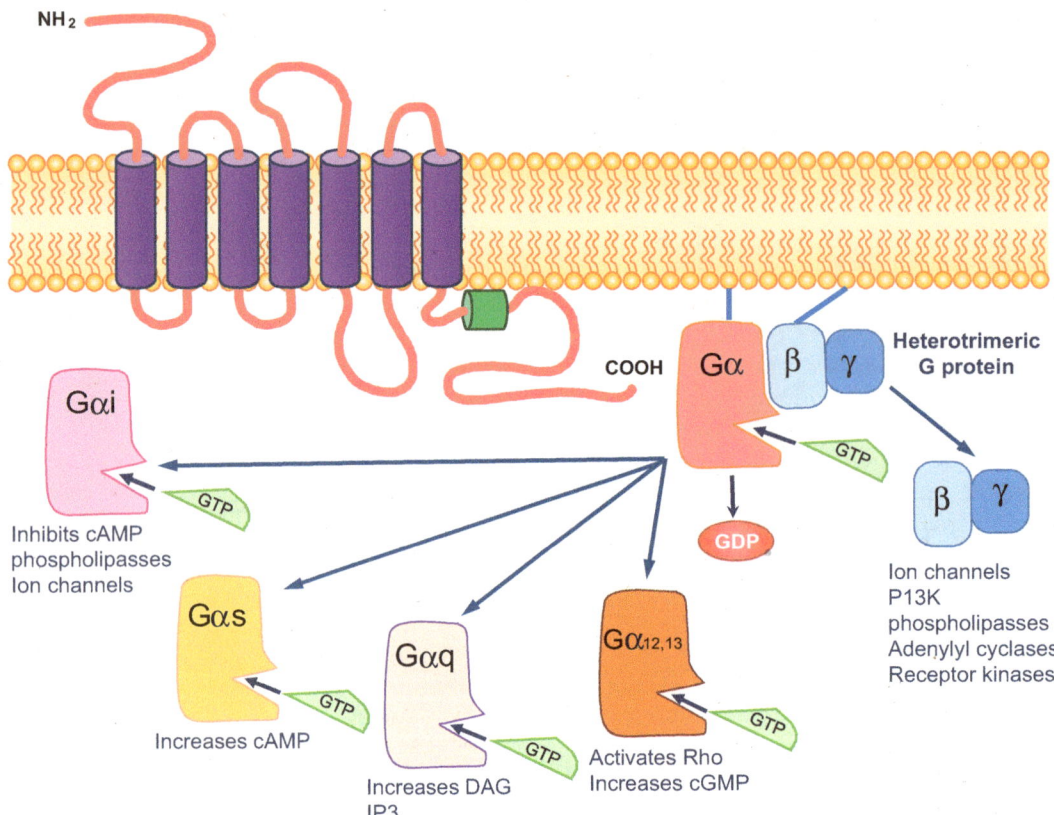

Figure 2.3 The signaling cascade which is initiated upon binding of an agonist to a GPCR. The exchange of GDP by GTP results in the dissociation of the Gα subunit from the βγ subunit. Upon binding to GTP, the alpha subunit becomes activated and can alter downstream effectors. Each of the four Gα subunit types has specific sets of second messengers. The dissociated β/γ subunit can independently activate a variety of effectors.

Cholera toxin (CTX) is a prime example of an effector that blocks GTP hydrolysis and causes serious pathological conditions. CTX is a protein complex that is secreted by the bacterium *Vibrio cholerae*. It is responsible for the massive watery diarrhea, which is the major health outcome of cholera infection. CTX acts by inducing chemical modifications that prevent GTP hydrolysis, resulting in continuous and unregulated activation of Gαs (**Figure 2.4**). This, in turns causes increased AC activity and the elevation of

Figure 2.4 Effects of cholera on intestinal cells, resulting in diarrhea. Cholera toxin, a protein complex produced by the *Vibrio cholera* bacterium, causes massive diarrhea by interacting with G proteins. The toxin prevents GTP hydrolysis, causing persistent activation of adenylate cyclase and marked elevation of cAMP. In intestinal cells, high cAMP increases chloride influx through the cystic fibrosis conductance regulator (CTFR). This is followed by enhanced water secretion into the intestinal lumen. The net result is a large volume of watery diarrhea.

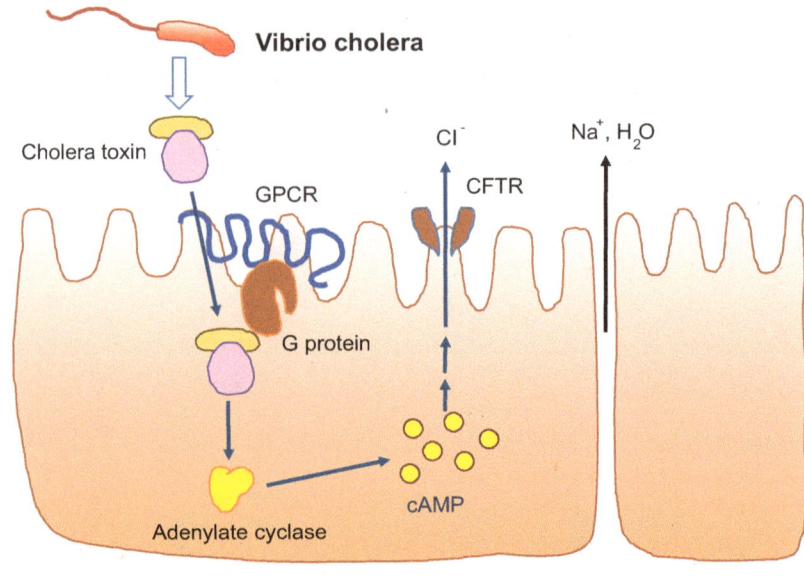

intracellular cAMP levels to more than 100-fold over normal levels. Within the gastrointestinal (GI) tract, the downstream mediators of cAMP stimulate an influx of chloride ions via the cystic fibrosis transmembrane conductance regulator (CFTR), resulting in greatly enhanced water secretion into the intestinal lumen. The overall outcome is a rapid fluid loss from the intestine, up to 2 L/h. If not treated by fluid and electrolyte replacements, cholera infection can lead to high morbidity and mortality due to severe dehydration.

The GPCRs vary in their ability to couple to and activate distinct G proteins and downstream signaling pathways [9]. Some receptors can activate only one class of G proteins and generate a single class of intracellular signals, while others are more promiscuous and can couple to several classes of G proteins and activate multiple intracellular signals. Furthermore, the formation of receptor homo- or heterodimers (as discussed below) results in variable combinations of signaling events. Collectively, the ability of the GPCRs to signal is regulated at several levels. These include the number of receptors that reside on the cell surface at any given time, alterations in signaling efficiency of the receptors, and formation of functional linkages with different effectors.

The G protein signaling systems are not constant but, rather, display a memory of prior activation or signaling tone. For instance, a strong activation of a particular receptor often results in its reduced responsiveness to subsequent stimulations (desensitization) by the same ligand, whereas low activation can lead to increased responsiveness (sensitization). Consequently, a given dose of an agonist or a specific drug may generate distinctly different responses, depending on the prior activation state of the system. Such regulatory features are important for preventing receptor overstimulation and also for allowing a linear response range. Receptors also differ in their basal or constitutive (agonist-independent) status.

Historically, GPCRs were viewed as existing at an equilibrium between an active and an inactive state. However, accumulating evidence indicates that they actually exist in multiple conformational states, with each conformation conferring different downstream effects [10]. These findings gave rise to the theoretical framework for the search of pathway-specific biased ligands capable of selective activation of only G protein- or β-arrestin-mediated signaling. The enhanced selectivity of the biased ligands offered great opportunities for the pharmaceutical industry to expand their repertoire of desirable and more effective medicines.

2.2.4 General characteristics and regulation of the DARs

Recent years have brought remarkable advances in the knowledge of the genetics, chemistry, pharmacology, neurobiology, and therapeutic applications of the DARs. As presented in **Table 2.2**, the five human *DAR* genes are localized in different chromosomes and differ at the level of genomic structure by the presence or absence of introns in their coding regions. Whereas

Table 2.2 Genetic, structural and pharmacological properties of DAR subtypes					
Gene symbol	*DRD1*	*DRD2*	*DRD3*	*DRD4*	*DRD5*
Chromosomes	5q35.1	11q23.1	3q13.3	11p15.5	4p16.1
No. of exons	2	8	7	4	1
No. of introns	None	6	5	3	None
Splice variants	None	Yes	Yes	Yes	None
No. of amino acids	446	414, 443[a]	400	387–515	477
G proteins[b]	$G\alpha_s$, $G\alpha_{olf}$	$G\alpha_i$, $G\alpha_o$,	$G\alpha_i$, $G\alpha_O$	$G\alpha_i$, $G\alpha_O$	$G\alpha_s$, $G\alpha_q$

a: D2S and D2L, respectively.
b: Major, but other G proteins are coupled to these receptors.

DRD2, DRD3, and *DRD4* have 6, 5, and 3 introns, respectively, *DRD1* and *DRD5* have no introns. Among the DARs, only D2R-like have functional splice variants. Alternative splicing of 87 base-pairs between introns 4 and 5 of *DRD2* generates two receptor variants–short (D2S) and long (D2L)—which differ by 29 residues in the ICL3 and exhibit distinct signaling, physiological, and pharmacological properties [9]. D3R has several variants, most of which are nonfunctional, while D4R has few polymorphic variants that differ by the number of 16-amino acid sequence repeats in the ICL3 [3].

The transcriptional regulation of the DAR genes has been covered in a comprehensive review [11], and some of this information is briefly summarized here. The gene structure of the two DAR classes are dissimilar not only with respect to the organization of their coding regions, but also in their transcriptional regulatory regions. The 5'-regulatory regions of *DRD2* and *DRD3* contain noncoding exon(s), several kilobases upstream from their coding exons, while those of the *DRD1*-like have only one noncoding exon and which is separated by a small intron from the coding exon. The promoters of the *DAR* genes lack functional TATA and CAAT boxes, are GC rich, and have several consensus-binding sites for the transcription factor Sp1. The regulatory region of both *DRD2* and *DRD3* contain an initiator-like element, indicating transcription initiation from this position, which is under strong negative regulation as has been found in mammalian cell cultures. The 5'-flanking regions of the *DRD3* and *DRD5* have a much lower GC content than those in the other *DAR* genes. The transcription of the *DAR* genes is regulated by cell-type–specific nuclear factors. Some studies have identified several transcription factors (DNA binding proteins) that regulate the *DAR* genes, with more data generally available on *DRD1* and *DRD2* than on the other receptors.

Another level of DAR regulation is provided by **microRNAs (miRNAs)**. These noncoding RNAs exert their regulatory translational repression and degradation of mRNA through the RNA-induced silencing complex (RISC). A core component of the RISC complex is the Argonauts (Ago) miRNA binding proteins, particularly Ago2, which generates selective miRNAs from their precursors and mediates miRNA-dependent degradation and translational repression of target mRNAs. Several recent reports have examined the involvement of different miRNAs in DAR regulation. In one study, the 3'UTR of the *DRD2* in human dopaminergic neurons was targeted by miR-326 and miR-9. Overexpression of these MiRs resulted in marked reductions of both receptor mRNA and receptor protein synthesis [12]. In another study, inhibition of endogenous miR-142-3p in a mouse CAD catecholaminergic neuronal cell line increased the D1R protein expression levels [13]. As the field of noncoding RNA-mediated translational suppression mechanisms becomes more prominent in the catecholamine research arena, additional knowledge of the overall regulation of DAR expression should be forthcoming.

Posttranslational modifications such as phosphorylation, glycosylation, and palmitoylation play important roles in the dynamics of the DARs and ultimately determine their efficacy. Several consensus phosphorylation sites, located at the cytoplasmic tail as well as ICL3, are involved in homologous and heterologous desensitization [14]. In addition, glycosylation sites at the N-terminus and ECL1 are important for the maturation and intracellular transport of some of the receptors [15]. In most receptors, the cytoplasmic tail has a site of palmitoylation, i.e., a covalent lipid modification of the side chain of Cys residues with the 16-carbon fatty acid palmitate. Palmitoylation facilitates the anchoring of the receptors to the plasma membrane and plays a role in receptor oligomerization [14].

The determination of the exact structure of the ligand-binding pocket within the DARs has long been a challenging endeavor. Scientists typically resolve the chemical structure of a protein by means of X-ray crystallography. For that, various methods are used to induce the protein to condense into a tightly packed crystal lattice. Once this is achieved, X-rays are delivered to the crystal and the structure of the protein is resolved from the diffraction patterns. However, getting the DAR proteins to crystalize with a bound ligand had been problematic for many years. This difficulty stems from the fact that membrane-embedded receptors are notoriously difficult proteins for crystallization because of their low abundance and hydrophobicity. In fact, of the five DAR subtypes, only the D3R has been successfully crystallized.

In addition to X-ray crystallography, methods such as affinity labeling, site-directed mutagenesis, *in silico* receptor homology modeling, and virtual ligand docking have been useful and rather informative with respect to the elucidation of the ligand-binding site. Such approaches have identified certain amino acid residues in helices 3, 5 and 6 as being involved in the binding of DA ligands. Given that the DAR ligands are charged water-soluble molecules, they require a water-accessible-binding site. Their binding to the receptor involves electrostatic interactions between the protonated amine of the ligand and a conserved Asp in TM3. A cluster of three Ser residues in TM5 form hydrogen bonds with catechol hydroxyls, while a cluster of three aromatic residues in TM6 interacts with the aromatic ring of the ligands.

A comparison of the structure, second messengers, downstream effectors, and central nervous system (CNS) distribution of D1-like and D2-like dopamine receptors is presented in **Figure 2.5**. The DARs have a very broad spectrum of neurologic, psychologic, endocrine, renal, cardiovascular, metabolic, immune and oncogenic actions. A search in the PubMed database for the term "dopamine receptors," included in either the title or the abstract, brings up more than 15,000 entries, underscoring the all-encompassing physiological/pathological significance of these receptors.

Within the brain, the DARs regulate voluntary movements, reward, sleep pattern, working memory, feeding, attention, cognitive functions, olfaction, vision, and reproductive behavior. In the periphery, DARs are involved in several aspects of the general operation of the sympathetic nervous system (SNS), and participate, often independent of the SNS, in the control of cardiovascular, renal, gastrointestinal, immune and reproductive functions. DARs are also associated with the malignant transformation of several peripheral

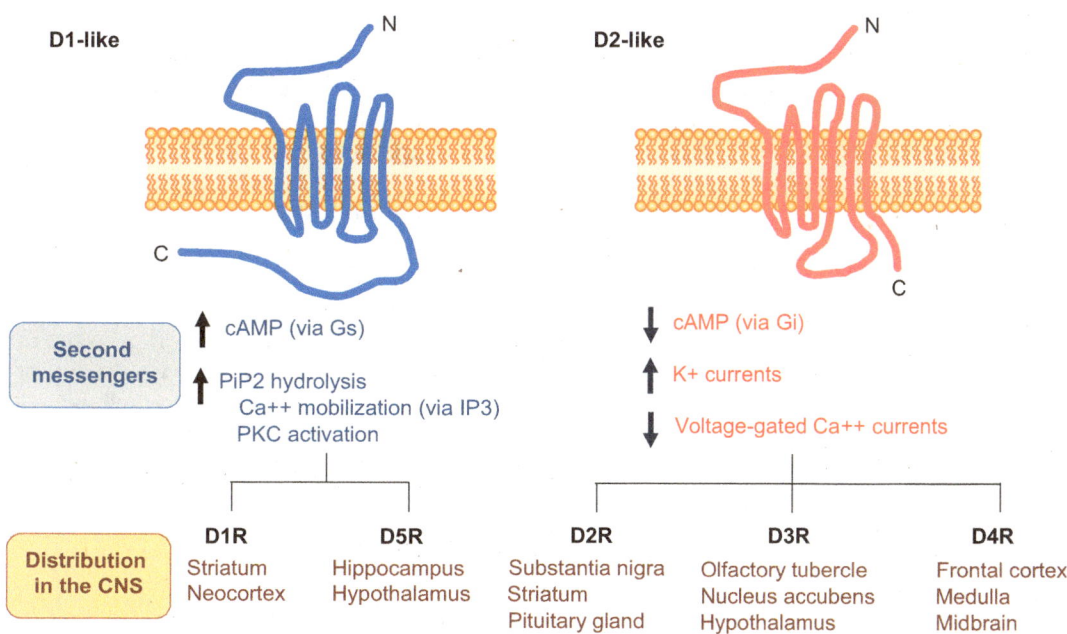

Figure 2.5 Structure, second messengers, and CNS distribution of D1-like and D2-like dopamine receptors. D1-like receptors have a longer C-terminal tail and a smaller third intracellular loop that links to G protein. The D1-like are generally associated with increased cAMP via activation of Gs, while the D2-like are primarily associated with inhibition of cAMP via Gi. The major distribution of the five DARs in the central nervous system and pituitary gland are shown.

organs and tissues. These topics are covered in great detail in subsequent chapters throughout the book.

There are several reports on putative associations between **polymorphisms** in the *DAR* genes and addiction, substance abuse, alcoholism, pathological gambling, and bulimia nervosa [16]. However, a comprehensive review of the existing literature did not arrive at definitive conclusions as to whether polymorphisms or mutations of certain DARs are associated with specific neuropsychiatric disorders [17]. This is not surprising given the complexity of brain disorders that involve not only many anatomical sites but also interactions among multiple neurotransmitters. The relative simplicity of the peripheral DAR system suggests that variations in DAR gene sequences may be predictive of differences in efficacy or side effects of peripherally acting drugs.

2.2.5 Specific properties and brain distribution of D1-like receptors

Although D1R and D5R share significant structural and pharmacologic characteristics, they exhibit distinct tissue distribution and have some discrete, though sometimes overlapping, functions. D5R was initially identified as an isoform of D1R and was therefore named D1B. Following its cloning, D5R was reclassified as an independent receptor. The human D1R and D5R proteins are composed of 466 and 477 residues, respectively. The hydrophobic seven membrane-spanning helices of these receptors are bundled to form a ligand-binding pocket which is lined with charged amino acids that confer ligand specificity for each receptor [18]. The amino acid sequence in the hydrophobic core is 75% and 52% homologous among D1R-like and D2R-like, respectively. The high homology in the hydrophobic core among D1R and D5R explains the fact that currently there are no ligands that clearly distinguish between the two receptors.

The ICL3 within the receptor structure is critical for the interactions with G proteins. D1-like receptors have a shorter ICL3 and a longer C-terminal tails than D2-like (**Figure 2.5**). DA binds to D1R at lower affinity than to D5R, apparently due to the longer C-terminal tail. Overall, D5Rs are distinguished from D1Rs by having higher constitutive (i.e., ligand-independent) activity, increased affinity and higher potency for agonists, decreased affinity for antagonists, and lower agonist-mediated maximal activation [19].

The prevailing dogma holds that D1-like receptors are associated with stimulatory actions through the activation of the Gαs/olf family of G proteins, which increase AC and stimulate cAMP production. However, in several peripheral tissues (both normal and malignant), D1R is actually linked to the GC/cGMP/protein kinase G (PKG) signaling pathway, where its activation can result in inhibition of gene expression and suppression of cellular growth as well as induction of apoptosis [20,21]. Both D1R and D5R are palmitoylated at cysteine residues located at the C-terminal tail region. Both receptors are also N-glycosylated at asparagine residues, a modification that supports efficient plasma membrane localization of several GPCRs. However, chemical or mutational inhibition of glycosylation had no discernable effects on the proper targeting of D1R to the cell surface, whereas glycosylation appears to be essential for the surface trafficking and ligand binding of D5R.

In the absence of specific ligands for D1R and D5R, immunohistochemistry could not be used for the purpose of mapping their distribution within the brain. By default, *in situ* hybridization has been employed as the method of choice. However, interpretation of such data on the brain distribution of these receptors should be made with caution because mRNA abundance does not necessarily reflect the actual level of the encoded protein or the number of ligand-binding sites. Moreover, receptors are synthesized in the neuronal cell body, followed by their delivery by axoplasmic transport to distal

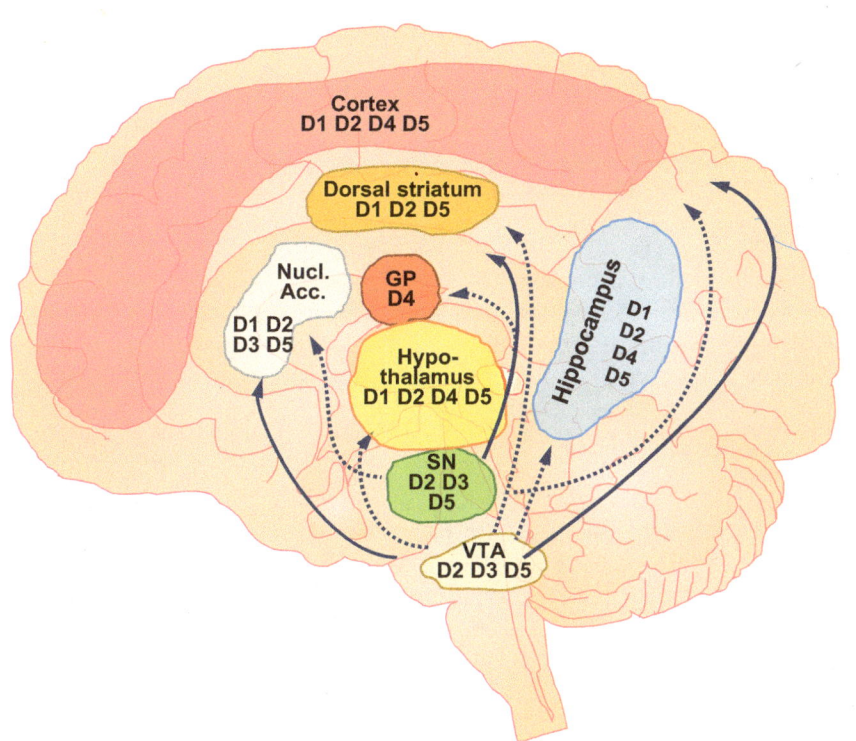

Figure 2.6 Distribution of the DAR in the brain in relation to the dopaminergic projections. Dopaminergic neurons from the ventral tegmental area (VTA) project to the limbic area [nucleus accumbens (Nucl. Accu.), hippocampus and amygdala] through the mesolimbic pathway and to cortical regions through the mesocortical pathway. Dopaminergic neurons from the substantia nigra (SN) project to the globus pallidus (GP) and dorsal striatum via the nigrostriatal pathway. Expression of the various DARs are shown in the respective regions. A fourth dopaminergic projection, the tuberoinfundibular/tuberohyphophysial pathway (not shown) originates in the arcuate nucleus of the hypothalamus and innervates the median eminence of the hypothalamus and the posterior pituitary gland and is primarily associated with D2R. Redrawn and modified from [85].

neuronal projections. *In situ* hybridization can only detect mRNA within the cell body and, therefore, data obtained with this technique do not reflect the distribution of the receptors in axons or dendrites.

The localization of the five DAR subtypes within the human brain is schematically illustrated in **Figure 2.6**, in anatomical relationships to the four dopaminergic neuronal projections: mesolimbic, mesocortical, nigrostriatal, and tuberoinfundibular. A detailed review of the brain dopaminergic projections is provided in **Chapter 3**. D1R is the most highly expressed DAR in the brain [9]. It has been detected at high density in the dorsal and ventral striatum, nucleus accumbens, and the olfactory tubercle. Lower levels of D1R expression are found in the basolateral amygdala, cerebral cortex, septum, thalamus and hypothalamus. Unlike D2R-like which are present at both presynaptic and postsynaptic localizations, D1R and D5R are found only postsynaptically and are not known to function as autoreceptors. The overall importance of D1R for a significant number of brain functions is highlighted by the observations that of all the individual DAR knockouts in mice, the D1R knockout (D1R KO) has the most severe phenotype, including spatial learning deficits, hyperactivity, and abnormal memory retention [23].

D5R has a lower abundance in the brain than D1R. Only limited D5R mRNA levels are seen in the dorsal striatum, nucleus accumbens, and olfactory tubercle, with little or no mRNA found in the hippocampus, the lateral mammillary nucleus, the parafascicular nucleus of the thalamus, and the anterior pretectal nuclei. However, upon a more careful examination, low levels of D5R mRNA were detected throughout the striatum and cortex in both humans and monkeys [24]. D5R KO mice are viable, fertile, and have normal reflexes, but have altered startle reflex, prepulse inhibition and some problems with exploratory locomotion behavior. They also have hypertension because of increased sympathetic tone and altered renal functions [25].

Overall, it has been difficult to assign functions that are unique to D5R and clearly distinguished from those of D1R. Such discrimination awaits the successful development of receptor subtype-specific agonists and antagonists. The expression of D1R and D5R in peripheral organs is covered in detail in **Chapter 5**.

2.2.6 Specific properties and brain distribution of D2-like receptors

The D2-like receptors are composed of D2R, D3R and D4R and include two D2 receptor isoforms, different isoforms of D3R, and polymorphic forms of D4R [9,11]. The coding regions of D2, D3 and D4 contain 6, 5 and 3 introns, respectively, with D3 lacking the fourth intron of D2, and D4 lacking the third and fourth introns of D2. The D2-like receptors are involved in both pre- and postsynaptic inhibition. In general, this class of receptors is associated with the regulation of mood and emotional stability in the limbic system, the control of movement in the basal ganglia, the control of cognition and memory in the prefrontal cortex, and the regulation of hormone synthesis and release from the pituitary gland.

D2Rs are composed of two nearly identical isoforms, D2 long (D2L or $D2_{444}$) and D2 short (D2S or $D2_{415}$), which are generated by alternative splicing of the *Drd2* gene. D2L differs from D2S by an additional 29 amino acids encoded by exon 6. This insertion is localized within the third cytoplasmic loop of the receptor, the region which interacts with G proteins. Consequently, this insertion accounts for the differential interactions of D2L and D2S with G proteins, the activation of distinct downstream signaling pathways, and their diversity of functions [26]. Analyses of D2L KO mice indicate that D2L is primarily involved in the control of postsynaptic functions, whereas D2S is involved in the control of neuron firing and DA release.

The search for cells that express only one D2R isoform has not been successful. The prevailing concept holds that both isoforms are present at each location that expresses D2R, albeit at a highly variable ratio. The two isoforms have a similar affinity for DA and both activate the canonical signaling pathways. Studies with cultured pituitary cells have shown that estrogen, through its α receptors (ERα), affects the regulation of D2R splicing, which generates the two isoforms. It is unclear, however, as to which factor(s) are involved in D2R splicing in other brain areas where the two isoforms are co-expressed.

Within the brain, D2Rs are found at the highest level in medium- and large-sized striatal neurons, the olfactory tubercle, retina, and the basal ganglia (caudate and putamen) as well as in the nucleus accumbens, ventral tegmental area, and the substantia nigra (**Figure 2.6**). Both D2R mRNA and protein are found at lower concentrations in the septal region, amygdala, hippocampus, hypothalamus, thalamus, cerebellum and cerebral cortex. Limited co-expression of D1R and D2R is seen in medium spiny neurons and in some of the pyramidal neurons in the prefrontal cortex.

The pituitary gland is not a bona fide part of the brain but is intimately associated with the ventral part of the hypothalamus (the median eminence) through neuronal connections, i.e., the tuberohypophysial dopaminergic pathway, which originates in the arcuate nucleus of the hypothalamus and innervates the neural lobe of the pituitary. In addition, DA reaches the anterior lobe through a specialized vasculature, the hypophysial portal system, which connects the median eminence with the anterior lobe. Both systems, neuronal and vascular, are described in greater detail in **Chapter 3**.

Figure 2.7 shows expression of D2R in all three lobes of the pituitary gland—anterior, intermediate and neural—and its localization, at different expression levels, in the various pituitary cell types. Within the anterior lobe, D2Rs are primarily expressed in lactotrophs, where D2R mediates the DA-induced inhibition of prolactin (PRL) release and suppression of lactotroph proliferation [27]. Lower levels of D2R are present in other pituitary cell types, i.e., somatotrophs, gonadotrophs and corticotrophs, where D2R are involved, to a limited extent, in the regulation of the production and release of their respective hormones [28]. The D2R are highly expressed in the melanotrophs of the intermediate lobe of the pituitary, where they are involved in the regulation of the production and release of proopiomelanotropin

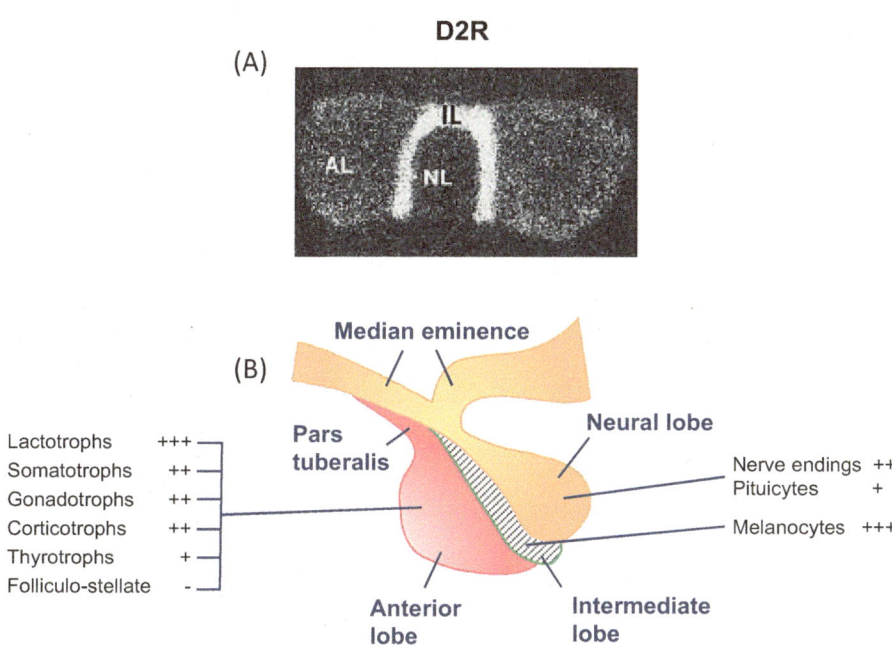

(A)

D2R

(B)

Median eminence

Pars tuberalis

Neural lobe

Lactotrophs   +++
Somatotrophs   ++
Gonadotrophs   ++
Corticotrophs   ++
Thyrotrophs   +
Folliculo-stellate -

Nerve endings ++
Pituicytes   +

Melanocytes +++

Anterior lobe

Intermediate lobe

Figure 2.7 Expression of D2R in the rat pituitary gland. As determined by autoradiography, D2R is highly expressed in the intermediate lobe (IL), with a lower expression in the anterior (AL) and neural (NL) lobes (A). The relative expression of D2R by the various cell types within the three lobes of the pituitary gland is shown in (B).

(POMC)-derived hormones [29]. In the neural lobe, which is composed primarily of nerve endings (vasopressin, oxytocin and dopamine), few glial cells, classified as pituicytes, express low levels of D2R. The distribution and functions of D2R in other peripheral tissues are covered in subsequent chapters.

D3R and D2R share 78% amino acid homology within their transmembrane-spanning domains. The *DRD3* gene can undergo alternative splicing, but most of the detected truncated mRNA species are not translated into functional receptor proteins [30]. The encoded D3R protein comprises 400 residues and contains several posttranslational modifications such as glycosylation, phosphorylation and palmitoylation sites. Compared to D2R, the distribution of D3R is more restricted (**Figure 2.6**). The highest level of D3R expression is found in the limbic area, nucleus accumbens, the olfactory tubercle and the island of Csalleja, with a poor expression in the dorsal striatum, and no D3R mRNA detectable within the pituitary gland. Later studies have found co-distribution of mRNAs for D3R and D2R in medium-sized cells within the striatum.

Despite the similarity in the binding sites of D2R and D3R, a number of agonists and antagonists that specifically bind to each receptor have been developed. Currently, there is considerable interest in the development of additional D3R-selective antagonists/partial agonists, mostly for treating substance abuse disorders but also for treating psychosis-related syndromes. Because D3R expression is more restricted to the limbic system, D3R-selective antagonists/partial agonists could, in theory, attenuate some psychotic symptoms and/or drug-seeking behavior and relapse without inducing the undesirable motor side effects that are frequently associated with the currently available D2R antagonists [31].

The *DRD4* gene is composed of five coding exons and generates a protein composed of 387 residues [32]. The nucleotide sequence of the *DRD4* gene contains a large number of polymorphisms, primarily found in exon 3, which is the region that encodes the IC3 domain. The length of the polymorphism varies from 2,916 amino acids to 11,916 amino acids. In polymorphic forms, a 48-bp sequence exists as a 2- to 11-fold variable number of tandem repeats. Functional studies revealed that a seven-repeat allele is slightly different from the two- and four-repeat alleles in terms of cAMP activity, suggesting that it may be associated with the response to DA-mediated antipsychotics.

D4R receptors are widely expressed in the brain, especially in the hippocampus (CA1, CA2, CA3, and dentate gyrus), frontal cortex, entorhinal cortex, caudate putamen, nucleus accumbens, olfactory tubercle, cerebellum, supraoptic nucleus, and substantia nigra pars compacta (**Figure 2.6**). At the subcellular level, D4R is located primarily at the cell periphery, but not in a certain population of neurons with clear cytoplasmic localization. Immunohistochemical studies found that D4R is localized in dendritic shafts and spines (postsynaptically) within the striatum, with projections to the substantia nigra.

The D4R plays multiple roles in the brain, including mediation of cortico–striatal neurotransmission by controlling the activity of aspartate and glutamate receptors. It is also involved in phospholipid methylation and affects the kinetics of ion channels, which are important for synaptic strength and the modulation of neuronal firing activity. D4R has been linked to many neuropsychological disorders, including schizophrenia, Parkinson's disease, bipolar disorder, addictive behaviors, and eating disorders. Mice with *DRD4* gene knockout have lowered responses to novel stimuli but an enhanced response to methamphetamine and cocaine.

2.2.7 DAR oligomerization, desensitization, and constitutive activity

The concept of DAR **oligomerization** was initially met with skepticism, given some predisposition for misinterpreting data that are based on receptor overexpression, shortcomings of co-immunoprecipitation methods, and/or the use of agonists/antagonists of unclear selectivity. However, biophysical approaches such as fluorescent resonance energy transfer (FRET), bioluminescent resonance energy transfer (BRET), and atomic force microscopy (AFM), with and without radiolabeling and genetic manipulations, have shown that DAR heterodimerization is a more prevalent phenomenon than was previously thought and has substantial physiological, pathological, and therapeutic implications [33].

The DARs can form four types of oligomers: (1) homodimers, e.g., D1R-D1R; (2) heterodimers with another member of the DAR subfamily, e.g., D1R-D2R; (3) heterodimers with another GPCR, e.g., D1R-A1R (adenosine A1 receptor), or (4) heterodimers with a receptor from a structurally different class, e.g., DAR-GABA$_A$ (receptor for GABA), or DAR/SSTR1 (somatostatin receptor 1). Dimers between DAR subfamily members, as well as with adrenergic and serotonergic receptors and also with receptors for adenosine, somatostatin, and prostaglandin, are highly relevant to the actions of peripheral DA actions because most of these receptors are expressed in the same organs that express DARs. Hence, such heterodimers have the potential to influence both the functions and therapeutic responses of peripheral DARs.

Several paradigms have been proposed to address the mechanism of multimer formation [34,35]. One model stipulates that homodimers are preassembled intracellularly before they reach the cell membrane. A second model assumes that heterodimers are generated by a free-floating, "collision-coupling" type of interaction within the cell membrane. A third model implicates a more regulated heterodimer formation through an association with chaperones that partition the receptor partners into membrane microdomains such as lipid rafts or caveolae.

Computational models have focused on putative motifs within the structure of the receptors that may participate in multimer formation [36]. Two patterns of association that involve the TM helices, have been proposed. One, termed contact dimerization, entails the packing of two different TM bundles with separate binding sites through their interactions at lipid-facing interfaces. Another model, termed domain-swapped dimerization, assumes an interpenetration of the TM bundles, with TMs from two different receptors forming interlacing units.

Receptor **desensitization** is defined as a time-dependent diminished responsiveness to a bound agonist that affects the dynamics, plasticity, and the overall activity of a given receptor. Desensitization protects against receptor overstimulation and enables the integration of a biological signal through feedback from second messengers [37]. Therapeutically, however, desensitization is viewed as a considerable impediment because it ultimately limits the efficacy of some drugs. Two types of desensitization modes have been recognized: homologous and heterologous [7]. Homologous desensitization refers to loss of response to an agonist that acts upon a specific receptor subtype, whereas heterologous desensitization refers to diminished responsiveness to a ligand resulting from input by signaling component(s) downstream of the receptor [38].

GPCR desensitization was initially attributed to the phosphorylation of ligand-occupied receptors by second messenger–dependent protein kinases such as protein kinase A (PKA), or protein kinase C (PKC) [39,40]. Following the discovery of the GRKs and β-arrestins, a more comprehensive model of desensitization has emerged [7]. As illustrated in **Figure 2.8**, this model posits that occupancy of the receptor by an agonist promotes the coupling of the receptor to G proteins and induces the activation of proximal effectors such as GRK. Upon receptor phosphorylation at specific sites on the ICLs and C-terminus by GRKs, the receptor becomes a high-affinity target for β-arrestins. Binding of β-arrestins promotes the receptor uncoupling from the G proteins in spite of continuous occupancy by the agonist. The phosphorylated receptor/β-arrestin complex is then targeted to clathrin-coated pits, followed by endocytosis and internalization. Once internalized, the receptors can either undergo dephosphorylation and recycle back to the plasma membrane, or they can be translocated to the lysosomes for degradation. It has been proposed that the sequence of residues that GRK phosphorylates

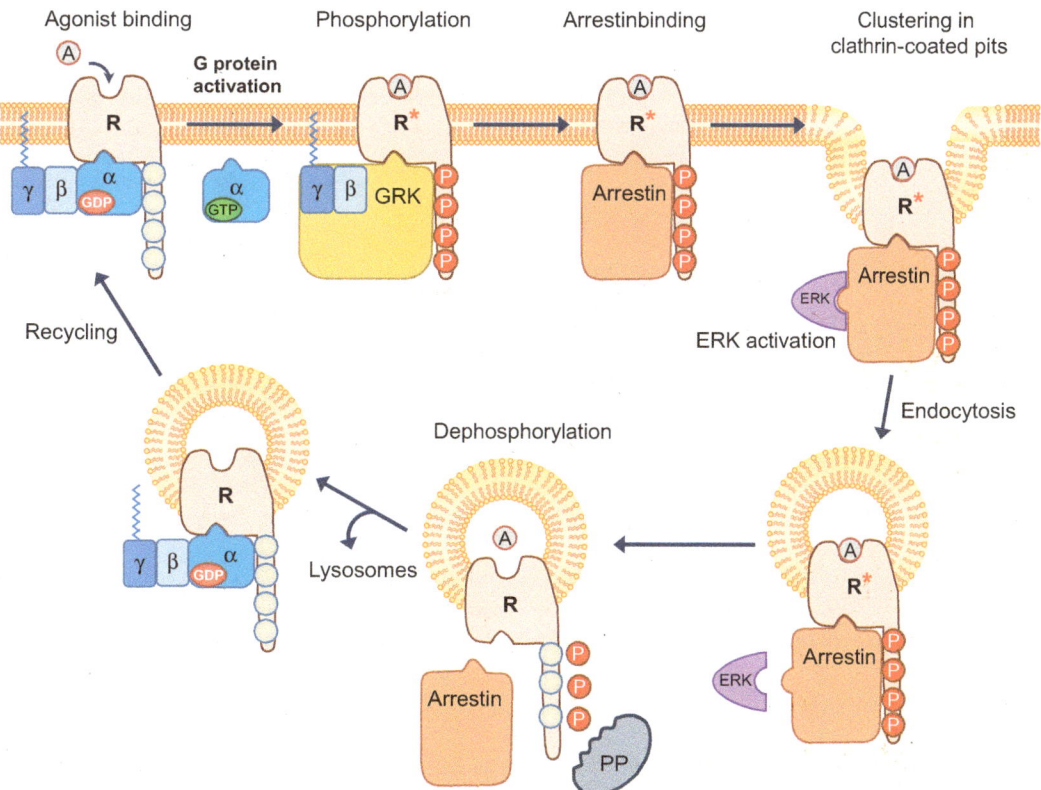

Figure 2.8 A model of desensitization of a GPCR (R) through a cycle of phosphorylation and dephosphorylation. Binding of an agonist (A) activates the receptor (R*) and promotes the dissociation of the Gα subunit from the Gβγ subunit. The G protein-coupled receptor kinases (GRKs) then phosphorylate the receptor at specific sites. This facilitates the binding of arrestins to the receptor, activation of ERK (extracellular receptor kinase) and resulting in its uncoupling from the G proteins in spite of a continuous occupancy by the agonist. The phosphorylated receptor/arrestin complex is targeted to clathrin-coated pits, leading to its internalization by endocytosis. Within the cytoplasm, the receptor/arrestin complex undergoes dephosphorylation and dissociation from arrestins. Dephosphorylated receptors can translocate to the lysosomes for degradation or be recycled back to the plasma membrane. Redrawn and modified from [39].

on the receptor may regulate how β-arrestin functions following receptor binding and may determine whether the receptor is recycled or degraded.

Both GRKs and β-arrestins have multiple functions beyond their roles in receptor desensitization. There are seven GRKs, which are subdivided into four subfamilies: (1) GRK1-like (1 and 7), (2) GRK2-like (2 and 3), (3) GRK4-like (4 and 5), and (4) GRK6. GRK1 and GRK7 are limited to the regulation of visual opsins, while GRK4 has some expression in the cerebellum, kidney and testes. GRK2, the most widely studied family member, phosphorylates non-receptor substrates and interacts with D1R- and D2R-initiated signaling cascades [41]. Both GRK2 and GRK4 have been implicated in the desensitization of renal D1R, one of the best studied peripheral DARs [42]. The renal dopaminergic system is covered in **Chapter 7**.

The β-arrestins also act as G protein-independent transducers [43]. They are subdivided into two groups: the visual arrestins (arrestin 1 and arrestin 4), which are expressed in the rod and cone cells of the eye, and the nonvisual arrestins (arrestin 2 and arrestin 3), which are ubiquitously expressed. As such, they can serve as scaffolding proteins that interact with multiple partners and affect receptor signaling, trafficking, and ubiquitination [10]. Unlike the substantial knowledge on the role of arrestins in peripheral adrenergic receptor (AR) functions, there are no published reports on their activity in peripheral cells that specifically express endogenous DARs.

Constitutive receptor activity is defined as activation of a given receptor in the absence of agonists. This concept has been demonstrated for many GPCRs and has been accepted as one of the principal aspects of their signaling. Many *in vitro* assays that have measured intracellular cAMP, GTPγ binding, and/or thymidine incorporation have established the constitutive activation of both D1-like and D2-like receptors [44]. As determined by mutagenesis studies, constitutive activity of D1-like receptors is governed by contributions from the third ECL, the third ICL., and the C-terminal tail and is regulated by several kinases. Among the DARs, the D5R has the highest basal signaling, while there are only limited data on constitutive activity of D2-like receptors. A major caveat is that constitutive activity is difficult to demonstrate *in vivo* because of the constant presence of endogenous DA.

2.3 CANONICAL VERSUS NONCANONICAL SIGNALING BY DAR SUBTYPES

Over the last decade, the knowledge of the DAR signaling cascade has greatly expanded [44,45], leading to an extensive reassessment of the mechanism of signaling by the DARs. The classical view stipulated a "linear" response, whereby an agonist-activated receptor is coupled to a specific G protein, resulting in predictable stimulation or inhibition of the intracellular responses. However, recent data revealed much more complex responses by typical agonists, involving not only G proteins but also effectors such as β-arrestins and GRKs. Consequently, new terms have been proposed to define the noncanonical signaling of the DARs as "collateral efficacy," "functional selectivity," "inverse agonism" or "biased agonism" [46,47]. Regardless of the terminology being used, G proteins occupy a central, albeit not an exclusive, position, as the regulators of DAR-driven signaling cascade. Although most of activities of the DARs are channeled through the G proteins, some activation occurs through G protein-independent pathways, such as those mediated by β-arrestins.

2.3.1 The consequences of DAR coupling to G proteins

The general association of GPCRs with the heterotrimeric G proteins was introduced at the beginning of this chapter (**Table 2.2**). The DAR-associated Gα subunits are divided into several classes: Gαs/olf and Gαio (associated with stimulation and inhibition of AC, respectively), $G\alpha q_{/11}$ (associated with the activation of phospholipase C; PLC), and $G\alpha_{12/13}$ (associated with the activation of RhoA as well as with guanylate cyclase; GC). Other effectors/second messengers reported to be activated by DAR agonists have not been clearly assigned to specific G protein subunits or act independent of the G proteins [9]. Both the α and $\beta\gamma$ subunits can activate multiple cell-specific effectors that include AC, GC, phosphodiesterases (PDEs), PLC, nitric oxide synthase (NOS), transporters and various ion channels. These, in turn, regulate second messengers such as cAMP, cGMP, diacylglycerol (DAG), inositol triphosphate (IP3), NO, arachidonic acid, and a variety of ions, ultimately leading to an integrated physiological response. The G protein-induced signal is terminated by the RGS proteins, which enhance the intrinsic GTPase activity of the $G\alpha$ subunit [48]. Upon hydrolysis of the bound GTP to GDP, the α and $\beta\gamma$ subunits reassociate, and the inactive G protein becomes linked again to the receptor.

2.3.2 The adenylate cyclase/cAMP/PKA signaling pathway

Adenylyl cyclase (AC), also known as adenylate cyclase, is an enzyme that plays key regulatory roles in mediating the actions of multiple receptors, including DARs. The AC catalyzes the conversion of ATP to cAMP, a second messenger that serves as a regulatory signal by controlling cAMP-dependent kinases, a variety of transcription factors, and many ion transporters. There are nine mammalian transmembrane ACs, with a tenth, a "soluble" form (sAC), having distinct catalytic and regulatory properties [49]. Membrane-bound ACs are grouped into four categories, based on their regulatory properties. Group I consists of Ca^{2+}-stimulated AC 1, 3 and 8; group II consists of $G\beta\gamma$-stimulated AC 2, 4 and 7; group III comprises $Gi\alpha/Ca^{2+}$-inhibited AC 5 and 6, while group IV contains forskolin-insensitive AC 9. The sAC is not regulated by G proteins but, rather, it is stimulated by calcium and bicarbonate.

The mammalian transmembrane ACs have a similar topology, which comprises a variable N-terminus and two repeats of a membrane-spanning domain (C1a and C2a), followed by a cytoplasmic domain. The C1a and C2a subdomains are homologous and form the active site of the enzyme. The two cytoplasmic domains (C1 and C2) form an intramolecular "dimer" at their interface, thereby creating a site that is primed for bidirectional regulation. In the absence of activators, the relative affinity between the C1 and C2 domains is weak, while forskolin or Gs each increases their affinity by 10-fold. Forskolin binds at the interface, while Gs binds at a cleft on the opposite side of the catalytic site. Synergistic activation by both regulators results in a 100-fold increase in affinity in the active site.

Increased cAMP levels activate PKA, which acts upon several targets, including cAMP response element binding protein (CREB) and DARPP-32 (32-kDa dopamine and cAMP-regulated phosphoprotein). PKA is a tetramer, composed of two regulatory (R) and two catalytic (C) subunits. Binding of four cAMP molecules to the R subunit promotes conformational changes that result in the dissociation of the C monomers. Upon ATP binding, the C monomers become activated and can phosphorylate cytoplasmic and nuclear proteins that contain the appropriate consensus sequence. CREB is a transcription factor that is activated by PKA-induced Ser133 phosphorylation and regulates the transcription of numerous genes [50]. DARPP-32 is a multifunctional phosphoprotein with a protein phosphatase 1 (PP1) inhibitory function. DARPP-32 becomes activated upon PKA- and cyclin-dependent kinase 5-induced Thr34 phosphorylation [51]. As discussed in **Chapters 11 and 12**, DARPP-32 play a critical role in tumorigenesis and is overexpressed in breast, prostate, colon, and stomach cancers.

Collectively, the above signaling proteins act as amplification mechanisms for the D1R/PKA axis, and are down-regulated as a consequence of

D2R activation. Both D1R-like and D2R-like also modulate, in opposite directions, K$^+$, Na$^+$ and Ca^{2+} ion channels, whose functions are especially critical for DA neurotransmission in the CNS but are also important at some peripheral sites, including the kidney, intestine and exocrine pancreas, as is discussed in later chapters [16].

2.3.3 Cross-talk of multiple signaling pathways

Figure 2.9 captures the complexity of the combinatorial DAR signaling. **Panel A** depicts the canonical signaling pathways by D1-like and D2-like receptors, which are linked to Gs/olf and Gi/o respectively, and stimulate and inhibit AC. Dimerization and oligomerization increase the complexity of DAR signaling by acquiring pharmacological and functional properties that are distinct from those of the individual monomers [16]. As depicted in **Figure 2.9, panel B**, D1R-D2R heterodimers have been reported to be linked to the Gq subunit, which regulates phospholipase C (PLC), a class of membrane-associated enzymes that cleave phospholipids. The 13 mammalian PLCs are classified into six isotypes (β, γ, δ, ε, ζ, η) according to structure, with each having unique as well as overlapping controls over its expression and subcellular distribution. The regulators of each PLC vary, but typically include heterotrimeric G protein subunits, protein tyrosine kinases, small G proteins, Ca^{2+}, and phospholipids. Activation of PLC stimulates the

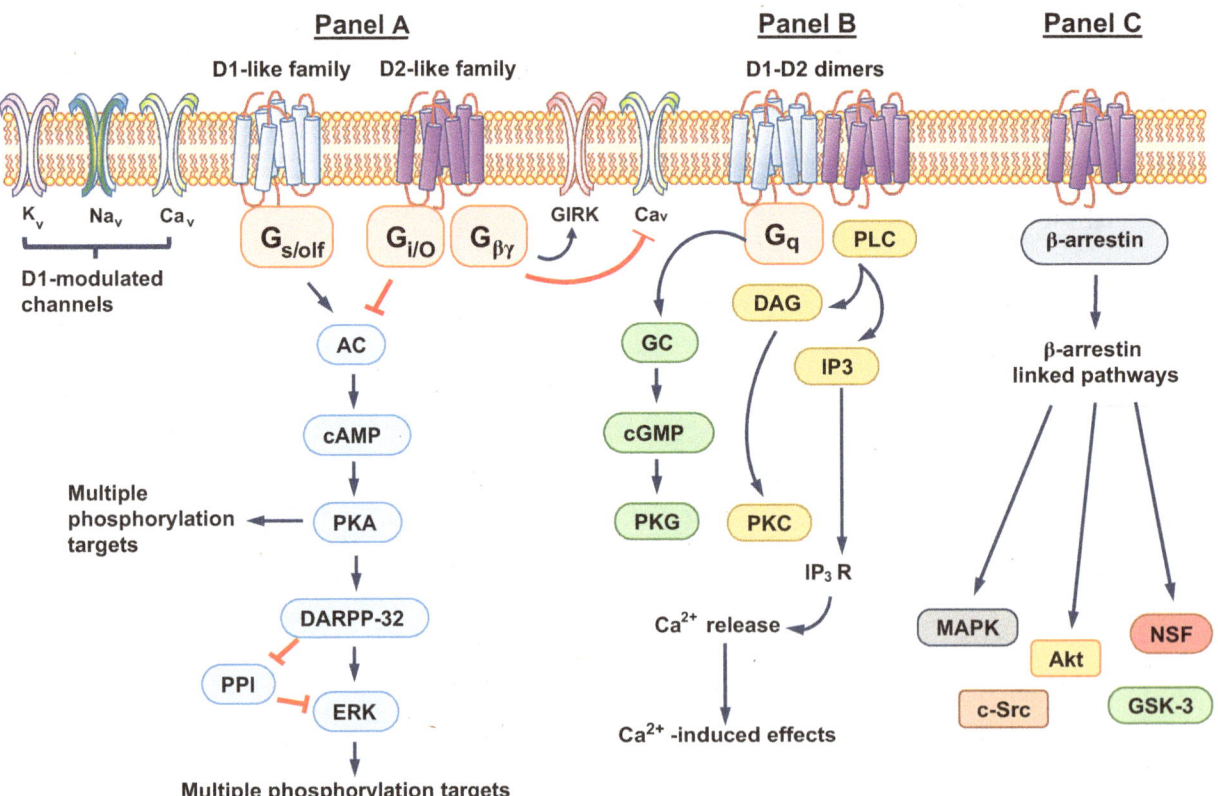

Figure 2.9 Canonical and non-canonical signaling pathways activated by D1- and D2-like dopamine receptors. D1-like and D2-like receptors signal through Gs/olf and Gio, which respectively stimulate and inhibit adenylate cyclase (AC) and alter intracellular cAMP levels (**Panel A**). Downstream effectors include PKA and 32-kDa DA- and cAMP-regulated phosphoprotein (DARPP-32), each with multiple targets. DARPP has protein phosphatase 1 (PPI) inhibitory functions. D1R also modulates voltage-gated K^+_v, Na^+_v, and Ca$^{2+}_v$ channels, while the dissociated Gβγ subunit modulates G protein-gated, inwardly rectifying K$^+$ (GIRK) and Ca$^{2+}_v$ channels. **Panel B:** D1-D2 heterodimers can couple to phospholipase C (PLC), which activates diacylglycerol (DAG), protein kinase C (PKC), and inositol triphosphate (IP3), which regulates Ca^{2+} release. Heterodimers are also linked to Gq subunit which activates guanylate cyclase (GC), increases cGMP and activates protein kinase G (PKG). **Panel C:** G protein-independent actions of β-arrestin include MAPK, Akt, c-Src, N-ethylmaleimide-sensitive factor (NSF), and glycogen synthase kinase-3 (GSK-3). Redrawn and modified from [16].

production of inositol triphosphate (IP3) and diacylglycerol (DAG), which regulate intracellular calcium and activate PKC. Further permutations of the signaling pathways can result from DAR heterodimerization with nonhomologous receptors.

Figure 2.9, **panel A**, also shows that in response to D2R-like activation, the dissociated βγ subunits can suppress N-type calcium channels [52] and can stimulate G protein-gated inwardly rectifying potassium channels (GIRKs) [53]. The G protein-gated ion channels are primarily found in CNS neurons, where they affect the flow of potassium (K^+), calcium (Ca^{2+}), sodium (Na^+), and chloride (Cl^-) ions across the plasma membrane and are involved in maintaining the electrochemical gradient across the cells. However, G protein-gated ion channels are also important at many peripheral sites, including kidney, intestine and exocrine pancreas, as discussed in later chapters. In addition, the RGS and GRK proteins contribute directly and indirectly to the amplification or diminution of signaling by the various DARs. Finally, as shown in **Figure 2.9**, **Panel C**, β-arrestins exert G protein-independent actions on noncanonical pathways [43] that include Akt, MAPK, c-Src, Mdm2 (mouse double minute homolog), NSF (N-ethylmaleisensitive factor) and GSK (Glycogen Synthase Kinase).

2.3.4 Role of the cGMP pathway

The **cGMP pathway** deserves special attention in this book because it is a major pathway that is activated by D1R in breast cancer [21,54] and other malignancies. The cGMP is generated from GTP by two distinct guanylate cyclases: particulate (pGC) and soluble (sGC). The pGCs are transmembrane receptors that primarily bind to natriuretic peptides, while the cytosolic heterodimeric sGC serves as the main target of NO [55,56]. As shown in **Figure 2.10**, several drugs, including YC-1, BAY 41-2272 and Riociguat directly stimulate sGC, while ODQ is a selective inhibitor of sGC. Once elevated, cAMP or cGMP are rapidly hydrolyzed by phosphodiesterases (PDEs), a superfamily that comprises 11 members that differ in structure, catalytic properties, and subcellular localization [57]. Based on substrate specificity, the PDEs are grouped into three classes: PDE 4, 7, and 8, which selectively hydrolyze cAMP, PDE 5, 6, and 9, which are specific for cGMP, and the remainder have dual activity.

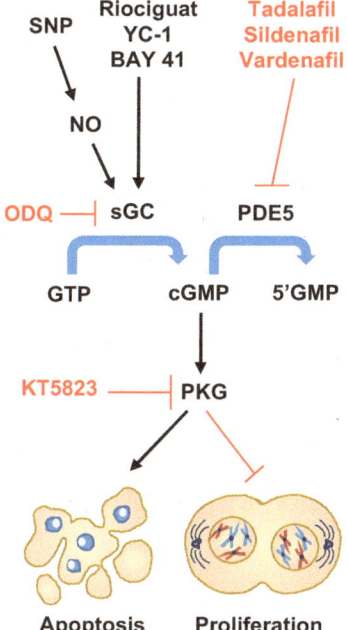

Figure 2.10 Characteristics of the guanylate cyclase/cGMP/PKG pathway. Soluble guanylate cyclase (sGC) is activated by nitric oxide (NO) and by three drugs: Riociguate, YC-1 and Bay 41-2271 (BAY 41), and it is inhibited by ODQ. The activated sGC converts GTP to cGMP while phosphodiesterase 5 (PDF5) selectively hydrolyzes cGMP to 5'GMP. PDE5 inhibitors include Tadalafil (Cialis), Sildenafil (Viagra) and Vardenafil (Levitra). Elevated cGMP activates protein kinase G (PKG), which is inhibited by KT5823 and in many systems can induce apoptosis and/or inhibit cell proliferation. SNP: sodium nitroprusside, which often serves as a nitric oxide (NO) donor.

Several **PDE5 inhibitors**, e.g., sildenafil (Viagra), tadalafil (Cialis), and vardenafil (Levitra) are used to treat erectile dysfunction and have an excellent safety record [58]. Cialis is the longest acting of the three drugs and has been recently on clinical trials to treat head and neck cancer [59]. The main targets of cGMP are two kinases—PKG-I and PKG-II [60]—both of which are inhibited by KT5823. The PKG-I gene is expressed as cytosolic PKG-Iα or PKG-Iβ isoform, while the PKG-II gene is expressed as a membrane-associated PKG-II protein. The kinetics, localization and substrates of the PKG enzymes differ. The PKGs phosphorylate many downstream effectors, some of which overlap with those that are targeted by PKA, while others are distinct. In many, but not all, cell types, activated PKG leads to the suppression of cell proliferation and/or apoptosis.

2.4 DAR AGONISTS AND ANTAGONISTS

2.4.1 Overview of DAR-selective ligands

The well-recognized association of brain DA with many neurological, cognitive, movement, and other pathological conditions has fostered vast investment by the pharmaceutical industry in the development of DA-altering drugs. Nonetheless, in spite of major advances in molecular genetics and neuropharmacology of the DARs, there remain critical needs for solving their three-dimensional structure, for identifying specific motifs within the receptors that confer ligand selectivity, and for a better understanding of the spectrum of drug actions that result from receptor heterodimerization [61].

The generation of new DAR-selective ligands by the pharmaceutical industry continues to be mostly empirical, is often unpredictable, and has only modestly benefited from computer-aided drug–design applications. Some caveats in the testing of new DAR-targeting drugs include (1) the use of naïve cells for receptor overexpression; (2) an overreliance on a single biochemical parameter, i.e., cAMP levels; (3) a single time point analysis that overlooks receptor dynamics and desensitization; (4) *in vivo* studies whereby the outcome is often obscured by interactions among multiple systems, drug availability, and drug metabolism; and (5) the unpredictability of the capacity of any newly developed drug to penetrate the **blood–brain barrier (BBB)**.

Drugs that alter DARs comprise the largest class of pharmaceuticals. These drugs can be divided into the following categories: (1) therapeutic drugs for the treatment of a wide variety of brain neurological disorders, (2) drugs that do not penetrate the brain, for treating peripheral DA-related disorders, (3) drugs that are primarily employed as receptor probes and radioligands in basic research, and (4) diagnostic drugs that are especially suited for positron emission tomography (PET) or single photon emission computerized tomography (SPECT) imaging. In addition to the drugs that directly bind DARs, many others act by affecting DA synthesis, reuptake, or metabolism [62] and will be briefly discussed in subsequent chapters. The therapeutic potential of agonists and antagonists acting upon central and peripheral DARs is presented in **Table 2.3**.

To encompass the full range of drug actions at the DARs, an updated drug definition was proposed [63]. According to this terminology, an agonist is defined as a substance that binds to a specific receptor and stimulates the signaling pathway known to be associated with it. A partial agonist causes a less than maximal response but can also act as a partial antagonist. An antagonist has no effects on its own but, rather, blocks an agonist-induced signaling. A somewhat confusing term is inverse agonist, defined as a ligand that binds to a receptor and inhibits agonist-independent (constitutive) signaling. These concepts are schematically presented in **Figure 2.11**.

Table 2.3 Therapeutic potential of antipsychotics in central and peripheral locations

Generic name	Trade name	DAR[1]	5HTR[2]	HA[3]	AR[4]	MR[5]
Typical						
Haloperidol	Haldol	2, 3–5	1, 2, 7	1		
Chlorpromazine	Thorazine	2, 3–5	1, 2, 6, 7	1	2	
Atypical						
Olanzapine	Zyprexia	1, 2, 4	1, 2, 6, 7	1	1	1
Quetiapine	Seroquel	1–3	1, 2, 7	1		
Risperidone	Risperdal	1–5	1, 2, 5, 7	1	1, 2	
Ziprasidone	Geodon	1–5	1, 2, 6, 7	1	1	1
Aripiprazole	Abilify	1–3	1, 2, 7	1	1	

1: Dopaminergic receptors; 2: Serotonergic receptors; 3: Histaminergic receptors; 4: Adrenergic receptors; 5: Muscarinic receptors; the numbers underneath each of these receptors designate receptor subtypes.

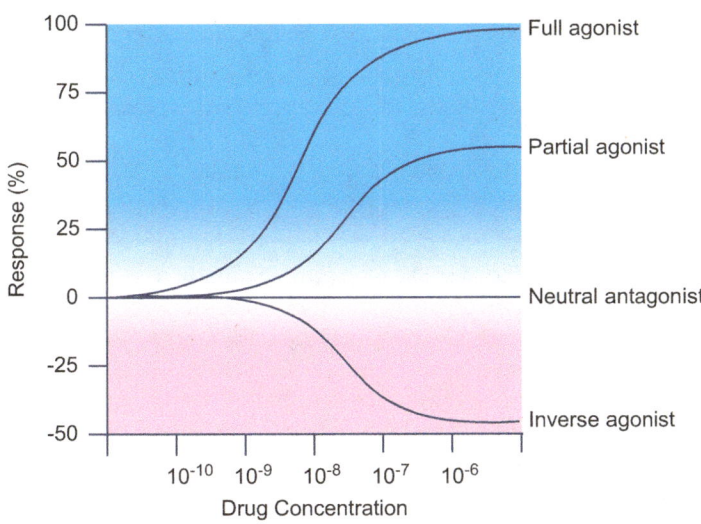

Figure 2.11 Conceptual characterization of the actions of various agonists and antagonists. Full and partial agonists elicit dose-dependent graded responses. An inverse agonist binds to the same receptor as does an agonist but induces the opposite response. A prerequisite for a response to an inverse agonist is that the receptor has constitutive (intrinsic or basal) activity in the absence of a ligand. A neutral antagonist has no activity in the absence of an agonist or inverse agonist but can block the activity of either. By definition, the efficacy of a full agonist is 100%, a neutral antagonist has 0% efficacy, and an inverse agonist has <0% (i.e., negative) efficacy.

2.4.2 D2R-altering drugs

Drugs that target D2R constitute the largest category of therapeutic dopaminergic agents. D2R agonists are used to treat Parkinson's disease, drug abuse, sexual dysfunction, and restless leg syndrome, while D2R antagonists have been primarily employed in the treatment of different psychoses. The DA hypothesis of schizophrenia, formulated over 60 years ago, stipulated that positive symptoms of the disease (e.g., disorganized thoughts, delusion, and hallucination) are caused by hyperactivity of certain brain dopaminergic systems [64]. The serendipitous discovery that D2R antagonists such as chlorpromazine and haloperidol ameliorated some of the positive symptoms of psychosis, initiated an era of neuroleptic ("tranquilizing") drug discovery [65]. However, it was soon realized that the early neuroleptics, which were later renamed "first generation antipsychotics," caused severe side effects such as Parkinsonism, tardive dyskinesia, and elevated serum PRL levels, all of which are typical of D2R blockade. These findings provided strong incentives to search for drugs with lesser extrapyramidal side effects and culminated in the development of the second generation or "atypical antipsychotics" (AAPs) medications [66].

2.4.3 Atypical antipsychotics

The last two to three decades have witnessed an explosive use of AAPs, albeit not always fully justified, for a growing list of mental disorders: schizophrenia, bipolar disorder, mania, attention deficit disorder, major depression, posttraumatic stress disorder, and autism [67]. Imaging methods such as PET and SPECT showed that the density and activity of D2R are often increased in the basal ganglia and prefrontal cortex of patients with schizophrenia and substance abuse disorders, providing a more concrete support for the continuous use of D2R blocking agents in certain patient subpopulations. Similar to the early neuroleptics, all AAPs antagonize D2R (and other DARs with lesser affinity), but they also bind at variable affinities to a myriad of serotonergic, adrenergic, muscarinic, and histaminergic receptors [68,69], as listed in **Table 2.4**.

Although AAPs have made valuable improvements in the cognitive functions of psychoses, the overall treatment efficacy has been less than satisfactory, with unmet needs for developing drugs that ameliorate impairments in working memory, attention, and social cognition. Totally unexpected, and of considerable concern, were the disturbing findings that many of the most commonly prescribed AAPs caused substantial weight gain and serious metabolic disorders such as diabetes, hyperlipidemia and hypertension [70,71]. Given the presence of DARs in peripheral organs that regulate metabolism, the AAP-induced metabolic dysfunctions likely occur as a result of a combined central and peripheral actions, as discussed in greater detail in **Chapter 8**.

2.4.4 Drug selectivity

The development, structure, pharmacology, and therapeutic efficacy of selective drugs for DAR subtypes have been covered in several extensive reviews [61,72]. In spite of a certain overlap in the binding affinities of various drugs to the three D2R-like, some antagonists, mostly derived from a piperazine structure, show higher selectivity for D3R or D4R over D2R. Nonetheless, the development of potent and highly selective D3R or D4R agonists has been less successful. Few promising D3R agonists include tetralin analogs, while effective D4R agonists include modified acetamides and benzamides.

Given the high homology in the binding pocket of D1R and D5R, currently available drugs do not discriminate well between the two receptors. Introduced more than 20 years ago, SCH-23390 has served as a prototypical D1R selective antagonist, while other antagonists, representing derivatives of divergent chemicals, have been coming through the pipeline [61]. Some D1R agonists, e.g., SKF83959, function as full agonists for D1R, but as partial agonists at D1R/D2R heterodimers, underlying the difficulty in assigning a definitive ligand selectivity when heterodimers are involved.

Table 2.4 Receptors targeted by typical and atypical antipsychotics

Receptor subfamily	Location	Action	Therapeutic potential
Central			
D1, D2 (agonism)	Substantia nigra/striatum	Motor control	Parkinson's disease
D1, D2 (antagonism)	Limbic cortex and associated structure	Information processing	Schizophrenia
D2 (agonism)	Anterior pituitary	Inhibition of PRL release	Hyperprolactinemia
Peripheral			
D1 (agonism)	Blood vessels	Vasodilation	Congestive heart failure
D1 (agonism)	Proximal renal tubules	Natriuretic	Heart failure
D2 (agonism)	Sympathetic nerve endings	Decrease release	Hypertension

2.4.5 Peripheral dopaminergic altering drugs

As illustrated in **Figure 2.12**, typical brain capillaries differ from most peripheral capillaries by having tight junctions between the endothelial cells that line the vessel's lumen, thereby forming the BBB. Regions within the brain which contain fenestrated capillaries and therefore lack the BBB are the area postrema, the median eminence of the hypothalamus, the posterior pituitary gland, and the pineal gland. The BBB excludes from the brain 100% of large-molecule neuro-therapeutics, and over 98% of all small-molecule drugs [73]. Small molecules cross the BBB in pharmacologically significant amounts if their molecular mass is less than 400–500 Da and the drug forms less than 8–10 hydrogen bonds with solvent water. It is surprising, therefore, that almost all of the clinically useful DAR altering drugs penetrate the brain, and only a very few, categorized as peripherally active drugs, do not.

Two peripheral D2R antagonists that do not penetrate the brain are **domperidone** [74] and **metoclopramide** [75]. Both are considered prokinetic drugs because they are defined by their ability to increase GI motility and have been used in clinical practice to treat several peripheral DA-associated disorders. Domperidone, which is chemically derived from butyrophenones, is a selective D2R and D3R antagonist with no significant interactions with D1R. It regulates the motility of gastric and small intestinal smooth muscle, as well as esophageal motor functions. Domperidone also has antiemetic activity because it blocks the DARs in the chemoreceptor trigger zone. The most severe adverse effect of domperidone is a prolongation of cardiac QT interval, which increases the risk of life-threatening arrhythmias.

Metoclopramide, a derivative of benzamides, binds to D2R at nanomolar affinity, has lower binding affinity to D1R, and also acts as a mixed 5-HT3R antagonist/5-HT5R agonist. Metoclopramide is used to treat nausea and vomiting that are associated with uremia, migraine headache, radiation sickness, effects of chemotherapy, labor, infection, and emetogenic drugs. Adverse side effects of metoclopramide include akathisia (restlessness), and focal dystonia, reflecting some blockade of central DARs.

Fenoldopam, a benzazepine derivative, is a peripheral D1R agonist. Administered parenterally, fenoldopam acts as a vasodilator in the peripheral arteries, and as a diuretic in the kidneys. Fenoldopam has been approved by the FDA in 1997 for in-hospital, short-term management of severe

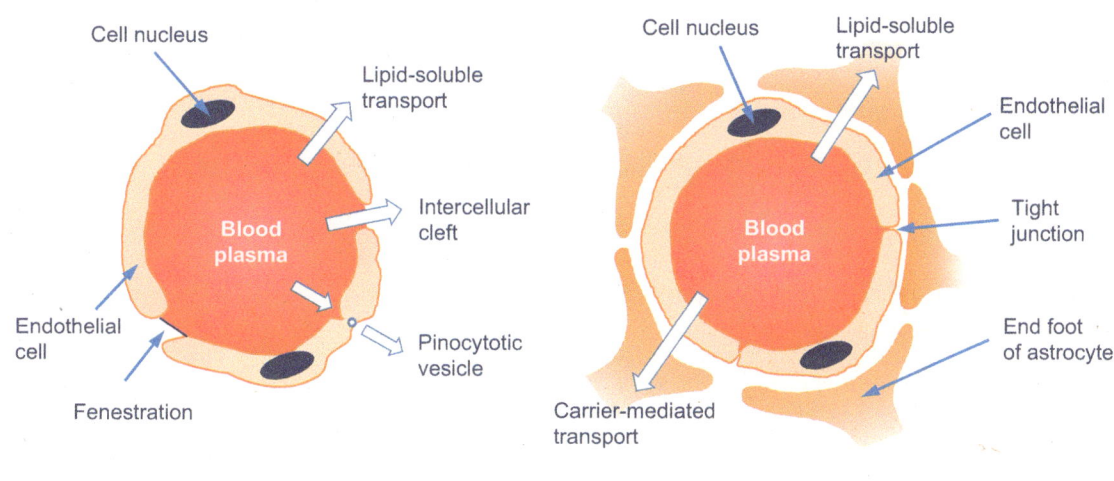

Peripheral capillary   **Brain capillary**

Figure 2.12 Structural and functional differences between peripheral and brain capillaries. Endothelial cells that line the lumen of peripheral capillaries have fenestrations and intercellular clefts that enable the passage of many types of molecules from the circulation to surrounding tissues. The endothelial cells in brain capillaries have tight junctions, thereby forming the blood brain barrier. Various molecules can still get entry to the brain through lipid soluble transport or carrier-mediated transport.

hypertension, when a rapid, but a reversible, reduction of blood pressure is required [76]. Hypertensive emergencies that use fenoldopam include accelerated hypertension, hypertensive encephalopathy, acute left ventricular failure, acute aortic dissection, pheochromocytoma crisis, interaction between tyramine-containing foods or drugs and monoamine oxidase inhibitors, eclampsia, drug-induced hypertension, and occasionally intracranial hemorrhage.

In contrast to DA, fenoldopam has no effect on beta AR, although it has some alpha-1 and alpha-2 AR antagonist activity. Adverse effects of fenoldopam include headache, nausea, hypotension, reflex tachycardia, and increased intraocular pressure. In hypertensive patients, fenoldopam rapidly decreased blood pressure, increased renal blood flow, and maintained or improved the glomerular filtration rate. In normotensive volunteers, fenoldopam increased renal blood flow without significant effects on systemic blood pressure or heart rate. Of great interest is our recent report [21] that fenoldopam, acting via the sGC/cGMP pathway, was a potent inducer of apoptosis in cultured breast cancer cells, and a robust suppressor of growth of breast cancer xenografts in immune-deficient mice.

2.5 DOPAMINE RECEPTOR KNOCKOUT MICE

Several lines of transgenic mice, each lacking a specific DAR subtype, were generated during the 1990s. The expectation was that gene deletion will ultimately clarify the functional specificity of each DAR because pharmacological approaches have been hampered by lack of highly selective agonists and antagonists. However, as discussed in a comprehensive review in 2004 [77], the phenotypic analysis of various DAR KO mice provided inconsistent conclusions. Some data reinforced the prevailing dogmas on the functions of the DAR subtypes, while others were more controversial. The latter was attributed to two main reasons. First, some of the variability of reported data were due to differences among laboratories in the use of mice with diverse genetic backgrounds, as well as to a lack of standardization of behavioral testing methods. Second, because of brain plasticity during embryonic development, the total deletion of a specific DAR can induce compensatory mechanisms, which obscure its normal functions throughout postnatal life. Site-specific and/or time-dependent conditional gene inactivation could overcome some of these deficiencies.

Knockout of any one DAR subtype is compatible with life, while differentially displaying motor, cognition, reward, and other behavioral dysfunctions. By far, most of the studies with DAR-deficient mice focused on the potential neurological disturbances. Here we review some of the studies that describe the phenotypes of D1R- and D2R-knockedout mice which have also examined the consequences of receptor deficiency on peripheral DA targets.

2.5.1 Phenotypes of individual DAR-KO mice

D1R KO mice are characterized by growth retardation and low survival. These were attributed to a loss of fine motor control in food seeking because such malfunctions could be corrected once the mutant mice were provided with an easy access to palatable food [78]. D1R KO mice had higher blood pressure than wild-type mice, which was ascribed to a defect in sodium transport in the proximal renal tubules [79]. D5R KO mice are viable and healthy, and do not display the growth retardation seen in D1R KO mice [80]. They also develop hypertension, which appeared to be caused by increased sympathetic tone due to a CNS defect.

Using the D5R KO mice, another group found overexpression of sodium transporters and sodium channels in distal renal nephrons, implicating a critical role for D5R in renal salt homeostasis [81]. The same group also reported that D3R KO mice, which otherwise appeared normal, also developed renal hypertension. Such hypertension was minimized by treating the mice with a D1R agonist, underscoring the complex interactions among the DAR subtypes in the kidney. D2R KO mice also have high blood pressure and susceptibility to sodium-related hypertension [82]. Collectively, studies with the various DAR KO mice revealed that DARs play critical roles in renal functions [83], as is discussed in more detail in **Chapter 7**.

The reports on DAR expression on dendritic cells (DCs), which play fundamental roles in the induction of the antitumor adaptive immune response, prompted Figueroa et al. [84] to use the D3R KO mice to determine the regulation of DCs functions by DA. Their data showed that D3R-deficiency in DCs enhanced the expansion of cytotoxic T-lymphocytes (CTLs) *in vivo* and induced a stronger antitumor immunity. Coculture experiments showed that D3R-deleted bone marrow-derived DCs potentiated antigen cross-presentation and CTLs activation. These findings implicated D3R as a potential new therapeutic target for strengthening antitumor immunity.

2.5.2 D2R deletion and pituitary functions

Extensive *in vitro* and *in vivo* studies have established that DA, acting primarily via D2R, has a strong impact on multiple functions of the pituitary gland [27,28]. The effects of D2R deletion on pituitary functions have been covered in an extensive review by Cristina et al. [85]. Briefly, such mice have chronic hyperprolactinemia, which was more pronounced in females than in males; a moderate decrease in αMSH (melanocyte stimulating hormone) content; and some growth retardation, presumably due to suppression of the GH-IGF (growth hormone-insulin-like growth factor) axis. Within the pituitary, the lactotrophs had altered structure, and the number of somatotrophs, gonadotrophs and thyrotrophs was moderately reduced. Females with D2R deletion developed pituitary hyperplasia at 8 months of age, and highly vascularized adenomas by 16 months. The increase in angiogenesis was attributed to an overexpression of vascular endothelial growth factor (VEGF) in follicular stellate cells by an unidentified paracrine factor from the lactotrophs. Another study, using prolactinomas from D2R KO mice, reported increased number of cells defined as "side populations," which represent tumor stem cells [86], leading the authors to suggest that such cells are the precursors of pituitary tumor pathogenesis.

2.6 SYNOPSIS

This chapter underscores the difficulties in predicting the overall actions of dopamine at any given cellular target. This complexity results from the facts that (1) dopamine can activate five different receptors that often act in opposite directions, with the ratio of receptor expression in each cell likely affecting the overall balance of the response; (2) the receptors can homo-or heterodimerize and, consequently, become linked to a variety of G proteins that activate a multitude of signaling pathways; (3) the responses to DA can be amplified as well as altered by G protein-associated kinases that also have receptor-independent actions; and (4) the response to receptor activation can change because of time-dependent receptor desensitization. Drugs that target the DARs have many therapeutic applications in terms of brain-associated motor control and information processing disorders, and also for selected endocrine, cardiovascular and renal disorders. DAR KO mice are viable and fertile but have cognitive and movement dysfunctions as well as some peripheral deficiencies that include hypertension, immune alterations, altered pituitary functions, and mild growth disturbances.

REFERENCES

1. Limbird LE. The receptor concept: A continuing evolution. *Mol Interv.* 2004;4(6):326–336.

2. Kebabian JW, Petzold GL, Greengard P. Dopamine-sensitive adenylate cyclase in caudate nucleus of rat brain, and its similarity to the "dopamine receptor." *Proc Natl Acad Sci USA.* 1972;69(8):2145–2149.

3. Missale C, Nash SR, Robinson SW, Jaber M, Caron MG. Dopamine receptors: From structure to function. *Physiol Rev.* 1998;78(1):189–225.

4. Park JY, Lee SY, Kim HR, Seo MD, Chung KY. Structural mechanism of GPCR-arrestin interaction: Recent breakthroughs. *Arch Pharm Res.* 2016;39(3):293–301.

5. Chung KY. Structural aspects of GPCR-G protein coupling. *Toxicol Res.* 2013;29(3):149–155.

6. Schioth HB, Fredriksson R. The GRAFS classification system of G-protein coupled receptors in comparative perspective. *Gen Comp Endocrinol.* 2005;142(1–2):94–101.

7. Gainetdinov RR, Premont RT, Bohn LM, Lefkowitz RJ, Caron MG. Desensitization of G protein-coupled receptors and neuronal functions. *Annu Rev Neurosci.* 2004;27:107–144.

8. Sprang SR. Invited review: Activation of G proteins by GTP and the mechanism of Galpha-catalyzed GTP hydrolysis. *Biopolymers.* 2016;105(8):449–462.

9. Beaulieu JM, Gainetdinov RR. The physiology, signaling, and pharmacology of dopamine receptors. *Pharmacol Rev.* 2011;63(1):182–217.

10. Bologna Z, Teoh JP, Bayoumi AS, Tang Y, Kim IM. Biased G protein-coupled receptor signaling: New player in modulating physiology and pathology. *Biomol Ther.* 2017;25(1):12–25.

11. D'Souza UM. Gene and promoter structures of the dopamine receptors. in *The Dopamine Receptors*, Neve K (ed.), Springer, New York, 2010; pp. 23–46.

12. Shi S, et al. MicroRNA-9 and microRNA-326 regulate human dopamine D2 receptor expression, and the microRNA-mediated expression regulation is altered by a genetic variant. *J Biol Chem.* 2014;289(19):13434–13444.

13. Tobon KE, Chang D, Kuzhikandathil EV. MicroRNA 142-3p mediates post-transcriptional regulation of D1 dopamine receptor expression. *PLoS One.* 2012;7(11):e49288.

14. Ng GY, et al. Desensitization, phosphorylation and palmitoylation of the human dopamine D1 receptor. *Eur J Pharmacol.* 1994;267(1):7–19.

15. Karpa KD, Lidow MS, Pickering MT, Levenson R, Bergson C. N-linked glycosylation is required for plasma membrane localization of D5, but not D1, dopamine receptors in transfected mammalian cells. *Mol Pharmacol.* 1999;56(5):1071–1078.

16. Ledonne A, Mercuri NB. Current concepts on the physiopathological relevance of dopaminergic receptors. *Front Cell Neurosci.* 2017;11:27–36.

17. Wong AH, Buckle CE, Van Tol HH. Polymorphisms in dopamine receptors: What do they tell us? *Eur J Pharmacol.* 2000;410(2–3):183–203.

18. Civelli O, Bunzow JR, Grandy DK. Molecular diversity of the dopamine receptors. *Annu Rev Pharmacol Toxicol.* 1993;33:281–307.

19. Tiberi M, Caron MG. High agonist-independent activity is a distinguishing feature of the dopamine D1B receptor subtype. *J Biol Chem.* 1994;269(45):27925–27931.

20. Borcherding DC, et al. Dopamine receptors in human adipocytes: Expression and functions. *PLoS One.* 2011;6(9):e25537.

21. Borcherding DC, et al. Expression and therapeutic targeting of dopamine receptor-1 (D1R) in breast cancer. *Oncogene.* 2016;35(24):3103–3113.

22. Brichta L, Greengard P, Flajolet M. Advances in the pharmacological treatment of Parkinson's disease: Targeting neurotransmitter systems. *Trends Neurosci.* 2013;36(9):543–554.

23. Glickstein SB, Schmauss C. Dopamine receptor functions: Lessons from knockout mice [corrected]. *Pharmacol Ther.* 2001;91(1):63–83.

24. Kumar U, Patel SC. Immunohistochemical localization of dopamine receptor subtypes (D1R-D5R) in Alzheimer's disease brain. *Brain Res.* 2007;1131(1):187–196.

25. Moraga-Amaro R, et al. Dopamine receptor D5 deficiency results in a selective reduction of hippocampal NMDA receptor subunit NR2B expression and impaired memory. *Neuropharmacology.* 2016;103:222–235.

26. Radl D, et al. Differential regulation of striatal motor behavior and related cellular responses by dopamine D2L and D2S isoforms. *Proc Natl Acad Sci USA.* 2018;115(1):198–203.

27. Ben-Jonathan N. Dopamine: A prolactin-inhibiting hormone. *Endocr Rev.* 1985;6(4):564–589.

28. Ben-Jonathan N, Hnasko R. Dopamine as a prolactin (PRL) inhibitor. *Endocr Rev.* 2001;22(6):724–763.

29. De Souza EB. Serotonin and dopamine receptors in the rat pituitary gland: Autoradiographic identification, characterization, and localization. *Endocrinology.* 1986;119(4):1534–1542.

30. Shafer RA, Levant B. The D3 dopamine receptor in cellular and organismal function. *Psychopharmacology.* 1998;135(1):1–16.

31. Moritz AE, Free RB, Sibley DR. Advances and challenges in the search for D2 and D3 dopamine receptor-selective compounds. *Cell Signal.* 2018;41:75–81.

32. Vallone D, Picetti R, Borrelli E. Structure and function of dopamine receptors. *Neurosci Biobehav Rev.* 2000;24(1):125–132.

33. Missale C, Fiorentini C, Collo G, Spano P. The neurobiology of dopamine receptors: Evolution from the dual concept to heterodimer complexes. *J Recept Signal Transduct Res.* 2010;30(5):347–354.

34. Fuxe K, et al. Diversity and bias through receptor-receptor interactions in GPCR heteroreceptor complexes: Focus on examples from dopamine D2 receptor heteromerization. *Front Endocrinol.* 2014;5:1–10.

35. George SR, Kern A, Smith RG, Franco R. Dopamine receptor heteromeric complexes and their emerging functions. *Prog Brain Res.* 2014;211:183–200.

36. Lukasiewicz S, Faron-Gorecka A, Dobrucki J, Polit A, Dziedzicka-Wasylewska M. Studies on the role of the receptor protein motifs possibly involved in electrostatic interactions on the dopamine D1 and D2 receptor oligomerization. *FEBS J.* 2009;276(3):760–775.

37. Ferguson SS. Evolving concepts in G protein-coupled receptor endocytosis: The role in receptor desensitization and signaling. *Pharmacol Rev.*2001;53(1):1–24.

38. Kelly E, Bailey CP, Henderson G. Agonist-selective mechanisms of GPCR desensitization. *Br J Pharmacol.* 2008;153(Suppl 1):S379–S388.

39. Kliewer A, Reinscheid RK, Schulz S. Emerging paradigms of G protein-coupled receptor dephosphorylation. *Trends Pharmacol Sci.*2017;38(7):621–636.

40. Benovic JL, et al. Phosphorylation of the mammalian beta-adrenergic receptor by cyclic AMP-dependent protein kinase: Regulation of the rate of receptor phosphorylation and dephosphorylation by agonist occupancy and effects on coupling of the receptor to the stimulatory guanine nucleotide regulatory protein. *J Biol Chem.* 1985;260(11):7094–7101.

41. Evron T, Daigle TL, Caron MG. GRK2: Multiple roles beyond G protein-coupled receptor desensitization. *Trends Pharmacol Sci.* 2012;33(3):154–164.

42. Armando I, Konkalmatt P, Felder RA, Jose PA. The renal dopaminergic system: Novel diagnostic and therapeutic approaches in hypertension and kidney disease. *Transl Res.* 2015;165(4):505–511.

43. Reiter E, et al. Beta-arrestin signalling and bias in hormone-responsive GPCRs. *Mol Cell Endocrinol.* 2017;449:28–41.

44. Zhang B, Albaker A, Plouffe B, Lefebvre C, Tiberi M. Constitutive activities and inverse agonism in dopamine receptors. *Adv Pharmacol.* 2014;70:175–214.

45. Beaulieu JM, Espinoza S, Gainetdinov RR. Dopamine receptors: IUPHAR Review 13. *Br J Pharmacol.* 2015;172(1):1–23.

46. Boyd KN, Mailman RB. Dopamine receptor signaling and current and future antipsychotic drugs. *Handb Exp Pharmacol.* 2012(212):53–86.

47. Urban JD, et al. Functional selectivity and classical concepts of quantitative pharmacology. *J Pharmacol Exp Ther.* 2007;320(1):1–13.

48. Sjogren B. The evolution of regulators of G protein signalling proteins as drug targets: 20 years in the making—IUPHAR Review 21. *Br J Pharmacol.* 2017;174(6):427–437.

49. Sadana R, Dessauer CW. Physiological roles for G protein-regulated adenylyl cyclase isoforms: Insights from knockout and overexpression studies. *Neurosignals.* 2009;17(1):5–22.

50. Shaywitz AJ, Greenberg ME. CREB: A stimulus-induced transcription factor activated by a diverse array of extracellular signals. *Annu Rev Biochem.* 1999;68:821–861.

51. Svenningsson P, et al. DARPP-32: An integrator of neurotransmission. *Annu Rev Pharmacol Toxicol.* 2004;44:269–296.

52. Zamponi GW, Snutch TP. Decay of prepulse facilitation of N type calcium channels during G protein inhibition is consistent with binding of a single Gbeta subunit. *Proc Natl Acad Sci USA.* 1998;95(7):4035–4039.

53. Peng L, Mirshahi T, Zhang H, Hirsch JP, Logothetis DE. Critical determinants of the G protein gamma subunits in the Gbetagamma stimulation of G protein-activated inwardly rectifying potassium (GIRK) channel activity. *J Biol Chem.* 2003;278(50):50203–50211.

54. Windham PF, Tinsley HN. cGMP signaling as a target for the prevention and treatment of breast cancer. *Semin Cancer Biol.* 2015;31C:106–110.

55. Pyriochou A, Papapetropoulos A. Soluble guanylyl cyclase: More secrets revealed. *Cell Signal.* 2005;17(4):407–413.

56. Sharma RK, Duda T. Membrane guanylate cyclase, a multimodal transduction machine: History, present, and future directions. *Front Mol Neurosci.* 2014;7:56–68.

57. Azevedo MF, et al. Clinical and molecular genetics of the phosphodiesterases (PDEs). *Endocr Rev.* 2014;35(2):195–233.

58. Kouvelas D, Goulas A, Papazisis G, Sardeli C, Pourzitaki C. PDE5 inhibitors: *In vitro* and *in vivo* pharmacological profile. *Curr Pharm Des.* 2009;15(30):3464–3475.

59. Califano JA, et al. Tadalafil augments tumor specific immunity in patients with head and neck squamous cell carcinoma. *Clin Cancer Res.* 2015;21(1):30–38.

60. Wolfertstetter S, Huettner JP, Schlossmann J. cGMP-dependent protein kinase inhibitors in health and disease. *Pharmaceuticals.* 2013;6(2):269–286.

61. Zhang A, Neumeyer JL, Baldessarini RJ. Recent progress in development of dopamine receptor subtype-selective agents: potential therapeutics for neurological and psychiatric disorders. *Chem Rev.* 2007;107(1):274–302.

62. Scatena R, et al. An update on pharmacological approaches to neurodegenerative diseases. *Expert Opin Investig Drugs.* 2007;16(1):59–72.

63. Strange PG. Antipsychotic drug action: Antagonism, inverse agonism or partial agonism. *Trends Pharmacol Sci.* 2008;29(6):314–321.

64. Baumeister AA, Francis JL. Historical development of the dopamine hypothesis of schizophrenia. *J Hist Neurosci.* 2002;11(3):265–277.

65. Shen WW. A history of antipsychotic drug development. *Compr Psychiatry.* 1999;40(6):407–414.

66. Kapur S, Mamo D. Half a century of antipsychotics and still a central role for dopamine D2 receptors. *Prog Neuropsychopharmacol Biol Psychiatry.* 2003;27(7):1081–1090.

67. Farah A. Atypicality of atypical antipsychotics. *Prim Care Companion J Clin Psychiatry.* 2005;7(6):268–274.

68. Richtand NM, et al. Dopamine and serotonin receptor binding and antipsychotic efficacy. *Neuropsychopharmacology.* 2007;32(8):1715–1726.

69. Seeman P. Atypical antipsychotics: Mechanism of action. *Can J Psychiatry.* 2002;47(1):27–38.

70. Coccurello R, Moles A. Potential mechanisms of atypical antipsychotic-induced metabolic derangement: Clues for understanding obesity and novel drug design. *Pharmacol Ther.* 2010;127(3):210–251.

71. Nasrallah HA. Atypical antipsychotic-induced metabolic side effects: Insights from receptor-binding profiles. *Mol Psychiatry.* 2008;13(1):27–35.

72. Li P, Snyder GL, Vanover KE. Dopamine targeting drugs for the treatment of schizophrenia: Past, present and future. *Curr Top Med Chem.* 2016;16(29):3385–3403.

73. Pardridge WM. The blood-brain barrier: Bottleneck in brain drug development. *NeuroRx.* 2005;2(1):3–14.

74. Reddymasu SC, Soykan I, McCallum RW. Domperidone: Review of pharmacology and clinical applications in gastroenterology. *Am J Gastroenterol.* 2007;102(9):2036–2045.

75. Parkman HP, et al. Clinical response and side effects of metoclopramide: Associations with clinical, demographic, and pharmacogenetic parameters. *J Clin Gastroenterol.* 2012;46(6):494–503.

76. Murphy MB, Murray C, Shorten GD. Fenoldopam: A selective peripheral dopamine-receptor agonist for the treatment of severe hypertension. *N Engl J Med.* 2001;345(21):1548–1557.

77. Holmes A, Lachowicz JE, Sibley DR. Phenotypic analysis of dopamine receptor knockout mice; recent insights into the functional specificity of dopamine receptor subtypes. *Neuropharmacology.* 2004;47(8):1117–1134.

78. Drago J, et al. Altered striatal function in a mutant mouse lacking D1A dopamine receptors. *Proc Natl Acad Sci USA.* 1994;91(26):12564–12568.

79. Albrecht FE, et al. Role of the D1A dopamine receptor in the pathogenesis of genetic hypertension. *J Clin Invest.* 1996;97(10):2283–2288.

80. Yang Z, Sibley DR, Jose PA. D5 dopamine receptor knockout mice and hypertension. *J Recept Signal Transduct Res.* 2004;24(3):149–164.

81. Wang X, et al. Upregulation of renal D5 dopamine receptor ameliorates the hypertension in D3 dopamine receptor-deficient mice. *Hypertension.* 2013;62(2):295–301.

82. Ueda A, et al. Disruption of the type 2 dopamine receptor gene causes a sodium-dependent increase in blood pressure in mice. *Am J Hypertens.* 2003;16(10):853–858.

83. Zeng C, et al. Dysregulation of dopamine-dependent mechanisms as a determinant of hypertension: Studies in dopamine receptor knockout mice. *Am J Physiol Heart Circ Physiol.* 2008;294(2):H551–H569.

84. Figueroa C, et al. Inhibition of dopamine receptor D3 signaling in dendritic cells increases antigen cross-presentation to CD8+ T-cells favoring anti-tumor immunity. *J Neuroimmunol.* 2017;303:99–107.

85. Cristina C, et al. Dopaminergic D2 receptor knockout mouse: An animal model of prolactinoma. *Front Horm Res.* 2006;35:50–63.

86. Mertens F, et al. Pituitary tumors contain a side population with tumor stem cell-associated characteristics. *Endocr Relat Cancer.* 2015;22(4):481–504.

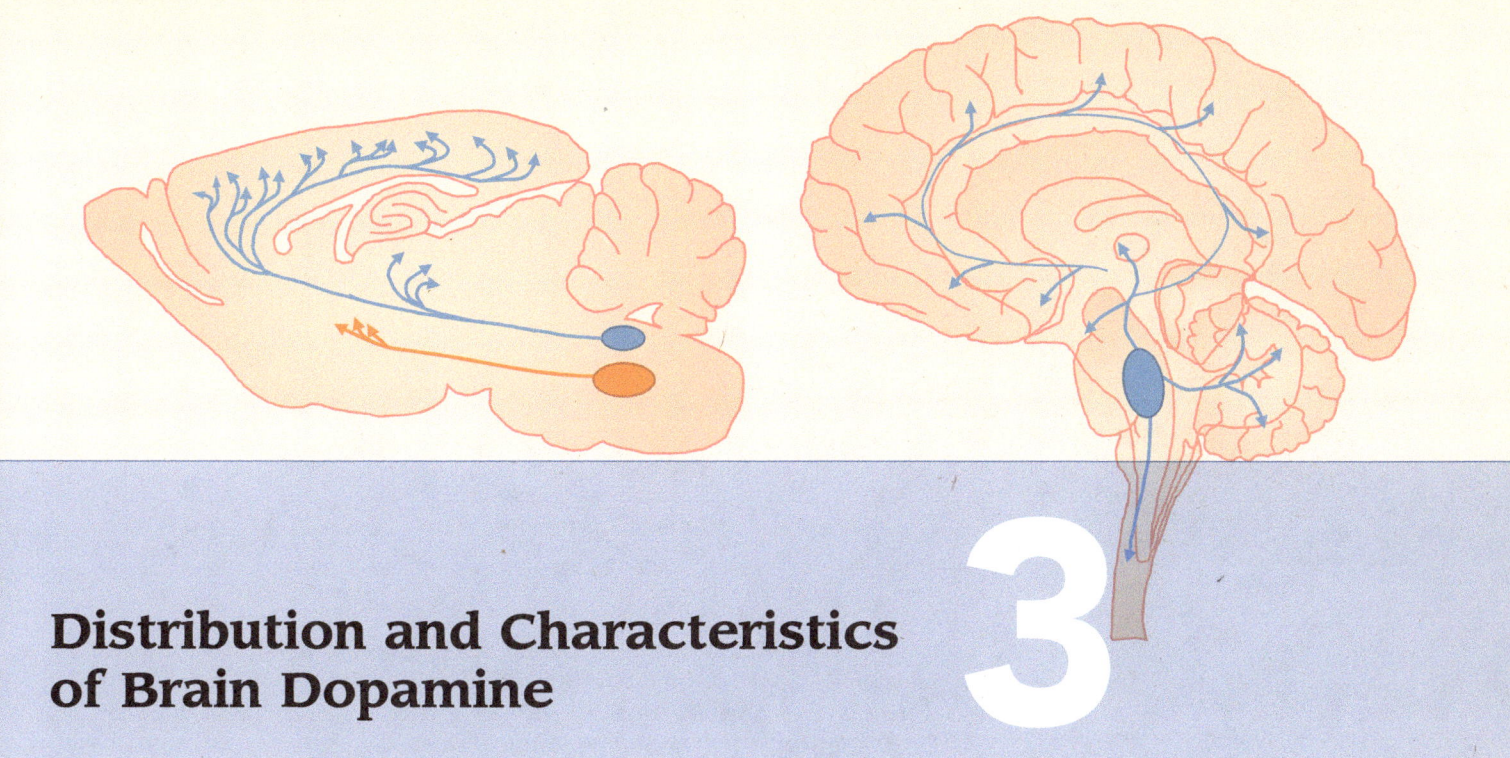

Distribution and Characteristics of Brain Dopamine

3

3.1 INTRODUCTION

When judged by abundance or by quantity alone, the brain dopaminergic neurons are not very impressive. Compared with glutamate, which is excitatory in >90% of all synapses in the human brain, or with gamma-aminobutyric acid (GABA), which is inhibitory at about 90% of the synapses that do not use glutamate, the number of dopaminergic synapses amount to less than 0.001% of the total synapses within the human brain. By some estimates, there are 400,000–600,000 dopamine (DA)-producing neurons in the human brain, constituting a miniscule fraction of the 100 billion neurons, and many more neuroglia, which populate the brain. Thus, the importance of the brain dopaminergic systems does not reside in their sheer number, or in simple excitatory or inhibitory synaptic activity, but in their critical association with a broad range of neurological functions and the wide variety of disorders that result from their dysfunction.

Given the prominence of DA as a neurotransmitter, it is not surprising that a substantial body of the research and clinical literature on the brain has been focused on DA. **Figure 3.1** presents the number of research/clinical publications, listed in PubMed with DA in their title, which have been published per decade since the 1950s. As of to-date, there are over 47,000 publications with a focus on DA, amounting to about 1,000 publications on DA each year since 2000. Moreover, several books dedicated to DA have been published in the last 20–30 years.

In spite of the very large database on DA, the endocrine functions of brain DA have received only a sporadic and an incomplete coverage. The objective of this chapter is to outline the distribution, structure and general features of the brain dopaminergic neurons, while emphasizing those areas of the brain and adjacent glands (i.e., pituitary and pineal) that are associated with the endocrine functions of brain DA. The information in this chapter serves as a background for **Chapter 4**, which focuses on the multiple endocrine functions of brain DA.

The first section provides historical perspectives on the discovery of the catecholamines and the increased knowledge on the anatomical distribution and projections of noradrenergic and adrenergic neurons. This is

Figure 3.1 The number of research/clinical publications with a focus on dopamine. Publications listed in PubMed with "dopamine" in their title that have been published per decade since the discovery of DA in the 1950s.

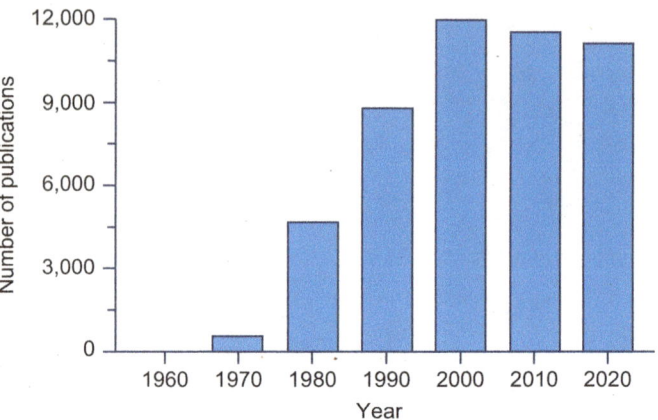

followed by a detailed analysis of the localization, distribution, and specific features of dopaminergic neurons. We then review the structure and major functions of the hypothalamus, with a focus on the location, functions, and DA content of the hypothalamic nuclei that are primarily associated with neuroendocrine regulation. Finally, we describe the neural and vascular connections between the hypothalamus and the pituitary gland.

3.2 NORADRENERGIC AND ADRENERGIC NEURONS

3.2.1 Historical perspectives

The catecholamines were discovered in a reverse order of their position in the biosynthetic pathway: L-Tyrosine → L-Dopa → Dopamine → Norepinephrine → Epinephrine. The order of discoveries was presumably due to the early recognition of the biological prominence of the catecholamines in different organs, together with the relative ease of experimental manipulations with these organs at the times of their discovery: epinephrine (Epi) or adrenaline, as a major product of the adrenal gland, norepinephrine (NE) or noradrenaline, as a principle neurotransmitter in the sympathetic nervous system, and DA as the dominant catecholamine in the brain.

Studies that began in the 1890s and continued into the 1930s identified vasoactive substances in extracts of the adrenal gland, using the frog heart and rodent blood pressure as experimental models. Once their physiological functions were recognized, it did not take long before the isolation, purification, and eventual chemical synthesis of both Epi and NE. On the other hand, DA was recognized as an independent neurotransmitter only in the 1950s.

By the early 1960s, methods such as fluorimetry and gas chromatography were successfully employed to determine the concentrations of the three catecholamines in extracts of various brain regions. These analyses, in combination with observations on the effects of drugs such as reserpine (which depletes catecholamines), or phenothiazines (which block catecholamine actions), were instrumental in documenting the global distribution of the catecholamines in different brain areas. Thereafter, mapping of the catecholaminergic neurons within the brain progressed in tandem with the establishment of histochemical methods for their visualization and the continuous improvements in both resolution and specificity of these methods [1].

The first breakthrough for mapping the biogenic amine neurons came with the development of the formaldehyde histofluorescence technique. This method was based on the finding that exposure of brain tissue sections to formaldehyde vapor induced the cyclization of both catecholamines and indoleamines and formation of intense fluorescent products.

The introduction of the more sensitive glyoxylic acid histochemistry, and subsequent application of antisera against catecholamine biosynthetic enzymes in immunocytochemical preparations, paved the way for a better distinction between the different catecholamines and a more accurate neuronal mapping. Once the dopamine transporter (DAT) was found to be highly specific for dopaminergic neurons, this property was exploited in immunocytochemical applications for increased specificity of the distribution of dopaminergic pathways throughout the brain.

In the 1960s, Dahlstrom and Fuxe published the first detailed map of catecholaminergic and serotonergic cell bodies in the rat brain (reviewed by [1]). Moving from the caudal to the rostral locations in the brain stem, they identified 12 catecholamine cell groups, designated **A1 to A12**, and 9 indoleamine cell groups, designated B1 to B9. In subsequent years, additional DA cell groups were identified more rostrally and were named **A13 to A16**. With the aid of immunohistochemistry for phenylethanolamine *N*-methyltransferase (PNMT; the enzyme that converts NE to Epi), adrenaline cell groups were identified in the caudal brain stem and were designated C1 and C2.

In addition to visualizing the brain catecholamines on microscopic slides, *in vivo* methods such as selective brain lesions, iontophoresis, pharmacological manipulations, anterograde and retrograde fiber tracings, pushpull cannulation, and *in vivo* imaging all contributed in various ways to increased precision of the neuronal mapping. The ultrastructure of the catecholaminergic neurons was subsequently elucidated with the use of electron microscopy.

The rat brain was used for most of the mapping of the catecholaminergic neurons, supported by few studies that have used cats, rabbits, and monkeys. The distribution of the catecholaminergic neurons in these species was generally corroborated upon studying human postmortem pathological specimens. Based on the classification that was established in the rat brain, 18 distinct catecholamine cell groups are now recognized. Of these, 7 groups are noradrenergic (A1–A7), 9 groups are dopaminergic (A8–A16), and 2 are adrenergic (C1–C2), as is illustrated in **Figure 3.2**.

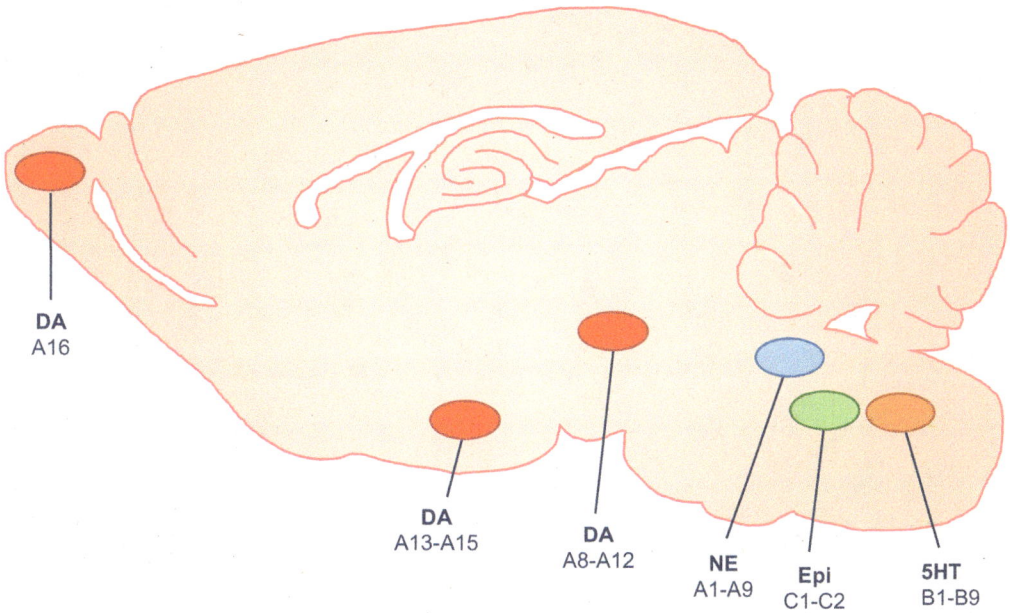

DA
A16

DA
A13-A15

DA
A8-A12

NE
A1-A9

Epi
C1-C2

5HT
B1-B9

Figure 3.2 Locations of biogenic amine neuronal cell bodies in the rat brain. Shown are serotonin (5HT), epinephrine (Epi), norepinephrine (NE), and dopamine (DA) neuronal cell bodies grouped into the A, B, and C cell groups according to the conventional map of catecholaminergic and serotonergic neurons. See text in **Section 3.2** for other explanations.

3.2.2 Distribution and major functions of the brain noradrenergic neurons

Several reviews cover the anatomy and physiology of the brain noradrenergic and adrenergic neurons [2–5], and they are briefly summarized in the next two sections. Noradrenergic neurons are relatively few in number, and their cell bodies are confined to two small clusters within the brain stem: the **locus ceruleus (LC)**, encompassing the A4, A5, and A6 groups, and a more diffuse group, located within the lateral tegmental area (LTA), which encompass A1, A2, and A7 (**Table 3.1**). The A1 and A2 groups have been categorized by some researchers as subceruleus.

The brain stem is located between the spinal cord and the diencephalon and is divided into three major structures: the medulla oblongata (myelencephalon), the pons with the cerebellum (metencephalon), and the midbrain (mesencephalon). Although the pons and cerebellum together constitute the metencephalon, only the pons, but not the cerebellum, is considered as a part of the brainstem. The transition zone between the medulla and the spinal cord is located at the level of the foramen magnum and is also at the level of the pyramidal decussation.

The LC is well conserved across the mammalian species and is the largest group of noradrenergic neurons in the brain. It is a bilateral structure that sits near the wall of the fourth ventricle and the mesencephalic trigeminal nucleus in the pons. The proximity of the LC to the fourth ventricle provides a good access for the noradrenergic neurons to the cerebrospinal fluid (CSF) and also for a potential exposure to regulatory hormones and toxic chemicals. The LC has been most extensively studied in the rat, where it contains about 1,600 medium-sized neurons. It is also known as the pigmented nucleus of the pons because of melanin granules, which give it a blue color under the microscope. The neuromelanin, which is formed by the polymerization of NE, is analogous to the DA-derived black neuromelanin within the substantia nigra.

The body of the LC can be subdivided into three components, populated by neurons of varying morphology: anterior pole, compact core, and posterior pole. The anterior pole comprises large multipolar cells that stain intensely for dopamine beta-hydroxylase (DBH), while the posterior pole mostly contains fusiform neurons, a spindle-like cell shape that is wide in the middle and tapers at both ends. The dorsal portion of the compact core contains densely packed fusiform neurons with smaller somata, while the ventral portion has larger multipolar neurons. The LC receives afferent input, primarily sensory, from several brain regions, including the hypothalamus, cingulate gyrus, and amygdala. The cerebellum and afferents from the raphe nuclei (the origin of the serotonergic neurons) also send projections to the LC.

In spite of the rather small "footprint" of the noradrenergic cell bodies, they send very long and widespread projections to most brain areas, where they exert powerful effects on their targets through alpha- and beta-adrenergic receptors. **Figure 3.3** compares the distribution of the noradrenergic neurons in human and rat brains. Along their length, the noradrenergic neurons have many varicosities that are capable of releasing their product in

Table 3.1 Noradrenergic neurons in the mammalian brain

Cell group	Nucleus of origin	Projections
Locus ceruleus (A4, A5, A6)	Locus coeruleus	Spinal cord, brainstem, cerebellum, hypothalamus, thalamus, basal telencephalon, isocortex
Lateral tegmental area (A1, A2, A7)	Dorsal motor vagus, nucleus tractus solitaries, lateral tegmentum	Spinal cord, brainstem, hypothalamus, basal telencephalon

Source: Moore, R.Y. and Bloom, F.E., *Annu. Rev. Neurosci.*, 2, 113–168, 1979.

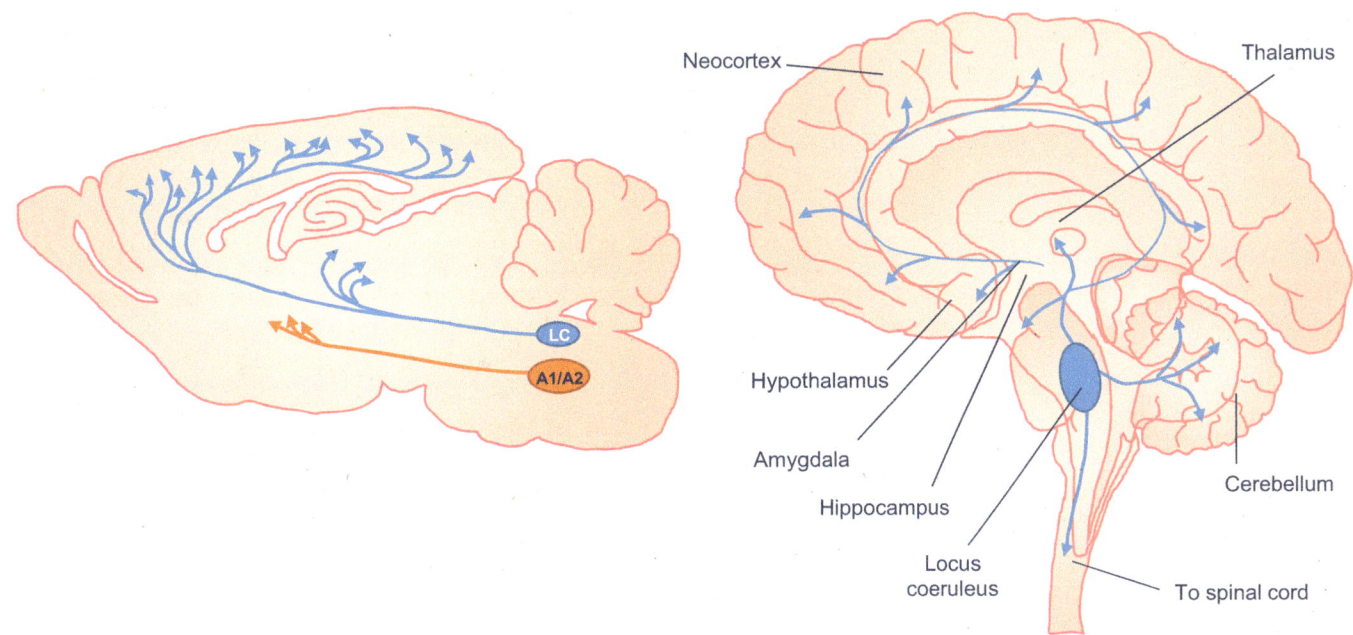

Figure 3.3 Distribution and projections of noradrenergic neurons in the human (right) and rat (left) brains. Both the locus ceruleus (LC) and subceruleus (A1 and A2) are shown in the rat brain. The noradrenergic cell bodies, located in the brain stem, send very long axons to most brain areas as well as to the spinal cord.

the absence of synapses. In fact, only about 5%–10% of the terminals that can be labeled by uptake of [3][H]-NE make typical synaptic contacts. The dorsal noradrenergic bundle, which originates in the LC, projects to several cortical areas, the hippocampus, amygdala, and other forebrain areas. This pathway provides most of the NE input to the forebrain and plays important roles in functions such as attention, arousal, waking, learning, and memory.

Other noradrenergic projections innervate the thalamus, cerebellum and tectum, while descending fibers project to more caudal regions as well as to the spinal cord. Within the spinal cord, the heaviest noradrenergic innervation is found in the thoracic and upper lumbar segments. The rest of the spinal cord has only scattered noradrenergic innervation in the dorsal and ventral horns. Noradrenergic fibers from the superior cervical ganglion, derived from a sympathoadrenal lineage, also innervate the pineal gland [6]. The extensive noradrenergic innervation of the hypothalamus is covered in detail in **Section 3.4.3**.

3.2.3 Distribution and major functions of the brain adrenergic neurons

The mapping of adrenergic innervation within the brain became possible upon the introduction of anti-PNMT antibodies in immunohistochemical preparations. The adrenergic cell groups (C1 and C2) are located in the medulla, close to, and slightly overlapping with, the noradrenergic A1 and A2 cell groups. Few additional PNMT-positive cells have been detected in amacrine cells of the retina. The majority of the ascending adrenergic fibers go to the hypothalamus, as is described in **Section 3.4.3**. Low PNMT activity is detected in the substantia nigra, stria terminalis, and septum, while only a few adrenergic terminals are found in the cerebellum. PNMT activity is not detectable in the cerebral cortex, caudate, amygdala, hippocampus, nucleus accumbens, and olfactory tubercle. Within the spinal cord, only the sympathetic lateral column has significant adrenergic innervation.

Unlike the wealth of information on the functions of brain dopaminergic and noradrenergic neurons, there is little knowledge on the physiological functions of the adrenergic neurons. Nonetheless, the anatomic sites with the highest density of PNMT-containing terminals in the rat brain provide some clues as to the putative actions of Epi. For example, PNMT-containing neurons in the caudal C1, C2, and the dorsal C3 cell groups are strongly and selectively activated by glucose deprivation, suggesting that the brain adrenergic neurons are involved in mediating glucose homeostasis [7]. On the other hand, it has been argued that given that Epi is co-localized with NE in most neurons, it should be considered a metabolite of NE rather than a bona fide brain neurotransmitter [8].

3.3 DOPAMINERGIC NEURONS

3.3.1 Classification of the brain dopaminergic neurons

The brain dopaminergic neuronal systems are more complex in their anatomy, more diverse in their localization, more versatile in their functions, and more numerous than the other two catecholaminergic systems [1,9]. There are also many interactions, at various anatomical sites, between the dopaminergic and the noradrenergic pathways that affect a number of complex functions, often attributed only to DA.

By a general consensus, there are three main dopaminergic pathways: **mesolimbic, mesocortical, and nigrostriatal**. According to some classifications, however, the mesolimbic and mesocortical projections are viewed as a single pathway, named mesocorticolimbic, with two branches. It is also important to note that there are significant interspecies differences between rodents and primates, both in terms of the sites of origin and the termination fields of the dopaminergic pathways [5].

In addition to the three aforementioned dopaminergic pathways, there are several smaller dopaminergic neuronal systems that either have no axons, e.g., retinal and olfactory bulb, extend to very short distances, e.g., **incerto-hypothalamic (IHDA), tuberoinfundibular (TIDA), and tuberohypophysial (THDA)**, or originate from the sympatho-adrenal system, e.g., fibers from the superior cervical ganglion to the pineal gland. Table 3.2 describes the nuclei of origin and projections of all the dopaminergic systems in the mammalian brain.

The following section briefly reviews the mesocortical, mesolimbic, and nigrostriatal dopaminergic pathways, which are primarily associated with cognitive, reward, learning, executive, and motor functions. This is followed by a short discussion of the retinal, olfactory, and superior cervical ganglionic

Table 3.2 Dopaminergic neuronal systems in the mammalian brain		
System	**Nucleus of origin**	**Projections**
Mesocorticolimbic	Ventral tegmental area substantia nigra pars compacta	Isocortex, olfactory bulb, septal area, nucleus accumbens
Nigrostriatal	Ventral tegmental area, substantia nigra pars compacta	Caudate-putanem, globus pallidus
Incerto-hypothalamic (IHDA)	Zona incerta, posterior hypothalamus	Dorsal hypothalamic areas, septum, lateral hypothalamic area
Tuberoinfundibular (TIDA)	Arcuate, periventricular, and paraventricular nuclei,	Median eminence of the hypothalamus
Tuberohypophysial (THDA)	Arcuate and periventricular nuclei	Neurointermediate lobe of the pituitary
Olfactory bulb	Periglomerular cells	Glomerular cells (mitral)
Retinal	Amacrine and Interflexiform cells of the retina	Inner and outer plexiform layers of the retina
Pineal	Superior cervical ganglion	Pinealocytes

Source: Moore, R.Y. and Bloom, F.E., *Annu. Rev. Neurosci.*, 1, 129–169, 1978.

dopaminergic neurons. **Section 3.4.3** covers the IHDA, an intrahypotha-lamic dopaminergic network that is associated with major neuroendocrine functions, while **Section 3.4.5** covers both the TIDA and THDA dopaminer-gic pathways, which originate in the hypothalamus and project to the pitui-tary gland.

3.3.2 Mesocortical, mesolimbic, and nigrostriatal dopaminergic pathways

In rodents, the cell bodies of the mesocortical and mesolimbic pathways are located in the ventral tegmental area (VTA), while in primates, they arise from both the substantia nigra pars compacta and the VTA (**Figure 3.4**). The dopaminergic somata in these pathways receive input information from most regions of the brain, enabling them to carry out multiple integrative functions. For example, the VTA receives afferent input from the central gray, isthmic and midbrain reticular formation, lateral hypothalamus, basal fore-brain, and the locus coeruleus [9]. The nigra pars compacta receives afferent input from the globus pallidus and neostriatum via the pars reticulate, along with input from the midbrain raphe, the central nucleus of the amygdala, the prefrontal cortex, and the lateral habenula. In humans, the dopaminergic cell bodies of the pars compacta are colored black because of neuromelanin, which is formed by the polymerization of DA.

Ascending dopaminergic fibers of the mesocortical pathway innervate a large number of subcortical and cortical structures, including wide areas of the neocortex. Within the neocortex, the motor cortical, prefrontal (espe-cially orbitofrontal) and anterior cingulate areas are among the most densely innervated with dopaminergic terminals, while there is only paucity of dopa-minergic input to sensory areas such as the primary visual cortex. This pat-tern of innervation is in accordance with the general appreciation that the main role of brain DA is in the regulation of motivational and reward-related, motor and cognitive functions, rather than in basic sensory processes [5].

Dysfunctions of the mesocortical dopaminergic pathway are associated with addiction, attention deficit hyperactivity disorder (ADHD) and the negative symptoms of schizophrenia. The latter symptoms are defined as a diminished

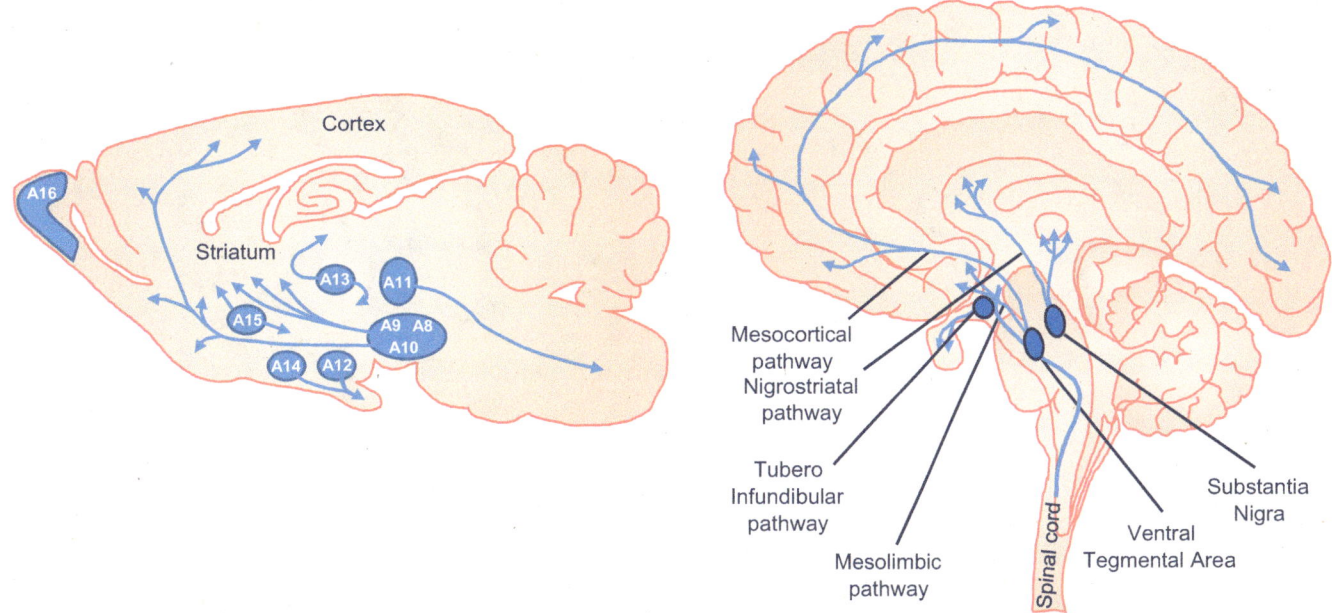

Figure 3.4 Distribution and projections of dopaminergic neurons in the human (right) and rat (left) brains. The four major dopaminergic pathways: mesolimbic, mesocortical, nigrostriatal, and tuberoinfundibular are shown in the human brain. More details of the classification of the dopaminergic cell body groups are shown in the rat brain. See text for more detail.

capacity to experience pleasure (anhedonia), a decreased social affiliation (asociality), a lack of motivation or drive (apathy), a decreased expression of emotion (flat or blunted affect), and a diminished speech capacity (alogia).

The mesolimbic dopaminergic pathway transmits DA from the VTA to the ventral striatum [10]. In histological preparations, the striatum is seen as stripes of gray and white matter and hence its name. In primates, the striatum is divided into a ventral sector, which consists of the nucleus accumbens and the olfactory tubercle, and a dorsal sector, which comprises the caudate nucleus and putamen. The major cells that populate the striatum are medium spiny GABAergic neurons that express DA receptors and are intermixed with inhibitory cholinergic interneurons. The ventral striatum, especially the nucleus accumbens, is associated with reward-related cognition, pleasure, and positive reinforcement. Dysfunctions of the mesolimbic dopaminergic system result in disorders similar to those listed above for the mesocortical system.

The nigrostriatal pathway connects the nigra pars compacta to the nucleus accumbens and putamen within the dorsal striatum [9]. It is by far the most abundant dopaminergic innervation, comprising as much as 70% of all brain dopaminergic neurons. This pathway is primarily involved in the regulation of movement, as part of a system called the basal ganglia motor loop. The number of the nigrostriatal neurons is significantly reduced with age and is also subjected to damage by neurotoxins such as 1-methyl-4-phenyl-1,2,3,6-tetrahydropyridine (MPTP) [11].

Loss of neurons in the substantia nigra or impairments of their functions are the main causes of **Parkinson's disease (PD)**, a neurodegenerative disease characterized by movement disorders such as stooped posture, stiffness, slowing of movement, and trembling (**Figure 3.5**). In advance stages, PD can progress to dementia and even to death. When administered systemically, DA does not cross the blood–brain barrier (BBB). Therefore, the main pharmacological intervention in patients with PD is the administration of L-Dopa, usually in combination with inhibitors of peripheral L-Dopa decarboxylase such as carbidopa or benserazide. Such a treatment provides symptomatic relief at the earlier stages of the disease, when sufficient number of dopaminergic neurons are still functional, but does not prevent the

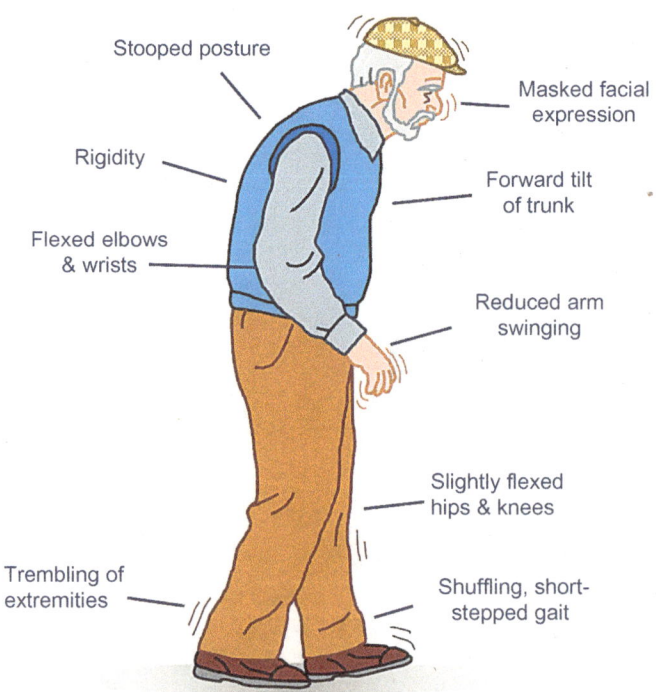

Figure 3.5 Physical disorders associated with Parkinson's disease. Illustrated are movement disorders, facial expression, and overall posture appearance of a patient with advanced Parkinson's disease.

progressive neuronal loss. Deep brain stimulation with implanted electrodes has been quite effective in many PD patients at the early stages of the disease.

Early clinical studies with intra-striatal transplants of fetal mesencephalic tissue in PD patients have provided proof-of-principle for the cell replacement strategy [12]. The grafted dopaminergic neurons reinnervated the striatum, restored regulated DA release and movement-related frontal cortical activation, and gave rise to a significant symptomatic relief. In the most successful cases, patients were able to withdraw L-Dopa treatment after transplantation and to resume an independent life. However, the following problems have been linked to the use of fetal tissue: (1) lack of sufficient amounts of tissue for transplantation for the large number of patients, (2) variability in the functional outcome among patients, and (3) signs of troublesome dyskinesias in a significant number of patients after transplantation. Consequently, neural tissue transplantation remains at an experimental stage in the treatment of Parkinson's disease.

Another approach has been the use of adult mesenchymal stem cells (MSCs). These have shown a remarkable therapeutic power in animal models of PD, given their differentiation competence, migratory capacity, and the production of bioactive molecules [13]. The results of MSC therapy in animal models and some clinical trials suggested that this form of cellular therapy may slow progression of the disease and could promote neuro-regeneration. However, further research is needed to address the limitations of this transplantation prior to eventual clinical application.

3.3.3 Retinal, olfactory, and pineal dopaminergic neurons

The retina of the eye has two types of DA-producing cells: **amacrine cells (ACs)**, which lack axons (anaxonic), and **inteplexiform cells (IPC)**, which have multiple processes [14,15]. The AC are a diverse class of intrinsic interneurons of the inner retina (**Figure 3.6**). They receive synaptic input from the bipolar cells as well as from other amacrines and in turn provide input to the ganglion cells and feedback information to the bipolar cells. These cells release DA into the extracellular milieu and are especially active during the daylight hours, becoming silent at night. The retinal DA enhances the activity of cone cells (which are responsible for color vision and spatial acuity) while suppressing rod cells (which are responsible for vision at low light levels). Consequently, DA increases retinal sensitivity to color and contrast during bright light conditions but at the cost of reduced sensitivity when the light is dim. The circadian rhythm of retinal DA levels is independent of input from the suprachiasmatic nucleus, the master circadian pacemaker, and depends on the actions of locally produced melatonin and GABA. Patients with Parkinson's disease have reduced retinal DA levels and suffer from a number of visual dysfunctions. Impaired color and contrast discrimination has been considered as preclinical signs of Parkinson's disease.

The DA neurons in the **olfactory bulb (OB)** belong to the A16 cell group (see **Figure 3.2**) and represent the major DA system within the forebrain [16]. Tyrosine hydroxylase (TH)-positive cells are mostly localized in the glomerular cell layer, accounting for nearly 10% of all juxtaglomerular cells. These cells have synaptic contacts with the afferent olfactory receptor neuron terminals and/or with external tufted cells and participate in the early steps in odor information processing that occur in the input layer of the OB.

Most neurogenesis in the mammalian brain is completed during embryonic life. However, few areas, including the hippocampal dentate granule cells and interneurons in the OB, continue to generate new neurons throughout postnatal life [17]. The embryonic and postnatal neurogenesis produce functionally distinct subpopulations of dopaminergic neurons in the OB. Large, axon-bearing DA neurons are exclusively produced during the early embryonic stages, while the small DA-positive anaxonic cells are generated through adult neurogenesis.

Figure 3.6 Diagram of retinal cells and their synaptic organization. Shown are the dopamine (DA)-producing neurons: amacrine cells (AC) and interprexiform cells (IPC). Shown are the main photoreceptors, rods (R) and cones (C) and their interconnections to horizontal cells (HC), bipolar cells (BC), and ganglion cells (GC). See **Section 3.3** for more detail. (Redrawn and modified from Popova, E., *J. Comp. Physiol. A Neuroethol. Sens. Neural. Behav. Physiol.*, 200, 333–358, 2014.)

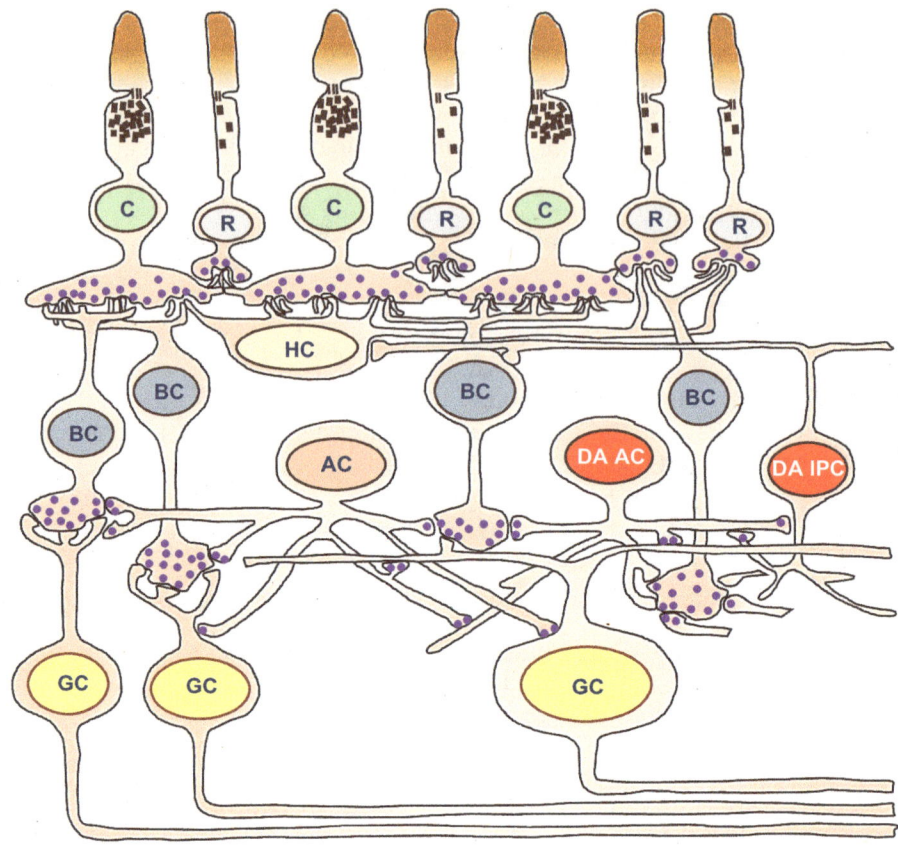

Olfaction plays important roles in the neuroendocrine regulation of reproduction. These functions are particularly important in rodents, where olfactory cues can advance and delay puberty, can suppress or facilitate the estrous cycles, and can also cause an early termination of pregnancy. Such cues also facilitate sexual behavior and inform mate selection in several species. Although olfactory cues are less critical in humans, the ability to smell is important for the perception of taste and can alert the individuals to dangers like a gas leak, fire, or rotten food. Moreover, congenital anosmia is associated with **Kallmann's syndrome**, defined as idiopathic hypogonadotropic hypogonadism. These issues are discussed in more detail in **Chapters 4 and 5**.

The **pineal gland** is a very small organ, shaped like a pine cone (hence its name) and is located near the center of the brain. The main function of the pineal gland is to translate the rhythmic cycles of night and day, encoded by the retina, into hormonal signals that are transmitted to the rest of the neuronal system. The pineal gland secretes a single hormone, **melatonin**, whose release is regulated by light. In humans, melatonin has two main functions: to help control the circadian rhythm and to regulate certain reproductive hormones.

As illustrated in **Figure 3.7**, the pineal gland of mammals is innervated by sympathetic nerve fibers that originate in the superior cervical ganglia and release both DA and NE [18]. Because peripheral denervation of the pineal results in a complete depletion of TH-positive cells, it supports an external, rather than a local, source of DA. Pineal DA and dihydroxyphenylacetic acid (DOPAC) levels increased significantly during the dark phase, with DA levels reaching a peak just prior to the time of peak melatonin production. DA was found to suppress both the production and release of melatonin by activating D4Rs that are hetero-dimerized with α1AB adrenergic receptors [19].

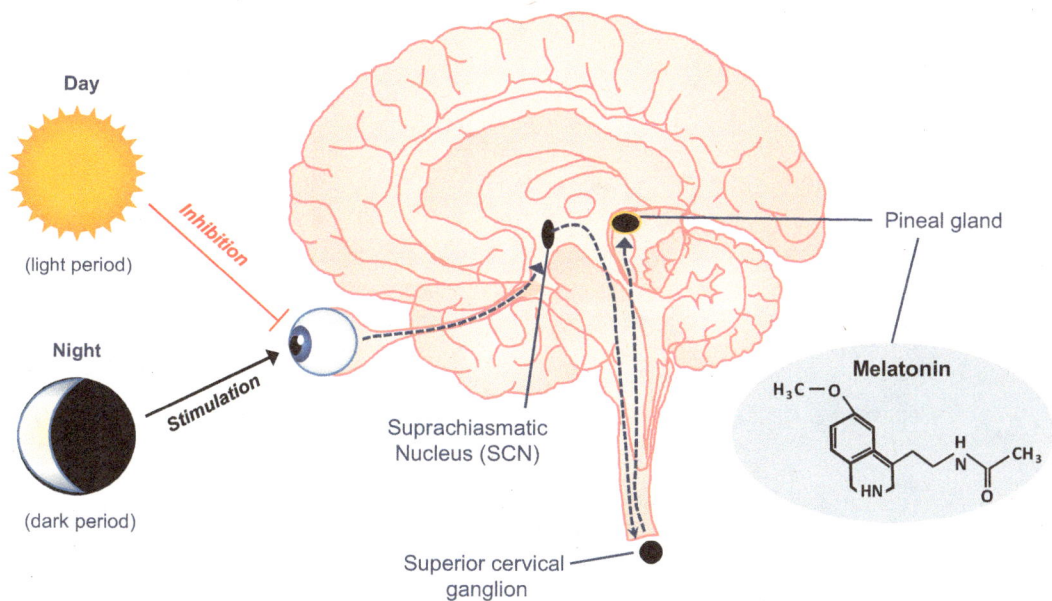

Figure 3.7 Neural connections that regulate the activity of the pineal gland. Photic information from the retina reaches the suprachiasmatic nucleus (SCN) via the retino-hypothalamic tract. After integration, the SCN sends rhythmic information to the SCN via the superior cervical ganglion. The release of melatonin, the major product of the pineal, is increased at night and reduced at daytime. The changing levels of melatonin, together with the neuronal output from the pineal to many brain regions, regulate circadian rhythmicity throughout the body.

3.4 THE HYPOTHALAMUS: STRUCTURE AND FUNCTIONS

3.4.1 Anatomy of the hypothalamus

The hypothalamus, where DA occupies a prominent position as a major neurotransmitter, is the center of neuroendocrine regulation [20,21]. A comprehensive review, entitled "Functional Anatomy of the Hypothalamus and Pituitary," by Lechan and Toni, covers this topic. This review is a chapter in *Endotex*, a thorough, constantly updated, free Web textbook, written by 400 experts and oriented toward physicians caring for patients with endocrine diseases. The book is downloadable at https://www.ncbi.nlm.nih.gov/books/NBK279126/. Below we summarize selected relevant information from this chapter.

The hypothalamus is located below the thalamus and above the pituitary gland (**Figure 3.8**). It is situated in the ventral diencephalon and is composed of many fiber tracts and discrete nuclei that are positioned symmetrically around the third ventricle. In midsagittal section, the human hypothalamus is bound anteriorly by the lamina terminalis, posteriorly by the posterior commissure at the caudal limit of the mammillary body, and superiorly by the hypothalamic sulcus. The lateral boundaries on each side of the hypothalamus include the internal capsule, cerebral peduncle, and subthalamus. Ventrally, the hypothalamus forms the floor of the third ventricle. Its inferior surface, called the tuber cinereum, contains the median eminence, one of seven circumventricular areas of the brain that have permeable capillaries and are not protected by the BBB.

The hypophysiotropic hormones of the hypothalamus (also called releasing and inhibiting hormones) collect in the median eminence and are secreted from there into the hypophyseal portal system, which vascularizes the pituitary gland [22]. The pars nervosa, a component of the posterior pituitary gland, forms a continuance with the median eminence via the pituitary stalk, or infundibulum. The stalk passes through the dura mater of the diaphragm sellae and carries dopaminergic and magnocellular axons down to the posterior pituitary. A detailed description of the neural and vascular

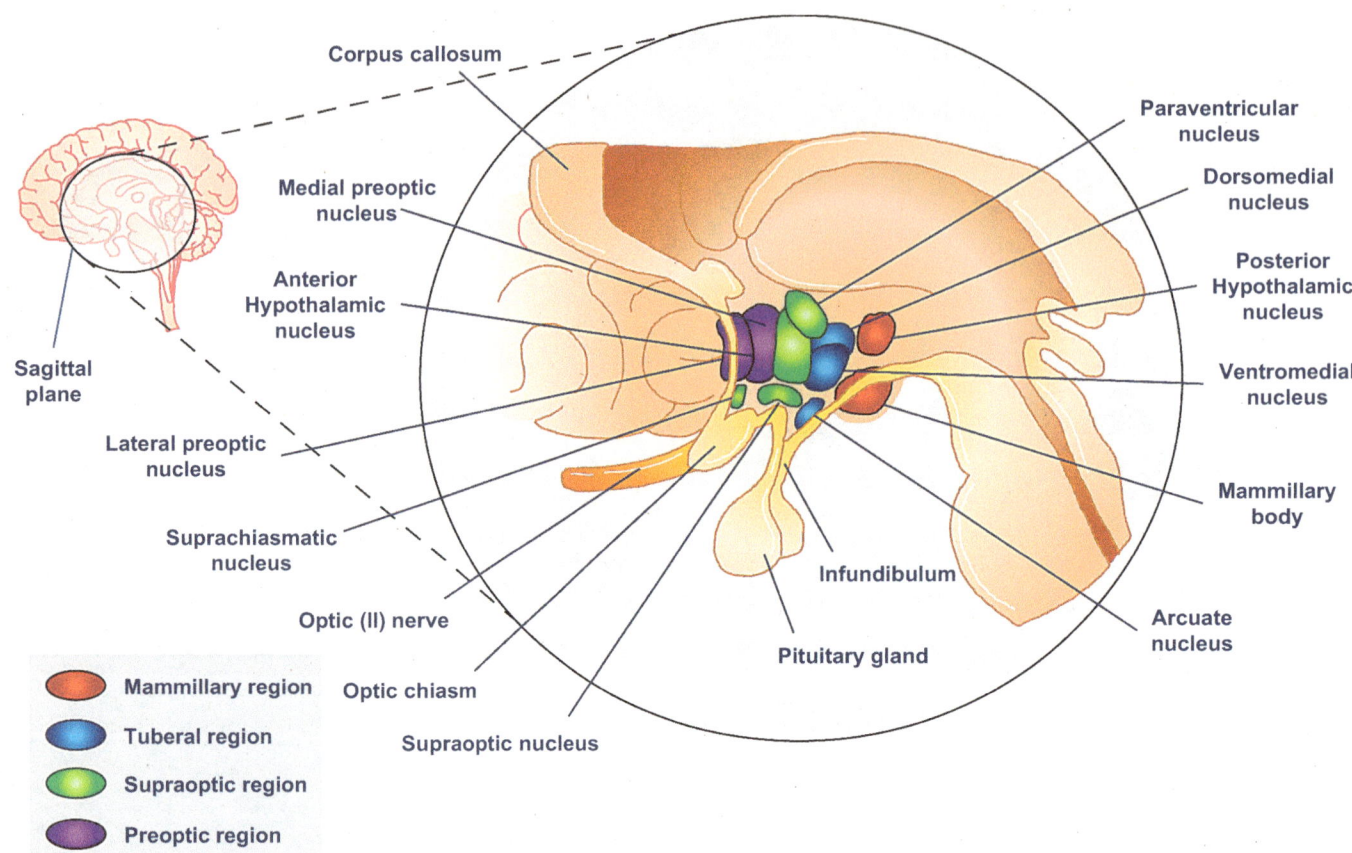

Figure 3.8 Major nuclei associated with neuroendocrine functions in the human hypothalamus. As seen in a sagittal plane, the nuclei are grouped by color into four regions: preoptic, supraoptic, tuberal and mammillary.

connections between the hypothalamus and the pituitary is presented in **Sections 3.4.5** and **3.4.6**.

From rostral to caudal direction, the hypothalamus is divided into three regions, each of which receives a different blood supply: (1) the anterior (chiasmatic) region, which extends between the lamina terminalis and the anterior infundibular process; (2) the median (tubular) region, which proceeds to the anterior column of fornix; and (3) the posterior (mammillary) region, which extends to the caudal mammillary bodies. Morphologically and functionally, each of these regions is further subdivided into a lateral region, occupied primarily by the bidirectional median forebrain bundle that connects the hypothalamus with other parts of the brain, and the medial and periventricular regions that contain the majority of the hypothalamic nuclei.

The hypothalamus is interconnected with many parts of the brain, particularly with the brain stem reticular formation and the areas that regulate the autonomic nervous system. It also constitutes a functional part of the limbic system by having connections to the amygdala and septum. Most nerve fibers within the hypothalamus are bidirectional. Projections to the caudal areas go through the medial forebrain bundle, the mammilotegmental tract and the dorsal longitudinal fasciculus, while projections to the rostral areas are carried out by the mammilothalamic tract, the fornix and the terminal stria.

3.4.2 The hypothalamic nuclei: Locations and major neuroendocrine functions

The hypothalamus is composed of many nuclei with diverse functions (see **Figure 3.8**). Each nucleus contains a mixture of several neuronal types that work in concert to regulate multiple functions that affect homeostasis.

Endocrine functions that are regulated by the hypothalamus include metabolism, water balance, satiety, reproduction, circadian rhythm, growth control, stress responses, and sexual and parental behaviors. The hypothalamic nuclei are divided by size into two main types: magnocellular and parvocellular. The magnocellular are neuroendocrine neurons, among the largest in the brain, which synthesize and secrete **oxytocin** and **arginine vasopressin (AVP)**, also called antidiuretic hormone (ADH). The parvocellular are much smaller neuroendocrine neurons that synthesize and release a variety of hypothalamic neurohormones. Both types of nuclei act as mediators between the central nervous system (CNS) and the endocrine system and play critical roles in the regulation of pituitary functions.

Several of the hypothalamic nuclei are classified as sexually dimorphic. Few nuclei differ in structure between males and females, e.g., the sexually dimorphic nucleus in the preoptic area, which is larger in males, while most others differ only in function. The most common functional differences among the sexes are in the control of reproduction and sexual behavior, which are determined by the distribution and/or responsiveness of sex steroid receptors in certain hypothalamic nuclei. In addition, the pulsatile pattern of growth hormone (GH) secretion is also sexually dimorphic in terms of pulse regularity, amplitude of the diurnal rhythm, and the magnitude of basal GH release. Sexual dimorphism is determined early in life as a consequence of a timely exposure of the fetus to male and female sex steroids and is irreversible [23].

Most, but not all, of the hypothalamic nuclei are associated with neuroendocrine functions, in which DA plays either a major or a contributory role. A brief description of these nuclei and the neurohormones and neuromodulators that they produce is presented below. A more extensive review of the direct involvement of some hypothalamic nuclei in the regulation of pituitary functions is provided in **Chapter 4**.

For a systematic description of the nuclei, it is convenient to divide the hypothalamus into three zones—periventricular, medial, and lateral—and then review the various nuclei in each zone by their location, from the rostral to the caudal direction.

As depicted in **Table 3.3**, the periventricular zone contains four of the most important nuclei for neuroendocrine regulation: periventricular, suprachiasmatic nucleus (SCN), paraventricular nucleus (PVN), and arcuate. The periventricular nucleus is a cluster of small neurons, located at the wall of the third ventricle, which extend from the rostral through the caudal zones. Neurons in the rostral area produce somatostatin and thyroid releasing hormone **(TRH)**, while those in the intermediate area also produce leptin, gastrin, and neuropeptide Y. In humans and primates (but not in rodents), the periventricular neurons in the intermediate zone also produce gonadotropin-releasing hormone **(GnRH)**. The caudal region of the periventricular nucleus is primarily associated with sympathetic nervous system regulation. Given its proximity to the third ventricle, the periventricular nucleus does not have an effective BBB, enabling its exposure to feedback regulation by peripheral hormones, as discussed in **Chapter 4**.

Table 3.3 List of the hypothalamic nuclei by zones	
Zones	**Nuclei (listed in a rostral to caudal order)**
Periventricular	Periventricular, suprachiasmatic nucleus (SCN), paraventricular nucleus (PVN), arcuate
Medial	Medial preoptic, anterior hypothalamic, dorsomedial (DMH), ventromedial (VMH), premammillary, mammillary, posterior hypothalamic
Lateral	Lateral preoptic, lateral hypothalamic, supraoptic (SON)

The **SCN** is among the smallest of the hypothalamic nuclei. It sits above the optic chiasma and serves as the mammalian master circadian pacemaker. Multiple lines of evidence have shown that the SCN is necessary for many circadian phenomena, that it is rhythmic *in vivo*, and that it has an autonomous circadian clock that maintains rhythmicity when isolated or when used in transplantation [24]. Individual cells within the SCN can function independently, while forming a unified circadian network through intercellular coupling.

The SCN receives input from the photosensitive ganglion cells in the retina through the retino-hypothalamic tract. It synchronizes the information on the solar day-night cycle and through its efferent connections affects the many circadian clocks that exist in various cells and organs throughout the body. When categorized by its neuropeptide content, the SCN can be divided into two regions: a ventral core and a dorsal shell. The ventral core, which receives the retinal input, contains vasoactive intestinal peptide (VIP) and gastrin-releasing peptide (GRP), while the shell primarily contains AVP.

The **PVN** is located adjacent to the third ventricle [21], and its main product is oxytocin. The PVN and supraoptic nuclei (SON), although discrete structures that are located at different hypothalamic regions, are often grouped together because both house the magnocellular neurons that innervate the posterior pituitary [25]. Each magnocellular neuron has one long axon that projects to the posterior pituitary, where it gives rise to many free neurosecretory nerve terminals without synaptic contacts. The magnocellular neurons also have several large dendrites that project ventrally, receive input from a number of neuronal systems, and can also release oxytocin locally. The parvocellular neurons within the PVN produce corticotropic release hormone (CRH) and TRH.

The **arcuate nucleus** is located near the third ventricle and above the median eminence. It is composed of a diverse population of neurons as well as nonneuronal cells such as astrocytes and tanycytes, which are specialized ependymal cells. The arcuate nucleus contains large amounts of DA and is the origin of the TIDA and THDA neurons that innervate the median eminence and the posterior pituitary, respectively, as discussed in **Sections 3.4.3** and **3.4.4**. The parvocellular neurons within the arcuate nucleus produce several releasing/inhibiting hormones that include GnRH, GH releasing hormone (GHRH), β-endorphin, melanocyte stimulating hormone (MSH) and somatostatin. In addition, the arcuate nucleus produces a variety of neuromodulators such as kisspeptin, NPY, substance P, Agouti-related peptide (AgRP) and CART (cocaine and amphetamine regulatory transcript). The arcuate nucleus has the highest concentrations of leptin receptors in the brain and as discussed in **Chapter 4**, it is the best characterized hypothalamic nucleus involved in energy homeostasis.

As illustrated in **Table 3.3**, the median zone of the hypothalamus contains the following nuclei: medial preoptic, anterior hypothalamic, dorsomedial, ventromedial, premammilary, mammillary, and posterior hypothalamic. Of these, only the medial preoptic, dorsomedial, and ventromedial nuclei have significant neuroendocrine functions and are covered below. The other nuclei are generally associated with thermoregulation, behavior, and a few other functions.

The **medial preoptic nucleus** represents the largest collection of neurons in the preoptic area. It is bounded laterally by the lateral preoptic nucleus, and medially by the preoptic periventricular nucleus. The nucleus can be divided into subnuclei, with the central portion in the rat and few other species larger in males than in females and has been named a sexually dimorphic nucleus. Such gender-dependent structural differences have not been consistently observed in humans, although there are reports on homologous regions in the human medial preoptic area that exhibit sexual dimorphism in structure [26].

The **sexually dimorphic nucleus** in rodents has gender-related different expression levels of sex steroid receptors and is linked to male sexual behavior. Lesions in this region in several species eliminate male copulatory behavior and inhibit sexual desire. The medial preoptic nucleus contains GnRH, somatostatin, β-endorphin, and substance P. In addition to sexual and parenting behavior, the preoptic nucleus is associated with the regulation of cardiovascular functions, body temperature, fluid balance, and water intake.

The dorsomedial hypothalamic nucleus (DMH) is located above the ventromedial nucleus and below the caudal part of the PVN. It receives information from neurons and hormones involved in feeding regulation, body weight and energy consumption, and it passes this information on to brain regions involved in sleep and wakefulness regulation, body temperature and corticosteroid secretion. Lesions in the DMH neurons in rats prevent food entrainment of wakefulness, locomotor activity, and core body temperature, verifying its role in oscillation between feeding and the circadian rhythm. Such lesions also caused a weakened level of response to feeding stimulation by insulin.

The ventromedial nucleus sits close to the base of the diencephalon, adjacent to the third ventricle, and above the median eminence and pituitary complex [27]. It is a bilateral cell group with an elliptical shape, composed of several subdivisions—anterior, dorsomedial, ventromedial and central—which differ anatomically, neurochemically and behaviorally. The VMH is highly conserved across species and has served as a model for studying the neuronal organization into nuclei during embryogenesis. The VMH is involved with energy balance, feeding and obesity and has been well recognized as an integral part of the complex hypothalamic circuitry that regulates satiety. Several neuropeptides, which are either produced locally, or act upon their respective receptors, i.e., neuropeptide Y (NPY), ghrelin, leptin, glucagon-like peptide 1, insulin, urocortin, and CRH, regulate energy balance. A subset of neurons within the VMH also has a high expression of estrogen receptors and play a role in the regulation of female sexual behavior, specifically of the lordosis response, an arching of the back that is a posture assumed by some female mammals during mating.

As presented in **Table 3.3**, the lateral zone contains the lateral preoptic nucleus, lateral hypothalamic nucleus, and supraoptic nucleus. The preoptic nucleus has already been described above. The **lateral hypothalamic nucleus**, or as it is often called the lateral hypothalamic area (LHA), is an essential partner in the circuit that regulates food intake and body weight. Two neuronal pathways that produce the neuropeptides melanin-concentrating hormone (MCH) and orexins (ORX) are localized in the LHA and provide monosynaptic projections to the cerebral cortex and autonomic preganglionic neurons. Both MCH and ORX neurons regulate the cognitive and autonomic aspects of food intake and body weight regulation. In addition, neurons in this area are associated with arousal, the reduction of pain perception, and the control of body temperature.

The **SON**, the second magnocellular nucleus, is situated adjacent to the optic chiasm and its main product is vasopressin. Parvocellular neurons within the SON produce neuroactive substances such as dynorphin, cholecystokinin (CCK) and CART. Because of their large size, abundance of neurosecretory products, long axons, and lack of synaptic contacts, the magnocellular neurons have served for many years as an excellent model system in neuroscience. Much information on electrical excitability, expression of neuroactive substances, axoplasmic transport, and the mechanism of release has been obtained from studying the magnocellular neurons of the PVN and SON under both *in vivo* and *in vitro* conditions [25].

3.4.3 Catecholaminergic innervation of the hypothalamus

Although the hypothalamus receives some DA from collateral fibers of the midbrain dopaminergic pathways, most hypothalamic DA is locally produced by two pathways: the TIDA and the **IHDA** neurons. The perikarya of the IHDA are located at the A13 DA cell group in the zona incerta, situated medially to the internal capsule and ventrally to the mammillothalamic tract of the thalamus [28]. These neurons send diffuse efferent projections to the anterior hypothalamic area, lateral hypothalamus and lateral preoptic area. Lesser density of dopaminergic neurons from the IHDA is found in the supraoptic, ventromedial and arcuate nuclei, and only a few immunopositive DA fibers are seen in the dorsomedial nucleus or the paraventricular nucleus. The A13 cell group also sends dopaminergic efferent projections to several other brain regions, especially those that are associated with defensive behavior, and include the central amygdala and the ventromedial hypothalamus.

The perikarya of the **TIDA** neurons are located in the arcuate, periventricular, and paraventricular nuclei and the medial preoptic-septal region. Their terminals are found in the **median eminence**, a region that contains one of the highest concentrations of DA in the brain. DA terminals are especially abundant in the external zone of the median eminence, where they comprise as much as a third of all terminals [29]. The TIDA terminals contain small dense core vesicles and converge on the primary capillary plexus of the hypophysial portal vessels. Some terminals contact the basement membrane of perivascular spaces, while others terminate on various neurosecretory cells or are situated in close proximity to ependymal cells. Except for tanycytes and some astrocytes, the median eminence contains a negligible number of neuronal perikarya and a dearth of classical synapses. The TIDA system represents genuine neurosecretory neurons whose product is released into perivascular spaces surrounding the capillary loops and is carried by the portal blood to the anterior pituitary.

The relative concentration of DOPAC is lower in the median eminence than in other brain areas rich in DA. DOPAC levels generally represent the amount of DA that is released and then recaptured by nerve terminals. Upon comparing DA uptake in several brain regions, it was found that the TIDA neurons lack a high-affinity transport system for DA [30]. Although immunoreactive dopamine transporter (DAT) has been detected in the median eminence [31], there has been no confirmation of its functionality. Therefore, the general consensus is that DA that is released within the median eminence is quickly transported by the hypophysial portal blood away from the terminals and little to no DA is available for reuptake. Notably, this is also the case with the majority of the peripheral DA producing cells, most of which do not have adjacent synapses. Such a feature also explains the resistance of the TIDA neurons to destruction by the neurotoxin 6-hydroxydopamine, which requires an active uptake mechanism.

The **THDA** neurons have cell bodies in the rostral arcuate and periventricular nuclei and terminals in the neural and intermediate lobes (ILs) of the posterior pituitary. Because the THDA neurons do not directly play a role within the hypothalamus itself, they are covered in detail in **Section 3.4.6**.

Table 3.4 presents the DA and NE content of selected hypothalamic nuclei and the median eminence of the rat. The DA concentration in the median eminence (65 ng/mg protein) is exceptionally high [32]. Practically all hypothalamic nuclei contain some DA, which is most highly concentrated in the arcuate nucleus, the paraventricular, the suprachiasmatic and the ventromedial nuclei. Some DA is also detected in the medial forebrain bundle at the posterior hypothalamic level. On the other hand, nuclei within the posterior hypothalamus—the premammillary nuclei, the caudal subdivisions of the arcuate nucleus, and the posterior hypothalamic nucleus—contain only low concentrations of DA.

Table 3.4 Dopamine and norepinephrine content of hypothalamic nuclei (listed alphabetically) and the median eminence of the rat		
Nucleus/area	Dopamine (ng/mg protein)	Norepinephrine (ng/mg protein)
Anterior hypothalamic	5	16
Arcuate	28	36
Dorsomedial	9	21
Median eminence	65	29
Paraventricular	10	51
Periventricular	7	34
Premammilary	3	16
Suprachiasmatic	12	40
Supraoptic	4	24
Ventromedial	10	38

Source: Palkovits, M. et al., *Brain Res.*, 77, 137–149, 1974.

The hypothalamus is extensively innervated by noradrenergic fibers coming from both the LC and the LTA. To identify the origin and distribution of the noradrenergic neurons in the rat, loss of NE in specific hypothalamic regions was determined after surgical transection of the lower brain stem or following electrolytic lesions of NE-containing cell groups [33]. As evident in **Table 3.4**, except for the median eminence, larger amounts of NE than DA are present throughout the hypothalamic nuclei, with the highest levels of NE found in the paraventricular, suprachiasmatic and ventromedial nuclei [32]. Notably, NE levels showed rhythmic variations in the suprachiasmatic nucleus (SCN), the central pacemaker of the circadian timing system, [34]. This circadian rhythm is driven by an endogenous pacemaker, independent of external light. The authors proposed that NE in the SCN provides a route through which brainstem activity feedbacks onto the circadian pacemaker in the SCN.

The greatest concentrations of Epi within the hypothalamus are found in the arcuate and paraventricular nuclei as well as the median eminence [35]. Intermediate levels are seen in the periventricular, dorsomedial and ventromedial nuclei, and the retrochiasmatic area, where the adrenergic neurons are interspersed with fibers of the supraoptic decussations. Low density of Epi was noted in the remaining hypothalamic nuclei. The dense adrenergic innervation of the PVN suggested an association with oxytocin secretion, while that in the DMH pointed toward a potential involvement of Epi in food and water intake. Interestingly, the adrenergic innervation of the arcuate nucleus, along with the observation that Epi was more effective than DA or NE in triggering ovulation in proestrus pentobarbital blocked rats, implied that gonadotrophin secretion is also influenced by the brain adrenergic neurons [36].

3.5 THE HYPOTHALAMO-PITUITARY COMPLEX: NEURAL AND VASCULAR CONNECTIONS

3.5.1 Functional anatomy of the hypothalamo-pituitary complex

The **pituitary gland** is suspended from the median eminence by the infundibulum (pituitary stalk) and is nested within the sella turcica of the sphenoid bone of the skull. The median eminence is one of the seven areas of

Figure 3.9 Diagram of the median eminence at the borders of the third ventricle (III). Shown are organization of three major zones: ependymal zone (E), zona interna (ZI), and zona externa (ZE). The portal capillaries within the zona externa are contacted by axon terminals of the tuberoinfundibular system and by processes of tanycytes, specialized ependymal cells. The neurohypophysial fiber tract courses through the zona interna. (Redrawn and modified from Lechan and Toni, Endotext: https://www.ncbi.nlm.nih.gov/books/NBK279126/.)

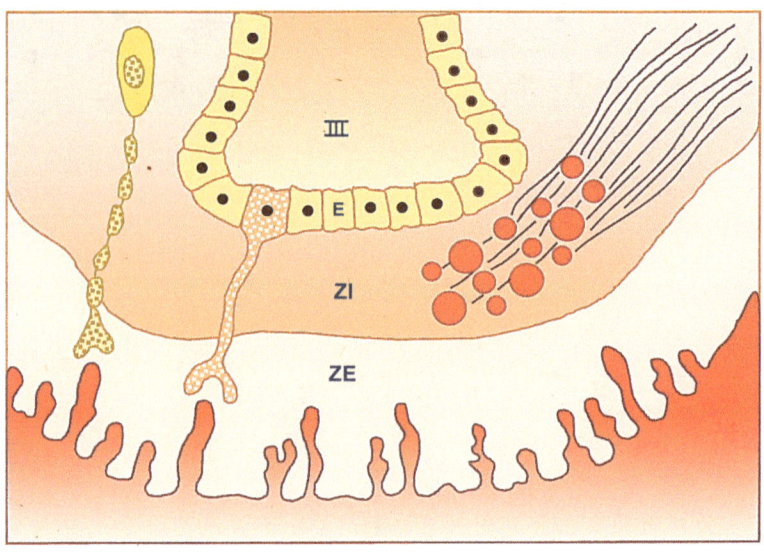

the brain called circumventricular organs, which have fenestrated capillaries and are therefore devoid of the BBB [37]. As illustrated in **Figure 3.9**, the median eminence is composed of three distinct zones: ependymal, internal, and external. The ependymal zone forms the floor of the third ventricle and has several specialized features, including tight junctions between adjacent cells and highly specialized cells, the tanycytes. **Tanycytes** are glial-like cells that extend protrusions and microvilli into the cerebrospinal fluid (CSF) at their ventricular surface and long cytoplasmic processes into the body of the median eminence. Tanycytes are exposed to CSF from the third ventricle and also have an access to circulating hormones and various metabolites through the fenestrated capillaries [38]. Some of the tanycytes act as conduits for trafficking certain molecules into the brain parenchyma, while others act as neural stem/progenitor cells that supply the postnatal and adult hypothalamus with new neurons. During embryonic development, tanycytes also serve as a scaffolding for axons that enter the median eminence, guiding them to their ultimate destination in the external zone.

The internal zone (zona interna) of the median eminence lies directly below the ependymal zone and is primarily composed of unmyelinated axons of the magnocellular neurons that project to the posterior pituitary. The internal zone also contains axons of the hypothalamic TIDA neurons as they descend into the external zone. The external zone (zona externa) is located beneath the internal zone and contains the primary portal capillaries and cytoplasmic extensions of the tanycytes. It also has many unmyelinated axons and axon terminals of the TIDA neurons, characterized by dense-core vesicles ranging from 50 to 130 nm in diameter. The close proximity of many of the axon terminals to the portal capillaries enables the release of stored hypothalamic releasing/inhibiting hormone into pericapillary spaces and from there by diffusion via the fenestrated endothelium into the long portal vessels leading to the pituitary gland.

3.5.2 Embryonic development and gross anatomy of the pituitary gland

The human pituitary gland is a size of a pea and weighs ~500 mg, while the rat pituitary gland weighs 12–15 mg. The pituitary consists of two lobes, anterior and posterior that arise from distinct parts of embryonic tissue. The posterior pituitary (PP) or neural lobe (NL) has its embryological origin in nervous tissue, given that it develops from a down-growth of the diencephalon that forms the floor of the third ventricle. On the other hand, the

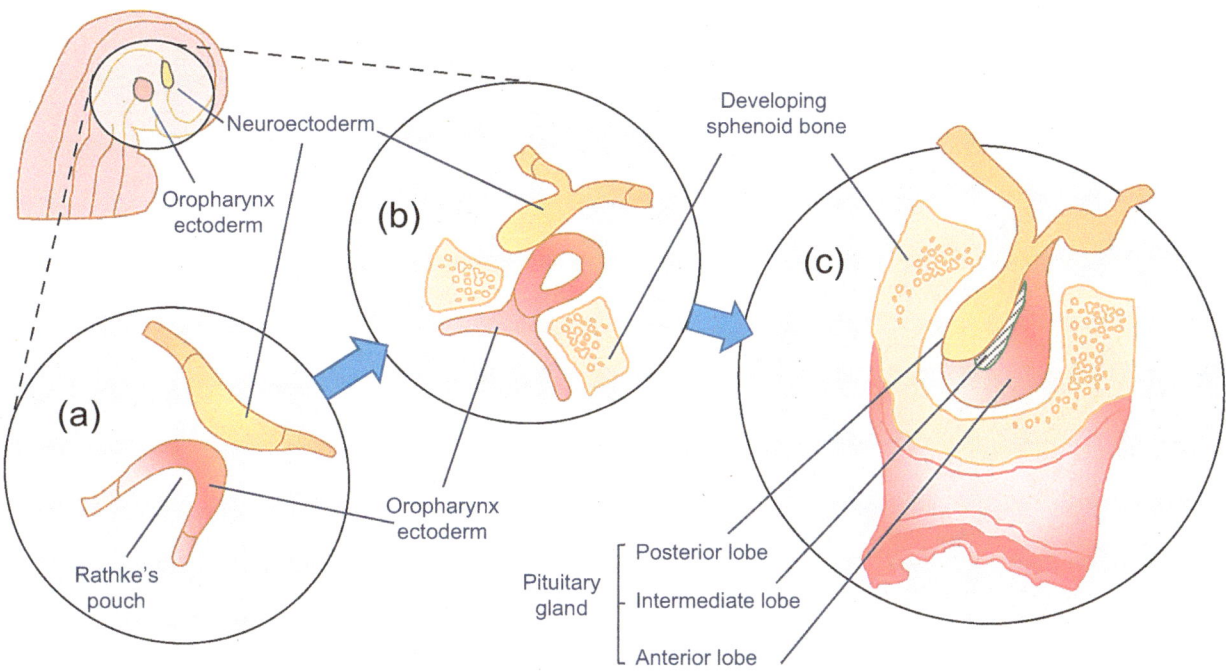

Figure 3.10 Diagram of the embryonic development of the pituitary gland. Panel A: Infundibulum and Rathke's pouch develop from the neural ectoderm and oral ectoderm, respectively. Panel B: Rathke's pouch constricts at the base and separates from the oral epithelium. Panel C: The mature pituitary gland is made of three components: anterior lobe or pars distalis, neural lobe or pars nervosa, and intermediate lobe or pars intermedia, which is only a remnant in humans.

anterior lobe (AL), also called the pars distalis or anterior pituitary (AP), is derived from an up-growth of the oral ectoderm within the primitive oral cavity, called the Rathke's pouch (**Figure 3.10**). The pituitary organogenesis is a highly complex and tightly regulated process that depends on several transcription factors such as PROP1, PIT1 (POU1F1), HESX1, LHX3, and LHX4 [39]. Mutations in these genes can result in different combinations of hypopituitarism that can be associated with structural alterations of the CNS, causing the congenital form of panhypopituitarism. Detailed information of the cellular composition and hormone synthesis and regulation by the pituitary gland is presented in **Chapter 5**.

Many mammals have a distinct IL that is interposed between the AL and NL. In humans, however, the IL is well developed in the fetus but undergoes involution to colloid-filled cysts that mingle with the neural lobe in the adult. Neither the role of the IL in fetal human development nor the reasons for its involution are known. Unless otherwise specified below, when presenting evidence obtained from animal studies, the term posterior pituitary refers to a combined neural and IL. Another small region, the pars tuberalis, extends upward from the pars distalis and joins the infundibular stalk arising from the posterior lobe (pars nervosa). The function of the pars tuberalis is poorly understood.

3.5.3 The hypothalamo-pituitary complex: Neural connections and DA concentrations

Two main neuronal pathways connect the hypothalamus with the pituitary gland. One tract is composed of magnocellular neurons (neurohypophysial tract) and carries oxytocin and vasopressin and innervates the neural lobe. The other tract is composed of dopaminergic neurons and innervates the NL as well as the IL in nonhuman species who have a distinct IL. The AP does not receive any direct innervation from the hypothalamus although it contains few nerve fibers primarily associated with local blood vessels.

Figure 3.11 Diagram of the dopaminergic innervation and vascular connections of the rat hypothalamo-pituitary complex. The periventricular-hypophysial tract (PHDA) innervates the intermediate lobe (IL). The tuberohypophysial tract (THDA) innervates the neural lobe (NL) as well as the IL. The tuberoinfundibular tract (TIDA) has terminals in the external zone of the median eminence. The hypophysial portal vasculature with a primary plexus in the median eminence is connected through the long portal vessels to the secondary capillary plexus in the anterior lobe (AL). The short portal vessels connect the NL and IL, while the IL is avascular.

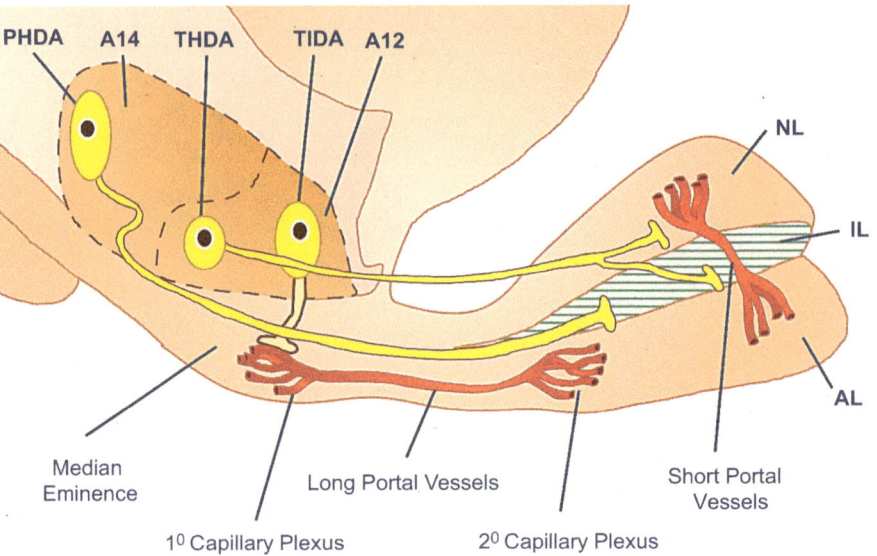

Based on anatomical and functional studies, two dopaminergic systems that regulate prolactin (PRL) were initially identified: the TIDA and the THDA. Further refinements in neuronal tracing techniques revealed that most of the THDA neurons projecting to the neurointermediate lobe actually originate from the A14 cells in the periventricular nucleus and therefore were termed the periventricular-hypophysial dopaminergic (PHDA) system (**Figure 3.11**). Because of their lower abundance and heterogeneous distribution, the hypothalamic dopaminergic neurons are not as well characterized as their nigrostriatal counterparts.

The **neurohypophysial magnocellular** neurons are characterized by axonal dilatation, also called Herring bodies, which are composed of a large number of neurosecretory granules measuring 200–350 nm in diameter [40]. In addition to vasopressin and oxytocin, these neurons produce and transport other peptides to the neural lobe, including dynorphin, enkephalin, galanin, cholecystokinin, TRH, VIP, GnRH, neuropeptide Y, substance P, CRH, and endothelin. Presumably, some of the hypothalamic hormones and neuromodulators can reach the anterior pituitary not only via the portal vasculature but also by diffusion from the neural lobe to the anterior love through the avascular IL.

Information on the physical and functional association of DA with the pituitary gland is based on two comprehensive reviews [41,42]. DA projections to the intermediate and neural lobes originate from the anterior and central portions of the arcuate nucleus, respectively. In the neural lobe, the dopaminergic nerve endings are found in close proximity to pituicytes (specialized glial cells of the neural lobe), magnocellular axon terminals, and precapillary spaces. In the avascular IL, they make close contact with the melanotrophs. In spite of the general notion that the IL is composed of a homogeneous population of cells, two distinct subpopulations of melanotrophs: **proopiomelanocortin (POMC)**-expressing, and non-POMC-expressing, have been identified and which differ in secretory activity, staining properties and receptor expression. The IL also has other cell types, including marginal, folliculo stellate and interstitial cells.

Folliculostellate cells (FS cells) are star-shaped follicle-forming cells that are found in both the anterior and ILs [43]. They are non-endocrine, agranular cells that are positive for S-100 protein, which serves as a marker for these cells. The functions ascribed to FS-cells include formation of an extensive three-dimensional network, scavenger activity by engulfing degenerated cells, paracrine regulation of endocrine cells by producing various growth factors and cytokines, and intercellular communication via their long cytoplasmic processes and gap junctions.

The concentration of DA in the rat neurointermediate lobe is similar to that in the medial basal hypothalamus but is lower than that in the median eminence [44]. Although the dopaminergic terminals in the neural lobe are neurosecretory, many terminals in the IL form synaptic-like contacts with the melanotrophs and suppress their proliferation as well as inhibit the release of β-endorphin and α-MSH (melanocyte-stimulating hormone). Notably, synaptic contacts between neurons and nonneuronal cells are uncommon. The presence of synapses, as well as the detection of spontaneous electrical activity in the melanotrophs, are more typical of the axon-less, catecholamine-producing adrenal chromaffin cells than most endocrine cells. Although the exact cellular origin of the melanotrophs is unclear, by analogy to chromaffin cells, they may arise from neuro-ectodermal progenitors that have lost their axons and migrated to an ectopic site.

The posterior pituitary express TH and DDC, but not DA-β-hydroxylase [44]. Isolated rat posterior pituitaries can synthesize DA *de novo* from tritiated tyrosine, with negligible production of NE. The source of these biosynthetic enzymes appears to be central, i.e., by transport from the THDA pathway rather than local. This is based on the findings that within seven days after pituitary stalk transection, DA concentrations in the posterior pituitary is reduced by more than 80% [42]. A portion of local NE, however, originates from the superior cervical ganglia. There are also diurnal variations in the concentrations of DA and NE in the posterior pituitary, with the highest levels observed during daylight hours.

There is significant genetic, biosynthetic, and functional diversity among the TIDA and THDA systems [44]. For example, DA synthesis in the posterior pituitary increases significantly after dehydration, whereas that in the median eminence remains unaffected. Injection of DA agonists increases, while injection of DA antagonists decreases, Dopa accumulation in the posterior lobe but not in the median eminence. This suggests autoreceptor regulation in the posterior pituitary but its absence in the median eminence. Indeed, well-defined DA receptors are present in the IL but not in the median eminence.

Additional evidence on the diversity among the two dopaminergic systems comes from studies that have used the neurotoxin monosodium glutamate (MSG), which causes selective destruction of retinal and arcuate nuclear neuronal perikarya. MSG reduces DA in the medial basal hypothalamus by more than 60%, but it does not affect that in the posterior pituitary [45]. The unchanged levels of NE in the two sites are consistent with its origin in extrahypothalamic neurons. In addition, it has been reported [46] that DA levels in the median eminence of female Snell dwarf mice are markedly reduced, while DA concentrations in the posterior pituitary are unaffected. Collectively, the above observations suggest that the perikarya and/or axon terminals of the TIDA and THD systems have different susceptibilities to pharmacological agents and to genetic defects.

3.5.4 The hypothalamo-pituitary complex: Vascular connections and DA levels in portal blood

Information on the pituitary vasculature is based on a review by Ben-Jonathan et al. [47]. Given the lack of direct innervation of the anterior pituitary, the **hypophysial portal vasculature** serves as the sole functional link between the brain and the anterior pituitary. By definition, portal vasculature refers to a system whereby a capillary bed pools into another capillary bed through veins, without first going through the heart. Both capillary beds and the blood vessels that connect them are considered as part of the portal venous system (**Figure 3.12**). The two best characterized portal systems in

Figure 3.12 Schematic illustration of the components of portal vasculature. By definition, a portal vasculature is composed of two capillary beds connected by venules and from there a direct connection into arterioles and the arterial circulation.

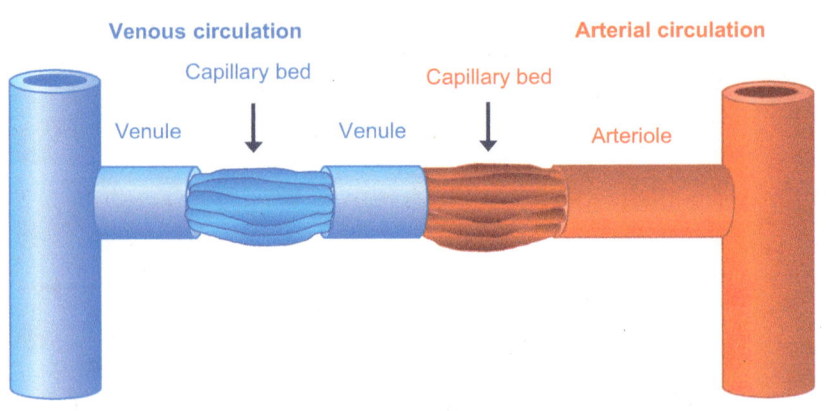

the body are the portal vasculature of the liver and the hypophysial portal vasculature. In the latter, the primary capillary bed is located in the median eminence, while the secondary capillary beds are located at both the anterior and neural lobes.

The capillaries in the median eminence, pituitary stalk and neural lobe are supplied by the superior, middle and anterior hypophysial arteries, respectively. Blood reaches the anterior lobe from the median eminence via the long portal vessels, which run along the pituitary stalk, and from the neural lobe via the short portal vessels, which bridge the avascular cleft of the pars intermedia. The secondary capillary plexus within the anterior pituitary is linked to the systemic venous circulation by the Y-shaped pituitary veins. **Figure 3.13** shows a schematic of the vascular and neural connections the hypothalamo-pituitary complex in humans.

Blood flows within the portal vasculature in different directions, depending upon the state of vasoconstriction in the various vascular beds. Such versatility of blood flow is well suited for exchanging information between the different components of the hypothalamo-pituitary system. Thus, the significant plasticity of the portal vasculature allows the delivery of hypothalamic or posterior pituitary agents to the anterior pituitary and also provides a route for transporting pituitary hormones to the hypothalamus for feedback information.

Studies in the mid-to-late 1970s established DA as the physiological inhibitor of PRL synthesis and release [48–50], followed later by increased evidence on the functions of DA as the regulator of lactotroph proliferation and electrical activity, as discussed in **Chapter 5**. As a mandatory criterion for the validation of a substance as a hypophysiotropic hormone, its presence in the hypophysial portal blood must be demonstrated. To this end, a variety of surgical techniques have been developed for portal blood collection in rats, sheep, and monkeys. In spite of some drawbacks resulting from anesthesia and surgical stress, this approach has been indispensable for studying many aspects of the hypothalamo-pituitary interactions.

Higher concentrations of DA in portal than in arterial plasma of rats was first reported by us and others in the 1970s [48,50]. This was followed few years later by similar findings in monkeys [51], and sheep [52]. Most information on the presence of DA in hypophysial portal blood under various physiological conditions and its relationship to the regulation of anterior pituitary hormone release have been made using the rat as the experimental model. Unfortunately, the mouse is too small for this surgical procedure. Hence, significant amounts of important information that could come from studying different genetic mouse models cannot be materialized.

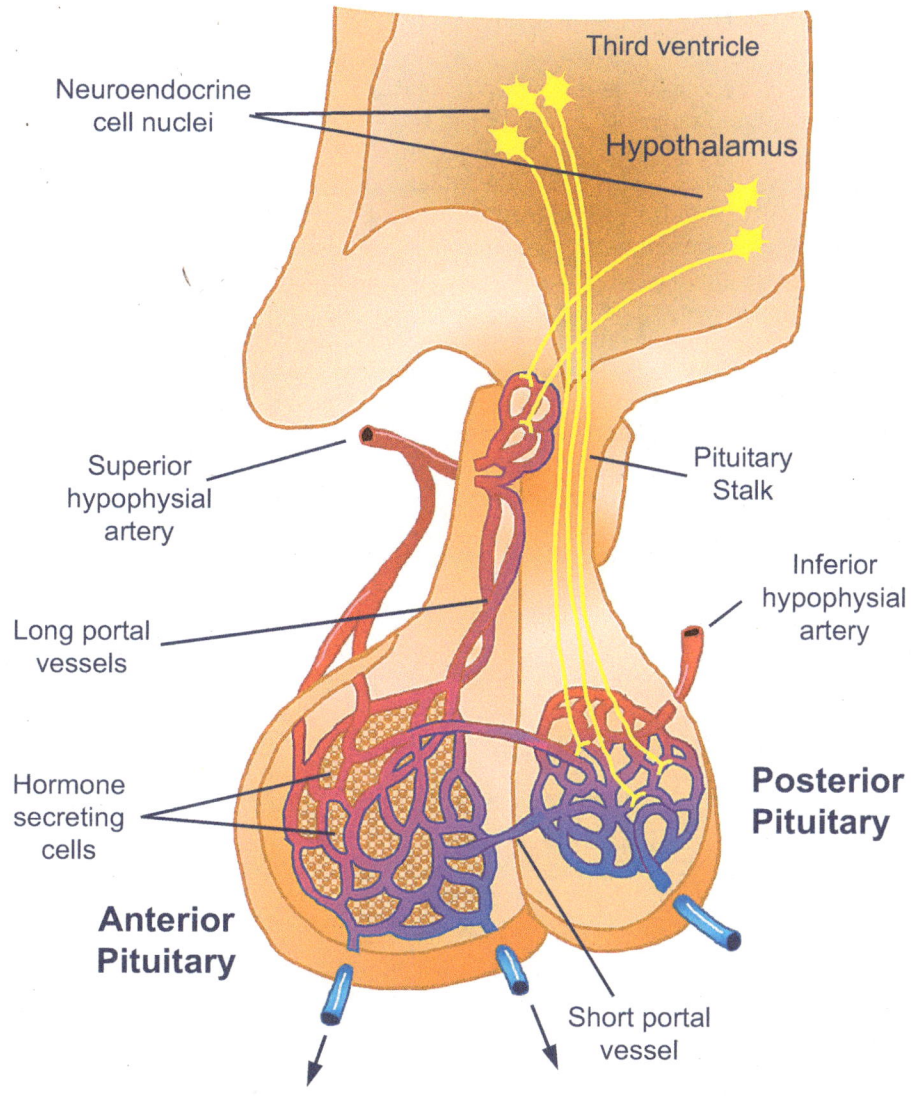

Figure 3.13 Neural and vascular connections of the human hypothalamo-pituitary complex. The hypophysial portal vasculature shows the primary and secondary capillary plexuses in the median eminence and anterior lobe, respectively, and the short portal vessels connecting the posterior and anterior pituitaries. Magnocellular neurons carrying primarily vasopressin and oxytocin connect the hypothalamus with the posterior pituitary. The TIDA and THDA dopaminergic tracts have terminals in the posterior pituitary and median eminence, respectively.

As previously reviewed [41,44], DA that is secreted into portal blood is primarily derived from a newly synthesized pool rather than from a storage pool. Administration of catecholamine enzyme blockers causes a prompt reduction in DA levels in portal blood and a subsequent rise in circulating PRL. Of the total amount of DA synthesized in the median eminence, only 15%–20% is released into portal blood. Therefore, DA turnover rates, which are based on the whole pool of DA in the median eminence, do not necessarily reflect the changes in the small fraction which is released into portal blood.

Unlike the long portal vessels that lie on the surface of the stalk, the short portal vessels bridging the neural and anterior lobes are extremely minute and technically inaccessible for cannulation. Therefore, a direct determination of DA levels in these vessels is virtually impossible, and indirect methods must be employed. One such method is posterior pituitary lobectomy (LOBEX). The rationale for this approach is that the removal of the PP should result in a rise in plasma PRL level which is proportional to the DA output from the gland. Indeed, acute LOBEX in rats induces a 2- to 3-fold elevation of plasma PRL levels, without having an effect on plasma LH levels [53]. Furthermore, intracarotid injections of DA immediately reverse the LOBEX-induced rise in PRL.

3.6 SYNOPSIS

The term catecholamines is derived from their structure, which incorporates an amine side chain and a catechol (dihydroxyphenyl) ring. The catecholamine family comprises DA, NE, and Epi. Although they differ only in a hydroxyl and a methyl group, respectively, they have a dissimilar distribution within the brain, activate different sets of receptors, and exert distinct functions. Over the last 50–60 years, the anatomy and physiology of the brain catecholaminergic neurons have been documented in great detail.

In addition to regulating important neurological functions such as locomotion, cognition and reward, the brain dopaminergic neurons are a critical component of the neuroendocrine functions of the brain. Three dopaminergic pathways, the IHDA, TIDA, and THDA, originate in the hypothalamus, supply DA to most hypothalamic nuclei and link the nervous system to the endocrine system via the pituitary gland. Within the pituitary gland, only the posterior pituitary (both neural and ILs) is directly connected to the hypothalamus through a nerve tract, whereas the anterior lobe is not innervated but, rather, receives hypothalamic information through the specialized hypophysial portal vasculature.

REFERENCES

1. Bjorklund A, Dunnett SB. Dopamine neuron systems in the brain: An update. *Trends Neurosci.* 2007;30(5):194–202.

2. Moore RY, Bloom FE. Central catecholamine neuron systems: Anatomy and physiology of the norepinephrine and epinephrine systems. *Annu Rev Neurosci.* 1979;2:113–168.

3. Waterhouse BD, Navarra RL. The locus coeruleus-norepinephrine system and sensory signal processing: A historical review and current perspectives. *Brain Res.* 2019;1709:1–15.

4. Mejias-Aponte CA. Specificity and impact of adrenergic projections to the midbrain dopamine system. *Brain Res.* 2016;1641(Pt B):258–273.

5. Chandler DJ, Waterhouse BD, Gao WJ. New perspectives on catecholaminergic regulation of executive circuits: Evidence for independent modulation of prefrontal functions by midbrain dopaminergic and noradrenergic neurons. *Front Neural Circuits.* 2014;8:53–65.

6. Fernandez-Espejo E, Armengol JA, Flores JA, Galan-Rodriguez B, Ramiro S. Cells of the sympathoadrenal lineage: Biological properties as donor tissue for cell-replacement therapies for Parkinson's disease. *Brain Res Brain Res Rev.* 2005;49(2):343–354.

7. Ritter S, Llewellyn-Smith I, Dinh TT. Subgroups of hindbrain catecholamine neurons are selectively activated by 2-deoxy-D-glucose induced metabolic challenge. *Brain Res.* 1998;805(1–2):41–54.

8. Mefford IN. Are there epinephrine neurons in rat brain? *Brain Res.* 1987;434(4):383–395.

9. Moore RY, Bloom FE. Central catecholamine neuron systems: Anatomy and physiology of the dopamine systems. *Annu Rev Neurosci.* 1978;1:129–169.

10. Van den Heuvel DM, Pasterkamp RJ. Getting connected in the dopamine system. *Prog Neurobiol.* 2008;85(1):75–93.

11. Stark AK, Pakkenberg B. Histological changes of the dopaminergic nigrostriatal system in aging. *Cell Tissue Res.* 2004;318(1):81–92.

12. Lindvall O. Developing dopaminergic cell therapy for Parkinson's disease—Give up or move forward? *Mov Disord.* 2013;28(3):268–273.

13. Mendes FD, et al. Therapy with mesenchymal stem cells in Parkinson Disease: History and perspectives. *Neurologist.* 2018;23(4):141–147.

14. Jackson CR, et al. Retinal dopamine mediates multiple dimensions of light-adapted vision. *J Neurosci.* 2012;32(27):9359–9368.

15. Popova E. Role of dopamine in distal retina. *J Comp Physiol A Neuroethol Sens Neural Behav Physiol.* 2014;200(5):333–358.

16. Bonzano S, Bovetti S, Gendusa C, Peretto P, De MS. Adult born olfactory bulb dopaminergic interneurons: Molecular determinants and experience-dependent plasticity. *Front Neurosci.* 2016;10:189–198.

17. Galliano E, et al. Embryonic and postnatal neurogenesis produce functionally distinct subclasses of dopaminergic neuron. *Elife.* 2018;7:e32373.

18. Hernandez G, et al. Tyrosine hydroxylase activity in peripherally denervated rat pineal gland. *Neurosci Lett.* 1994;177(1–2):131–134.

19. Moujir F, Santana C, Hernandez FJ, Reiter RJ, Abreu P. Daily time course of the contents in monoamines and their metabolites in the pineal gland of Syrian hamster. *Neurosci Lett.* 1997;223(2):77–80.

20. Biran J, Tahor M, Wircer E, Levkowitz G. Role of developmental factors in hypothalamic function. *Front Neuroanat*. 2015;9:47–63.

21. Swaab DF, et al. Functional neuroanatomy and neuropathology of the human hypothalamus. *Anat Embryol*. 1993;187(4):317–330.

22. Porter JC, Mical RS, Ben-Jonathan N, Ondo JG. Neurovascular regulation of the anterior hypophysis. *Recent Prog Horm Res*. 1973;29:161–198.

23. Ciofi P, Lapirot OC, Tramu G. An androgen-dependent sexual dimorphism visible at puberty in the rat hypothalamus. *Neuroscience*. 2007;146(2):630–642.

24. Mohawk JA, Takahashi JS. Cell autonomy and synchrony of suprachiasmatic nucleus circadian oscillators. *Trends Neurosci*. 2011;34(7):349–358.

25. Brown CH. Magnocellular neurons and posterior pituitary function. *Compr Physiol*. 2016;6(4):1701–1741.

26. Yang T, Shah NM. Molecular and neural control of sexually dimorphic social behaviors. *Curr Opin Neurobiol*. 2016;38:89–95.

27. McClellan KM, Parker KL, Tobet S. Development of the ventromedial nucleus of the hypothalamus. *Front Neuroendocrinol*. 2006;27(2):193–209.

28. Eaton MJ, Wagner CK, Moore KE, Lookingland KJ. Neurochemical identification of A13 dopaminergic neuronal projections from the medial zona incerta to the horizontal limb of the diagonal band of Broca and the central nucleus of the amygdala. *Brain Res*. 1994;659(1–2):201–207.

29. Ajika K, Hokfelt T. Projections to the median eminence and the arcuate nucleus with special reference to monoamine systems: Effects of lesions. *Cell Tissue Res*. 1975;158(1):15–35.

30. Demarest KT, Moore KE. Lack of a high affinity transport system for dopamine in the median eminence and posterior pituitary. *Brain Res*. 1979;171(3):545–551.

31. Revay R, Vaughan R, Grant S, Kuhar MJ. Dopamine transporter immunohistochemistry in median eminence, amygdala, and other areas of the rat brain. *Synapse*. 1996;22(2):93–99.

32. Palkovits M, Brownstein M, Saavedra JM, Axelrod J. Norepinephrine and dopamine content of hypothalamic nuclei of the rat. *Brain Res*. 1974;77(1):137–149.

33. Palkovits M, et al. Noradrenergic innervation of the rat hypothalamus: Experimental biochemical and electron microscopic studies. *Brain Res*. 1980;191(1):161–171.

34. Cagampang FR, Okamura H, Inouye S. Circadian rhythms of norepinephrine in the rat suprachiasmatic nucleus. *Neurosci Lett*. 1994;173(1–2):185–188.

35. Fuller RW. Pharmacology of brain epinephrine neurons. *Annu Rev Pharmacol Toxicol*. 1982;22:31–55.

36. Kalra SP. Catecholamine involvement in preovulatory LH release: Reassessment of the role of epinephrine. *Neuroendocrinology*. 1985;40(2):139–144.

37. Knigge KM, Scott DE. Structure and function of the median eminence. *Am J Anat*. 1970;129(2):223–243.

38. Goodman T, Hajihosseini MK. Hypothalamic tanycytes-masters and servants of metabolic, neuroendocrine, and neurogenic functions. *Front Neurosci*. 2015;9:e387.

39. de Moraes DC, Vaisman M, Conceicao FL, Ortiga-Carvalho TM. Pituitary development: A complex, temporal regulated process dependent on specific transcriptional factors. *J Endocrinol*. 2012;215(2):239–245.

40. Stopa EG, LeBlanc VK, Hill DH, Anthony EL. A general overview of the anatomy of the neurohypophysis. *Ann N Y Acad Sci*. 1993;689:6–15.

41. Ben-Jonathan N. Dopamine: A prolactin-inhibiting hormone. *Endocr Rev*. 1985;6(4):564–589.

42. Saavedra JM. Central and peripheral catecholamine innervation of the rat intermediate and posterior pituitary lobes. *Neuroendocrinology*. 1985;40(4):281–284.

43. Devnath S, Inoue K. An insight to pituitary folliculo-stellate cells. *J Neuroendocrinol*. 2008;20(6):687–691.

44. Ben-Jonathan N, Hnasko R. Dopamine as a prolactin (PRL) inhibitor. *Endocr Rev*. 2001;22(6):724–763.

45. Heiman ML, Ben-Jonathan N. Increase in pituitary dopaminergic receptors after monosodium glutamate treatment. *Am J Physiol*. 1983;245(3):E261–E265.

46. Morgan WW, Bartke A, Pfeil K. Deficiency of dopamine in the median eminence of Snell dwarf mice. *Endocrinology*. 1981;109(6):2069–2075.

47. Ben-Jonathan N, Arbogast LA, Hyde JF. Neuroendocrine regulation of prolactin release. *Prog Neurobiol*. 1989;33(5–6):399–447.

48. Ben-Jonathan N, Oliver C, Weiner HJ, Mical RS, Porter JC. Dopamine in hypophysial portal plasma of the rat during the estrous cycle and throughout pregnancy. *Endocrinology*. 1977;100(2):452–458.

49. MacLeod RM, Lehmeyer JE. Studies on the mechanism of the dopamine-mediated inhibition of prolactin secretion. *Endocrinology*. 1974;94(4):1077–1085.

50. Gibbs DM, Neill JD. Dopamine levels in hypophysial stalk blood in the rat are sufficient to inhibit prolactin secretion *in vivo*. *Endocrinology*. 1978;102(6):1895–1900.

51. Neill JD, Frawley LS, Plotsky PM, Tindall GT. Dopamine in hypophysial stalk blood of the rhesus monkey and its role in regulating prolactin secretion. *Endocrinology*. 1981;108(2):489–494.

52. Thomas GB, et al. Concentrations of dopamine and noradrenaline in hypophysial portal blood in the sheep and the rat. *J Endocrinol*. 1989;121(1):141–147.

53. Peters LL, Hoefer MT, Ben-Jonathan N. The posterior pituitary: Regulation of anterior pituitary prolactin secretion. *Science*. 1981;213(4508):659–661.

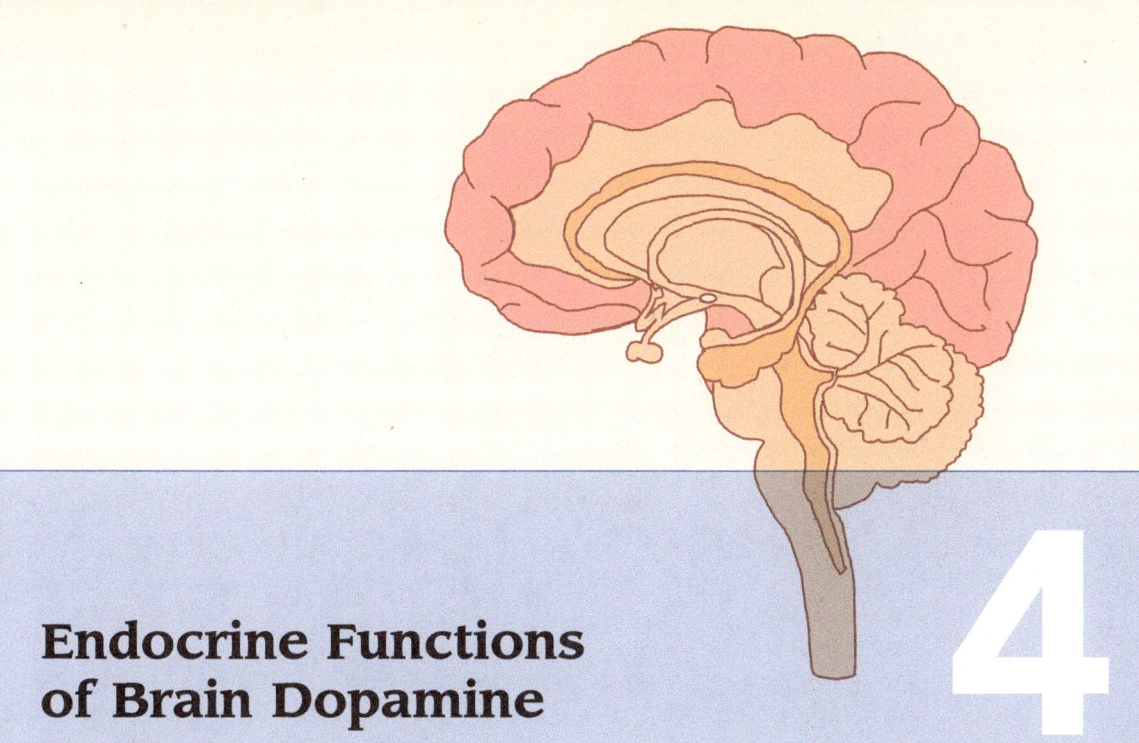

Endocrine Functions of Brain Dopamine

4

4.1 INTRODUCTION

Given the prominence of dopamine (DA) as a critical neurotransmitter in health and disease, there is a vast literature on its actions within the brain, especially in conjunction with neurological and psychiatric disorders. Yet, much less attention has been given to the neuroendocrine functions of brain DA, which is the subject of this chapter. To this end, we have selected four major functions for which there is substantial evidence on the involvement of brain DA: circadian rhythms, stress response, food intake, and reproduction/sexual behavior.

Circadian clocks generate self-sustaining, cell-autonomous oscillations with a time period of approximately 24 h. The features of a circadian clock in all organisms include its persistence under constant conditions, an oscillation that is temperature compensated, and an entrainment to external input. The neural circuitry that regulates the circadian rhythms includes the retina, suprachiasmatic nucleus, and the pineal gland. These structures operate under bidirectional neural and hormonal control to and from various organs and affect multiple physiological and endocrine functions.

Stress is a state of threatened homeostasis caused by extrinsic adverse forces (stressors) and is counteracted by an intricate repertoire of physiologic and behavioral responses aiming at maintaining/reestablishing the optimal body equilibrium. Stress affects two major systems in opposite ways: It activates the hypothalamo-pituitary adrenal (HPA) axis (fight and flight response), while suppressing the hypothalamo-pituitary gonadal (HPG) axis.

Obesity and its opposite disorder anorexia are pervasive problems of humans in modern times. There is clearly a need for better understanding of the mechanisms that regulate appetite versus energy expenditure because the balance of the two processes ultimately determine body weight. Within the brain, the hypothalamus serves as a relay station that receives information on the state of energy fluxes by pancreatic, adipose, stomach and intestinal hormones and coordinates endocrine, autonomic, and behavioral responses that regulate feeding behavior and appetite around an optimal set point.

Reproduction fulfills two broad functions. One is to ensure the optimal conditions for the conception and rearing of the offspring of each individual, and another is the preservation of the species. More than any other physiological functions, each species has developed specific patterns of reproductive cycles, sexual behavior, as well as length of gestation and lactation, that are best suited for its social structure and living environment. The neuroendocrine system controls and coordinates both the physiological and the behavioral aspects of reproduction.

4.2 CIRCADIAN RHYTHMS

4.2.1 Circadian rhythms and their impact on health and disease

Living organisms have an internal clock that helps the body adapt to the external environment. Throughout the 24-h day, the clock regulates many physiological, endocrine, and mental processes, including sleep, body temperature, metabolism, blood pressure, hormone release and alertness. Circadian rhythms are endogenously generated ~24-h biological rhythms that are organized in two levels: a molecular level represented by the clock genes, and a systemic regulatory level represented by the neuroendocrine networks. The circadian oscillation is synchronized by external environmental cycles, primarily the light/dark cycle of the geophysical day and night. External rhythmic events that can synchronize biological rhythms are called zeitgebers or synchronizers. "**Entrainment**" is defined as a synchronization between oscillators with different but similar periods that occurs when one of the oscillators imposes its period on the other.

The importance of the circadian rhythm for multiple organisms, from flies to man, was recognized by the granting of the 2017 Noble Prize in Physiology and Medicine to three U.S. scientists, Michael Rosbash, Jeffrey C. Hall, and Michael W. Young. Working with fruit flies in the 1980s, they discovered a gene named *period*, which produces the PER protein. PER was found to fluctuate over the 24-h sleep-wake cycle, building up overnight and decreasing during daytime. In 1994, Michael Young discovered a second clock gene, *timeless*, encoding the TIM protein, which is required to maintain a self-sustaining circadian rhythm. Upon binding of TIM to PER, the two proteins can enter the cell nucleus and block the activity of *period*, thereby constituting an inhibitory feedback loop.

In subsequent years, various proteins that regulate the circadian rhythms in mammals have been isolated and characterized. Although an in-depth discussion on the molecular mechanisms that govern circadian rhythms is beyond the scope of this chapter, the interested reader can find excellent information in two comprehensive reviews [1,2]. The short background information presented below is largely based on those reviews.

The molecular mechanism of the endogenous circadian clock in mammals comprises feedback loops of cyclic gene products that apply negative and positive transcriptional regulation of "clock" genes and proteins. The cell-autonomous circadian clocks are generated by a feedback loop composed of the clock genes "circadian locomotor output cycles kaput" (Clock) and "brain and muscle ARNT-like 1" (Bmal1), which produce the proteins CLOCK and BMAL1, respectively. These transcription factors heterodimerize and bind to promotor elements to up-regulate the expression of period (Per1–3) and cryptochrome (Cry1–2). Subsequently, PER and CRY proteins form heterodimers that interact with the BMAL1:CLOCK complex to repress their own transcription.

The clock genes cycle in opposite phases and define the daily variations in physiological functions that shape the circadian rhythm. During the night, the PER-CRY complex is degraded, and BMAL1:CLOCK can start a new cycle of transcription. The entire cycle takes about 24 h to complete.

The BMAL1:CLOCK heterodimer also activates the transcription of other clock genes, such as reverse erythroblastosis virus (Rev-Erb) a and b, which repress Bmal1 gene expression. Several kinases and phosphatases that regulate the speed, precision, and functions of the circadian clock also contribute to circadian oscillations. Other important regulatory factors include ubiquitination, acetylation/deacetylation and methylation as well as input from microRNAs and several RNA-binding protein complexes.

The body's circadian rhythm is closely linked to health and disease. Every day, the normal sleep-wake cycle is regulated by the intrinsic clocks that increase the release of **melatonin**, the "sleep" hormone, in the evening and elevate the serum levels of **cortisol**, the "stress and alertness" hormone, upon awakening in the morning. The explosion in the use of melatonin to regulate sleep in otherwise healthy individuals has led to the spending of more than $400 million by U.S. consumers on melatonin supplements in 2018. Synthetic melatonin is made in factories that are not regulated by the U.S. Food and Drug Administration. Most commercial products are offered at dosages that can cause blood melatonin levels to rise much above the endogenously produced levels in the body. A typical dose of 1–3 mg can elevate blood melatonin levels by as much as 10–20 times above normal. For melatonin to be helpful, the correct dosage, method, and the time of day it is taken must be appropriate to the sleep problem. Taking it at the "wrong" time of the day may reset the biological clock in an undesirable direction [3].

As shown in **Figure 4.1**, circadian changes in humans encompass a wide spectrum of physiological, endocrine and mental functions, including blood pressure, body temperature, alertness, mood, and coordination. These changes are driven by interactions between central neural and neuroendocrine control mechanisms, intrinsic peripheral clocks in different organs, local cytokines, and circulating hormones. Short-term misalignments between the internal clock and the external environment, as occur following a jetlag, can impose acute disruptive effects on the sleep-wake rhythm. Prolonged or

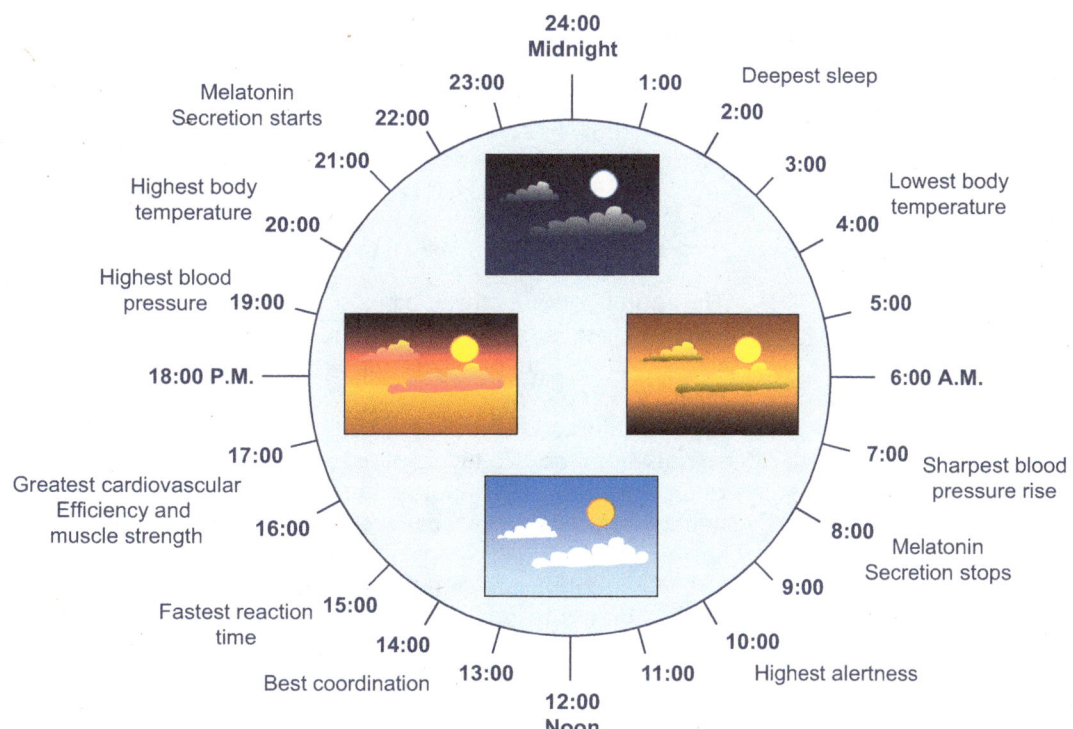

Figure 4.1 Changes in circadian rhythm in humans. Shown are circadian changes in physiological, hormonal, and behavioral parameters. (Redrawn and modified from Joe Cohen, Circadian Rhythms, selfhacked.com.)

chronic disruptions of the circadian rhythms are linked to diverse pathogenic processes such as sleep disorders, depression, metabolic syndrome, obesity, cardiovascular diseases and even carcinogenesis. Notably, the molecular clock is also linked to the control of energy balance, as evident by the fact that both *Clock* mutant and *PER* mutant mice develop obesity.

Mammals have peripheral clocks with the capacity to generate endogenous rhythms in nearly all organs and tissues. The **suprachiasmatic nucleus (SCN)** of the hypothalamus serves as the ultimate pacemaker, or master clock, of the circadian rhythm. Under normal conditions, the various peripheral clocks are entrained to the light-dark cycle by an integrated input from the master clock and its associated neuroendocrine networks. Some peripheral clocks can be synchronized by feeding rhythms as the dominant signals. Hence, irregular feeding rhythms may lead to uncoupling of peripheral clocks from the SCN master clock, with adverse effects on fitness and even health.

4.2.2 The neurocircuitry that controls circadian rhythms and the role of dopamine

The brain circuitry that regulates the circadian rhythms is composed of three major components: the retina, the SCN, and the pineal gland. In addition, modulating functions are provided by some hypothalamic nuclei, the striatum, and several circulating hormones. Accumulating evidence has established a significant involvement of DA in the regulation of each site within the core circadian neurocircuitry. In addition, DA synthesis and release, which are altered by the circadian oscillations, have a direct impact on the diurnal of release of pituitary hormones such as prolactin (PRL) and an indirect modulatory effect on the circadian rhythmicity of several other hormones, as detailed below.

The critical role of DA in affecting the circadian rhythms is highlighted by the recognition that alterations of circadian rhythms are among the most debilitating non-motor symptoms in **Parkinson's Disease (PD)**. As was discussed in **Chapter 3**, PD is a neurodegenerative disease, characterized by the loss of dopaminergic neurons in the nigrostriatal and mesolimbic pathways, including the ventral tegmental area (VTA). The most common circadian abnormality in PD patients is alteration of the sleep-wake cycle. Comprehensive information on the role of DA as a modulator of circadian rhythms at the central nervous system can be found in an extensive review [4] and is briefly summarized below.

4.2.3 Dopamine and retinal rhythmicity

The retina contains three types of photoreceptors: rods, cones and intrinsically photosensitive retinal ganglion cells (ipRGCs). Rods are responsible for vision at low light levels (scotopic vision). Cones are active at higher light levels (photopic vision), are capable of color vision, and are responsible for high spatial acuity. Ganglion cells collect electrical messages of the visual signals from the two layers of neurons and serve as the final neuronal output of the retina [5]. The mammalian retina is not only a light-sensing tissue that conveys photic information to the brain, but it also has an intrinsic circadian system.

The retinal cells and pigment epithelium have circadian oscillators that are integrated through neural synapses, electrical coupling (gap junctions), and released neurochemicals. Ganglion cells produce the photopigment melanopsin, which plays a non-image-forming role in the setting of the circadian rhythms. Fibers from the ganglion cells form the retino-hypothalamic tract that innervates the SCN. These fibers contain two substances, pituitary adenylate cyclase activating polypeptide (PACAP) and glutamate, both of which regulate, through their respective receptors, the circadian rhythms of the SCN.

Depending upon the species, DA is produced in amacrine and/or inter-plexiform cells of the retina [6]. A description of the DA-producing cells in the retina is presented in **Chapter 3**, and an illustration of their cell-to-cell connections is shown in **Figure 3.6**. The retinal dopaminergic neurons fire spontaneously at a modest rate, which is modulated by excitatory glutamatergic and inhibitory glycinergic and GABAergic synaptic inputs to their dendrites. Endogenous retinal DA release is low during darkness and increases during exposure to constant light. Acting via D2R, DA regulates the rhythmic transcription of melanopsin, thereby influencing the light entrainment signals sent to the SCN. Acting via D1R, DA is involved in the coupling (opening) and uncoupling (closing) of gap junctions, which are important for communication between the different retinal layers.

A rise of DA preceded the rise in PER2, and the elimination of DA from the brain, or the blockade of one of its receptors, results in decreased PER2, which can be reversed by administration of DA agonists. It has been proposed that retinal circadian activity is divided into two phases: (1) a DA-independent phase at low light intensities, and (2) a DA-dependent phase at high light intensities. Both phases have an impact on signaling the night-day transitions to the SCN.

The retina produces its own melatonin that acts as an important local endogenous circadian signal. Retinal melatonin has reciprocal and antagonistic relationship with DA. For example, daily injections of melatonin into retinas of mice unable to synthesize melatonin, induced circadian rhythms of DA release, suggesting that the normal synthesis of retinal DA is controlled by the rhythmic release of melatonin from the photoreceptors. Another study used mice with a selective disruption of retinal tyrosine hydroxylase (TH) resulting in retina-specific DA deficiency [7]. These mice had a significant disruption in their circadian rhythm, contrast sensitivity and visual acuity, all of which were rescued by the delivery of D1R or D4R agonists.

4.2.4 Dopamine and the suprachiasmatic nucleus

The SCN, located in the hypothalamus above the optic chiasma, is a small bilateral nucleus composed of about 10,000 neurons. Neurons of the SCN can maintain their own rhythm in the absence of external cues [8]. The SCN processes and interprets the photic information obtained from the retina and passes it on to the pineal gland and to other regions in the central nervous system. The neural pathway that extends from the SCN to the pineal gland involves a multi-synaptic connection via the **superior cervical ganglion**. See **Chapter 3** for a description of the structure of the SCN and the pineal gland, and **Figure 3.7** for a diagram of their neural and endocrine connections.

The fetal SCN begins to oscillate before birth and is entrained to the maternal circadian rhythm [9]. In hamsters, D1R was first detected in the fetal SCN on embryonic day (E) 15, the day before birth in this species, and its expression persists through adulthood. Although TH-immunoreactive fibers coursing below the fetal SCN are first seen on day E15, TH-immunoreactive cells and fibers are detected in the SCN only by postnatal day 5. The authors postulated that the expression of D1R in the fetal SCN is consistent with their role in entraining the fetal circadian pacemaker to maternal cues. However, the receptor-transmitter mismatch between D1R and TH-positive fibers in the fetal SCN suggests that DA acts as a paracrine or a humoral signal in this entrainment. These findings are supported by another study [10], reporting that exposure to a D1R agonist (SKF-38393) during gestation suppresses the development of afferent connections to the fetal SCN. This deficiency causes changes in the initial response of the circadian timing system to light, as judged by a delay in light-induced *c-fos* expression in the SCN.

Studies with D1R-deficient adult mice have revealed that D1R-mediated DA transmission within the SCN is an integral component that determines the rate of entrainment following changes in the light cycle [11]. After identifying a direct connection from DA neurons of the VTA to the SCN, it was found that specific activation of these neurons accelerates the entrainment to a light cycle shift. Moreover, D1R null [D1R(-/-) or *Drd1*-knockout (KO)] mice have a slow rate of photo-entrainment in response to a phase shift in the light cycle, which can be rescued by re-expression of *Drd1* exclusively within the SCN.

There is also evidence for functional dopaminergic systems in circadian-driven brain sites in primates [12]. In both baboons and humans, *in situ* hybridization has revealed differential expression of D1R and D5R in several hypothalamic nuclei. D1R was expressed in the SCN, supraoptic nucleus (SON), and paraventricular nucleus (PVN), while D5R was expressed only in the SON and PVN. In newborn baboons, injection of the D1/D5 dopamine receptor agonist SKF 38393 at night increases the uptake of 2-deoxy-glucose (a marker of heightened neuronal activity) in the SCN, SON, and PVN. By contrast, *c-fos* mRNA expression is induced in the SON and PVN but not in the SCN. These data show that functional D1R and D5R are present in the primate hypothalamus and that acute stimulation of these receptors influences SCN, SON, and PVN activity.

4.2.5 Effects of DA on the pineal: Regulation of melatonin synthesis and release

The pineal gland is a small structure located on the epithalamus. Its main secretory product is melatonin, which is synthesized from tryptophan by four enzymatic steps in the indoleamine biosynthetic pathway (**Figure 4.2**). Dietary tryptophan is absorbed into the bloodstream from the gut and is circulated through the body. In cells that express tryptophan hydroxylase (TPOH), tryptophan is hydroxylated to 5-hydroxy-tryptophan in the mitochondria, followed by decarboxylation by aromatic amino acid decarboxylase (AADC; the same enzyme that converts L-Dopa to DA) to serotonin (5-hydroxy-tryptamine or 5-HT). Serotonin is the final product of the indoleamine pathway in many brain and peripheral locations, where it functions as a neurotransmitter and/or a hormone. In the pineal gland, however, serotonin is further acetylated

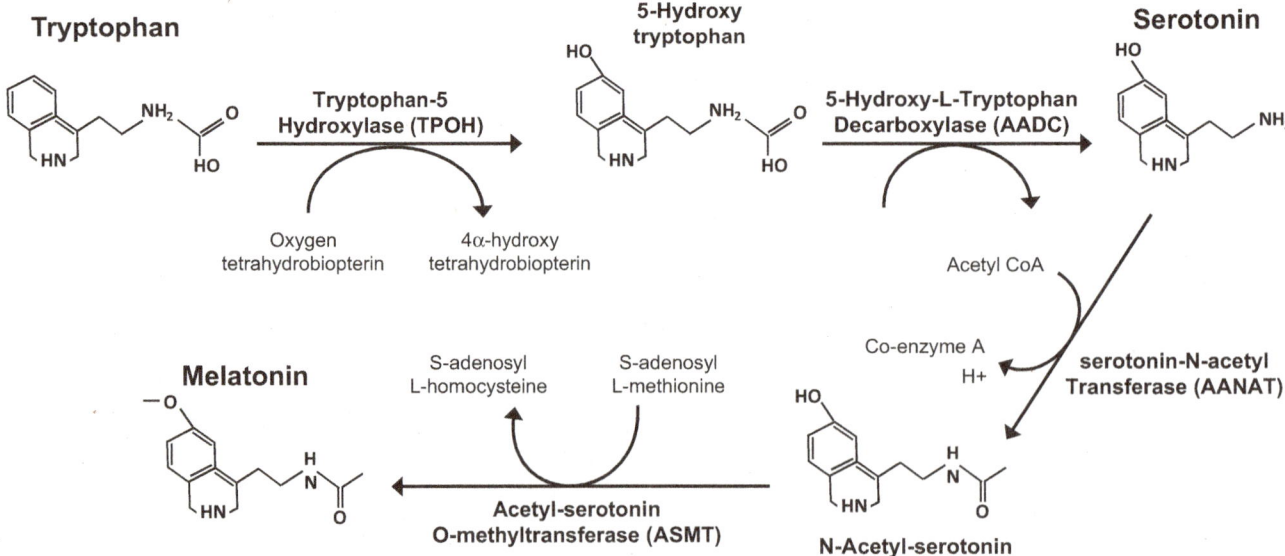

Figure 4.2 Biosynthesis of melatonin. The indoleamine biosynthetic pathway leading to the production of melatonin from tryptophan via serotonin as an intermediate.

on its free amine by serotonin N-acetyl transferase (AANAT), followed by O-methylation on the hydroxyl group by acetyl-serotonin-O-methyl transferase (ASMT), generating melatonin. Among the four biosynthetic enzymes in the pineal, the activities of TPOH and AANAT increase at night, while those of AADC and HIOMT run continuously.

Melatonin is produced at night and is secreted from the pineal into the cerebrospinal fluid (CSF) and blood circulation [13]. In continuous darkness, melatonin production in the pineal persists, while an acute light exposure at night rapidly stops its production. In the systemic circulation, melatonin undergoes first-pass metabolism in the liver, and its half-life in human serum is less than 1 h. Because melatonin is not stored in the pineal but is released as soon as it is made, serum melatonin levels closely reflect the rate of its synthesis in the pineal. In rats, the concentration of melatonin in the pineal increases by a factor of 10 from day to night, and by a factor of 5–10 in the blood. In addition to the pineal, melatonin is produced by other sites, including the retina, where it serves as a local cytokine rather than a secretable hormone.

As discussed in a recent review [3], melatonin is an amphiphilic compound, i.e., it has a polar, water-soluble group attached to a nonpolar, water-insoluble hydrocarbon chain. As such, melatonin can cross the cell, organelle, and nuclear membranes and directly interact with intracellular molecules in a non–receptor-mediated actions. Melatonin is an effective antioxidant that protects lipids, protein, and DNA from oxidative damage. The receptor-mediated actions of melatonin entail its binding to two receptors, MT_1 and MT_2, which are high-affinity G protein-coupled receptors encoded by the MTNR1A (human chromosome 4q35.1) and MTNR1B (human chromosome 11q21-q22) genes.

Both short- and long-term effects of melatonin depend on the target organs, local concentration of the hormone, type of cellular receptors and signaling system, duration of the signal, and the affinity and desensitization of the different receptor subtypes. For example, the immediate effects of melatonin depend on the phase of the melatonin circadian production cycle as follows: (1) rising evening phase, (2) phase of daily nocturnal peak, and (3) dawn falling phase at the end of the night. These changes ultimately determine the concentration of extracellular melatonin at any given time and the duration of its interaction with its targets and their sensitivity [3]. **Figure 4.3** depicts the light input into the pineal via the SCN and superior cervical ganglion (SCG) and broad effects of circulating melatonin on multiple systems, including body temperature, metabolic parameters, serum hormones, blood pressure and immune functions.

The use of immunofluorescence for DA and immunocytochemistry for the catecholamine biosynthetic enzymes have documented the catecholaminergic innervation of the pineal gland of many mammalian species: rats, rabbits, ferrets, ground squirrels, golden hamsters, and bovine. In some species, both TH and dopamine beta-hydroxylase (DBH) were detected in the pineal, indicating production of both DA and norepinephrine (NE), while in others, DBH was undetectable, indicating a predominance of DA. Some of the TH-positive fibers innervating the pineal originate from the superior cervical ganglion, but pinealocytes are also capable of local DA synthesis. For example, electrical stimulation of rat pineal glands incubated with ^{14}C-tyrosine enhanced the formation of ^{14}C-dopamine [14]. Compared to its basal levels at midday, the content of endogenous DA in the pineal at midnight increased by 450%, and that of NE by only 50%.

Depending upon the species, DA has been reported to alter melatonin production in the pineal. In the rat pineal, both N-acetyl transferase and melatonin content exhibit the expected rise 8 h (0400 hours) after the onset of darkness. Pineal DA and 3,4,dihydroxyphenylacetic acid (DOPAC; a major DA metabolite) levels are highest during the dark phase, reaching a peak at 0200 hours, just before the rise in melatonin production. However,

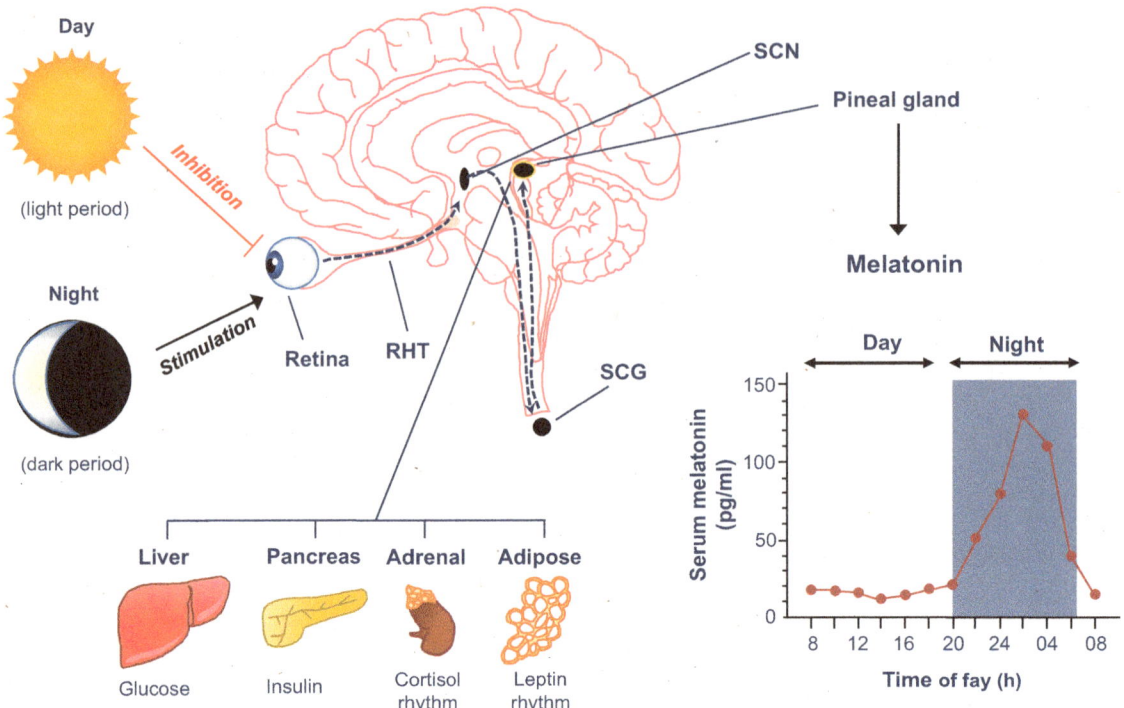

Figure 4.3 Entrainment of melatonin synthesis by the light/dark periods. Shown are the neural circuitry associated with the generation of melatonin, the circadian profile of serum melatonin concentrations, and the main target organs/hormones affected by circulating melatonin. SCN: suprachiasmatic nucleus; SGG: superior cervical ganglion; RHT: retinohypothalamic tract.

no significant difference has been found in NE and TH activity levels during the 24 h period in the hamster. These data suggest a potential role of pineal DA in the induction of melatonin synthesis, and indicate that catecholamines act differently on pineal glands from different species.

Among the five DARs, D4R is the dominant receptor in the rat pineal gland. The expression level of *Drd4* mRNA in pinealocytes is greater than that in most other tissues, and it is under photoneural control. In one study, DA caused a dramatic decrease in melatonin production in the pineal gland by activating heterodimers between the adrenergic receptors α_{1B} or β_1 and the D4R [15]. The authors postulated that such heteromers provide a rapid feedback mechanism for the neuronal hormone system to modulate circadian-controlled outputs.

4.2.6 Effects of dopamine on the circadian rhythms of circulating hormones

In humans, melatonin, PRL, growth hormone (GH), adrenocorticotropic hormone (ACTH), cortisol, and leptin all exhibit diurnal variations in their serum levels (**Table 4.1**). During nighttime sleep, melatonin, GH, PRL and leptin are elevated and ACTH and cortisol are suppressed, while serum cortisol levels rise upon waking in the morning. A direct involvement of DA in the regulation of the circadian rhythm is best established for melatonin and PRL, while there is only limited evidence that either brain or peripheral DA is a critical factor in the control of diurnal GH, cortisol and leptin release.

DA is the primary physiological inhibitor of PRL synthesis and release [16,17]. When DA input from the median eminence to the anterior pituitary is up, PRL release is down, and when DA input is down, PRL release is up. The ability of DA to confer circadian rhythmicity on PRL is not ubiquitous but, rather, depends on the species and the physiological conditions. Female rats have long served as an experimental model for studying the circadian-inducing capability of DA on PRL release [18]. During early pregnancy, PRL

Table 4.1 Circadian rhythms of selected circulating hormones

Time/hormone	Night	Morning	Noon	Afternoon	Evening
Melatonin	↑↑↑	↓	←→	←→	←→
GH	↑	←→[1]	←→	←→	←→
ACTH	↓	↑↑	↑	←→	←→
PRL	↑	←→	←→	↑[2]	←→
Leptin	↑	↓	←→	←→	↑
Cortisol	↓	↑↑	↑	←→	←→

1: Pulsatile release throughout the day.
2: Afternoon surge only during proestrus or early pregnancy (rodents).

secretion occurs in a semi-circadian rhythm, characterized by a large nocturnal surge and a slightly smaller diurnal surge (**Figure 4.4**). Both surges are necessary for the initiation and maintenance of luteal function. During this time, the tuberoinfundibular dopaminergic (TIDA) neurons exhibit a daily pattern of activity that is inversely related to PRL secretion. Thus, DA levels in hypophysial portal blood, TH activity in the median eminence, and expression of Fos-related antigens in dopaminergic neurons are all reduced during the PRL surges and are elevated in the inter-surge intervals when PRL is low. The PRL surges begin to reduce in magnitude on day 8 of pregnancy and are completely eliminated by day 10.

The physiological trigger for the initiation of the semi-circadian rhythm of PRL is uterine cervical stimulation during mating [19]. Such daily PRL surges can be reproduced either by a sterile mating or by an artificial stimulation of the cervix, resulting in a condition called pseudopregnancy, which lasts for about 10–11 days (**Figure 4.4**). When the cervically stimulated animals are placed in a 12:12 light/dark environment with lights on at 0600 hours,

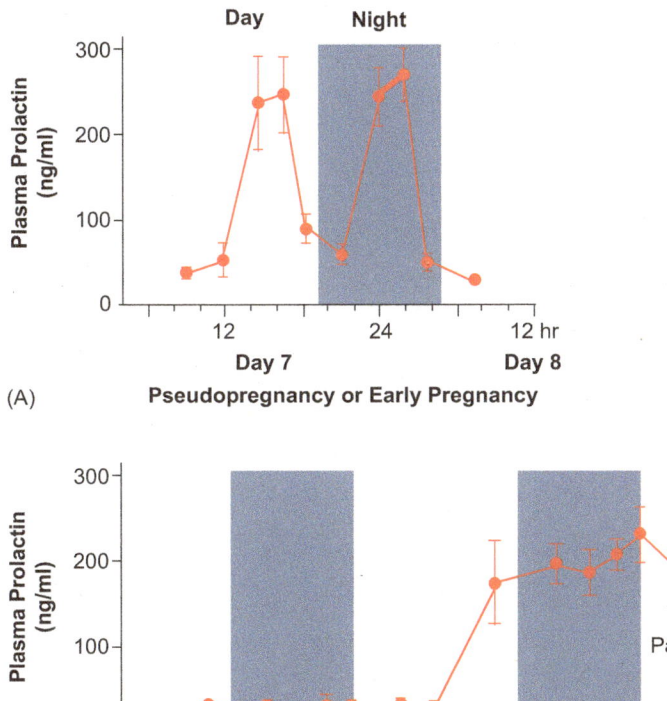

Figure 4.4 Semi-circadian profile of serum prolactin during early (A) and late (B) rat pregnancy. Redrawn and modified from Andrews, Z.B. et al., *Endocrinology*, 142, 2719–2724, 2001.

a nocturnal PRL surge occurs at about 0300 hours and a diurnal surge at 1700 hours [19]. This pattern of PRL secretion is observed for several days without reapplication of the stimulus and is independent of the time of day at which the stimulus was applied, indicating that it is driven by an endogenous circadian rhythm entrained by the master clock. Indeed, the TIDA neurons in the arcuate nucleus display rhythmic release of DA, which is out of phase with PRL release. However, DA is not the only substance responsible for the circadian rhythm of PRL and other neuroactive substances, e.g., oxytocin, vasopressin, gamma-aminobutyric acid (GABA), and feedback regulation by PRL on the TIDA neurons have all been implicated. Although a robust circadian pattern of PRL release is not seen in humans, PRL release is generally high at nighttime and low in daytime.

Three other human pituitary hormones: ACTH, thyroid-stimulating hormone (TSH) and GH also show diurnal changes in their secretion. ACTH release is stimulated by hypothalamic corticotropin-releasing hormone (CRH) and is inhibited by adrenal corticosteroids, while TSH is stimulated by thyrotropin releasing hormone (TRH) and is inhibited by thyroid hormones. GH release is regulated by two hypothalamic substances: growth hormone releasing hormone (GHRH) and somatostatin (SSI), and two peripheral hormones: insulin-like growth factor-1 (IGF-1), and GH releasing peptide (ghrelin). Although some studies have implicated DA in the control of the diurnal variations in the release of these pituitary hormones, especially by affecting the corresponding releasing/inhibiting hormones, the data are inconsistent and are not as clear as is the case for PRL.

Leptin release from adipose tissue is rhythmic on a daily basis (**Table 4.1**). The leptin rhythm is regulated by intrinsic clocks in adipocytes and by timing cues coming from the master SCN clock through autonomous nervous signals and neuroendocrine rhythms [i.e., adrenal glucocorticoids (GCs) and pineal melatonin]. In addition, feeding and the metabolic status markedly affect the shape of rhythmic leptin release by the adipocytes. In one study, short-term bromocriptine treatment has been shown to lower circulating leptin levels in obese women, suggesting that dopaminergic neurotransmission is involved in the control of leptin release in humans [20]. Presumably, the DA agonist could affect leptin release by acting centrally and/or by activating D2Rs expressed in human adipocytes [21].

4.3 THE NEUROENDOCRINE STRESS RESPONSE

4.3.1 Time-related, multi-facetted activation of the stress response

As discussed in a comprehensive review [22], stress is defined as a response to any external demand for change. Whether the applied stress is physical (i.e., bodily harm), emotional (i.e., verbal abuse), or perceived (i.e., fear and anxiety), it can have two opposing consequences on the well-being of the individual. One is positive, by empowering the organism to cope with, and adapt to, unexpected challenges in order to survive, the so-called "fight or flight" response. Another is negative: It is associated with certain disturbances in both behavior and normal bodily functions.

In humans, physical or mental stress can result in depression, anxiety, post-traumatic stress disorder (PTSD) as well as in drug and alcohol addiction. In addition to altering behavior and promoting neuropsychiatric disturbances, stress can interfere with normal cardiovascular, immune, endocrine, and reproductive functions and can cause some illness. **Figure 4.5** shows an example of the outcome of psychological and physical stress, which activates the hypothalamo-pituitary adrenal axis (HPA),

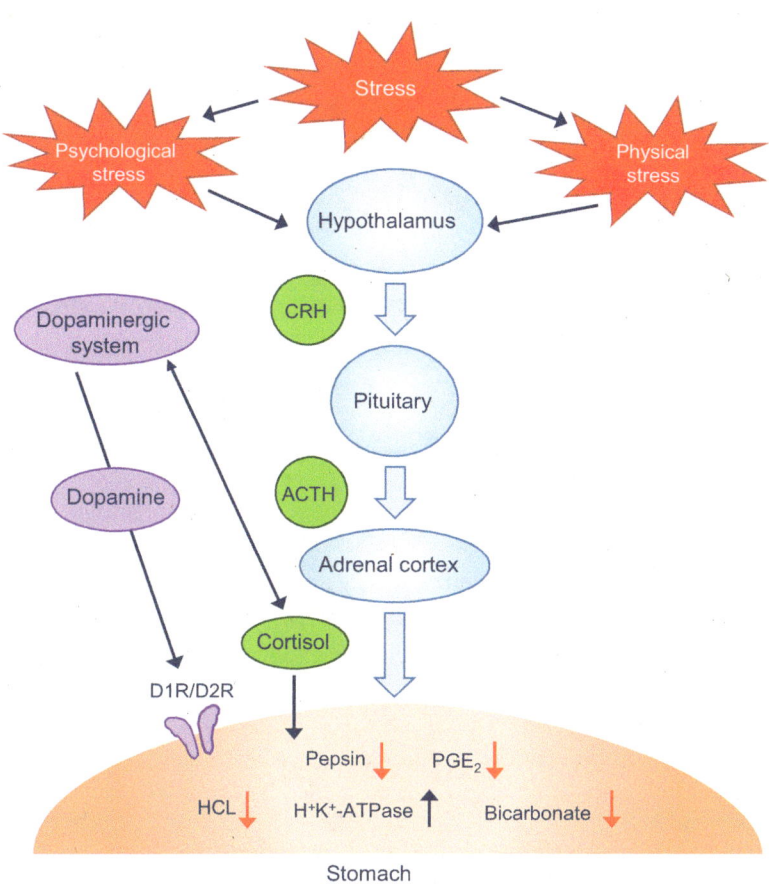

Figure 4.5 Activation of the hypothalamo-pituitary-adrenal axis (HPA) that generates gastric ulcers. Activation of the HPA by a variety of stressful stimuli and the dopaminergic system is involved in the induction of gastric ulcers. ACTH: adrenocorticotropic hormone; CRH: corticotropin-releasing hormone. (Redrawn and modified from Rasheed, N. and Alghasham, A., *Adv. Pharmacol. Sci.*, 2012, 1–12, 2012.)

and together with the dopaminergic system enhances the development of gastric ulcer, presumably by increasing the release of acidic digestive juices [23].

As depicted in **Table 4.2**, a typical stress response encompasses multiple central and peripheral changes that interact in a time-dependent manner [22]. **A first-wave** of response, occurring within few seconds, involves enhanced release of monoamine neurotransmitters, DA, NE and 5-HT in several brain sites, as well as increased CRH release from the hypothalamus, followed by stimulation of ACTH release from the pituitary. In parallel, the release of hypothalamic gonadotropin-releasing hormone (GnRH) is reduced, leading to a decreased secretion of the pituitary gonadotropins: luteinizing hormone (LH) and follicle-stimulating hormone (FSH). Additional changes that occur within a short timeframe involve the secretion of opiates, PRL, GH, glucagon, arginine-vasopressin (AVP) and renin. The first-wave stress response in the periphery includes rapid increases in NE release by activated sympathetic preganglionic neurons in the spinal cord, which in turn stimulate the splanchnic nerve that innervates the adrenal medulla.

The second-wave stress response, starting within several minutes, involves increased peripheral secretion of GCs, resulting from the activation of the HPA-axis, increased production and release of NE and epinephrine (Epi) from the adrenal medulla, and reduced secretion of gonadal steroids, i.e., ovarian estrogens (E2) and testicular androgens, resulting from the lower circulating levels of the gonadotropins. This neurohormonal stress response has an additional endocrine component in the form of glucagon, which together with GCs enhances the release of glucose, amino acids and fatty acids, as a coordinated catabolic response to stress.

Table 4.2 Rapid first wave and delayed second wave of the stress response

Site	First wave (seconds to minutes)	Second wave (minutes to hours)
Brain/pituitary	↑DA, NE, 5-HT	
	↑CRH/ACTH	
	↓GnRH/LH, FSH	↑PRL, AVP, Renin
	↑Opioids	
Periphery	↑Sympathetic activity	
	↑Adrenal medulla	↑NE, Epi,
	↑Adrenal cortex	↑GCs
	↓Gonads	↓E2, Testosterone
Feedback on brain	↑NE/Epi via vagus	↑NE from locus coeruleus
		↑GCs genomic actions

Source: Stockhorst, U. and Antov, M.I., *Front Behav. Neurosci.*, 9, 359–372, 2015.
ACTH: adrenocorticotropic hormone; AVP: arginine-vasopressin; CRH: corticotropin-releasing hormone; DA: dopamine; E2: estrogen; Epi: epinephrine; FSH: follicle-stimulating hormone; GC: glucocorticoids; GnRH: gonadotropin-releasing hormone; 5-HT: 5-hydroxy-tryptamine (serotonin); LH: luteinizing hormone; NE: norepinephrine; PRL: prolactin.

The first- and second-wave stress changes also differ in the onset and duration of their effects. In the brain, increased NE, DA, 5-HT, and CRH levels usually occur within seconds, and their effects subside quickly and rarely outlast the duration of the stressor. The concentrations of GCs in the brain reach peak levels only about 20 min after onset of the stressor. Genomic actions of GCs (and other steroids) take longer to manifest, with the slower genomic GC actions actively reversing and normalizing the rapid effects of the various first-wave stress mediators.

The brain also receives feedback from stress-induced changes in the periphery. Although peripheral catecholamines cannot cross the blood–brain barrier (BBB), they can bind to adrenergic receptors on the vagus nerve, which transmits the impulse to the nucleus of the solitary tract. From there, the signals reach and activate the locus coeruleus (LC), which in turn increases NE levels throughout the brain, including the basolateral amygdala where NE binds to adrenergic G protein-coupled receptors and initiates the cAMP signal cascade. The GCs, which readily cross the BBB, exert negative feedback at both hypothalamic and pituitary sites to eventually reduce CRH and ACTH secretion.

As shown in **Figure 4.6**, the primary brain sites involved in the perception of stress and the generation of the stress response are (1) the hypothalamus; (2) the limbic structures, especially the amygdalar nuclei, which are involved in anxiety and fear and the bed nucleus of the stria terminalis that are responsive to sustained threats; and (3) the orbitofrontal cortex, which responds to novel aversive stimuli [24].

4.3.2 Central dopamine and the stress response: Interactions with the HPA axis

As was discussed in detail in **Chapters 2** and **3**, the dopaminergic neurons in the substantia nigra, VTA, and hypothalamus give rise to three main pathways: the nigrostriatal, the mesolimbocortical (MLC), and the TIDA, respectively. Accordingly, DA receptors are mainly localized in the striatum, the limbic system, the brain cortex, and the basal hypothalamus. Although several neurotransmitter systems throughout the brain are activated by stressful stimuli, the MLC dopaminergic neurons appear to be particularly responsive

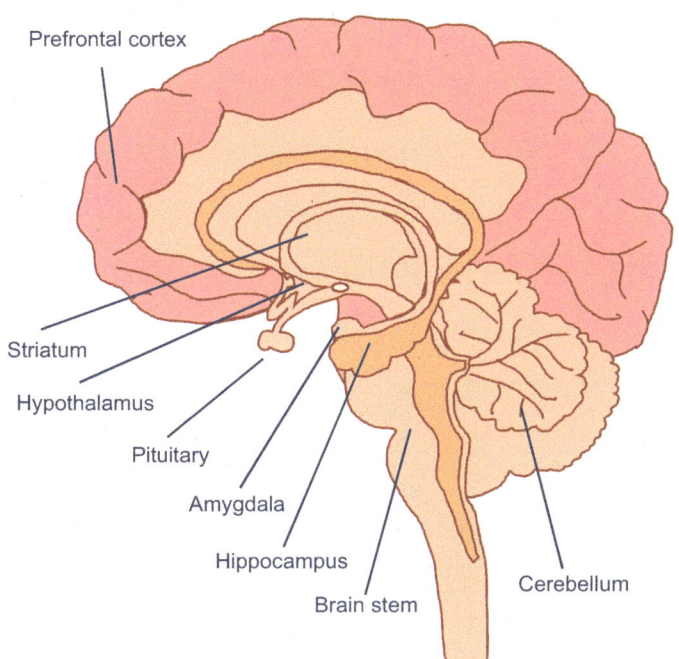

Figure 4.6 Primary brain structures associated with the perception of stress and activation of the stress response.

to the effects of stress because they are activated by low intensity stressors that do not produce detectable changes in most ascending catecholaminergic pathways [25]. The MLC dopaminergic neurons are unique in that they have only a very low density of autoreceptors with autoregulatory capabilities, underlying the higher rates of burst firing and DA turnover in these neurons than in other midbrain dopaminergic projections. In addition, the MLC DA neurons are modulated by many substances, including GABA, serotonin, excitatory amino acids, substance P, opiates, and the noradrenergic systems. Any or all of these factors can contribute to a selective activation of the MLC DA neurons by stressors, resulting in widespread effects on higher brain functions that are altered by stress such as fear, cognition, memory, self-control, motivation, and drive.

Of particular relevance to our discussion is the involvement of DA with the neuroendocrine axis of the stress response. A well-accepted paradigm is that the release of ACTH and GCs in response to systemic and emotional stressors results from the activation of different brain pathways that converge into the parvocellular component of the PVN of the hypothalamus. CRH-producing neurons are primarily localized in the PVN although they are also found at lower levels in the medial preoptic nucleus, bed nucleus of the stria terminalis, olfactory bulb, pontine gray, tegmental reticular nucleus, external cuneate nucleus, and midline thalamus [26]. Neurons from the PVN send axons to the external zone of the median eminence where they release CRH and other ACTH secretagogues into the pituitary portal blood and thus regulate pituitary functions.

The effects of CRH are mediated by binding to and activating G protein-coupled receptors of two subtypes, CRH R_1 and CRH R_2. CRH R_1 is found throughout the cortex, forebrain, cerebellum, mesencephalon and pons, as well as in the pituitary corticotrophs and melanotrophs, while CRH R2 has a more restricted distribution. The hypothalamic regions with the highest levels of CRH R1 are the supraoptic nucleus, magnocellular PVN, arcuate nucleus, and SCN. Double-labeled immunofluorescence has showed co-localization of CRH R_1 in the arcuate nucleus with both neuropeptide Y (NPY) and TIDA neurons.

Multiple studies have shown that brain DA is an important activator of the HPA axis in response to stress [22]. First, D2R agonists increase plasma corticosterone levels, likely acting via central mechanisms given that this increase was not blocked by the peripheral D2R antagonist domperidone, which does not cross the BBB. Second, peripheral administration of both D1 and D2 agonists induced c-fos expression (a marker of neuronal activation) and increased CRH mRNA levels in the PVN. Third, lesions of the dopaminergic neurons in the VTA with 6-hydroxydopamine partially reduced both basal and restrained stress levels of corticosterone in rats, suggesting a stimulatory role of DA. However, this type of neurotoxic lesions of the VTA resulted not only in depletion of DA but also in impaired release of other neurotransmitters and neuromodulators.

Application of immobilization (IMO) stress to experimental animals has served as a good experimental model for testing the effects of a severe stressor characterized by a prolonged post-stress activation of the HPA axis. This model allows the study of the putative roles of DA not only in the initial response to a stressor but also in the dynamics of the recovery phase after termination of exposure [27]. This is particularly important in the case of the HPA axis because the area under the curve of the GCs response to stressors, which includes the post-stress period, is critical for understanding the physiological and pathological impact of the stressors on the organism.

In one study [27], IMO was applied to rats together with the administration of selective D1 and D2 receptor antagonists (SCH23390 and eticlopride, respectively). The DA antagonists reduced the elevation of ACTH and corticosterone in response to acute IMO, particularly during the post-IMO period. In another study, chronic restraint stress in rats induced DA neuronal loss in the substantia nigra, VTA and arcuate nucleus [28]. The reduction in the arcuate nucleus was prolonged and most severe, with the number of DA neurons decreasing by 11% at the 2nd week, by 38% at the 4th week, and by 56% at the 8th week. Lower decreases were seen in the VTA. Collectively, these studies reveal differences in the responsiveness of the DA neurons to acute and chronic stressors and show that a prolonged stress can result in neuronal loss within several DA systems.

4.3.3 Circulating dopamine and the stress response

The involvement of peripheral DA with the stress response has not received the attention it deserves. As discussed in more detail in **Chapter 6**, basal circulating DA level in humans is very low (~0.1 nmol/L) and requires special effort for its analysis. In addition, more than 95% of serum DA in humans (but not in rodents) is in the form of dopamine sulfate (DA-S), which is undetectable by routine analytical methods [29]. These limitations explain why analysis of serum DA has been generally overlooked when studying serum catecholamines, with most investigators resorting to the measurement of urinary DA despite major drawbacks. Early studies found that high stress events such as off-road motor races, resulted in a dramatic elevation in urinary DA concentrations. Sleep deprivation of 24 h has also showed increased urinary DA and NE, but not Epi. Increased urinary DA reflects an enhanced release of DA from peripheral sources because the BBB prevents central DA from reaching the systemic circulation.

A recent study by the U.S. Navy evaluated neuroendocrine and physical performance responses in sailors and Marines undergoing a Survival, Evasion, Resistance, and Escape (SERE) training course [30]. Blood samples were obtained from the recruited men at baseline (T1), stress (T2), and recovery (T3) time points and were analyzed for plasma catecholamines, testosterone, and NPY. The serum concentrations of DA, NE, Epi and cortisol increased 2- to 3-fold at T2, with a concomitant reduction in testosterone. NPY concentrations did not increase at T2, and decreased significantly at T3. Notably, there was a significant elevation of DA under resting (i.e., in the absence of exercise) conditions, possibly due to an "anticipation," a rise which was not fully recovered at T3. Unlike NE and Epi, serum DA concentrations remained elevated at T2 and 24 h later at T3, indicating a relatively long recovery time that is not complete from a homeostatic regulatory perspective. The prolonged elevation and the delayed recovery of DA deviates from the typical pattern of stress-induced NE and Epi release. In response to stress, NE and Epi are released within seconds, but they also rapidly return to normal resting concentrations once the sympathetic drive is reduced.

Whereas the above studies demonstrated that peripheral DA responds strongly to stressful stimuli, the specific targets of circulating DA and the physiological parameters that it alters are presently unknown. The other two catecholamines regulate a wide variety of physiological effects, from increases in heart rate and blood pressure to energy release and blood flow to the skeletal muscle, all as part of the "fight or flight" response. However, the effects of DA on any of these parameters are mild at best, and in some vascular beds, DA reduces, rather than elevates, blood pressure. One potential target for the stress-elevated peripheral DA is the activation of the HPA axis, by acting on the pituitary and/or the adrenals. In the human pituitary, D2Rs are expressed at moderate levels in ACTH-producing corticotrophs, while D1R-like and D2R-like receptors are expressed in all three zones of the normal human adrenal cortex [31]. Future studies that address these issues should be undertaken.

4.3.4 Dopamine and the stress response: Interactions with sexually dimorphic hormones

Gender is an important determinant of human health, with a clear pattern for sex-specific prevalence of various mental and physical disorders. Men have higher susceptibility to infectious diseases, hypertension, aggressive behavior, and drug abuse, while conditions such as autoimmune diseases, chronic pain, depression, and anxiety disorders are more common among women. Some of these gender differences emerge during the reproductive years and gradually diminish after menopause, suggesting the involvement of specific gender-related hormones.

Epidemiological studies report higher prevalence of stress-related disorders such as acute stress disorder and PTSD in women than in men following exposure to trauma.

Because of the complexity of social and cultural factors that influence the response to stress in humans, studies of gender differences in animal models of acute and chronic stress have generated important information. Despite some limitations, such animal models can provide a good approximation of certain aspects of this complex responsiveness and offer the opportunity for distinguishing between biological and sociocultural factors that enter into the equation when studying humans. The section below provides information on interactions between DA and hormones with a clear pattern of sexual dimorphism in both secretion and functions. These hormones include (1) two ovarian hormones, estrogen and progesterone; (2) a testicular hormone, testosterone; (3) four anterior pituitary hormones, LH, FSH, PRL and GH; and (4) a posterior pituitary hormone, oxytocin (OT).

The HPA response patterns differ markedly between males and females in both animal and human studies [32]. Although basal level and response to stress is higher in female rodents, the situation is more complex in humans. Men have a relatively higher secretion of ACTH with comparable total cortisol levels under basal conditions. However, the sensitivity of the adrenal cortex to ACTH in women appears to be higher than in men given that there were no gender differences at the pituitary level upon a challenge with a synthetic CRH. It thus appears that female sex hormones attenuate the sympathoadrenal and HPA responsiveness, leading to a more sluggish cortisol feedback on the brain and less or delayed containment of the stress response. Animal and human studies have suggested that *in utero* resetting of the HPA may be an important factor. A sexually dimorphic response to programming of the HPA axis has been found in animal studies, with female rats being more sensitive than males to activation of the HPA axis following prenatal stress.

Estrogen and progesterone modulate DA activity in the striatum and nucleus accumbens (NAc) under several conditions [33]. For example, there are estrous cycle-dependent variations in the basal extracellular concentration of striatal DA, in amphetamine (AMPH)-stimulated DA release, and in striatal DA-mediated behaviors. Ovariectomy attenuates basal extracellular DA, AMPH-induced striatal DA release, and behaviors mediated by the striatal DA system. Estrogen rapidly and directly acts on the striatum and NAc via a G protein-coupled membrane estrogen receptor (GPER) to enhance DA release and DA-mediated behaviors. In male rats, estrogen does not affect striatal DA release, and the removal of testicular hormones is also without effect. It has been proposed that estrogen induces rapid changes in neuronal excitability by acting on membrane receptors located in striatal GABAergic neurons and DA terminals. This modulates the dopaminergic terminal excitability, resulting in enhanced DA release.

Whereas stress activates the HPA axis, it concomitantly suppresses the hypothalamo-pituitary gonadal (HPG) axis (**Figure 4.7**). The main components of the HPG axis are (1) hypothalamic kisspeptin, which regulates reproductive events by interacting with the GnRH neurons, and GnRH; (2) pituitary LH, FSH and PRL; and (3) the gonadal sex steroids: ovarian estrogen and progesterone and testicular testosterone, which provide feedback regulation to both the hypothalamic and pituitary reproductive hormones.

The actions of DA on stress-induced suppression of the HPG axis are exerted at several levels, both direct and indirect. One level is at the arcuate **kisspeptin** neurons, which co-express D2R, the DA receptor subtype responsible for the seasonal inhibition of GnRH pulses in some species. Expression

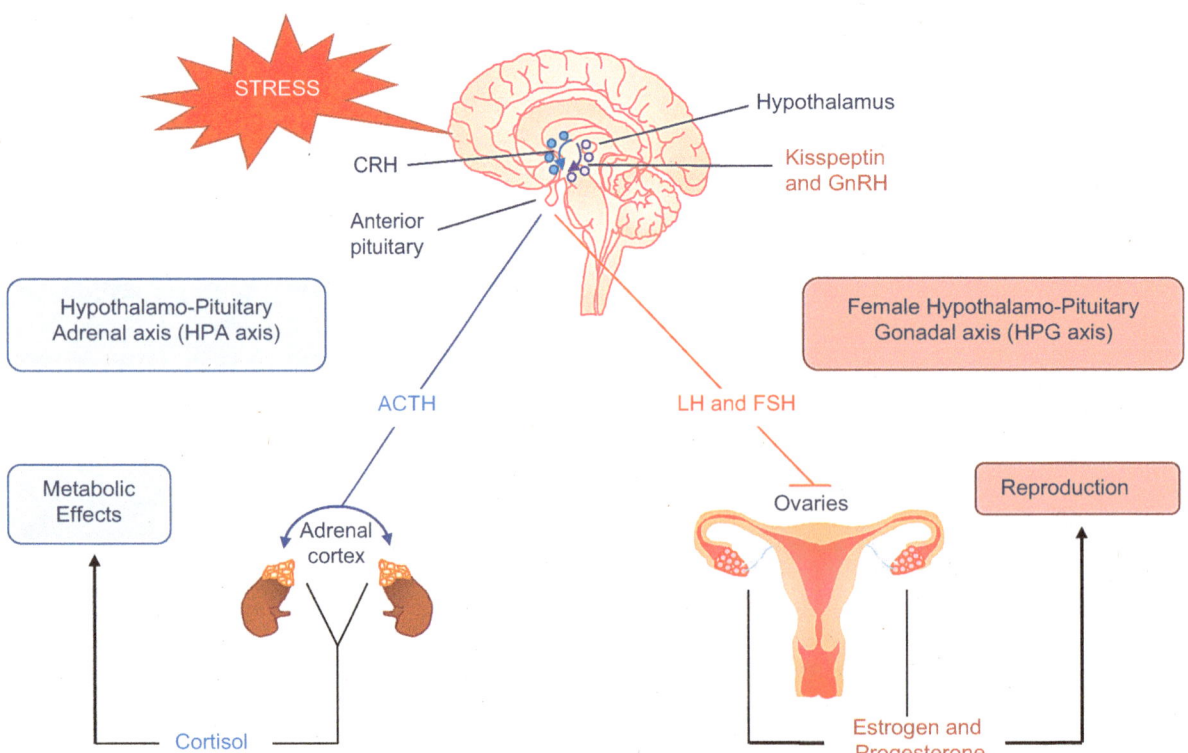

Figure 4.7 Opposite effects of stress on two regulatory systems. Stress causes stimulation of the hypothalamo-pituitary adrenal axis (HPA) while inhibiting the hypothalamo-pituitary-gonadal (HPG) axis. ACTH: adrenocorticotropic hormone; CRH: corticotropin-releasing hormone; FSH: follicle-stimulating hormone; GnRH: gonadotropin-releasing hormone; LH: luteinizing hormone.

of kisspeptin and kisspeptin receptor mRNA is down-regulated by stressors such as restraint, hypoglycemia, and lipopolysaccharides, suggesting that kisspeptin signaling plays a critical role in the transduction of stress-induced suppression of reproductive processes. A second level is at the **GnRH neurons**. DA potently suppresses the electrical activity of GnRH neurons in both males and female rats, and a third of GnRH neurons increase basal firing rate after the administration of DA receptor antagonists, demonstrating the ability of DA for suppressing GnRH [34]. A third level is indirect, via increased **PRL release**. Clinical data show that hyperprolactinemia is a frequent cause of reproductive dysfunction leading to infertility in both males and females. The suppressive affects of PRL on reproduction occur at multiple sites: the hypothalamus, the pituitary, and the gonads. Serum PRL crosses the BBB and affect the *Kiss1*-expressing neurons where it suppresses GnRH release and, thus, the release of the gonadotropins (see further discussion in **Section 4.5**).

PRL release is dramatically affected by a wide variety of stress stimuli, including ether stress, restraint stress, thermal stress, hemorrhage, surgery, social conflict, and even academic stress in humans [35]. Given that DA is the primary physiological inhibitor of PRL, it follows that DA is a major factor behind the stress-mediated effects on PRL secretion. In some cases, this has indeed been confirmed by the use of DA antagonists, which have prevented stress-induced alterations in PRL. A modest increase in TIDA neuronal activity accompanies restraint stress in estrogen-treated, but not in cycling or lactating rats. The general consensus is that in addition to DA, stress-induced PRL release involves other substances known to regulate PRL release, including serotonin, histamine, β-endorphin, OT, and vasopressin. Notably, the physiological importance to the organism of the PRL-secretory response to stress is elusive.

The pattern of **GH secretion** in both rodents and humans is also sex dependent. In adult male rats, GH is secreted in regular pulses every 3.5–4 h, while its secretion in females is continuous at low amplitude [36]. In adult men, there are 8–10 pulses of GH secretion per day lasting a mean of 96 min, with 128 min between each pulse. The diurnal secretory pattern of GH in humans is fully developed only after puberty, with a major peak seen at late night/early morning that is associated with the REM sleep, and a number of GH peaks during the light hours of the day. In women, there is a major peak in the early morning, but otherwise the GH secretory pattern is irregular. The sex-dependent GH secretion patterns are considered a major factor in establishing and maintaining sexual dimorphism in hepatic gene transcription and linear body growth. In both men and women, several types of stress stimuli increase GH release, including hypoglycemia, exercise and surgery as well as some pathological states such as anorexia nervosa and diabetes mellitus.

The involvement of DA in the stress-induced GH release is based on several lines of evidence. For example, the GHRH neurons in the periventricular and infundibular regions of the hypothalamus have a dense innervation of dopaminergic fibers. Perifusion of hypothalamic slices from male rats with DA causes a significant increase in GHRH release with the opposite effect on somatostatin [37]. The continuous infusion of DA into normal healthy men resulted in increased mean GH secretion, comparable to that observed with GHRH. When given together, DA and GHRH have additive effects on GH secretion, and similarly, the DA agonist bromocriptine augments the effects of GHRH [38].

OT is another sexually dimorphic hormone that is strongly affected by stress. OT is made in the hypothalamus and is released to the general circulation from its storage in the posterior pituitary. OT has a classical role in endocrine regulation, where it acts as an important mediator during parturition through activation of uterine contractions, and during lactation, through the milk ejection reflex. Beyond its involvement in peripheral endocrine functions, OT acts within the brain as a key substrate for a range of social behaviors, including social bonding, parental and sexual behavior, as well as in the responses to stress, anxiety and aggression. The differential effects of OT on males and females appears to be due to the actions of estrogen, which stimulates OT release from the hypothalamus and promotes OT receptor binding in the amygdala, while testosterone directly suppresses OT in mice [39].

The PVN contains two subpopulations of OT neurons: (1) magnocellular neurons, which project to the neural lobe, where OT is secreted into the peripheral circulation; and (2) parvocellular neurons, which project toward the brainstem, spinal cord, and supraoptic nucleus to release OT in a somato-dendritic manner. OT fibers from the PVN also project to the nucleus and VTA, adjacent to dopaminergic neurons. Exposure to various stressors such as immobilization, shaker, social defeat, and forced swimming, which increase the release of OT into the peripheral circulation of male and female rats, also activates central release of OT, although the peripheral and central release of OT do not necessarily coincide [40]. Interactions between the neural circuits that control OT and DA are complex and bidirectional, i.e., OT causes the release of DA, while activation of the dopaminergic systems increase the release of OT under certain conditions.

4.4 FOOD INTAKE AND METABOLIC HOMEOSTASIS

4.4.1 Control of body weight and the consequences of dysregulation of food intake

The control of body weight by the brain relies upon the detection and integration of signals that reflect energy stores and fluxes and upon interactions with different peripheral hormonal as well as with social, emotional, and circadian factors [41]. Body weight can remain rather stable over time by matching food intake to energy expenditure, i.e., the sum of excreted energy, heat loss, and physical work. Deviations from this equilibrium will result in weight gain or loss.

Appropriate food intake has important implications for the daily well-being of the individual, and its dysregulation can result in several unfavorable clinical conditions. Enhanced appetite and increased food intake can lead to **obesity**, which had become a serious global health problem due to its increasing prevalence and associated comorbidities. Obese individuals are at increased risk for developing type 2 diabetes, cardiovascular disease, and certain types of cancer [42].

Dysregulation of appetite also lies at the root of **anorexia** nervosa and **bulimia** nervosa. Both are eating disorders characterized by loss of self-control in eating behavior and disturbed emotions, which affect 2%–3% of young women [43]. Anorexia is a serious eating disorder, with the highest mortality rate among psychiatric disorders. It is exemplified by chronic self starvation, amenorrhea, and severe weight loss due to the reductions of both fat mass and lean body mass. Bulimia is an eating disorder in which the subject engages in recurrent binge eating. To compensate for the high intake of food and to prevent weight gain, this is usually followed by the induction of vomiting, use of laxatives, enemas, diuretics, excessive exercising, or fasting, resulting in dysregulation of the endogenous endocrine axes.

4.4.2 The hypothalamus: A major relay station that controls feeding behavior

Early studies identified the hypothalamus as a critical feeding center, based on the induction of a significant increase in feeding behavior by lesions in the ventromedial hypothalamus (VMH) of the rat, whereas lesions in the ventrolateral hypothalamus led to reduced feeding behavior and malnutrition.

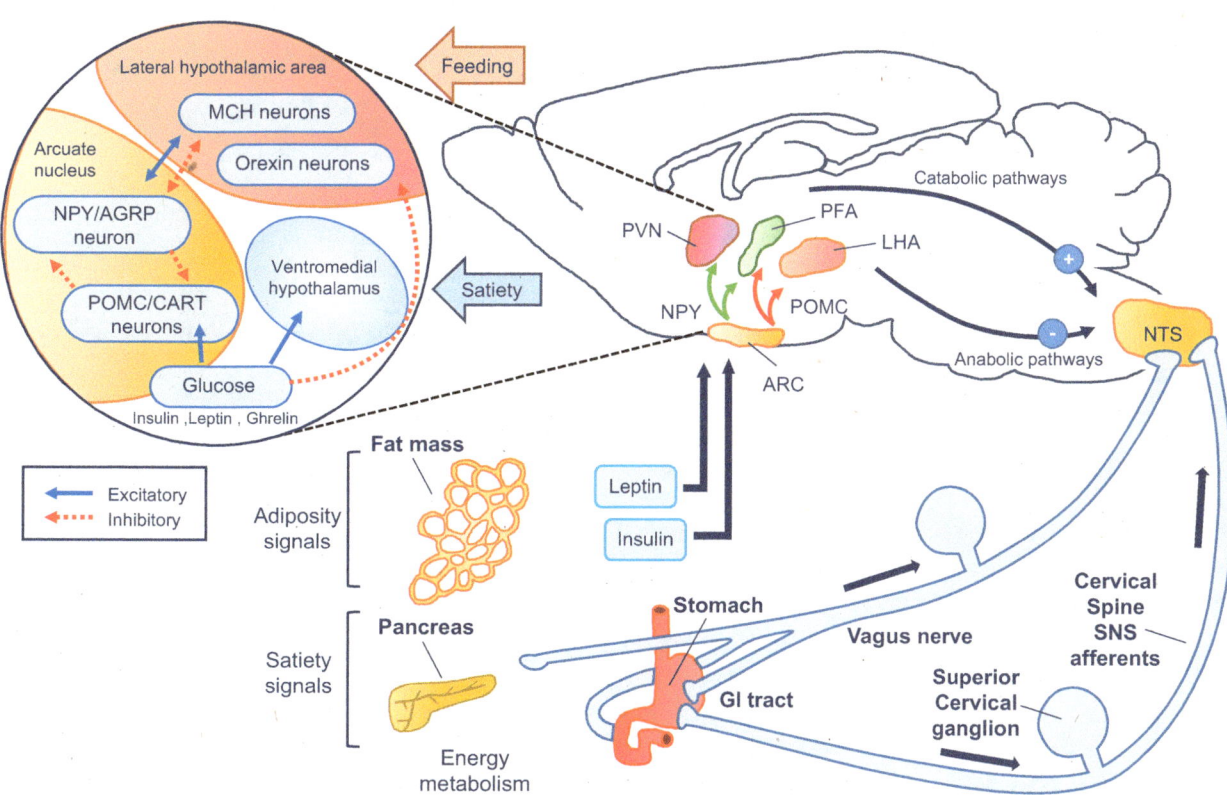

Figure 4.8 Regulation of appetite and feeding behavior. The hypothalamus is the major integrator of peripheral signals and coordinator of autonomic, endocrine and behavioral responses in the regulation of appetite and feeding. Leptin and insulin, the adiposity signals, interact with the brain at the hypothalamic arcuate nucleus (ARC). Satiety signals, generated in the gastrointestinal (GI) tract, provide information via sensory axons in the vagus and sympathetic (SNS) nerves into the nucleus of the solitary tract (NTS) in the brain stem. AgRP: agouti related peptide; CART: cocaine- and amphetamine-regulated transcript; LHA: lateral hypothalamic area; MCH: melanin concentrating hormone; NPY: neuropeptide Y; PFA: periformical area; POMC: proopiomelanocortin; PVN: paraventricular nucleus. See text for additional explanations. (Redrawn and modified from Woods, S.C., *Am. J. Physiol. Gastrointest. Liver Physiol.*, 286, G7–13, 2004.)

Figure 4.8 illustrates the central role of the hypothalamus as an integrator of information on the nutritional status of the body and as a coordinator of endocrine, autonomic, and behavioral responses that regulate feeding behavior and appetite.

The main factors in the hypothalamic neurocircuitry associated with food intake are proopiomelanocortin (POMC) and its derivative alpha melanocyte stimulating hormone (αMSH) and cocaine- and amphetamine-related transcript (CART), NPY, agouti-related peptide (AgRP), orexins, melanin concentrating hormone (MCH), and the melanocortin receptors 3 and 4 (MCR 3&4). The major peripheral hormones associated with the regulation of body weight are adipose-derived leptin, pancreas-derived insulin and amylin, stomach-derived ghrelin, and intestinal-derived **Cholecystokinin** (CCK) and peptide tyrosine tyrosine (PYY). Few other factors, e.g., glucagon-like peptide 1 (GLP-1) and pancreatic polypeptide (PP), have also some contributions to the control of body weight but will not be covered here. **Table 4.3** presents a list of major factors that regulate appetite and body weight.

The involvement of DA in the control of food intake and body weight regulation has been underscored by the phenotype of the D2R KO mice [D2R(-/-)]. Compared with wild-type littermates, the D2R KO mice show reduced food intake and low body weight, together with increased basal energy expenditure level. Although the D2R null mice do not provide a clue as to which brain dopaminergic system is associated with food intake, multiple studies, detailed below, have shown that both the hypothalamic and the mesocorticolimbic dopaminergic systems are actively engaged in the control of food intake and feeding behavior.

Table 4.3 Major positive and negative regulators of appetite and body weight and their site of origin (listed alphabetically)

Orexigenic	Site of origin	Anorexigenic	Site of origin
AgRP	Brain	Amylin	Pancreas
β-endorphin	Brain and pituitary	α-MSH	Brain and pituitary
Dynorphin A	Brain and pituitary	CART	Brain
GABA	Brain	CCK	Intestine
Galanin	Brain and periphery	Dopamine	Brain
Ghrelin	Stomach	GLP-1	Intestine
MCH	Brain	Insulin	Pancreas
Norepinephrine	Brain	Leptin	Adipose
NPY	Brain	Peptide YY	Brain
Orexin	Brain	Serotonin	Brain

AgRP: agouti related peptide; CART: cocaine- and amphetamine-related transcript; CCK: cholecystokinin; GABA: gamma-aminobutyric acid; GLP-1: glucagon-like peptide 1; MCH: melanin concentrating hormone; NPY: neuropeptide Y; α-MSH: alpha melanocyte stimulating hormone.

The **arcuate nucleus** of the hypothalamus is a prime example of the convergence of multiple peripheral and central factors acting together in the regulation of food intake. Neurons within the arcuate nucleus express the orexigenic AgRP and NPY, and the anorexigenic POMC. Both AgRP/NPY and POMC neurons express receptors for, and are targeted by, leptin, the satiety hormone, and ghrelin, the hunger hormone. Leptin inhibits the activity of AgRP/NPY neurons and stimulates the activity of POMC neurons, leading to decreased feeding and increased energy expenditure via the stimulation of the MC3R and MC4R by αMSH, a cleaved product of POMC. On the other hand, ghrelin activates the AgRP/NPY cells and inhibits the POMC cells, thereby stimulating the release of AgRP, a potent MC3R and MC4R antagonist, resulting in increased appetite. The AgRP/NPY neurons can also inhibit the POMC neurons themselves through NPY and GABA. Many of the dopaminergic neurons within the arcuate nucleus either co-express or have synaptic connections with most of the factors listed above, enabling direct interactions of DA with orexigenic/anti-orexigenic activities. In fact, the anorectic peptides leptin and CART peptide inhibit hypothalamic dopamine release.

The **ventromedial nucleus** of the VMH is heavily involved in the regulation of food intake and also has interactions with the dopaminergic neurons. The use of *in vivo* microdialysis showed that DA release from the VMH correlated with meal size and post-meal intervals [44]. Overnight food-deprived rats received a single VMH injection of sulpiride, a D2R antagonist or saline as control. Food intake after sulpiride injection was greater in obese rats but was not different in lean rats. The authors suggested that down-regulation of D2R in the VMH induces sensitization of a behavior aimed at having large meals. Thus, low D2R expression may be causal for the exaggerated DA release observed in obese rats during food ingestion and for the reduced satiety feedback effect of DA.

The sections below review selected properties and activities of the main peripheral factors that regulate food intake and their reciprocal interactions with DA. Additional information can be found in two extensive reviews [41,45]. A good discussion on the control of appetite and body weight is also found in a comprehensive review: "Functional Anatomy of the Hypothalamus and Pituitary," by Lechan and Toni. It is a chapter in *Endotex*, a Web textbook by the Endocrine Society, which can be downloaded at https://www.ncbi.nlm.nih.gov/books/NBK279126/.

4.4.3 Leptin, a major suppressor of appetite

Leptin is a 16-kDa protein hormone composed of 167 amino acids and which is transcribed from the *Ob* (*lep*) gene, located on the long (q) arm of chromosome 7 in humans (7q32.1). Leptin is produced predominantly, but not exclusively, in white adipose tissue and its circulating levels are proportional to the amount of stored fat. Leptin binds to membrane receptors belonging to the cytokine type 1 receptor family and activates the Janus kinase/signal transducers and activators of transcription (JAK/STAT3) pathway [46]. Other signaling pathways, including ERK, STAT5 and PI3K, are also activated by leptin in various tissues. Six types of leptin receptors (ObRa, ObRb, ObRc, ObRd, ObRe, and ObRf) have been identified, only four of which are found in humans (ObRa–ObRd). Leptin receptors are expressed in different organs such as kidney, liver, heart, gastrointestinal tract, ovaries, testes, spleen, and pancreas and throughout the brain.

The discovery of leptin came about as a result of a long and persistent pursuit of the biological causes of extreme obesity in mice [47]. A strain of mice, now known as obese, or ob/ob, was discovered in 1950 at the Jackson Laboratories, Bar Harbor Maine. These mice are massively obese, weighing as much as three times more than normal mice, and they have an insatiable appetite. Douglas Coleman, a researcher at the Jackson's laboratories, hypothesized that a circulating satiety factor was lacking in this mouse and predicted that the factor acts on the hypothalamus to modulate food intake. His hypothesis was based on published reports that a discrete lesion in the VMH nucleus in a 28-year-old woman was associated with increased food consumption and profound obesity, while bilateral lesions of the lateral hypothalamus in animals resulted in anorexia and weight loss. Thus, the concept of a hypothalamic ventromedial nucleus satiety center and lateral hypothalamic orexigenic center that is influenced by peripheral signals was developed and dominated the thinking about the hypothalamic control of feeding for decades.

Jeffrey Friedman, a molecular geneticist at Rockefeller University, New York, set out to find how a defect in a presumably just one gene in the Jackson's obese mouse could have such profound effects on both the animal's body weight and feeding behavior. Using positional cloning (a novel technique at the time), he began searching for the responsible gene in the late 1980s. It took him eight years, and in 1994 he successfully cloned the *ob* gene in mice and its homolog in humans. He subsequently purified the gene product and called the protein hormone leptin. It was soon realized that the ob/ob mouse has a mutation in the *ob* gene, resulting in an absolute leptin deficiency. The fat-reducing property of the newly synthesized leptin was verified by its ability to reduce the voracious feeding in the ob/ob obese mice and to reverse their obesity. The discovery of leptin revolutionized the thinking on the mechanisms governing appetite and satiety and established a new concept that fat is an endocrine organ.

Leptin can penetrate the BBB via a saturable transport through brain endothelia that express ObRb. Within the brain, leptin acts at multiple sites to suppress food intake and to increase energy expenditure. Leptin resistance, a state characterized by the inability of leptin to induce anorectic and metabolic effects in spite of hyperleptinemia, is a common condition in various chronic diseases, including some forms of obesity.

Expression of leptin receptors is highest in the hypothalamus, where they are found in the arcuate nucleus, the PVN, and the dorsomedial, lateral and ventromedial regions. The major site of leptin's action is the arcuate nucleus, where it alters the activities of two groups of neurons with opposing functions: α-MSH-producing neurons that co-express CART, and AgRP-producing neurons that co-express NPY. During fasting, when circulating leptin levels are low, the expression of α-MSH and CART, which promote weight loss and energy expenditure, is reduced while the expression of AgRP and NPY, which promote weight gain and reduce energy expenditure, is increased.

Several studies showed multiple interactions between leptin and the hypothalamic and mesocorticolimbic dopaminergic systems. At the hypothalamic level, the presence of D2R in leptin-sensitive, STAT3-positive cells in the arcuate nucleus was revealed by double immunofluorescence histochemistry [48]. Leptin injections induced STAT3 phosphorylation in D2R-expressing hypothalamic neurons, while D2R stimulation by the agonist quinpirole suppressed leptin-induced STAT3 phosphorylation. Treatment with D2R agonists and antagonists modulated the leptin-induced food intake and changes in body weight in wild-type mice but not in D2R(-/-) mice. The blockade of D2R, but not D1R receptor, attenuated the acute hypophagic effect of leptin in fasted mice. In addition, leptin dose-dependently inhibited depolarization-induced DA release from hypothalamic synaptosomes (nerve endings). It has also been reported that VTA dopaminergic neurons express the leptin receptors and respond to leptin with activation of the JAK-STAT pathway and a reduction in their firing rate [49].

4.4.4 Pancreatic hormones: Insulin and amylin

Comprehensive information on the pancreatic hormones that participate in the regulation of feeding can be found in an extensive review [41]. Anatomically, the pancreas is a heterogeneous gland with both exocrine and endocrine functions. The exocrine pancreas secretes pancreatic juice into the duodenum, helping with the breakdown of carbohydrates, proteins and lipids in ingested food that enters the duodenum from the stomach. The endocrine pancreas is composed of islets that are dispersed within the exocrine pancreas. The islets contain four types of cells: (1) A-cells that secrete glucagon, (2) B-cells that secrete insulin and amylin, (3) D-cells that secrete somatostatin, and (4) F-cells that secrete PP. Both insulin and glucagon regulate glucose homeostasis. Insulin is the primary determinant of glucose removal from the blood and its uptake into the cells, and it is also the suppressor of glucose secretion by the liver. Glucagon is a primary stimulant of glucose production and secretion by the liver.

Insulin is a 5.8-kDa protein hormone composed of 51 amino acids. The mature hormone is organized as a dimer consisting of an A-chain and a B-chain linked by disulfide bonds. The insulin receptor (IR) is a transmembrane receptor belonging to the large class of tyrosine kinase receptors, and it can be activated by insulin, IGF-I and IGF-II. Ligand binding triggers auto-phosphorylation of some tyrosine residues, with each subunit phosphorylating its partner. This generates a binding site for the insulin receptor substrate (IRS-1), which becomes phosphorylated and activates PI3K, Akt and other kinases.

Plasma insulin levels are low during fasting (basal condition) and increase during and immediately after meals (prandial condition) or glucose administration (stimulated condition). Basal, prandial and stimulated insulin levels are in direct relationship to the state of obesity, with leaner individuals having lower plasma insulin levels than obese individuals. Thus, similar to leptin, the amount of insulin in the circulation is directly proportional to white fat, enabling it to convey important signals on the degree of adiposity to insulin-sensitive tissues [50].

Circulating insulin can permeate the BBB by a receptor-mediated transport through capillary endothelial cells. Insulin receptors are expressed throughout the brain, including DA-containing neurons in the arcuate nucleus and the VTA [51]. Within the brain, insulin induces both short- and long-term suppressive effects on food intake and body weight. Intra-cerebroventricular (ICV) or hypothalamic delivery of insulin inhibits food intake (anorexigenic effect) and produces a loss of body weight in both rodents and primates. The interactions of insulin with orexigenic and anorexigenic neuropeptides in the control of feeding behavior and energy homeostasis helps to maintain the body weight around a set point. Based on the similarities in the brain

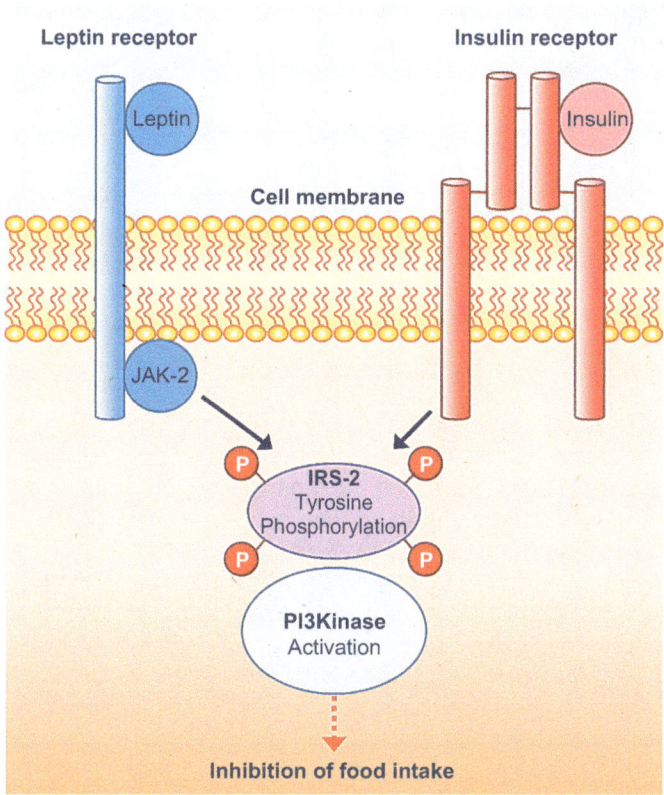

Figure 4.9 Cross talk between leptin and insulin receptors. Activation of both receptors results in the convergence of their signaling in the control of feeding behavior. IRS-2: insulin receptor substrate 2.

actions of insulin and leptin and the cross-talk between their signaling, it has been suggested that convergent mechanisms mediate the cellular responses to insulin and leptin in hypothalamic neurons (**Figure 4.9**).

Several lines of evidence show that insulin alters the activity of the VTA dopaminergic neurons and their output, depending upon the length of exposure to insulin and the state of obesity [51]. Insulin administered ICV had synergistic actions when paired with D2R antagonism to reduce sucrose consumption. Chronic hyperinsulinemia, either by a continuous ICV insulin infusion or in the hyperinsulinemic obese Fa/Fa Zucker rats, results in elevated DA transporter (DAT) mRNA in the midbrain. In addition, insulin has been shown to increase the spike frequency in about half of VTA/Substantia nigra neurons. Using fast-scan cyclic voltammetry, another study has showed that insulin can reduce evoked somato-dendritic DA in the VTA, an effect abolished by a DA transporter inhibitor [51]. The authors concluded that insulin depresses DA concentration in the VTA via increased reuptake of DA through DAT. In addition, insulin-mediated decrease of DA in the VTA may suppress salience of food once satiety is reached.

Amylin is a 37-amino-acid peptide belonging to the calcitonin gene-related peptide (CGRP) family. It is co-secreted with insulin from pancreatic beta cells in response to prandial stimuli and in direct proportion to the meal size. Circulating amylin can cross the BBB, but amylin-immunoreactive neurons have also been detected in the hypothalamus [52]. With respect to the feeding behavior, amylin inhibits food intake following both systemic or ICV administration at low doses. Interactions between amylin and DA are supported by the demonstration that amylin inhibited the release of DA, but not NE or serotonin, from hypothalamic synaptosomes. The authors concluded that the anorectic effects of amylin involve inhibition of the release of DA in the hypothalamus. Another study reported that amylin acts in the VTA to reduce palatable food intake [53]. Although a direct interaction of amylin with DA was not tested in this study, based on the well-established role of DA in reward-related processes, the authors proposed that the effects of amylin are mediated by the blunting of DA.

4.4.5 Gastrointestinal hormones involved in the regulation of food intake

Ghrelin is a 28-amino acid peptide, initially purified from the rat stomach as an endogenous ligand of the GH secretagogue receptor. Subsequent research has established ghrelin as a major appetite-stimulating hormone. The metabolic effects of ghrelin are opposite to those of leptin, as it stimulates food intake and decreases energy expenditure [54]. Ghrelin is produced by endocrine cells in gastric oxyntic glands. Its serum levels are high when nutrient availability is low, as occurs during fasting, and are low when energy supply is sufficient, as occurs following food consumption. Thus, circulating ghrelin levels are inversely correlated with the body mass index, i.e., they are upregulated in undernourished states, such as anorexia nervosa, and are downregulated in obesity. Given that its levels rise just before food consumption in both rodents and humans, Ghrelin appears to serve as a cue for meal initiation.

Ghrelin binds to GHSR1a, a seven-transmembrane receptor that is coupled to Gαq/11 and is linked to the activation of phospholipase Cγ and Ca^{2+} release. In the brain, these receptors are highly expressed in the hypothalamus, specifically in the arcuate and ventromedial nuclei, but are also found in other areas, including the VTA, hippocampus, and substantia nigra [55]. The widespread expression of the ghrelin receptor in the brain supports the notion that ghrelin fulfills other functions in the brain, including the promotion of cell survival and neuroprotection.

Many studies have shown that DA, primarily that present within the mesocoticolimbic system, is the main mediator of the actions of ghrelin on both motivation and reward that accompany food intake [54]. For example in one study, ghrelin administered directly to the VTA increased food seeking activities in rats, while its orexigenic action was blunted by the administration of a ghrelin antagonist in the VTA. In addition, palatable food-related behaviors were impaired in VTA-lesioned rats. In another study, local injection of ghrelin into the VTA induced DA release in the NAc (as measured by microdialysis) and increased food consumption [56]. The authors concluded that peripherally administered ghrelin activates GHSRs in the VTA and induces bimodal effects on mesolimbic DA, depending on the food-consumptive states.

CCK is an octapeptide that is synthesized in neuroendocrine I cells scattered along the proximal two-thirds of the small intestine. CCK is released in response to fatty acids and amino acids that enter the gut and binds to two G protein-coupled receptors: GPR40 and CaSR [57]. CCK has multiple physiological actions, including (1) stimulation of pancreatic exocrine secretion and gall bladder contraction, which are critical for the digestion of these nutrients; (2) regulation of gastric emptying and bowel transit to titrate the delivery of nutrients, and (3) inhibition of food intake by acting on the brain.

Following its release from the small intestine, CCK acts directly on vagal afferent neurons that terminate in the nucleus tractus solitarius (NTS) and activates ascending pathways that control ingestive behavior (**Figure 4.8**). In addition, microinjections of CCK into the brain identified six hypothalamic sites (anterior hypothalamus, dorsomedial hypothalamus, lateral hypothalamus, paraventricular nucleus, supraoptic nucleus, and ventromedial nucleus) and two hindbrain sites (NTS and fourth ventricle) that significantly suppressed food intake during the first hour postinjection [58]. There is evidence that CCK interacts with DA in the control of food intake. For example, injection of a DA receptor agonist (apomorphine) or CCK inhibited food intake and these effects were reversed by pretreatment with a DA receptor antagonist, cis-flupentixol [59]. Blockade of cholecystokinin-A receptors, by treatment with L-364,718, but not cholecystokinin-B receptors, by treatment with L-365,260, blocked the inhibitory effect of CCK on food intake but did not affect the inhibitory effect of apomorphine.

PYY is a 36 amino acid peptide hormone. Together with PP and NPY, PYY comprises the PP family of peptides [60]. There are two forms of PYY, which differ by two amino acids. PYY1–36 is the form released from enteroendocrine L-cells in the colon and ileum in response to nutrient signals. In the blood, PYY1–36 is rapidly converted to PYY3–36 by cleavage of the two N-terminal amino acids. PYY3–36, which is the major circulating form, exerts different and even opposite biological functions than PYY1–36. The distinct biological functions exerted by the two forms result from their different binding affinities for the five Y receptor subtypes: Y1, Y2, Y4, Y5 and Y6. All are inhibitory G protein-coupled receptors that reduce c-AMP and calcium mobilization. Whereas PYY1–36 has similar affinities for the Y1 and Y2 receptor, PYY3–36 is a high-affinity Y2 receptor ligand. PYY effectively crosses the BBB and acts on hypothalamic and extrahypothalamic sites to increase or reduce food intake, depending on the receptor being targeted. Overexpression of PYY protects against diet-induced obesity, and PYY3–36 administration reduces food intake in both obese and non-obese subjects [60].

4.4.6 Involvement of dopamine in the control of feeding behavior

A well-accepted paradigm posits that food intake is determined by a complex interplay of circulating signals of energy homeostasis with brain circuits that encode the diverse behavioral repertoire required to acquire and consume food. Two extensive reviews [45,61] cover this topic. To define the role of DA in feeding behavior, an elegant genetic approach was employed in which the ability to synthesize DA was selectively eliminated by generating DA-deficient (DD) mice [62]. Such mice grow normally for the first week but then start showing signs of bradykinesia and hypophagia and die of starvation by 3 to 4 weeks. The DD mice can be rescued by daily injections of L-Dopa, which generates hyperlocomotion and hyperphagia lasting for 8–10 h after the injection. In the same study, intraventricular injection of PYY increased the amount of time that both DD and control mice spent near food, but DD mice did not eat much when DA levels were low (27 h after last L-Dopa treatment). A viral-mediated gene transfer was then used to restore DA synthesis in specific brain locations in the DD mice. When DA was restored in the central caudate putamen, but not in the NAc, DD mice fed adequately and maintained body weight for many months without L-Dopa treatment. ICV administration of PYY led to marked increase in food intake in the DD restored mice. The authors concluded that the ability of PYY to stimulate feeding is partially restored by DA signaling in the caudate-putamen. However, this study did not identify which form of PYY actually exerted the orexigenic action.

The consumption of food, especially when sated, is often driven by its rewarding properties, which involve activation of the mesolimbic DA pathway. DA neurons of the VTA, the origin of the mesocoticolimbic dopaminergic neurons, have been implicated in the incentive, reinforcing and motivational aspects of food intake. The VTA dopaminergic neurons express the ObR and respond to leptin by an activation of the JAK-STAT pathway and a reduction in their firing rate. In a study, direct administration of leptin to the VTA caused decreased food intake, while long-term RNAi-mediated knockdown of the ObR in the VTA led to increased food intake and lower sensitivity to palatable food [49]. Collectively, these studies have established that interactions between VTA dopaminergic systems and leptin signaling are important in the control of food intake.

CART was discovered as a novel up-regulated mRNA in a drug abuse paradigm. Transcription of the gene results in two alternatively spliced mRNAs of different length that produce pro-peptides of different lengths, called proCART 1–89 and proCART 1–102. Both pro-peptides are found in rats, while only proCART 1–89 is found in humans. Posttranslational processing by prohormone convertases results in two biologically active CART peptides, whose names CART 55–102 and CART 62–102 are derived from the long form of proCART. Subsequent studies have established that the CART peptides act as inhibitors of feeding. CART peptide-containing projections were traced from the lateral hypothalamus to the VTA and the dorsomedial substantia nigra, where they synapse onto dopaminergic dendrites [63]. The authors concluded that the significant behavioral states influenced by CART peptides, including feeding and locomotion, is mediated by direct and/or indirect modulation of VTA dopaminergic neuronal activity.

Two other neuropeptides, **orexin** and **MCH**, have similar activities in promoting consumption of palatable or caloric food and in being stimulated by intake of this food. They appear to do so by interacting with limbic DA. There are two orexins, orexin A (OX-A) and orexin B (OX-B), also called hypocretin 1 and hypocretin 2, which are cleaved from a 130-amino acid precursor. Neurons containing OX reside exclusively in the hypothalamus, spanning the dorsomedial hypothalamic nucleus through the perifornical area and into the lateral hypothalamic area [64]. MCH was isolated from a salmon pituitary as an antagonist of αMSH-induced skin darkening, and it was later recognized for its role in stimulating feeding. Neurons containing MCH are distinct from, but adjacent to, those containing OX, lying predominantly in the lateral hypothalamus but also in the perifornical area and zona incerta. Projections from OX- and MCH-containing neurons terminate in many of the same brain areas, including the locus coeruleus, hippocampus, thalamus, NAc, VTA, amygdala, cortex, and various nuclei of the hypothalamus [64].

As illustrated in **Figure 4.10**, ventral tegmental DA interacts with multiple neuronal projections from hypothalamic regions associated with the control of food intake and appetite. The receptors for OX and MCH are also located in these same brain areas. Based on several lines of evidence, the authors concluded that OX contributes to the rise in accumbal DA that normally occurs prior to meal consumption, while MCH contributes to the fall in DA observed during the feeding bout.

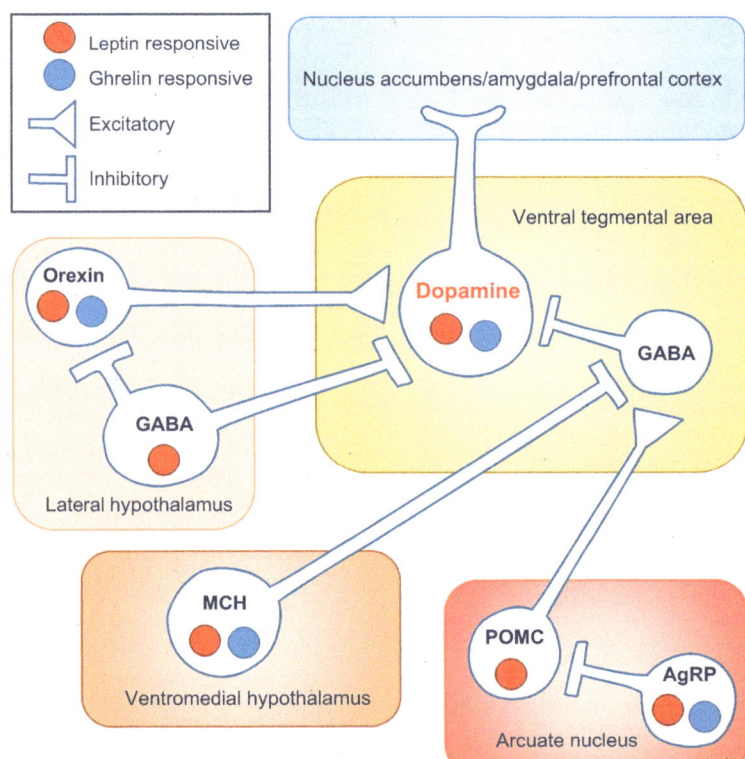

Figure 4.10 Involvement of dopamine in the regulation of feeding behavior. Direct and indirect interactions between ventral tegmental dopaminergic neurons and orexigenic/anorexigenic neurons from several of hypothalamic areas are involved in the regulation of feeding behavior. AgRP: agouti related peptide; GABA: gamma-aminobutyric acid; MCH: melanin concentrating hormone; POMC: proopiomelanocortin. (Redrawn and modified from van Zessen, R et al., *Proc. Nutr. Soc.*, 71, 435–445, 2012.)

4.5 NEUROENDOCRINE REGULATION OF REPRODUCTION AND SEXUAL/MATERNAL BEHAVIOR

The ultimate function of the reproductive system is to ensure the survival of the species by preserving the integrity of each individual and its offspring. Production of viable offspring requires the timely generation and delivery of functional gametes, the successful fertilization and implantation, pregnancy that supports optimal fetal development, well-timed parturition, and provision of milk for nutrition of the neonate. Although reproductive success is not essential for the life of the individual, it is crucial for survival of the species. To this end, each species has evolved different patterns of reproductive cycles, sexual behavior, as well as length of gestation and lactation, that are best suited for its social structure and living environment. The neuroendocrine system controls and coordinates most of the physiological, and all of the behavioral, aspects of reproduction.

As illustrated in **Figure 4.11**, key hormones in the hypothalamo-pituitary-gonadal (HPG) axis are (1) hypothalamic kisspeptins and GnRH; (2) pituitary LH, FSH, and PRL; and (3) male and female gonadal steroids, testosterone and estrogen/progesterone, respectively. All the reproductive hormones are functionally interconnected via positive and negative feedback regulatory loops that homeostatically regulate reproductive functions. These mechanisms are aimed at ensuring optimal reproductive fecundity, i.e., the control of spermatogenesis in the male and the regulation of the ovarian cycle, ovulation, fertilization, pregnancy, parturition, and lactation in the female. Not less important for reproductive success is the regulation of sexual behavior, which brings the two sexes together at the optimal time for conception, and maternal behavior, which ensures the appropriate growth and well-being of the young.

Given the complexity of the HPG axis and the multiple components involved, genetic factors or environmental insults can cause acute or permanent alterations in gonadal hormone synthesis, changes in the timing of puberty, loss or reduced fertility, and dysregulation in fetal and neonatal

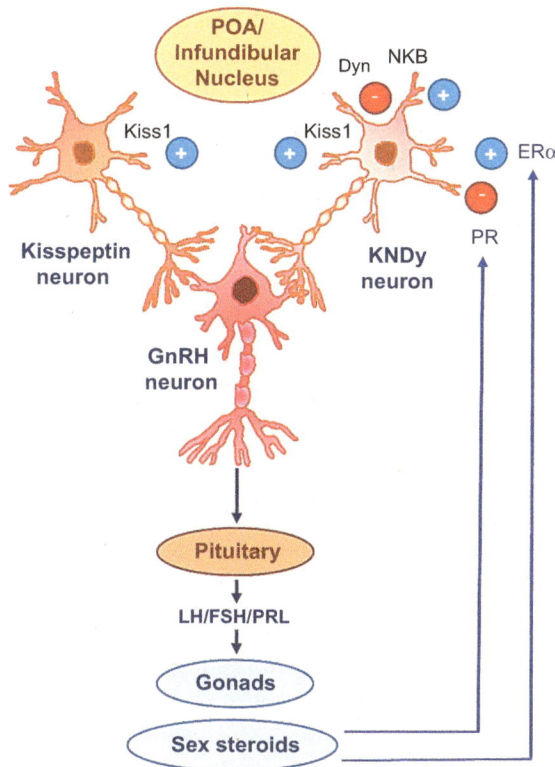

Figure 4.11 The regulation of the hypothalamo-pituitary gonadal axis in females. Hypothalamic and pituitary hormones and feedback loop by ovarian steroids are associated with the regulation of the hypothalamo-pituitary-gonadal axis in the female. See text for further explanations. estrogen receptor alpha; Dyn: dynorphin; FSH: follicle-stimulating hormone; kisspeptin 1; NKB: neurokinin B; PR: progesterone receptor; PRL. (Redrawn and modified from Marques et al., *Endotex*, a Web textbook by the Endocrine Society: https://www.ncbi.nlm.nih.gov/books/NBK279070/.)

development. A good review by Marques et al., entitled: "Physiology of GNRH and Gonadotropin Secretion" is included as a chapter in *Endotex*, a Web textbook by the Endocrine Society, which can be downloaded at https://www. ncbi.nlm.nih.gov/books/NBK279070/.

4.5.1 Synthesis of GnRH, origin and migration of GnRH neurons, and genetic dysfunctions

GnRH is a linear decapeptide with two protected termini: pyroglutamic acid at the N-terminus and carboxyamide at the C-terminus. It is a product of the *GNRH1* gene, located on chromosome 8p11.2. Similar to other neuropeptides, GnRH is synthesized as part of a large prohormone that is cleaved enzymatically and is further modified within the secretory granules. Cleavage of the prohormone generates the GnRH decapeptide and a 56-amino acid GnRH-associated protein (GAP), which is secreted together with GnRH. The function of GAP is unknown but it inhibits PRL release in some species. A second human gene, named *GnRH-II*, was later cloned and mapped to chromosome 20p13. The most prominent difference in tissue distribution of GnRH-I and GnRH-II in humans is that expression of GnRH-I is confined to the brain, whereas GnRH-II is expressed at the highest level outside the brain, i.e., kidney, bone marrow and prostate [65].

GnRH neurons are unusual in that they are derived from progenitors in the epithelium of the olfactory placode [66]. During embryonic life, the nascent GnRH neurons migrate along the vomeronasal axons, across the cribiform plate and into the mediobasal hypothalamus where migration ceases and the neurons detach from their axonal guides. Mature GnRH neurons are distributed throughout the anteroventral periventricular and preoptic area of the hypothalamus. Multiple factors are involved in the embryonic migration of the GnRH neurons, ranging from transcription factors to a variety of transmembrane tyrosine kinases or G protein-coupled receptors and their ligands, to extracellular matrix proteins. Some of the factors influence the movement of the GnRH neurons indirectly, by altering the pace or targeting of the olfactory system. Loss of the GnRH neurons or their misdirection along the migration route results in failure of sexual maturation in mice and man [67].

The importance of olfactory bulb development to reproduction in humans is exemplified by the phenotype of **Kallmann's syndrome** (KS), a genetic disorder classified as hypogonadotropic hypogonadism. Such patients have poorly defined secondary sexual characteristics, are infertile, and are at increased risk of developing osteoporosis. They are distinguished by having either a reduced, or a total loss of the sense of smell. Anosmia in KS patients is due to a perturbed olfactory bulb development, whereas the observed infertility is due to an impaired maturation or defective migration of the GnRH neurons. Mutations in prokineticin genes (*PROK1* and *PROK2*) lead to hypogonadotropic hypogonadism without anosmia, suggesting that factors other than suboptimal migration can also lead to functional deficiencies in GnRH. **Table 4.4** summarizes the major features of Kallmann's syndrome and common treatment options.

4.5.2 Kisspeptins, GnRH pulsatility, and the role of dopamine

Kisspeptins are a group of peptides ranging from 10 to 54 amino acids that are derived from a single precursor. They bind to and activate the G protein-coupled receptor Kiss1R with similar efficacy [68]. The hypothalamic kisspeptin neurons play a key role in regulating the activity of GnRH neurons [69]. In both rodents and sheep, the kisspeptin neurons in the preoptic area are sexually dimorphic, being more numerous in females than males. There are two major groups of kisspeptin cell bodies: a large number in the arcuate

Table 4.4 Main features and treatment of Kallman's Syndrome

Characteristics	Rare disorder, familial and sporadic forms
Potential causal genes	KAL1, FGFR1, FGF8, CHD7, PROKR2, PROK2
Gender distribution	1 in 30,000 males; 1 in 120,000 females
Common clinical deficits	Anosmia, absence of GnRH
Occasional clinical features	Facial asymmetry, color blindness, renal anomalies
Clinical features (men)	Microphallus, cryptorchidism, delay in puberty
Clinical features (women)	Delay in puberty, anovulation
Hormone evaluation	Low LH, low FSH, low gonadal steroids
Treatment (men or women)	Secondary sex characteristics and fertility can be achieved by treatment with gonadotropins

FSH: follicle-stimulating hormone; GnRH: gonadotropin-releasing hormone; LH: luteinizing hormone.

nucleus, and a smaller number in the periventricular area of the third ventricle of rodents and the preoptic area of non-rodents [70].

Both sets of neurons project to GnRH cell bodies, which contain the kisspeptin receptor Kiss1R. Kisspeptin neurons in the arcuate nucleus also co-express neurokinin B and the endogenous opioid peptide, dynorphin, and are named kisspeptin/neurokinin B/dynorphin (KNDy) neurons. The kisspeptin neurons in the rostral periventricular area play an essential role in enabling ovulation in rodents by activating the GnRH neurons, while those in the arcuate nucleus are involved in the regulation of pulsatile GnRH release and also mediate the negative feedback of the sex steroids on the HPG axis given that GnRH neurons themselves do not express receptors for gonadal steroids (**Figure 4.12**).

The GnRH neurons extend processes to the median eminence and, upon the appropriate stimuli, they release GnRH into the hypophysial portal capillaries of the median eminence. The portal blood carries GnRH into the

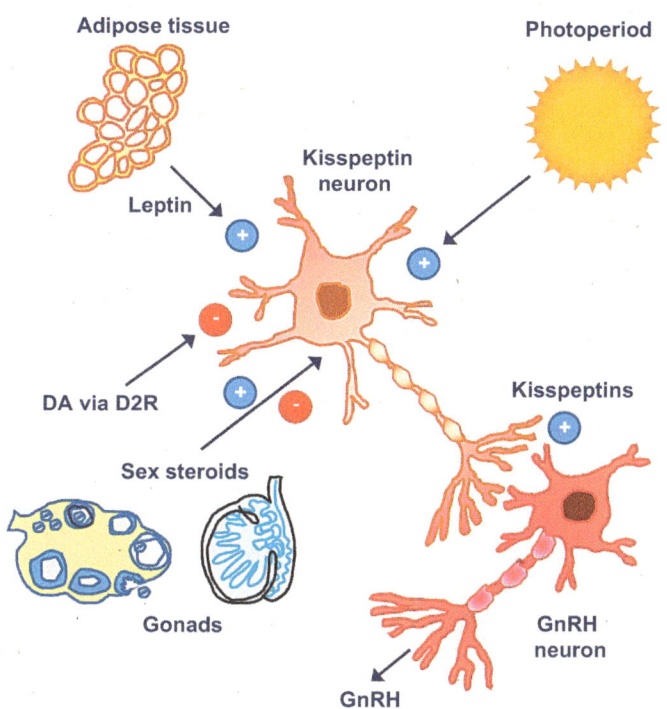

Figure 4.12 The regulation of GnRH release. The kisspeptin neurons integrate most of the external and internal information which regulates GnRH release. Feedback control by the sex steroids as well as dopamine, metabolic and environmental signals have direct effects on kisspeptin neurons. DA, dopamine; GnRH: gonadotropin-releasing hormone. (Redrawn and modified from Sonigo, C. and Binart, N., *Ann. Endocrinol. Paris*, 73, 448–458, 2012.)

anterior pituitary where it binds to its receptor (GnRHR), a member of the seven-transmembrane, G protein-coupled receptor superfamily (GPCR), and stimulates the production and release of LH and FSH. GnRH has a short half-life of 2–4 min due to rapid cleavage by peptidases. As a result of the rapid degradation and massive dilution, the peripheral circulation does not contain biologically active concentrations of GnRH.

The pattern of GnRH secretion is of utmost importance for the optimal performance of the reproductive system. In females, two distinct modes of GnRH release are recognized: pulsatile and surge. The pulsatile mode refers to an episodic release of GnRH, with distinct pulses of GnRH secretion into the portal circulation and undetectable GnRH concentrations during inter-pulse intervals. Pulses of LH occur approximately every hour in follicular-phase females, and every 2 to 3 h in luteal-phase females and males. The surge mode occurs during the preovulatory phase, in which the presence of GnRH in portal blood is high and persistent. Studies with rodents and other species identified a subpopulation of arcuate kisspeptin neurons as the putative GnRH pulse generator [71]. However, little is known about the mechanisms of synchronization and the afferent hormonal and transmitter modulation that are required for establishing the normal patterns of LH pulsatility.

The pulsatile release of GnRH and LH is essential for fertility in all mammals. This notion is based on the findings that a constant stimulation of the pituitary gonadotropes with GnRH, both under *in vivo* and *in vitro* conditions, suppresses LH secretion. This suppression occurs because of down-regulation and desensitization of the GnRH receptors in response to continuous exposure to GnRH. In patients with functional hypothalamic amenorrhea, the ovulatory cycle can be restored by pumps that deliver GnRH pulses. Recently, some success was achieved with a constant infusion of kisspeptin, which presumably restored pulsatile GnRH release.

DA is involved in the regulation of GnRH release both directly, by acting on the GnRH neurons, and indirectly, by affecting the kisspeptin neurons. Studies with immortalized GnRH cells (GT1-1) revealed that DA stimulated GnRH release via D1R receptors that are positively coupled to adenylate cyclase [72]. In another study, a 14-day treatment with bromocriptine, a DA agonist, increased by 70% GnRH mRNA expression, while the antagonist haloperidol caused a reduction. Since hypophysectomy did not alter the effects of these drugs, it was concluded that their action is not mediated by variation in pituitary hormone secretion.

The association of DA with the kisspeptin neurons has been well documented in sheep. Unlike rodents, which breed continuously throughout the year, ewes are seasonal breeders. During the nonbreeding season (anestrus), there is a dramatic increase in the ability of estradiol to inhibit the pulsatile release of GnRH [73]. Several lines of evidence showed that (1) most KNDy neurons co-express D2R, which mediates the actions of DA on GnRH pulse frequency in anestrus ewes; (2) estradiol increases the percentage of KNDy neurons containing D2R; (3) microinjections of the D2R antagonist, pimozide, into the arcuate nucleus increase LH pulse frequency in anestrus ewes; and (4) the stimulatory actions of intramuscular injections of pimozide were blocked by ICV infusions of a Kiss1R antagonist. Based on these data, it was proposed that the increased ability of estradiol to inhibit GnRH pulse frequency in anestrus ewes is due to higher DA release from A15 neuronal projections onto the KNDy neurons as well as the increased responsiveness of these neurons to the inhibitory effect of DA.

4.5.3 Reproductive functions of PRL and reciprocal interactions with dopamine

As discussed in our 2008 review [74], PRL is one of the most versatile hormones in the mammalian endocrine repertoire, affecting reproductive, metabolic, immune, behavioral, osmoregulatory, and oncogenic functions. Among the many functions ascribed to PRL, its involvement with reproduction has been best characterized. However, in many reviews that cover reproduction, PRL often does not receive an appropriate recognition. One of the difficulties in placing the reproductive functions of PRL in perspective is that its spectrum of activities varies markedly with the species being studied. Whereas PRL is essential for the initiation and maintenance of lactation in all mammals, its roles in other reproductive functions differ from one species to another.

The sources of PRL and the control of its production and release are also dissimilar among species. For example, in humans, in addition to the pituitary, PRL is produced by multiple tissues where it is regulated in a cell-specific manner and acts as a cytokine. With few exceptions, PRL production in other animals is restricted to the pituitary, with PRL acting only as a classical circulating hormone. DA of hypothalamic origin is well recognized as a potent physiological inhibitor of pituitary PRL production and release in all species. This aspect of DA-PRL interactions is covered in **Chapter 5**, while here the focus is on interactions between DA and PRL that occur within the brain and have an impact on reproduction.

Unlike most pituitary hormones, PRL lacks a specific endocrine gland as a target organ that can provide feedback information to the pituitary. Instead, PRL regulates its own secretion by acting in a short-loop feedback mechanism on the hypothalamic dopaminergic neurons [75]. One question is how does a relatively large protein of 199 amino acids cross the BBB to gain access to the dopaminergic neurons? The arcuate nucleus and median eminence have an incomplete BBB by virtue of having fenestrated endothelial cells (see **Chapter 3**, **Section 3.5.5**). However, this is not the only route by which PRL gains access to hypothalamic DA because systemic administration of PRL activates neurons throughout the hypothalamus, not only in the arcuate nucleus. Moreover, PRL levels in the cerebrospinal fluid parallel the changes in its circulating levels. Instead, there is evidence that PRL crosses the BBB through a saturable, carrier-mediated transport system, which is independent of the PRL receptor (PRLR) given that it also occurs in PRLR KO mice.

In rodents, concomitant with the LH preovulatory surge, PRL exhibits a proestrus surge. This surge does not affect ovulation *per se* but, rather, is essential for the support of the corpus luteum (luteotropic action), which promotes the production of progesterone needed for the maintenance of gestation. The situation in humans is quite different. PRL does not exhibit a mid-cycle surge and is not luteotropic. With the exception of lactation, PRL does not have discernible effects on reproductive processes under normal conditions. However, hyperprolactinemia is a major cause of infertility in both men or women, as well as a cause for impotence in men, suggesting that PRL contributes, in a subtle manner, to optimal reproduction in humans. Hyperprolactinemia in other species also results in anovulation and other reproductive disturbances.

PRLR immunoreactive neurons in the rat brain are found in the cerebral cortex (pyramidal cell layer), septal nuclei, and the amygdaloid complex [76]. A dense staining of PRLR is seen in the substantia nigra, habenula and the paraventricular thalamic nucleus. Immunostaining was also seen in the choroid plexus and the subcommissural organ. Within the hypothalamus, PRLRs are found in the suprachiasmatic, supraoptic, paraventricular, dorsomedial and arcuate nuclei; a similar distribution of the PRLR is seen in human and monkey brains.

As summarized in a 2015 review [75], exogenous PRL administration to rodents stimulated hypothalamic DA synthesis and turnover in the arcuate nucleus and median eminence and promoted DA secretion into portal blood. In contrast, hypoprolactinemia, induced by the administration of DA agonists, resulted in the suppression of DA secretion, indicating that basal activity of these neurons depends upon the endogenous circulating levels of PRL.

The time course of PRL action on the TIDA neurons has a "rapid" component of increased activity observed 2–4 h after PRL treatment and a delayed component seen 12 h after the treatment.

4.5.4 Hyperprolactinemia and the suppression of GnRH release

Hyperprolactinemia suppresses ovulation in rodents and causes infertility in humans by affecting three potential target sites: hypothalamus (GnRH release), pituitary (gonadotropin release) and gonads (sex steroid release). There is evidence that PRL can affect all three sites, but the suppression of hypothalamic GnRH release by PRL appears to be the most dominant mechanism. For example, hyperprolactinemia in both men and women is associated with a marked reduction in both the frequency and amplitude of LH pulses. Restoration of fertility can be achieved either by normalizing serum PRL concentrations or by a pulsatile delivery of GnRH, suggesting a PRL-induced suppression of GnRH [77].

Several animal models have been used to determine the mechanism by which hyperprolactinemia suppresses reproductive processes [75]. One model is the lactating rat. During lactation, circulating PRL is intermittently elevated in response to the suckling stimulus, resulting in a state of anovulation. Indeed, breast feeding has been used by some women as a natural method of contraception. The second model is the cycling female rat treated with PRL either by repeated injections or by continuous infusion. Evidence obtained with both animal models shows that elevated PRL suppresses GnRH release by acting primarily on the kisspeptin neurons rather than on the GnRH neurons themselves. Indeed, only a few GnRH neurons in mice express PRLR, suggesting an action of PRL upstream of the GnRH neurons.

Lactation is associated with a period of infertility in most mammalian females, including women. It has been argued that although hyperprolactinemia is generally viewed as a pathological condition with adverse consequences, the situation is quite different during lactation, when hyperprolactinemia is of great physiological importance. An inhibitory action of PRL on fertility during lactation allows the mother to focus her energy on feeding the offspring, rather than investing precious resources in another pregnancy. In humans, this function serves as a critical regulator of population growth, by spacing the timing of births to allow the mother to ration her metabolic investment across sequential pregnancies.

Induction of hyperprolactinemia in normal cycling female mice has caused anovulation, reduced GnRH and gonadotropin secretion, and diminished kisspeptin expression. A daily intraperitoneal administration of kisspeptin restores gonadotropin secretion and ovarian cyclicity, suggesting that kisspeptin neurons play a major role in anovulation resulting from hyperprolactinemia. These data, however, do not clarify whether the PRL-induced activation of the kisspeptin-GnRH axis is direct, or indirect, by way of activating the dopaminergic system. The interactions between elevated PRL, the kisspeptin-GnRH neurons, as well as the ovaries, are illustrated in **Figure 4.13**.

4.5.5 Regulation of sexual and maternal behavior

Unlike research on reproductive processes, where parameters such as hormone levels, ovulation, spermatogenesis, or lactation can be analyzed in experimental animals with relative ease, studies on sexual behavior, using laboratory animals, have met with difficulties, especially if the goal is to better understand human sexuality. It is not surprising, therefore, that many publications with titles like "What can animal models tell us about human sexual response?" [78], or "Assessment of sexual behavior in rats: The potentials and pitfalls" [79], have dominated the literature of experimental research on sexual behavior. It is broadly acknowledged that human behavior is much

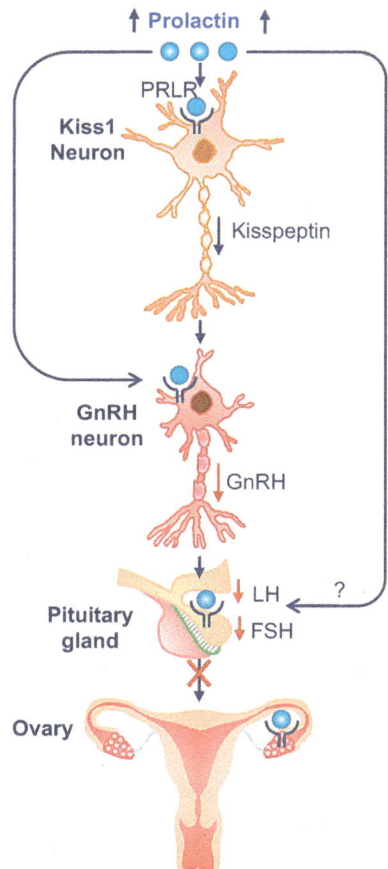

Hypogonadism, Infertility, Amenorrhea

Figure 4.13 Suppression of the reproductive axis by hyperprolatinemia. The prolactin (PRL) receptors (PRLR) are expressed on all potential regulatory sites: kisspeptin and GnRH neurons, pituitary gonadotrophs and steroid-producing ovarian cells. FSH: follicle-stimulating hormone; GnRH: gonadotropin-releasing hormone; Kiss1: kisspeptin 1; LH: luteinizing hormone; PRL: prolactin. (Redrawn and modified from Kaiser, U.B., *J. Clin. Invest.*, 122, 3467–3468, 2012.)

more complex than any animal model can reproduce given that family, social, cultural influences, and personal beliefs, which are so important in the clinical practice, cannot be incorporated into animal models [80]. Similarly, maternal behavior is species specific, is expressed under different physiological/environmental conditions, and differs substantially between rodents and humans.

In spite of the above caveats, the general consensus is that animal models are crucial for understanding some of the complexities of sexual behavior, potential causes of sexual dysfunctions, and the intricacies of maternal care. A major advantage that can be provided by animal models, and cannot be replicated when studying human sexual behavior, is the ability to genetically alter critical components such as sex steroids, kisspeptin, or GnRH and their receptors by knockout or conditional expression in animal models.

The hypothalamus is the primary site that integrates incoming sensory and peripheral signals and coordinates the behavioral response. Other regions, including the VTA, periaqeductal gray and prefrontal cortex, also contribute to reproductive behaviors [81]. Lesions in the medial preoptic area (MPOA) of the hypothalamus prevent key elements of **sexual behavior** in both males and females, and disrupt maternal behavior in females. The MPOA contains neurons that are responsive to sex steroids and is sexually dimorphic, i.e., it is larger and has more neurons in males than in female rodents. Additional studies have showed that individual neurons in the MPOA fire during sexual stimulation.

The output of the MPOA include a major projection to the VMN, where lesions have prevented both male (mounting) and female (lordosis) sexual behavior. The two nuclei have projections to the lateral periaqueductal gray, which mediates motor and autonomic patterns associated with sexual behavior. The PVN also plays modulatory roles, especially with respect to penile erection in males and the onset of maternal behavior and responses to social stimuli in females. Other important hypothalamic nuclei include the VMN, particularly in the context of female sexual behavior, and the SON in terms of sexual interactions.

Among the neuromodulators involved in reproductive/maternal behaviors, DA and OT stand out as key players [81]. Major sources of DA that mediate behaviors are the incerto-hypothalamic system, which innervates several hypothalamic nuclei involved in social interaction and behavioral responses, and the TIDA, which regulates the release of PRL, itself involved in sexual and maternal behavior. As demonstrated by agonist, antagonist and DA release studies, the action of brain DA in female rodents is associated with lordosis, a quantifiable marker of sexual receptivity. In males, DA agonists elicit penile erection and accelerate sexual interaction when paired with a mate, and these effects are blocked by central DA antagonist administration. Such treatments are most effective when given directly into the PVN or the MPOA. It was also reported that a D1R antagonist given systemically has decreased intromission frequency in paired sexually naïve males, together with attenuated copulation-induced Fos expression in medial preoptic neurons, suggesting a role for endogenous DA in this region at intromission.

As reviewed in [81], neuroanatomical, pharmacological and behavioral evidence reveal an interdependent relationship between DA and OT neurons in eliciting sexual behavior. Although OT is released into the blood and increases during sexual arousal and copulation in rodents and humans, it is the OT action within the brain that plays an important role in behaviors associated with sexual encounter. In males, central OT is acutely implicated in both penile erection and ejaculation mechanisms, while in females, OT is especially associated in proceptive behavior and lordosis.

Another important interaction that affects sexual behavior in both sexes is between DA and estradiol [82]. Estradiol acts in concert with hypothalamic DA, through the activation of similar intracellular signaling pathways, to stimulate female sexual behavior. Some of the effects of estradiol are mediated by its direct action on the mesolimbic dopaminergic neurons. Notably, the estradiol–DA interactions also influence male sexual behavior. Although testosterone is regarded as the primary male sex hormone, in rodents, testosterone action on the brain induces male sexual behavior largely because it is aromatized to estradiol. For example, peripheral testosterone injection in male castrates restores sexual behavior only in the presence of aromatase, and estradiol can restore sexual behavior when aromatase has been inhibited. Moreover, male mice with knockout of the aromatase enzyme do not initiate copulation.

The neuroendocrine control of **maternal behavior** in mammals is covered in several reviews [83–85]. The progression through pregnancy, parturition, lactation, and weaning is interspersed by physiological and hormonal changes, some of which are abrupt, e.g., parturition, while others are gradual, e.g., pregnancy and weaning. The synchronization of maternal behavior with parturition and lactation ensures that the mother responds to needs of the young at the appropriate time. Around the time of birth, changes in hormones such as estrogen, progesterone, PRL, and OT trigger a cascade of neurological adaptations that result in a stereotypic behavior.

Although some parameters associated with maternal behavior can be studied in humans, many others have been examined in great detail in animal models. Maternal behavior in rodents is basically composed of four parameters: retrieval behavior, nest building, nursing behavior, and pup grooming. Under well-controlled experimental paradigms, behavioral patterns in response to various situations or treatments can be quantified. Animal models have major advantages in terms of genetic manipulations, control of variables, short life cycle, and low cost [85].

Studies with rodents have shown that DA affects the expression of typical maternal behavior in mothers who have given birth, as well as in non-mothers who exhibit maternal behaviors because of repeated exposure to young. Most research has focused on the mesolimbic DA projections from the VTA to the NAc in the ventral striatum [86]. DA is released in the NAc during maternal behavior, and the increased DA release is associated with stronger maternal responses. The medial preoptic area (MPOA) is important for the motivational aspects of maternal behavior. MPOA lesions reduce NAc activity during maternal behavior. Moreover, inhibition of D1R in both the NAc and the MPOA disrupts maternal behavior, whereas stimulation of D1R in the NAc facilitates maternal behavior. DA seems to be especially important for active, goal-directed maternal behaviors, such as pup retrieval, licking, and grooming, and appears to be less important for more passive behaviors such as nursing.

Finally, DA has an indirect effect on maternal behavior through its control of PRL release. Both PRL and placental lactogen, acting through the PRLR, are heavily involved in the induction of maternal behavior [84]. Evidence has been derived from multiple experimental paradigms, including hypophysectomy, gonadectomy and replacement therapy. For example, to determine the central site of action of PRL, gonadectomized, steroid-treated, bromocriptine injected virgin female rats were infused with PRL or placental lactogen into the brain. Infusion into the MPOA stimulated a rapid onset of maternal behavior toward foster pups. The same dose of PRL infused into the lateral vehicles failed to stimulate maternal behavior, indicating that the actions of PRL in the MPOA is site specific. The potential role of PRL acting on the VMH and PVN, which express the PRLR, has not been well established. The authors concluded that these studies clearly demonstrate a role for lactogenic hormones in combination with estrogens and progesterone in the induction of maternal behavior.

4.6 SYNOPSIS

DA from all regions of the brain is associated with four major categories of neuroendocrine-regulated functions: circadian rhythms, stress response, food intake/metabolic homeostasis, and reproduction/sexual and maternal behavior. For each of these functions, background detailing their overall management has been first presented, followed by a summary of the available data on the involvement of dopamine in their operation. Some of the presented evidence is strong and convincing, while in other cases it is indirect or incomplete. The multiplicity of dopamine action is best explained by considering the variety of neurotransmitters, hormones and neuromodulators that regulate DA production and release, as well as the five different receptors with complementing and/or opposing actions through which dopamine exerts its influence.

REFERENCES

1. Merbitz-Zahradnik T, Wolf E. How is the inner circadian clock controlled by interactive clock proteins?: Structural analysis of clock proteins elucidates their physiological role. *FEBS Lett.* 2015;589(14):1516–1529.

2. Golombek DA, Bussi IL, Agostino PV. Minutes, days and years: Molecular interactions among different scales of biological timing. *Philos Trans R Soc Lond B Biol Sci.* 2014;369(1637):20120465.

3. Cipolla-Neto J, Amaral FGD. Melatonin as a hormone: New physiological and clinical insights. *Endocr Rev.* 2018;39(6):990–1028.

4. Korshunov KS, Blakemore LJ, Trombley PQ. Dopamine: A modulator of circadian rhythms in the central nervous system. *Front Cell Neurosci.* 2017;11:91–104.

5. La MC, Ross-Cisneros FN, Sadun AA, Carelli V. Retinal ganglion cells and circadian rhythms in Alzheimer's disease, Parkinson's disease, and beyond. *Front Neurol.* 2017;8:162–185.

6. Popova E. Role of dopamine in distal retina. *J Comp Physiol A Neuroethol Sens Neural Behav Physiol.* 2014;200(5):333–358.

7. Jackson CR, et al. Retinal dopamine mediates multiple dimensions of light-adapted vision. *J Neurosci.* 2012;32(27):9359–9368.

8. Ramkisoensing A, Meijer JH. Synchronization of biological clock neurons by light and peripheral feedback systems promotes circadian rhythms and health. *Front Neurol.* 2015;6:128–139.

9. Strother WN, Norman AB, Lehman MN. D1-dopamine receptor binding and tyrosine hydroxylase-immunoreactivity in the fetal and neonatal hamster suprachiasmatic nucleus. *Brain Res Dev Brain Res.* 1998;106(1–2):137–144.

10. Ferguson SA, Rowe SA, Krupa M, Kennaway DJ. Prenatal exposure to the dopamine agonist SKF-38393 disrupts the timing of the initial response of the suprachiasmatic nucleus to light. *Brain Res.* 2000;858(2):284–289.

11. Grippo RM, Purohit AM, Zhang Q, Zweifel LS, Guler AD. Direct midbrain dopamine input to the suprachiasmatic nucleus accelerates circadian entrainment. *Curr Biol.* 2017;27(16):2465–2475.

12. Rivkees SA, Lachowicz JE. Functional D1 and D5 dopamine receptors are expressed in the suprachiasmatic, supraoptic, and paraventricular nuclei of primates. *Synapse.* 1997;26(1):1–10.

13. Zisapel N. Melatonin-dopamine interactions: From basic neurochemistry to a clinical setting. *Cell Mol Neurobiol.* 2001;21(6):605–616.

14. Racke K, Krupa H, Schroder H, Vollrath L. *In vitro* synthesis of dopamine and noradrenaline in the isolated rat pineal gland: Day-night variations and effects of electrical stimulation. *J Neurochem.* 1989;53(2):354–361.

15. Gonzalez S, et al. Circadian-related heteromerization of adrenergic and dopamine D(4) receptors modulates melatonin synthesis and release in the pineal gland. *PLoS Biol.* 2012;10(6):e1001347.

16. Ben-Jonathan N. Dopamine: A prolactin-inhibiting hormone. *Endocr Rev.* 1985;6(4):564–589.

17. Ben-Jonathan N, Hnasko R. Dopamine as a prolactin (PRL) inhibitor. *Endocr Rev.* 2001;22(6):724–763.

18. Smith MS, Neill JD. Termination at midpregnancy of the two daily surges of plasma prolactin initiated by mating in the rat. *Endocrinology.* 1976;98(3):696–701.

19. Andrews ZB, Kokay IC, Grattan DR. Dissociation of prolactin secretion from tuberoinfundibular dopamine activity in late pregnant rats. *Endocrinology.* 2001;142(6):2719–2724.

20. Kok P, et al. Activation of dopamine D2 receptors lowers circadian leptin concentrations in obese women. *J Clin Endocrinol Metab.* 2006;91(8):3236–3240.

21. Borcherding DC, et al. Dopamine receptors in human adipocytes: Expression and functions. *PLoS One.* 2011;6(9):e25537.

22. Stockhorst U, Antov MI. Modulation of fear extinction by stress, stress hormones and estradiol: A review. *Front Behav Neurosci.* 2015;9:359–372.

23. Rasheed N, Alghasham A. Central dopaminergic system and its implications in stress-mediated neurological disorders and gastric ulcers: Short review. *Adv Pharmacol Sci.* 2012;2012:1–12.

24. Papathanassoglou ED, Giannakopoulou M, Mpouzika M, Bozas E, Karabinis A. Potential effects of stress in critical illness through the role of stress neuropeptides. *Nurs Crit Care.* 2010;15(4):204–216.

25. Horger BA, Roth RH. The role of mesoprefrontal dopamine neurons in stress. *Crit Rev Neurobiol.* 1996;10(3–4):395–418.

26. Peng J, et al. A quantitative analysis of the distribution of CRH neurons in whole mouse brain. *Front Neuroanat.* 2017;11:63–76.

27. Belda X, Armario A. Dopamine D1 and D2 dopamine receptors regulate immobilization stress-induced activation of the hypothalamus-pituitary-adrenal axis. *Psychopharmacology.* 2009;206(3):355–365.

28. Sugama S, Kakinuma Y. Loss of dopaminergic neurons occurs in the ventral tegmental area and hypothalamus of rats following chronic stress: Possible pathogenetic loci for depression involved in Parkinson's disease. *Neurosci Res.* 2016;111:48–55.

29. Goldstein DS, et al. Sources and physiological significance of plasma dopamine sulfate. *J Clin Endocrinol Metab.* 1999;84(7):2523–2531.

30. Szivak TK, et al. Adrenal stress and physical performance during military survival training. *Aerosp Med Hum Perform.* 2018;89(2):99–107.

31. Boschetti M, et al. Role of dopamine receptors in normal and tumoral pituitary corticotropic cells and adrenal cells. *Neuroendocrinology.* 2010;92(Suppl 1):17–22.

32. Verma R, Balhara YP, Gupta CS. Gender differences in stress response: Role of developmental and biological determinants. *Ind Psychiatry J.* 2011;20(1):4–10.

33. Becker JB. Gender differences in dopaminergic function in striatum and nucleus accumbens. *Pharmacol Biochem Behav.* 1999;64(4):803–812.

34. Brown RS, Herbison AE, Grattan DR. Effects of prolactin and lactation on A15 dopamine neurones in the rostral preoptic area of female mice. *J Neuroendocrinol.* 2015;27(9):708–717.

35. Freeman ME, Kanyicska B, Lerant A, Nagy G. Prolactin: Structure, function, and regulation of secretion. *Physiol Rev.* 2000;80(4):1523–1631.

36. Steyn FJ, Ngo ST. Endocrine rhythms of growth hormone release: Insights from animal studies. *Best Pract Res Clin Endocrinol Metab.* 2017;31(6):521–533.

37. Kitajima N, et al. Effects of dopamine on immunoreactive growth hormone-releasing factor and somatostatin secretion from rat hypothalamic slices perifused *in vitro*. *Endocrinology.* 1989;124(1):69–76.

38. Vance ML, et al. Role of dopamine in the regulation of growth hormone secretion: Dopamine and bromocriptine

augment growth hormone (GH)-releasing hormone-stimulated GH secretion in normal man. *J Clin Endocrinol Metab.* 1987;64(6):1136–1141.

39. Scott N, Prigge M, Yizhar O, Kimchi T. A sexually dimorphic hypothalamic circuit controls maternal care and oxytocin secretion. *Nature.* 2015;525:519–522.

40. Winter J, Jurek B. The interplay between oxytocin and the CRF system: Regulation of the stress response. *Cell Tissue Res.* 2019;375(1):85–91.

41. Woods SC, Lutz TA, Geary N, Langhans W. Pancreatic signals controlling food intake; insulin, glucagon and amylin. *Philos Trans R Soc Lond B Biol Sci.* 2006;361:1219–1235.

42. Xu Y, O'Malley BW, Elmquist JK. Brain nuclear receptors and body weight regulation. *J Clin Invest.* 2017;127(4):1172–1180.

43. Kontis D, Theochari E. Dopamine in anorexia nervosa: A systematic review. *Behav Pharmacol.* 2012;23(5–6):496–515.

44. Fetissov SO, Meguid MM, Sato T, Zhang LH. Expression of dopaminergic receptors in the hypothalamus of lean and obese Zucker rats and food intake. *Am J Physiol Regul Integr Comp Physiol.* 2002;283(4):R905–R910.

45. Ferrario CR, et al. Homeostasis meets motivation in the battle to control food intake. *J Neurosci.* 2016;36(45):11469–11481.

46. Barrios-Correa AA, Estrada JA, Contreras I. Leptin signaling in the control of metabolism and appetite: Lessons from animal models. *J Mol Neurosci.* 2018;66(3):390–402.

47. Feve B, Bastard JP. From the conceptual basis to the discovery of leptin. *Biochimie.* 2012;94(10):2065–2068.

48. Kim KS, et al. Enhanced hypothalamic leptin signaling in mice lacking dopamine D2 receptors. *J Biol Chem.* 2010;285(12):8905–8917.

49. Hommel JD, et al. Leptin receptor signaling in midbrain dopamine neurons regulates feeding. *Neuron.* 2006;51(6):801–810.

50. Figlewicz DP, Benoit SC. Insulin, leptin, and food reward: Update 2008. *Am J Physiol Regul Integr Comp Physiol.* 2009;296(1):R9–R19.

51. Mebel DM, Wong JC, Dong YJ, Borgland SL. Insulin in the ventral tegmental area reduces hedonic feeding and suppresses dopamine concentration via increased reuptake. *Eur J Neurosci.* 2012;36(3):2336–2346.

52. Brunetti L, et al. Effects of ghrelin and amylin on dopamine, norepinephrine and serotonin release in the hypothalamus. *Eur J Pharmacol.* 2002;454(2–3):189–192.

53. Mietlicki-Baase EG, et al. Amylin receptor activation in the ventral tegmental area reduces motivated ingestive behavior. *Neuropharmacology.* 2017;123:67–79.

54. Al MO, Nogueiras R, Dieguez C, Girault JA. Ghrelin and food reward. *Neuropharmacology.* 2019;148:131–138.

55. Frago LM, Chowen JA. Involvement of astrocytes in mediating the central effects of ghrelin. *Int J Mol Sci.* 2017;18(3):2–19.

56. Kawahara Y, et al. Peripherally administered ghrelin induces bimodal effects on the mesolimbic dopamine system depending on food-consumptive states. *Neuroscience.* 2009;161(3):855–864.

57. Dockray GJ. Cholecystokinin. *Curr Opin Endocrinol Diabetes Obes.* 2012;19(1):8–12.

58. Blevins JE, Stanley BG, Reidelberger RD. Brain regions where cholecystokinin suppresses feeding in rats. *Brain Res.* 2000;860(1–2):1–10.

59. Bednar I, Forsberg G, Linden A, Qureshi GA, Sodersten P. Involvement of dopamine in inhibition of food intake by cholecystokinin octapeptide in male rats. *J Neuroendocrinol.* 1991;3(5):491–496.

60. Stadlbauer U, Woods SC, Langhans W, Meyer U. PYY3-36: Beyond food intake. *Front Neuroendocrinol.* 2015;38:1–11.

61. Waterson MJ, Horvath TL. Neuronal regulation of energy homeostasis: Beyond the hypothalamus and feeding. *Cell Metab.* 2015;22(6):962–970.

62. Hnasko TS, Szczypka MS, Alaynick WA, During MJ, Palmiter RD. A role for dopamine in feeding responses produced by orexigenic agents. *Brain Res.* 2004;1023(2):309–318.

63. Dallvechia-Adams S, Kuhar MJ, Smith Y. Cocaine- and amphetamine-regulated transcript peptide projections in the ventral midbrain: Colocalization with gamma-aminobutyric acid, melanin-concentrating hormone, dynorphin, and synaptic interactions with dopamine neurons. *J Comp Neurol.* 2002;448(4):360–372.

64. Barson JR, Morganstern I, Leibowitz SF. Complementary roles of orexin and melanin-concentrating hormone in feeding behavior. *Int J Endocrinol.* 2013;2013:1–10.

65. Cheng CK, Leung PC. Molecular biology of gonadotropin-releasing hormone (GnRH)-I, GnRH-II, and their receptors in humans. *Endocr Rev.* 2005;26(2):283–306.

66. Wierman ME, Kiseljak-Vassiliades K, Tobet S. Gonadotropin-releasing hormone (GnRH) neuron migration: Initiation, maintenance and cessation as critical steps to ensure normal reproductive function. *Front Neuroendocrinol.* 2011;32(1):43–52.

67. Biehl MJ, Raetzman LT. Developmental origins of hypothalamic cells controlling reproduction. *Semin Reprod Med.* 2017;35(2):121–129.

68. Beltramo M, Dardente H, Cayla X, Caraty A. Cellular mechanisms and integrative timing of neuroendocrine control of GnRH secretion by kisspeptin. *Mol Cell Endocrinol.* 2014;382(1):387–399.

69. Sonigo C, Binart N. Overview of the impact of kisspeptin on reproductive function. *Ann Endocrinol.* 2012;73(5):448–458.

70. Lehman MN, Hileman SM, Goodman RL. Neuroanatomy of the kisspeptin signaling system in mammals: Comparative and developmental aspects. *Adv Exp Med Biol*. 2013;784:27–62.

71. Herbison AE. The gonadotropin-releasing hormone pulse generator. *Endocrinology*. 2018;159(11):3723–3736.

72. Yoshida H, Paruthiyil S, Butler P, Weiner RI. Role of cAMP signaling in the mediation of dopamine-induced stimulation of GnRH secretion via D1 dopamine receptors in GT1–7 cells. *Neuroendocrinology*. 2004;80(1):2–10.

73. Nestor CC, et al. Regulation of GnRH pulsatility in ewes. *Reproduction*. 2018;156(3):R83–R99.

74. Ben-Jonathan N, Lapensee CR, Lapensee EW. What can we learn from rodents about prolactin in humans? *Endocr Rev*. 2008;29(1):1–41.

75. Grattan DR. 60 years of neuroendocrinology: The hypothalamo-prolactin axis. *J Endocrinol*. 2015;226(2):T101–T122.

76. Roky R, et al. Distribution of prolactin receptors in the rat forebrain. Immunohistochemical study. *Neuroendocrinology*. 1996;63(5):422–429.

77. Kaiser UB. Hyperprolactinemia and infertility: New insights. *J Clin Invest*. 2012;122(10):3467–3468.

78. Pfaus JG, Kippin TE, Coria-Avila G. What can animal models tell us about human sexual response? *Annu Rev Sex Res*. 2003;14:1–63.

79. Heijkoop R, Huijgens PT, Snoeren EMS. Assessment of sexual behavior in rats: The potentials and pitfalls. *Behav Brain Res*. 2018;352:70–80.

80. Ventura-Aquino E, Paredes RG. Animal models in sexual medicine: The need and importance of studying sexual motivation. *Sex Med Rev*. 2017;5(1):5–19.

81. Baskerville TA, Douglas AJ. Interactions between dopamine and oxytocin in the control of sexual behaviour. *Prog Brain Res*. 2008;170:277–290.

82. Stolzenberg DS, Numan M. Hypothalamic interaction with the mesolimbic DA system in the control of the maternal and sexual behaviors in rats. *Neurosci Biobehav Rev*. 2011;35(3):826–847.

83. Levy F. Neuroendocrine control of maternal behavior in non-human and human mammals. *Ann Endocrinol*. 2016;77(2):114–125.

84. Bridges RS. Neuroendocrine regulation of maternal behavior. *Front Neuroendocrinol*. 2015;36:178–196.

85. Pires GN, Tufik S, Giovenardi M, Andersen ML. Maternal behavior in basic science: Translational research and clinical applicability. *Einstein*. 2013;11(2):256–260.

86. Henschen CW, Palmiter RD, Darvas M. Restoration of dopamine signaling to the dorsal striatum is sufficient for aspects of active maternal behavior in female mice. *Endocrinology*. 2013;154(11):4316–4327.

87. Woods SC. Gastrointestinal satiety signals I. An overview of gastrointestinal signals that influence food intake. *Am J Physiol Gastrointest Liver Physiol*. 2004;286(1):G7–13.

88. van Zessen R, van der Plasse G, Adan RA. Contribution of the mesolimbic dopamine system in mediating the effects of leptin and ghrelin on feeding. *Proc Nutr Soc*. 2012;71(4):435–445.

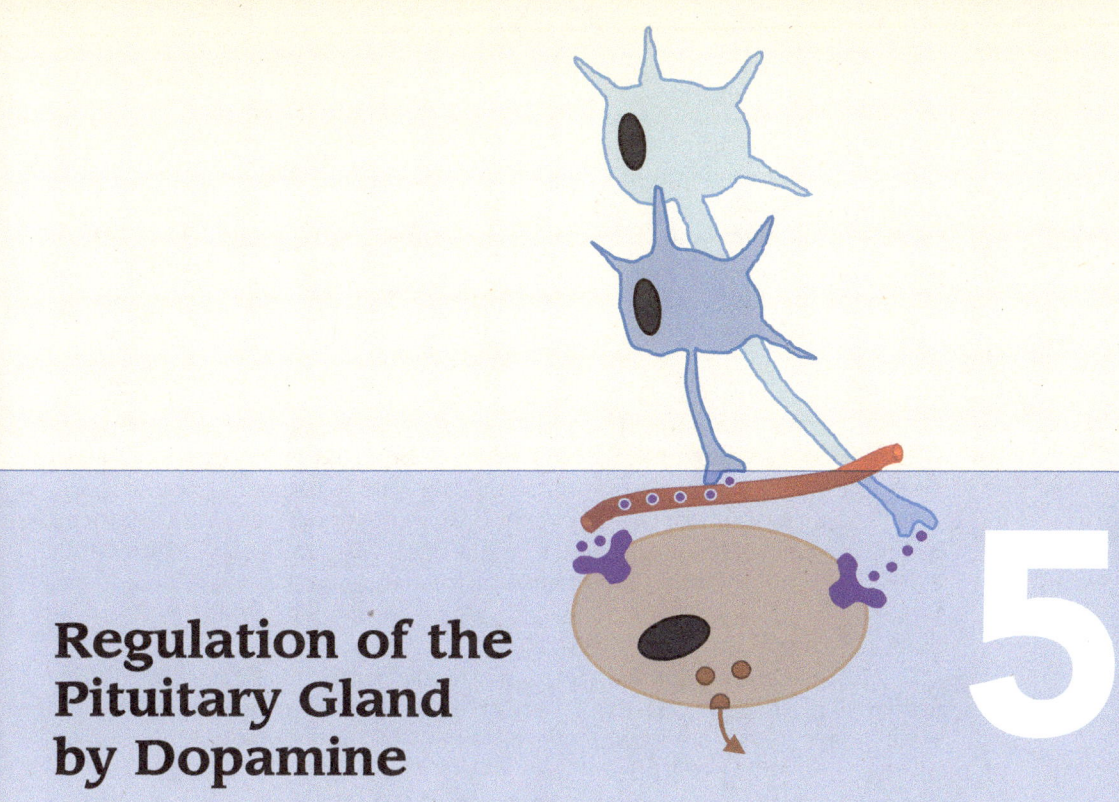

Regulation of the Pituitary Gland by Dopamine

5

5.1 INTRODUCTION

The subject of this chapter is the involvement of dopamine (DA) with the neuroendocrine regulation of the pituitary hormones. The various hormones are broadly grouped by their anatomical localization, i.e., the neural, intermediate, and anterior lobes. The first section covers the ontogeny, cellular composition and innervation of the neural lobe (NL), followed by a description of the synthesis, release, receptors, major functions and dysfunctions of **vasopressin (AVP) and oxytocin (OT)**, the two major hormones of the NL. Evidence for the involvement of DA in the regulation of each of these hormones is presented and evaluated. The next section is organized by the same format and covers β-endorphin and **α-melanocyte stimulating hormone (α-MSH)**, the two major hormones of the intermediate lobe (IL). The largest portion of the chapter is dedicated to the anterior pituitary and its six hormones. First, the ontogeny and composition of the various anterior pituitary cell types are reviewed. This is followed by a discussion on the six hormones, which are grouped by their common functional attributes. These include: prolactin (PRL) and growth hormone (GH), the somatolactogenic hormones, luteinizing hormone (LH) and follicle-stimulating hormone (FSH), the reproductive hormones, and adrenocorticotropic hormone (ACTH) and thyroid stimulating hormone (TSH), the stress and metabolic hormones.

5.2 NEURAL LOBE HORMONES: VASOPRESSIN AND OXYTOCIN

5.2.1 Ontogeny and composition of the posterior pituitary (neurointermediate lobe)

The mature posterior pituitary gland in humans consists of distal axons of hypothalamic neurons and their terminals, specialized supportive astroglia known as **pituicytes**, which surround the terminals, and capillaries that originate from branches of the carotids. During embryogenesis, the gland is formed by evagination of neural tissue from the floor of the third ventricle.

After its downward migration, the neurohypophysial tissue becomes encapsulated by the ascending oral ectodermal cells of Rathke's pouch, which eventually form the anterior pituitary. In humans, this development is completed by the end of the first trimester, and the two major neurohypophysial hormones, **arginine vasopressin (AVP)**, also called antidiuretic hormone (ADH), and **OT**, are already detectable in the neurohypophysial tissue at this time [1].

The IL in humans differentiates from the rostral part of the neural tissue during fetal life and actively releases α-MSH. However, toward the end of gestation, the human IL undergoes involution to a group of colloid-filled cysts that mingle with the NL in the adult [2]. The IL does not exist as a discrete pituitary lobe postnatally, except during pregnancy, when it undergoes hyperplasia [3]. Neither the mechanisms underlying the pregnancy-related hyperplasia, nor the exact functions of the IL in pregnant women are known. Whales, dolphins and elephants also lack a distinct IL, while most other mammalian species have a well-defined IL.

The posterior pituitary [neurointermediate lobe (NIL)] receives three neuronal projections: (1) the magnocellular neurons **(neurohypophysial tract)**, which extend from the supraoptic (SON) and paraventricular nuclei (PVN), carry AVP and OT, respectively, and innervate the NL; (2) the **tuberohypophysial dopaminergic neurons (THDA)**; which originate in the arcuate nucleus and innervate the NL; and (3) **the periventricular hypothalamic dopaminergic (PHDA)** neurons, which originate in the periventricular nucleus, and innervate the IL in species that have a distinct gland [4]. In addition to their release into the systemic circulation, the secretory products of these neurons can be delivered directly to the anterior lobe (AL, or AP) via the short portal vessels that connect the two lobes. See a detailed description of the neurovascular connections of the hypothalamo-pituitary complex in **Chapter 3**, and illustration in **Figure 3.11**.

During embryogenesis, the magnocellular neurons originate from the neuroepithelium of the third ventricle and migrate, first to the SON and then to the PVN and accessory nuclei. PreproAVP gene expression and proAVP synthesis are detected in the newly formed neurons soon after their arrival to the magnocellular nuclei. The number of AVP-immunoreactive neurons increases postnatally. However, this increase is unlikely due to formation of new neurons but can be explained by expression of AVP in initially "silent" neurons or by increased AVP synthesis that makes the cells detectable by immunocytochemistry [1]. In addition to AVP and OT, the magnocellular neurons co-express endogenous opioids such as dynorphins, endorphins and enkephalins, as well as other neuropeptides, including galanin and corticotropic releasing hormone (CRH), all of which affect the activity of the vasopressin and/or OT neurons [5].

The main functions of AVP are antidiuresis and blood pressure control, whereas the major functions of OT are uterine contraction and milk ejection. The phylogeny of the neuro-hypophysial hormones is traced to vasotocin, which is found in the primitive lamprey [6]. Two distinct lineages, one involved in reproduction (OT-like peptides) and the other in osmoregulation (vasopressin-like peptides), have developed by gene duplication and point mutations. All these peptides, which remained highly conserved through evolution, are nonapeptides with an intramolecular disulfide bridge and an amidated C-terminus. In homologous peptides from 50 vertebrate species, six of the nine amino acids are invariant. As illustrated in **Figure 5.1**, the OT-like peptides include OT, mesotocin (MT), and isotocin (IT) all of which have substitutions in positions 4 and/or 8. Within the vasopressin-like peptides, which include AVP and arginine vasotocin (AVT), substitutions occur in positions 3 and 8.

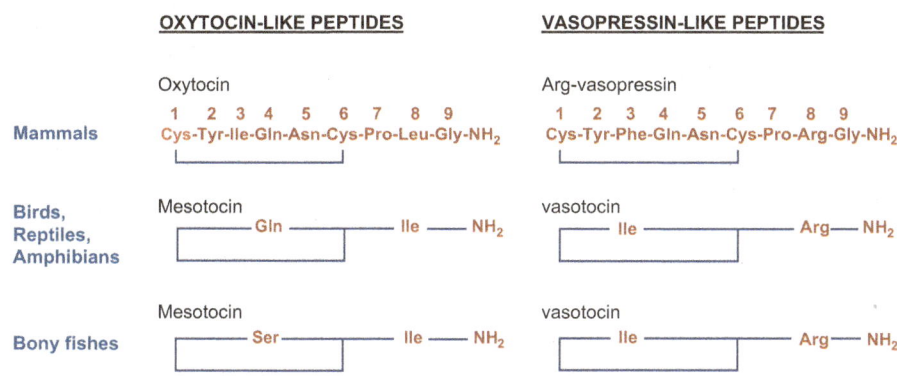

Figure 5.1 Comparison of the structures of oxytocin-like and vasopressin-like peptides in different species. Species-dependent alterations in oxytocin-like peptides occur in residues 4 and 8, while those in vasopressin-like peptides occur in residues 3 and 8.

The pituicytes, identified by their expression of glial fibrillary acidic protein (GFAP), are resident astroglia within the NL. They have an intimate relationship with the axonal terminals of the neurohypophysial hormones and exhibit morphological plasticity. Ultrastructural studies have revealed that when hormone demand is low, the pituicytes engulf the neurosecretory processes and interpose their processes between the secretory endings and the basal lamina [7]. Because any secreted hormone must pass through the basal lamina into the perivascular spaces before reaching the fenestrated capillaries, the interpositions by the pituicytes form physical and chemical barriers to hormones entering the circulation. During prolonged physiological stimulations such as during dehydration and lactation, the processes of the pituicytes retract from the basal lamina through rearrangement of the cytoskeleton and permit closer neural contacts. Studies with isolated neurohypophyses and cultured adult rat pituicytes have shown that they undergo morphological changes in response to osmotic stimuli or receptor-mediated activation of adenylate cyclase, reflecting the processes which occur *in vivo*.

5.2.2 Synthesis, transport, and processing of the neurohypophysial hormones

Each of the OT and AVP genes contains three exons and two introns. Both genes are located on the same chromosomal locus (chromosome 2 in mice and chromosome 20 in humans) but are transcribed in the opposite directions [8]. The domain which separates the OT and VP genes is relatively short and is called the "intergenic region" (IGR). The IGR in the rat and human is about 10–11 kbp in length, whereas that in the mouse is only 3.6 kbp. More than half of the rat IGR is represented by a long interspersed repeated DNA element that is completely missing in the mouse and human IGRs. The high sequence conservation found both upstream and downstream of these genes suggests that these domains contain regulatory DNA sequences.

As shown in **Figure 5.2**, each of the nonapeptides is synthesized as part of larger precursor protein that undergoes posttranslational processing, cleavage and modification, forming the mature nonapeptide [9]. The precursors have a multi-domain structure consisting of a signal peptide, followed by the nine-amino acid hormone, and a cysteine-rich domain called neurophysin, which is connected to the hormone moiety by the amino acids glycine-lysine-arginine residues (GKRs). The GKR is a site of a multistep enzymatic process of cleavage of the hormone from neurophysin and the formation of the mature nonapeptide by amidation of the position-9 amino acid glycine. At the C-terminus, the AVP prohormone also contains a 39-amino acid

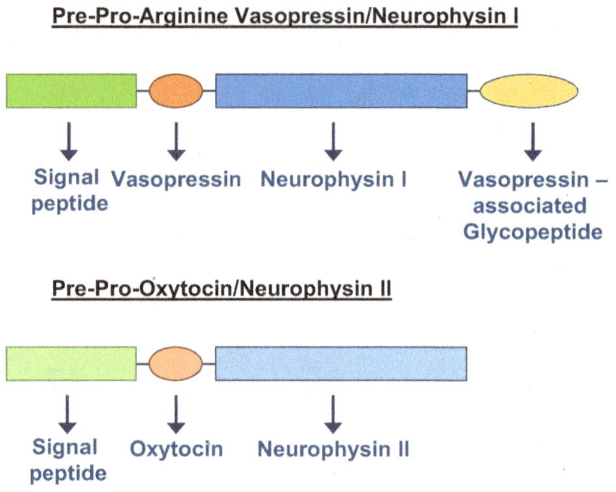

Figure 5.2 Processing of the neurohypophysial preprohormones. Cleavage of pre-pro-arginine vasopressin and pre-pro-oxytocin proteins yields different bioactive peptides and carrier proteins.

Pre-Pro-Arginine Vasopressin/Neurophysin I

Signal peptide · Vasopressin · Neurophysin I · Vasopressin – associated Glycopeptide

Pre-Pro-Oxytocin/Neurophysin II

Signal peptide · Oxytocin · Neurophysin II

glycopeptide, named vasopressin-associated glycopeptide (VAG), which is absent in the OT prohormone.

The regulated secretory pathway is shared by excitable cells that include neurons, neuroendocrine, and endocrine cells. It entails the sorting of the newly made proteins from the endoplasmic reticulum to the Golgi apparatus, where they are packed within secretory granules. The latter are either transported toward the plasma membrane in a cytoskeleton-dependent manner to be released by exocytosis or are transported down the axon toward the nerve terminals, from where they are released, as is the case with the magnocellular neurons.

The signal peptide is removed from the prohormone during the process of translation in the endoplasmic reticulum, while the rest of the processing occurs within the secretory vesicles. The precursor proteins of the nonapeptides are packaged in large dense core vesicles that are transported to the posterior pituitary by axoplasmic flow. During transport, the AVP prohormone is enzymatically cleaved at both the vasopressin-neurophysin and neurophysin-glycopeptide processing sites by PC3, a subtilisin-like prohormone convertase [10].

Neurophysin I is associated with AVP, while **neurophysin II** is associated with OT. Both are carrier proteins that after cleavage are non-covalently linked to their respective peptides within the secretory granules. The self-association of the peptide-neurophysin complex plays critical roles in the sorting of the precursor protein to the regulated secretory pathway. The neurophysins are also responsible for the protection of the peptides, which are present at high concentrations within the secretory granules, against disulfide exchange and degradation. During exocytosis, the neurophysins are released together with their respective peptides from the nerve terminals into the circulation and are detectable in the cerebrospinal fluid. Although there is significant information on the structure and function of the neurophysins as carrier proteins, it is presently unknown whether circulating neurophysins have independent functions as hormones or if they have specific targets in the periphery.

Both nonapeptides are also produced outside the brain, i.e., in female reproductive tissues like the ovary, corpus luteum, follicular fluid and uterus. Other peripheral tissues that produce the nonapeptides include the testes, adrenal, thymus, and pancreas, where they modulate local actions by acting as paracrine/autocrine factors rather than as circulating hormones [9]. Interactions between DA and peripherally produced OT and AVP are covered in **Chapters 6, 7 and 9**.

5.2.3 Vasopressin: Receptors, functions, and regulation

AVP, also known as the **ADH**, binds to two major receptors: V1R and V2R (for review [11]). In humans, the two receptors have a high degree of sequence homology, with about 102 invariant amino acids among the 370–420 residues. The V1Rs are further divided into two subclasses, V1aR and V1bR. V1aR is found on vascular smooth muscle cells, where it mediates vasoconstriction and enhances prostaglandin release, and is also found in the liver, kidneys and platelets. In addition, the V1aRs are expressed in many brain nuclei, where they are involved in the regulation of several types of social behaviors. V2Rs are found on the distal convoluted tubules and medullary collecting ducts of the nephrons and mediate the antidiuretic actions of AVP. A third receptor, the V3R (also called V1bR) is located in the anterior pituitary and the pancreatic isles, where it respectively facilitates the release of ACTH and intervenes in insulin secretion.

The AVP receptors are members of the G protein-coupled receptor (GPCR) superfamily. The receptors consist of seven hydrophobic transmembrane alpha helices, joined by alternating intracellular and extracellular loops, an extracellular N-terminal domain, and a cytoplasmic C-terminal domain. Similar to other GPCRs, activation of the vasopressin receptors by a bound ligand induces conformational changes in the relative orientation of the transmembrane domains, enabling the binding of G proteins to certain intracellular loops. The V1a and V1b receptors activate phospholipases via Gq/11, while the V2 receptor activates adenylyl cyclase and increases intracellular cAMP by interacting with the G protein subtype Gs. See **Chapter 2** for a detailed discussion on the overall structure and functions of GPCRs.

Dysfunctions of AVP result in **diabetes Insipidus (DI)**, a rare disease with about 20,000 cases in the United States per year. The disorder is characterized by the production of very large volumes of urine (polyuria) and excessive drinking (polydipsia). This condition is also exemplified by the Brattleboro rat, a strain of laboratory rat that descended from a litter born in West Brattleboro, Vermont, in 1961. After noticing that the litter of 17 pups drank and urinated excessively, the researchers proceeded to breed these rats. They then found that these rats have congenital DI due to a deletion of one nucleotide within the second exon of the AVP gene. The mutation resulted in an open reading frame with altered C-terminus that lacks a stop codon for terminating translation. The new reading frame also caused a loss of the single glycosylation site. In the absence of AVP, the Brattleboro rat cannot concentrate its urine and compensates for the massive fluid loss by drinking very large volumes of water. Given the inability of homozygous Brattleboro rats to synthesize AVP, neurophysin, and VAG, they have served as an excellent model for studying DI, as well as the physiological importance of each of the products of the AVP gene [12]. Both production and release of OT are normal in these rats, and they have no problems either with delivery or with lactation [13].

In humans, DI disrupts the normal life of affected individuals because of the continuous passing of very large volumes of urine and because of the increased thirst. There are two major causes of DI: central (neurogenic) and peripheral (nephrogenic). Neurogenic DI is due to impaired secretion of AVP, resulting either from a congenital inability to produce AVP or from a traumatic brain injury, a pituitary surgery, or tumors. Only in rare cases, is central DI caused by a genetic problem. Nephrogenic DI is due to failure of the kidneys to respond to AVP, either because of mutated V1Rs or because of uncoupling of the G proteins from the receptor, and it is usually inherited [14].

Past treatment for central DI consisted of administration of posterior pituitary extracts that contain vasopressin. In later years, desmopressin (DDAVP), a synthetic analog of vasopressin, was found to have benefits over natural vasopressin, e.g., ease of administration and stability at room

Figure 5.3 Effects of DDAVP infusion on water consumption in Brattleboro rats. The continuous infusion of desmopressin (DDAVP) caused rapid suppression of high water consumption in female Brattleboro rats. The termination of infusion resulted in a fast resumption of high water consumption. (Redrawn and modified from Hyde, J.F. et al., *Endocrinology*, 125, 35–40, 1989.)

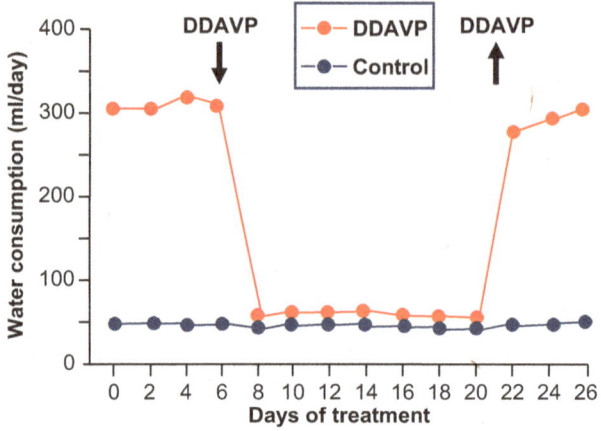

temperature. Desmopressin can be taken as a nasal spray, tablets or injections, and it also has some benefits to adults who have problems with nighttime urination (nocturia). **Figure 5.3** demonstrates the rapid effect of desmopressin (DDAVP) on water consumption in Brattleboro rats, and the immediate return to polyuria upon discontinuation of treatment [13].

Desmopressin is ineffective in nephrogenic DI, and patients with this condition must be provided with adequate amounts of water to drink to prevent dehydration. A diet low in salt and proteins is recommended because it reduces urine output. Nonsteroidal anti-inflammatory drugs and thiazide diuretics are sometimes used to treat this disorder. The two drugs act by different mechanisms to increase the amounts of sodium and water reabsorbed by the kidney, resulting in a decreased urine volume.

The major factors that affect AVP release are changes in blood osmolality and/or blood volume. Osmoreceptors, located in the brain circumventricular organs [which are outside the blood–brain barrier (BBB)], detect changes in plasma osmotic pressure and transmit the signals to the hypothalamus to activate the synthesis and release of AVP from the SON and posterior pituitary, respectively. Baroreceptors, located in the heart, detect changes in blood volume and send afferent signals to both the SON and PVN to increase the synthesis and release of AVP. The secretion of AVP is also stimulated by pain and by drugs such as nicotine, morphine and barbiturates. In trauma situations, significant amounts of AVP are released to counteract blood loss. The result is constriction of the smooth muscles of blood vessels and a rise in arterial blood pressure. In addition, the thirst sensation after acute alcohol intake is caused by the suppression of AVP release by ethanol, resulting in diuresis and enhanced thirst. The effects of DA on the release of the posterior pituitary neuropeptides are discussed in **Section 5.2.5**.

5.2.4 Oxytocin: Receptors, physiology, and regulation

Only one type of OT receptor (OTR), belonging to the Rhodopsin-type GPCR family, has been identified to date [15,16]. The coupling of OTR to G proteins has been best studied in the human myometrium. As shown in **Figure 5.4**, Gaq activation by a bound ligand stimulates phospholipase C-β, leading to increased inositol 3 phosphate (IP3) and diacylglycerol (DAG) levels. These changes lead to increased Ca^{2+} release from the sarcoplasmic reticulum, and activation of protein kinase C (PKC) as well as the Ca-calmodulin complex, resulting in the stimulation of myometrial contractions.

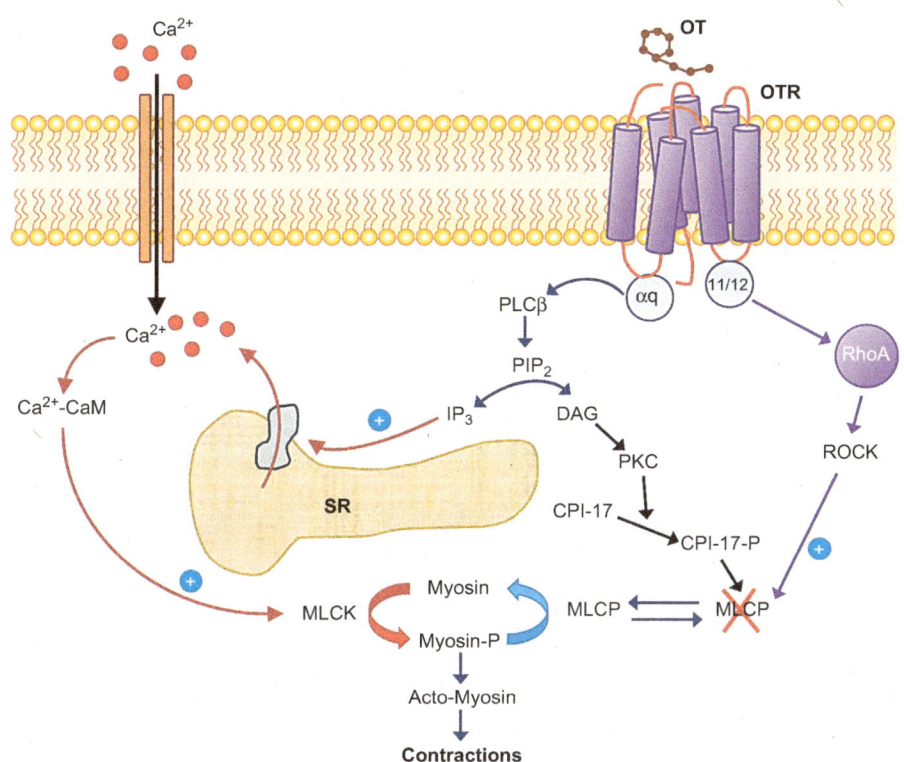

Figure 5.4 Induction of myometrial contractions by oxytocin (OT). Upon binding of oxytocin (OT) to OT receptor (OTR), the activated Gaq subunit increases phospholipase Cβ (PLCβ), which stimulates the generation of inositol-tri-phosphate (IP3) and diacylglycerol (DAG) via hydrolysis of phosphatidylinositol-4-5-bisphosphate (PIP2). IP3 increases the release of Ca^{+2} from the sarcoplasmic reticulum (SR), while DAG activates protein kinase C (PKC). The overall increase in Ca^{2+} levels activates the calcium-calmodulin (Ca^{2+}-CaM) complex, which in turn activates myosin light-chain kinase (MLCK). This promotes myosin activity and stimulates myometrial contractions. Activation of the G11/12 subunit leads to stimulation of RhoA and Rho kinase (ROCK), which regulate phosphatase (MLCP). (Redrawn and modified from Kim, S.H. et al., *Mol. Cell Endocrinol.*, 449, 56–63, 2017.)

OT has multiple central and peripheral actions that include reproduction, parturition, maternal behavior, social behavior, lactation, erectile dysfunction, and ejaculation. During pregnancy, in response to the high levels of estrogens, OT induces the release of prostaglandins. The two hormones acting together bring about thinning and dilation of the uterine cervix before parturition. During the process of delivery, the two hormones cause uterine contractions, applying pressure that enhances the baby's descent in the pelvis. After delivery, OT continues to cause myometrial contractions, helping to constrict the uterine blood vessels and prevent excessive blood loss. A second important peripheral action of OT is on the contraction of the myoepithelial cells in the lactating mammary gland, where OT is responsible for milk ejection during breastfeeding.

The two main stimuli for OT release are vagino-cervical distension such as occurs during delivery and suckling during lactation. OT release in response to cervical stimulation represents a classical neuroendocrine reflex, defined as a self-sustaining positive feed-forward cycle of neurohormone release and its effects on peripheral targets. The OT reflex is initiated by an increased pressure on the cervix or vaginal walls, which stimulates local somato-sensory neurons with synapses in the dorsal horn of the spinal medulla. Ascending axonal connections in the anterolateral columns of the spinal cord transfer the stimulus to OT neurons within the hypothalamic SON and PVN and from there to their terminals in the neurohypophysis. Increased plasma OT levels, amplified by a locally released OT-mediated positive feedback, promote additional uterine contractions via OTR and increased pressure on the cervix. Mating, sexual stimulation, and various forms of stress also stimulate OT release.

OT is also produced by some parvocellular neurons that project to other parts of the brain and spinal cord. By acting on receptors localized in the hypothalamus, prefrontal cortex, amygdala, and hippocampus, OT is involved in many types of behavior, including sexual and maternal behavior, and social bonding (see **Chapter 4** for discussion on sexual and maternal behavior). Because of these actions, the popular literature has often termed OT the "love" or the "feeling good" hormone. The short portal vessels that connect the NL to the anterior pituitary provide a conduit for OT to participate in the control of anterior pituitary hormones via OTRs, which are expressed in a number of pituitary cell types. In the anterior pituitary, OT is involved in the direct regulation of two hormones, PRL and ACTH, while having indirect effects on gonadotroph functions.

5.2.5 Effects of dopamine on the neurohypophysial hormones

Several lines of evidence demonstrate that brain DA is involved in blood pressure regulation by interacting with AVP. In conscious rats, chemical stimulation of the ventral tegmental area (VTA; the origin of the mesolimbic/mesocortical dopaminergic pathway) by a local administration of substance P analog, causes a prolonged increase in blood pressure and heart rates. These increases are completely blocked by treatment with D1 or D2 receptor antagonists, as well as by pretreatment with vasopressin V-1 receptor antagonists [17]. An association between DA, AVP and blood pressure regulation has also been observed in studies that used D5R-knockout (KO) mice [18]. The disruption of DA transmission through D5R results in hypertension, apparently due to the activation of several receptors in the central nervous system (CNS), including OTR, V1R, and non-N-methyl D-aspartate receptors.

Interactions between DA and AVP also occur at the NL. A double-labeling technique has shown that dopaminergic nerve endings containing dense core vesicles 60–100 nm in diameter and small clear vesicles are always found in close proximity to vasopressin neurosecretory endings, suggesting interactions between the two systems [19]. In cultures of isolated rat neurohypophysial tissues, DA and selected agonists increased the release of both AVP and OT, and this was blocked by DA receptor antagonists [20]. These data suggest that the dopaminergic control of neurohypophysial hormone secretion also occurs at the level of the posterior pituitary, independent of the hypothalamus.

Most evidence shows excitatory roles for DA on OT release. For example, basal release of OT in nonsuckled, lactating rats increases after intravenous administration of the D1R agonist SKF 38393. This effect, as well as the suckling-induced OT release, are prevented by treatment with the D1R antagonist SCH 23390. These data suggest that DA exerts a stimulatory influence over OT secretion during lactation through an action at the DAR [21]. These findings were supported by another report showing that DA and apomorphine increased OT synthesis in incubated neurohypophysial tissues, an effect which was blocked by administration of galanin together with DA [22].

Bidirectional interactions between OT and central dopaminergic neurons also occur during a variety of social interactions. OT promotes prosocial behavior through a direct effect on the VTA dopaminergic neurons associated with pleasure and reward [23]. A model of the various stimuli that increase OT release and their effects on both behavioral and endocrine parameters is shown in **Figure 5.5**A, and interactions between OT and DA neurons are shown in **Figure 5.5**B. Although some studies have reported inhibitory effects of DA over OT release, this may be due either to the involvement of different DA receptors and/or to an indirect effect of PRL, which is stimulated by OT. Nonetheless, it is of concern that in many of the above studies, both DA and its agonists were used at micromolar concentrations, much beyond the range of their binding affinities to the various receptors.

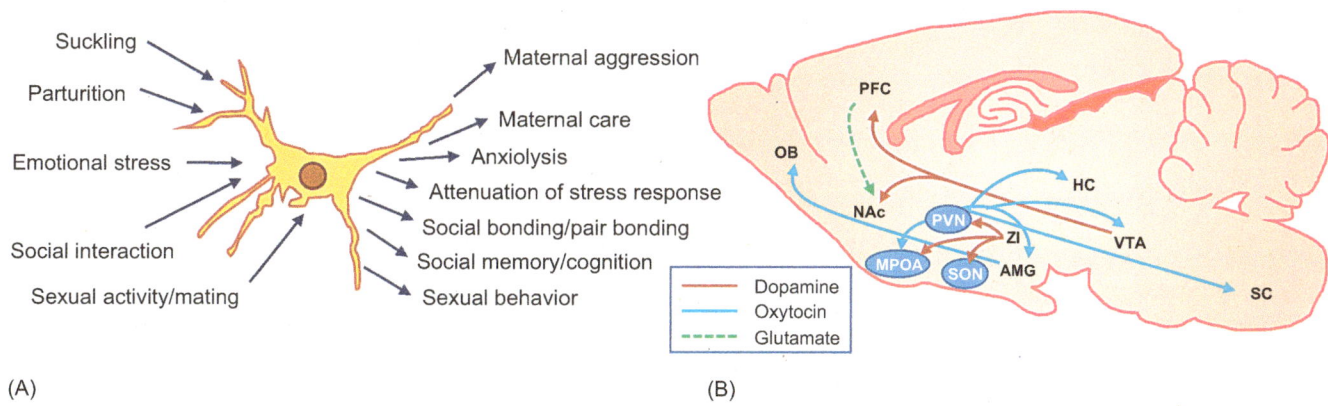

Figure 5.5 The multiple systems that are affected by oxytocin and its interactions with dopamine. Panel A shows the various stimuli that affect brain oxytocin release and the major systems that are affected by oxytocin. (Redrawn and modified from Neumann, I.D., *J Neuroendocrinol.*, 20, 858–865, 2008.) Panel B shows the major dopaminergic pathways within the rat brain and their relationship to oxytocin neuron populations. AMG, amygdala; HC, hippocampus; MPOA, medial preoptic area; NAc, nucleus accumbens; OB, olfactory bulbs; PFC, prefrontal cortex; PVN, paraventricular nucleus; SC, spinal cord; SON, supraoptic nucleus; VTA, ventral tegmental area; ZI, zona incerta. (Redrawn and modified from Baskerville, T.A. and Douglas, A.J., *CNS Neurosci. Ther.*, 16, e92–e123, 2010.)

5.3 INTERMEDIATE LOBE HORMONES: β-ENDORPHIN AND α-MSH

5.3.1 Structure, innervation and regulation of the intermediate lobe

The IL of the pituitary serves as a classical model of an interface between the CNS and endocrine effectors. In laboratory rats and mice, the IL and its hypothalamic innervation are an easily accessible peripheral neuroendocrine system with many similarities to more complex systems within the CNS. Thus, the ability to isolate and examine both the innervation and its targets at the cellular and organ levels provides an excellent opportunity for studying neuroendocrine and neurotrophic interactions [24].

In contrast to most endocrine glands which are well vascularized, the IL contains very few, if any, blood vessels, and its products reach the two other pituitary lobes by diffusion. The IL is richly supplied by nerve fibers originating from the hypothalamus. In mammals, the IL is innervated by DA, norepinephrine (NE), gamma-aminobutyric acid (GABA), and serotonin neurons, while in lower vertebrates, e.g., fish and amphibians, the IL is also innervated by peptidergic fibers, which participate in the regulation of the secretory activity of the melanotrophs [25]. Like other regions of the pituitary, both nerve terminals in the IL and the melanotrophs are outside the BBB and are accessible to input from circulating regulatory molecules.

The rodent IL consists of 10–15 layers of densely arranged cells separated into lobules by strands of connective tissue. The principal cell type is the **melanotroph**, a large polyhedral cell with a smooth ovoid nucleus. The melanotrophs secrete **α-MSH, β-endorphin**, and several other peptide derivatives of a common 241-amino acid precursor protein named proopiomelanocortin (**POMC**). Histological studies have revealed heterogeneity in the staining properties of individual melanotrophs, which correlates with different levels of POMC-mRNA expression [26].

The regulation and functions of the melanotrophs have been extensively studied in amphibians, where α-MSH stimulates pigment dispersion in dermal melanophores and is responsible for the skin darkening during the process of background adaptation. In humans, α-MSH is produced in many extra-pituitary sites and binds to the melanocortin 1 receptor. It stimulates melanogenesis, or the production of melanin by the melanocytes in both skin and hair [27]. In addition to its skin-darkening action, α-MSH has anti-inflammatory and antimicrobial effects. It also contributes to innate immunity and participates in the control of food intake and body weight. Within the hypothalamus, α-MSH suppresses appetite and is involved in sexual arousal. See more discussion on α-MSH, appetite and sexual behavior in **Chapter 4** and on its functions within the skin in **Chapter 11**.

5.3.2 Synthesis and processing of POMC and its derivatives

POMC is synthesized by both corticotrophs of the AL and melanotrophs of the IL, but the generation of biologically active peptides through proteolytic cleavage differs among the two cell types (**Figure 5.6**). The major products released by the corticotrophs are **ACTH**, a 39-amino acid polypeptide and **β-endorphin**, a 31-amino acid polypeptide, while the end-products of POMC processing in the melanotrophs are β-endorphin and α-MSH, a 13-amino acid peptide with an acetylated N-terminus and an amidated C-terminus [2].

β-endorphin is also produced in many brain areas, including the medial basal hypothalamus, thalamus, amygdala, periaqueductal gray, the inferior colliculus, the raphe nucleus, locus ceruleus, several regions of the reticular formation, and the solitary tract. Primarily released in response to pain, β-endorphin reduces the perception of pain by acting on central and peripheral opiate receptors. Whereas it has similar analgesic properties as morphine and codeine, β-endorphin does not lead to addiction or dependence as can be caused by these drugs. The euphoria, or sensation of well-being that often accompanies exercise of sufficient intensity and duration is also attributed to increased β-endorphin release.

5.3.3 Regulation of intermediate lobe hormones by dopamine

Studies on the distribution of axon terminals within the IL during development and in adult rats found both synaptic and non-synaptic interactions between dopaminergic terminals and melanotrophs. Several studies have shown that DA has major inhibitory effects on the production and processing of POMC, as well as on the release of its derived peptides from the IL (for review [26]). For example, the regulation of POMC mRNA expression in rats was altered

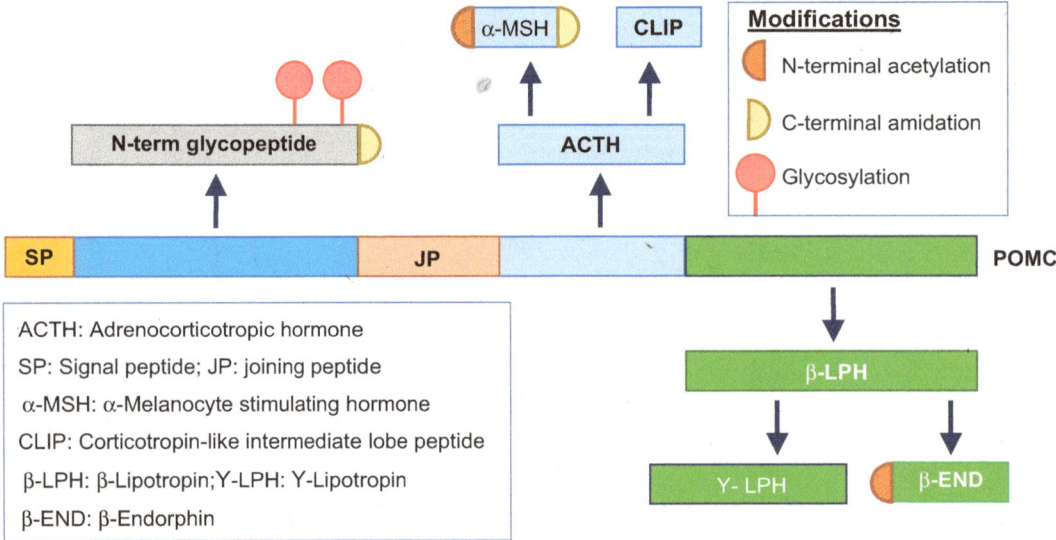

Figure 5.6 Proteolytic cleavage of proopiomelanocortin (POMC) precursor in corticotrophs and melanotrophs. The major peptides secreted by the corticotrophs are ACTH and β-endorphin, while the melanotrophs primarily secrete α-melanocyte stimulating hormone (α-MSH). See text for additional explanations. POMC: proopiomelanocortin.

by injections of haloperidol, a DA antagonist, or bromocriptine, a DA agonist. In addition, the maintenance of high dopaminergic activity in rats by DA agonists prevented stress-released melanotrophic peptides, while a blockade of DA transmission augmented this response, especially in the presence of β-3 adrenergic agonists. These data suggested that reduced DA tone, coupled with enhanced β-3 adrenergic activity, is needed to produce the full effects of stress-induced release of melanotrophic peptides from the IL. Whereas a DA agonist, ergocryptine, inhibited basal α-MSH release from cultured fetal and early neonatal rat melanocytes, it had no effect on the release of POMC-derived peptides from the corticotrophs, indicating intrinsic differences in the responsiveness of the two POMC-producing cell types [28].

5.4 COMPOSITION AND ONTOGENY OF THE ANTERIOR PITUITARY

Three reviews serve as a background for the next two sections. One is a chapter entitled "Functional Anatomy of the Hypothalamus and Pituitary" in *Endotex*, a Web textbook by the Endocrine Society, downloadable at https://www.ncbi.nlm.nih.gov/books/NBK279126/. A second review is in *Science* [29], and a third review is in *Endocrine Reviews* [30].

5.4.1 Anterior pituitary structure, cell types, and dopamine receptors

The anterior pituitary is composed of nests of cuboidal hormone-producing cells located adjacent to venous sinusoids. The sinusoids, which are part of the secondary capillary plexus of the hypophysial portal, are lined with fenestrated epithelium that collects the secretory products of the pituitary cells. As previously discussed, the pituitary gland lies outside the BBB. Thus, in addition to the releasing/inhibiting hormones that reach the anterior pituitary from the hypothalamus via the hypophysial portal vasculature, peripheral hormones and modulators can reach the anterior pituitary from the systemic circulation and provide feedback information.

The long-held dogma that the mammalian anterior pituitary is devoid of innervation has been challenged by several reports. In fact, peptidergic nerve fibers were detected in anterior pituitaries of humans, monkeys, dogs and rats [31]. Large numbers of nerve fibers containing substance P (SP), calcitonin gene-related peptide (CGRP), and galanin (GAL), were seen in the human pituitary stalk. These nerve fibers run along the pituitary stalk and enter the anterior pituitary; some GAL-immunoreactive fibers reach the center of the human anterior pituitary. Older publications from the 1990s reported expression of tyrosine hydroxylase (TH) in anterior pituitaries from rats and humans. Yet, there was no follow up on these findings, and the identity of the cells that express TH and whether the enzyme was functional and locally generated DA remain unknown.

As listed in **Table 5.1**, the anterior pituitary contains five discrete hormone producing cell types, each of which produces its respective hormone(s): somatotrophs (GH), lactotrophs (PRL), corticotrophs (ACTH), thyrotrophs (TSH), and gonadotrophs (both LH and FSH). The older classification of the anterior pituitary cells was based on their staining properties: acidophils for somatotrophs and lactotrophs, basophils for thyrotrophs and gonadotrophs, and chromophobes for corticotrophs. Following the introduction of specific immunocytochemical methods for each of the pituitary hormone, this staining classification became obsolete. All secretory anterior pituitary cells also exhibit spontaneous and extracellular calcium-dependent electrical activity, but differ with respect to the patterns of firing and associated calcium signaling and hormone secretion.

Table 5.1 Anterior pituitary hormones

Name	Name	Structure	Secretory cells	Target	Main effects
Adrenocorticotropic hormone	ACTH	Polypeptide	Corticotrophs	Adrenal gland	Secretion of glucocorticoids, mineralocorticoids, androgens
Thyroid stimulating hormone	TSH	Glycoprotein	Thyrotrophs	Thyroid gland	Secretion of thyroid hormones
Follicle-stimulating hormone	FSH	Glycoprotein	Gonadotrophs	Gonads	Growth of reproductive tract
Luteinizing hormone	LH	Glycoprotein	Gonadotrophs	Gonads	Sex hormone production
Growth hormone	GH	Polypeptide	Somatotrophs	Liver, adipose	Promotes growth; carbohydrate and lipid metabolism
Prolactin	PRL	Polypeptide	Lactotrophs	Breast	Milk production

ACTH: adrenocorticotropic hormone; FSH: follicle-stimulating hormone; GH: growth hormone; LH: luteinizing hormone; PRL: prolactin; TSH: thyroid stimulating hormone.

Figure 5.7 Anterior pituitary hormones. Shown are the six protein hormones of the anterior pituitary, their target organs/ glands and their primary functions. ACTH: adrenocorticotropic hormone; FSH: follicle-stimulating hormone; GH: growth hormone; LH: luteinizing hormone; PRL: prolactin; TSH: thyroid stimulating hormone.

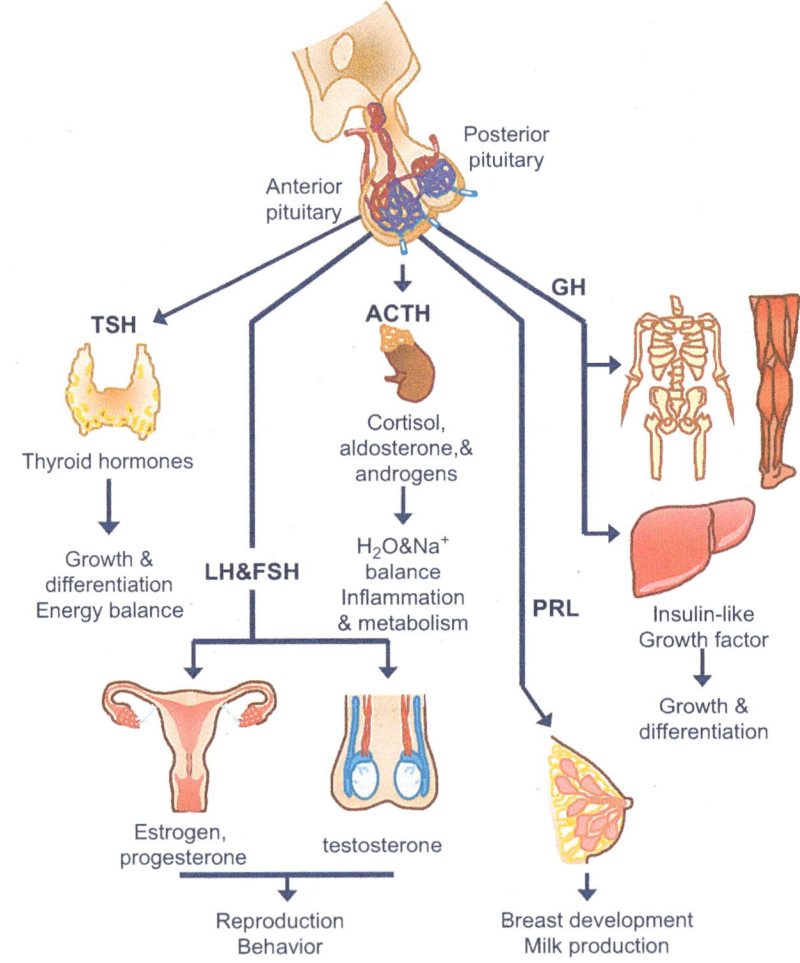

Figure 5.7 shows the primary targets of each pituitary hormone: thyroids for TSH, male and female gonads for LH and FSH, adrenals for ACTH, and multiple targets for PRL and GH. A detailed discussion on each hormone, its regulation, release, and feedback from its targets is presented in the subsequent sections. In addition to the classical protein hormones, the anterior pituitary synthesizes a wide variety of peptides, growth factors, cytokines, binding proteins and neurotransmitters that carry out paracrine and/or autocrine functions in the control of pituitary secretion and/or cell proliferation.

The relative expression of DA receptors (DARs) in the human pituitary is D2R>>>D4R>>D5R>D1R, while D3R is undetectable [32]. Clinically nonfunctioning pituitary tumors (non-hormone secreting) showed a predominance of D2R, low expression of D4R, and undetectable D1R, D3R and D5R [33]. On the other hand, a recent study reported expression of D5R in nonfunctioning human pituitary adenomas as well as in rat (GH3) and mouse (MMQ) lactotroph/somatotroph cell lines [34]. In non-tumorous rat [35–37] and human [32,38] anterior pituitaries, D2R was primarily expressed in the lactotrophs. Both long and short D2R isoforms were found in the normal human pituitary [39]. D2R is expressed at variable levels in other hormone-producing cells, including gonadotrophs, corticotrophs, thyrotrophs and somatotrophs [38,40].

Folliculostellate cells are non-endocrine, star-shaped cells, devoid of secretory granules that are dispersed among of endocrine cells of the anterior pituitary. They are glial-like cells derived from neuroectodermal origin and are identified by several markers: vimentin, S-100 protein, and GFAP. The cells are organized into follicles, possess phagocytic properties, and communicate with each other and with endocrine cells through gap-junctions. Their long processes generate a three-dimensional (3D) network, used for the transmission of signals throughout the pituitary and which help in the coordination of its functions.

Folliculostellate cells produce signaling proteins such as vascular endothelial growth factor (VEGF), epidermal growth factor (EGF), fibroblast growth factor (FGF), nerve growth factor (NGF), insulin-like growth factor (IGF), and transforming growth factors (TGF-α and -β). They also express cytokines such as interleukin-6 and leukemia inhibitory factor (LIF) and produce nitric oxide (NO), **Pituitary adenylate cyclase-activating polypeptide (PACAP)** and follistatin. It has been reported that folliculostellate cells specifically influence the function of gonadotrophs and lactotrophs. There are no published records whether folliculostellate cells express TH or DA receptors.

Stem cells that do not produce hormones have also been identified in the adult anterior pituitary. The cells are identified by expression of established stem cell markers and are located primarily in the marginal zone around the pituitary cleft. The pituitary stem cells are presumably involved in cell regeneration after pituitary injury and have a role in pituitary tumor formation. In one study [41], progenitor/stem cells were isolated from nonfunctioning human pituitary tumors (NFPTs). Approximately 70% of the NFPTs formed spheres that co-express the stem cell markers DAX1, SF1, and ERG1. Upon *in vitro* incubation, D2R as well as somatostatin (SSTR2) agonists inhibited their proliferation, suggesting the maintenance of anti-proliferative effects of DA and somatostatin. This finding is supported by another study [42], reporting on co-expression of D2R and somatostatin receptors in stem cells isolated from nonfunctioning human pituitary adenomas. The authors indicated that stem cells with similar properties also exist in murine models of pituitary adenomas.

5.4.2 Ontogeny of anterior pituitary cell lineage

The embryonic development of the hypothalamo-pituitary complex has been primarily studied in rodents, using a combination of morphological, biochemical and genetic approaches. Information on the process of ontogeny has also benefitted from the availability of naturally occurring mutations and transgenic animal models. For an in-depth coverage of pituitary cell differentiation, the interested reader is referred to two reviews [29,43].

The ontogeny of the anterior pituitary depends upon a progressive cascade of activated extrinsic or intrinsic transcription factors and signaling molecules. The initial extrinsic phase of murine pituitary development comprises signals emanating from both the ventral diencephalon and the oral ectoderm. As illustrated in **Figure 5.8**A at mouse embryonic day (E) 6.5–7, the anterior

Figure 5.8 Ontogeny of the anterior pituitary in the mouse embryo. Panel A depicts the time course of development of the pituitary gland. Shown are the various transcription factors that affect the differentiation of the neural and/or oral ectoderm and their time of activation from day 6.5 to 10.5 of embryonic life. E: embryonic day. See text for additional explanations. (Redrawn and modified from de Moraes, D.C. et al., *J. Endocrinol.*, 215, 239–245, 2012.) Panel B shows the pituitary-specific transcription factors involved in the development of the anterior pituitary from Rathke's pouch. Thyrotrophs, lactotrophs and somatotrophs are derived from a common lineage, determined by Prop-1 and Pit-1. Corticotrophs and gonadotrophs originate from independent lineages. See text for other explanations. ACTH: adrenocorticotropic hormone; FSH: follicle-stimulating hormone; GH: growth hormone; LH: luteinizing hormone; PRL: prolactin; TSH: thyroid stimulating hormone. (Redrawn and modified from Cohen, L.E. and Radovick, S., *Endocr. Rev.*, 23, 431–442, 2002.)

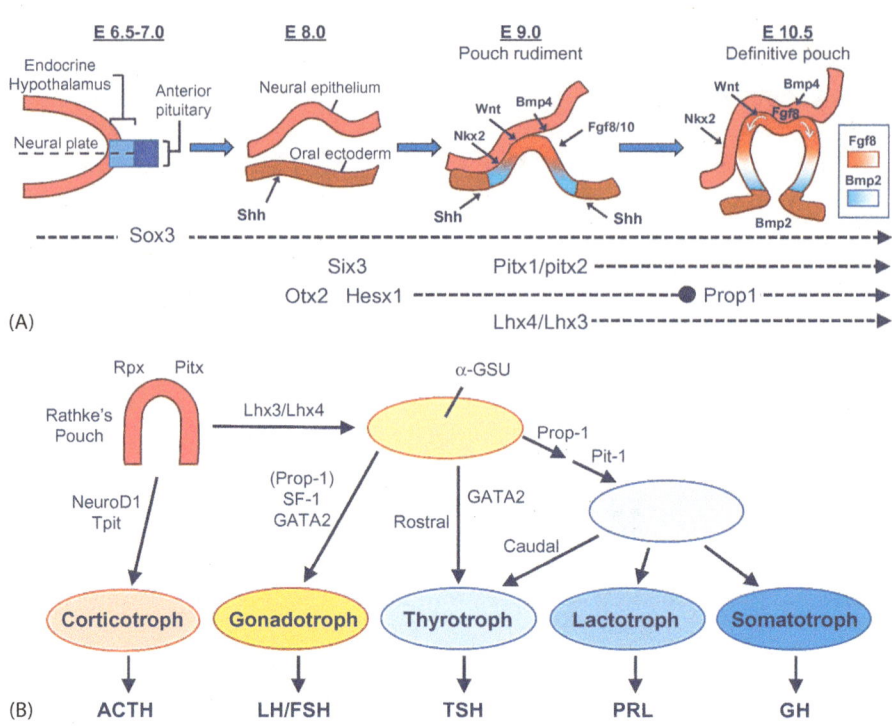

portion of the neural plate is destined to give rise to the primordial pituitary, while the adjacent midline region will become the endocrine hypothalamus [43]. At E8, the oral ectoderm starts to proliferate in response to Shh, Six3, Otx2 and Hex1 and participates in midline formation. Proliferation continues at E9 in response to Bmp4, Fgf8, Wnt2 and Nkx2 coming from the neural epithelium. At the same time, the oral ectoderm begins to invaginate upward and to form a rudimental **Rathke's pouch**, which expresses Lhx3/4 and Pitx1/2. At the edge of the pouch, Bmp2 makes contact with the oral ectoderm and antagonizes Fgf2, which is expressed by the neural epithelium. Subsequently, an Bmp2–Fgf8 ventral–dorsal gradient is established that determines the activation of specific genes in each cell group according to their localization within the pouch.

In parallel with the invagination of the oral ectoderm, pituitary precursor cells proliferate and migrate. The Wnt and Shh pathways regulate proliferation, while the Bmp and Fgf pathways participate in both proliferation and cellular migration. The formation of Rathke's pouch is complete at E10.5, and the pituitary precursor cells start expressing specific factors that determine their patterns of differentiation. Activation of distinct target genes occurs in response to an established dorsal–ventral gradient of Fgf8 and a ventral–dorsal gradient of Bmp2. Thus, depending on its location, each cell has a distinct starting point within the differentiation process (**Figure 5.8**A). For example, ventral cells express the transcription factors Isl1 and Gata2, while the dorsal cells express Pax6, Tpit and Prop1.

Pituitary organogenesis in humans progresses along the same lines but at a different timescale (**Table 5.2**). It begins during week 4 of fetal development, when a thickening of cells in the oral ectoderm form the hypophysial placode, giving rise to Rathke's pouch. The pituitary organizer is a domain within the ventral diencephalon that expresses Bmp4, Fgf8, and Fgf10, which induce the formation of the pituitary precursor, Rathke's pouch, from the oral ectoderm. The Wnt signaling pathway regulates this pituitary organizer such that loss of Wnt5a leads to an expansion of the pituitary organizer and

Table 5.2 Timeline of embryonic development of the human pituitary	
Time of gestation	**Processes**
Week 4	Hypophysial pouch, Rathke's pouch, diverticulum from roof of the mouth
Week 5	Elongation, contacts infundibulum, diverticulum of prosencephalon
Week 6	Connecting stalk between pouch and oral cavity degenerates
Week 10	GH and ACTH detectable
Week 16	Adenohypophysis fully differentiated
Week 20–24	GH levels peak, then decline

ACTH: adrenocorticotropic hormone; GH: growth hormone.

an enlargement of Rathke's pouch. Wnt signaling is classified into canonical signaling, which is mediated by β-Catenin, and noncanonical signaling, which operates independently of β-Catenin.

During weeks 6–8 of human gestation, Rathke's pouch constricts at its base and eventually separates from the oral epithelium. At the same time, a downward extension of the ventral diencephalon forms the posterior lobe, and the two nascent lobes connect to form the composite structure of the adult pituitary. Cells of the anterior wall of Rathke's pouch undergo extensive proliferation to form the anterior lobe, while the posterior wall proliferates more slowly to form the vestigial IL. Cell patterning and terminal differentiation occurs within the anterior lobe, forming the five specialized endocrine cell types of the pituitary gland.

As illustrated in **Figure 5.8**B, distinct pituitary cell types emerge from a common primordium in response to opposing signaling gradients that originate from organizing centers. Terminal cell type differentiation requires selective gene activation and long-term active repression. These are mediated by cell type-specific and promoter-specific recruitment of coregulatory complexes. Thyrotrophs, lactotrophs and somatotrophs are derived from a common lineage, determined by the transcription factors Prop-1 and Pit-1, while independent lineages lead to the formation of corticotrophs and gonadotrophs. Overall, pituitary organogenesis is a complex and tightly regulated temporal and spatial process that depends on interactions between transcription factors such as Prop1, Pit1 (Pou1f1), Hesx1, Lhx3, and Lhx4. Mutations in these genes can result in various forms of hypopituitarism and can be associated with structural alterations that cause congenital forms of panhypopituitarism.

5.5 SOMATOLACTOGENIC HORMONES: PRL AND GH

PRL and GH share a similar tertiary protein structure and utilize conserved, single-pass transmembrane receptors belonging to the type 1 cytokine receptor superfamily. The two hormones are single-chain polypeptides comprising 191–199 residues with a molecular mass of 22–23 kDa. They have two to three disulfide bridges whose location is conserved across species [44]. Both hormones, as well as their receptors, emerged by gene duplications in early vertebrate evolution. The two hormones are involved in multiple physiologic processes, including development, growth, reproduction, and metabolism, with few overlapping functions. Despite their many similarities, GH and PRL are produced by distinct anterior pituitary cells, and differ in the controls of their expression, synthesis, release, and actions.

5.5.1 Structure and properties of somatotrophs and lactotrophs

Somatotrophs first appear at E15.5 in the mouse, while lactotrophs mostly develop postnatally. Both cell types (as well as thyrotrophs) require *Pit1*, the Pou homeodomain transcription factor, for their differentiation and hormone expression. *Pit1* gene mutations have been associated with dwarfism and hypothyroidism in both mice and humans. By light microscopy, both cell types are identified as acidophils. Somatotrophs have centrally located nuclei and a diffuse cytoplasmic staining for GH and are located predominantly in the lateral wings of the anterior pituitary. Lactotrophs are polygonal cells scattered throughout the gland. The main cell type in the adult human anterior pituitary are the somatotrophs, constituting approximately 50% of the total cell population, while the number of lactotrophs varies with the physiological state [45]. Whereas lactotrophs constitute only 9% of the pituitary cell population in men and in nulliparous women, they go up to 30% of the total anterior pituitary cell population in multiparous women.

Mammosomatotrophs (also named somatolactotrophs) resemble somatotrophs in appearance but contain both GH and PRL by immunohistochemistry. They often have irregular, elongated and pleomorphic large granules typical of PRL secretion. The relative abundance and functions of the mammosomatotrophs have been debated. It has been argued that they serve as a reserve pool of transitional cells that offer plasticity to the pituitary, enabling it to respond to altered physiological states. In some cases, transformation is going only in one direction, while in other cases, there is a reciprocal transformation between the two cell types.

In spite of their terminal differentiation, pituitary cells in adult pituitaries continue to undergo mitosis, which is augmented under some conditions. For instance, during human pregnancy, the pituitary gland increases in size, with most of the added volume due to lactotroph hyperplasia. To this end, preexisting lactotrophs proliferate and somatotrophs are recruited to switch from GH to PRL production. Reversible transdifferentiation also occurs in the pituitary, whereby cells are recruited from heterologous cell types, with GH-producing cells having the capacity to transdifferentiate to gonadotrophs. The high proliferative potential of the lactotrophs accounts for the higher incidence of lactotroph tumors (prolactinomas) than other types of pituitary tumors.

Somatotrophs and lactotrophs are excitable cells that express a plethora of voltage-gated Na^+, K^+, and Ca^{2+} channels that generate sporadic or rhythmic electrical activity. The spontaneous plateau-bursting action potentials occur without coupling to calcium release from intracellular stores. This electrical activity generates calcium signals of sufficient amplitude to keep a steady hormone release [46]. Both spontaneous electrical activity and basal hormone secretion can be further amplified by activation of $G_{q/11}$- and G_s-coupled receptors and inhibited by $G_{i/o/z}$-coupled receptors. The voltage-gated Ca^{2+} influx that is coupled to this electrical activity plays an essential role in the stimulus-secretion coupling in both cell types.

5.5.2 PRL: Structure, synthesis, regulation, and functions

Most information on PRL is derived from our previous review [44]. hPRL is a single chain protein hormone composing 199 residues. It assumes a nonconventional "up-up-down-down" four-helical bundle topology that is a common feature of the hematopoietic cytokines. The α-helices are stabilized by three disulfide bonds (C4–11, C58–174, and C191–199). In addition to the pituitary lactotrophs, PRL in humans is also produced in multiple extrapituitary sites, including the brain, decidua, breast, myometrium, prostate and immune cells, where it is regulated in a cell-specific manner and acts as a cytokine. Consequently, even after hypophysectomy, humans are not totally deprived of endogenous PRL. With few exceptions, the expression of PRL in other animals is confined to the pituitary, with PRL acting as a classical circulating hormone.

The human *PRL* gene is located on chromosome 6 and shares 40% homology with GH and placental lactogen (PL), the three hormones that are classified as lactogens. The *PRL* gene contains six exons and is regulated by a proximal promoter composed of a 2- to 2.5-kb sequence at the 5′ flanking region. Eight cis-acting elements within the proximal promoter bind Pit-1, a homeobox transcription factor that confers tissue- and cell-specific PRL expression. Although Pit-1 is expressed primarily in the pituitary gland, it is not restricted to lactotrophs and requires interactions with other transcription factors to confer lactotroph phenotype. The first exon (1a) has a transcriptional start site 5.8 kb upstream of the pituitary start site (**Figure 5.9**). In extrapituitary sites such as decidua, myometrium and lymphoid cells, exon 1a is spliced to exon 1b, generating an mRNA transcript that is 150 bp larger than the pituitary counterpart in the 5′ untranslated region; both pituitary and extrapituitary sites produce an identical PRL protein.

Pituitary PRL is subjected to multiple regulators that can be classified into four broad categories: endocrine, paracrine, juxtacrine and autocrine. Endocrine regulators originate from the hypothalamus and the gonads and reach the lactotrophs through the blood. Paracrine factors reach the lactotrophs by diffusion from neighboring pituitary cells. Juxtacrine interactions emanate from the extracellular matrix of adjacent cells. Autocrine agents are synthesized by the lactotrophs themselves. Hence, the overall PRL production and secretion reflects a balance between local and distant releasing and inhibiting factors.

Pituitary PRL gene expression is affected by multiple hormones, neurotransmitters and growth factors. Compounds that bind to G protein-linked receptors, e.g., thyrotropin releasing hormone (TRH), vasoactive intestinal peptide (VIP) and DA, activate protein kinase A, protein kinase C and/or calcium/calmodulin-dependent pathways. These are mediated by a variety of transcription factors that bind to consensus sequences within the PRL promoter. Estrogens, on the other hand, diffuse into the nucleus, where they bind to their receptors. The activated estrogen receptor (ER) acts as transcription factor by binding to an estrogen response element (ERE) located within the

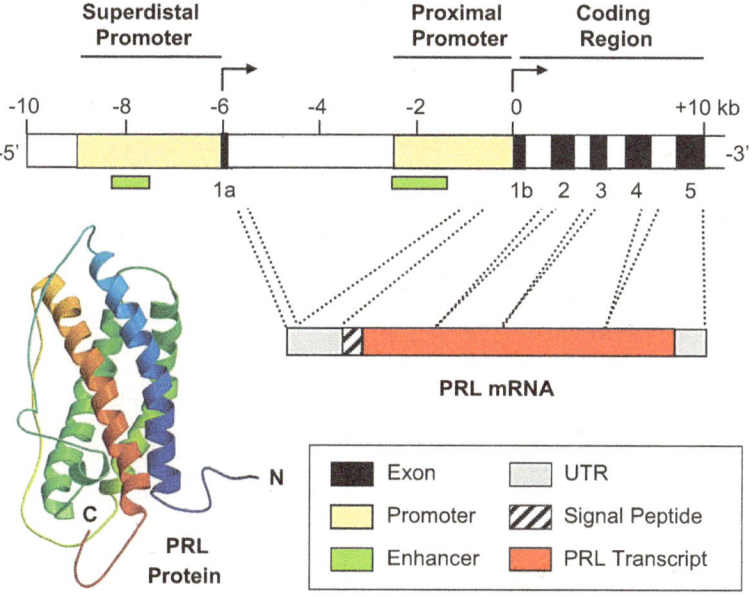

PRL mRNA

PRL Protein

Exon	UTR
Promoter	Signal Peptide
Enhancer	PRL Transcript

Figure 5.9 The human PRL gene, mRNA and protein. The prolactin (PRL) gene is composed of five exons and its expression in the pituitary is regulated by a proximal promoter. The first exon (1a) has a start site located 5.8 kb of the pituitary stat site and is used as an alternative promoter in extrapituitary sites. In this case, exon 5a is spliced to exon 5b and generates an mRNA transcript that is 150 bp larger than the pituitary transcript. Both transcripts, however, generate an identical PRL protein, a single-chain polypeptide that assumes an up-up-down-down four-helical configuration. UTR: untranslated region.

distal region of the proximal PRL promoter. Among growth factors, insulin and EGF stimulate, whereas transforming growth factor β (TGFβ) suppresses PRL gene expression. Many of the growth factors bind to transmembrane receptors with intrinsic tyrosine kinase activity and exert pleiotropic actions such as stimulation of PRL gene transcription, increases in hormone storage, and alterations of lactotroph morphology.

Posttranslational modification of PRL includes glycosylation, phosphorylation and proteolytic cleavage. hPRL is N-glycosylated on Asn 31 via an Asn-X-Ser consensus sequence. The carbohydrate moiety contains fucosylated and partially sialylated complex oligosaccharides. It has been reported that glycosylated and nonglycosylated PRL utilize different routes of sorting and release, with glycosylated PRL constitutively secreted, whereas the release of non-glycosylated PRL involves a storage step. This modification is especially relevant to the release of glycosylated PRL in extrapituitary sites that lack secretory granules. The serum levels of glycosylated hPRL vary during pregnancy, lactation, hyperprolactinemia, and in some disease stages, and it is also abundant in human milk and amniotic fluid. Glycosylated PRL has reduced receptor binding affinity and mitogenic activity, thereby diminishing PRL actions at target tissues. Yet, glycosylation may alter proteolytic cleavage of PRL, regulate its distribution, or delay its clearance.

Phosphorylated PRL has been characterized in bovine and rat pituitaries, but not in humans. The major phosphorylation site in bovine PRL is Ser 90, which is conserved in PRL, GH, and PL of most species. The addition of a bulky, negatively charged side chain to Ser 90 may disrupt hormone folding, reducing its receptor binding and impairing its biological activity. A 16-kDa PRL variant, which acts as an antiangiogenic factor, is formed by cleavage at 145–149, followed by the reduction of the interchain disulfide bond.

DA is a potent physiological inhibitor of PRL. The pituitary lactotroph is unique in its capacity for high constitutive secretion of PRL, but its activity is tonically suppressed by DA, which reaches the pituitary from the hypothalamus as well as from the NL (**Figure 5.10**). The involvement of posterior pituitary DA in the regulation of PRL release has been demonstrated by the

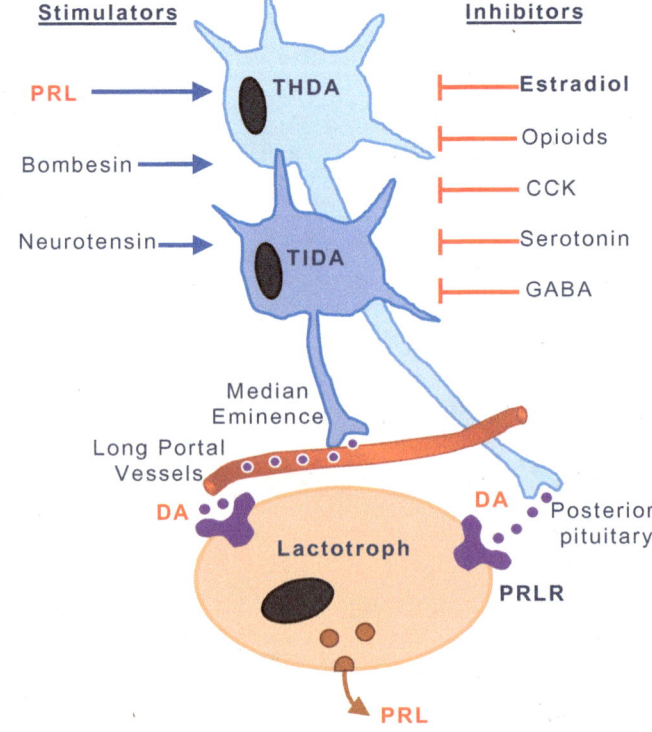

Figure 5.10 The regulation of the pituitary lactotrophs and PRL production and release by DA. Both the tuberoinfundibular (TIDA) and tuberohypophysial (THDA) dopaminergic neurons have their cell bodies in the arcuate nucleus of the hypothalamus. The TIDA neurons terminate in the median eminence and release dopamine (DA) into hypophysial portal blood. The THDA neurons have terminals in both the neural and intermediate lobes of the pituitary (posterior pituitary), and the DA released from these terminals reaches the lactotrophs by diffusion. The two dopaminergic systems are regulated by multiple substances that provide stimulation or inhibition over DA synthesis and release. CCK: cholecystokinin; GABA: gamma-aminobutyric acid; PRL: prolactin; PRLR: prolactin receptor.

rise in PRL release following posterior pituitary lobectomy [47]. In the lacto-trophs, DA binds to D2R and exerts multiple actions, including lowering of intracellular calcium and cAMP levels, inhibition of PRL gene expression and release, and suppression of cell proliferation. Deficiency in, or unresponsive-ness to, DA are major causes for the development of prolactinomas, the most common pituitary adenomas (covered below and also in **Chapter 12**). PRL itself, acting via a short loop negative feedback, is the primary regulator of the hypothalamic dopaminergic system. Whereas a singular releasing factor for PRL has not been identified, several neuropeptides, including TRH and VIP, are capable of acute stimulation of PRL release under some conditions (**Figure 5.10**).

The effects of DA on both lactotrophs and PRL are revealed by the phe-notype of D2R KO mice [48]. These mice have chronic hyperprolactinemia and pituitary hyperplasia and are growth retarded due to alterations in the GH-IGF-I axis. There are fewer somatotrophs, gonadotrophs and thyrotrophs in the knockout mice and the lactotrophs do not show the typical dense secretory granules but, rather, appear degranulated. PRL levels are higher in female than in male knockouts, and pituitary hyperplasia is observed at 8 months only in females. After 16 months of age, highly vascularized ade-nomas develop, especially in females. Prominent vascular channels in the hyperplastic and adenomatous pituitaries, as well as extravasated red blood cells not contained in capillaries, has also been a common finding.

PRL is one of the most versatile hormones, fulfilling more than 100 func-tions and surpassing the number of known actions of all other pituitary hor-mones combined. PRL functions are associated with reproduction, lactation, growth/development, osmoregulation, metabolism, immune regulation, brain functions, and behavior. The effects of PRL on reproduction involves multiple systems and tissues that differ in importance in a species-specific manner.

The principal target for PRL is the mammary gland, where PRL promotes the growth and differentiation of the lobuloalveolar structures and is essen-tial for the initiation and maintenance of lactation and the production of milk proteins and lipids in all mammals. In some species, however, continuous lactogenesis is supported by GH rather than by PRL. In rodents, PRL has both luteotropic and luteolytic actions on the ovary and supports progesterone production after embryo implantation. In these species, PRL plays a major role in modulating the reproductive cycle and is crucial for the maintenance of pregnancy and lactation. Although rodents have a significant preovulatory rise in PRL, this does not occur in humans. The profile of PRL release during human pregnancy is also entirely different from that in rodents. It involves three independently regulated compartments that produce PRL—the mater-nal, fetal, and decidual compartments—where PRL rises at different times throughout pregnancy [44]. In humans, a total absence of PRL is very rare and is not associated with discernable clinical problems. On the other hand, excess release of PRL, or hyperprolactinemia, results in reproductive distur-bances, as discussed below.

PRL plays a major role in the control of water and electrolyte balance in fish and amphibians, with lesser osmoregulatory actions in birds and mammals. Although the control of development and body growth is usually ascribed to GH, there are some functional overlaps between GH and PRL, especially in lower vertebrates. PRL stimulates proliferation of the pancre-atic islets and increases insulin secretion, primarily during pregnancy. PRL also has well-established mitogenic, secretory and morphogenic effects on the prostate. In the immune system, PRL induces proliferation, differentiation and activation of various lymphoid cells, although transgenic animals lack-ing PRL or its receptor have little, if any, discernible immune disturbances. In the brain, PRL affects the production and release of several hypothalamic releasing and inhibiting hormones and has significant effects on maternal behavior. In the skin, PRL is involved in the regulation of hair growth cycles and keratinocyte proliferation. In adipose tissue, PRL participates in the regulation of lipid synthesis and breakdown and adipokine release.

As discussed in a recent review [49], **hyperprolactinemia** is the most common endocrine disorder of the hypothalamo-pituitary axis. A prolactinoma is the most common cause of chronic hyperprolactinemia, once pregnancy, primary hypothyroidism, and drugs that increase serum PRL levels have been ruled out. Among pituitary adenomas, prolactinomas are the most common, accounting for about 40% of cases. The magnitude of elevated serum PRL serves as a useful diagnostic tool for determining the etiology of hyperprolactinemia. PRL levels >250 ng/mL are suggestive of a prolactinoma, while most patients with pituitary stalk dysfunction, drug-induced hyperprolactinemia or systemic diseases have PRL levels <100 ng/mL.

The most prevalent clinical problems associated with prolactinomas are hypogonadism, and infertility in both genders, impotence in men, and reduced sexual drive in women. Patients with macro-prolactinomas can also have visual field defects, headaches and neurological disturbances due to mass tumor effects. The goals of therapy are the normalization of PRL levels so as to restore eugonadism and the reduction of tumor mass to relieve neurological problems. Given their association with minimal morbidity, DA agonists have replaced surgery as the mainstay of treatment for prolactinomas. Cabergoline is a DA agonist with more specific D2R binding capacity than bromocriptine, and is currently the drug of choice for the treatment of PRL- and GH-secreting pituitary adenomas. The serum half-life of cabergoline is about 65 h, allowing oral administration just once or twice per week. This, and its better tolerability in comparison with bromocriptine enhances patient compliance. After two years of successful treatment, DA withdrawal is usually considered for microprolactinomas and in some cases of macroadenomas. About 10% of prolactinomas are resistant to DA agonist therapy. DA-resistant prolactinomas are more aggressive and tend to be large, invasive, hyperangiogenic tumors with high mitotic indices, making their management by surgery, radiosurgery, or medical therapies challenging.

5.5.3 GH: Structure, synthesis, and regulation

A large portion of the background information below is based on a chapter by Gunawardane et al. entitled "Normal physiology of growth hormone in adults," in *Endotex*, a Web textbook. The chapter can be downloaded at https://www.ncbi.nlm.nih.gov/books/NBK279056/.

Human GH (hGH, also named somatotropin) is a 191-amino acid, single-chain polypeptide with two disulfide bonds and molecular weight of 22 kDa. Although hGH shares only 22% primary amino acid sequence homology with hPRL, the 3D topology of the two hormones is very similar, enabling the binding of hGH to both its cognate receptor (hGHR) and the hPRLR. On the other hand, hPRL binds only to hPRLR and not to the hGHR. hGH is a heterogeneous protein consisting of several isoforms. This heterogeneity is the consequence of multiple hGH genes, mRNA splicing, posttranslational modifications, and peripheral metabolism [50].

The GH locus, a 66-kb region of DNA, is located in chromosome 17q22-q24 and consists of five homologous genes that appear to have been duplicated from an ancestral GH-like gene. Of the five encoded genes, hGH-N (normal) is predominantly expressed in the pituitary somatotrophs, whereas the other four genes, the three chorionic somatomammotropins (hCS-L, hCS-A, and hCS-B) and hGH-V (variant), are expressed selectively in the placenta. In contrast, the mouse genome contains a single pituitary-specific GH gene and lacks any GH-related CS genes [51]. Given the significant differences among species with respect to the synthesis, pattern and regulation of release, as well as spectrum of actions, extrapolation of results obtained with non-primates to the characteristics of GH in humans should be done with caution.

hGH exerts its effects either directly, through its cognate receptor, or indirectly through the stimulation of insulin-like growth factor 1 (IGF-1), a 70-amino acid peptide, found in the circulation primarily bound to transport proteins named IGF binding proteins (IGFBP 1-6). IGF-1 is synthesized in the liver and also in peripheral tissues such as bone, myoblasts, muscle, erythroid precursors, ovary, and kidney and the CNS. IGF-1 also provide feedback inhibition to pituitary GH synthesis and release.

hGH is both physiologically and clinically a very important protein hormone. It reaches its highest serum levels during puberty, when it affects somatogenic growth and helps to maintain multiple tissues and organs, general health and homeostasis. A reduction in the rate of hGH production begins in middle age, with the decline continuing to old age. The full manifestation of disorders of GH release and/or action depends upon the age at which they occur. Excessive GH is almost always caused by a benign GH-producing pituitary tumor (adenoma). Children with this condition develop great stature (**gigantism**), whereas a prolonged excessive GH level in adults produces **acromegaly**, characterized by swelling of the hands and feet and altered facial features. Patients with acromegaly also have organ enlargement and serious disorders such as high blood pressure, diabetes and heart disease.

GH deficiency (GHD) in children is defined as growth failure associated with inadequate GH production. It is commonly evaluated in children whose length or height remains below the normal range. Children diagnosed with GHD benefit from GH replacement therapy, which can improve the linear growth, but only until fusion of the growth plates at the end of their growth spurt (around 17–19 years of age). Recombinant GH therapy is given to prepubertal patients by daily subcutaneous injections. GH insufficiency in adults is associated with osteoporosis, sarcopenia, and cognitive dysfunction as well as with cardiac and metabolic diseases. Because of its reputation as a performance-enhancing agent, GH is widely abused in the sporting arena and is on the list of banned substances by the World Anti-Doping Agency. Unlike many substances of abuse such as synthetic anabolic steroids, GH is a naturally occurring substance. Therefore, the demonstration of exogenous GH administration in athletes must rely on detecting concentrations in excess of an established reference interval.

GH release is **pulsatile** in nature, characterized by highly ordered secretory events that are dispersed by periods of low or undetectable levels of GH during basal secretory periods (inter-pulse intervals). The amplitude, frequency, and regularity of the GH pulses, as well as the basal levels of GH secretion, vary between individuals, sexes and the physiological needs. The GH pulses in humans are ordered relative to the sleep-wake cycle, with maximal GH release occurring following the onset of slow-wave sleep.

As shown in **Figure 5.11**, GH release is regulated by a variety of central and peripheral factors. Two hypothalamic hormones, GHRH and somatostatin (SST) provide stimulation and inhibition, respectively. Two peripheral hormones, ghrelin, originating from the stomach, and IGF-1, coming primarily from the liver, provide stimulation and inhibition, respectively. Amino acids and free fatty acids also affect GH release.

GHRH was originally isolated from a pancreatic tumor taken from a patient with acromegaly and somatotroph hyperplasia. GHRH is derived from a 108-amino acid prepro-hormone, which gives rise to GBRH(l-40) and [1–44], both of which are found in the human hypothalamus. GHRH binds to a GPCR-type receptor, activates adenylate cyclase and stimulates transcription of the GH gene and GH release. **Somatostatin** [also called somatotropin release inhibitory factor (SRIF)] is derived from a 116-amino acid prohormone that

Figure 5.11 The regulation of growth hormone (GH) synthesis and release. Two hypothalamic peptides, growth hormone releasing hormone (GHRH) and somatostatin (SST) respectively stimulate and inhibit GH production and release from the somatotrophs. Positive regulation of GH comes from ghrelin, coming from the stomach, from circulating amino acids and from various stressors. Negative regulation of GH is provided by insulin-like growth factor-1 (IGF-1) primarily from the liver and from circulating free fatty acids (FFA). GH itself can reach the hypothalamus via a short loop mechanism and regulates the release of GHRH and SST.

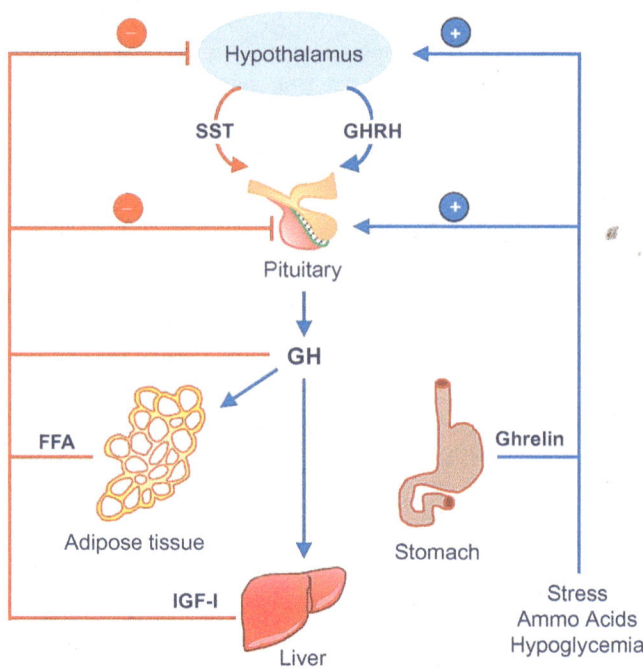

gives rise to two forms, somatoatatin-28 and -14, both of which are cyclic peptides due to an intramolecular disulfide bond. Somatostatin inhibits GH secretion directly from the somatotrophs and also reduces the impact of ghrelin by inhibiting its secretion from the stomach. There are five somatostatin receptor (SSTR) subtypes, with SSTR2 and 5 the most abundant in the pituitary.

Ghrelin is an orexigenic (appetite-stimulating) peptide derived from preproghrelin, a 117-amino acid peptide, by cleavage and n-octanoylation at the third residue, giving rise to a 28-amino acid active peptide. Ghrelin is the endogenous ligand of the GH secretagogue receptor (GHS-R) la, a member of the GPCR superfamily that is coupled to the phospholipase C-phosphoinositide pathway.

DA has both direct and indirect effects on somatotroph homeostasis and on GH synthesis and release. The direct effects of DA on GH release were examined using perifused rat anterior pituitary cells [52]. Pulses of DA caused a prompt rise in GH release, an effect which is reversible, concentration-dependent, and partially antagonized by metoclopramide, a DA antagonist. Pretreatment with DA followed by a GHRH challenge during a continuous DA perifusion has reduced the stimulatory effect of GHRH on GH release. These data suggest that DA has opposite actions at the somatotrophs: stimulation of basal GH release and inhibition of GHRH-induced GH secretion. Yet, the identity of the DA receptors that respond to DA or the overall expression/distribution of the various DAR in somatotrophs have not been reported.

Transgenic mice with a deleted dopamine transporter ($DAT^{-/-}$), which increases DA tone, have anterior pituitary hypoplasia and dwarfism [53]. The spatial distribution of both lactotrophs and somatotrophs is altered, their number is dramatically reduced, GH concentration is down, and the density of D2R is markedly reduced throughout the pituitary gland. Because $DAT^{-/-}$ mice have a 50% reduction in hypothalamic GHRH mRNA levels and a slight increase in mRNA for SST, it was concluded that the control of somatotroph development by DA occurred at the hypothalamic level and implicates GHRH.

The above observations are supported by the phenotype of D2R KO mice [54]. Absence of D2R alters the GHRH-GH-IGF-I axis and impairs body growth and somatotroph population. Although body weights were similar at birth, somatic growth was lesser in male D2R$^{-/-}$ mice from 1–8 months of age and in D2R$^{-/-}$ females during the first 2 months. In one study, the rate of skeletal maturation, and the weights of liver and white adipose tissue were decreased in knockout male mice even though food intake was unchanged. Serum GH concentration was significantly decreased during the first 2 months in mice of both sexes, and IGF-I and IGFBP-3 levels were lower. PRL was significantly higher in knockout mice, with females attaining higher levels than males. In addition, pituitaries from adult knockout mice had impaired basal GH release and a lower response to GHRH *in vitro*. The authors concluded that the D2R participates in GHRH/GH release in the first month of life.

Although DA agonists are less effective in decreasing GH hypersecretion in acromegaly than in subduing hyperprolactinemia, cabergoline has been beneficial in the treatment of acromegaly with concomitant PRL hypersecretion, and it is an acceptable agent in the medical treatment of GH hypersecretion [55]. The rationale is that up to 40% of GH-secreting adenomas co-secrete PRL, and D2R are expressed by some somatotrophs. Indeed, the effects of a novel somatostatin-DA chimera was recently tested in a small clinical trial [56]. The chimera showed a promising outcome in terms of its ability to reduce GH, IGF-1, and PRL concentrations, demonstrating enhanced efficacy of the somatostatin analog and DA agonist moieties.

5.6 REPRODUCTIVE HORMONES: LH AND FSH

The production of viable progeny depends on the optimal functioning of highly specialized organs, the gonads. In the male, the testes fulfill a dual function: spermatogenesis and steroidogenesis, both of which are under regulation by the gonadotropins from the anterior pituitary. The reproductive capability of the female, unlike that of the male, is intermittent, i.e., the production of ova, or mature germ cells, is not continuous. In contrast to the millions of spermatozoa produced daily by the male, only one or few mature ova are generated by the female at each reproductive cycle. The cyclic release of ova from the hormone-producing Graafian follicles is regulated by cyclic alterations in gonadotropin secretion, which is governed by complex positive and negative feedback mechanisms.

The human ovulatory cycle, or **menstrual** cycle, is characterized by a midcycle hormone-driven ovulation, followed by menses at the end of each cycle. Menses results from the withdrawal of hormonal support and shedding of the superficial layer of the uterine lining. Menstrual cycles begin during puberty, are interrupted during pregnancy and lactation, and cease at the time of menopause, typically between of 45 and 55 years of age. In contrast, sperm production in men continues throughout life, with a slight reduction to old age. Thus, during their lifetime, women are fertile for a much shorter time (30–40 years) than men (potentially up to 80 years).

Figure 5.12 compares the components of the hypothalamo-pituitary-gonadal (HPG) axes in the two sexes. The reproductive axis consists of three major sites and their associated hormones: the hypothalamus (kisspeptin and GnRH), the pituitary (LH and FSH), and the gonads (testes and androgens in males vs. ovaries and estrogen/progesterone in females). The gonadal steroids feedback information to both the hypothalamus and pituitary, affects the secondary sex characteristics, and regulate the structure and functions of the reproductive tract (epididymis, vas deferens, urethra and penis in males vs. oviducts, uterus, cervix and vagina in females). Although the various components of the reproductive tract do not produce hormones

Figure 5.12 The hypothalamic-pituitary-gonadal (HPG) axis in males and females. Gonadotropin-releasing hormone (GnRH), released from the hypothalamus into the hypophysial portal blood, stimulates the production and release of two gonadotropins: luteinizing hormone (LH) and follicle-stimulating hormone (FSH). Both hormones are identical in males and females. In males, the main targets of the gonadotropins are the testes, where they control both steroidogenesis and spermatogenesis. Testosterone provide feedback inhibition over the secretion of GnRH, LH and FSH. In females, the main targets of the gonadotropins are the ovaries, where they regulate both steroidogenesis and ovulation. Estrogen and progesterone provide negative and positive feedback to GnRH and the gonadotropins, depending upon the stage of the ovulatory cycle.

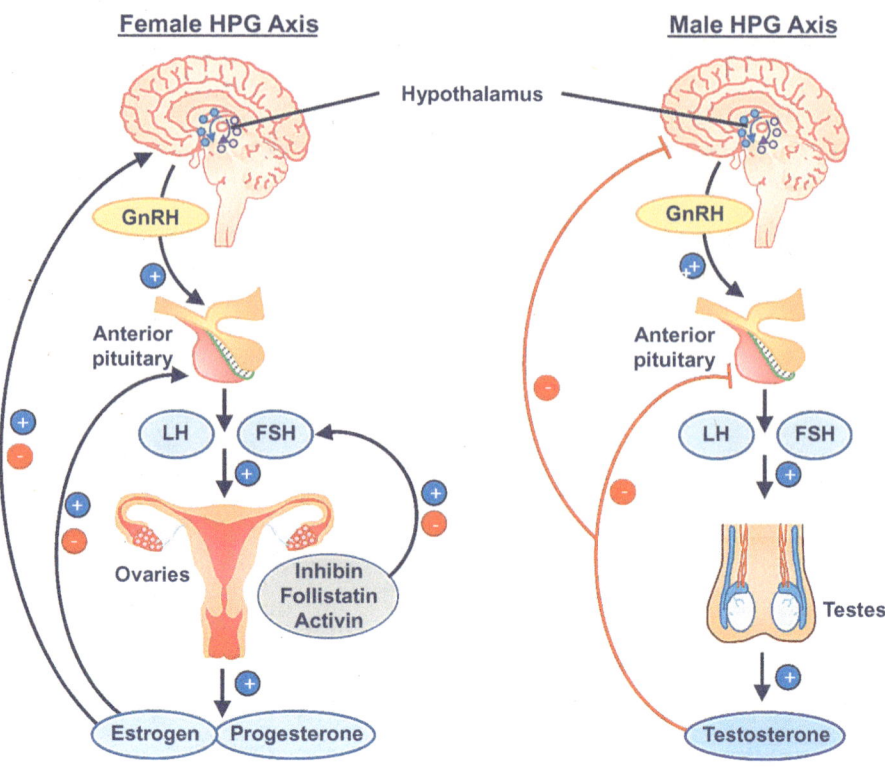

per se (with the exception of the uterus during pregnancy), they have several locally produced cytokines that can alter the efficacy of the sex steroids. The reproductive hormones of the hypothalamus are covered in **Chapter 4**, **Section 5.5**, while the gonads and the reproductive tracts are covered in **Chapter 10**. The focus here is on pituitary gonadotrophs, their hormones, LH and FSH, and their regulation by DA.

5.6.1 The gonadotrophs: Structure, properties, and secretory products

The gonadotrophs account for only 5%–10% of the total pituitary cell population. In spite of their relatively low number, they fulfill a fundamental role in the development and maintenance of fertility through the synthesis and secretion of LH and FSH [57,58]. Although named for their functions in females, LH and FSH in the male are identical in structure to their counterparts in the female. Each hormone is a glycoprotein composed of alpha and beta subunits. The alpha subunits in LH, FSH as well as TSH are identical, while the beta subunits differ and confer the biological activity of each hormone.

Given the low percentage of the gonadotrophs within the heterogeneous pituitary cell population, studies on their properties using primary pituitary cell cultures have been challenging. A major advantage has been the development of murine gonadotroph-derived cell lines, αT3-1 and LβT2 cells, which provide valuable tools for analyzing the molecular and cellular events underling the synthesis and secretion of LH and FSH. Yet, immortalized cell lines may have characteristics distinct from those of mature gonadotrophs or may represent or reflect the response of a subpopulation of this cell type. To overcome such limitations, researchers have recently developed novel strategies for identifying and purifying gonadotrophs from transgenic animals.

Gonadotrophs are found throughout the anterior pituitary, alone or in groups, often in close contact with blood capillaries [59]. They are oval in shape with prominent, often eccentrically located, nuclei. They have noticeable rough endoplasmic reticulum, well-developed Golgi complexes and many secretory granules, that stain strongly for both LH and FSH. An issue that has long been debated, but remained unsettled, is whether subpopulations of gonadotrophs are monohormonal, i.e., produce only LH or FSH. Neither number nor morphology of the gonadotrophs is constant, but change with physiologic conditions. Castration, for example, induces a 2- to 3-fold increase in the number of gonadotrophs, presumably due the removal of negative feedback by gonadal hormones and increased cell number. It also results in the appearance of "castration cells," hypertrophied gonadotrophs characterized by dilation of the endoplasmic reticulum and the presence of large vacuoles.

Along with thyrotrophs, gonadotrophs express the *Cga* gene, which encodes the alpha glycoprotein subunit. In addition to the genes encoding the beta subunits of the gonadotropins, *Fshb* and *Lhb*, gonadotrophs are defined by two other genes not expressed in other secretory pituitary cell types: the GnRH receptor (GnRHR) gene (*Gnrhr*) and dentin matrix protein 1 gene [60]. The gonadotrophs also secrete other soluble proteins, including chromogranins and secretogranins. Morphologically, gonadotropes contain two types of secretory granules: small granules containing secretogranin II, and large granules containing chromogranin A.

Similar to lactotrophs, gonadotrophs are excitable cells. They express numerous voltage-gated sodium (Na_V), calcium (Ca_V), potassium (K_V), and chloride (Cl_V) channels at the plasma membrane and fire spontaneous action potentials [60]. Gonadotrophs also express ligand-gated ion channels, and their activation by hypothalamic or intrapituitary ligands causes increased firing frequency and facilitates Ca^{2+} influx and hormone release. The main Ca^{2+}-mobilizing receptor is GnRHR. The involvement of calcium in the stimulation of hormone release was studied using simultaneous measurements of exocytosis and intracellular calcium concentration with fluorescent indicators in single pituitary gonadotrophs [61]. The results showed that the spontaneous calcium oscillations under resting conditions are not linked to regulated or basal exocytosis, while intracellular calcium mobilization is essential for GnRH-stimulated exocytosis.

5.6.2 GnRH and the GnRH receptor

GnRH was isolated and sequenced in 1977 by the Nobel laurates Roger Guillemin and Andrew Schally. It is a decapeptide with protected N- and C-termini and which has the following structure: pyro-Glu-His-Trp-Ser-Tyr-Gly-Leu-Arg-Pro-Gly-NH$_2$. Like several other proteins and peptides, GnRH is enzymatically processed from a larger precursor, located on chromosome 8. In mammals, the decapeptide end-product is synthesized from an 89-amino acid prohormone that is the target of various regulatory mechanisms of the HPG axis.

As described in **Chapter 4**, GnRH is produced by hypothalamic neurons in the preoptic area, is regulated by kisspeptin, and is transported by the hypophysial portal blood to the anterior pituitary, where it stimulates the release of both LH and FSH. GnRH is released intermittently in pulses. The variations in the frequency of pulsatile GnRH release have differential effects on the synthesis and release of the gonadotropins. For example, FSH is preferentially stimulated at low GnRH pulse frequencies, whereas LH is stimulated at high pulse frequencies. Likewise, low GnRH pulse frequencies favor *Fshb* gene expression, whereas high GnRH pulse frequencies favor *Lhb* gene expression [62]. Apart from the hypothalamus, GnRH is expressed in several peripheral tissues, including the gonads and the placenta, where it acts as a cytokine and is not released into the systemic circulation. Across the vertebrate species, there are as many as 23 naturally occurring GnRH analogs with multiple substitutions in their amino acid sequence, relative to the mammalian GnRH. The wide distribution of GnRH in various tissues suggests that it has acquired significant functions through phylogeny [63].

As indicated in a recent review [64], three GnRH receptors or receptor-like sequences have been identified, with distinct distributions and functions. The GnRHR type I is the predominant form in mammalian gonadotrophs. In some species, including humans, GnRHR-1 is also expressed in reproductive tissues (e.g., breast, endometrium, ovary, prostate) and in tumors derived from these tissues. GnRHR is a 60 kDa, G protein-coupled receptor, whose number in the gonadotrophs depends on the developmental and reproductive stage and determines their responsiveness to GnRH. Receptor number is regulated at the transcriptional level by estradiol (E2) and the pattern of GnRH pulsatility. Transcription of *Gnrhr* in cultured gonadotrophs occurs in the absence (basal) and presence (regulated) of GnRH stimulation.

Several studies reported that activated GnRHR interacts with Gsα protein, resulting in the activation of a variety of signaling pathways. Questions remain, however, which other G proteins are involved in GnRHR signaling in the gonadotroph, how each G protein contributes to signaling, and how the GnRH pulses are decoded to activate signaling cascades that preferentially induce FSH or LH production. Studies using LβT2 gonadotroph cells showed that both Gαq/11 and Gαs are involved in GnRHR signaling. Activation of Gaq/11 proteins by GnRHR increases phospholipase Cβ, which in turn cleaves phosphatidylinositol-4-5-bisphosphate (PIP2) into inositol tri-phosphate (IP3) and diacylglycerol (DAG). IP3 stimulates Ca^{2+} release from endoplasmic reticulum stores into the cytosol, whereas DAG activates protein kinase C (PKC). Some PKC isoforms are activated by calcium as well. Many investigators reported that GnRH induces increases in intracellular calcium levels and PKC activation, followed by the activation of mitogen-activated protein kinase (MAPK) cascades in gonadotrophs. The MAPKs can translocate to the nucleus and activate several transcription factors.

5.6.3 FSH: Structure, synthesis, regulation and functions

FSH is a 35.5 kDa glycoprotein hormone consisting of two subunits, alpha and beta, which are held together by non-covalent bonds [62]. Its overall structure is similar to those of LH, TSH and human chorionic gonadotropin (hCG), a product of the placenta during pregnancy. The alpha subunits of all four glycoprotein hormones are identical and consist of 96 amino acids, while the beta subunits vary. Both subunits are required for the biological activity of each hormone. FSH has a beta subunit of 111 amino acids that confers its specific biologic action and is responsible for interaction of the hormone with the FSH receptor (FSHR). The carbohydrate moiety of FSH is covalently bound to asparagine, and is composed of N-acetyl-galactosamine, mannose, N-acetyl-glucosamine, galactose, and sialic acid.

The regulation of FSH production and release is multifaceted. The major stimulator of FSH synthesis and release is hypothalamic GnRH, while gonadal steroids provide feedback information that can be inhibitory or stimulatory, depending on the physiological conditions. The gonads also produce three dimeric proteins—inhibin, activin, and follistatin—which exert a direct effect on FSH secretion from the gonadotrophs without affecting LH. Although the primary source of inhibin is the gonads of both sexes, both activin and follistatin are produced in extragonadal tissues, including the anterior pituitary, and can affect FSH through an autocrine/paracrine mechanism. Activin can influence the gonadotropins at many levels. First, it directly stimulates FSH biosynthesis and release from the gonadotrophs.

Second, it up-regulates GnRHR gene expression, leading to alterations in the synthesis and release of both gonadotropins in response to GnRH. Third, activin can stimulate GnRH release from GnRH neurons, thereby affecting both FSH and LH secretion. Inhibin and follistatin negatively affect these effects by preventing the binding of activin to its receptor at the cell membrane, and by blocking the activation of downstream signal transduction pathways.

A rise of FSH during the early follicular phase of the menstrual cycle stimulates the growth and maturation of antral ovarian follicles and increases E2 synthesis by granulosa cells. At midcycle, concomitant with the surge of LH, there is also a small rise in FSH although its physiological significance is unclear. FSH is clinically used for controlled ovarian stimulation in women undergoing assisted reproduction, i.e., *in vitro* fertilization of retrieved oocytes. FSH is also used to treat anovulatory infertility in women. The necessity for FSH in female reproduction is clearly demonstrated clinically and in animal models. Women with loss-of function mutations in the *FSHB* or *FSHR* genes have primary or secondary amenorrhea (an abnormal absence of menstruation), and the development of their follicles is arrested at the preantral stage. Similar phenotypes are observed in *Fshb*- and *Fshr*-deficient mice.

FSH is also used to treat hypogonadotropic hypogonadism in men. Although FSH targets the Sertoli cells to regulate spermatogenesis, its absolute necessity for male reproduction has been a matter of debate. Men homozygous for a certain inactivating mutation in the *FSHR* gene have variable oligozoospermia, or low sperm count. Similarly, male *Fshr*-deficient mice have reduced testicular size and low sperm counts but otherwise are fertile. These data suggest that FSH plays a role in maintaining normal spermatogenesis but may not be absolutely required for fertility in males. Still, men with inactivating mutations in the *FSHB* subunit gene, who are unable to produce the dimeric ligand, are azoospermic.

The FSHR is primarily expressed in the gonads [65]. In the testis, it is located in Sertoli cells where it supports spermatogenesis, while in the ovary, it is expressed in granulosa cells where it regulates growth and maturation of ovarian follicles and estrogen production. The FSHR belongs to the Rhodopsin subfamily of the GPCRs. The receptor is composed of an unusually large N-terminal extracellular domain where recognition and binding of its cognate ligand occur, and ends with an intracellular C-terminal tail, containing functional motifs that link the receptor to intracellular mediators. The FSHR activates a complex signaling network that involves several G protein subtypes (e.g., Gs, Gi, and Gq/11), interacts with other receptors (e.g., the IGF-1 receptor and the EGFR), and is also associated with β-arrestins and adaptor proteins. Downstream signaling includes a number of intertwined pathways such as PKA, PKB, PKC, PI3K and ERK1/2. The complex signaling network allows for fine-tuning regulation of the gonadotropic stimulus, where activation/inhibition of its multiple components varies, depending on the cell context, developmental stage of the host cells, and concentration of receptors and ligands [65].

5.6.4 LH: Structure, synthesis, regulation, and functions

LH is a dimeric protein composed of alpha and beta subunits. The alpha subunit is identical to those of the other glycoprotein hormones, while the beta subunit consists of 120 amino acids and has large homologies to the beta subunit of hCG, which contains an additional 24 amino acids. The two hormones also differ in the composition of their sugar moieties, which affects their bioactivity and rate of degradation. The biologic half-life of LH is 20 min, much shorter than that of FSH (3–4 h) or hCG (24 h). The amino acid sequences of the transmembrane domains of LHR and

FSHR are similar, whereas the extracellular and cytoplasmic domains are 50% identical. The LHRs bind both LH and hCG. In addition to the male and female gonads, LHR is expressed in the brain, placenta, and other extragonadal sites.

A sharp **preovulatory surge** in serum LH concentrations, and a smaller rise in FSH, occur in midcycle. During the surge, the concentration of LH rises to 10–20 times its resting level, and the duration of the surge is 36–48 h. Several factors act in coordination to generate the midcycle LH surge. Importantly, there is a dramatic switch from a negative to a positive feedback action of E2 at both the pituitary and hypothalamic level. The switch is triggered when a persistently increasing serum E2 concentration reaches a critical point and stimulate kisspeptin. In response to a kisspeptin-induced GnRH rise, the gonadotrophs become highly sensitive to GnRH stimulation due to increased number of GnRHR receptors; a small rise in progesterone levels in the late follicular phase may also have a triggering role.

The preovulatory LH surge is essential for rupture of the Graafian follicle and release of the mature egg. After ovulation, the egg enters the fallopian tube and travels to the uterus. The empty Graafian follicle undergoes structural and biochemical transformations (luteinization) and is converted into a corpus luteum. The corpus luteum is a temporary ovarian structure that produces high levels of progesterone, moderate levels of E2 and inhibin, and a small amount of estrogen. In case of conception, the corpus luteum is sustained, and it is essential for the maintenance of pregnancy. Otherwise, the corpus luteum decays within 10 days, and new corpora lutea are formed with each subsequent ovulation.

Beyond the events of the mid-cycle surge, the main function of LH is to increase the production of androgens by ovarian theca cells. The androgens, androstendione and testosterone, diffuse into the granulosa cells, where they are converted to E2 and estrone by aromatase (CYP19). Aromatase action, and therefore estrogen production, is controlled by FSH. Hence, the function of theca cells and granulosa cells are controlled by LH and FSH, respectively, according to the "two cell–two hormone" paradigm. In humans, LH stimulates the production of progesterone by the corpus luteum, while in rodents, this is done by PRL. In male mammals, LH stimulates androgen production by the testicular Leydig cells. Clinically, LH is used to support FSH therapy in the treatment of some fertility disorders. The longer-acting hCG is used to induce ovulation in women, to increase sperm count in men, and to treat boys with undescended testes.

5.6.5 Regulation of the hypothalamo-pituitary-gonadal axis by dopamine

DA is involved in the regulation of the HPG axis at all three levels: hypothalamus, pituitary, and gonads. At the hypothalamic level, DA interacts with both the kisspeptin/neurokinin B/dynorphin (KNDy)-kisspeptin neurons and the GnRH neurons. In seasonal breeders (e.g., ewes), KNDy neurons receive direct contact from dopaminergic terminals, and this input increases during anestrous, supporting the notion that DA inhibits GnRH secretion by suppressing the activity of arcuate nucleus KNDy neurons, thereby providing some explanation for the seasonal shift in E2-negative feedback in the sheep [66]. Furthermore, as many as 50% of GnRH neurons in rats express D1R and/or D2R, which mediate potent and direct hyperpolarization of GnRH neurons by DA [67]. In addition to the direct postsynaptic actions, DA exerts presynaptic modulation of the GABA/glutamate afferent inputs to GnRH neurons through D1R and/or D2R receptors. The effects of DA on the GnRH neurons are independent of sex or the estrous cycle, indicating that the suppressive role for DA within the GnRH neuronal network is unrelated to gonadal steroid modulation.

To examine the direct effects of DA on the gonadotrophs, monolayer cultures of LβT2 cells have been used [68]. This cell line was created from pituitary tumors in transgenic mice carrying the LHβ-subunit regulatory region. The cells express the gonadotropin α-subunit, the LHβ-subunit, and GnRHR, and respond to GnRH with dose-dependent increases in LH secretion, exhibiting functional characteristics consistent with those of normal pituitary gonadotrophs. The LβT2 cells express both the PRLR and D2R. PRL and the DA agonist bromocriptine (Br), alone or in combination, were found to block the alpha-subunit and LHβ-subunit mRNA responses to GnRH, whereas the LH secretory response was differentially affected by the treatments. GnRH dose-dependently stimulated LH release, with a 4- to 5-fold increase at 10 μM GnRH. Unexpectedly, PRL or Br stimulated basal LH release, with PRL, but not Br, enhancing the LH secretory response to GnRH, an effect which was completely blocked by Br. These results established direct actions of PRL and DA on the gonadotrophs and showed interactions between the two hormones in the regulation of gonadotropin secretion. Moreover, an uncoupling between LH synthesis and release in both basal and GnRH-stimulated responses to PRL and DA was clearly apparent.

Dopaminergic input to the fetal pituitary affects the proper functioning of the gonadotrophs postnatally [36]. PACAP is expressed at high level in the fetal pituitary but decreases profoundly between E19 and postnatal day 1 (PN1), with a further decrease from PN1 to PN4. In a study using single-cell RT-PCR of pituitary cell cultures from newborn rats, the gonadotrophs were positive for D2R and PACAP mRNAs. PACAP expression in E19 pituitary cultures was suppressed by the PACAP6-38 antagonist as well as by the D2R agonist bromocriptine. Increasing concentrations of bromocriptine inhibited PACAP-induced cAMP production, decreased PACAP promoter activity, and reduced PACAP mRNA levels in αT3-1 gonadotroph cells. In addition, blockade of DAR by injecting haloperidol into newborn rat pups partially reversed the developmental decline in pituitary PACAP mRNA that occurs between PN1 and PN4. These results show that DA receptor signaling regulates PACAP expression, supporting the notion that a rise in hypothalamic DA at birth abrogates cAMP signaling in fetal gonadotrophs to interrupt a feed-forward mechanism that maintains PACAP expression at high level in the fetal pituitary. The authors proposed that the perinatal decline in pituitary PACAP reduces pituitary follistatin, permitting GnRH receptors and FSH-β to facilitate activation of the neonatal gonad.

DA is also directly involved in the regulation of gonadal functions in both sexes. To this end, the source of DA is either local, through catecholamine biosynthesis, or through the systemic circulation. The distribution of DA receptors and the effects of DA on reproductive processes involving both the gonads and the lower reproductive tract are covered in **Chapter 10**.

5.7 STRESS AND METABOLIC HORMONES: ACTH AND TSH

As discussed in **Chapter 4** and depicted in **Figure 4.5**, ACTH plays a pivotal role in the overall homeostasis, as well as the stress response and is a major component of the **hypothalamo-pituitary-adrenal axis (HPA)**. ACTH is released from corticotrophs and its main target is the ACTH receptor, primarily found in the zona fasciculata of the adrenal cortex. Binding of ACTH to its receptor increases the production and release of glucocorticoids (cortisol in humans and corticosterone in rodents), which affect many central and peripheral sites as part of the stress response and fulfill important cardiovascular, metabolic, immunologic and homeostatic functions.

The hypothalamo-pituitary-thyroid axis (HPT) consists of hypothalamic TRH, pituitary TSH and the thyroid hormones, thyroxine (T_4), and triiodothyronine (T_3). TSH is a glycoprotein hormone made of two subunits with structural similarities to LH and FSH. TSH is produced by pituitary thyrotrophs and its main target is the TSH receptor, found primarily in the follicular cells of the thyroid gland. TSH stimulates production and secretion of the thyroid hormones, which affect almost every tissue in the body and act to increase basal metabolic rate, protein synthesis, neural maturation, and proper development and differentiation of multiple cells. Both thyrotrophs and TSH are primarily stimulated by hypothalamic TRH and inhibited by thyroid hormones.

5.7.1 Corticotrophs and ACTH

Several reviews [24,69,70] served as source of information for this section. Corticotrophs, which can be detected in fetal mice from E13.5, are the first endocrine cell type to terminally differentiate in the pituitary. By E15.5, the cells extend cytoplasmic projections, or cytonemes, that form contacts between apparently isolated cells, as well as with the perivascular space lining capillaries where the projection is frequently filled with secretory granules. This early organization is maintained postnatally and throughout adulthood. The relationships between the vasculature and endocrine cell networks reflects the importance of the dynamic secretion of the different hormones, i.e., the rapid release of ACTH in response to stress. In the adult pituitary, corticotrophs constitute about 20% of the total pituitary cell population.

The major secretory product of the corticotrophs is **ACTH**, a 39-amino acid peptide hormone, which is a cleaved product of the POMC precursor protein (see **Section 5.3.2** and **Figure 5.6**). POMC is encoded by a single copy gene on chromosome 2p13.3. It contains a 5′ promoter and three exons. POMC is a versatile multifunctional precursor protein that gives rise to an array of bioactive peptides with unique bioactivities. This versatility is reflected in the complexities of the promoter/enhancer structure of the POMC gene, which is suppressed by glucocorticoids (GCs), and is induced by CRH, AVP, and pro-inflammatory cytokines. The multiplicity of responsive elements in the proximal enhancer of the POMC gene contributes to corticotrophs function as integration centers, allowing them to make fine-tuned responses to environmental and psychic inputs. This fine-tuned integration is crucial for the animal's adaptation response to stressors. Corticotrophs also secrete β-lipoprotein, serving also as a precursor for the generation of β-endorphin by the corticotrophs. In contrast to the corticotrophs, the end-product of the POMC processing in the melanotrophs of the pars intermedia is α-MSH.

The most potent factor in the regulation of the corticotrophs is **CRH**, a 41-amino acid peptide derived from a 196-amino acid preprohormone. CRH is produced by neurons of the PVN and is released into hypophysial portal blood from axon terminals in the median eminence. In the corticotrophs, CRH binds to CRH-type 1 receptors and stimulates adenylyl cyclase. Elevated cAMP activates protein kinase A (PKA) which, in turn acts on voltage-operated Ca^{2+} channels, causing an influx of Ca^{2+} and the subsequent activation of calcium calmodulin kinase II (CaMKII). These signaling events lead to ACTH release and a concomitant activation of POMC gene expression. AVP, released from hypothalamic neurons or from the NL, acts synergistically with CRH to increase the electrical activity of the corticotrophs, increase intracellular calcium concentration, and enhance the secretion of ACTH.

The effects of GCs on the HPA axis is exerted at different sites and by various mechanisms. One integrative site is the hippocampus, which contains the highest concentration of glucocorticoid receptors in the CNS. Another site is the PVN, where the GCs inhibit the synthesis and release of both CRH and AVP. At the pituitary corticotrophs, GCs inhibit the transcription of POMC and antagonize CRH-stimulated ACTH release. They also repress cell cycle regulators such as L-Myc, N-Myc and E2F2 and suppress cortico-troph proliferation.

5.7.2 Major disorders associated with dysfunctions of the HPA axis

Given the multiple functions of GCs throughout the body, it is not surprising that excess or deficiency in ACTH and/or GCs result in several endocrine disorders that cause significant morbidity. **Figure 5.13** depicts typical changes in hormones and structures of the HPA axis that are associated with two major diseases: Cushing's syndrome and Addison's disease.

Cushing's syndrome (CS) results from extended exposure to excessive GCs from endogenous or exogenous sources [71]. The most common form of endogenous CS, occurring in ~80% of the cases, is ACTH-dependent, and is caused by an ACTH-secreting pituitary adenoma. The incidence of pituitary-dependent adrenal hyperplasia is three times greater in women than men. About 10% of the ACTH-dependent CS is due to ectopic (nonpituitary) ACTH secretion, and more rarely, to CRH production by benign or malignant neu-roendocrine tumors. About 15% to 20% of CS cases are ACTH-independent, caused either by adrenocortical hyperplasia, or by adrenocortical tumors that secrete excessive cortisol, causing suppression of ACTH. Exogenous (iatro-genic) CS, the most common form, is caused by excessive intake of GCs, used as anti-inflammatory or immunosuppressive medications.

As illustrated in **Figure 5.14**, a patient with Cushing's syndrome presents with multiple physical, medical, and mental changes. There is significant weight gain with uneven distribution of fat. The fat accumulates in the abdomen and torso, including the neck area, face, midline, and upper back, while the extremities are not affected, and may even become thinner. A round and red face, known as "moon face," is a typical physical feature. Skin problems include acne, infections, easy bruising, and long healing periods after injury or bruise. High blood pressure, which can lead to serious and even life-threatening complications, is also common. Because cortisol plays an important role in the normal muscle functioning, fatigue and muscle weakness are prevalent. Decreased sexual interest and low libido also afflict patients suffering from Cushing's syndrome. Male patients can have fertility problems and erectile dysfunction problems, while women may have imbalance of estrogen and progesterone and irregular menstrual periods. In addition, those suffering from this disease tend to have depression, anxiety and irritability.

Figure 5.13 The regulation of the hypothalamo-pituitary-adrenal axis (HPA) in humans. The middle panel shows the HPA under normal situation. Corticotropin releasing hormone (CRH) is released from the hypothalamus into hypophysial portal blood and stimulates the production and release of adrenocorticotropic hormone (ACTH). The main targets of ACTH are the adrenal glands, where it stimulates the production and release of cortisol. Cortisol, in turn provides negative feedback to both CRH and ACTH. The left panel shows the situation in Addison's disease. Destruction of the steroid-producing cells in the adrenal cortex causes a major deficiency in cortisol and consequently elevated ACTH levels, but little to no cortisol. The right panel shows the situation in Cushing's syndrome. A corticotroph adenoma produces and releases excessive levels of ACTH, resulting in elevated plasma cortisol levels and adrenal hyperplasia.

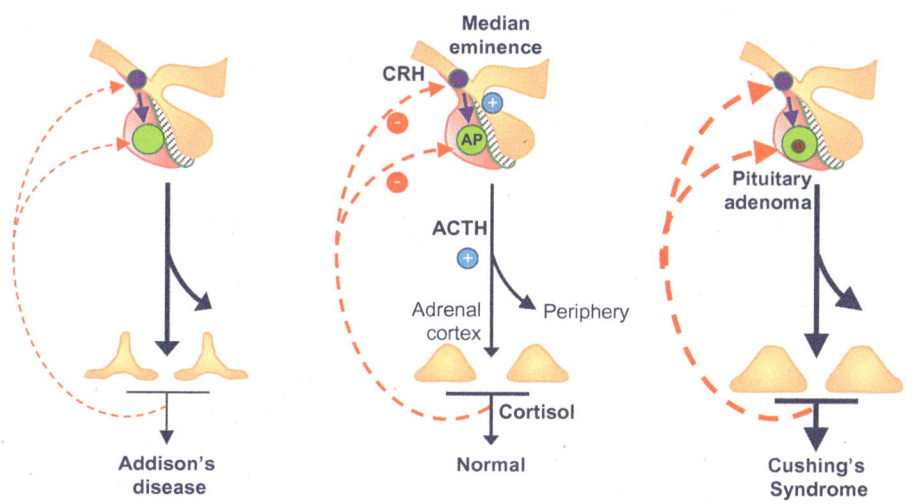

Figure 5.14 Physical appearance and major disturbances in patients with Cushing's or Addison's disease. See text for additional explanations.

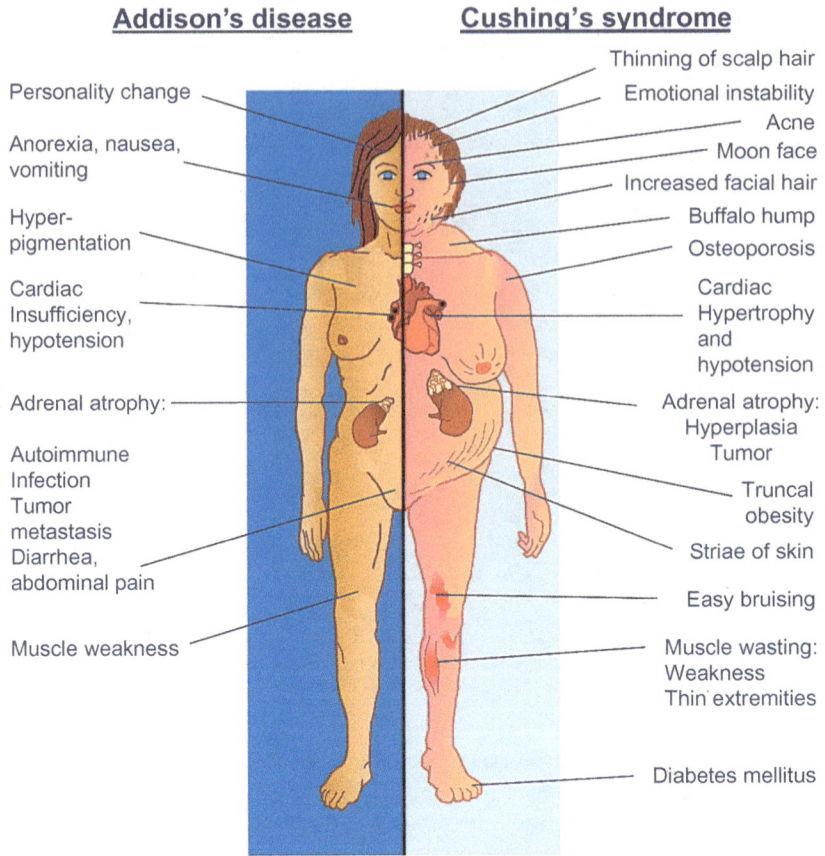

Addison's disease, also called primary adrenal insufficiency, is nearly the opposite of Cushing's syndrome. It is a rare condition that affects 1:100,000 people and can happen at any age to either men or women. It is a chronic condition caused by selective immune destruction of the steroid-producing cells in the adrenal cortex [72]. Autoimmune etiology accounts for 70% of the cases. Long-lasting infections such as tuberculosis, HIV, and some fungal infections can also harm the adrenal glands, as do cancer cells that spread to the adrenals. The adrenal glands can also be affected by "secondary adrenal insufficiency," caused by a reduction in hypothalamic CRH or pituitary ACTH. Another cause of secondary adrenal insufficiency is prolonged or improper use of steroid hormones such as prednisone. Less common causes include pituitary tumors and damage to the pituitary gland during surgery or radiation.

As illustrated in **Figure 5.14**, a patient with Addison's disease has several characteristic physical medical, and mental changes. There is weight loss, postural hypotension, muscular weakness and pains, fatigability, and an overall lack of energy. Patients have hyperpigmentation with some skin regions appearing as bronze color and also changes in the distribution of body hair. Such patients have loss of appetite, abdominal pains accompanied with nausea and vomiting, and frequent urination. Hypoglycemia is a common symptom in Addison's patients. Women have problems with their menstrual cycle, and there is lack of libido for both sexes.

People with Addison's disease can lead normal lives as long as they take their medications. Although patients survive with lifelong steroid replacement therapy, long-term outcome is compromised by complications associated with chronic steroid replacement, particularly vertebral fracture and type 2 diabetes. In addition to these complications, there is the ever-present risk of an unexpected adrenal crisis, which is inevitably fatal without prompt medical intervention. The average age of diagnosis for Addison's disease is 39 years, resulting in many years of intensive health care and reduced work capacity.

5.7.3 Involvement of dopamine in the regulation of the HPA axis

DA affects the HPA axis at both the hypothalamic and pituitary levels. Putative relationships between dopaminergic fibers and CRH-neurons were examined in male rats by means of immunocytochemistry, following the elimination of noradrenergic and adrenergic inputs to the hypothalamus [73]. The simultaneous immunocytochemical localization of DBH- or TH-immunoreactive fibers and CRH neurons in hypothalami from brainstem-lesioned, colchicine treated animals, revealed close distribution of catecholaminergic fibers and CRH neurons within the PVN. The authors concluded that the PVN receives selective dopaminergic innervation which likely influences the functions of the pituitary and adrenal glands via hypothalamic CRH.

Early studies found that D2R are expressed in more than 75% of cells within the human pituitary gland, indicating that they are not only expressed in lactotrophs and melanotrophs, which represent no more than 30% of the entire pituitary cell population. There were also reports on co-expression of somatostatin receptors and D2R in corticotroph adenomas. A recent study [40] undertook a closer look at the cellular distribution of D2R in human pituitaries. Using double immunohistochemistry, a strong D2R expression was found in corticotroph cell clusters lining the colloid-filled cysts, which represent the remnant melanotrophs of the residual pars intermedia. Variable D2R expression, ranging from weak to strong, was found in corticotroph cell clusters located between the cysts and belonging to the anterior lobe. The frequent expression of D2R in corticotroph adenomas associated with Cushing's disease, is supported by the findings that DA agonists such as cabergoline, are effective in controlling cortisol secretion in 30%–40% of patients with Cushing's disease after long-term treatment.

Another study used the murine pituitary corticotroph cell line AtT20 [74]. Application of 9-cis retinoic acid (RA) induced functional D2R in these cells and increased cell sensitivity to the DA agonist bromocriptine (Br). The combined administration of 9-cis RA and Br lowered the steady-state level of POMC, reduced ACTH release and suppressed cell viability more efficiently than either drug alone. The authors concluded that a combination of 9-cis RA and Br should be considered as a potential treatment for patients with ACTH-dependent Cushing's disease. A recent report indicated that dopastatin, a chimeric compound capable of binding to both SSTRs and D2R, showed antisecretory efficacy in human corticotroph tumors *in vitro* [75]. Yet, repeated dopastatin administration in human subjects caused accumulation of highly active dopaminergic metabolites that eventually blocked its action. Therefore, dopastatin's clinical development has been halted, but second-generation chimeras are under development.

5.7.4 Thyrotrophs and TSH

Thyrotrophs comprise about 5% of the adult anterior pituitary cell population. Most thyrotrophs are large stellate cells, filled with secretory granules, 100–150 nm in diameter, that are localized at the cell margins [76]. The cells are characterized by a dense arrangement of parallel arrays of rough endoplasmic reticulum. Clustered or isolated elongated TSH cells are also observed. Some cells with secretory granules 150–250 nm in diameter appear sporadically in the gland, and ultrastructurally resemble gonadotrophs. Rat thyrotrophs constitute an heterogeneous cell population in terms of morphology and ultrastructure. Hyperplastic pituitaries resulting from hypothyroidism, contain cells that stain for both TSH and PRL or TSH and GH derived from transdifferentiated somatotrophs or lactotrophs into thyrotrophs.

The glycoprotein hormone alpha subunit (αGSU) is the first pituitary hormone gene expressed during embryonic development, at mouse E10.5 [43]. Thyrotrophs are derived from two different populations. The first appears in the rostral tip of the developing pituitary by E12. This population is transient and is independent of Pit1 expression. The second arises by E15.5 and is PIT1-dependent. This population corresponds to the adult thyrotrophs, indicating that Pit1 is important for transactivating TSHβ, and for maintaining the cellular lineage. Thyrotroph embryonic factor (Tef) is expressed exclusively in the rostral portion of the developing pituitary, where the thyrotrophic precursors are located. Tef transactivates the *TSHβ* promoter, with Pitx1 and Pitx2 also contributing to thyrotroph differentiation. In humans, the fetal thyroid gland reaches maturity by gestational week 11–12, close to the end of the first trimester and begins to secrete thyroid hormones by about gestational week 16. During this period, an adequate supply of maternal thyroid hormones must be sustained to ensure normal neurological development.

The main target of TSH, whose expression is restricted to the pituitary thyrotrophs, is the follicular cells of the thyroid gland. The actions of TSH on thyroid cells are mediated by TSHR, a G protein–adenylyl cyclase–coupled receptor and include changes in thyroid cell morphology and growth, iodine metabolism, and stimulation of the production and release of T_3 and T_4. TSH is a glycoprotein consisting of alpha and beta subunits. The alpha-subunit, composed of 92 amino acids is thought to be the effector unit responsible for the stimulation of adenylate cyclase. The β subunit, composed of 118 amino acids, is unique to TSH, and determines its receptor specificity [77]. The rat and human TSH gene consists of three exons, whereas the mouse gene has five exons. The α-subunit is produced in excess of the beta-subunit and is rate limiting in the production of the non-covalent α-β dimer. Expression of the *TSHβ* gene is stimulated by the transcription factors Pit-1 and Gata2, whereas *TSHβ* gene expression is suppressed by thyroid hormones. TRH plays a dominant role in the stimulation of TSH synthesis.

TRH, a tripeptide with a structure of (pyro) Glu-His-Pro-NH2, was the first hypothalamic neurohormone to be isolated by the groups of Schally and Guillemin in 1969 (for reviews [78,79]). TRH is derived from a large precursor protein, prepro-TRH, by posttranslational processing by prohormone convertase enzymes. The rat prepro-TRH is a 29-kDa polypeptide composed of 255 amino acids. It contains an N-terminal leader sequence, five copies of the TRH progenitor sequence Gln-His-Pro-Gly flanked by paired basic amino acids, four non-TRH peptides lying between the TRH progenitors, an N-terminal flanking peptide, and a C-terminal flanking peptide. Rats and mice have five Gln-His-Pro-Gly TRH progenitor sequences, whereas humans have six TRH sequences. Serum thyroid hormone levels can affect the processing of pro-TRH by altering the prohormone convertases.

The highest concentrations of TRH are found in the PVN and the median eminence, from where TRH is secreted into hypophysial portal blood [80]. The hypothalamic TRH neurons receive afferents from multiple brain regions. Neurons from the arcuate nucleus transmit the nutritional status, those from the suprachiasmatic nucleus convey circadian cycle information, and neurons from the brain stem send information about external temperature. Stimuli that induce TRH–TSH release also coordinately increase *Trh* transcription.

The response of the thyrotroph to TRH is bimodal: First, it stimulates the release of stored hormone, and then it stimulates gene expression which increases TSH synthesis. TRH binds to TRHR1, a seven-transmembrane-spanning GPCR, and activates the Gq/11 protein and phospholipase C. The production of IP3 and DAG mobilizes intracellular calcium pools and activation of PKC. The interaction of TRH with TRHR1 induces rapid desensitization of the response, through receptor internalization in clathrin-coated vesicles and accumulation in endosomes. TRH stimulates not only the synthesis and release of TSH from thyrotrophs but also the release of PRL from lactotrophs, and in some species, also of GH from somatotrophs.

The regulation of TSH by **thyroid hormones** is fine tuned by their circulating levels. Thus, TSH synthesis and release are inhibited by high serum levels of T_3 and T_4 (hyperthyroidism) and are stimulated by low levels (hypothyroidism). Thyroid hormone action is mediated by thyroid hormone receptors (THR), encoded by two separate genes: *Thrα* and *Thrβ*, which generate several isoforms that are differentially expressed in a tissue-specific manner [81]. *Thrα1* and *Thrα2* are widely expressed. In contrast, *Thrβ1* is primarily expressed in the liver, whereas expression of *Thrβ2* is present mostly in the pituitary and hypothalamic TRH neurons.

5.7.5 Disorders of the hypothalamo-pituitary-thyroid axis

Humans have five major types of thyroid dysfunctions, each with its own symptoms. The five groups are (1) hypothyroidism, caused by having insufficient free thyroid hormones; (2) hyperthyroidism, caused by having excessive free thyroid hormones; (3) structural abnormalities of the thyroids, most commonly as a goiter (enlargement of the gland); (4) tumors, which can be benign or malignant; and (5) abnormal thyroid function tests without any clinical symptoms (subclinical hypothyroidism or subclinical hyperthyroidism).

In some types of thyroid disorders, such as subacute thyroiditis or postpartum thyroiditis, symptoms can subside after a few months and laboratory tests may return to normal. However, most thyroid diseases do not resolve on their own. Common hypothyroid symptoms include fatigue, low energy, weight gain, inability to tolerate cold, slow heart rate, dry skin and constipation. Common hyperthyroid symptoms include irritability, anxiety, weight loss, fast heartbeat, inability to tolerate heat, diarrhea, and enlargement of the thyroid. Tumors, often called thyroid nodules, can have many different symptoms ranging from hyperthyroidism to hypothyroidism to swelling in the neck and compression of the structures in the neck.

Diagnosis of potential thyroid dysfunctions starts with a history and physical examination. If dysfunction of the gland is suspected, laboratory tests can help support or rule out thyroid disease. As illustrated in **Figure 5.15**, a sensitive serum TSH assay can be initially used to indicate which additional diagnostic steps should be undertaken. If an autoimmune disorder of the thyroid such as Grave's disease is suspected, blood tests, looking for antithyroid autoantibodies, can be obtained. Ultrasound, biopsy and radioiodine scanning and uptake may be used to help with the diagnosis, particularly if a nodule is suspected. Treatment of thyroid disease varies, based on the

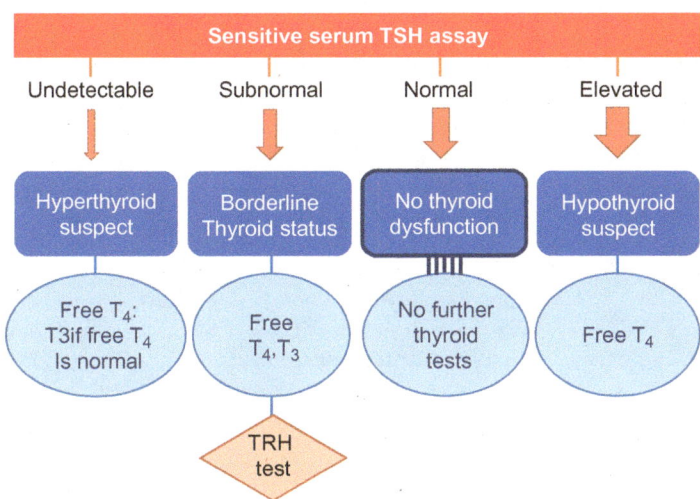

Figure 5.15 Common sequence of diagnostic tests for identifying thyroid disturbances. Changes in serum thyroid stimulating hormone (TSH) levels suggest potential alteration in the thyroid gland. These can be supported by analysis of circulating thyroid hormone (T_3 and T_4) levels and required additional tests such as biopsy, imaging, radioiodine uptake and tests for autoantibodies. TRH: thyrotropin releasing hormone.

disorder. Levothyroxine is the mainstay of treatment for hypothyroidism, while hyperthyroidism caused by Graves' disease can be managed with iodine therapy, antithyroid medication, or surgical removal of the thyroid gland. Thyroid surgery may also be performed to remove a thyroid nodule or to reduce the size of a goiter if it obstructs nearby structures or for cosmetic reasons.

5.7.6 Involvement of dopamine in the regulation of the HPT axis

The effects of DA and related compounds of TRH release were studied using rat hypothalamic fragments [82]. Whereas DA itself slightly inhibited TRH release, DA antagonists such as fluphenazine or alpha-flupentixol exhibited an inhibitory effect by themselves, so that the specific receptors involved in mediating the dopaminergic actions could not be characterized. Another study used rat striatal slices and nucleus accumbens fragments and found that DA agonists, acting via D2R, increased TRH release [83]. Although of interest, these sites are too remote from the hypophysiotropic areas and do not directly affect the HPT axis. Early studies used monolayer cultures of rat anterior pituitary cells and reported that DA inhibited basal and epinephrine-stimulated TSH secretion [84]. The data showed that epinephrine is a potential regulator of TSH secretion on its own and via its interactions with TRH, DA and somatostatin.

DA exerts inhibitory effects on the HPT axis in humans. Several studies [85–87] have reported that metoclopramide and domperidone, both peripheral DA antagonists, as well as sulpiride, an antidopaminergic agent, increased serum TSH levels in human volunteers and enhanced TRH-induced TSH release. The degree of the TSH response was significantly greater in females than males and was sustained for longer times. Support for an inhibitory role of DA comes from a study with freshly dispersed human pituitary cell cultures obtained within 24 h after accidental death [88]. Whereas epinephrine and serotonin stimulate TSH release, DA and somatostatin inhibit TRH-stimulated TSH release, showing that the control of TSH secretion by hypothalamic neuropeptides and biogenic amines in human pituitary is similar to that in the rat.

5.8 SYNOPSIS

The ontogeny, structure and functions of the various pituitary cells have been reviewed with an emphasis on the unique properties of their respective hormones and their impact on physiology and pathophysiology. DA is involved, directly or indirectly, in the regulation of both expression and release of virtually all hormones produced by the three lobes of the pituitary: neural, intermediate and anterior. Some of the regulation by DA occurs at the level of the hypothalamus, by affecting the production of the respective releasing or inhibiting hormones. In other cases, DA interacts directly with DAR expressed by the pituitary cells themselves and regulates their activity.

REFERENCES

1. Ugrumov MV. Magnocellular vasopressin system in ontogenesis: Development and regulation. *Microsc Res Tech.* 2002;56(2):164–171.

2. Saland LC. The mammalian pituitary intermediate lobe: An update on innervation and regulation. *Brain Res Bull.* 2001;54(6):587–593.

3. Abou Samra AB, Pugeat M, Dechaud H, Nachury L, Tourniaire J. Acute dopaminergic blockade by sulpiride stimulates beta-endorphin secretion in pregnant women. *Clin Endocrinol (Oxf).* 1984;21(5):583–588.

4. Moore RY, Bloom FE. Central catecholamine neuron systems: Anatomy and physiology of the dopamine systems. *Annu Rev Neurosci.* 1978;1:129–169.

5. Shuster SJ, Riedl M, Li X, Vulchanova L, Elde R. The kappa opioid receptor and dynorphin co-localize in vasopressin magnocellular neurosecretory neurons in guinea-pig hypothalamus. *Neuroscience.* 2000;96(2):373–383.

6. Acher R, Chauvet J. Structure, processing and evolution of the neurohypophysial hormone-neurophysin precursors. *Biochimie.* 1988;70(9):1197–1207.

7. Hatton GI. Pituicytes, glia and control of terminal secretion. *J Exp Biol.* 1988;139:67–79.

8. Young WS, III, Gainer H. Transgenesis and the study of expression, cellular targeting and function of oxytocin, vasopressin and their receptors. *Neuroendocrinology.* 2003;78(4):185–203.

9. Banerjee P, Joy KP, Chaube R. Structural and functional diversity of nonapeptide hormones from an evolutionary perspective: A review. *Gen Comp Endocrinol.* 2017;241:4–23.

10. Roberts MM, Robinson AG, Hoffman GE, Fitzsimmons MD. Vasopressin transport regulation is coupled to the synthesis rate. *Neuroendocrinology.* 1991;53(4):416–422.

11. Maybauer MO, Maybauer DM, Enkhbaatar P, Traber DL. Physiology of the vasopressin receptors. *Best Pract Res Clin Anaesthesiol.* 2008;22(2):253–263.

12. Bernal A, Mahia J, Puerto A. Animal models of central diabetes insipidus: Human relevance of acquired beyond hereditary syndromes and the role of oxytocin. *Neurosci Biobehav Rev.* 2016;66:1–14.

13. Hyde JF, North WG, Ben-Jonathan N. The vasopressin-associated glycopeptide is not a prolactin-releasing factor: Studies with lactating Brattleboro rats. *Endocrinology.* 1989;125(1):35–40.

14. Lu HA. Diabetes Insipidus. *Adv Exp Med Biol.* 2017;969:213–225.

15. Kim SH, Bennett PR, Terzidou V. Advances in the role of oxytocin receptors in human parturition. *Mol Cell Endocrinol.* 2017;449:56–63.

16. Jurek B, Neumann ID. The oxytocin receptor: From intracellular signaling to behavior. *Physiol Rev.* 2018;98(3):1805–1908.

17. Cornish JL, Wilks DP, Van den Buuse M. A functional interaction between the mesolimbic dopamine system and vasopressin release in the regulation of blood pressure in conscious rats. *Neuroscience.* 1997;81(1):69–78.

18. Yang Z, Sibley DR, Jose PA. D5 dopamine receptor knockout mice and hypertension. *J Recept Signal Transduct Res.* 2004;24(3):149–164.

19. Pelletier G. Identification of ending containing dopamine and vasopressin in the rat posterior pituitary by a combination of radioautography and immunocytochemistry at the ultrastructural level. *J Histochem Cytochem.* 1983;31(4):562–564.

20. Galfi M, et al. Effects of dopamine and dopamine-active compounds on oxytocin and vasopressin production in rat neurohypophyseal tissue cultures. *Regul Pept.* 2001;98(1–2):49–54.

21. Crowley WR, Armstrong WE. Neurochemical regulation of oxytocin secretion in lactation. *Endocr Rev.* 1992;13(1):33–65.

22. Galfi M, et al. Inhibitory effect of galanin on dopa mine induced increased oxytocin secretion in rat neurohypophyseal tissue cultures. *Regul Pept.* 2003;116(1–3):35–41.

23. Hung LW, et al. Gating of social reward by oxytocin in the ventral tegmental area. *Science.* 2017;357(6358):1406–1411.

24. Jenks BG. Regulation of proopiomelanocortin gene expression: An overview of the signaling cascades, transcription factors, and responsive elements involved. *Ann N Y Acad Sci.* 2009;1163:17–30.

25. Lamacz M, Tonon MC, Louiset E, Cazin L, Vaudry H. The intermediate lobe of the pituitary, model of neuroendocrine communication. *Arch Int Physiol Biochim Biophys.* 1991;99(3):205–219.

26. Chronwall BM, Bishop JF, Gehlert DR. Rat intermediate lobe in culture: A histological and biochemical characterization. *Peptides.* 1988;9(Suppl 1):169–180.

27. Catania A, Airaghi L, Colombo G, Lipton JM. Alpha-melanocyte-stimulating hormone in normal human physiology and disease states. *Trends Endocrinol Metab.* 2000;11(8):304–308.

28. Lugo DI, Pintar JE. Ontogeny of basal and regulated proopiomelanocortin-derived peptide secretion from fetal and neonatal pituitary intermediate lobe cells: Melanotrophs exhibit transient glucocorticoid responses during development. *Dev Biol.* 1996;173(1):110–118.

29. Scully KM, Rosenfeld MG. Pituitary development: Regulatory codes in mammalian organogenesis. *Science.* 2002;295(5563):2231–2235.

30. Cohen LE, Radovick S. Molecular basis of combined pituitary hormone deficiencies. *Endocr Rev.* 2002;23(4):431–442.

31. Liu S. Peptidergic innervation in pars distalis of the human anterior pituitary. *Brain Res.* 2004;1008(1):61–68.

32. Neto LV, et al. Expression analysis of dopamine receptor subtypes in normal human pituitaries, nonfunctioning pituitary adenomas and somatotropinomas, and the association between dopamine and somatostatin receptors with clinical response to octreotide-LAR in acromegaly. *J Clin Endocrinol Metab.* 2009;94(6):1931–1937.

33. Pivonello R, et al. Dopamine receptor expression and function in clinically nonfunctioning pituitary tumors: Comparison with the effectiveness of cabergoline treatment. *J Clin Endocrinol Metab.* 2004;89(4):1674–1683.

34. Leng ZG, et al. Activation of DRD5 (dopamine receptor D5) inhibits tumor growth by autophagic cell death. *Autophagy.* 2017;13(8):1–16.

35. Goldsmith PC, Cronin MJ, Weiner RI. Dopamine receptor sites in the anterior pituitary. *J Histochem Cytochem.* 1979;27(8):1205–1207.

36. Winters SJ, et al. Dopamine-2 receptor activation suppresses PACAP expression in gonadotrophs. *Endocrinology.* 2014;155(7):2647–2657.

37. Mansour A, et al. Localization of dopamine D2 receptor mRNA and D1 and D2 receptor binding in the rat brain and pituitary: An in situ hybridization-receptor autoradiographic analysis. *J Neurosci.* 1990;10(8):2587–2600.

38. Pivonello R, et al. Novel insights in dopamine receptor physiology. *Eur J Endocrinol.* 2007;156(Suppl 1):S13–S21.

39. Pivonello R, et al. Dopamine receptor expression and function in human normal adrenal gland and adrenal tumors. *J Clin Endocrinol Metab.* 2004;89(9):4493–4502.

40. Pivonello R, et al. Dopamine D2 receptor expression in the corticotroph cells of the human normal pituitary gland. *Endocrine.* 2017;57(2):314–325.

41. Peverelli E, et al. Dopamine receptor type 2 (DRD2) and somatostatin receptor type 2 (SSTR2) agonists are effective in inhibiting proliferation of progenitor/stem-like cells isolated from nonfunctioning pituitary tumors. *Int J Cancer.* 2017;140(8):1870–1880.

42. Wurth R, et al. Phenotypical and pharmacological characterization of stem-like cells in human pituitary adenomas. *Mol Neurobiol.* 2017;54(7):4879–4895.

43. de Moraes DC, Vaisman M, Conceicao FL, Ortiga-Carvalho TM. Pituitary development: A complex, temporal regulated process dependent on specific transcriptional factors. *J Endocrinol.* 2012;215(2):239–245.

44. Ben-Jonathan N, Lapensee CR, Lapensee EW. What can we learn from rodents about prolactin in humans? *Endocr Rev.* 2008;29(1):1–41.

45. Asa SL, Ezzat S. The pathogenesis of pituitary tumors. *Annu Rev Pathol.* 2009;4:97–126.

46. Stojilkovic SS. Pituitary cell type-specific electrical activity, calcium signaling and secretion. *Biol Res.* 2006;39(3):403–423.

47. Murai I, Ben-Jonathan N. Chronic posterior pituitary lobectomy: Prolonged elevation of plasma prolactin and interruption of cyclicity. *Neuroendocrinology.* 1986;43(4):453–458.

48. Cristina C, et al. Dopaminergic D2 receptor knockout mouse: An animal model of prolactinoma. *Front Horm Res.* 2006;35:50–63.

49. Vilar L, et al. Controversial issues in the management of hyperprolactinemia and prolactinomas: An overview by the neuroendocrinology department of the Brazilian society of endocrinology and metabolism. *Arch Endocrinol Metab.* 2018;62(2):236–263.

50. Harvey S, Baudet ML. Extrapituitary growth hormone and growth? *Gen Comp Endocrinol.* 2014;205:55–61.

51. Su Y, Liebhaber SA, Cooke NE. The human growth hormone gene cluster locus control region supports position-independent pituitary- and placenta-specific expression in the transgenic mouse. *J Biol Chem.* 2000;275(11):7902–7909.

52. Serri O, Deslauriers N, Brazeau P. Dual action of dopamine on growth hormone release *in vitro. Neuroendocrinology.* 1987;45(5):363–367.

53. Bosse R, et al. Anterior pituitary hypoplasia and dwarfism in mice lacking the dopamine transporter. *Neuron.* 1997;19(1):127–138.

54. Diaz-Torga G, et al. Disruption of the D2 dopamine receptor alters GH and IGF-I secretion and causes dwarfism in male mice. *Endocrinology.* 2002;143(4):1270–1279.

55. Racine MS, Barkan AL. Medical management of growth hormone-secreting pituitary adenomas. *Pituitary.* 2002;5(2):67–76.

56. de Boon WMI, et al. A novel somatostatin-dopamine chimera (BIM23B065) reduced GH secretion in a first-in-human clinical trial. *J Clin Endocrinol Metab.* 2019;104(3):883–891.

57. Bernard DJ, Fortin J, Wang Y, Lamba P. Mechanisms of FSH synthesis: What we know, what we don't, and why you should care. *Fertil Steril.* 2010;93(8):2465–2485.

58. Fischer R. Understanding the role of LH: Myths and facts. *Reprod Biomed Online.* 2007;15(4):468–477.

59. Ristic N, et al. Diet-induced obesity and ghrelin effects on pituitary gonadotrophs: Immunohistomorphometric study in male rats. *Cell J.* 2016;17(4):711–719.

60. Stojilkovic SS, Bjelobaba I, Zemkova H. Ion channels of pituitary gonadotrophs and their roles in signaling and secretion. *Front Endocrinol (Lausanne).* 2017;8:126.

61. Masumoto N, et al. Simultaneous measurements of exocytosis and intracellular calcium concentration with fluorescent indicators in single pituitary gonadotropes. *Cell Calcium.* 1995;18(3):223–231.

62. Stamatiades GA, Carroll RS, Kaiser UB. GnRH-A key regulator of FSH. *Endocrinology.* 2019;160(1):57–67.

63. Schneider F, Tomek W, Grundker C. Gonadotropin-releasing hormone (GnRH) and its natural analogues: A review. *Theriogenology.* 2006;66(4):691–709.

64. Stamatiades GA, Kaiser UB. Gonadotropin regulation by pulsatile GnRH: Signaling and gene expression. *Mol Cell Endocrinol.* 2018;463:131–141.

65. Ulloa-Aguirre A, Reiter E, Crepieux P. FSH receptor signaling: Complexity of interactions and signal diversity. *Endocrinology.* 2018;159(8):3020–3035.

66. Weems P, et al. Effects of season and estradiol on KNDy neuron peptides, colocalization with D2 dopamine receptors, and dopaminergic inputs in the ewe. *Endocrinology.* 2017;158(4):831–841.

67. Liu X, Herbison AE. Dopamine regulation of gonadotropin-releasing hormone neuron excitability in male and female mice. *Endocrinology.* 2013;154(1):340–350.

68. Henderson HL, Townsend J, Tortonese DJ. Direct effects of prolactin and dopamine on the gonadotroph response to GnRH. *J Endocrinol.* 2008;197(2):343–350.

69. Harno E, Gali RT, Coll AP, White A. POMC: The physiological power of hormone processing. *Physiol Rev.* 2018;98(4):2381–2430.

70. Le Tissier PR, et al. Anterior pituitary cell networks. *Front Neuroendocrinol.* 2012;33(3):252–266.

71. Guaraldi F, Salvatori R. Cushing syndrome: Maybe not so uncommon of an endocrine disease. *J Am Board Fam Med.* 2012;25(2):199–208.

72. Gan EH, Pearce SH. Management of endocrine dusease: Regenerative therapies in autoimmune Addison's disease. *Eur J Endocrinol.* 2017;176(3):R123–R135.

73. Liposits Z, Paull WK. Association of dopaminergic fibers with corticotropin releasing hormone (CRH)-synthesizing neurons in the paraventricular nucleus of the rat hypothalamus. *Histochemistry.* 1989;93(2):119–127.

74. Occhi G, et al. Activation of the dopamine receptor type-2 (DRD2) promoter by 9-cis retinoic acid in a cellular model of Cushing's disease mediates the inhibition of cell proliferation and ACTH secretion without a complete corticotroph-to-melanotroph transdifferentiation. *Endocrinology.* 2014;155(9):3538–3549.

75. Theodoropoulou M, Reincke M. Tumor-directed therapeutic targets in Cushing disease. *J Clin Endocrinol Metab.* 2019;104(3):925–933.

76. Yoshimura F, Nogami H, Yashiro T. Fine structural criteria for pituitary thyrotrophs in immature and mature rats. *Anat Rec.* 1982;204(3):255–263.

77. Bargi-Souza P, Goulart-Silva F, Nunes MT. Novel aspects of T3 actions on GH and TSH synthesis and secretion: Physiological implications. *J Mol Endocrinol.* 2017;59(4):R167–R178.

78. Joseph-Bravo P, Jaimes-Hoy L, Uribe RM, Charli JL. 60 years of neuroendocrinology: TRH, the first hypophysiotropic releasing hormone isolated: Control of the pituitary-thyroid axis. *J Endocrinol.* 2015;226(2):T85–T100.

79. Chiamolera MI, Wondisford FE. Minireview: Thyrotropin-releasing hormone and the thyroid hormone feedback mechanism. *Endocrinology.* 2009;150(3):1091–1096.

80. Oliver C, Ben-Jonathan N, Mical RS, Porter JC. Transport of thyrotropin-releasing hormone from cerebrospinal fluid to hypophysial portal blood and the release of thyrotropin. *Endocrinology.* 1975;97(5):1138–1143.

81. Chiamolera MI, et al. Fundamentally distinct roles of thyroid hormone receptor isoforms in a thyrotroph cell line are due to differential DNA binding. *Mol Endocrinol.* 2012;26(6):926–939.

82. Joseph-Bravo P, Charli JL, Palacios JM, Kordon C. Effect of neurotransmitters on the in vitro release of immunoreactive thyrotropin-releasing hormone from rat mediobasal hypothalamus. *Endocrinology.* 1979;104(3):801–806.

83. Przegalinski E, Jaworska L, Budziszewska B. The role of dopamine receptors in the release of thyrotropin-releasing hormone from the rat striatum and nucleus accumbens: An *in vitro* study. *Neuropeptides.* 1993;25(5):277–282.

84. Dieguez C, Foord SM, Peters JR, Hall R, Scanlon MF. Interactions among epinephrine, thyrotropin (TSH)-releasing hormone, dopamine, and somatostatin in the control of TSH secretion in vitro. *Endocrinology.* 1984;114(3):957–961.

85. Scanlon MF, et al. Dopamine is a physiological regulator of thyrotrophin (TSH) secretion in normal man. *Clin Endocrinol (Oxf).* 1979;10(1):7–15.

86. Delitala G, Devilla L, Canessa A, D'Asta F. On the role of dopamine receptors in the central regulation of human TSH. *Acta Endocrinol (Copenh).* 1981;98(4):521–527.

87. Krulich L. Neurotransmitter control of thyrotropin secretion. *Neuroendocrinology.* 1982;35(2):139–147.

88. Abrahamson MJ, Wormald PJ, Millar RP. Neuroendocrine regulation of thyrotropin release in cultured human pituitary cells. *J Clin Endocrinol Metab.* 1987;65(6):1159–1163.

89. Neumann ID. Brain oxytocin: A key regulator of emotional and social behaviours in both females and males. *J Neuroendocrinol.* 2008;20(6):858–865.

90. Baskerville TA, Douglas AJ. Dopamine and oxytocin interactions underlying behaviors: Potential contributions to behavioral disorders. *CNS Neurosci Ther.* 2010;16(3):e92–123.

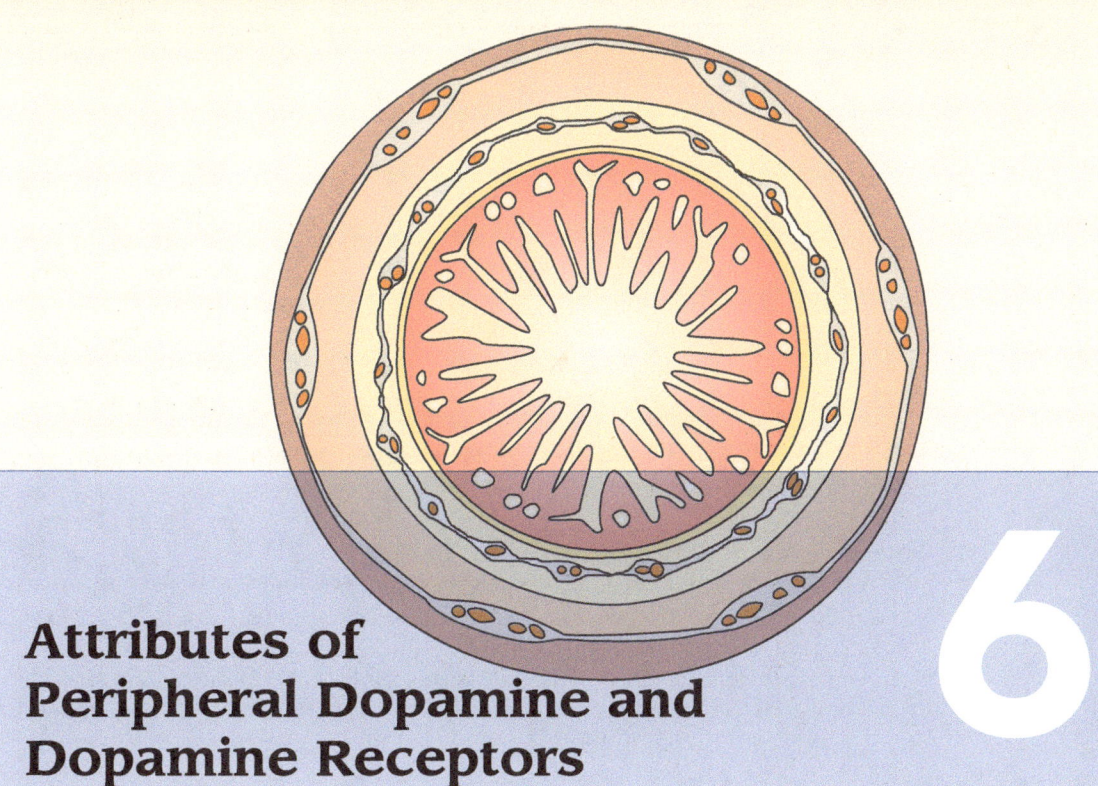

Attributes of Peripheral Dopamine and Dopamine Receptors

6.1 INTRODUCTION

The presence of dopamine (DA) in the systemic circulation has been known for decades. Yet, compared to the extensive interests in brain DA, little attention has been given to its peripheral counterpart, and its functions as a circulating hormone and a cellular modulator in health and disease. Similarly, there has been an intense interest in the control of circulating norepinephrine (NE) and epinephrine (Epi) but not in their sister molecule DA. This chapter begins by reviewing the available information on the putative sources of circulating DA and then discusses the unique aspects of its synthesis, storage, release, and metabolism. Unlike most other species, circulating DA in humans is primarily present in the form of **dopamine sulfate (DA-S)**, which is biologically inactive and has a significantly longer half-life in the circulation than unmodified DA. However, a specific enzyme, **arylsulfatase A (ARSA)**, is expressed and secreted by several peripheral tissues and can convert DA-S back to bioactive DA. The remainder of the sections provide concise information on the expression of various dopamine receptors (DARs) in peripheral tissues, which are grouped by common characteristics. (e.g., cardiovascular, immune, reproductive, etc.). Thus, this chapter summarizes both the presence of DA in the periphery and identifies its target organs and cells.

6.2 SOURCES OF CIRCULATING DOPAMINE

6.2.1 Biosynthesis, storage, release, and metabolism of peripheral catecholamines

The homeostasis of catecholamines is covered in great detail in **Chapter 1**. Here we recapitulate some of the salient features that are especially pertinent to peripheral catecholamines. The three catecholamines, DA, NE, and Epi, are synthesized by four sequential enzymes (**Figure 1.1**). The first enzyme, tyrosine hydroxylase (TH), converts tyrosine to dihydroxyphenylalanine (Dopa) by inserting an OH group on the ring. Tyrosine circulates at a concentration of 1–1.5 mg/dL and enters the cells by an active transport. TH serves as the rate-limiting step and is expressed in a number of peripheral organs and tissues. The second enzyme, Dopa decarboxylase (DDC; also known as aromatic L-amino acid decarboxylase), generates DA from Dopa by removing the carboxyl group from the side chain and is widely expressed. Dopa can also become available to catecholamine-producing cells from the circulation. Cells that express the third enzyme, dopamine β-hydroxylase (DBH), which inserts an OH group on the side chain, can synthesize NE as a final product. Expression of phenylethanolamine N-methyl-transferase (PNMT), which catalyzes the transfer of a methyl group from S-adenosylmethionine (SAM) to NE, enables the production of Epi. Within the adrenals, the high concentration of cortisol, which diffuses from the zona fasciculata of the cortex to the adrenal medulla, enhances PNMT expression. This accounts for the fact that in the normal human adrenal medulla, about 80% of the catecholamine content is Epi, while only 20% is NE.

NE and **Epi** are present at variable concentrations in human plasma, and their tissues of origin vary [1]. Serum NE originates from both sympathetic nerve endings and the adrenal medulla, while serum Epi originates exclusively from the adrenal medulla. Although PNMT is expressed in several extra-adrenal tissues, including the lung, kidney, pancreas, and some cancer cells, Epi is produced by these tissues in only small amounts that remain locally and contributes minimally to its circulating levels. The release of catecholamine by the adrenal medulla is stimulated by exercise, hemorrhage, ether anesthesia, surgery, hypoglycemia, anoxia, asphyxia, and psychological conditions such as threat or fear. Hypoglycemia and some stressful stimuli increase the rate of Epi secretion more than NE, while anoxia and asphyxia produce a greater increase in NE.

Changes in plasma NE levels have served as an index of sympathetic nerve activity. However, the relationship between circulating NE levels and sympathetic nerve traffic is complex given that plasma NE concentrations at any given time depend on a balance between the rate of NE release into plasma and the rate of its removal from plasma. This complexity does not invalidate the analysis of plasma NE levels in the diagnosis or assessment of drug effects, but interpretation should be made with caution.

Catecholamines are metabolized by two major enzymes, **monoamine oxidase (MAO)**, which carries out oxidative deamination of the side chain, and **catechol-O-methyltransferase (COMT)**, which transfers a methyl group to the catechol ring using SAM as a methyl donor (**Figure 1.4**). Deamination by MAO is generally responsible for the intracellular inactivation of catecholamines that occurs within neurons, while COMT is primarily responsible for the extracellular metabolism of secreted catecholamines. COMT is found in most peripheral tissues (e.g., blood cells, liver, kidney, and vascular smooth muscle). Peripheral NE is primarily metabolized to normetanephrine by COMT, while Epi is metabolized to metanephrine. Some of the metabolites also undergo conjugation by glucuronidases and sulfotransferases. The latter is especially important for the processing of peripheral DA and is discussed in more detail below.

6.2.2 The origin of circulating dopamine

The **blood–brain barrier (BBB)** prevents a bidirectional transport of DA between the brain and the systemic circulation, while the DA precursor, L-Dopa, can easily interchange between the two compartments. Consequently, DA that is present in the general circulation or in urine does not come from the brain but only from peripheral sources. Basal plasma DA levels in humans are very low (~0.1 nmol/L) and require special effort for their analysis. Except for the few reports on plasma DA levels under various physiological and pathological conditions, DA in the systemic circulation has been ignored in most clinical studies. The measurement of urinary DA and its metabolites has traditionally been used for some diagnostic purposes, but

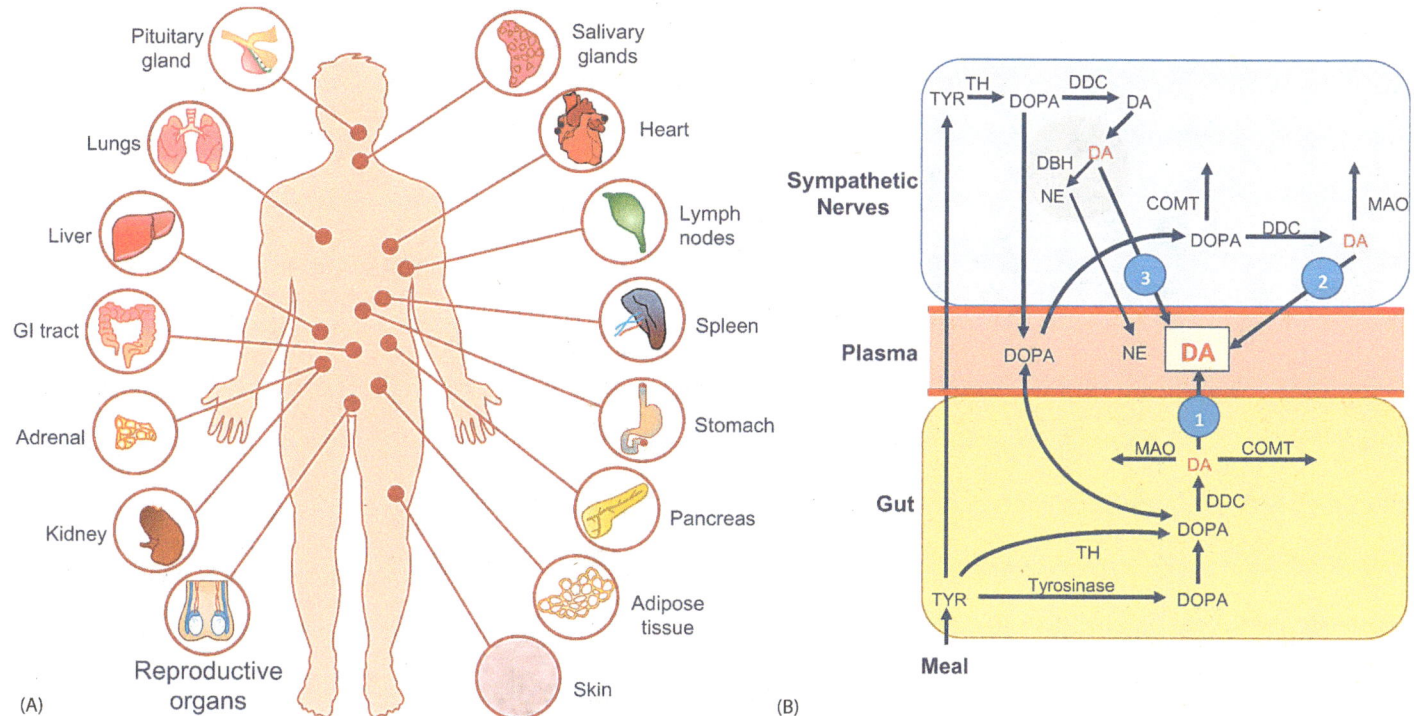

Figure 6.1 Locations of dopamine receptors in peripheral tissues and the sources of circulating dopamine. Panel A shows expression of dopamine receptors (DARs) in peripheral tissues throughout the human body. Panel B shows the various sources of dopamine (DA) in plasma: (1) from the gut, (2) from *de novo* synthesis in sympathetic nerves, and (3) from conversion of Dopa in sympathetic nerves and other cell types. Sympathetic nerves that express tyrosine hydroxylase (TH) can generate DA from tyrosine (TYR), available from the diet, or from Dopa, available from the diet as well as from blood. In the gut, either TH or tyrosinase can convert TYR to Dopa. COMT: catecholamine O-methyl transferase; DBH: dopamine beta-hydroxylase; DDC: Dopa decarboxylase; MAO: monoamine oxidase; NE: norepinephrine.

this alternative has several drawbacks, including the inability to assess rapid changes in DA release in response to various stimuli or medications.

As shown in **Figure 6.1**A, circulating DA originates from neuronal, nonneuronal, and dietary sources [2,3]. The major source of circulating DA is the gastrointestinal (GI) tract, with additional contributions coming from the coincidental release of DA with NE and DBH during exocytosis from the sympathoadrenal system, where DA serves as an intermediate of catecholamine biosynthesis. Tissues that are populated by DA-producing neuroendocrine cells, called amine precursor uptake and decarboxylation (**APUD**) cells [4], can also release their product into the circulation. In addition to *de novo* synthesis of DA from tyrosine, tissues that lack TH can take up Dopa from the circulation via carrier-mediated transport and convert it to DA by the ubiquitously expressed DDC (**Figure 6.1**B). Beyond the endogenous sources, DA can be obtained from the diet [5]. Certain foods (e.g., bananas) contain DA, while others (e.g., tuna and cereals) contain Dopa. Some foods also contain tyramine, which can be converted to DA by the cytochrome P450 enzyme complex.

Many peripheral nonneuronal cells express TH and/or DDC and are known for their ability to synthesize DA (**Figure 6.1**A). These include the anterior [6] and posterior [7] pituitaries, kidneys [8], salivary glands [9], GI tract, spleen, and pancreas [2], mesenteric lymph nodes, thymus, and lymphocytes [10], mast cells [11], bone marrow-derived mesenchymal stem cells [12], carotid body [13], keratinocytes [14], melanocytes [15], and adipocytes [16] as well as the gonads and most reproductive tract organs [17]. Yet, only a small fraction of the DA that is produced in these sites reaches the circulation because most of the released DA remains locally at high concentrations and acts as a paracrine/autocrine factor on adjacent DAR-expressing cells.

6.2.3 Special characteristics of peripheral dopamine

As discussed in **Chapter 1**, the synthesis, storage, release, reuptake and metabolism of DA are dynamic processes that differ in many respects between the "**closed**" **system** of the brain dopaminergic neurons, and the "**open-ended**" **system** of peripheral DA-producing cells (see **Figure 1.2**). Within the closed system of neuron/synapse/neuron, the concentration of the released DA can be as high as 5–10 µM, whereas the concentration of circulating DA reaching peripheral target cells does not exceed 20–30 nM. This large differential in the effective concentrations of DA impinging upon peripheral organs has not always been taken into consideration in many studies using cultured cells/organs, resulting in the use of pharmacological, rather than physiological, doses of DA.

All monoaminergic neurons, including catecholamines and indoleamines, have intrinsic processes that guard against neuronal overstimulation by the released neurotransmitter. They also have specific mechanisms for maintaining an adequate intraneuronal storage/secretory capacity capable of prompt responses to frequent stimuli. Except for the **tuberoinfundibular (TIDA)** neurons, which release DA into the hypophysial portal vasculature, all other brain dopaminergic neurons are linked via synapses to recipient neurons in a "closed" system configuration. The latter is briefly summarized below so as to compare it with the distinct properties of the "open-ended" system of peripheral DA.

In a typical dopaminergic neuron, DA is synthesized in the trans-Golgi network and is packed into secretory vesicles (**Figure 6.2**). Within the vesicles, the concentration of DA can be 100–1,000 times higher than its levels in the cytoplasma. The vesicles also provide protection for DA from degradation by mitochondrial MAO. DA loaded within the vesicles exists in a dynamic equilibrium, whereby a passive outward leakage of DA into the cytoplasm is counterbalanced by an inward active transport. This active transport is controlled by the vesicular monoamine transporter **(VMAT)**, which maintains an H⁺-electrochemical gradient between the cytoplasm and vesicles.

As shown in **Figure 6.2**, in a typical neuron, DA is released into the synaptic cleft by a calcium-mediated exocytosis and binds to DARs, which are localized either presynaptically (autoreceptors) or postsynaptically (recipient

Figure 6.2 Homeostasis of dopamine in a typical neuronal system. The synthesis, storage, release and actions of dopamine (DA) in neurons. Tyrosine hydroxylase (TH) is localized both in the cytoplasm and in storage vesicles, which also contain Dopa decarboxylase (DDC) and vesicular monoamine transporter (VMAT2). DA is released by a calcium-dependent mechanism into the synaptic space. From there, DA can either bind to postsynaptic or presynaptic dopamine receptors (DARs), can be taken back into the secreting cell by the DA transporter (DAT), or it can be degraded by COMT (catechol-O-methyltransferase). Intracellular DA that is not protected by storage vesicles can be degraded by mitochondrial monoamine oxidase (MAO).

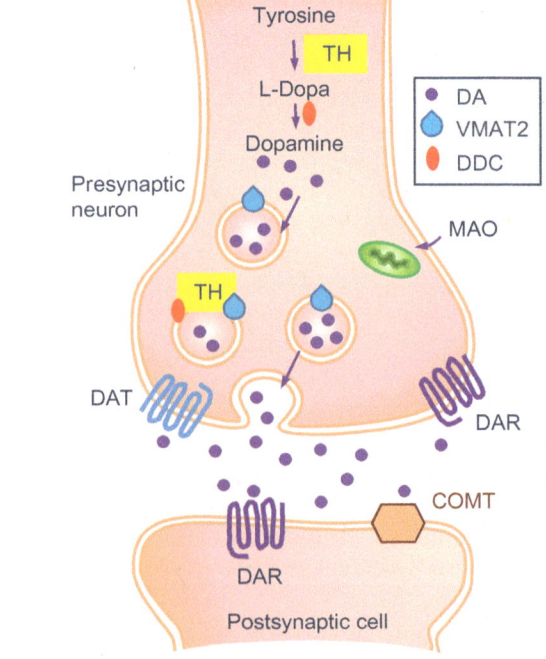

receptors). After triggering action potentials in the postsynaptic neurons and activating intracellular signaling pathways, DA rapidly disassociates from the receptors and can undergo one or more of the following actions: It can (1) be taken back ("reuptake") into the presynaptic terminal by the **DA transporter (DAT)**, (2) repackaged into storage vesicles by VMAT, (3) become deaminated by intracellular MAO to dihydroxyphenylacetic acid (**DOPAC**), and/or (4) become O-methylated by membranous or soluble COMT to **3-methoxytyramine**. Homovanylic acid (HVA) is the final product of O-methylation of DOPAC.

Although the behavior of peripheral DA-producing cells follows the general scheme described above, there are several distinct dissimilarities. DA synthesis in peripheral organs can be either independent or dependent on their sympathetic innervation. Thus, chemical sympathectomy with 6-hydroxydopamine (**6-OH-DA**) in rats can cause a marked decrease in DA in some peripheral organs, but its preservation in others, providing proof for a nonneuronal origin [18]. When DA synthesis is independent of sympathetic innervation, the question is whether DA in the producing cells is stored and released in a regulated manner from secretory vesicles or it passively diffuses out soon after synthesis. Unlike neural or endocrine cells, which have an abundance of secretory vesicles, most other tissues do not have a well-developed storage and secretory capacity for endocrine-like releasable factors. It has been proposed that in some peripheral organs, DAT, which usually brings DA into the cell, acts as an outward facing transporter and facilitates the diffusion of DA out of the cells [19].

Without the restricted space of a synapse, DA is released from peripheral nonneuronal cells into the extracellular space and diffuses away from the producing cells into the blood. DA can then reach its target cells via the circulation and can also become inactivated by metabolic enzymes located either at adjacent or at remote sites. Deamination of peripheral DA can be done by two isoenzymes, MAO-A and MAO-B, which are highly expressed in the liver, with a lower expression in the myocardium, lung, kidney, and duodenum [20]. O-methylation is carried out by COMT, also having the highest activity in the liver, followed by the kidney, stomach, and intestine [21].

6.3 UNIQUE CHARACTERISTICS OF DOPAMINE SULFATE

Investigators studying catecholamine have been faced with several dilemmas with respect to the production and functions of peripheral DA in humans [3]. One is the recognition that urinary excretion rates of DA and its metabolites exceed those of NE and its metabolites. Another is the fact that almost all DA circulates in a **sulfoconjugated form** as DA-S. These findings raise the following questions: (1) Where is DA-S coming from? (2) Why is it so abundant? (3) What might be its physiological functions? The following sections present the available information on the origin(s) of DA-S and its potential functions.

6.3.1 Sulfoconjugation of peripheral dopamine

The major metabolic inactivation of secreted peripheral DA in humans is sulfoconjugation. This reaction is carried out by the enzyme sulfotransferase type 1A3 (**SULT1A3**), which acts by transferring a sulfonate group (SO_3^{-1}) from a universal donor 3′-phospho-adenosine, 5′ phosphosulfate (PAP), to recipient molecules [22]. Sulfoconjugation is a rather common metabolic conversion, encompassing steroid and thyroid hormones, some neuroendocrine peptides, and glycoprotein hormones. Sulfoconjugation is known to alter the bioactivity, metabolic half-life, solubility, and/or receptor binding affinity of the affected hormones. Plasma DA is more than 95% sulfonated, and both the 3-O- and 4-O-sulfate isomers are present, with the 3-O-sulfate isomer being in a greater abundance by an order of magnitude [22].

Under basal conditions, serum DA-S at ~5 ng/mL exceeds by tenfold the basal levels of free DA (0.3 ng/mL), NE (0.2 ng/mL), or EPI (0.05 ng/mL) combined. Yet, the presence of DA-S in human serum has been overlooked by most investigators, likely because DA-S is undetectable by routine analytical methods and requires a special extraction method for its measurement. In addition, unlike NE and Epi, which have attracted a lot of attention because of their essential roles in the stress circuitry response, there has been a lack of understanding and only little appreciation for the roles played by circulating DA.

Sulfoconjugation is the major form of DA inactivation in human serum, while glucuronidation predominates in rats. No ortholog of SULT1A3 is known in rodents, emphasizing the greater importance of DA sulfoconjugation in humans than in laboratory animals [23]. This issue should be taken into account when using rodents to examine the physiology or pathophysiology of peripheral DA. The highest activity of SULT1A3 is found in the GI tract, with moderate enzyme activities found in the liver, lung, pancreas, and platelets [3]. SULT1A3 has high specificity toward catecholamines as well as toward catechol-estrogens [22]. Within the catecholamine family, the enzyme has higher affinity for DA than for NE or Epi because of a single amino acid substitution.

Ingestion of a meal after fasting in human volunteers induced a 50-fold rise in serum DA-S levels [3]. This and other data led the authors to conclude that serum DA-S is derived from sulfoconjugation of DA that is synthesized from L-Dopa in the GI tract, with both dietary and other endogenous sources contributing additional values to its circulating levels. A five- to tenfold elevation of serum DA-S levels was seen in chronic alcoholics and cocaine abusers although neither the causes for these rises nor the clinical implications of DA-S to these conditions are known. Serum DA-S levels are also increased in patients with renal hypertension, pheochromocytoma, and chronic renal insufficiency.

6.3.2 Functions of DA-S and the role of arylsulfatase A

DA-S does not bind DAR and is biologically inactive. To this end, it has been proposed that sulfoconjugation of DA serves two complimentary purposes [3]. One is to act as a gut–blood barrier to effectively detoxify (inactivate) DA derived from ingested food (as a DA and/or from L-Dopa) before it enters the blood stream. Another proposed function is to limit the activity of DA as an autocrine-paracrine factor in the GI and other DA-producing sites. However, as discussed below, it appears that the most critical function of DA-S is to serve as a stable reservoir of indolent DA that can be readily converted to authentic DA when the need arises.

Figure 6.3 shows that unlike the irreversible inactivation of DA by deamination, O-methylation or glucuronidation, sulfoconjugation is reversible, and DA-S can be converted back to bioactive DA by **ARSA**. Of particular importance is the fact that DA-S has a serum half-life of 4–5 h, as compared with 2–3 min for free DA. Thus, DA-S constitutes a relatively stable serum reservoir of inactive (and nondegradable) DA, which can be converted back to bioactive DA when necessary by cells that express ARSA. It is envisioned that ARSA is released from the lysosomes to the proximity of DAR expressing cells, where it can convert DA-S into a bioactive form that binds the receptors. The ability of some normal as well as malignant tissues to convert the inactive DA-S to the bioactive DA is fundamental for the overall functions of peripheral DA. The sulfation/desulfation reactions of DA can also be considered as a buffering system that reversibly deactivates DA when its concentrations are too high, while converting it back to bioactive DA when needed.

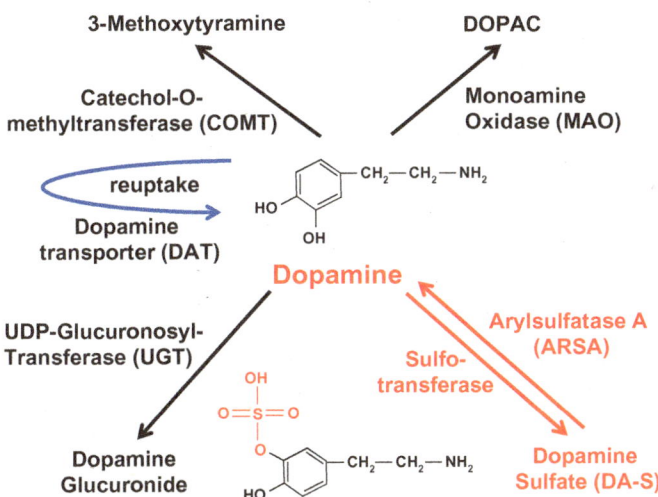

Figure 6.3 Dopamine (DA) metabolism. DA can be converted to 3-methoxytyramine by COMT or to dihydroxyphenylacetic acid (DOPAC) by MAO. DA conjugation to glucuronide is done by UDP-Glucuronosyl-Transferase (UGT), while sulfoconjugation is done by sulfotransferase. The latter reaction is reversible, and DA-S can be converted back to bioactive DA by ARSA.

ARSA is a member of the superfamily of sulfatases that catalyze the hydrolysis of sulfate ester bonds of a variety of substrates, ranging from sulfated proteoglycans to conjugated steroids and catecholamines [24]. Of the 17 members of the sulfatase family, ARSA and ARSC are of particular relevance for many endocrine and oncogenic functions. ARSC (also known as steroid sulfatase) is a microsomal enzyme that is configured as an integral membrane protein of the endoplasmic reticulum. Together with aromatase, ARSC maintains high estrogen levels in the breast and may have a role in the promotion of breast cancer.

ARSA is a soluble and secretable lysosomal enzyme with high specificity for sulfated cerebrosides and catecholamines. Mutations in ARSA can lead to metachromatic leukodystrophy, a rare demyelinating disease caused by a deficiency in ARSA that results in the accumulation of sulfatides in the central and peripheral nervous system. The disease is inherited as an autosomal recessive trait and leads to a progressive neurodegeneration. It is a fatal disease with very limited therapeutic options, primarily palliative [25].

As illustrated in **Figure 6.4**, human adipose tissue and adipocytes express ARSA with a robust enzymatic activity [26]. DA and DA-S were equally potent in their ability to increase lipolysis and also to suppress PRL release from the adipocytes, indicating that DA-S was indeed converted to

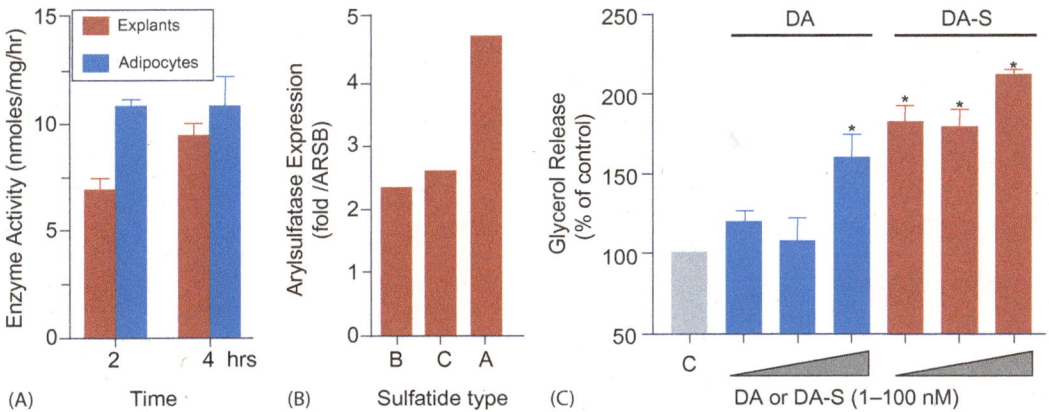

Figure 6.4 Expression of arylsulfatase A (ARSA) and proof of bioactivity of dopamine sulfate (DA-S) in human adipocytes. Panel A: Enzymatic activity of arylsulfatase in adipose explants and isolated mature adipocytes. Panel B: Relative expression of ARSA (A), ARSB (B) and ARSC (C) in adipose tissue. Panel C: Both dopamine (DA) and DA-S increase lipolysis in adipocytes. (Redrawn and modified from Borcherding, D.C. et al., *PLoS One*, 6; e25537, 2011.)

bioactive DA, enabling it to exert these actions. Notably, ARSA has been detected in serum from patients with a variety of cancers [27]. Given the renewed interest in the role of peripheral DA in pathophysiology and tumorigenesis, ARSA merits increased attention as a potential diagnostics tool for multiple disorders.

6.4 DOPAMINE RECEPTORS IN THE CARDIOVASCULAR, PULMONARY, AND RENAL SYSTEMS

The early identification of peripheral DAR subtypes relied on physiological and pharmacological approaches that lacked specificity and/or sensitivity. Following the cloning of the five DAR and the subsequent development of powerful research tools such as real-time polymerase chain reaction (RT-PCR), *in situ* hybridization, immunohistochemistry, Western blotting, and whole-body positron emission tomography (PET) imaging, a more comprehensive understanding of the distribution of specific DARs in the various components of the cardiopulmonary system has emerged.

6.4.1 The cardiovascular and renal systems

Administration of DA agonists or antagonists to laboratory animals and humans has caused changes in blood pressure [28]. In addition, all DAR knockout (KO) mice exhibit some increases in blood pressure [29]. These findings served as the backdrop for looking at the expression of DAR subtypes in all major organs that are closely involved in the regulation of blood pressure: heart, blood vessels, kidneys, and adrenals.

Various DAR subtypes are expressed in the heart [30], coronary vessels [31], aortic and mesenteric arteries [32], and the portal vein [33] but are absent in smaller vessels such as arterioles, capillaries, or venules. Immunoreactive D1R was localized in rat cardiomyocytes and in vascular smooth muscle cells (VSMC) of the coronary arteries but not in interstitial fibrotic tissue [34]. With the exception of D3R, all other DAR subtypes have been identified by immunocytochemistry and immunoblotting in the human heart, including the endocardium, myocardium, and epicardium [35]. D1R has been localized primarily in the epicardial layer of the human heart, while other DARs are found in the atrial and ventricular walls (**Figure 6.5**).

Endothelial cells line the inside of every blood vessel, forming a one-cell-thick layer, the endothelium, which is also found on the inner walls of the heart chambers and lymphatic vessels. In large vessels such as veins and arteries, the endothelium forms the blood vessel wall along the much thicker layers of muscle cells and elastic fibers. In capillaries, on the other hand, the endothelium makes up the entirety of the blood vessel wall. Using immunofluorescence and Western blotting, a recent paper [36] reported expression of D2R in a human vascular endothelial cell (HUVEC)-C endothelial cell line.

The kidneys provide effective mechanisms for the regulation of blood pressure by managing blood volume through the activity of the renin-angiotensin system. Several studies have examined the distribution of DARs in nephrons, local blood vessels, and other kidney substructures [28,37]. As shown in **Figure 6.6**, all DAR subtypes are expressed in different locations within the kidney. Along the nephron, the glomeruli express D3R and D4R, the proximal tubules express all DARs except for D2R, whereas D1R, D3R, and D5R are expressed in the medullary, but not in cortical, portions of the thick ascending limb of Henle. The collecting ducts (cortical and medullary) express all DARs except for D2R, while the macula densa and juxtaglomerular cells express D1R and D3R. The adventitia of the renal arteries and pre-junctional nerves express D2R, D3R, and D4R, while the tunica media expresses D1R, D3R, and D5R.

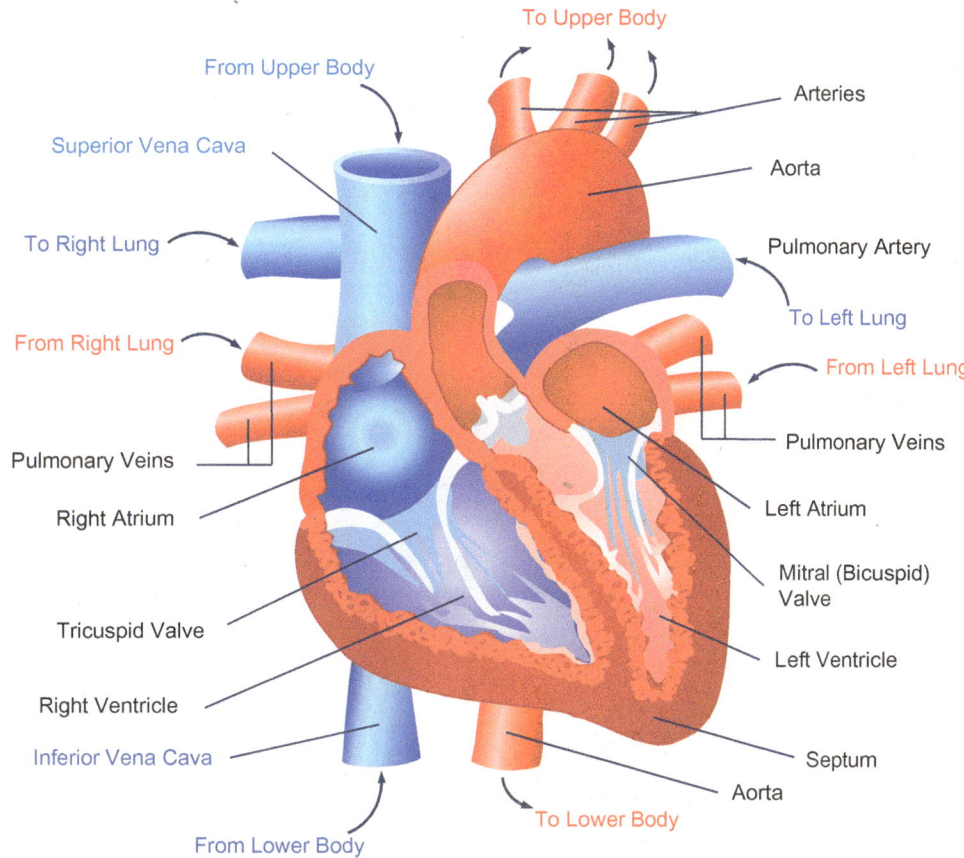

Figure 6.5 Structure of the human heart. Shown are the left and right ventricles, left and right atria, tricuspid and mitral valves. Major blood vessels connected to the heart include the aorta, pulmonary arteries, pulmonary veins, and superior and inferior vena cava. The coronary vessels are not shown. Various dopamine receptors (DARs) are expressed by all layers of the heart: endocardium, epicardium, and myocardium as well as in the major blood vessels and the coronary vessels. See text for details. (Redrawn and modified from: https://www.gettyimages.com/photos/stocktrek-images.)

Both the adrenal **cortex** and **medulla** secrete hormones that are involved in blood pressure regulation. Medullary NE and Epi have multiple actions on the cardiac muscle and blood vessels, while cortical aldosterone helps in the control of blood pressure by managing the balance of blood potassium, sodium, and pH. In addition, cortisol from the adrenal cortex works in conjunction with catecholamines to regulate the reaction to stress, glucose levels, and blood pressure. D2R in the human adrenal cortex was localized in the zona glomerulosa, which secretes aldosterone, and the zona reticularis, which secretes androgens [38]. Lower D2R expression was seen in the zona fasciculate, which produces glucocorticoids, and the medulla, which produces catecholamines. Another study reported that both D2R and D4R were more highly expressed in the normal human adrenal cortex than in aldosterone-producing adenomas [39]. The chromaffin cells of the rat adrenal medulla express D2R, which likely serves as autoreceptors, but do not express D1R.

6.4.2 The respiratory system

Accumulating evidence indicates that DA affects ventilation, bronchial diameter, pulmonary circulation, and lung water and mucus clearance and is also involved in the neuromodulation of chemosensory structures associated with respiration [40]. Through these complex mechanisms, DA can exert beneficial as well as unfavorable effects on respiration. The respiratory structures involved include airway, lungs, pulmonary blood vessels, and chemosensory organs.

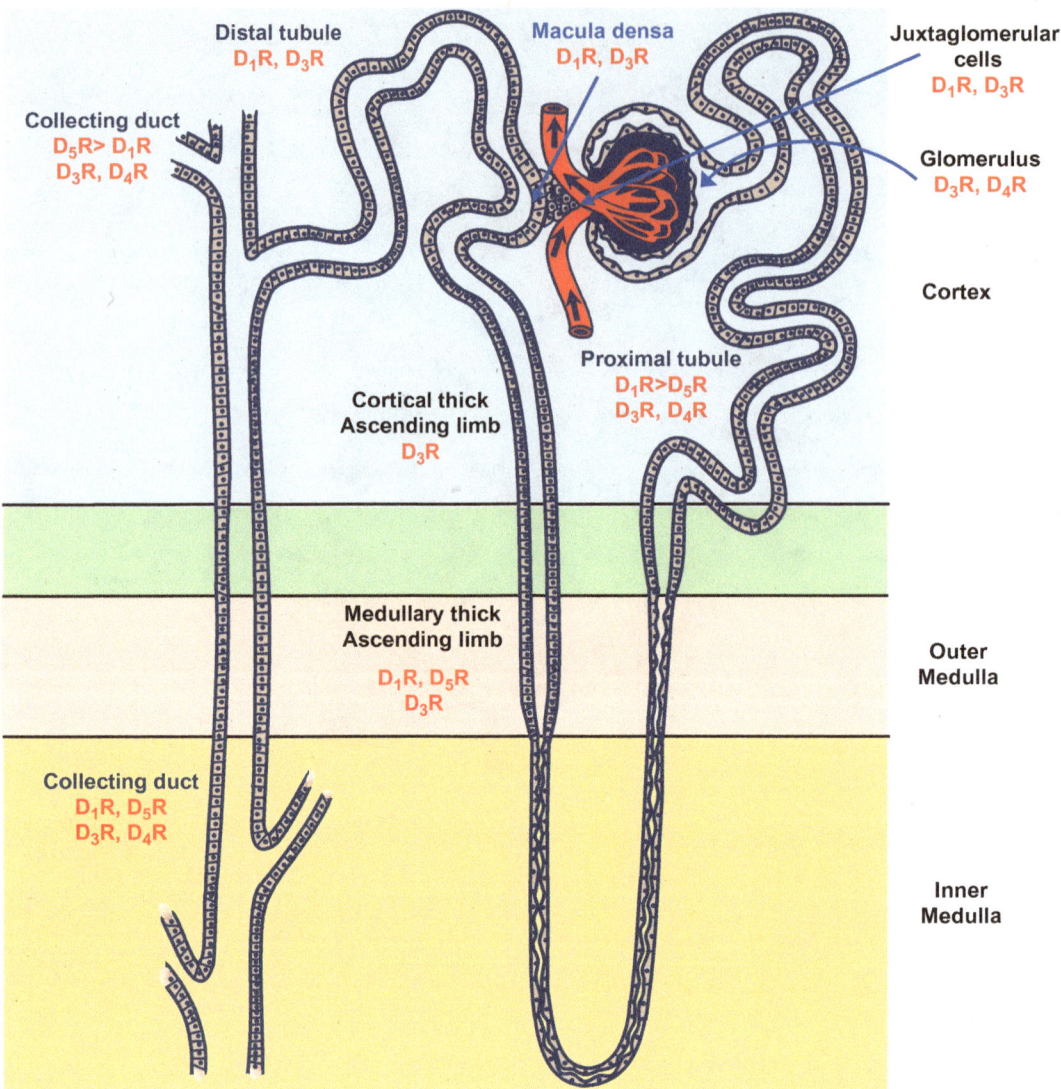

Figure 6.6 Distribution of specific dopamine receptors along the nephron. A nephron begins at the glomerulus and courses through the cortex, outer and inner medulla of the kidney toward the collecting duct. (Redrawn and modified from Zeng, C. et al., *Physiol. Genomics*, 19, 233–246, 2004.)

Expression of various DARs has been found in both the smooth muscle and the epithelial components of airway structures such as the pharynx, larynx, trachea, bronchi, and bronchioles [40,41]. The subcellular distribution of DAR subtypes in different layers of the pulmonary and bronchial arteries was also examined [42]. In large intrapulmonary artery branches, D1R was located within the tunica intima (the innermost layer which contains the endothelium) and the tunica media (which contains the smooth muscle cells), and it was also seen within the tunica media of medium-sized pulmonary artery branches [43]. On the other hand, D2R was seen within the tunica adventitia (the outmost layer made of connective tissue) of extrapulmonary arteries, and within large- and medium-sized intrapulmonary artery branches.

Both D1R and D2R are expressed in pulmonary alveolar cells [44]. Alveolar epithelial type 1 cells (AT1) are terminally differentiated squamous cells that cover 95% of the lung interior and are involved in solute transport. AT1 cells express both D1R and D2R, but only D1R is coupled to the amiloride-sensitive Na^+ channel activity that controls the level of alveolar fluid. Using whole-body PET scanning with a D1R radioactive analog has revealed high radiation absorbance in the lungs [45].

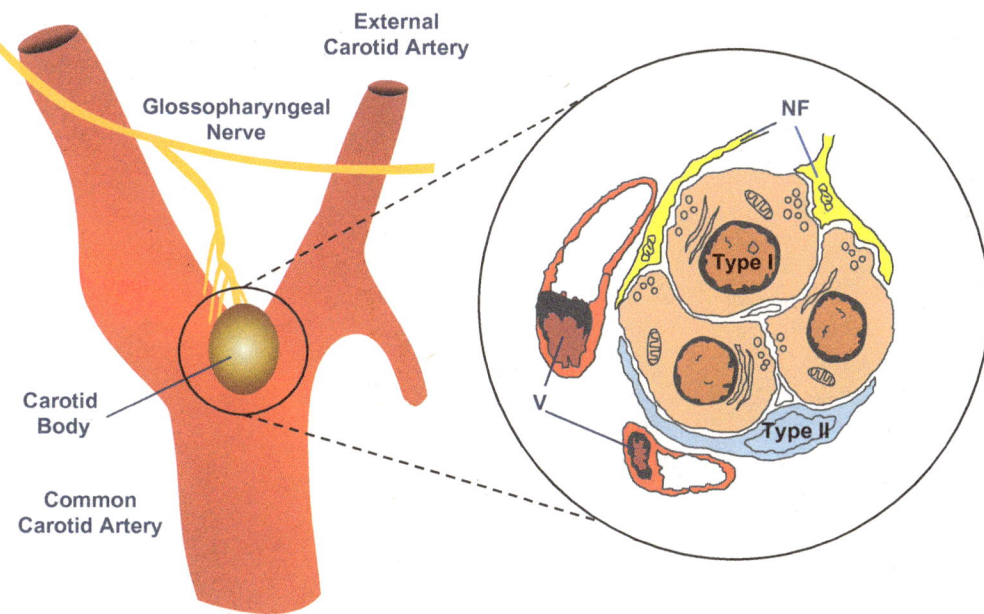

Figure 6.7 Diagram of the carotid body. The carotid body is located at the bifurcation of the carotid artery and is innervated by the glossopharyngeal nerve. Cross section of the carotid body shows the oxygen-sensing type I glomus cells, type II glial cells, afferent sensory nerve fibers (NF), and the surrounding blood vessels (V). (Redrawn and modified from Lopez-Barneo, J. et al., *J. Neural. Transm. (Vienna)*, 116, 975–982, 2009.)

The **carotid body** is a small cluster of chemoreceptors and supporting cells located near the bifurcation of the carotid artery (**Figure 6.7**). The carotid body detects changes in the composition of arterial blood (i.e., the partial pressure of arterial oxygen, carbon dioxide, and pH), and conveys this information to the respiratory regulatory centers in the brain. The carotid body is composed of two cell types: neurosecretory, oxygen-sensitive type I glomus cells, and glial type II cells. Glomus cells are closely associated with the afferent ending of the carotid sinus nerve, whose cell bodies are located in the petrosal ganglion [46]. D2Rs were identified in the afferent nerve terminals and also in the parenchyma and stroma of the carotid body [13,47]. In another study, D2Rs were co-localized with TH in type I glomus cells and their expression was found to increase during prenatal rat development, consistent with the maturation of hypoxic chemosensitivity [48]. Expression of any other DAR has not been reported.

6.5 DOPAMINE RECEPTORS IN THE DIGESTIVE SYSTEM AND IN ORGANS THAT REGULATE METABOLISM

Efficient conversion of food into energy and basic nutrients is contingent upon having a coordinated cooperation between a duct system that transports the food and accessory organs and glands that facilitate food processing. The primary organs that regulate metabolism include the brain (primarily the hypothalamus), endocrine pancreas, liver, and adipose tissue, with important contributions to metabolic regulation coming from the thyroids, pituitary, adrenals, kidneys, and skeletal muscle (**Figure 6.8**).

6.5.1 The digestive system

Food passes from the oral cavity into the GI tract, which is composed of the esophagus, stomach, small intestines, and large intestines. The digestive process is facilitated by the secretory activity of organs associated with digestion that include the salivary glands, pancreas, liver, and gallbladder.

Figure 6.8 The central role of the hypothalamus in the regulation of metabolic homeostasis. Within the hypothalamus, the major regions involved in metabolism are the lateral hypothalamic area (LHA), periventricular area (PVH), dorsal medial hypothalamus (DMH), ventral medial hypothalamus (VMH), and arcuate nucleus (ARC). The hormones and neuropeptide involved include: melanin concentrating hormone (MCH), orexins (OXs), agouti-related peptide (AgRP), neuropeptide Y (NPY), cocaine and amphetamine-related transcript (CART), proopiomelanocortin (POMC), and α-melanocyte stimulating hormone (αMSH). WAT: white adipose tissue; BAT: brown adipose tissue. (Redrawn and modified from Lopez, M., et al., *Trends Mol Med.*, 19, 418–427, 2013.)

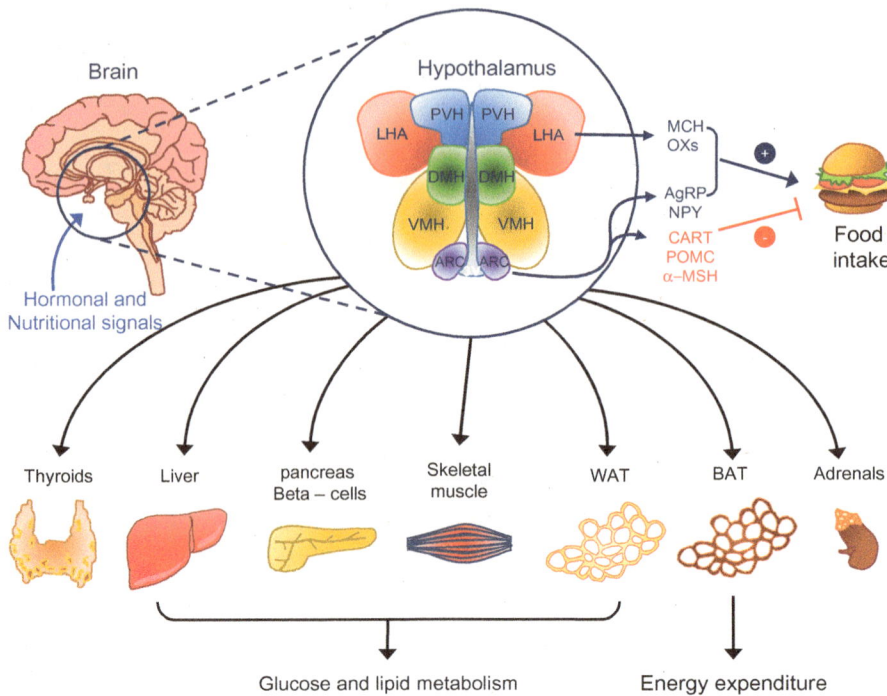

Immunohistochemistry and Western blotting of the rat salivary glands has revealed expression of DAR primarily in striated and excretory ducts [9]. The acinar cells of the submandibular gland have strong immunoreactivity for D2R, while cells of the convoluted granular tubules were negative for both D1-like and D2-like receptors. Acinar cells of the parotid glands have the highest expression of both D1R and D2R among other salivary glands. Because the localization of DA and dopaminergic markers did not correlate with neuron-specific enolase, the authors proposed that the catecholamine stores and dopaminergic markers are independent from glandular innervation.

An early study with human esophageal specimens employed DAR agonists and antagonists and cAMP accumulation as biochemical markers for confirming the presence of DAR [49]. In a later study [50], both D1R and D2R were detected in the lower esophageal sphincter but not in the esophageal body, supporting data by others that local DA was involved in the regulation of esophageal motility.

Using *in situ* hybridization, all five DAR subtypes were detected in both the stomach and the duodenum of rats, with D4R and D5R showing the highest expression [51]. In mice, immunoreactive D1R, D2R, and D5R were detected along the entire lower GI tract, from the stomach to the distal colon. An older study using radioactive DA found a single class of saturable DA binding sites in gastric and duodenal mucosa from human volunteers [52].

The **enteric nervous system (ENS)** of the gut is a subdivision of the autonomic nervous system (ANS). As illustrated in **Figure 6.9**, the ENS is composed of two ganglion-like nerve plexuses—the myenteric and submucosal—which are integrated within the gut wall [53]. Although the ENS is independent of the central nervous system (CNS), some of its properties resemble those of the CNS rather than those of the peripheral autonomic ganglia. The ENS, which has been called "the little brain within the gut," controls GI motility, mucosal blood flow, ion and water transport, and resorption. D2R expression, co-localized with TH, was detected in some myenteric neurons within the proximal and distal section of the rat colon [54]. Another study found that subsets of submucosal and myenteric neurons were immunoreactive for D1R, D2R, and D3R [55].

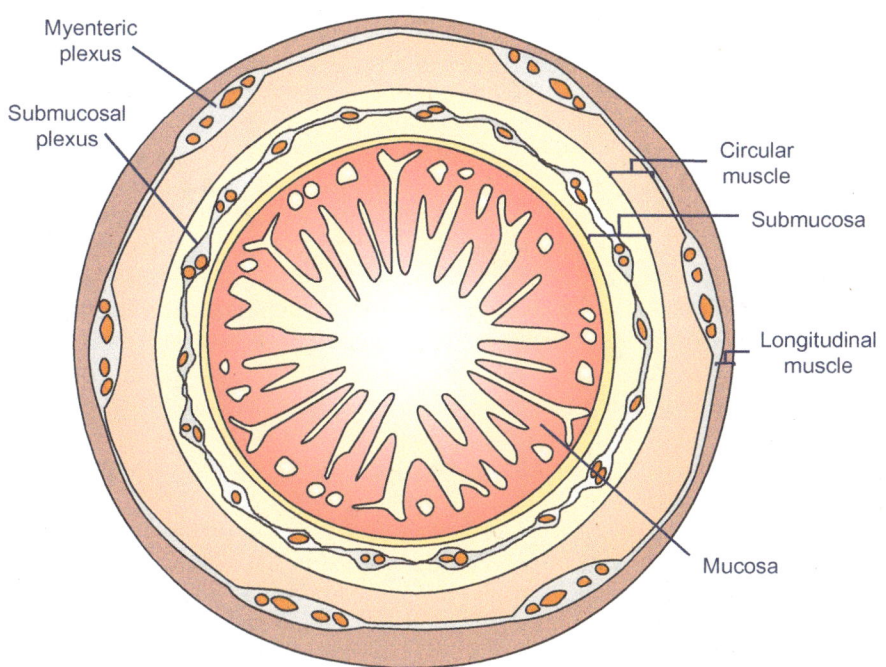

Myenteric plexus

Submucosal plexus

Circular muscle

Submucosa

Longitudinal muscle

Mucosa

Figure 6.9 Transverse section through the small intestine showing the structure of the enteric nervous system. Enteric neurons are organized in two main ganglia: an outer myenteric plexus located between the longitudinal and circular muscle layers, and an inner submucosal plexus residing within the submucosa. (Redrawn and modified from Heanue, T.A. and Pachnis, V., *Nat. Rev. Neurosci.*, 8, 466–479, 2007.)

The **exocrine pancreas** releases pancreatic juice which is composed of digestive enzymes secreted from exocrine acinar cells, and bicarbonate secreted from the epithelial cells lining small pancreatic ducts. Both play essential roles in the conversion of ingested food into fuel. Nutrients absorbed from the small intestine are processed by the liver. Bile produced in the liver plays an important role in digesting fat. The gallbladder concentrates the bile, stores it, and releases it into the small intestines.

The presence of DAR in the above organs has been detected by complementary methods. Whole-body PET biodistribution, using the D1R-specific ligand 11C-NNC 112, showed its highest retention in the gallbladder, liver, lungs, kidneys, and wall of the urinary bladder [45]. When PET imaging was done using D2R antagonists, the highest retention was found in the gallbladder wall, small intestine, liver, and urinary bladder wall. These data revealed widespread distribution of D1R and D2R in many components of the digestive system. Additional studies, using DA agonists and antagonists, have identified binding sites in canine pancreatic acini [56], while D2R was identified by RT-PCR and microarrays in human pancreatic tumors [57,58]. Expression of D5R was higher in human liver tumors than in adjacent liver tissue, while D1R was lower in hepatocellular carcinoma than in normal tissues. Other DARs were low or undetectable. All five DAR subtypes were expressed at the mRNA and protein levels in human HepG2 hepatocarcinoma cells [59] and in the normal rat liver [60].

6.5.2 Metabolic regulation

DAR expression in the brain, hypothalamus, and the pituitary gland is covered in **Chapters 4** and **5**, while those in the liver, adrenals, and kidneys are covered in the sections above. Although there are reports that DA affects some parameters in isolated skeletal muscles, these actions were attributed to the action of DA on adrenergic receptors, with no unequivocal demonstration of the presence of DAR in skeletal muscles. There are, at present, also no published reports on DAR expression in the thyroid gland. Hence, this section deals only with DAR that are expressed in the endocrine pancreas and adipose tissue.

The **endocrine pancreas** is a key player in metabolic regulation by secreting the blood sugar–lowering hormone insulin and its opposite hormone glucagon. All five DARs were expressed in different endocrine cells of the human pancreas [61]. D1R was present in insulin-secreting beta cells, and D2R was expressed in glucagon-secreting alpha cells, somatostatin-secreting delta cells, and pancreatic polypeptide producing cells. D4R was expressed in both beta and polypeptide cells, whereas D5R was expressed only in delta cells. Based on the morphological distribution of DAR, the authors suggested that specific DARs should be considered novel targets for clinical treatment of diabetes.

Adipose tissue is an active metabolic organ with a triple function: (1) homeostasis of lipid storage and utilization, (2) release of adipokines and cytokines which regulate the metabolic activity of numerous organs, and (3) thermogenesis. In 2011, we provided the first comprehensive analysis of DAR expression and functions in human adipose tissue [26]. DAR expression was analyzed by RT-PCR and Western blotting in adipose tissue explants, primary adipocytes, and two human adipocyte cell lines, LS14 and SW872. D1R was the most highly expressed, followed by D4R > D2R = D5R, while D3R was undetectable. When adipose tissue was separated into mature adipocytes and stromo-vascular cells (which contain preadipocytes, endothelial, and immune cells), D1R expression was fivefold higher in mature adipocytes, D4R was similar, while D2R was lower. During adipogenesis, D2R expression rose on day 3, while D4R was decreased and remained low throughout cell differentiation. D1R expression showed a biphasic response: a drop on day 3, followed by a rise on day 6. **Chapter 8** provides more detail on these findings.

Brown adipose tissue (BAT) is especially abundant in newborns and in hibernating mammals and is also present and metabolically active in adult humans. The main function of BAT is thermoregulation, accomplished through a non-shivering thermogenesis. Using DA agonists and antagonists, a 1992 report described the detection of a DAR of unspecified identity in rat BAT [62]. More recently, expression of D1R and D2R in SV-40 T immortalized murine brown adipocytes was confirmed using Western blotting [63].

6.6 DOPAMINE RECEPTORS IN HEMATOPOIETIC AND IMMUNE SYSTEMS

The synthesis and development of blood cells occur by the process of hematopoiesis. It begins during embryonic development and continues throughout adulthood to produce and replenish the blood system. Cellular blood components are derived from hematopoietic stem cells, residing mainly in the bone marrow. The immune system consists of lymphoid organs divided into the primary and secondary immune systems as well as the myeloid and lymphoid cells. The primary lymphoid organs are the bone marrow and thymus. These are the sites where hematopoiesis occurs and immature lymphocytes grow, develop, and differentiate. Secondary, or peripheral, lymphoid organs consist of the spleen and lymph node. These play important roles in antigen presentation and the adaptive immune response initiation.

6.6.1 Dopamine receptors in lymphoid organs

The **thymus gland** is the main organ of the lymphatic system. Its primary function is to promote the development of T-lymphocytes. Once mature, these cells leave the thymus and are transported via the circulation to the lymph nodes and spleen. The **spleen** synthesizes antibodies in its white pulp, while its red pulp serves as a major reservoir of monocytes, the progenitors of dendritic cells and macrophages.

Both thymus and spleen have a self-contained dopaminergic system, including locally produced DA, VMAT, and all five DAR subtypes [64]. Immunohistochemistry has revealed individual DAR immunoreactivity primarily in the thymic cortical–medulla transitional zone, and to a lesser extent in the medulla, but not in the cortex. In the spleen, immunoreactivity for VMAT-1, VMAT-2, and all five DARs was located primarily in the white pulp border and to a lesser extent in the white pulp. The authors concluded that DA, likely originating from immune cells and/or from the sympathetic neuroeffector plexus, is released into the lymphoid microenvironment and plays a role in the maturation and selection of lymphocytes and the activation of immune responses.

6.6.2 Dopamine receptors in peripheral blood lymphocytes

Peripheral blood lymphocytes (PBLs) are mature lymphocytes that circulate in the blood rather than being localized in organs such as the spleen and lymph nodes. The PBLs include T cells (which function in cell-mediated, cytotoxic adaptive immunity), B cells (active in the humoral, antibody-driven adaptive immunity), and natural killer (NK) cells (which function in cell-mediated, cytotoxic innate immunity).

A 2011 review [65] covers the DA system in PBLs. Using methods such as ligand binding, immunocytochemistry, RT-PCR, Western blotting, and flow cytometry, D2R, D3R, D4R, and D5R were detected in human PBLs, while D1R was undetectable. Significant decreases in most of the DARs were seen in patients with neurodegenerative diseases such as Parkinson's, Wilson's, and Alzheimer's diseases and amyotrophic lateral sclerosis, while increases in D3R, D4R, and D5R were seen in Tourette's syndrome and in migraine. The authors suggested that analysis of DARs and/or DATs in PBLs could serve as a diagnostic tool for identifying derangement of DA transmission in neuropsychiatric diseases as well as for monitoring the effects of pharmacological intervention.

As summarized in two reviews [66,67] and illustrated in **Figure 6.10**, other hematopoietic cells, including leukocytes, macrophages, platelets, mast cells, and NK cells, either make their own DA or respond to DA and its specific agonists, suggesting the presence of active DARs, albeit the exact identity of the receptors has not yet been determined in all cell types. More detail on DA functions in the hematopoietic and immune systems is provided in **Chapter 9**.

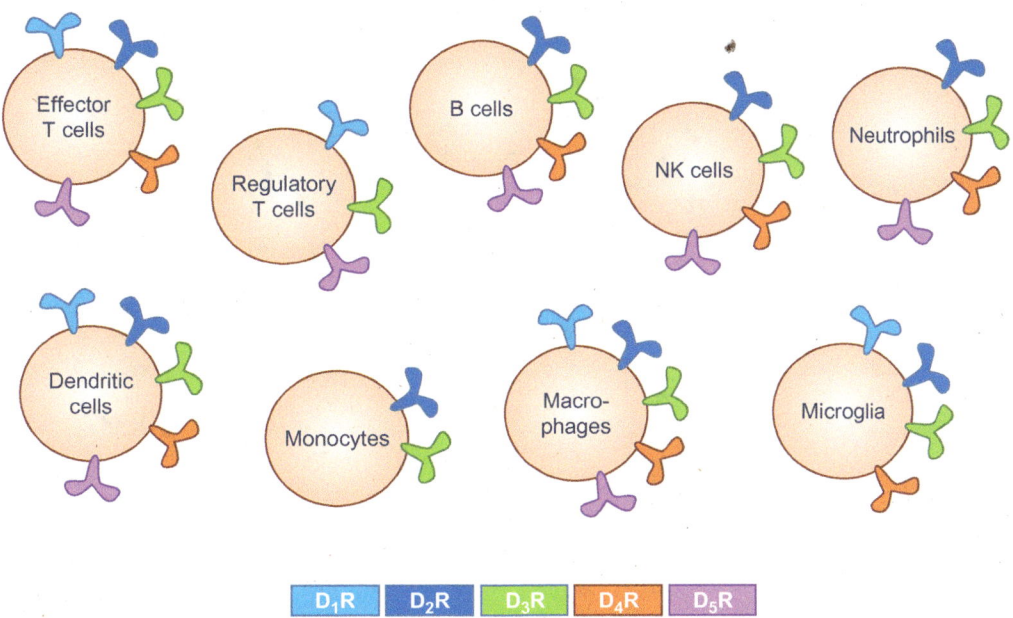

Figure 6.10 Expression of different dopamine receptor (DAR) subtypes in various immune cells. NK: natural killer. (Redrawn and modified from Levite, M. Dopamine in the immune system: Dopamine receptors in immune cells, potent effects, endogenous production and involvement in immune and neuropsychiatric diseases, in *Nerve Driven Immunity*, Levine M (ed), Springer, Vienna, 1–25, 2019.)

6.7 DOPAMINE RECEPTORS IN SKIN AND BONES

6.7.1 Skin structure

The skin and its accessory structures—hair, nails, and a variety of glands—make up the integumentary system and provide an overall protection to the underlying structures. The skin is composed of multiple layers of ectodermal tissues held together by connective tissue. As shown in **Figure 6.11**, the uppermost layer is the **epidermis**, organized as a stratified squamous epithelium composed of proliferating, basal, and terminally differentiated keratinocytes. The keratinocytes are the major cells that inhabit the epidermis, constituting 95% of its cellular content. Other cells within the epidermis include the mechanoreceptor Merkel cells, the pigment producing melanocytes, and the antigen-presenting dendritic Langerhans cells. The **dermis**, which lies under the epidermis, contains collagen fibers, blood vessels, nerve endings, sweat glands, sebaceous glands, fibroblasts, sense organs, smooth muscles, and hair follicles. The innermost layer is the **hypodermis**, consisting primarily of fat tissue (adipocytes), which functions as insulation and as an energy source.

6.7.2 Dopamine receptors in different skin components

An indirect evidence for the importance of the DA-DAR axis in the control of skin functions comes from non-motor dysfunctions in patients with Parkinson's disease. Skin-related symptoms include sialorrhea (excessive salivation), seborrhea (red, itchy skin rash), hyperhidrosis (excessive sweating), sensory dysfunction/denervation of the skin, and sympathetic changes. Yet, it is unclear whether some or all the above changes are due to central and/or local deficiency of the dopaminergic system.

Several DAR subtypes have been detected in various layers and cells within the skin. For example, in skin samples from the nude mouse, D2R immunoreactivity was detected in the basal layer of the epidermis, D4R was seen in the uppermost part of the epidermis, while D3R was undetectable [68]. In samples of human plantar skin (which covers the soles of the feet),

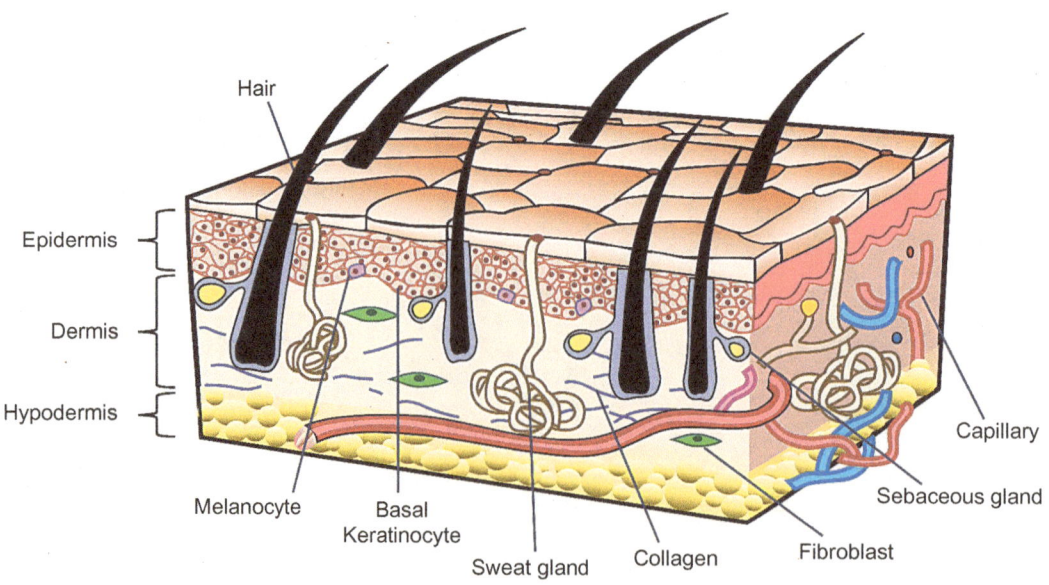

Figure 6.11 Structure of the three layers of the human skin and their components. The outermost epidermis contains keratinocytes and melanocytes. The major cellular components of the dermis are the fibroblasts. The hypodermis consists primarily of adipose tissue. The rich vascular system provides nutrients to all cells. Skin appendages such as hair follicles and sebaceous and sweat glands are located throughout the dermis and epidermis. Redrawn and modified from several sources.

D1R expression was abundant in the dermis layer, moderate in the lower epidermal layers, and low in subcutaneous tissue. On the other hand, D2R expression was prominent in the subcutaneous tissue near blood vessels [69]. The authors concluded that the specific distribution of D1 and D2 receptors in human plantar skin is related to the functions of a particular zone of the human skin that supports the weight of all the body.

Melanocytes are melanin-producing, neural crest-derived cells that are located at the bottom layer of the epidermis. Melanin is a dark pigment primarily responsible for skin color. Melanin is synthesized from tyrosine or L-Dopa by the enzyme tyrosinase, thus sharing a biosynthetic pathway with DA. Once synthesized, melanin is contained within special organelles, the melanosomes, which are transported to nearby keratinocytes to induce pigmentation that serves as a protection against ultraviolet radiation. Melanocytes also have a role in the immune system.

Vitiligo is an acquired skin disorder where melanocytes are selectively destroyed, causing white patches on the skin. A possible association between vitiligo and genes in the DA pathway was examined in punch biopsies from involved and uninvolved skin of vitiligo patients and from non-sun-exposed skin of control subjects [70]. As judged by both mRNA expression and protein levels, D1R and D5R were significantly higher in vitiligo skin compared to controls. The authors concluded that the DA pathway influences melanogenesis either directly or through the melanocortin pathway.

Using Western blotting and confocal microscopy, D1Rs were detected in adult murine dermal fibroblasts collected from normal and diabetic mice and in dermal fibroblasts collected from human subjects [71]. In microdissected human scalp hair follicles from women, expression of D1R was detected by RT-PCR, without a further confirmation by immunocytochemistry or Western blotting. The D1R transcripts were present in both human skin and hair follicles and persisted in organ culture [72].

6.7.3 Bone structure

Bone is a rigid structure that constitutes part of the vertebrate skeleton. Bones fulfill multiple functions, including support of the body to enable motility, protection for various organs, production of red and white blood cells (in the bone marrow), and storage of minerals. Bone is composed of three primary cell types: **osteoblasts**, **osteocytes**, and **osteoclasts**. Osteoblasts are bone-forming cells that generate a protein mixture known as osteoid, composed of type I collagen, which upon mineralization becomes bone. They also manufacture hormones such as prostaglandins and robustly produce alkaline phosphatase, which is involved in bone mineralization and the formation of matrix proteins. Osteocytes are mature bone cells, which originate from the osteoblasts and migrate into the bone matrix in spaces called lacunae. Osteocytes have many processes that contact osteoblasts and other osteocytes, forming a communication network. Their functions include the formation of bone, the maintenance of matrix, and calcium homeostasis. Osteoclasts are the cells responsible for bone resorption and remodeling. They are large, multinucleated cells located on bone surfaces in resorption pits. Because osteoclasts are derived from a monocyte stem-cell lineage, they are equipped with phagocytic-like mechanisms similar to circulating macrophages.

6.7.4 Dopamine receptors in bone cells

All five DARs were expressed in primary osteoclasts from female mice, and all but D3R were expressed in primary osteoblasts [73]. However, MC3T3-E1 cells, a pre-osteoblast cell line, had a robust expression of only D1R and D4R. Treatment of these cells with DA during differentiation caused suppression of osteoblast mineralization and osteoblast marker gene expression. These data lead the authors to suggest that the effects of atypical antipsychotics such as risperidone, which cause reduced bone mineral density and increased fracture risk, could be due, in part, to direct effects of the drugs on bone cells such as osteoclasts via their DARs. This notion is supported by another report that D2R agonists suppressed LPS-induced osteoclast formation in murine bone marrow culture *ex vivo*. This indicates that the dopaminergic signaling plays an important role in bone homeostasis via direct effects on osteoclast differentiation and further suggests that the clinical use of neuroleptics is likely to affect bone mass [74]. On the other hand, opposite effects by DA on MC3T3-E1 cells has been reported in another study, confirming expression of all five DARs in these cells [75]. In this study, DA treatment increased both cell proliferation and osteogenic mineralization. The discrepancy between the two studies could be explained by the difference in abundance of D1R-like vs. D2R-like in their cell cultures.

6.8 DOPAMINE RECEPTORS IN MALE AND FEMALE REPRODUCTIVE SYSTEMS

6.8.1 Overall features of the reproductive system

The reproductive axis in both genders comprises the hypothalamus, pituitary, gonads, and the reproductive tract, all of which are functionally connected via feed-forward and feed-back interactions between central and peripheral hormones. Expression of DARs in the hypothalamus and pituitary gland is covered in **Chapters 4** and **5**. To recapitulate this information, reproduction in both males and females is centrally controlled by gonadotropin-releasing hormone (GnRH)–producing neurosecretory neurons with cell bodies in the preoptic area of the hypothalamus and axon terminals in the median eminence. GnRH is released from these terminals in a pulsatile manner into the hypophysial portal blood and is delivered to the anterior pituitary. Within the pituitary, GnRH binds to GnRH receptors on the gonadotrophs and regulates the release of the gonadotropins, luteinizing hormone (LH) and follicle-stimulating hormone (FSH). Both hormones differentially promote sexual differentiation, affect secondary sex characteristics, gonadal development, gametogenesis, and regulate the fertility in both sexes. In addition to the gonadotropins, prolactin (PRL) exerts diverse actions in females during the reproductive cycle, pregnancy, and lactation, and its pathological overproduction interferes with many reproductive processes in both genders. Below we focus on DAR expression in the gonads and the reproductive tracts of both males and females.

6.8.2 The male lower reproductive system

The male lower reproductive system includes the testes (gonads), epididymis, vas deferens, ejaculatory duct, urethra, seminal vesicles, prostate, bulbourethral glands, and penis. Different DAR subtypes are expressed in testes from several species. Within the seminiferous tubules of the rat testis, D2R is expressed in all pre- and post-meiotic germ cells, with the highest expression in spermatogonia, the undifferentiated male germ cells [76]. The same investigators also detected D2R protein in isolated spermatogenic cells and spermatozoa from rats, mice, bulls, and humans. Confocal microscopy of mature spermatozoa has showed that D2R was primarily localized in the flagellum, but also in the acrosome of the sperm head, except in human spermatozoa [76]. As illustrated in **Figure 6.12**, D2R was localized in the acrosome, mid piece, and flagellum in boar sperm and was activated during *in vitro* capacitation [77]. There are no published data on expression of DAR other than D2R in the testis.

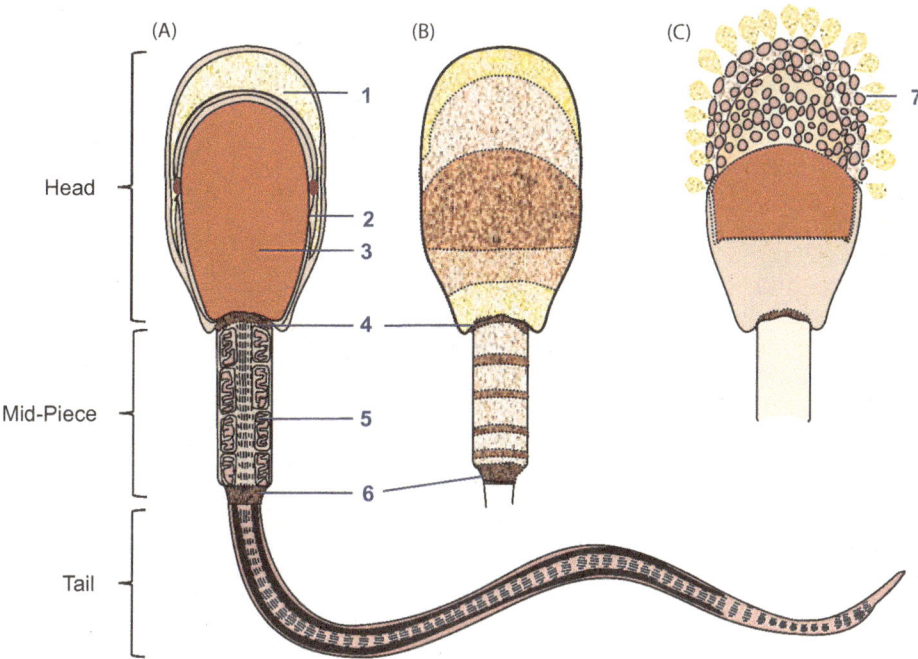

Figure 6.12 Schematic representation of the boar sperm cell. Panel A: Sectional view of a sperm cell. Solid lines represent membrane bilayers: 1, acrosome; 2, nuclear envelope; 3, nucleus; 4, posterior ring and neck; 5, mitochondria; 6, annular ring; Panel B: Surface view of the sperm head and midpiece with the subdomains. Panel C: The acrosome reaction. 7, mixed vesicles that are formed during the acrosome reaction via multiple fusions between the plasma membrane and the outer acrosomal membrane. (Redrawn and modified from Gadella, B.M. et al., *Int J Dev Biol.*, 52, 473–480, 2008.)

There is only scant information on DAR expression in the male extra-gonadal reproductive tract. An early study used a combination of radio-receptor binding and autoradiography to examine the localization of the DA agonist 3H-dihydroergotoxine in rat male sex organs. The drug was bound to smooth muscle cells of the vas deferens, seminal vesicles, and prostate gland, as well as to glandular tissue of the seminal vesicles and prostate gland [78]. These findings were supported by a later study showing the presence of D1R and D2R transcripts in human and rat seminal vesicles [79]. Both D2R [80] and D4R [81] were expressed in the guinea pig vas deferens, while D2R was detected in vas deferens from mice but not rats [82]. The human corpus cavernosum penis is a highly vascularized tissue which fills with blood during erection. D1R was twofold more abundant than D2R in this tissue, and both were primarily localized in the smooth muscle component of this structure [83].

6.8.3 The female reproductive system

The female lower reproductive system is composed of the ovaries, oviducts (Fallopian tubes), uterus, cervix, and vagina. The placenta, which accommodates the developing fetus, and the mammary glands, which provide nutrients to the newborn, are integral components of the female reproductive system. As illustrated in **Figure 6.13**, immunoreactive D1R in the human ovary was localized in granulosa cells of large antral follicles and in the corpus luteum [84]. Another study found that isolated human granulosa cells expressed all DARs except for D3R [85]. In the rat ovary, stronger D1R and D2R signals were seen in the thecal and interstitial cells of the corpus luteum than in granulosa cells [85]. In the equine ovary, D2R was high in the corpus luteum and ovarian cortex, low in granulosa/theca, and undetectable in ovarian medulla.

D1R expression was uniformly distributed throughout the equine corpus luteum and was undetectable elsewhere [86]. In pseudopregnant rabbits, DA, D1R, and D3R were expressed in luteal cells and exerted a dual

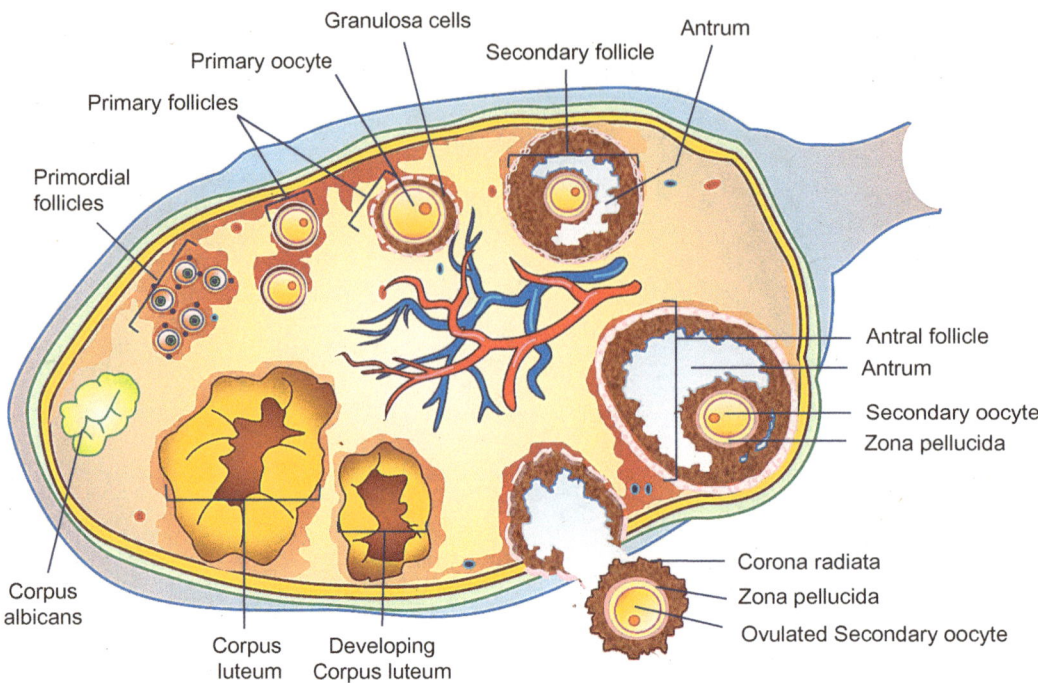

Figure 6.13 Schematic presentation of the human ovary. Shown are follicles and enclosed oocytes at different stages of development from primordial to antral follicles. Following ovulation, the oocyte is released into the body cavity and the empty follicle is transformed into a corpus luteum. (Redrawn and modified from several sources.)

modulatory function in the lifespan of CL (i.e., the DA/D1R was luteotropic), while the D3R was luteolytic [87]. Collectively, the distribution of DAR is both cell-type and receptor-type selective in different species. Although there are several reports on expression of DAR in *Xenopus* oocytes, there are no published records on their expression in human oocytes.

There is a dearth of published data on DAR expression in various components of the lower female reproductive tract. A single report described a moderate expression of the long and short D2R isoforms in pig oviducts and uterus [88]. In the human placenta, D2R expression was seen in the cytotrophoblasts, syncytial trophoblasts, vascular endothelial cells, and fibroblasts in the chorionic villi [89]. In the rat placenta, expression of D1R and D2R was evident in the spongiotrophoblast and giant cells of the junctional zone [90]. There are no reports on DAR expression in mammary glands of any species, with their expression seen in human breast tumors, but not in adjacent normal breast tissue [91], as discussed in **Chapter 12**.

6.9 SYNOPSIS

The homeostasis of central and peripheral DA differs in several respects. First, there is no cross transfer between the brain and DA in the systemic circulation, with the latter originating from both neuronal and nonneuronal tissues, as well as from some food items. Second, although the biosynthetic pathways for the two DA systems are similar, they differ primarily in the mechanisms that govern the release, reuptake, and metabolic inactivation of DA. Third, a unique aspect of peripheral DA is its inactivation by sulfoconjugation, which prolongs its serum half-life but renders it biologically inactive. Tissues that express ARSA can respond to circulating DA-S by converting it to free DA. The ability of normal and malignant cells to respond to DA-S is paramount to the pathophysiology of peripheral DA.

Outside the CNS, DA functions primarily as a local paracrine messenger. In blood vessels, it inhibits NE release and acts as a vasodilator (at physiological concentrations). In the kidneys, it increases sodium excretion and urine output. In the pancreas, it reduces insulin production. In the digestive system, it reduces GI motility and protects intestinal mucosa. In the immune system, it reduces the activity of lymphocytes. With the exception of blood vessels, DA in each of these peripheral systems is synthesized locally and exerts its effects near the cells that release it. The peripheral actions of DA can be executed via variable levels of expression of the five DARs in various organs and systems that include cardiopulmonary, digestive, metabolic hematopoietic, immune, dermal, and reproductive functions.

REFERENCES

1. Goldstein DS, Eisenhofer G, Kopin IJ. Sources and significance of plasma levels of catechols and their metabolites in humans. *J Pharmacol Exp Ther.* 2003;305(3):800–811.

2. Eisenhofer G, et al. Substantial production of dopamine in the human gastrointestinal tract. *J Clin Endocrinol Metab.* 1997;82(11):3864–3871.

3. Goldstein DS, et al. Sources and physiological significance of plasma dopamine sulfate. *J Clin Endocrinol Metab.* 1999;84(7):2523–2531.

4. Rubi B, Maechler P. Minireview: New roles for peripheral dopamine on metabolic control and tumor growth: Let's seek the balance. *Endocrinology.* 2010;151(12):5570–5581.

5. Naila A, Flint S, Fletcher G, Bremer P, Meerdink G. Control of biogenic amines in food—Existing and emerging approaches. *J Food Sci.* 2010;75(7):R139–R150.

6. Jaubert A, Drutel G, Leste-Lasserre T, Ichas F, Bresson-Bepoldin L. Tyrosine hydroxylase and dopamine transporter expression in lactotrophs from postlactating rats: Involvement in dopamine-induced apoptosis. *Endocrinology.* 2007;148(6):2698–2707.

7. Arbogast LA, Ben-Jonathan N. Tyrosine hydroxylase in the stalk-median eminence and posterior pituitary is inactivated only during the plateau phase of the preovulatory prolactin surge. *Endocrinology.* 1989;125(2):667–674.

8. Zhang MZ, et al. Intrarenal dopaminergic system regulates renin expression. *Hypertension.* 2009;53(3):564–570.

9. Tomassoni D, et al. Dopamine, vesicular transporters, and dopamine receptor expression in rat major salivary glands. *Am J Physiol Regul Integr Comp Physiol.* 2015;309(5):R585–R593.

10. Qiu YH, Peng YP, Jiang JM, Wang JJ. Expression of tyrosine hydroxylase in lymphocytes and effect of endogenous catecholamines on lymphocyte function. *Neuroimmunomodulation.* 2004;11(2):75–83.

11. Ronnberg E, Calounova G, Pejler G. Mast cells express tyrosine hydroxylase and store dopamine in a serglycin-dependent manner. *Biol Chem.* 2012;393(1–2):107–112.

12. Huang Y, Chang C, Zhang J, Gao X. Bone marrow-derived mesenchymal stem cells increase dopamine synthesis in the injured striatum. *Neural Regen Res.* 2012;7(34):2653–2662.

13. Iturriaga R, Alcayaga J, Gonzalez C. Neurotransmitters in carotid body function: The case of dopamine—Invited article. *Adv Exp Med Biol.* 2009;648:137–143.

14. Sivamani RK, et al. Acute wounding alters the beta2-adrenergic signaling and catecholamine synthetic pathways in keratinocytes. *J Invest Dermatol.* 2014;134(8):2258–2266.

15. Marles LK, Peters EM, Tobin DJ, Hibberts NA, Schallreuter KU. Tyrosine hydroxylase isoenzyme I is present in human melanosomes: A possible novel function in pigmentation. *Exp Dermatol.* 2003;12(1):61–70.

16. Vargovic P, et al. Adipocytes as a new source of catecholamine production. *FEBS Lett.* 2011;585(14):2279–2284.

17. Mayerhofer A, Frungieri MB, Bulling A, Fritz S. Sources and function of neuronal signalling molecules in the gonads. *Medicina (B Aires).* 1999;59(5Pt 2):542–545.

18. Kawamura M, et al. Differential effects of chemical sympathectomy on expression and activity of tyrosine hydroxylase and levels of catecholamines and DOPA in peripheral tissues of rats. *Neurochem Res.* 1999;24(1):25–32.

19. Soares-da-Silva P, Vieira-Coelho MA, Serrao MP. Uptake of L-3,4-dihydroxyphenylalanine and dopamine formation in cultured renal epithelial cells. *Biochem Pharmacol.* 1997;54(9):1037–1046.

20. Saura J, et al. Localization of monoamine oxidases in human peripheral tissues. *Life Sci.* 1996;59(16):1341–1349.

21. Mannisto PT, Kaakkola S. Catechol-O-methyltransferase (COMT): Biochemistry, molecular biology, pharmacology, and clinical efficacy of the new selective COMT inhibitors. *Pharmacol Rev.* 1999;51(4):593–628.

22. Strott CA. Sulfonation and molecular action. *Endocr Rev.* 2002;23(5):703–732.

23. Dajani R, et al. X-ray crystal structure of human dopamine sulfotransferase, SULT1A3. Molecular modeling and quantitative structure-activity relationship analysis demonstrate a molecular basis for sulfotransferase substrate specificity. *J Biol Chem.* 1999;274(53):37862–37868.

24. Ghosh D. Human sulfatases: A structural perspective to catalysis. *Cell Mol Life Sci.* 2007;64(15):2013–2022.

25. Sevin C, Aubourg P, Cartier N. Enzyme, cell and gene-based therapies for metachromatic leukodystrophy. *J Inherit Metab Dis.* 2007;30(2):175–183.

26. Borcherding DC, et al. Dopamine receptors in human adipocytes: Expression and functions. *PLoS One.* 2011;6(9):e25537.

27. Laidler P, Kowalski D, Silberring J. Arylsulfatase A in serum from patients with cancer of various organs. *Clin Chim Acta.* 1991;204(1–3):69–77.

28. Zeng C, et al. Functional genomics of the dopaminergic system in hypertension. *Physiol Genomics.* 2004;19(3):233–246.

29. Zeng C, et al. Dysregulation of dopamine-dependent mechanisms as a determinant of hypertension: Studies in dopamine receptor knockout mice. *Am J Physiol Heart Circ Physiol.* 2008;294(2):H551–H569.

30. Yan H, et al. D2 dopamine receptor antagonist raclopride induces non-canonical autophagy in cardiac myocytes. *J Cell Biochem.* 2013;114(1):103–110.

31. Tonnarini G, et al. Dopamine receptor subtypes in the human coronary vessels of healthy subjects. *J Recept Signal Transduct Res.* 2011;31(1):33–38.

32. Zeng C, et al. Dopamine D1 receptor augmentation of D3 receptor action in rat aortic or mesenteric vascular smooth muscles. *Hypertension.* 2004;43(3):673–679.

33. Ricci A, Collier WL, Amenta F. Pharmacological characterization and autoradiographic localization of dopamine receptors in the portal vein. *J Auton Pharmacol.* 1994;14(1):61–68.

34. Matsumoto T, et al. Type 1A dopamine receptor expression in the heart is not altered in spontaneously hypertensive rats. *Am J Hypertens.* 2000;13(6Pt 1):673–677.

35. Cavallotti C, Mancone M, Bruzzone P, Sabbatini M, Mignini F. Dopamine receptor subtypes in the native human heart. *Heart Vessels.* 2010;25(5):432–437.

36. Niewiarowska-Sendo A, Kozik A, Guevara-Lora I. Influence of bradykinin B2 receptor and dopamine D2 receptor on the oxidative stress, inflammatory response, and apoptotic process in human endothelial cells. *PLoS One.* 2018;13(11):e0206443.

37. Choi MR, et al. Renal dopaminergic system: Pathophysiological implications and clinical perspectives. *World J Nephrol.* 2015;4(2):196–212.

38. Pivonello R, et al. Dopamine receptor expression and function in human normal adrenal gland and adrenal tumors. *J Clin Endocrinol Metab.* 2004;89(9):4493–4502.

39. Wu KD, et al. Expression and localization of human dopamine D2 and D4 receptor mRNA in the adrenal gland, aldosterone-producing adenoma, and pheochromocytoma. *J Clin Endocrinol Metab.* 2001;86(9):4460–4467.

40. Ciarka A, Vincent JL, van de Borne P. The effects of dopamine on the respiratory system: Friend or foe? *Pulm Pharmacol Ther.* 2007;20(6):607–615.

41. Matsuyama N, et al. The dopamine D1 receptor is expressed and induces CREB phosphorylation and MUC5AC expression in human airway epithelium. *Respir Res.* 2018;19(1):53.

42. Ricci A, Mignini F, Tomassoni D, Amenta F. Dopamine receptor subtypes in the human pulmonary arterial tree. *Auton Autacoid Pharmacol.* 2006;26(4):361–369.

43. Kobayashi Y, Ricci A, Amenta F, Cavallotti C, Hattori K. Localization of dopamine receptors in the rabbit lung vasculature. *J Vasc Res.* 1995;32(3):200–206.

44. Helms MN, et al. Dopamine activates amiloride-sensitive sodium channels in alveolar type I cells in lung slice preparations. *Am J Physiol Lung Cell Mol Physiol.* 2006;291(4):L610–L618.

45. Cropley VL, et al. Whole-body biodistribution and estimation of radiation-absorbed doses of the dopamine D1 receptor radioligand 11C-NNC 112 in humans. *J Nucl Med.* 2006;47(1):100–104.

46. Bairam A, Carroll JL. Neurotransmitters in carotid body development. *Respir Physiol Neurobiol.* 2005;149(1–3):217–232.

47. Czyzyk-Krzeska MF, Lawson EE, Millhorn DE. Expression of D2 dopamine receptor mRNA in the arterial chemoreceptor afferent pathway. *J Auton Nerv Syst.* 1992;41(1–2):31–39.

48. Gauda EB, Northington FJ, Linden J, Rosin DL. Differential expression of a(2a), A(1)-adenosine and D(2)-dopamine receptor genes in rat peripheral arterial chemoreceptors during postnatal development. *Brain Res.* 2000;872(1–2):1–10.

49. Missale G, et al. Evidence for the presence of both D-1 and D-2 dopamine receptors in human esophagus. *Life Sci.* 1990;47(5):447–455.

50. Sigala S, et al. Different neurotransmitter systems are involved in the development of esophageal achalasia. *Life Sci.* 1995;56(16):1311–1320.

51. Deng X, Zheng Z, Ye S. Localization and expression of dopamine receptors in stomach and duodenum in rats. *Zhonghua Yi Xue Za Zhi.* 1997;77(2):103–105.

52. Hernandez DE, Mason GA, Walker CH, Valenzuela JE. Dopamine receptors in human gastrointestinal mucosa. *Life Sci.* 1987;41(25):2717–2723.

53. Wedel T, et al. Organization of the enteric nervous system in the human colon demonstrated by wholemount immunohistochemistry with special reference to the submucous plexus. *Ann Anat.* 1999;181(4):327–337.

54. Colucci M, et al. Intestinal dysmotility and enteric neurochemical changes in a Parkinson's disease rat model. *Auton Neurosci.* 2012;169(2):77–86.

55. Li ZS, Schmauss C, Cuenca A, Ratcliffe E, Gershon MD. Physiological modulation of intestinal motility by enteric dopaminergic neurons and the D2 receptor: Analysis of dopamine receptor expression, location, development, and function in wild-type and knock-out mice. *J Neurosci.* 2006;26(10):2798–2807.

56. Vayssette J, Vaysse N, Ribet A. Dopamine receptors in pancreatic acinar cells from dog. *Eur J Pharmacol.* 1986;122(3):321–328.

57. O'Toole D, et al. The analysis of quantitative expression of somatostatin and dopamine receptors in gastro-entero-pancreatic tumours opens new therapeutic strategies. *Eur J Endocrinol.* 2006;155(6):849–857.

58. Jandaghi P, et al. Expression of DRD2 is increased in human pancreatic ductal adenocarcinoma and inhibitors slow tumor growth in mice. *Gastroenterology.* 2016;151(6):1218–1231.

59. Xu JJ, et al. Dopamine D1 receptor activation induces dehydroepiandrosterone sulfotransferase (SULT2A1) in HepG2 cells. *Acta Pharmacol Sin.* 2014;35(7):889–898.

60. Shao X, et al. Exogenous dopamine induces dehydroepiandrosterone sulfotransferase (rSULT2A1) in rat liver and changes the pharmacokinetic profile of moxifloxacin in rats. *Drug Metab Pharmacokinet.* 2015;30(1):97–104.

61. Zhang Y, et al. Pancreatic endocrine effects of dopamine receptors in human islet cells. *Pancreas.* 2015;44(6):925–929.

62. Nisoli E, Tonello C, Memo M, Carruba MO. Biochemical and functional identification of a novel dopamine receptor subtype in rat brown adipose tissue. Its role in modulating sympathetic stimulation-induced thermogenesis. *J Pharmacol Exp Ther.* 1992;263(2):823–829.

63. Kohlie R, et al. Dopamine directly increases mitochondrial mass and thermogenesis in brown adipocytes. *J Mol Endocrinol.* 2017;58(2):57–66.

64. Mignini F, Tomassoni D, Traini E, Amenta F. Dopamine, vesicular transporters and dopamine receptor expression and localization in rat thymus and spleen. *J Neuroimmunol.* 2009;206(1–2):5–13.

65. Buttarelli FR, Fanciulli A, Pellicano C, Pontieri FE. The dopaminergic system in peripheral blood lymphocytes: From physiology to pharmacology and potential applications to neuropsychiatric disorders. *Curr Neuropharmacol.* 2011;9(2):278–288.

66. Basu S, Dasgupta PS. Dopamine, a neurotransmitter, influences the immune system. *J Neuroimmunol.* 2000;102(2):113–124.

67. Levite M. Dopamine in the immune system: Dopamine receptors in immune cells, potent effects, endogenous production and involvement in immune and neuropsychiatric diseases. in *Nerve Driven Immunity,* Levine M (ed), Springer, Vienna, 2019;1–25.

68. Fuziwara S, Suzuki A, Inoue K, Denda M. Dopamine D2-like receptor agonists accelerate barrier repair and inhibit the epidermal hyperplasia induced by barrier disruption. *J Invest Dermatol.* 2005;125(4):783–789.

69. Tammaro A, et al. Dopaminergic receptors in the human skin. *J Biol Regul Homeost Agents.* 2012;26(4):789–795.

70. Reimann E, et al. Expression profile of genes associated with the dopamine pathway in vitiligo skin biopsies and blood sera. *Dermatology.* 2012;224(2):168–176.

71. Chakroborty D, et al. Activation of dopamine D1 receptors in dermal fibroblasts restores vascular endothelial growth factor-A production by these cells and subsequent angiogenesis in diabetic cutaneous wound tissues. *Am J Pathol.* 2016;186(9):2262–2270.

72. Langan EA, et al. Dopamine is a novel, direct inducer of catagen in human scalp hair follicles in vitro. *Br J Dermatol.* 2013;168(3):520–525.

73. Motyl KJ, et al. A novel role for dopamine signaling in the pathogenesis of bone loss from the atypical antipsychotic drug risperidone in female mice. *Bone.* 2017;103:168–176.

74. Hanami K, et al. Dopamine D2-like receptor signaling suppresses human osteoclastogenesis. *Bone.* 2013;56(1):1–8.

75. Lee DJ, et al. Dopaminergic effects on in vitro osteogenesis. *Bone Res.* 2015;3:15020.

76. Otth C, et al. Novel identification of peripheral dopaminergic D2 receptor in male germ cells. *J Cell Biochem.* 2007;100(1):141–150.

77. Ramirez AR, et al. The presence and function of dopamine type 2 receptors in boar sperm: A possible role for dopamine in viability, capacitation, and modulation of sperm motility. *Biol Reprod.* 2009;80(4):753–761.

78. Amenta D, Cavallotti C, De RM, Ferrante F, Amenta F. Autoradiographic localization of the dopaminergic agonist 3H-dihydroergotoxine within the male reproductive system. *Funct Neurol.* 1987;2(2):207–216.

79. Hyun JS, et al. Localization of peripheral dopamine D1 and D2 receptors in rat and human seminal vesicles. *J Androl.* 2002;23(1):114–120.

80. Morishita H, Katsuragi T. Existence of postsynaptic dopamine D2 receptor as an enhancer of contractile response in vas deferens. *Eur J Pharmacol.* 1998;344(2–3):223–229.

81. Morishita H, Shibata K, Sakata N, Kita S, Katsuragi T. A new approach to finding specific dopamine D4 receptor agonists. *Eur J Pharmacol.* 2005;516(2):145–150.

82. Martin PL, Kelly M, Cusack NJ. (–)-2-(N-propyl-N-2-thienylethylamino)-5-hydroxytetralin (N-0923), a selective D2 dopamine receptor agonist demonstrates the presence of D2 dopamine receptors in the mouse vas deferens but not in the rat vas deferens. *J Pharmacol Exp Ther.* 1993;267(3):1342–1348.

83. d'Emmanuele di Villa Bianca R, et al. Peripheral relaxant activity of apomorphine and of a D1 selective receptor agonist on human corpus cavernosum strips. *Int J Impot Res.* 2005;17(2):127–133.

84. Mayerhofer A, et al. Functional dopamine-1 receptors and DARPP-32 are expressed in human ovary and granulosa luteal cells in vitro. *J Clin Endocrinol Metab.* 1999;84(1):257–264.

85. Rey-Ares V, et al. Dopamine receptor repertoire of human granulosa cells. *Reprod Biol Endocrinol.* 2007;5:40–50.

86. King SS, et al. Dopamine receptors in equine ovarian tissues. *Domest Anim Endocrinol.* 2005;28(4):405–415.

87. Parillo F, et al. Evidence for a dopamine intrinsic direct role in the regulation of the ovary reproductive function: In vitro study on rabbit corpora lutea. *PLoS One.* 2014;9(8):e104797.

88. Xu HP, et al. Molecular cloning, expression and variation analyses of the dopamine D2 receptor gene in pig breeds in China. *Genet Mol Res.* 2011;10(4):3371–3384.

89. Kim HJ, Koh PO, Kang SS, Paik WY, Choi WS. The localization of dopamine D2 receptor mRNA in the human placenta and the anti-angiogenic effect of apomorphine in the chorioallantoic membrane. *Life Sci.* 2001;68(9):1031–1040.

90. Kim MO, et al. Colocalization of dopamine D1 and D2 receptor mRNAs in rat placenta. *Mol Cells.* 1997;7(6):710–714.

91. Borcherding DC, et al. Expression and therapeutic targeting of dopamine receptor-1 (D1R) in breast cancer. *Oncogene.* 2016;35(24):3103–3113.

92. Lopez-Barneo J, et al. The neurogenic niche in the carotid body and its applicability to antiparkinsonian cell therapy. *J Neural Transm (Vienna).* 2009;116(8):975–982.

93. Lopez M, Alvarez CV, Nogueiras R, Dieguez C. Energy balance regulation by thyroid hormones at central level. *Trends Mol Med.* 2013;19(7):418–427.

94. Heanue TA, Pachnis V. Enteric nervous system development and Hirschsprung's disease: Advances in genetic and stem cell studies. *Nat Rev Neurosci.* 2007;8(6):466–479.

95. Gadella BM, Tsai PS, Boerke A, Brewis IA. Sperm head membrane reorganisation during capacitation. *Int J Dev Biol.* 2008;52(5–6):473–480.

Renal, Cardiovascular, and Pulmonary Functions of Dopamine

7

7.1 INTRODUCTION

The **kidneys** participate in whole-body homeostasis by regulating extracellular fluid volume, blood pressure, acid-base balance, and electrolyte concentration. In fulfilling these functions, the kidneys act in concert with other organs through interactions of locally produced hormones such as renin and angiotensin II, with circulating hormones such as aldosterone, antidiuretic hormone (**ADH** or vasopressin, **AVP**), and atrial natriuretic peptides (**ANP**). The first two sections of this chapter summarize the available information on the association of dopamine (DA) with two main functions of the kidney: regulation of blood pressure and control of electrolyte balance.

DA exerts multiple effects of the cardiovascular system, including direct effects on the heart, and actions on major vascular beds and on endothelial cells. The fourth and fifth sections of this chapter analyze the dose-dependent actions of DA on the cardiovascular system, which can be rather dissimilar. In spite of the traditional use of DA in clinical practice during cardiovascular emergencies, its beneficial therapeutic effects have been controversial and highly debated. The reasons for this uncertainty are discussed and evaluated in this chapter.

The last sections of this chapter focus on the role of DA in the physiology and pathophysiology of respiration. DA is involved in several aspects of pulmonary functions, including the cardiopulmonary response to exercise, the reduction of edema-associated vascular permeability in the lung, and oxygen-sensing by the carotid body.

7.2 RENAL FLUID HEMODYNAMICS AND HYPERTENSION

7.2.1 Kidney development and gross anatomy

The urinary system includes the kidneys, renal pelvis, ureters, bladder, and urethra (**Figure 7.1**). The kidney is a vital organ with considerable cellular complexity and functional diversity. Its main functions include the balancing the body's fluids, the removal of waste products and drugs, the release of

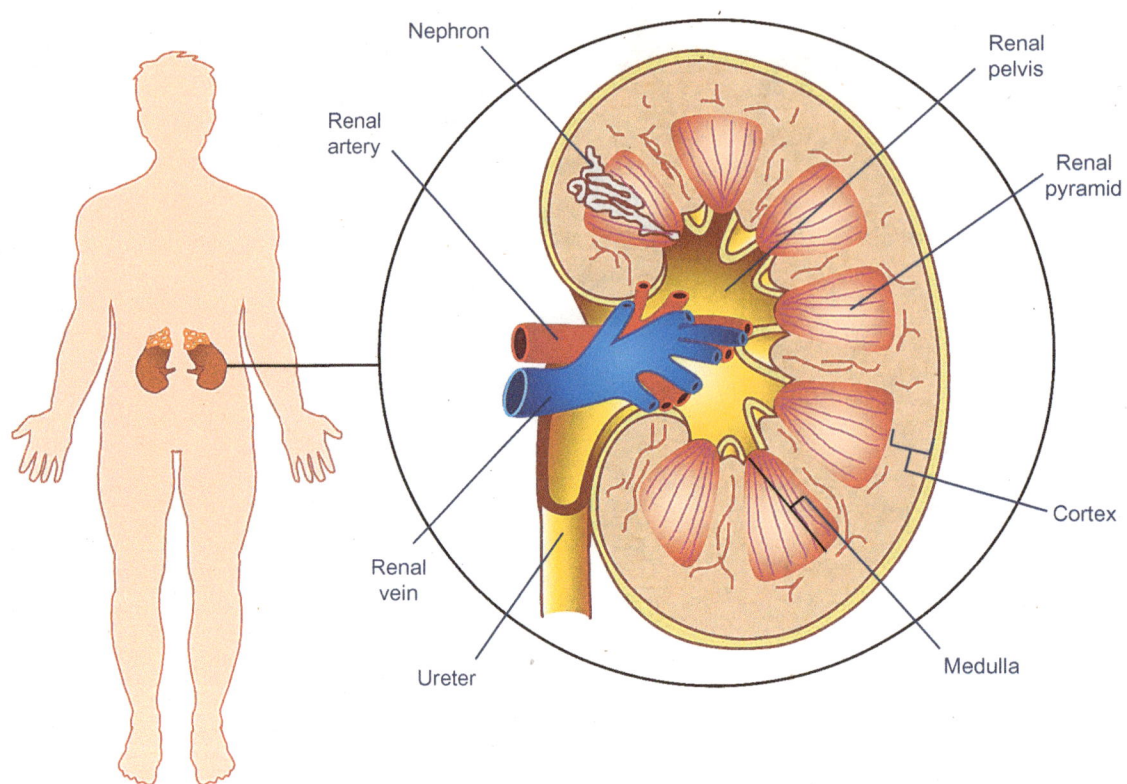

Figure 7.1 Diagram of a human kidney in cross section. Shown are the three major parts (pelvis, medulla, and cortex), location of a nephron, the pyramid structure composed of the collecting ducts, the ureter, and the renal vein and arteries.

hormones that regulate blood pressure, and the control of production of red blood cells. The remainder of the structures of the urinary system serve as the conduits for the transport, storage, and release of urine.

Embryonic development of the kidney depends on time-related reciprocal inductive signals between mesenchymal and epithelial progenitor tissues [1,2]. The urinary system is a component of the "urogenital system," and is anatomically associated with genital development. Kidney development in humans begins early during embryonic life (weeks 5–6 of gestation), and progresses through three developmental stages: pronephros, mesonephros, and metanephros, to the end of gestation. The early patterning of the kidney region depends on interactions between the *Pax/Eya/Six* genes, with essential roles also played by *lim1, Odd1*, and *Wnt* genes. Ureteric bud outgrowth and branching morphogenesis are controlled by the Ret/Gdnf pathway, which is under positive and negative regulation by a variety of factors. After birth, the kidneys continue to mature, and by 2 years of age, they are similar to the adult kidneys in their capacity for regulating body water and for ensuring waste elimination. During childhood, the kidneys grow in size and reach a near-adult size of 10 cm in diameter by 12 years of age; the bladder also continues to grow up to this age.

The kidneys in humans are located in the abdominal cavity on each side of the spine. Blood enters the kidneys through the renal arteries, which are branches of the abdominal aorta, and leaves via the renal veins, which drain into the inferior vena cava. Despite their relatively small size, the kidneys receive as much as 20% of the cardiac output, at a flow rate of about 1 L/min. This disproportionally large blood flow enables the kidneys to carry out a massive filtration endeavor that generates up to 180 L of filtrate each day. Once filtered, processed and concentrated, urine is moved from the kidney through the ureters into the bladder, where it is stored until being emptied upon the relaxation of the urethral sphincter.

The kidneys receive autonomic nerve input from both the parasympathetic and sympathetic nervous systems via the renal plexus. The sympathetic supply originates from spinal cord levels T11–L3, while the vagus nerve provides the parasympathetic innervation. The renal hilum serves as the entry and exit site for all structures that service the kidneys: blood vessels, nerves, lymphatics, and ureters.

A cross section of the kidney shows an outer region, the renal cortex, and an inner region, the renal medulla (**Figure 7.1**). Connective tissue extensions, designated the renal columns, radiate downward from the cortex through the medulla, revealing the renal papillae, which are bundles of collecting ducts (CDs) that transport urine made by the nephrons for excretion. The renal columns divide the kidney into 8–10 lobes and provide a supportive framework for blood vessels that enter and exit the cortex.

7.2.2 The nephron: Structure–function relationship

The adult human kidney contains 0.8–1.5 million nephrons, each of which constitute the basic structural and functional unit of the kidney. **Figure 7.2** shows the structure of a single nephron. A **nephron** is composed of a corpuscle, located in the kidney cortex, which specializes in filtration, and a long tubule, which winds through the medulla and specializes in reabsorption and secretion. The corpuscle is the initial filtration unit of the nephron and is composed of a capillary tuft, the glomerulus, which is enclosed within the **Bowman's capsule**. The glomerulus receives blood supply from an afferent arteriole with a wider diameter than the efferent artery. This narrowing facilitates the generation of hydrostatic pressure as a driving force for efficient filtration of the blood. Podocytes within the Bowman's capsule wrap around the capillaries and carry out the filtration process. The filtrate then moves into the tubule, where it is further processed by sequential reabsorption and secretion steps along the length of tubule, as described below.

As illustrated in **Figure 7.2**, the tubule originating from the Bowman's capsule is structurally divided into the following segments: (1) proximal convoluted tubule (PCT); (2) proximal straight tubule; (3) loop of Henle, consisting

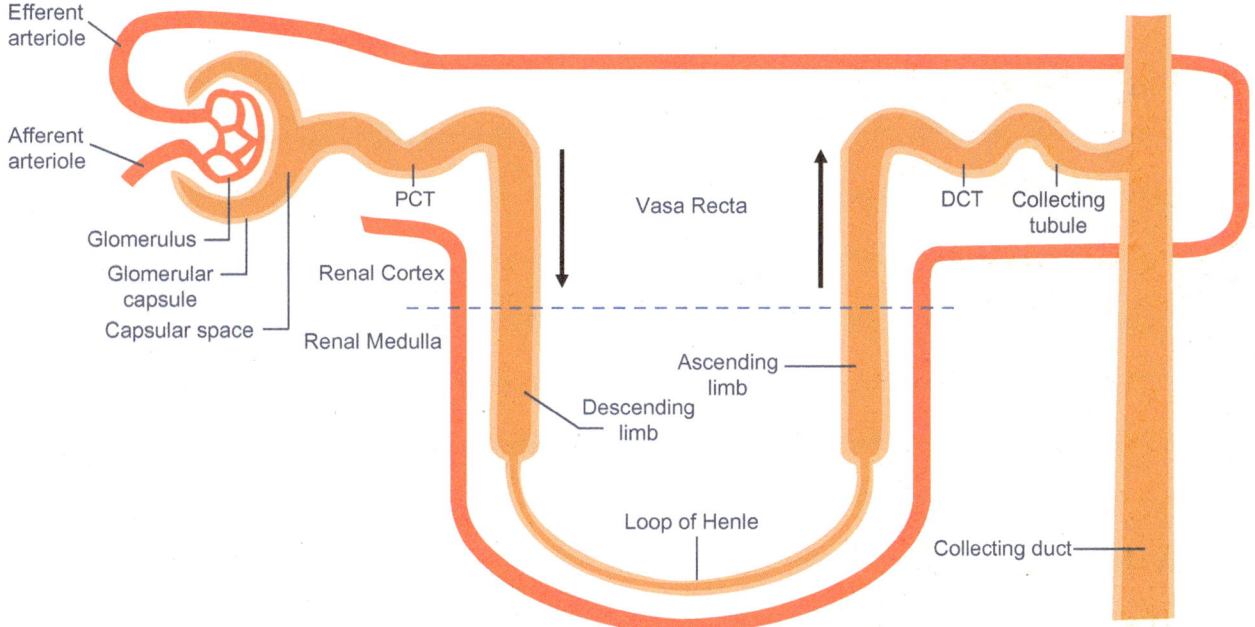

Figure 7.2 Schematic diagram of a nephron. The glomerulus, enclosed inside the Bowman's capsule, is located within the renal cortex, while the long tubule crosses into the medulla and back into the cortex. Shown are the descending limb, loop of Henle, ascending limb, and a collecting duct. DCT: distal convoluted tubule; PCT: proximal convoluted tubule.

of a descending thin limb, an ascending thin limb, and an ascending thick limb; (4) distal convoluted tubule (DCT); and (5) CDs. Each CD drains several nephrons and all CDs coalesce into the ureters, which transport the urine into the bladder.

Based on the location of their glomeruli in the renal cortex, three groups of nephrons are distinguished: superficial, midcortical, and juxtaglomerular. The juxtaglomerular nephrons, which lie in the cortex next to the medulla, differ from the other nephrons by having a longer loop of Henle, a larger glomerulus, and a distinct post-glomerulus blood supply. They are adjacent to the **justaglomerulus apparatus (JGA)**, which contains three types of cells: macula densa (acting as chemoreceptors), messengial cells (contractile elements), and juxtagranular cells. The latter are modified vascular smooth muscle cells with an epithelial appearance and are the site of renin synthesis and release [3].

7.2.3 Urine formation and composition

The process of urine formation is divided into four functional steps: glomerular filtration, tubular reabsorption, tubular secretion, and water conservation (**Figure 7.3**). Glomerular filtration occurs as blood passes into the glomerulus and produces a plasma-like filtrate devoid of proteins. The filtrate is captured by the Bowman's capsule and is funneled into the renal tubule. The filtration membrane allows water and soluble components such as small solutes and waste to pass but blocks the transport of blood cells and large proteins, which remain in the blood.

As the filtrate moves along the nephron, the cells lining the tubule selectively and actively take substances from the filtrate and move them out of the tubule back into the blood in the process of reabsorption. These substances include water, sodium, chloride, bicarbonate, glucose, and amino acids. If not reclaimed by the tubule cells, these physiologically important

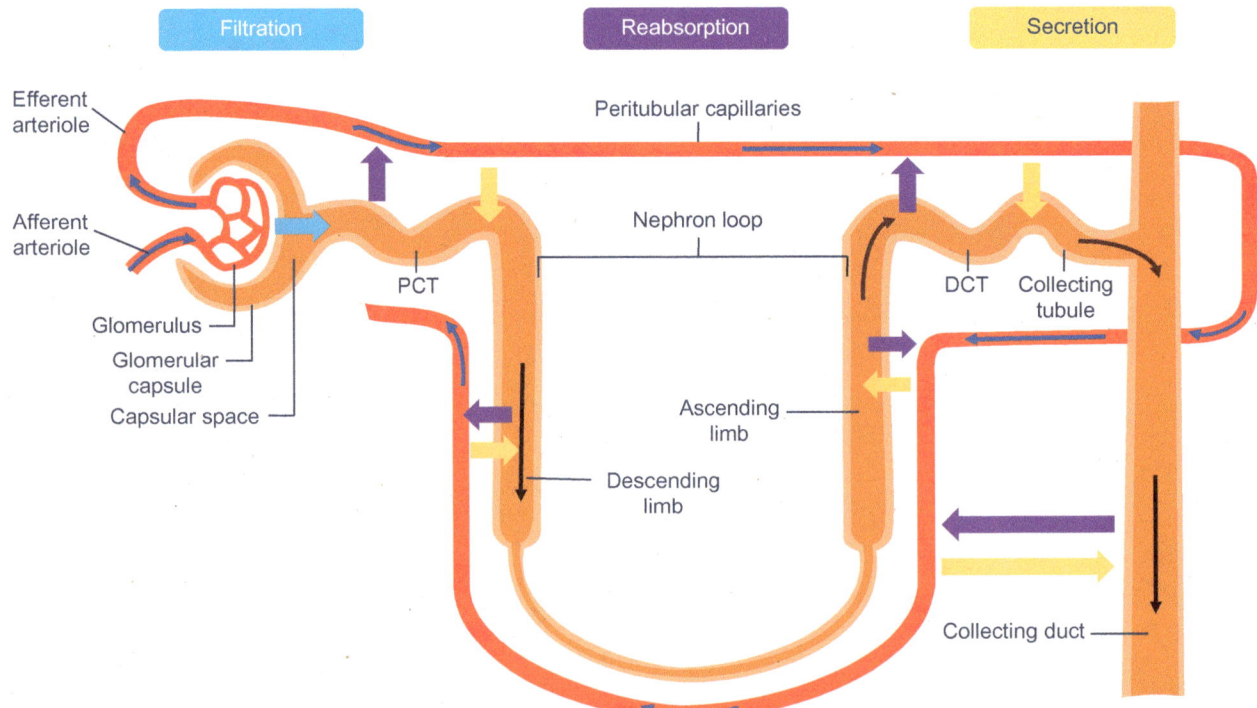

Figure 7.3 The different functions executed by a nephron. Shown are the various functions—filtration, reabsorption, and secretion—which occur along the nephron. Black arrows show the direction of movement of filtrate and blue arrows show the direction of blood flow. See text for other explanations. DCT: distal convoluted tubule; PCT: proximal convoluted tubule.

molecules would be lost in the urine. The tubular cells are very efficient and can reabsorb all of the glucose and amino acids and up to 99% of the water and important ions present in the glomerular filtrate.

Tubular secretion occurs at both the PCTs and DCTs. It represents the transfer of materials from the peritubular capillaries to the tubular lumen, and it is the opposite process of reabsorption. Secretion removes substances from blood that are either too large to be filtered (e.g., antibiotics, toxins) or are in excess in the blood (e.g., H^+, K^+). Secretion is done by an active transport as well as by passive diffusion. In fact, only a few substances are secreted, and those are typically waste products. The PCTs are responsible for secreting organic acids such as creatinine and other bases into the filtrate and efficiently regulate the pH of the filtrate by exchanging hydrogen ions for bicarbonate ions. Water reabsorption occurs along the length of the tubules and is stimulated by ADH in the CDs (see discussion below).

Urine is about 95% water and 5% waste products. Nitrogenous wastes excreted in urine include urea, creatinine, ammonia, and uric acid. Ions such as Na^+, K^+, H^+, and Ca^{2+} are also excreted. In the clinic, pathological examination of urine specimens is usually done by three procedures: visual examination, chemical examination (dipstick), and microscopic examination [4]. Parameters observed during gross examination include urine color, clarity vs. turbidity, and the presence of blood (hematuria). A number of medical conditions such as urinary tract infections, blood diseases, effects of drugs, and even some malignancies (e.g., melanoma) can specifically change some of these parameters and can be diagnosed during the gross examination.

The chemical examination of urine specimens measures the pH, specific gravity (reflecting the hydration status), and the presence of glucose and/or ketones (diabetes), nitrites (bacterial infections), bilirubin (liver disease) as well as elevated protein levels (proteinuria). After urine sedimentation, the microscopic examination looks for white blood cells (infection), red blood cells (kidney disease or blood disorders), bacteria or yeast (infections), epithelial cells, and casts (particles formed from coagulated proteins secreted by tubular cells). Any deviation from normal is indicative of additional diagnostic analyses for suspected diseases.

7.2.4 Endocrine regulation of blood pressure by the kidney

Two hormonal systems, **ADH** [5] and the **renin-angiotensin-aldosterone** system (RAAS) [6], participate in the regulation of blood pressure by the kidney, and are covered here because they have direct or indirect interactions with the renal dopaminergic system.

ADH, or AVP, is a nine-amino acid peptide produced by the hypothalamus and secreted from the posterior pituitary (see details on ADH synthesis and release in **Chapter 5**). Following its binding to V2 receptors in the CDs, ADH increases local water permeability, resulting in water reabsorption and the excretion of a more concentrated urine (antidiuresis). This occurs through interactions of ADH with aquaporin-2 [7]. The aquaporins (AQPs) are membrane proteins that form channels and transfer water and small solutes across cellular membranes. AQP2 is expressed in principal cells of the connecting tubules and CDs, where it is localized in storage vesicles in a basal state. Upon ADH stimulation, AQP2 is phosphorylated by ADH-regulated cAMP/protein kinase A (PKA) and is translocated to the apical plasma membrane. Driven by an osmotic gradient, pre-urinary water passes the membrane through the AQP2 channel and leaves the tubule on the basolateral side. When water homeostasis in the body is restored, ADH levels decline, and AQP2 is internalized from the plasma membrane into the storage compartment, rendering the plasma membrane watertight again.

Other actions of ADH in the kidney include increasing the permeability of the inner medullary portion of the CDs to urea by regulating the expression of cell surface urea transporters and by increasing calcium concentrations (**Figure 7.4**). ADH also acutely stimulates sodium absorption across the ascending loop of Henle, adding to the countercurrent multiplication that aids in proper water reabsorption later in the DCTs and CDs.

The RAAS is involved in several aspects of renal functions, including the regulation of blood pressure, the control of fluid and electrolyte balance, and natriuresis (**Figure 7.5**). Renin is not a hormone but, rather, an aspartyl protease that serves as the rate-limiting step of the RAAS. It is synthesized as an enzymatically inactive proenzyme that is constitutively secreted from several tissues. However, only renin-expressing cells in the JGA of the kidney can generate active renin from pro-renin, which is stored in vesicles and is released as an active renin into the circulation upon demand. Acute release of renin from the JGA is controlled by cAMP and is stimulated in response to decreased extracellular fluid volume by calcium-signaling pathways, prostaglandins, and nitric oxide (NO). Longer-lasting challenges of renin secretion lead to changes in the number of renin-producing cells in preglomerular vascular smooth muscle or extraglomerular mesangial cells.

Renin enzymatically converts circulating angiotensinogen, made by the liver, into angiotensin I. The inactive angiotensin I is then converted into active angiotensin II by **angiotensin-converting enzyme (ACE)**. ACE is produced in the lungs but binds to the surface of endothelial cells in the afferent arterioles and the glomerulus. Angiotensin II acts systemically as a potent vasoconstrictor, and it also causes constriction of both

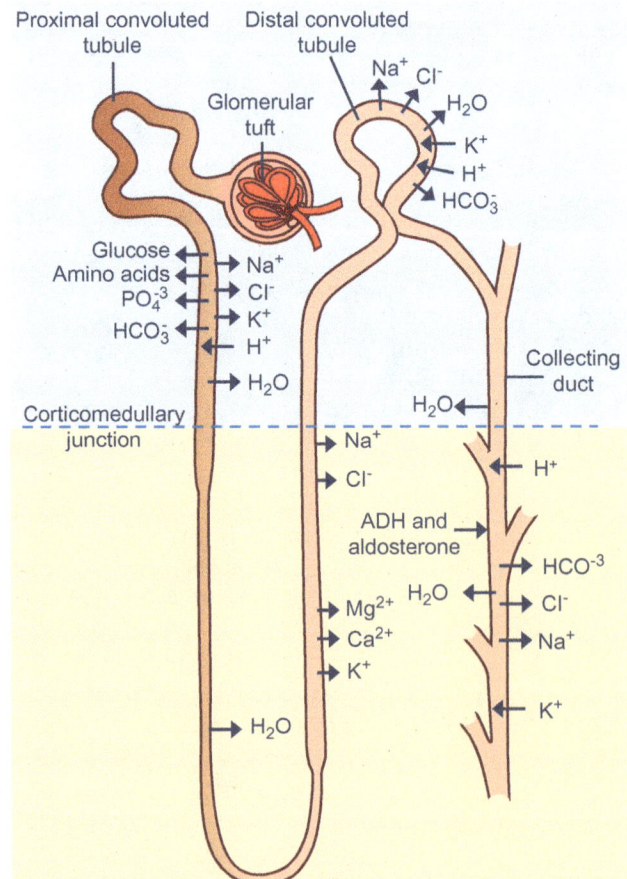

Figure 7.4 Exchange of water and electrolytes between blood and filtrate in different segments of the nephron. The hormonal control of water and electrolytes by antidiuretic hormone (ADH) and aldosterone occur primarily in the collecting ducts.

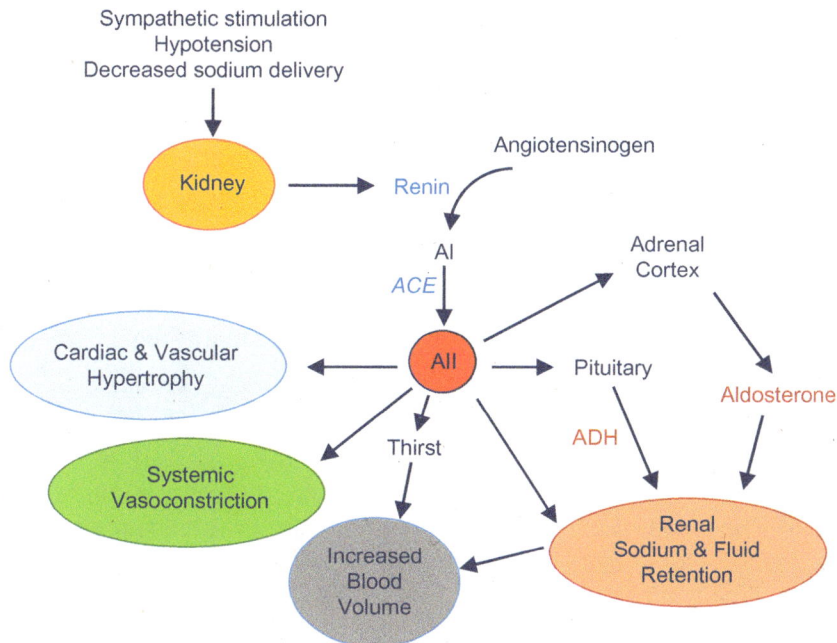

Figure 7.5 Central role of the renin–angiotensin–aldosterone system in the regulation of blood pressure and fluid homeostasis. The kidney-derived renin converts angiotensinogen to angiotensin I (AI). The inactive AI is converted to an active angiotensin II (AII) by angiotensin converting enzyme (ACE). Both the pituitary antidiuretic hormone (ADH) and adrenal aldosterone contribute in many ways to renal fluid hemodynamic and sodium retention. (Redrawn and modified from Ravid, M. *Diabetes Care*, 32, S410–S413, 2009.)

the afferent and efferent arterioles of the glomerulus. In cases of blood loss or dehydration, angiotensin II reduces both glomerular filtration rate (**GFR**), and renal blood flow, thus limiting fluid loss and preserving blood volume. The release of angiotensin II is stimulated by decreases in blood pressure, underlying its primary role in the preservation of adequate blood pressure.

Aldosterone, which is considered the "salt-retaining hormone," is released from the adrenal cortex in response to angiotensin II, or directly in response to increased plasma K^+ (**Figure 7.5**). Aldosterone promotes Na^+ reabsorption by the nephron and increases water retention. At the same time that aldosterone causes increased recovery of Na^+, it also causes loss of K^+. The dual effect of aldosterone on two minerals, together with its origin in the adrenal cortex form the basis for its designation as a mineralocorticoid hormone. Collectively, renin has an immediate effect on blood pressure that is due to angiotensin II–stimulated vasoconstriction and a prolonged effect that is mediated through Na^+ recovery by aldosterone. Notably, progesterone is an ovarian steroid that is structurally similar to aldosterone. Progesterone binds to the aldosterone receptor and weakly stimulates Na^+ reabsorption and increased water recovery. This process is unimportant in men but may cause increased water retention in women during periods of the menstrual cycle when progesterone levels are elevated.

When the RAAS is abnormally active, blood pressure rises. Many drugs that interrupt different steps in this system are used to lower blood pressure. Among these are ACE inhibitors and angiotensin receptor blockers, serving as one of the primary approaches to control high blood pressure, heart failure, kidney failure, and harmful effects of diabetes.

7.2.5 Dopamine: Roles in essential hypertension and renal hemodynamics

Long-term regulation of blood pressure rests on both renal and cardiovascular mechanisms. Essential hypertension, also called primary hypertension or idiopathic hypertension, is a form of high blood pressure without clearly identifiable causes. Essential hypertension affects ~25% of the adult population, and is a common risk factor for stroke, myocardial infarction, heart failure, and kidney failure [8,9]. This is a heterogeneous disease, caused by interactions of genetic and environmental factors.

Among the major causes of essential hypertension are abnormalities in the renal regulation of ion transport and fluid homeostasis, intrinsic and extrinsic to the kidney. Hormones, humoral factors, and neuronal activity, including the RAAS, ADH, and the sympathetic nervous systems (catecholamines), are preeminent in promoting high blood pressure. Yet, hypertension is caused not only by activation of pro-hypertensive factors, but also by defects or dysfunctions in antihypertensive systems serving as counter-regulatory mechanisms. Aberrations in counter-regulatory pathways such as the renal dopaminergic system, are intimately involved in the pathogenesis of essential hypertension [10,11].

The kidney has the complete bioenzymatic and receptor machinery necessary for establishing a self-contained dopaminergic system [9]. DA is formed in the kidney independent of innervation. Both tyrosine hydroxylase (TH) and Dopa decarboxylase (DDC) are expressed in kidney cells [12]. The major source of tubular DA, however, is L-Dopa, which is freely filtered into the glomerulus, serving as a prohormone (**Figure 7.6**). In the tubules, L-Dopa can be taken up by sodium-independent and -dependent transporters. It is then rapidly decarboxylated to DA by DDC, whose activity is regulated by hormones such as angiotensin II or ANP [13]. Mediated by organic cationic transporters (OCTs), the newly formed DA can leave the cells through the apical border. Circulating DA can also enter into the proximal tubular cells through local OCTs (**Figure 7.6**). In addition to DA production in the tubular epithelial cells, TH-positive sympathetic nerve varicosities are localized in close apposition to the epithelial cells of cortical collecting ducts (CCDs), contributing DA to the modulation of fluid and electrolyte transport in the CDs [14].

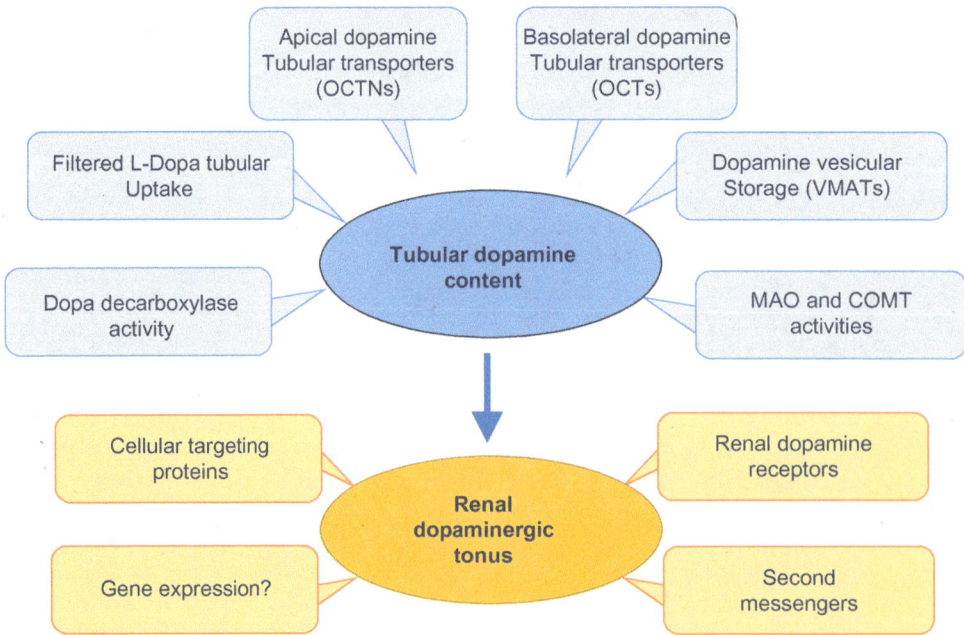

Figure 7.6 Synthesis and actions of dopamine in the renal tubules. (Redrawn and modified from Choi, M.R. et al., *World J. Nephrol.*, 4(2), 196–212, 2015.) COMT: catechol-O-methyltransferase; MAO: monoamine oxidase; OTC: organic cation transporter; VMAT: vesicular monoamine transporter. See text for other explanations.

As discussed in **Chapter 6, Section 6.4.1**, and illustrated in **Figure 6.5**, all five types of DA receptors (DARs) are expressed along the various segments of the nephron, from the glomerulus to the tubules, to the CDs. The D1-like receptors are present on smooth muscle cells of renal blood vessels, the juxtaglomerular apparatus, and the renal tubules, while the D2-like receptors exist on the glomeruli, postganglionic sympathetic nerve terminals, zona glomerulosa cells of the renal cortex, and the renal tubules [15,16].

At the glomeruli, DA augments **diuresis** and **natriuresis** by increasing water and sodium filtration. Binding of DA to D1R in mesangial and podocyte cells results in increased cAMP levels, inhibition of angiotensin II actions, and induction of depolarization and relaxation of the podocytes [15]. DA also decreases water permeability in the CDs by lowering cAMP production. In the CCDs, this effect is mediated by D4R, whereas in the inner medullary CDs, the inhibitory effect of DA is mediated by α2-adrenoceptors (ARs). DA also antagonizes the actions of ADH by promoting internalization of AQP2 from the plasma membrane into intracellular storage vesicles, albeit by an unclear mechanism [17].

The antihypertensive effect of DA in the kidney is also exemplified by its interactions with losartan, a nonpeptide angiotensin II receptor antagonist with high affinity and selectivity for the AT1 receptor. Losartan blocks the vasoconstrictor and aldosterone-secreting effects of angiotensin II by inhibiting its binding to the AT_1 receptor. In cultured rat proximal tubule cells that express endogenous AT1R and D1R, losartan causes D1R activation [18]. Moreover, administration of a D1R antagonist to rats with renal hypertension significantly attenuates the antihypertensive effect of losartan. These data indicate that losartan exerts its strong antihypertensive effects by inhibiting AT1R signaling, and this is reinforced by the enhancement of D1R signaling.

An intact renal DA system is necessary for maintaining several aspects of renal hemodynamics, from fluid balance to the control of blood pressure within the normal range [19,20]. Renal DA production in response to sodium loading is often impaired in humans with **essential hypertension**, and it may contribute to the progression of the disease. It has also been reported that renal DA or DAR are defective in human essential hypertension and in animal models of genetic hypertension [11]. This is supported by the development of hypertension in transgenic mice with disrupted DAR subtypes [21,22]. Hypertension may also be associated with abnormal post-translational modification of DAR or with alterations in their signaling pathways [23]. Peripheral D1R agonists such as fenoldopam are currently used as a first-line treatment in patients with hemodynamic instability and who are unresponsive to fluid therapy [24].

In spite of the well-defined antihypertensive actions of DA, the benefits of treating renal failure with intravenous infusion of DA itself have been inconsistent and controversial. The lack of consistency is due to the variable dose-dependent effects of DA on different organs. At a low dose of 1–2 µg/kg/min, DA acts predominantly on D1R at the renal, mesenteric, cerebral and coronary vessels, resulting in selective vasodilation. At the intermediate dose of 5–10 µg/kg/min, DA also activates β-1 adrenergic receptors (AR) in the heart, increasing heart rate (chronotropic effect) and stroke volume (inotropic effect) and resulting in increased cardiac output. At the high doses of >10 µg/kg/min, DA binds to α-1 AR, causing vasoconstriction and increased vascular resistance [9,25]. Undoubtedly, a better clinical prospective for treating specific renal dysfunctions with DA awaits the development of more selective dopaminergic drugs as well as targeted, rather than systemic, delivery.

7.3 RENAL NATRIURESIS, OXIDATIVE STRESS, AND DIABETIC NEPHROPATHY

7.3.1 Central role of the kidney in keeping electrolyte balance and osmolarity

Electrolytes are defined as minerals that carry an electric charge when dissolved in liquid such as blood. The major blood electrolytes—sodium, potassium, chloride, and bicarbonate—are involved in the regulation of nerve and muscle function and in the maintenance of acid-base balance and water balance. Electrolytes maintain normal fluid levels in various compartments because the amount of fluid that a compartment contains depends on the concentration of electrolytes in it. If the electrolyte concentration is high, fluid moves into that compartment by osmosis, while if the electrolyte concentration is low, fluid moves out of that compartment. To adjust fluid levels, the body can actively move electrolytes in or out of cells. Hence, having an **electrolyte balance** is very important in maintaining fluid balance among the compartments.

The kidneys play a critical role in keeping physiologically suitable electrolyte concentrations. This is done by filtering electrolytes and water from the blood, returning some to the blood and excreting any excess into the urine. Overall, the kidneys maintain a balance between the daily consumption and excretion of electrolytes and water. An electrolyte imbalance can result from dehydration or overhydration (e.g., intravenous fluid administration), effects of drugs, or certain heart, kidney, or liver disorders.

In addition to regulating the total fluid volume, the **osmolarity** (the amount of solute per unit volume) of bodily fluids must be tightly regulated. Extreme variations in osmolarity cause cells to shrink or swell, damaging or destroying cellular structures and disrupting normal cellular functions. The regulation of osmolarity is achieved by balancing the intake and excretion of sodium with that of water. Sodium is by far the major solute in the extracellular fluids and therefore it effectively determines the osmolarity of extracellular fluids.

An important concept is that the regulation of osmolarity must be integrated with the regulation of volume because changes in water volume alone will have either diluting or concentrating effects on bodily fluids (**Table 7.1**). For example, when a person becomes dehydrated, he or she loses proportionately more water than solute (sodium), resulting in increased osmolarity of bodily fluids. In this situation, the body must conserve water (e.g., by increased drinking or by antidiuresis) but not sodium, to lower the rise in osmolarity. On the other hand, a loss of large amount of blood from trauma

Table 7.1 Changes in urine volume and osmolarity under various conditions			
	Osmolarity		
Volume	**Decrease**	**No change**	**Increase**
Increase	Drinking large amounts of water	Ingestion of isotonic saline	Ingestion of hypertonic saline
No change	Replacement of sweat loss with plain water	Normal volume and osmolarity	Eating salt without drinking water
Decrease	Incomplete compensation for dehydration	Hemorrhage	Dehydration (sweat loss or diarrhea)

or surgery is associated with loses of sodium and water that are proportionate to the composition of bodily fluids. In this situation, the body should conserve both water and sodium (e.g., intravenous administration of saline).

As noted above, ADH plays a major role in lowering blood osmolarity by increasing water reabsorption in the kidneys, helping to dilute bodily fluids (**Figure 7.5**). To prevent osmolarity from decreasing below normal, the kidneys have other regulated mechanisms for reabsorbing sodium in the distal nephron. This is primarily done by aldosterone, which increases sodium reabsorption. The adrenal cortex directly senses plasma osmolarity. When osmolarity increases above normal, aldosterone secretion is inhibited. The lack of aldosterone causes less sodium to be reabsorbed in the distal tubule. At the same time, ADH secretion increases to conserve water, complementing the effects of low aldosterone levels to decrease the osmolarity of bodily fluids. The net effect is a decrease in the amount of urine excreted, with an increase in the osmolarity of the urine and the lowering of blood osmolarity.

7.3.2 Endocrine regulation of natriuresis by the kidney

Salt (NaCl) intake is an important determinant of blood pressure. Salt sensitivity of blood pressure refers to the responses to changes in dietary salt intake that produce meaningful increases or decreases of blood pressure. The underlying mechanisms that promote salt sensitivity are complex, ranging from genetic to environmental influences. Salt sensitivity represents a major hypertension phenotype given that about 50% of people with essential hypertension are salt-sensitive and as many as 25% of normotensive individuals are also salt-sensitive [10]. Salt sensitivity is an insidious "silent killer" because even in individuals with normal blood pressure, it can lead to cardiovascular morbidity and mortality and is associated with other diseases (e.g., asthma, gastric carcinoma, osteoporosis, and renal dysfunction).

Natriuresis is defined as the process of sodium excretion into urine by the action of the kidneys. Natriuresis lowers sodium concentration in the blood and also tends to lower blood volume because osmotic forces drag water out of the blood into the urine along with sodium. Many diuretic drugs take advantage of this mechanism and are used to treat medical conditions involving excess blood volume such as hypernatremia and hypertension. Several hormones—renin–angiotensin–aldosterone (discussed above), atrial natriuretic peptides [13], and endothelin-1 [26]—participate in the regulation of natriuresis by the kidney and are discussed below because they have direct or indirect interactions with the renal dopaminergic system.

ANP is a 28-amino acid peptide synthesized and stored in the atrial myocytes. ANP is released in response to stretching of the cardiac wall (e.g., increased blood volume) or after stimulation with endothelin, cytokines, or α-adrenergic agents [13]. The actions of ANP include inhibition of sodium channels and the sodium pump in the inner medullary CD cells, inhibition of aldosterone release from the adrenal cortex, inhibition of renin release (which ultimately reduces aldosterone release), and increasing GFR. ANP inhibits angiotensin II-dependent sodium and water reabsorption in the proximal tubules, and decreases water reabsorption by the distal and collecting tubules. The inhibitory effects of ANP are mediated by a cyclic GMP/protein kinase G (PKG)-dependent mechanism. All of these actions contribute to increased loss of sodium in the urine.

Endothelin-1 (ET-1) is a 21-amino acid peptide known as the most potent vasoconstrictor in humans. ET-1 is produced by almost every cell type in the kidney, albeit the majority of its production occurs in endothelial and principal cells of the inner medullary CDs [26]. The similar distribution of ET_A and ET_B receptors suggests an autocrine function. Several *in vitro* studies have shown inhibitory effects of ET-1 on sodium reabsorption in the isolated inner medullary CD and in CCD that involved decreases in Na/K-ATPase and epithelial sodium channel (ENaC) activities, respectively [27]. A deletion of the ET_B receptor-encoding gene, *Ednrb*, in CD cells, or a pharmacological inhibition of ET_A and ET_B receptors in isolated CCD augments ENaC activity in mice. In addition, inactivation of the ET_A receptor in mouse CD alters the normal response to ADH. Finally, renal ET-1 expression is increased on a high-sodium diet.

7.3.3 Involvement of dopamine with natriuresis

Significant evidence indicates that DA is a major regulator of salt reabsorption by the proximal tubules [9,10,22]. Based on the manipulations of DA biosynthetic and metabolizing enzymes, knockout of DAR subtypes, and treatments with DA altering drugs, renal DA was estimated to account for ~50% of the rise in sodium excretion following increased NaCl intake [28]. DA exerts natriuretic effects by increasing the GFR and by directly modulating sodium reabsorption by inhibiting the activities of Na/K-ATPase and Na^+/H^+ exchangers [11,29]. Because the dopaminergic regulation of sodium excretion is mediated by tubular, rather than by hemodynamic mechanisms, systemic administration of DA or dopaminergic agonists do not fully mimic the renal autocrine/paracrine function of DA [25]. Nonetheless, in cases of hypertensive crisis and acute kidney damage in patients, the action of intrarenal DAR has been leveraged by infusion of DA, as well as by the use of the peripheral D1-like agonist fenoldopam.

The mammalian proximal tubule absorbs about half of the isotonic glomerular filtrate [30]. The role of DA in the regulation of sodium transport by the proximal tubule is depicted in **Figure 7.6**. DA acutely inhibits three Na^+ transporters in the proximal tubule: the apical membrane Na^+/H^+ exchanger, Na^+/inorganic phosphate cotransporter, and the basolateral Na/K-ATPase. Apical membrane Na^+/H^+ exchange activity is responsible for all of the transcellular NaCl absorption in the proximal tubule via coupling with parallel Cl^-/base$^-$ exchange. It lowers luminal Cl^- and enhances paracellular Cl^- absorption. The proximal tubule apical membrane Na^+/H^+ exchange activity is encoded primarily by *NHE3*. Its importance in maintaining extracellular fluid volume is underscored by the hypovolemia and hypotension seen in NHE3-deficient rodents.

Several studies with rodents reinforce the close association of the renal dopaminergic system with salt sensitivity and natriuresis. For example, high salt sensitivity observed in the C57Bl/6J strain of mice has been attributed to both defective renal D1R functions and decreased renal DA production [31]. In addition, a deletion of any of the DAR subtypes results not only in elevated blood pressure but also in impaired pressure natriuresis [22]. This impairment is due either to the direct effect of various DAR on natriuresis, or it results from their interactions with other G protein-coupled receptors.

There are several types of interactions between DA, ANP, and ET-1 in the regulation of natriuresis. For example, OCT-dependent tubular uptake of DA was increased by ANP through the activation of natriuretic peptide receptor-A (NPR-A) receptor via PKG as its signaling pathway. This effect was reflected by an increase in urinary DA excretion, natriuresis, diuresis, and decreased Na/K-ATPase activity [32]. In addition, DA and ANP act together in a concerted manner to promote sodium excretion, especially through the overinhibition of Na/K-ATPase activity.

Another study used the spontaneously hypertensive (SHR) and normotensive Wistar-Kyoto (WKY) rats as the experimental model for examining interactions between ET-1 and DA in the control of natriuresis [33]. This was done by a selective infusion of different reagents into the right kidney of anesthetized rats. The D3R agonist (PD128907) caused natriuresis in WKY rats, which was partially blocked by the ET_B receptor antagonist. In contrast, PD128907 blunted sodium excretion in the SHRs. Additional *in vitro* studies lead the authors to conclude that D3R physically interact with proximal tubule ET_B receptors and that the blunted natriuretic effect of DA in SHR is due, in part, to abnormal $D3/ET_B$ receptor interactions.

7.3.4 Renal dopamine and oxidative stress

Oxidative stress causes an imbalance between the production of **reactive oxygen species (ROS)** and the system's ability to detoxify reactive intermediates or repair the resulting damage [9,19]. ROS are generated by enzymes such as NADPH oxidases, which are membrane-bound enzyme complexes facing the extracellular space. The enzymes catalyze the production of a superoxide free radical by transferring one electron from NADPH to oxygen: $O_2 + e^- \rightarrow {}^{\bullet}O_2^-$. During the process, O_2 is transported from the extracellular space to the cell interior and an H^+ is exported. Superoxide $[{}^{\bullet}O_2^-]$ is the precursor of most reactive oxygen species. Dismutation of a superoxide by superoxide dismutase (SOD) enzymes, produces hydrogen peroxide $[H_2O_2]: 2H^+ + {}^{\bullet}O_2^- + {}^{\bullet}O_2^- \rightarrow H_2O_2 + O_2$. Hydrogen peroxide in turn may be partially reduced to hydroxyl radical (${}^{\bullet}OH$) or fully reduced to water: $H_2O_2 + e^- \rightarrow HO^- + {}^{\bullet}OH\, 2\,H^+ + 2e^- + H_2O_2 \rightarrow 2\,H_2O$. Other reactions can convert hydrogen peroxide to hypochlorous acid (HClO), which is very reactive and can cause significant oxidative damage. Oxidative stress, through the overproduction of peroxides and free radicals, can cause apoptosis and damage various components of the cell, including proteins, lipids, and DNA. Positive effects of ROS include the induction of host defense genes and the mobilization of ion transport systems.

The Nox family of NADPH oxidases include seven Nox homologs, four of which (Nox1, Nox2, Nox4, and Nox5) are found in the vasculature and the kidney, where they constitute the major sources of ROS. The enzymes are dormant in resting cells but rapidly become activated by stimuli such as bacterial infections and some cytokines. A careful regulation of the activity of these enzymes is crucial for maintaining a healthy level of ROS in the body. Increased Nox activity boosts the production of ROS that participate in the pathogenesis of several disorders. The association of these enzymes with hypertension is supported by the increases in Nox1, Nox2, and Nox4 in several tissues in rats with spontaneous hypertension or with angiotensin II-induced hypertension. In addition, protein expression of the Nox5 gene is greater in renal proximal tubular cells from hypertensive than normotensive humans and may account for the increased oxidative stress in renal proximal tubule cells from the hypertensive patients.

The renal dopaminergic system provides a protective mechanism for the kidney by acting as a negative regulator of ROS [19]. As summarized in Table 7.2, the D1, D2, and D5 receptors can exert anti-oxidant effects via direct and indirect inhibition of pro-oxidant enzymes such as NADPH oxidase and also through the stimulation of anti-oxidant enzymes such as SOD and heme-oxygenase (HO), which can indirectly inhibit NADPH oxidase activity. D1R inhibits NADPH oxidase activity via PKA and protein kinase C (PKC) cross-talk and stimulates SOD, glutathione peroxidase, and glutamyl cysteine transferase activities. D5R, via inhibition of phospholipase D2, reduces NADPH oxidase activity, and increases the expression of another anti-oxidant enzyme, HO-1. Finally, the stimulation of renal D2R increases the expression of endogenous anti-oxidants such as Parkinson protein 7 (PARK7 or DJ-1), paraoxonase 2, and HO-2, all of which can inhibit NADPH oxidase activity.

7.3.5 Dopamine and diabetic nephropathy

Diabetes is the most prevalent cause of end-stage kidney disease. Diabetes is associated with elevated GFR, enhance tubular reabsorption, reduced sodium delivery to the macula densa, and progressive glomerular and

Table 7.2 Dopamine receptors (DARs) regulate the production of reactive oxygen species by inhibiting pro-oxidant and stimulating anti-oxidant enzymes

DAR type	Pro-oxidant enzymes (inhibition)	Anti-oxidant enzymes (stimulation)
D1R	NADPH oxidase, via PKA/PKC	SOD, glutathione peroxidase, glutamyl cysteine transferase, HO-1
D2R	NADPH oxidase	DJ-1, PON2, HO-2, glutathione, catalase, SOD
D5R	NADPH oxidase via PLD2	SOD, glutathione peroxidase, glutamyl cysteine transferase, HO-1

Modified from Cuevas S. et al., *Int. J. Mol. Sci.*, 14, 17553–17572, 2013.
DJ-1: Parkinson protein 7; HO-1: heme oxygenase; PKA: protein kinase A; PKC: protein kinase C; PON2: paraoxonase 2; PLD2: phospholipase D; SOD: superoxide dismutase.

tubero-interstitial injuries [9]. In addition to structural abnormalities, diabetes nephropathy is characterized by alterations in the renin–angiotensin system and the renal cyclooxygenase-2 (COX-2) enzyme, which catalyzes the formation of prostaglandins.

Clinical studies have documented an impairment of the renal dopaminergic system in diabetic patients, which likely contributes to diabetes nephropathy. In one report [34], the urinary DA excretion, representing local kidney production, was lower in type 2 diabetic patients than in controls and was decreased in insulin-treated patients compared to patients not treated with insulin. Urinary DA excretion correlated positively with sodium excretion in non-insulin treated patients and in controls but not in insulin-treated patients. In contrast to the findings in healthy volunteers, an intravenous sodium load failed to increase the DA excretion in type 2 diabetic patients, despite similar increments in sodium excretion. A low-dose DA infusion caused significantly lower natriuretic responses in insulin-treated type 2 diabetic patients compared to controls, but not in non-insulin-treated patients. These findings suggest that patients with type 2 diabetes have a derangement of the renal dopaminergic system that is accentuated by insulin treatment.

Several studies using experimental animals confirm and extend the above findings. One study used catechol-O-methyltransferase (COMT)-deficient mice, which have increased renal DA production due to decreased DA metabolism, and proximal tubule DDC-deficient mice (ptDDC$^{-/-}$), which have decreased tubular DA due to lack of synthesis [35]. Compared with wild-type diabetic mice, the COMT$^{-/-}$ mice had decreased hyperfiltration, decreased macula densa COX-2 expression, decreased albuminuria, decreased glomerulopathy, and lower expression of markers of inflammation, oxidative stress, and fibrosis. These differences were also seen in diabetic mice with a transplanted kidney from COMT$^{-/-}$ mice. In contrast, the diabetic ptDDC$^{-/-}$ mice had increased nephropathy. The authors concluded that the intrarenal dopaminergic system plays important roles in the development and progression of diabetic kidney injury and that decreased renal DA production has important consequences in the pathogenesis of diabetic nephropathy.

In addition to DA biosynthesis, expression and functions of the DA receptors are also affected by diabetes. Obese Zucker rats, serving as a model of obesity-related diabetes, are hyper-insulinemic, hyperglycemic, and hypertensive compared with lean rats [36]. DA produced a concentration-dependent inhibition of Na/K-ATPase activity in the proximal tubules of lean rats, but its inhibitory effect was reduced in obese rats. The D1-like receptors in the basolateral membranes of obese rats showed a 45% decrease in B(max) without a change in K(d) compared with lean rats. DA and SKF 38393 (a D1/D5 agonist) failed to stimulate G proteins in obese rats, suggesting a receptor–G protein coupling defect. It was concluded that a decrease in D1-like receptor binding sites and diminished activation of G proteins, resulting from defective coupling to the receptor, led to the reduced inhibition by DA of Na/K-ATPase activity in the proximal tubules of obese Zucker rats. Such a defect in renal DAR function may contribute to sodium retention and development of hypertension in obese rats.

7.4 CARDIAC FUNCTIONS

The cardiovascular system is composed of the heart and the circulatory system. The heart, composed of ventricles and atria, works as a pump that pushes blood to organs, tissues, and cells. The blood delivers oxygen and nutrients to every cell in the body and removes the carbon dioxide and waste products made by those cells. The muscle mass of the heart, the myocardium, shares structural and functional characteristics of both smooth and skeletal muscle. The cardiomyocytes are rather small cells that form the highly branched network of the functional syncytium, which acts together mechanically and electrically. The atrial and ventricular muscle tissue are structurally similar but differ in their electrical properties. In addition to these, the heart has conducting tissue (Purkinje fibers) which is adapted for rapid and efficient conduction of action potential as well as sinoatrial and atrioventricular nodes that are involved in the initiation and conduction of the heartbeat.

7.4.1 Homeostasis of dopamine in the myocardium

There is only sparse and incomplete information on *de novo* synthesis of DA in the heart. Studies with chick embryos (a model of maternal-independent embryonic development) revealed expression of TH in the developing heart during a restricted time of development [37]. Overexpression of TH or application of L-Dopa- or DA-induced cardiac differentiation, while TH caused disruption of cardiac morphogenesis. Comparable findings were observed in mice with TH inactivation [38]. The absence of TH resulted in a mid-gestational lethality, with about 90% of the mutant embryos dying between embryonic days 11.5 (E11.5) and E15.5 because of cardiovascular failure. Administration of L-Dopa to the pregnant females resulted in a complete *in utero* rescuing of the mutant mice.

There are no published records on TH expression in adult myocardial cells, and neither the reasons for, nor the mechanism of, TH silencing in the myocytes after birth are known. Nonetheless, DA can be available to the heart from several external sources. One is via the circulation as free DA or as conjugated DA sulfate (DA-S). Given the wide tissue distribution of arylsulfatase A, which deconjugates DA-S, the heart could utilize this mechanism to gain access to large amounts of circulating DA. A second source is via decarboxylation of circulating L-Dopa, by DDC, which is widely expressed in most tissues. An additional source of DA are ganglionated plexi (GP), consisting of conglomerations of autonomic ganglia on the epicardial surface of the heart and which express catecholamine biosynthesis markers [39].

The myocardium expresses extraneuronal monoamine transporters (EMTs) and organic cation transporters OCT1 and OCT2 [40] as well as major catecholamine metabolizing enzymes such as COMT and monoamine oxidase (MAO) [41]. The distribution of DARs in the heart is described in **Chapter 6**. With the exception of D3R, all DAR subtypes were identified in different regions of the human heart, including the endocardium, myocardium, and epicardium [42]. D1R was localized primarily in the epicardial layer of the human heart, while other DARs were found in the atrial and ventricular walls. In sum, the heart possesses all the components of a functional dopaminergic system: biosynthesis, metabolism, uptake, and receptors.

7.4.2 Actions of DA on the heart under pathophysiological conditions

The different pharmacological actions of DA are effected through dose-dependent activation of dopaminergic, α-adrenergic, and β-adrenergic receptors. Activation of β_1, β_2, and α_1 adrenergic receptors in the heart results in increased heart rate and contractility, while activation of presynaptic α_2 receptors has an inhibitory effect on norepinephrine (NE) release from nerve terminals.

Cardiac ischemia is characterized by reduced blood supply to the heart, resulting from blockade of the coronary arteries. Ischemic heart disease is a leading cause of death in Western countries and a major cause of hospitalization. The putative role of D1R in cardiac ischemia was examined using cardiomyocytes from neonatal rats incubated under ischemia-mimetic conditions, which increased D1R expression [43]. Treatment with SKF-38393 (D1R agonist) significantly increased both oxidative stress and expression of apoptotic markers, while SCH-23390 (D1R antagonist) had no significant effects. The authors concluded that D1R enhances apoptosis in cardiomyocytes in simulated ischemia conditions through the mitochondrial and death receptor pathways. Using a similar perfusion model of cardiomyocytes, another investigation focused on the role of D2R in ischemia [44]. Ischemia/reperfusion increased the expressions of D2R, an effect which was further enhanced by post-conditioning. Additional experiments showed that D2R activation provided cardio protection by promoting translocation of PKC to the particulate fraction. Based on these results it appears that the two receptors act in opposite ways: D1R enhances the effects of ischemia, while D2R is protective.

DA has been used in patients with cardiac dysfunction for more than five decades. A review of multiple trials with DA had evaluated mortality data, serious adverse events, myocardial infarction, arrhythmias, and renal replacement therapy [45]. The report indicated that the evidence in favor of treatment with DA in critically ill patients with cardiac dysfunction is inconclusive and of low quality, leading the authors to conclude that the use of DA for cardiac dysfunction can neither be recommended nor refuted.

Acute heart failure (AHF) is a life-threatening medical condition, where urgent diagnostic and treatment methods are of key importance. The cornerstone of AHF management is identifying the precipitating factors and specific phenotype. The main pathophysiology of AHF is congestion, both systemic and inside organs such as lung, kidney, or liver. Cardiac output is often preserved in AHF, except in a few cases of advanced heart failure, while renal dysfunction is rather prevalent. Given its ionotropic properties, DA has been widely used as the first line of defense in patients with hypotensive cardiogenic shock. The benefits of treatment with a low-dose DA for AHF was examined by reviewing multiple trials [46]. No differences were found between the DA treatment group and placebo. The authors concluded that there is no justification for a routine use of low-dose DA in non-hypotensive patients with AHF, while further studies are needed to define the role of low-dose DA in patients with AHF and hypotension.

7.5 BLOOD PRESSURE REGULATION IN VARIOUS VASCULAR BEDS

The circulation system is functionally divided into two main vascular beds: **pulmonary** and **systemic**. The pulmonary circulation moves blood from the heart to the lungs, where it absorbs oxygen and releases carbon dioxide. The oxygenated blood then flows back to the heart and from there to the rest of the body via the systemic circulation. The vascular system is composed of three major types of blood vessels: arteries, which carry blood away from the heart; capillaries, responsible for exchange of water and chemicals between blood and tissues; and veins, which carry blood from the capillaries back toward the heart. As shown in **Figure 7.7**, the various types of blood vessels differ in diameter and in cytoarchitecture. Blood vessels do not actively engage in the transport of blood given that they have no appreciable peristalsis. Instead, blood is propelled through the arteries and arterioles by pressure generated by the heartbeat. However, arteries and, to a limited degree, veins can regulate their inner diameter by contraction of the vascular smooth muscle cells (VSMCs). Local vasoconstriction and vasodilation can then change the blood flow to downstream organs.

7.5.1 Effects of DA on various circulation systems

Although DAR subtypes are expressed in all major parts of the cardiovascular system (see **Chapter 6**), their roles in the regulation of its functions have been understudied, especially when compared with the extensive research on NE and epinephrine (Epi). As elsewhere, vascular DAR belong to two classes. The D1-like activate adenylyl cyclase with a consequent increase in intracellular cAMP, while D2-like inhibit adenylyl cyclase and result in activation of K^+ channels, inhibition of Ca^{2+} channels, and the exchange of

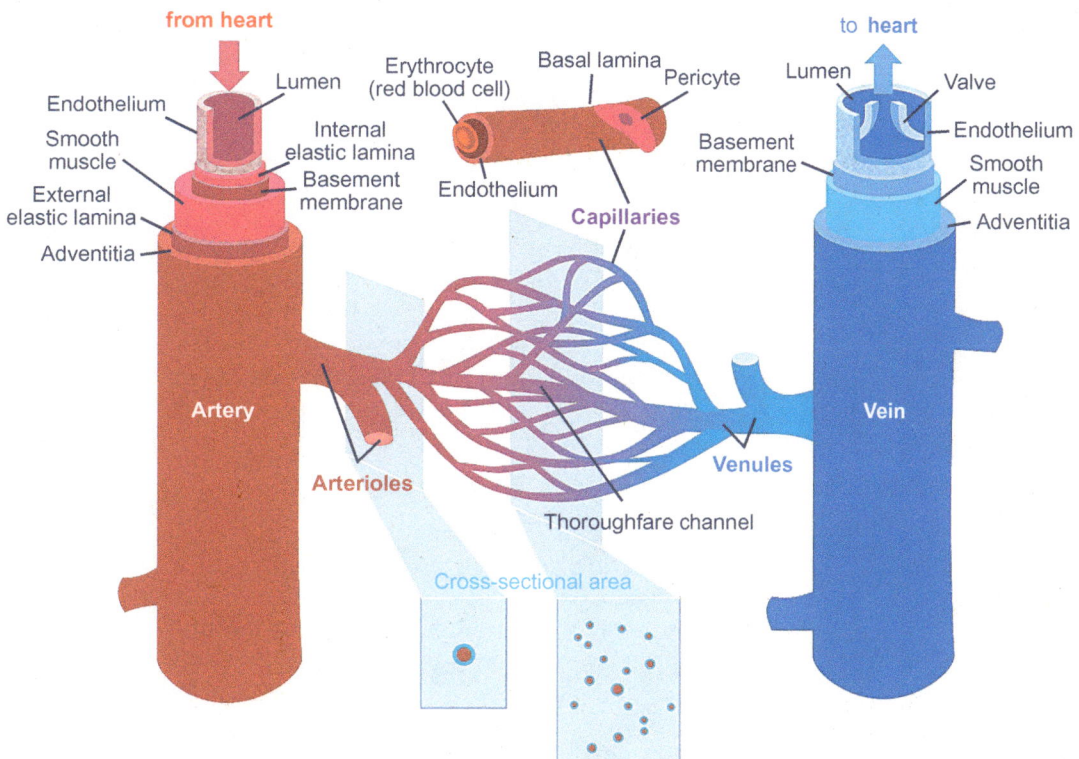

Figure 7.7 Cytoarchitecture of the various blood vessels connecting arteries and veins. The relative dimensions of the smaller blood vessels, from arteries to capillaries to venules are shown in the inserts.

phosphatidyl inositol. In general, D1-like receptors provoke renal, coronary, and mesenteric vasodilatation, while the stimulation of D2R-like is often associated with NE release from sympathetic nerve endings. Activation of postsynaptic α_1 and α_2 adrenergic receptors by DA mediates vasoconstriction, while activation of β_2 receptors mediates vasodilation.

Importantly, DA influences the various receptors in a dose-dependent manner. Low doses of DA act primarily on D1R, resulting in a selective vasodilatation in the renal, mesenteric, cerebral, and coronary beds. Vasodilatation can be balanced by an increase in stroke volume so that the arterial pressure remains relatively stable. Higher doses of DA activate not only dopaminergic, but also the β-adrenergic receptors, resulting in increased cardiac output due to combined increases in stroke volume and heart rate. Thus, doses of 5–10 μg/kg/min have predominant β-adrenergic effects and start to have α-adrenergic effects, resulting in combined increases in cardiac output and blood pressure. Higher doses of DA provoke a predominantly α-adrenergic activation, with vasoconstriction and increased systemic vascular resistance. The dose-dependent effects of DA, however, are not entirely predictable because of significant inter-individual variability in enzymatic DA inactivation. In practical terms, DA should be titrated to a desired physiologic effect because weight-based administration of DA can achieve quite different concentrations among individuals.

As documented by *in vivo* and *in vitro* studies, activation of D1R in resistance arteries by DA at low concentrations normally causes dilation. Treatment of dogs and rats with fenoldopam, a peripheral D1R agonist, decreased blood pressure without activating central DAR [47]. In another study, fenoldopam reduced mean arterial pressure, lowered arterial and venous resistance, and increased cardiac output in rats [48]. At the highest dose, fenoldopam also increased heart rates and reduced the mean circulatory filling pressure, the driving force of venous return. Chronic inhibition of dopamine beta-hydroxylase (DBH), which converts DA to NE, decreased urinary DA and NE in SHR rats, and caused a marked reduction in their blood pressure [49]. The authors postulated that the observed hypotension was due to the combination of reduced circulating NE and increased DA; the treatment itself did not prevent development of the disease in this genetic animal model of hypertension.

In one study, DA infusion in healthy human volunteers elicited a biphasic effect on mean arterial blood pressure [50]. At low doses, DA decreased diastolic pressure, indicating a depressor effect, whereas at high doses, a pressor effect was also evident, as judged by the rise in the systolic pressure. The authors suggested that the increase in cardiac output, caused by low DA concentration, is secondary to a decline in peripheral vascular resistance, independent of the effects of β-1 AR on cardiac contractility and heart rate. In another study, patients with episodes of orthostatic hypotension and hypovolemia were examined for the potential involvement of catecholamines [51]. Such episodes were accompanied by moderate increases in plasma NE and Epi, and a five-fold rise in DA-S, indicating an extraneuronal discharge of DA. The authors speculated that the hypovolemia, which might have been caused by increased fluid transfer from blood to urine, resulted from direct or indirect actions of circulating DA-S.

Septic shock is a potentially fatal medical condition that occurs when sepsis, which is organ injury or damage in response to infection, leads to a dangerously low blood pressure and abnormalities in cellular metabolism. The primary infection is caused by bacteria, but it can also be due to fungi, viruses and parasites. Infection can be localized in the lungs, brain, urinary tract, skin, or abdominal organs. Septic shock can cause multiple organ failure and death. Vasoactive medications in the treatment of septic shock include inotropes, vasopressors, vasodilators, and inodilators [52]. DA and NE are most often used together as a vasoconstrictor in the initial phase of septic shock. DA in a large dose activates α1-AR and has potent vasoconstriction effects. By having equivalent effects on the heart and the vasculature, DA increases cardiac output, mean arterial pressure and heart rate. NE has a great potency in increasing blood pressure via α1R, but its inotropic effect is not as potent as that of DA. Given inconclusive data on mortality rates following the use of DA, the question whether DA is beneficial or harmful in the treatment of septic shock or as a cardiac protector is currently unresolved [52].

Vasodilation is brought about by relaxation of the VSMC that reside in the walls of large veins and arteries and smaller arterioles (**Figure 7.7**). Application of DA or its agonists to arterial strips from human mammary [53], radial [54], or coronary [55] arteries caused relaxation, but only when the opposing contractile effects of α-AR were blocked. It was therefore concluded that the differential affinity of DA for DAR subtypes, or for α-AR, ultimately determines whether certain arteries respond to DA by contraction or by relaxation.

The findings that activation of either D1R or D2R similarly cause vasodilation have been enigmatic, until a plausible explanation was provided by a systematic analysis of receptor localization in different vascular beds [56]. The authors stipulated that post-junctional D1R in some arteries induce direct vasodilation, while pre-junctional D2R on postganglionic nerve terminals in other arteries, indirectly induce vasodilation by inhibiting NE release.

Increased VSMC proliferation can cause hypertrophy and hyperplasia of blood vessel walls, resulting in increased vascular resistance and exacerbation of the pathogenesis of hypertension. Studies with rat aortic-derived VSMCs showed that the proliferative effects of NE were attenuated by D1R or D3R agonists, which by themselves had no effect on VSMC proliferation [57]. A combined stimulation of D1R and D3R activities had additive inhibitory effects. Inhibition by D3R and D1R occurred via PKA and PKC activation, respectively. Other studies found that DA antagonized the induction of VSMC proliferation by insulin [58], neuropeptide Y [59], angiotensin II [60], and IGF-1 [61] in several vascular beds.

7.5.2 Dopamine and angiogenesis

Angiogenesis is the formation of new blood vessels from preexisting ones, as opposed to vasculogenesis, which is the generation of new blood vessels and which occurs primarily during embryonic development. Angiogenesis involves migration, growth, and differentiation of endothelial cells lining the inside wall of blood vessels. Stimulation of angiogenesis can be therapeutic in ischemic heart disease, peripheral arterial disease, and wound healing. Decreasing or inhibiting angiogenesis can be therapeutic in cancer, ophthalmic conditions, rheumatoid arthritis, and other diseases. Specific cases of DA effects on angiogenesis in different diseases are presented in respective chapters later in the book.

Hypoxia is a potent stimulus for angiogenesis. Most types of parenchymal cells (myocytes, hepatocytes, neurons, and astrocytes) respond to a hypoxic environment by secreting a key proangiogenic growth factor, vascular endothelial growth factor (VEGF). Members of the VEGF family stimulate cellular responses by binding to VEGFRs, which are tyrosine kinase receptors located on the cell surface. Binding of VEGF to the receptor induces receptor dimerization and activation through transphosphorylation. VEGF-A (commonly referred to as VEGF) binds to VEGFR receptor (VEGFR)-1 (flt-1) and VEGF-2 (kdr/flk-1). The latter mediates most of the known cellular responses to VEGF, while the function of VEGF-1 are less well defined.

VEGF is a potent cytokine with a dual function: increased vascular permeability and stimulation of angiogenesis [62]. Several experimental models have shown that DA suppresses vascular permeability by interacting with VEGF or its receptor. In one study, a murine model of lipopolysaccharide (LPS)-induced acute lung injury was used [63]. Pretreatment of the LPS-treated mice with DA reduced pulmonary edema by decreasing serum VEGF and VEGFR2 phosphorylation. The inability of a D2R agonist to reduce pulmonary edema in D2R$^{-/-}$ mice indicated that DA acted via D2R to inhibit pulmonary edema-associated vascular permeability. In another study, mice were inoculated intraperitoneally with ovarian carcinoma cells, which secrete VEGF and cause microvascular hyper-permeability [64]. Acting via D2R, DA reduced ascites fluid accumulation by enhancing the endocytotic internalization of VEGFR-2, leaving less VEGFR-2 on the surface of the cell membrane for interaction with VEGF.

In women undergoing *in vitro* fertilization (IVF), inadvertent overstimulation of ovarian follicular growth can lead to a life-threatening **ovarian hyperstimulation syndrome (OHSS)**. OHSS is characterized by enlarged ovaries and significant accumulation of ascites fluid, resulting in abdominal bloating, high risk of venous thromboembolism, decreased organ perfusion, and respiratory distress [65,66]. Most cases are mild, but moderate or severe OHSS occurs in up to 8% of IVF cycles. Prophylactic treatment with DA agonists such as cabergoline, quinagolide or bromocriptine have reduced the incidence but not the severity of OHSS, without compromising pregnancy outcomes. As inferred from *in vivo* [67] and *in vitro* [68] experimental models, the alleviation of OHSS by D2R activation results from the inhibition of ovarian VEGF production via D2R-mediated posttranscriptional mechanism.

The ability of DA to inhibit angiogenesis is beneficial for the suppression of tumor growth [69] but is disadvantageous during ischemic healing, which necessitates new blood vessel formation. The role of DA during ischemic healing was studied using wild-type and D2R knockout (D2R KO) mice with a unilateral hind limb-induced ischemia [70]. Recovery from ischemia was much faster in D2R KO than in control mice, indicating that DA impairs the healing process. DA antagonized angiotensin II-mediated angiogenesis by suppressing the expression of its receptor in endothelial cells. The authors concluded that DA prolongs the time of post-ischemic recovery, warranting the use of pharmacological intervention for blocking DA actions.

7.6 RESPIRATION AND OXYGEN-SENSING

Respiration is the movement of oxygen from the outside environment to the cells within tissues throughout the body, and the transport of carbon dioxide in the opposite direction. As shown in **Figure 7.8**, the respiratory tract is divided into two main components: the upper respiratory tract, which includes the nasal and oral cavities, pharynx and larynx, and the lower respiratory tract, which includes the trachea, primary bronchi and lungs. Air is breathed in through the nose or the mouth. A mucous membrane within the nasal cavity acts as a filter and traps pollutants and other harmful substances found in the air. From there, air moves into the pharynx, a passage in the intersection between the esophagus and larynx. A flap of cartilage, the epiglottis, allows air to pass through and the same time closes the opening of the esophagus to prevent food from moving into the airway.

The larynx houses the vocal folds and manipulates pitch and volume, which is essential for phonation. From the larynx, air moves into the trachea and down to the right and left primary bronchi. Each of these bronchi branch into secondary (lobar) bronchi that branch into tertiary (segmental) bronchi that branch into smaller airways called bronchioles that eventually

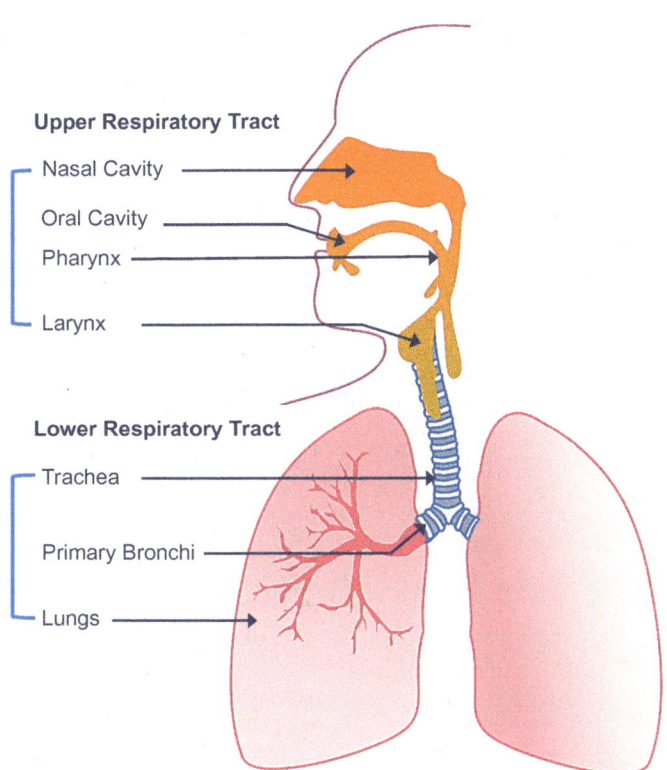

Upper Respiratory Tract

- Nasal Cavity
- Oral Cavity
- Pharynx
- Larynx

Lower Respiratory Tract

- Trachea
- Primary Bronchi
- Lungs

Figure 7.8 Diagram of the human respiratory system. Shown are the upper respiratory tract, which includes the nasal cavity, pharynx, and larynx, and the lower respiratory tract, which includes the trachea, primary bronchi, and lungs.

connect with tiny specialized structures called alveoli that function in gas exchange. More details on structure of the lungs are presented below in **Section 7.7**.

7.6.1 Dopamine homeostasis in the upper respiratory tract

Older studies have examined for *de novo* synthesis of DA in the different components of the upper respiratory tract as well as for the potential availability of DA from various sources. Two populations of cells were identified within paraganglia in the human larynx [71]. One population contained chromogranin and TH, representing neurosecretory catecholamine secreting cells, while a second population had choline acetyltransferase-positive cells, representing parasympathetic input. Most of the cells in the paraganglia of the recurrent and superior laryngeal nerves showed intense TH immunoreactivity, only a few exhibited DBH immunoreactivity, and none displayed phenyl-ethanolamine-N-methyltransferase (PNMT) immunoreactivity. The local arterioles were supplied with plexuses of nerve fibers showing TH and DBH immunoreactivity. The results indicate that DA is the major catecholamine in the laryngeal nerve paraganglia, while fewer cells produce NE. Based on the application of sympathectomy and surgical excision of the superior cervical ganglion [72], it was concluded that DA in the laryngeal nerves is stored independently of the sympathetic nerve fibers, likely in the paraganglia, while NE fibers originate from the supercervical ganglion. The ubiquitous presence of DDC ensures availability of DA via the conversion of circulating L-Dopa. Although there are few reports on the presence of NE in the trachea and bronchi, there are no published records on the presence of DA in these tissues.

Human airway smooth muscles express both D1R and D2R [73]. D1R modulates airway smooth muscle tone through cAMP production, which would favor airway relaxation in asthmatics. D1R, but not D2R, is also expressed in human airway epithelial cells, where it is involved in mucus overproduction, which could worsen airway obstructive symptoms. D1R and

D2R have also been identified in lung arteries [74], where they are found within the tunica adventitia of human pulmonary arteries and within the tunica media of large and medium intrapulmonary arteries of the rabbit.

7.6.2 Dopamine involvement in upper airway pathophysiology

The control of the bronchial diameter is complex and involves both neural and humoral factors. Thus, adrenergic and cholinergic systems are both involved in this regulation, as well as inhibitory and excitatory non-adrenergic, non-cholinergic nerves. Adrenergic control is carried out by NE released from nerve endings and Epi released from the adrenal medulla. Acting via β_2-AR, Epi and NE can produce bronchodilation, but they can also produce bronchoconstriction by stimulating α_1 and α_2 AR if pathological disturbances exist in the airways. Non-adrenergic/non-cholinergic nervous control of airway smooth muscle is exerted by neurotransmitters such as substance P, neurokinin, and neuropeptide Y.

There is growing evidence that DA is involved in the regulation of airway diameter (for review, see [74]). Rat tracheal smooth muscle cultures were used as a model to evaluate the role of dopaminergic and adrenergic receptors [75]. Contraction was induced by various doses of acetylcholine or by electric field stimulation. A low concentration of DA elicited a small initial contraction, followed by a marked relaxation. Cholinergic-induced contraction was completely reversed by DA. The biphasic dopaminergic response was not blocked by alpha- or beta-adrenergic antagonists, or by a D2R antagonist. A D1R antagonist produced a sustained increase of basal tone but did not block the initial dopaminergic contraction and partially inhibited the bronchodilator effect of DA. It was concluded that DA-induced relaxation in rat trachea is mediated by D1R and that adrenergic receptors are not involved. However, the intrinsic contractile activity of the D1R antagonist on airway smooth muscle deserves further research.

Asthma is a long-term inflammatory disease of the conducting zone of airways, especially the bronchi and bronchioles, causing increased contractability of surrounding smooth muscles. Asthma is characterized by variable and recurring symptoms, reversible airflow obstruction, and easily triggered bronchospasms. The disease is caused by a combination of genetic and environmental factors, including exposure to air pollution and allergens. Other potential triggers are medications such as aspirin. Although there is no cure for asthma, its symptoms can be alleviated by avoiding triggers and by the use of inhaled corticosteroids, long-acting beta agonists and anti-leukotriene agents.

DA inhibits bronchial constriction induced by histamine in normal and asthmatic human subjects [75]. A low dose of inhaled DA-induced bronchial dilatation in patients with acute asthma attack. However, DA did not modify the resting bronchial tone in normal subjects or in asthmatics without an acute attack. Additionally, D2R blockade with metoclopramide did not modify resting bronchial tone, suggesting that the DA effect is mediated by D1R.

7.6.3 Dopamine and carotid body functions

The carotid body is the main peripheral chemoreceptor that coordinates cardio-ventilatory functions [76,77]. See discussion on the carotid bodies in **Chapter 6** and their illustration in **Figure 6.7**. The carotid body is a small cluster of chemoreceptors located near the bifurcation of the carotid artery. It detects changes in the composition of arterial blood flowing through it, in particular the partial pressure of oxygen and carbon dioxide, but is also sensitive to changes in pH and temperature. The carotid body contains the most vascularized tissue in the human body, and its main function is to sense changes in arterial oxygen tension and conveys this information to respiratory centers in the brain to elicit hyperventilation in hypoxia.

The carotid body is made up of two types of cells, called glomus cells. Glomus type I cells are neuroendocrine cells derived from the neural crest that release a variety of neurotransmitters. Glomus type II cells resemble glia, express the glial marker S-100, and act as supporting cells and also as stem cells. Glomus type I cells are synaptically connected to nerve terminals of the petrosal ganglion neurons. In response to hypoxia, hypercapnia, or acidosis, the glomus cells release one or more transmitters, which act upon nerve terminals of chemosensory neurons and increase their afferent discharge [77]. The responsiveness of the carotid body to acute hypoxia relies on the inhibition of O_2-sensitive K^+ channels in glomus cells, which leads to cell depolarization, Ca^{2+} entry, and release of transmitters that activate afferent nerve fibers. Although this model of O_2-sensing is generally accepted, the molecular mechanisms underlying K^+ channel modulation by O_2 tension are unknown.

DA is the predominant catecholamine synthesized in the carotid body, with a 5:1 ratio of DA to NE. DA is taken up and is stored in dense core vesicles of glomus cells. DAR are located pre- and postsynaptically on the glomus cells. Studies with rats showed that hypoxia increases TH activity in the carotid body, resulting in increased DA synthesis and release. The mode of action of DA within the carotid body is species specific. For example, DA plays an inhibitory role in cats [78], as was found by examining cats before and after exposure to a simulated altitude of 14,000 feet for 2 days [79]. Treatment with domperidone, a peripheral D2R antagonist, showed that decreased DA activity accounted for the increased chemosensitivity of acclimatization, supporting the actions of DA as an inhibitory transmitter. However, DA appears to act as an excitatory transmitter in the rabbit carotid body [80]. Studies in healthy humans have demonstrated that intravenous low-dose DA was associated with decreased carotid sinus drive and diminished ventilatory response to hypoxia, an effect which was abolished by the DAR blocker haloperidol [81].

7.7 PULMONARY VENTILATION AND PATHOPHYSIOLOGY

The lungs are located in the thoracic cavity and are protected from physical damage by the rib cage. The diaphragm at the base of the lungs is a sheet of skeletal muscle that separates the lungs from the stomach and intestines and is the main muscle of respiration involved in breathing. The lungs are encased in two membranes: the inner visceral pleura, which covers the surface of the lungs, and the outer parietal pleura, which is attached to the inner surface of the thoracic cavity. The pleurae cavity contains pleural fluid, which decreases the amount of friction that lungs experience during breathing. The overall lung structure is depicted in **Figure 7.9**. Within the lung, the bronchi branch into secondary (lobar) and then into tertiary (segmental) bronchi. These lead into smaller airways called bronchioles that eventually connect with tiny specialized structures called alveoli. The alveoli are only one cell thick, allowing easy passage of oxygen and carbon dioxide (CO_2) between the alveoli and surrounding capillaries.

The alveolar compartment is lined with membranous pneumocytes, known as type I alveolar epithelial cells (AT1), and granular pneumocytes, called type II alveolar epithelial cells (ATII). ATI cells are extremely thin squamous cells that cover 90%–95% of the alveolar surface. They are involved in the process of gas exchange between the alveoli and blood. ATII are responsible for epithelial repair upon injury and for ion transport and are very active

Figure 7.9 Diagram and cross section of part of a human lung. Shown are the extensive capillary beds that surround the alveoli and how air flows into the alveoli.

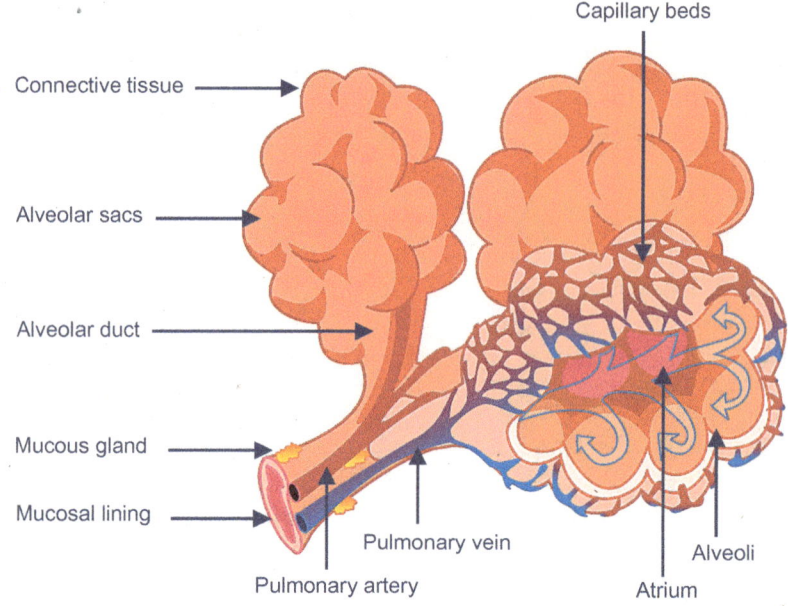

immunologically, contributing to lung defense by secreting antimicrobial factors. ATII cells also secrete cytokines and chemokines, which are involved in activation and differentiation of immune cells and are able to present antigen to specific T cells. Another cell type that is important in lung defense is the pulmonary macrophage.

7.7.1 *De novo* dopamine synthesis in the lung

DA in the rat lung is much lower than NE, and its levels are significantly reduced following sympathectomy, indicating neuronal rather than local origin [82]. L-Dopa levels, on the other hand, are maintained at much higher relative concentrations than those of TH activity and the concentrations of DA or NE after sympathectomy, suggesting the L-Dopa serves as a significant source for local DA synthesis. Indeed, another study reported that ATII cells express DDC, which converts L-Dopa to DA, and that they produce DA [83]. Notably, rats fed a 4% tyrosine-supplemented diet for 24 h had increased urinary DA levels, while benserazide, which inhibits the conversion of L-Dopa to DA, prevented the increase in DA in the tyrosine-fed rats. There was no increase in other catecholamines in the urine of tyrosine-fed rats (neither Epi nor NE). As illustrated in **Figure 7.10**, the authors postulated that tyrosine-supplemented diet increases L-Dopa production in chromaffin tissue (which has TH) and that the newly synthesized L-dopa is released into the circulation and taken up by the alveolar epithelial cells to produce DA. The same study also presented evidence that increasing the production of DA (by feeding rats a 4% enriched tyrosine diet) increased Na/K-ATPase α_1 and β_1 subunit abundance in alveolar type II cells and increased Na/K-ATPase activity and protein abundance at the basolateral membranes of peripheral lung tissue.

7.7.2 Actions of dopamine in the lung

There is evidence that DA affects ventilation, pulmonary circulation, neuromodulation of sensory pulmonary nerves, and lung water clearance. Through complex mechanisms, DA exerts beneficial as well as detrimental effects

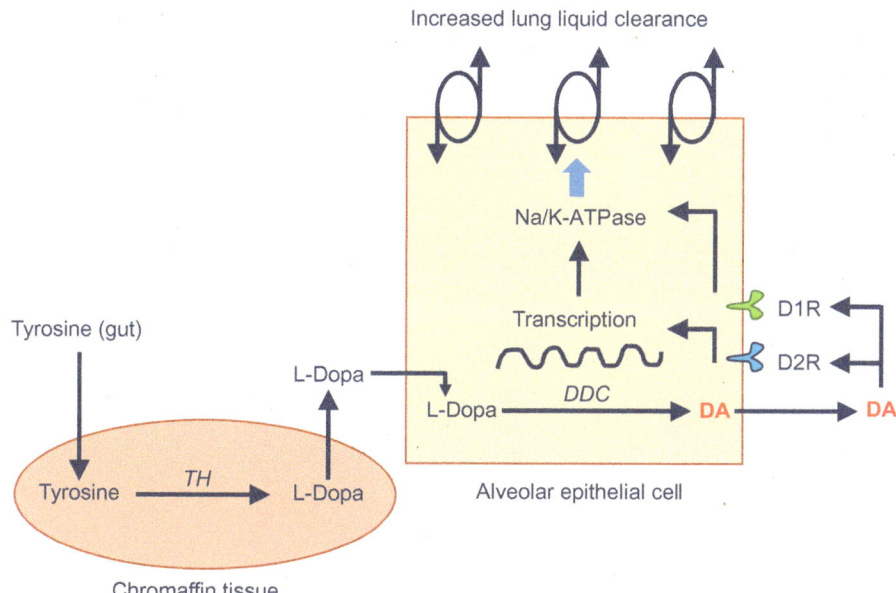

Figure 7.10 L-Dopa generation in chromaffin tissue by a tyrosine-enriched diet and the effects of DA on the alveoli. Circulating L-Dopa, released from chromaffin cells, is taken up by the alveolar cells and is converted to DA. Local DA is involved in the upregulation of alveolar epithelial Na/K-ATPase and increased liquid clearance. (Redrawn and modified from Adir, Y. et al., *Am. J. Respir. Crit. Care Med.*, 169, 757–763, 2004.)

on respiration. The beneficial effects of DA on the lung occur by decreasing edema formation and by improving respiratory muscle function. Yet, DA can also be harmful in cases of difficult weaning from mechanical ventilation. In particular, the effects of DA on respiration can be detrimental in seriously ill patients with a limited breathing reserve. This detrimental effect has been shown in patients with heart failure, where low-dose DA decreased minute ventilation during normoxic breathing, an effect that did not occur in age-matched control subjects [74]. Heart failure patients often have only mild arterial blood oxygen desaturation as a result of low cardiac output and pulmonary vascular congestion.

Chronic obstructive pulmonary disease (COPD), previously called chronic bronchitis or emphysema, is characterized by long-term breathing problems and poor airflow. The main symptoms include shortness of breath and cough with sputum production. COPD is a progressive disease that typically worsens over time. Tobacco smoking is the most common cause of COPD, with factors such as air pollution and genetics playing a smaller role. Based on the observation that DAR activation inhibits rapidly adapting stretch receptors in the lungs, DA administration was tested in animals and yielded positive results [74]. However, a randomized, placebo-controlled human study with DA produced disappointing results with respect to ventilation parameters. Moreover, sibenadet, a dual D2R-β_2 AR agonist, did not alleviate breathlessness, cough, or excess sputum in patients with COPD.

Pulmonary Edema entails fluid accumulation in the tissue and air spaces of the lung, leading to impaired gas exchange and may cause respiratory failure. Pulmonary edema is due either to failure of the left ventricle of the heart to remove blood adequately from the pulmonary circulation (cardiogenic pulmonary edema) or to injury to the parenchyma or vasculature of the lung (non-cardiogenic pulmonary edema). Treatment is focused on three aspects: first, improving respiratory function; second, treating the underlying cause; and third, avoiding further damage to the lung. Pulmonary edema, especially acute, can lead to fatal respiratory distress or cardiac arrest due to hypoxia. It is a cardinal feature of congestive heart failure. The rate of edema resolution in most patients with acute respiratory distress syndrome (ARDS) is markedly impaired, a finding that correlates with higher mortality [84]. Several mechanisms impair the resolution of alveolar edema in ARDS, including cell injury from unfavorable ventilator strategies or pathogens, hypoxia, cytokines, and oxidative stress. In patients with severe ARDS, alveolar epithelial cell death is a major mechanism that prevents the resolution of lung edema.

Mechanistically, pulmonary edema can result from a change in the hydrostatic and/or oncotic pressure gradients across the pulmonary circulation or from increased alveolar permeability [85]. The primary mechanism for removal of alveolar edema fluid depends on active ion transport across the alveolar epithelium. Thus, pulmonary edema is cleared from the alveoli by an active Na^+ transport, with Na^+ moving vectorially across the alveolar epithelium, entering the cell via apical amiloride-sensitive Na^+ channels and extruding from the cell via the basolaterally located Na/K-ATPase. Water follows the Na^+ gradients isosmotically, resulting in alveolar fluid reabsorption [86]. The resolution of alveolar edema across the normally tight epithelial barrier can be up-regulated by cAMP-independent mechanisms, including DA, glucocorticoids, thyroid hormone, and growth factors.

As summarized in a review [86], the function of Na/K-ATPase in the lung can be regulated by short- or long-term mechanisms. Several studies have demonstrated that DA increases alveolar fluid reabsorption in normal lungs and in animal models of lung injury by regulating Na/K-ATPase function. The short-term effects of DA in alveolar ETII cells cause translocation of preformed Na/K-ATPase pumps from intracellular pools (i.e., late endosomal compartment) to the cell plasma membrane. This rapid recruitment is regulated by the actin cytoskeleton and the microtubuli. Disruption of the microtubular transport system by colchicine blocks the stimulatory effects of DA on Na/K-ATPase translocation and impairs the lung's ability to clear edema in normal and injured lungs. Further studies have shown that Na/K-ATPase activity increased two-fold in ATII cells following 15 min of incubation with DA or the D1R agonist fenoldopam but not with quinipirole, the D2R agonist. DA increased Na/K-ATPase activity in alveolar epithelial type II cells by recruiting sodium pumps into the plasma membrane from an intracellular compartment via a novel protein kinase C-dependent pathway.

7.8 SYNOPSIS

This chapter reviews the functions of dopamine in three important systems: renal, cardiovascular and pulmonary. In the kidney, locally produced DA is closely associated with many of the renal functions, including ion transport, natriuresis and the maintenance of fluid hemodynamics. Renal DA is also involved with several kidney disorders such as essential hypertension and diabetic nephropathy. In the cardiovascular system, DA is involved in cardiac ischemia, in blood pressure regulation in various vascular beds, and in angiogenesis under several physiological and pathological conditions. In the respiratory system, DA is involved in upper airway pathophysiology, in carotid body oxygen-sensing functions and in several aspects of pulmonary ventilation and alveolar edema.

REFERENCES

1. Little MH, McMahon AP. Mammalian kidney development: Principles, progress, and projections. *Cold Spring Harb Perspect Biol.* 2012;4(5):1–18.

2. Dressler GR. The cellular basis of kidney development. *Annu Rev Cell Dev Biol.* 2006;22:509–529.

3. Kurtz A. Control of renin synthesis and secretion. *Am J Hypertens.* 2012;25(8):839–847.

4. He YM, Yao SW, Huang YJ, Liang BS, Liu HY. Investigating the recheck rules for urine analysis in children. *Genet Mol Res.* 2016;15(2):15–21.

5. Bankir L, Bouby N, Ritz E. Vasopressin: A novel target for the prevention and retardation of kidney disease? *Nat Rev Nephrol.* 2013;9(4):223–239.

6. Simoes E Silva AC, Flynn JT. The renin-angiotensin-aldosterone system in 2011: Role in hypertension and chronic kidney disease. *Pediatr Nephrol.* 2012;27(10):1835–1845.

7. Takata K, Matsuzaki T, Tajika Y, Ablimit A, Hasegawa T. Localization and trafficking of aquaporin 2 in the kidney. *Histochem Cell Biol.* 2008;130(2):197–209.

8. Zeng C, et al. Dysregulation of dopamine-dependent mechanisms as a determinant of hypertension: Studies in dopamine receptor knockout mice. *Am J Physiol Heart Circ Physiol.* 2008;294(2):H551–H569.

9. Choi MR, et al. Renal dopaminergic system: Pathophysiological implications and clinical perspectives. *World J Nephrol.* 2015;4(2):196–212.

10. Armando I, Konkalmatt P, Felder RA, Jose PA. The renal dopaminergic system: Novel diagnostic and therapeutic approaches in hypertension and kidney disease. *Transl Res.* 2015;165(4):505–511.

11. Banday AA, Lokhandwala MF. Dopamine receptors and hypertension. *Curr Hypertens Rep.* 2008;10(4):268–275.

12. Taveira-da-Silva R, da Silva SL, Vieyra A, Einicker-Lamas M. L-Tyr-induced phosphorylation of tyrosine hydroxylase at Ser40: An alternative route for dopamine synthesis and modulation of Na+/K+-ATPase in kidney cells. *Kidney Blood Press Res.* 2019;44(1):1–11.

13. Choi MR, Rukavina Mikusic NL, Kouyoumdzian NM, Kravetz MC, Fernandez BE. Atrial natriuretic peptide and renal dopaminergic system: A positive friendly relationship? *Biomed Res Int.* 2014;2014:20–30.

14. Loesch A, Unwin R, Gandhi V, Burnstock G. Sympathetic nerve varicosities in close apposition to basolateral membranes of collecting duct epithelial cells of rat kidney. *Nephron Physiol.* 2009;113(3):15–21.

15. Hussain T, Lokhandwala MF. Renal dopamine receptors and hypertension. *Exp Biol Med (Maywood).* 2003;228(2):134–142.

16. Zeng C, et al. Functional genomics of the dopaminergic system in hypertension. *Physiol Genomics.* 2004;19(3):233–246.

17. Boone M, Deen PM. Physiology and pathophysiology of the vasopressin-regulated renal water reabsorption. *Pflugers Arch.* 2008;456(6):1005–1024.

18. Li D, et al. Binding of losartan to angiotensin AT1 receptors increases dopamine D1 receptor activation. *J Am Soc Nephrol.* 2012;23(3):421–428.

19. Cuevas S, Villar VA, Jose PA, Armando I. Renal dopamine receptors, oxidative stress, and hypertension. *Int J Mol Sci.* 2013;14(9):17553–17572.

20. Zeng C, Zhang M, Asico LD, Eisner GM, Jose PA. The dopaminergic system in hypertension. *Clin Sci (Lond).* 2007;112(12):583–597.

21. Jose PA, Eisner GM, Felder RA. Renal dopamine receptors in health and hypertension. *Pharmacol Ther.* 1998;80(2):149–182.

22. Wang X, et al. Dopamine, kidney, and hypertension: Studies in dopamine receptor knockout mice. *Pediatr Nephrol.* 2008;23(12):2131–2146.

23. Felder RA, Eisner GM, Jose PA. D1 dopamine receptor signalling defect in spontaneous hypertension. *Acta Physiol Scand.* 2000;168(1):245–250.

24. Landoni G, et al. Beneficial impact of fenoldopam in critically ill patients with or at risk for acute renal failure: A meta-analysis of randomized clinical trials. *Am J Kidney Dis.* 2007;49(1):56–68.

25. Schenarts PJ, et al. Low-dose dopamine: A physiologically based review. *Curr Surg.* 2006;63(3):219–225.

26. De MC, Speed JS, Kasztan M, Gohar EY, Pollock DM. Endothelin-1 and the kidney: New perspectives and recent findings. *Curr Opin Nephrol Hypertens.* 2016;25(1):35–41.

27. Tokonami N, et al. Endothelin-1 mediates natriuresis but not polyuria during vitamin D-induced acute hypercalcaemia. *J Physiol.* 2017;595(8):2535–2550.

28. Zeng C, Yang Z, Asico LD, Jose PA. Regulation of blood pressure by D5 dopamine receptors. *Cardiovasc Hematol Agents Med Chem.* 2007;5(3):241–248.

29. Zhang LN, et al. Crosstalk between dopamine receptors and the Na(+)/K(+)-ATPase (review). *Mol Med Rep.* 2013;8(5):1291–1299.

30. Bacic D, et al. Dopamine acutely decreases apical membrane Na/H exchanger NHE3 protein in mouse renal proximal tubule. *Kidney Int.* 2003;64(6):2133–2141.

31. Escano CS, et al. Renal dopaminergic defect in C57Bl/6J mice. *Am J Physiol Regul Integr Comp Physiol.* 2009;297(6):R1660–R1669.

32. Kouyoumdzian NM, et al. Atrial natriuretic peptide stimulates dopamine tubular transport by organic cation transporters: A novel mechanism to enhance renal sodium excretion. *PLoS One.* 2016;11(7):e0157487.

33. Zeng C, et al. Renal D3 dopamine receptor stimulation induces natriuresis by endothelin B receptor interactions. *Kidney Int.* 2008;74(6):750–759.

34. Segers O, Anckaert E, Gerlo E, Dupont AG, Somers G. Dopamine-sodium relationship in type 2 diabetic patients. *Diabetes Res Clin Pract.* 1996;34(2):89–98.

35. Zhang MZ, et al. Intrarenal dopamine inhibits progression of diabetic nephropathy. *Diabetes.* 2012;61(10):2575–2584.

36. Hussain T, Beheray SA, Lokhandwala MF. Defective dopamine receptor function in proximal tubules of obese Zucker rats. *Hypertension.* 1999;34(5):1091–1096.

37. Lopez-Sanchez C, et al. Tyrosine hydroxylase is expressed during early heart development and is required for cardiac chamber formation. *Cardiovasc Res.* 2010;88(1):111–120.

38. Zhou QY, Quaife CJ, Palmiter RD. Targeted disruption of the tyrosine hydroxylase gene reveals that catecholamines are required for mouse fetal development. *Nature.* 1995;374(6523):640–643.

39. Hoover DB, et al. Localization of multiple neurotransmitters in surgically derived specimens of human atrial ganglia. *Neuroscience.* 2009;164(3):1170–1179.

40. Schomig E, Lazar A, Grundemann D. Extraneuronal monoamine transporter and organic cation transporters 1 and 2: A review of transport efficiency. *Handb Exp Pharmacol.* 2006(175):151–180.

41. Fujii T, Yamazaki T, Akiyama T, Sano S, Mori H. Extraneuronal enzymatic degradation of myocardial interstitial norepinephrine in the ischemic region. *Cardiovasc Res.* 2004;64(1):125–131.

42. Cavallotti C, Mancone M, Bruzzone P, Sabbatini M, Mignini F. Dopamine receptor subtypes in the native human heart. *Heart Vessels.* 2010;25(5):432–437.

43. Li HZ, et al. Effect of dopamine receptor 1 on apoptosis of cultured neonatal rat cardiomyocytes in simulated ischaemia/reperfusion. *Basic Clin Pharmacol Toxicol.* 2008;102(3):329–336.

44. Gao J, et al. Involvement of dopamine D2 receptors activation in ischemic post-conditioning-induced cardioprotection through promoting PKC-epsilon particulate translocation in isolated rat hearts. *Mol Cell Biochem.* 2013;379(1–2):267–276.

45. Hiemstra B, et al. Dopamine in critically ill patients with cardiac dysfunction: A systematic review with meta-analysis

and trial sequential analysis. *Acta Anaesthesiol Scand.* 2019;63(4):424–437.

46. Torres-Courchoud I, Chen HH. Is there still a role for low-dose dopamine use in acute heart failure? *Curr Opin Crit Care.* 2014;20(5):467–471.

47. Hahn RA, Wardell JR, Jr., Sarau HM, Ridley PT. Characterization of the peripheral and central effects of SK&F 82526, a novel dopamine receptor agonist. *J Pharmacol Exp Ther.* 1982;223(2):305–313.

48. Ng SS, Pang CC. In vivo venodilator action of fenoldopam, a dopamine D(1)-receptor agonist. *Br J Pharmacol.* 2000;129(5):853–858.

49. Igreja B, Wright LC, Soares-da-Silva P. Sustained high blood pressure reduction with etamicastat, a peripheral selective dopamine beta-hydroxylase inhibitor. *J Am Soc Hypertens.* 2016;10(3):207–216.

50. Olsen NV. Effects of dopamine on renal haemodynamics tubular function and sodium excretion in normal humans. *Dan Med Bull.* 1998;45(3):282–297.

51. Kuchel O, Leveille J. Idiopathic hypovolemia: A self-perpetuating autonomic dysfunction? *Clin Auton Res.* 1998;8(6):341–346.

52. Zhang Z, Chen K. Vasoactive agents for the treatment of sepsis. *Ann Transl Med.* 2016;4(17):333.

53. Teisman AC, Buikema H, van Veldhuisen DJ, de ZD, van Gilst WH. Direct vasodilating effects of the new dopaminergic agonist Z1046 in human arteries. *J Cardiovasc Pharmacol.* 2000;35(4):581–585.

54. Katai R, et al. The variable effects of dopamine among human isolated arteries commonly used for coronary bypass grafts. *Anesth Analg.* 2004;98(4):915–920.

55. Toda N, Enokibori M, Matsumoto T, Okamura T. Responsiveness to dopamine of isolated epicardial coronary arteries from humans, monkeys, and dogs. *Anesth Analg.* 1993;77(3):526–532.

56. Tayebati SK, Lokhandwala MF, Amenta F. Dopamine and vascular dynamics control: Present status and future perspectives. *Curr Neurovasc Res.* 2011;8(3):246–257.

57. Li Z, et al. Inhibitory effect of D1-like and D3 dopamine receptors on norepinephrine-induced proliferation in vascular smooth muscle cells. *Am J Physiol Heart Circ Physiol.* 2008;294(6):H2761–H2768.

58. Zeng C, et al. D1-like receptors inhibit insulin-induced vascular smooth muscle cell proliferation via down-regulation of insulin receptor expression. *J Hypertens.* 2009;27(5):1033–1041.

59. Zhou Y, et al. Inhibitory effect of D1-like dopamine receptors on neuropeptide Y-induced proliferation in vascular smooth muscle cells. *Hypertens Res.* 2015;38(12):807–812.

60. Yu C, et al. Activation of the D4 dopamine receptor attenuates proliferation and migration of vascular smooth muscle cells through downregulation of AT1a receptor expression. *Hypertens Res.* 2015;38(9):588–596.

61. Kou X, et al. Dopamine d(1)-like receptors suppress proliferation of vascular smooth muscle cell induced by insulin-like growth factor-1. *Clin Exp Hypertens.* 2014;36(3):140–147.

62. Breen EC. VEGF in biological control. *J Cell Biochem.* 2007;102(6):1358–1367.

63. Vohra PK, et al. Dopamine inhibits pulmonary edema through the VEGF-VEGFR2 axis in a murine model of acute lung injury. *Am J Physiol Lung Cell Mol Physiol.* 2012;302(2):L185–L192.

64. Basu S, et al. The neurotransmitter dopamine inhibits angiogenesis induced by vascular permeability factor/vascular endothelial growth factor. *Nat Med.* 2001;7(5):569–574.

65. Tang H, Mourad S, Zhai SD, Hart RJ. Dopamine agonists for preventing ovarian hyperstimulation syndrome. *Cochrane Database Syst Rev.* 2016;11:CD008605.

66. Kasum M, et al. Dopamine agonists in prevention of ovarian hyperstimulation syndrome. *Gynecol Endocrinol.* 2014;30(12):845–849.

67. Ferrero H, et al. Dopamine receptor 2 activation inhibits ovarian vascular endothelial growth factor secretion in an ovarian hyperstimulation syndrome (OHSS) animal model: Implications for treatment of OHSS with dopamine receptor 2 agonists. *Fertil Steril.* 2014;102(5):1468–1476.

68. Ferrero H, et al. Dopamine agonist inhibits vascular endothelial growth factor protein production and secretion in granulosa cells. *Reprod Biol Endocrinol.* 2015;13:104–112.

69. Sarkar C, Chakroborty D, Basu S. Neurotransmitters as regulators of tumor angiogenesis and immunity: The role of catecholamines. *J Neuroimmune Pharmacol.* 2013;8(1):7–14.

70. Sarkar C, Ganju RK, Pompili VJ, Chakroborty D. Enhanced peripheral dopamine impairs post-ischemic healing by suppressing angiotensin receptor type 1 expression in endothelial cells and inhibiting angiogenesis. *Angiogenesis.* 2017;20(1):97–107.

71. Dahlqvist A, Forsgren S. Expression of catecholamine-synthesizing enzymes in paraganglionic and ganglionic cells in the laryngeal nerves of the rat. *J Neurocytol.* 1992;21(1):1–6.

72. Dahlqvist A, Pequignot JM, Hellstrom S. Sympathectomy provides evidence of dopamine storage in rat laryngeal nerve paraganglia. *Acta Physiol Scand.* 1989;135(2):189–195.

73. Matsuyama N, et al. The dopamine D1 receptor is expressed and induces CREB phosphorylation and MUC5AC expression in human airway epithelium. *Respir Res.* 2018;19(1):53–65.

74. Ciarka A, Vincent JL, van de Borne P. The effects of dopamine on the respiratory system: Friend or foe? *Pulm Pharmacol Ther.* 2007;20(6):607–615.

75. Cabezas GA, Velasco M. DA1 receptors modulation in rat isolated trachea. *Am J Ther.* 2010;17(3):301–305.

76. Fitzgerald RS, Eyzaguirre C, Zapata P. Fifty years of progress in carotid body physiology—Invited article. *Adv Exp Med Biol.* 2009;648):19–28.

77. Iturriaga R, Alcayaga J, Gonzalez C. Neurotransmitters in carotid body function: The case of dopamine—Invited article. *Adv Exp Med Biol.* 2009;648):137–143.

78. Tatsumi K, Pickett CK, Weil JV. Decreased carotid body hypoxic sensitivity in chronic hypoxia: Role of dopamine. *Respir Physiol.* 1995;101(1):47–57.

79. Tatsumi K, Pickett CK, Weil JV. Possible role of dopamine in ventilatory acclimatization to high altitude. *Respir Physiol.* 1995;99(1):63–73.

80. Gomez-Nino A, Dinger B, Gonzalez C, Fidone SJ. Differential stimulus coupling to dopamine and norepinephrine stores in rabbit carotid body type I cells. *Brain Res.* 1990;525(1):160–164.

81. Bainbridge CW, Heistad DD. Effect of haloperidol on ventilatory responses to dopamine in man. *J Pharmacol Exp Ther.* 1980;213(1):13–17.

82. Kawamura M, et al. Differential effects of chemical sympathectomy on expression and activity of tyrosine hydroxylase and levels of catecholamines and DOPA in peripheral tissues of rats. *Neurochem Res.* 1999;24(1):25–32.

83. Adir Y, et al. Augmentation of endogenous dopamine production increases lung liquid clearance. *Am J Respir Crit Care Med.* 2004;169(6):757–763.

84. Matthay MA. Resolution of pulmonary edema. Thirty years of progress. *Am J Respir Crit Care Med.* 2014;189(11):1301–1308.

85. Dada LA, Sznajder JI. Mechanisms of pulmonary edema clearance during acute hypoxemic respiratory failure: Role of the Na,K-ATPase. *Crit Care Med.* 2003;31(4 Suppl):S248–S252.

86. Adir Y, Sznajder JI. Regulation of lung edema clearance by dopamine. *Isr Med Assoc J.* 2003;5(1):47–50.

87. Ravid M. Dual blockade of the renin-angiotensin system in diabetic nephropathy. *Diabetes Care.* 2009;32(Suppl 2):S410–S413.

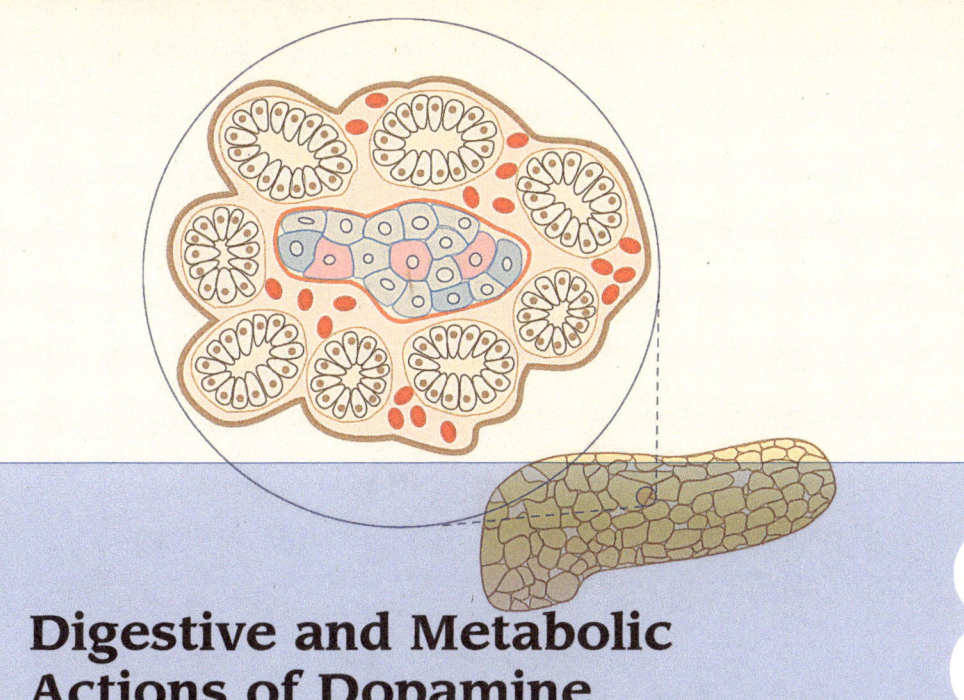

Digestive and Metabolic Actions of Dopamine

8

8.1 INTRODUCTION

The function of the gastrointestinal (GI) tract is **digestion and absorption** of ingested nutrients. To this end, the GI tract employs the processes of secretion, motility, and absorption, all of which must be integrated and well-coordinated. **Figure 8.1** shows the overall structure of the GI tract, and divides its functions into two complementary categories: (1) **motility**, or the control of food movement throughout the system, and (2) **digestion**, or the breakdown of food, secretion of digestive juices by accessory organs and internal secretory cells, and absorption of the processed nutrients. The GI tract also produces and secretes a number of **hormones**, some of which act locally to assist in digestion, while others have several peripheral as well as central targets. A list of GI-derived hormones is presented in **Table 8.1**.

Throughout its length, the GI tract operates as a hollow muscular tube with a similar basic histology among the various components. However, each of its components has distinct structural modifications that are needed in order to carry out its functional specialties. In a cross section from the lumen out, the GI tract is composed of four concentric layers: mucosa, submucosa, muscular layer, and adventitia or serosa. The mucosa is the innermost layer and comes in direct contact with the digested food, whereas the muscular layer is engaged in the transfer of food by contractions, relaxation, and peristalsis.

The second part of this chapter focuses on the involvement of dopamine (DA) with metabolic homeostasis in health and disease. To this end, features of the major organs that regulate metabolism are first briefly delineated. Next, two of the most common metabolic-related disorders, obesity and diabetes, are discussed, with an emphasis on the potential roles of DA. Finally, induction of obesity by antipsychotics, especially those that alter the dopaminergic system, is reviewed.

Figure 8.1 The human gastrointestinal system. Functions associated with motility are designated on the left, while those associated with digestion are shown on the right.

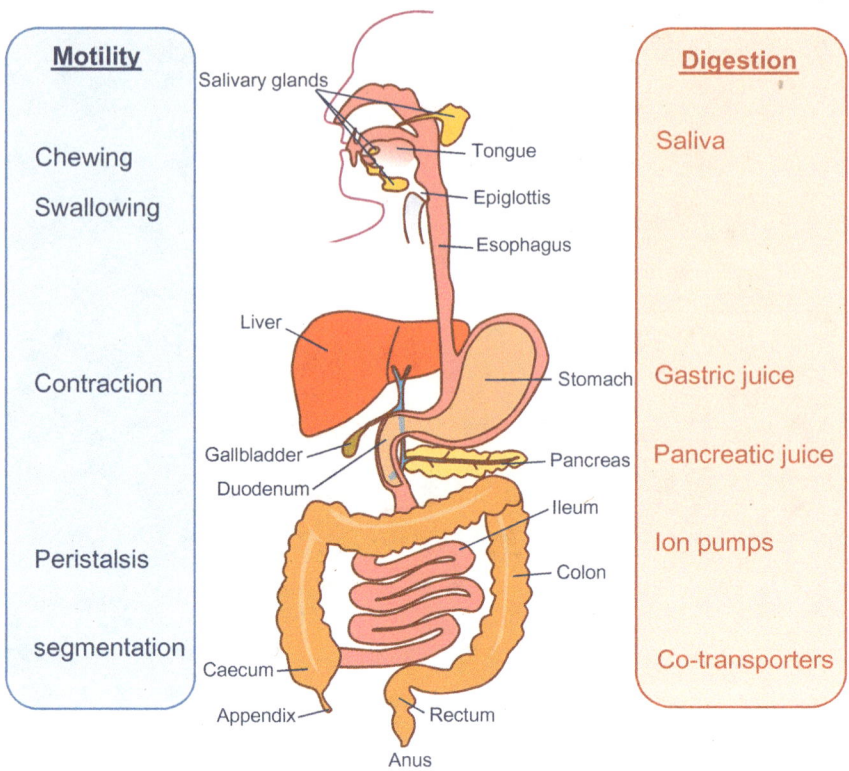

Table 8.1 Most important hormones secreted by the gastrointestinal tract				
Hormone	**Secreted by**	**Target(s)**	**Releasing stimuli**	**Actions**
Gastrin	G cells of stomach	Histamine secreting and parietal cells	Peptides in lumen	Increases acid secretion in stomach and gastric motility
Cholecystokinin (CCK)	Endocrine cells of small intestine	Gall bladder; pancreas; gastric muscle	Partially digested proteins and fatty acids in duodenum	Gallbladder contraction; inhibits gastric emptying; stimulates secretion of pancreatic enzymes
Secretin	Endocrine cells of small intestine	Pancreas; stomach	Acid and partially digested proteins and fatty acids in duodenum and small intestine	Stimulates bicarbonate secretion and pepsin release; inhibits gastric acid secretion, bile ejection
Motilin	Endocrine cells of upper small intestine	Smooth muscle of duodenum	Fasting	Stimulates gastric contractions

8.2 GASTROINTESTINAL SYSTEM: MOTILITY

Motility throughout the digestive system necessitates coordinated contractions and relaxation of the smooth muscle within the various segments of the GI tract [1]. Contractions produce propulsive forces that move the digesta along the GI tract and assist in the mixing of the ingested foodstuff with digestive enzymes by bringing the nutrients into close contact with the mucosa for efficient digestion and absorption. Relaxation/contraction of certain smooth muscles allows sphincters to open and close for a regulated passage of foodstuff and also enables the ingested material to be contained and processed within reservoirs such as the stomach.

8.2.1 Pharynx and esophagus

Passage of food through the GI tract begins at the mouth, where **chewing** initiates mechanical digestion. This is followed by **swallowing**, which pushes the food through the pharynx and into the esophagus. Eating and swallowing represent complex neuromuscular activities, consisting of three components: oral, pharyngeal, and esophageal, each of which is controlled by a different neurological mechanism. Parkinson's disease (PD) patients, as well as patients treated with neuroleptics, often experience impairments of chewing and swallowing [2]. Such disorders have been generally attributed to the degeneration of nigrostriatal dopaminergic neurons or to drug actions on these neurons, while overlooking the potential involvement of the local DA–DA receptors (DAR) axis. Therefore, the relevant information on the roles of local DA is presented and emphasized in the following sections.

The **pharynx** is a funnel-shaped tube connected to the posterior end of the mouth. It contains a flap of tissue, the **epiglottis**, which acts as a switch to route food into the esophagus and air into the larynx. As many as 60%–70% of late-stage PD patients report symptoms of **dysphagia**, a medical term that describes difficulty in swallowing. There are two types of dysphagia: oropharyngeal, which designates the difficulty in starting a swallow, and esophageal, which indicates a sensation of food being stuck in the neck or chest. Such problems severely reduce the quality of life for PD patients and constitute a major medical challenge for the physician. Aspiration pneumonia, a complication of chronic dysphagia, has a four times higher incidence in PD patients than in the general population and is a leading cause of death in long-term studies. The neurological basis for dysphagia in PD is poorly understood, owing in part to the lack of appropriate animal models for specifically addressing this problem.

A recent study [3] reported that administration of rotenone, a commercial pesticide, to rats caused nigrostriatal DA loss, **alpha synuclein** (α-syn) aggregation, polyubiquitin formation, and evidence of inflammation and microglial activation. Given that this treatment reproduced many of the neurodegenerative and behavioral phenotypes of PD, it offered a good experimental model for analyzing the involvement of DA in swallowing. Rotenone had dose-dependent effects on the timing and amplitude of chewing, as well as on the coordination of chewing and swallowing. However, in spite of an interesting set of data, the study fell short of resolving the question whether the swallowing problems were due to a deficiency in the central vs. the local dopaminergic systems.

Evidence that activation of pharyngeal D1R stimulates the swallowing reflex was provided in a study that treated guinea pigs with a D1R antagonist with and without substance P [4]. Substance P is released from vagal sensory nerves in the pharynx and mediates both swallowing and coughing reflexes. The blockade of D1R decreased the number of swallows and reduced substance P content in the laryngeal and pharyngeal mucosa, suggesting functional interactions between substance P and DA. Indeed, it was subsequently found that tyrosine hydroxylase (TH)-positive sympathetic nerve endings innervate the dorsal root ganglia and sensory ganglia of the glossopharyngeal nerve of guinea pigs that contained substance P.

The **esophagus** is a muscular tube with sphincters at both ends that leads from the pharynx to the stomach (**Figure 8.2**A). The upper esophageal sphincter (UES) is under conscious control and its main function is to keep food from going down the larynx during breathing, eating, belching, and vomiting. The lower esophageal sphincter (LES) is not under voluntary control and acts to prevent acid and stomach contents from moving backward into the mouth. Propulsion of food through the esophagus occurs through peristalsis, with a contraction occurring behind a bolus of food and a relaxation occurring ahead of the bolus of food (**Figure 8.2**B).

Studies with opossums found that intravenous administration of DA caused dose-dependent reductions in the lower esophageal sphincter pressure, as well as fewer contractions in the lowermost part of the body of the esophagus [5]. The onset of contractions in the esophageal body occurred 20 seconds after the onset of the sphincter response, indicating some synchronization. Because vagal stimulation also caused contraction in the body and relaxation of the sphincter, the authors proposed that the vagal effect on the esophagus is mediated, in part, by DARs.

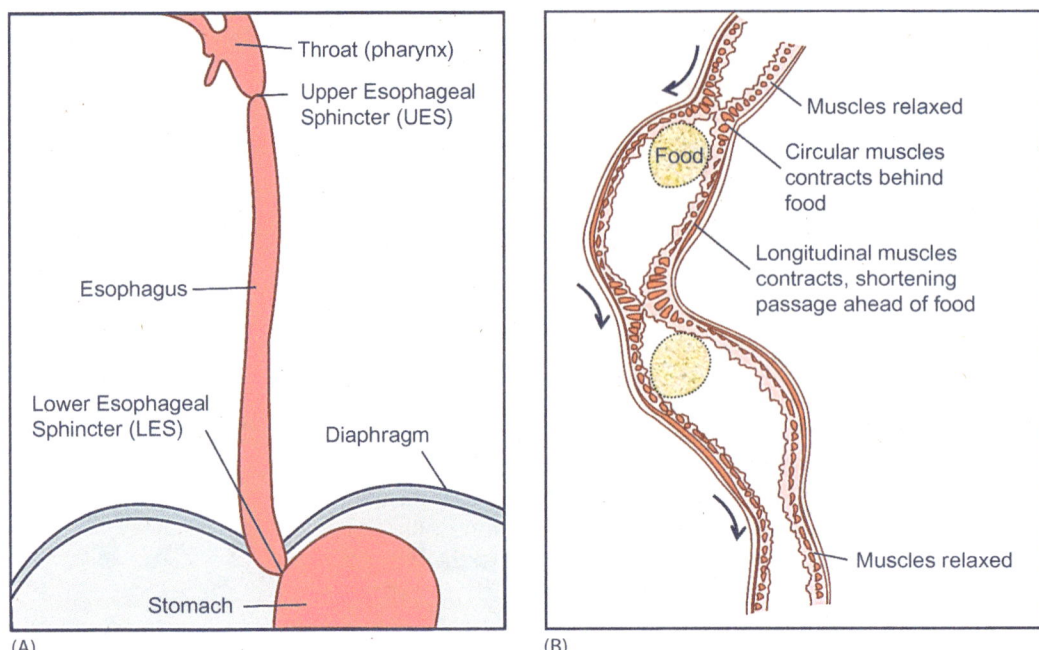

Figure 8.2 Panel A shows the human esophagus, which connects the pharynx to the stomach, with the upper and lower esophageal sphincters. Panel B is a diagram of the propulsion of food through the esophagus by coordinated muscle contraction and relaxation.

In humans, both D1R-like and D2R-like are located in the LES but not in the esophageal body [6]. Surgical specimens of human esophagus were treated with fenoldopam (a peripheral D1R agonist) or bromocriptine (a D2R agonist), followed by cAMP analysis [7]. The data showed that D2R activation decreased LES activity, while D1R activation caused an increase. In addition, the local esophageal DA system was dysregulated in LES **achalasia**, a neuromuscular disorder characterized by an absence of peristalsis in the esophageal body and a failure of the LES to relax in response to swallowing. In clinical trials, domperidone, a peripheral D2R antagonist, provided short-term relief of symptoms to patients with dyspepsia or dysfunctional gastroesophageal reflux.

Gastroesophageal reflux disease (**GERD**), or acid reflux disease, is a recurrent condition where acidic gastric juices leak upward into the esophagus [8]. The most common symptoms are an acidic taste in the mouth, regurgitation, and heartburn. The pathogenesis of GERD is multifactorial, involving lower esophageal sphincter relaxation or pressure abnormalities. As a result, reflux of acid, bile, pepsin, and pancreatic enzymes occurs, leading to esophageal mucosal injury. Other factors contributing to the pathophysiology of this disease include hiatal hernia, impaired esophageal clearance, delayed gastric emptying, and impaired mucosal defensive factors. Treatments for GERD include changes in food choices, medications, and surgery in extreme cases. Initial treatment is commonly done with proton-pump inhibitors such as omeprazole to neutralize the acid. Metoclopramide, a peripheral D2R antagonist, has been used alone or in combination with antacids to treat GERD for some years. However, use of metoclopramide has declined in recent years because of concerns with adverse effects such as PD-like movement disorders.

8.2.2 Stomach and duodenum

The **stomach** is a hollow muscular organ located between the esophagus and the small intestines. It is divided into three main parts: fundus, body, and pyloric region (pyloric sphincter). Like the rest of the alimentary canal, the stomach's wall is made of four concentric layers, but with structural adaptations in the mucosa and muscularis to enable the unique functions of the organ. In addition to the typical circular and longitudinal smooth muscle layers, the muscularis has an inner oblique smooth muscle layer. Consequently, in addition to moving the food through the lumen, the stomach can vigorously churn the food and mechanically break it down into smaller particles. Stomach functions include food storage, initial digestion of proteins, production of chyme (a semifluid mass of partly digested food), and the killing of bacteria by the strong acidity. The pyloric sphincter controls the passage of chyme from the stomach into the duodenum, where peristalsis takes over to move it through the rest of the intestines.

Immunostaining and *in situ* hybridization histochemistry of the rat stomach has showed the presence of TH, DA transporter (DAT), and vesicular monoamine transporters (VMAT) in the acid-producing parietal cells of the mucosa. Both TH enzyme activity and DA levels were unchanged after chemical sympathectomy, indicating independence of sympathetic innervation [9]. Active reuptake and storage of DA showed a regulated DA release from parietal cells, likely acting via D1R, the most abundant DAR in both the gastric and duodenal epithelia. The authors concluded that gastric epithelia has self-contained, functional DA-producing neuroendocrine cells and that DA has important roles in several gastric functions. Other investigators, using *in situ* hybridization, reported detection of mRNAs of all five DAR subtypes in both the muscular and mucosal layers of rat stomach [10].

In humans, TH, DA, and DA sulfate (DA-S) were detected in stomach samples from normal patients [11], while high-affinity D2R were detected in normal, benign, and malignant stomach tissues [12]. Both the density and affinity of the DA binding sites were similar in normal tissue and benign tumors, whereas the density of the receptors was decreased in the malignant tissue. **Gastroparesis**, a disorder which slows or stops food movement from the stomach to the small intestine, is common in diabetes and after some surgical procedures. Antidopaminergic drugs, such as domperidone and metoclopramide [13], have been effective in alleviating gastroparesis, although they also incur some undesirable side effects.

As illustrated in **Figure 8.3**, GI dysfunction is the most common non-motor symptom of PD [14]. Patients often have early satiety and nausea

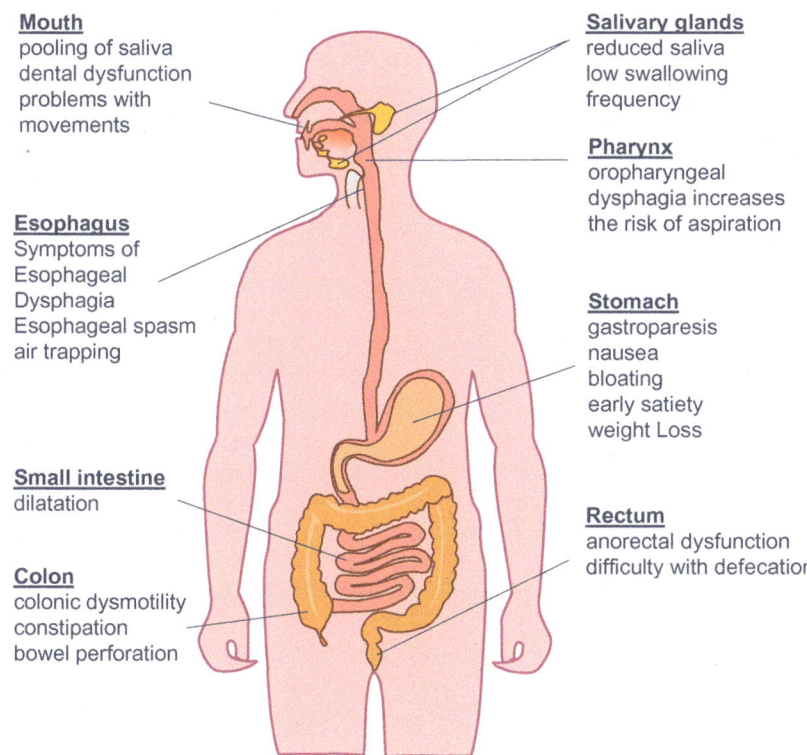

Mouth
pooling of saliva
dental dysfunction
problems with
movements

Salivary glands
reduced saliva
low swallowing
frequency

Pharynx
oropharyngeal
dysphagia increases
the risk of aspiration

Esophagus
Symptoms of
Esophageal
Dysphagia
Esophageal spasm
air trapping

Stomach
gastroparesis
nausea
bloating
early satiety
weight Loss

Small intestine
dilatation

Rectum
anorectal dysfunction
difficulty with defecation

Colon
colonic dysmotility
constipation
bowel perforation

Figure 8.3 Common gastrointestinal dysfunctions in Parkinson's disease. (Redrawn and modified from Pfeiffer, R.F. *Lancet Neurol.*, 2, 107–116, 2003.)

symptoms resulting from delayed gastric emptying, bloating from poor small bowel coordination, as well as constipation and defecation dysfunctions because of impaired colonic transit. Dopaminergic-related disturbances in gut motility have also been observed in transgenic mice lacking D2R [15]. These mice are smaller in size than wild-type littermates in spite of eating significantly more. They have greater defecation frequency, larger water content and mass of their stool, and significant decreases in total GI and colonic transit times. The higher motility in the absence of D2R indicated a physiologically relevant inhibitory effect of DA on gut motility.

The **duodenum** is a C-shaped structure, about 25 cm long in humans. Anatomically, it is considered part of the intestine while functionally it shares several characteristics with the stomach. The duodenum begins at the pyloric sphincter of the stomach and ends at the jejunum (**Figure 8.1**). Within the duodenum, chyme is mixed with bile from the gallbladder and digestive juices from the pancreas. Secretion of bicarbonate increases the pH at the duodenal mucosal surface, a process that provides protection against acid/pepsin injury and maintains the mucosal integrity. DA is produced in the human duodenum by locally expressed TH as well as by decarboxylation of L-Dopa [11]. D1R, D2R, and D3R were identified in the mouse duodenum [15].

Incubation of longitudinal smooth muscle strips from different sections of the rat intestine with DA have revealed region-specific effects of DA on smooth muscle motility [16]. In the duodenum, DA induced D1R- and D2R-dependent contractions, whereas in more distal parts, it caused alpha and beta adrenoceptor-dependent relaxation. Similar results were obtained when mechanically ventilated critically ill patients were infused with low-dose DA and showed adverse effects on gastroduodenal motility [17].

8.2.3 Small and large intestine

The **intestine** (also called the bowel, or gut) in humans extends from the pyloric sphincter of the stomach to the anus and consists of two main segments: the small intestine and the large intestine. The small intestine is subdivided into the duodenum, jejunum, and ileum, whereas the large intestine is subdivided into the cecum, ascending, transverse, descending and sigmoid colon, rectum, and anal canal. During embryonic development, the primitive gut is patterned into three segments: foregut, midgut, and hindgut, each of which gives rise to specific gut and gut-related structures. Components derived from the gut (i.e., the stomach and colon) develop as swellings or dilatations in cells of the primitive gut, while structures that are derived from the primitive gut, but are not part of the gut proper, generally develop as outpocketings of the primitive gut.

Immunoreactive D1R and D2R were detected in the distal colon of rats, including the myenteric plexus and smooth muscle [18]. Acute cold-restraint stress enhanced the contractions of the distal colon, which was more sensitive to DA than the control. Both DA content and D1R expression in the smooth muscle layer of the colon increased under the stress condition. Support for the role of local DA in gut motility comes from the use of **MPTP**, a selective DA neurotoxin, which generates a mouse model of PD [19]. MPTP caused a 40% reduction of DA neurons in the gut without altering the density of cholinergic neurons. Electrophysiological recording of contraction in isolated colons from MPTP-treated mice showed relaxation defects associated with the loss of DA. MPTP administration *in vivo* induced a transient increase in colon motility without changing gastric emptying or transit through the small intestine. In humans, DA inhibited motility in the upper gut but stimulated motility in the colon [20]. The contrasting effects of DA on intestinal motor activity confirms the presence of heterogeneous populations of DAR along the GI tract, and the indirect effects of DA on the ARs.

Irritable bowel syndrome (IBS) is a common disorder that affects the large intestine. Symptoms include cramping, abdominal pain, bloating, gas, and diarrhea, constipation or both. Almost 2 of 10 people suffer from IBS, with women being more affected than men. Causes include weak intestinal contractions that slow food passage, poorly coordinated signals between the brain and the intestines, inflammation or bacterial infection in the intestines, and changes in the intestinal microflora [21]. In a recent Polish study, serum and urinary serotonin and DA and their metabolites were analyzed in healthy controls, IBS patients with diarrhea (IBS-D), and IBS patients with constipation (IBS-C). Patients had their symptoms for at least 6 months and had stopped all medications 1 week before the test [22]. Patients with IBS-D had significantly higher blood levels of serotonin and urinary levels of its metabolite, 5-hydroxyindolacetic acid (5-HIAA) than healthy controls and patients with IBS-C, while those with IBS-C had higher plasma DA and urinary homovanylic acid (HVA) levels. It is unfortunate that blood levels of DA-S, the predominant circulating catecholamine in humans, were not analyzed in this study.

Prokinetic drugs enhance intestinal motility by increasing the frequency and/or the strength of contraction of different segments of the GI tract without disrupting their rhythm. Such drugs are used in the clinic to treat IBS, gastritis, GERD, gastroparesis (delayed gastric emptying), and functional dyspepsia (impaired digestion). The most widely prescribed prokinetic drugs are bethanechol (a muscarinic receptor agonist), metoclopramide and domperidone (peripheral D2R antagonists), and cisapride (a serotonergic 5-HT4 receptor agonist). Acting by enhancing the effects of acetylcholine and/or by blocking the inhibitory effects of DA [23], prokinetic drugs can relieve abdominal discomfort, bloating, constipation, heartburn, nausea, and vomiting.

8.3 GASTROINTESTINAL SYSTEM: DIGESTIVE FUNCTIONS

The digestive system is responsible for taking in foods and turning them into energy and nutrients that support the functions, growth, and repair of the body. The digestive activity can be divided into six processes: (1) ingestion, (2) mixing and moving of food and waste, (3) secretion of fluids and digestive enzymes, (4) breaking food down into components, (5) absorption of nutrients, and (6) excretion of waste. A list of digestive enzymes produced by the different components of the GI tract is presented in **Table 8.2**.

Each day, approximately 8 L of fluid pass through the human GI tract, with only 2 L coming from ingestion and the remainder coming from secretion by the GI system itself [24]. Of the latter, about half is secreted by the digestive glands—salivary glands, pancreas, liver, and gall bladder—and the other half is secreted by specialized epithelial cells within the digestive tract.

Table 8.2 Digestive enzymes of the gastrointestinal tract

Location	Enzyme name	Action
Salivary glands	Amylase; lipase	Starch; triglycerides
Stomach	Pepsin (pepsinogen); gastric lipase	Proteins; triglycerides
Pancreas	Amylase; lipase; phospholipase; trypsinogen; chymotrypsinogen	Starch; triglycerides; phospholipids; peptides
Intestinal epithelium	Enterokinase; sucrase; maltase; lactase; endopeptidases; aminopeptidase; carboxypeptidase	Activates trypsin; sucrose; maltose; lactose; interior and terminal peptide bonds; amide and carboxyl ends of peptides

Nearly all this fluid is normally reabsorbed, and feces contain significant amounts of fluid only when diarrhea occurs. Hence, the regulation of the secretory activity of different parts of the GI tract is important not only for the provision of appropriate nutrition but also for the maintenance of the correct fluid and electrolyte balance in the body.

8.3.1 Salivary glands

In humans, **saliva** is produced in three pairs of salivary glands—parotid, submandibular, and sublingual—whose secretory activities are controlled by autonomic sympathetic and parasympathetic innervation. Unlike the rest of the body, the two divisions of the autonomic system work in the same direction rather than antagonistically in the regulation of the salivary glands [25]. As shown in **Figure 8.4**A, the salivary glands are composed of two main cell types: acini, which produce the saliva, and ductal cells, which modify and deliver saliva to the mouth.

Between 0.5 and 1.5 L of saliva are produced in humans every day. In addition to its role in digestion, saliva provides a physical barrier against local irritants, lubricates and cleans the oral tissues, promotes mucosal repair, maintains a stable intraoral pH and electrolyte balance, prevents tooth demineralization, and contributes to the remineralization of dental hard tissues (**Figure 8.4**B). Saliva contains antibacterial, antiviral, and antifungal components that help to maintain the normal oral flora and provide protection. Saliva facilitates mastication and swallowing, initiates food processing, and affects the perception of taste. Saliva is also an easily accessible body fluid for the diagnosis of many local and systemic conditions [26]. Samples of human saliva contain low levels of DA and NE that do did not parallel their levels in plasma [27]. When tested volunteers expressed fear or anxiety, DA levels in their saliva increased tenfold. The mechanism by which stress stimulates DA release into saliva is unclear and merits additional investigation.

A review of the literature found many publications on DA as a key regulator of the salivary glands in ticks, whose saliva is the major route for

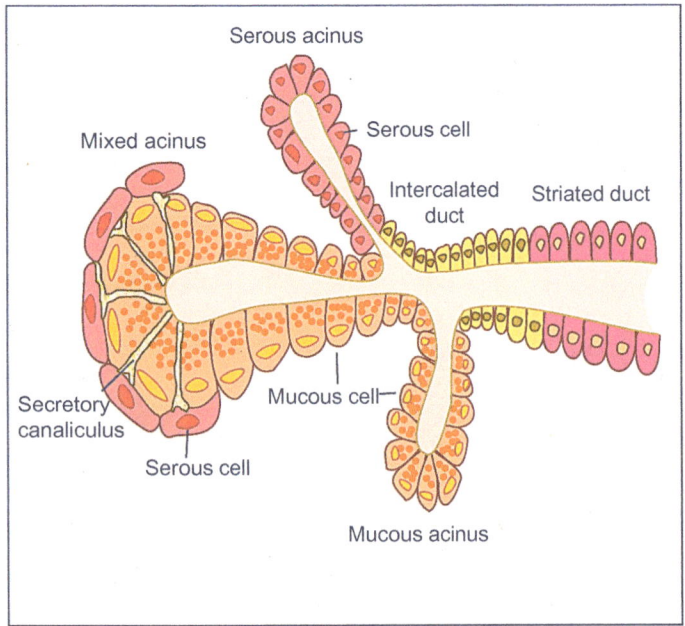

(A)

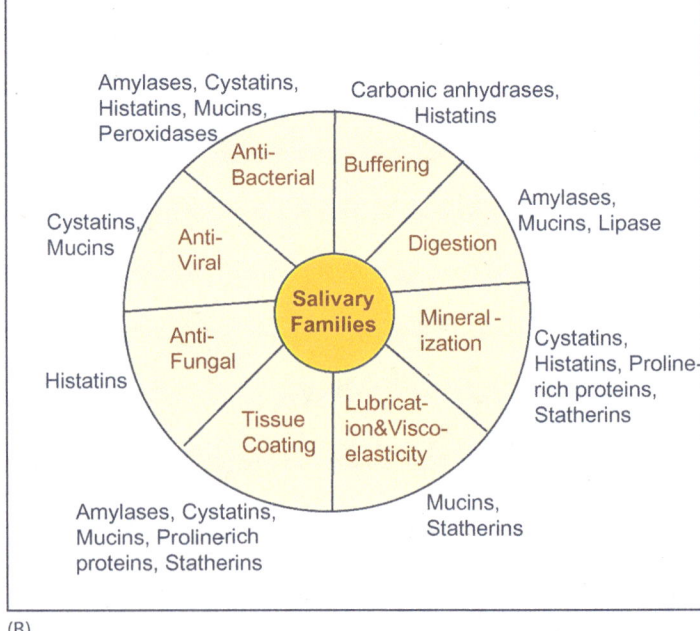

(B)

Figure 8.4 Panel A is a schematic representation of the different cellular components of a salivary gland. Panel B demonstrates the various functions attributed to saliva. (Redrawn and modified from Carpenter, G.H. *Annu. Rev. Food Sci. Technol.*, 4, 267–276, 2013.)

pathogen transmission in many mammals [28]. In rats, the only mammalian species studied in detail with respect to DA homeostasis, the salivary glands have a complete DA system, including all the biosynthetic and uptake enzymes, and several DAR subtypes [29]. Acinar cells of the parotid gland express the highest levels of D1R and D2R, while ductal cells show the highest expression of DA transporters. It has been proposed, albeit without actual experimental evidence, that the DA system regulates the quality, volume, and ionic concentration of saliva.

There is substantial information on the stimulatory actions of DA on **amylase** release from the salivary glands. Amylase is a glycoprotein enzyme that hydrolyzes dietary starch into disaccharides and trisaccharides and is the most abundant protein in saliva. Early studies, using incubated rat parotid glands with agonists and antagonists, found that both DA and NE stimulated amylase secretion [30]. The DA effects were primarily mediated through postsynaptic D1R-like receptors and to a lesser extent by presynaptic D1R-like, resulting in NE release. A later study [31] found that the DA-stimulated amylase release from rat parotid gland slices was blocked by a beta-adrenergic receptor (AR) antagonist, but not by D1R or D2R antagonists, indicating that the effects of DA are primarily mediated through beta-AR. The discrepancies among these reports likely reflect the use of different DA doses.

8.3.2 The stomach

The **stomach** is composed of several discrete layers: mucosa, submucosa, muscularis, and serosa. **Figure 8.5** shows the specialization of luminal cells of the stomach for particular secretory products: parietal cells for hydrochloric acid, chief cells for pepsinogen, mucus cells for bicarbonate-rich mucus, G cells for gastrin, D cells for somatostatin, X/A-like cells for ghrelin, and enterochromaffin cells for histamine and serotonin [24]. TH activity and DA have been detected in isolated parietal cells of the rat stomach [32]. High DA levels that persisted after chemical sympathectomy were present in gastric juice and the mucosa, confirming that local DA synthesis is independent of innervation. The mucosa also had significant expression of D5R.

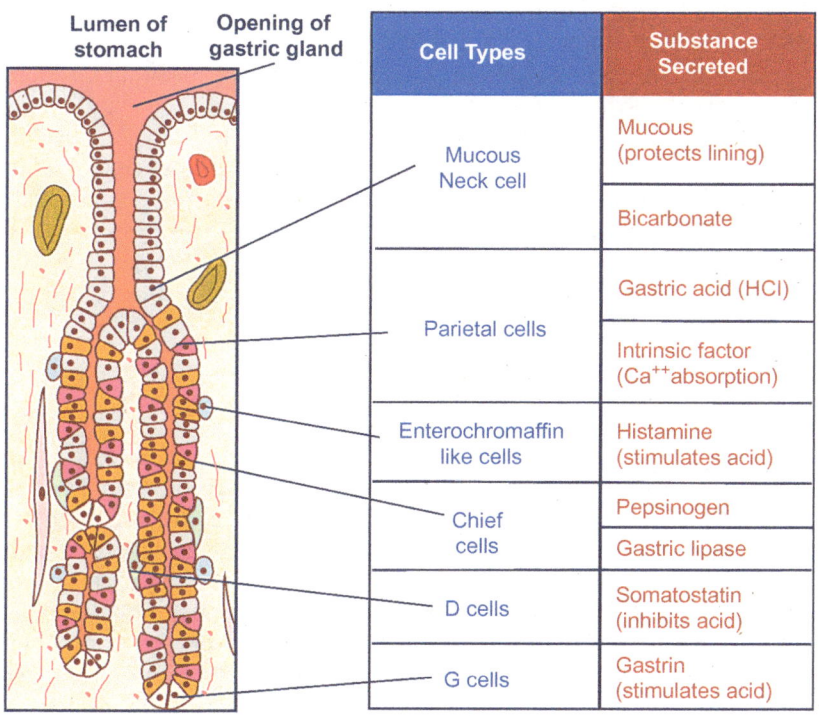

Cell Types	Substance Secreted
Mucous Neck cell	Mucous (protects lining)
	Bicarbonate
Parietal cells	Gastric acid (HCl)
	Intrinsic factor (Ca^{++} absorption)
Enterochromaffin like cells	Histamine (stimulates acid)
Chief cells	Pepsinogen
	Gastric lipase
D cells	Somatostatin (inhibits acid)
G cells	Gastrin (stimulates acid)

Figure 8.5 Different cell types within the lumen of the stomach, their secretory products and putative functions.

After a meal, the stomach can store food for several hours while the churning action of its longitudinal and oblique muscles physically breaks down the food. The powerful hydrochloric acid helps to convert the bolus into chyme. Food is digested in the stomach for several hours, during which time pepsin breaks down most of the proteins. Next, chyme is slowly transported from the pylorus sphincter into the small intestine where additional digestion occurs and nutrient absorption takes place. Unlike the intestines, the stomach has no absorptive functions of food except for alcohol and a few drugs.

The stomach walls are lined by a thick layer of mucus that prevents the stomach from digesting itself by the hydrochloric acid. When mucus is limited, an ulcer (erosion of tissue) may form. **Peptic ulcers** are open sores that develop on the inner lining of the stomach or duodenum. They affect as many as 5%–10% of people in Western societies, cause pain and indigestion, and can be transiently alleviated by antacid medications. The two main causes for gastric ulcers are long-term use of nonsteroidal anti-inflammatory drugs (NSAIDs) such as aspirin and ibuprofen and an infection with a specific bacterium, *Helicobacter pylori* [33]. *H. pylori* infection is a key factor in the etiology of various GI diseases, ranging from chronic active gastritis without clinical symptoms, to peptic ulceration, gastric adenocarcinoma, and gastric mucosa-associated lymphoid tissue lymphoma. Disease outcome is the result of a complex interplay between the host and the bacterium. Host immune gene polymorphisms and gastric acid secretion largely determine the bacterium's ability to colonize and act upon a specific gastric niche [34].

Several studies have shown that DA exerts gastro-protective effects. For example, D5R-deficient mice subjected to gastroduodenal lesions showed that the loss of D5R was associated with increased mucosal vulnerability to lesioning [35]. In rats with gastric mucosal injury, both central and peripheral D1R agonists provided gastro-protection [36]. In another study, stomach fluid was collected from rats subjected to manipulations that increase gastric acid secretion [37]. Treatment with quinpirole (a D2R agonist) and domperidone (a peripheral D2R antagonist) inhibited histaminergic-, pentagastrin- and cholinergic-stimulated gastric acid secretion. In the isolated perfused rat stomach, DA at 10 nM stimulated acid secretion [38], and at higher concentration stimulated the release of somatostatin [39]. Fasting for 4 days decreased the concentrations of Dopa and DA in the rat stomach, but neither the mechanism involved nor the physiological consequences of these changes were determined [40].

Based on evidence from these animal studies on the gastro protective effects of DA, selective D1R agonists have been proposed as useful adjuncts in the treatment of peptic ulcers [9]. Thus far, however, all published studies on the DA-DAR axis in the human stomach have focused on gastric cancer, and there are no clinical data to support potential beneficial effects of DA on ulcers. The effects of DA on gastric and liver cancers are discussed in **Chapter 13**.

8.3.3 Accessory digestive organs: Exocrine pancreas and gall bladder

The **pancreas** is composed of two structurally and functionally distinct glands that cohabit in one organ [41]. The bulk of the pancreas is made of exocrine cells that produce digestive enzymes, while the islets of Langerhans produce several hormones, primarily associated with glucose metabolism, as discussed later in this chapter. Digestive enzymes produced by the exocrine pancreas include amylase, lipase, phospholipase, trypsinogen, and chymotrypsinogen. The latter two are proteolytic enzymes that are synthesized as inactive zymogens to prevent autodigestion and are activated by cleavage in the intestines. The pancreatic exocrine cells release their secretory products into small ducts that join to form the main pancreatic duct running along the length of the pancreas. The pancreatic duct merges with the bile duct in the head of the pancreas, and the content of both is released directly into the duodenum.

Exocrine cells of the pancreas have a self-contained DA system, including TH, DA, and DAT [42], as well as several DAR subtypes [43]. The effects of DA on pancreatic secretion were examined in an early study in which pancreatic juice was collected from conscious dogs fitted with a pancreatic duct cannula [44]. Infusion of DA caused a secretory peak that lasted for 10 min, followed by a plateau. The stimulatory action of DA was blocked by domperidone, but not by alpha or beta-AR antagonists, suggesting activation of D2R-like receptors. The stimulatory effects of DA primarily affected water and bicarbonate secretion but not secretion of proteins (presumably enzymes). A later study using anesthetized dogs [45] found that D1R agonists stimulated the release of pancreatic juice that contained high bicarbonate and low proteins.

The **gallbladder** is a small pouch that sits below the liver and stores the bile produced by the liver. About 400–800 mL of bile is produced per day in adult humans. In response to signals (i.e., fat in chyme), the gallbladder squeezes the stored bile into the small intestine through the biliary duct. The main constituents of bile include bile acids (also called bile salts), phospholipids (mainly phosphatidylcholine), cholesterol, bilirubin, and inorganic salts. Bile salts act as surfactants and emulsify fats in the digest into micelles, making it easier for pancreatic-derived lipase to break down the triglycerides into fatty acids and monoglycerides, which can be absorbed by the villi on intestinal mucosal cells. Gallbladder inflammation can be caused by gallstones, excessive alcohol use, infections, or even tumors. The gallbladder itself is not essential for digestion given that its removal in healthy individuals causes no discernable health or digestion problems except for a small risk of diarrhea and fat malabsorption.

Whole-body positron emission tomography (PET) imaging in patients has showed high expression of D1R [46] and D2R [47] in the gallbladder. Infusion of high doses of DA (2.5 mg/kg/h) to rats fitted with a pancreatic and biliary fistula significantly suppressed bile volume [48]. Because this inhibitory effect was modified by neither propranolol nor haloperidol, it indicated a nonspecific effect, secondary to increased vascular resistance. Another study examined contractile effects of domperidone on gallbladder smooth muscle, using *in vivo* and *in vitro* models [49]. In the *in vivo* part, healthy men were given 20 mg domperidone or placebo tablets orally, and gallbladder emptying was monitored by ultrasonography. Domperidone caused a reduction of gallbladder volume. In the *in vitro* part, gallbladder strips from guinea pigs were field stimulated. Domperidone at 100 µM (but not at lower doses) caused contractions, while high doses of DA had no effects. Yet, the effect of domperidone appeared to be nonspecific because it was inhibited by atropine and abolished by tetrodotoxin. Collectively, much needs to be examined with respect to the actions of DA on the gallbladder.

8.3.4 Small and large intestines

The **small intestine** in humans is 20–25 feet in length and has a diameter of 1 inch. Extrinsic innervation to the small intestine is provided by parasympathetic nerve fibers from the vagus nerve and sympathetic nerve fibers from the thoracic splanchnic nerve. Intrinsic innervation comes from the enteric dopaminergic system, covered in detail in **Section 8.3.5**. The superior mesenteric artery is the main arterial supply to the intestine. Veins, running parallel to the arteries, drain into the superior mesenteric vein. Nutrient-rich blood from the small intestine is carried to the liver via the hepatic portal vein.

Table 8.3 Cells of the mucosa of the small intestine: Location and functions

Cell type	Location in the mucosa	Function
Absorptive	Epithelium/intestinal glands	Digestion and absorption of nutrients in chyme
Goblet	Epithelium/intestinal glands	Secretion of mucus
Paneth	Intestinal glands	Secretion of bactericidal enzyme lysozyme; phagocytosis
G cells	Intestinal glands of duodenum	Secretion of gastrin
I cells	Intestinal glands of duodenum	Secretion of cholecystokinin, which stimulates release of pancreatic juices and bile
K cells	Intestinal glands	Secretion of glucose-dependent peptide, which stimulates insulin release
M cells	Intestinal glands of duodenum and jejunum	Secretion of motilin, which accelerates gastric emptying, stimulates intestinal peristalsis and stimulates production of pepsin
S cells	Intestinal glands	Secretion of secretin

The main function of the small intestine is absorption. This process is facilitated by (1) deep ridges in the mucosa and submucosa that form circular folds, (2) small (0.5- to 1-mm long) hair-like vascularized projections called villi, and (3) brush border, or cylindrical microscopic apical surface extensions of the plasma membrane of the mucosa's epithelial cells, which are formed by microfilaments. All of the aforementioned modifications greatly increase the surface area of the small intestine, allowing more area for nutrients to be absorbed. **Table 8.3** presents the different cell types in the mucosa of the small intestine and their functions.

One of the most important functions of DA in the intestines is the regulation of **Na/K-ATPase** (NKA) pumps. These pumps are localized in the basolateral membrane of enterocytes of the small intestine and play a major role in nutrient transport by transferring K^+ ions into, and Na^+ out of the cell. The resulting gradient drives glucose transport into intestinal cells through a glucose–sodium cotransporter and also regulates the transport of amino acids and calcium. A recent comprehensive review [50] describes cross-talk between DAR and NKAs in several organs. Within the jejunum, NKA activity was inhibited by locally produced DA acting through D1R. Changes in jejunal NKA activity contributed to the maintenance of sodium homeostasis, especially when renal function was compromised. Another study used duodenal tissue from control and D5R-overexpressing transgenic mice, and found that DA enhanced epithelial permeability via a D5R mediated, cAMP-dependent pathway [51].

The **colon** assists in the maintenance of fluid balance in the body by absorbing the remaining water and electrolytes from indigestible food. The regulation of colonic functions by DA was demonstrated in studies employing trans-epithelial recording of isolated distal rat colonic segments [52]. The data showed that DA promoted epithelial Cl^- absorption coupled with HCO_3^- secretion. This was mediated by an apical electrogenic anion exchanger, SLC26A3, and required basolateral entry of HCO_3^- through Na^+/HCO_3^- cotransporter. The coupling of HCO_3^- secretion to Cl^- absorption suggested the DA not only regulates fluid transport, but also the luminal pH. The authors concluded that these findings have important implications to colon physiology, as well as to the pathophysiology of PD.

8.3.5 Enteric dopaminergic system and the gut microbiome

The **enteric nervous system (ENS)** is embedded in the lining of the GI system throughout its length from the esophagus to the anus and governs many of its functions [53]. Interactions between the ENS and central nervous system (CNS) are often described as the gut–brain axis. The gut–brain axis is

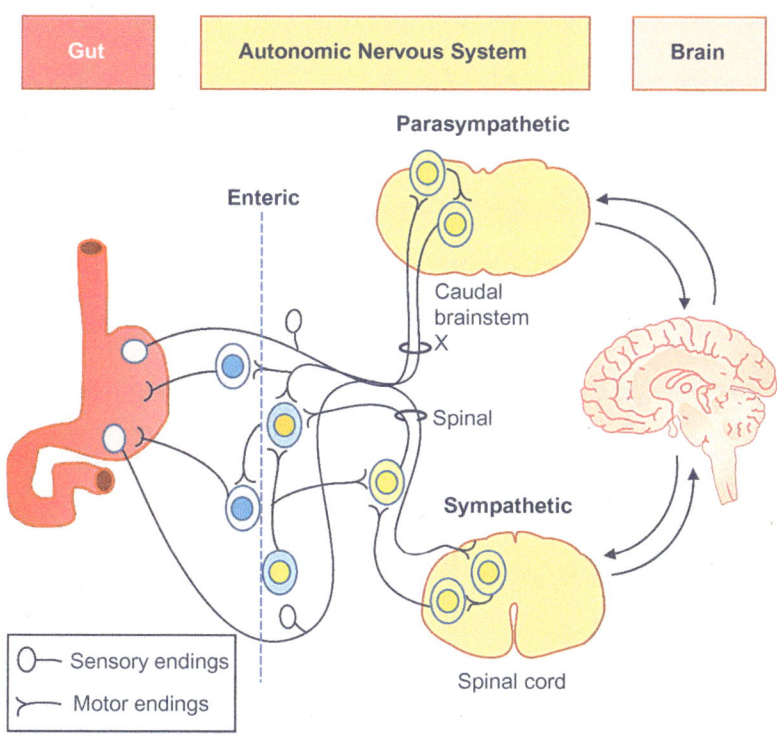

Figure 8.6 Shown is a simplified presentation of the sympathetic and parasympathetic (yellow) and enteric (blue) innervation of the stomach. The enteric nervous system is entirely contained within the gastrointestinal wall. For simplification, all nerve branches supplying the gut, pancreatic postganglionic neurons and the sacral parasympathetic system are not shown. (Redrawn and modified from Udit, S., and Gautron, L., *Front Neurosci.*, 7, 134–149, 2013.)

composed of a network of autonomic neurons that connect the caudal brainstem and spinal cord to the esophagus, GI tract, liver, and pancreas. Axons of these neurons travel through the vagus, splanchnic, mesenteric and pelvic spinal nerves to innervate the abdominal viscera. A diagram of part of the anatomy of the mammalian gut–brain axis, showing the enteric innervation of the stomach, is presented in **Figure 8.6**.

Although the general organization of the gut–brain axis appears simple relative to the CNS, the neurochemical, anatomical, and functional relationships between the different populations of gut–brain axis neurons can be rather complex. A better understanding of the anatomy and plasticity of the gut–brain axis should advance our grasp of autonomic neural circuits and chronic diseases that affect the gut–brain axis, including but not limited to, inflammatory bowel diseases, metabolic syndrome, visceral pain, and eating disorders [54].

Transcripts encoding TH and the DAT are present in the murine GI tract at the following relative abundance: small intestine >stomach or proximal colon >distal colon), and they are derived from intrinsic neurons [55]. TH protein was demonstrated by immunocytochemistry in neuronal perikarya (submucosal >>myenteric plexus; small intestine >stomach or colon). TH, DA, and DAT immunoreactivities were present in subsets of neurons (submucosal myenteric) in guinea pigs and mouse intestines. Surgical ablation of sympathetic nerves of the bowel did not affect DAT immunoreactivity and actually increased the numbers of TH-immunoreactive neurons, the expression of mRNA encoding TH and DAT, and enteric dihydroxyphenylacetic acid (DOPAC; the specific DA metabolite).

A recent investigation [56] examined for similarities between the central and the enteric dopaminergic systems in patients with PD. In addition to the degeneration of striatal DA neurons in PD, the presence of misfolded α-syn–containing aggregates is a primary pathological hallmark of PD in the CNS. This study identified phosphorylated alpha-synuclein (p-α-syn)–containing Lewy bodies in the gut enteric system in normal patients, while the total number and the area of p-α-syn–positive aggregates were increased in PD patients, as was the number of small- and large-sized aggregates. Increased expression of D1R, VIP, and serotonin receptor 3A was also observed in PD patients, while serotonin receptor 4 and muscarinic receptor 3 (M3R) were downregulated. The authors concluded that their findings support the hypothesis that the CNS pathology of increased p-α-syn in PD also applies to the ENS.

The human **microbiome** represents an ecological community of commensal, symbiotic, and pathogenic microorganisms such as bacteria, fungi, protozoa, and viruses that share the human body space [57]. It is estimated that the human microbiome, primarily that of the gut, consists of $\sim 10^{13}-10^{14}$ microorganisms, which is 10 times more than the total number of cells in the human body. The gut microflora benefits their host by protecting it against colonization by pathogens, assisting in the intake of nutrients from the diet, metabolizing certain drugs to functional forms, and helping in the absorption and distribution of fat. The gut microbiome is considered by some as a virtual organ in its own right because it produces an array of bioactive molecules that directly interact with the endocrine, nervous, and immune systems of their host. Although the functional significance of this activity is not yet fully understood, there is a growing appreciation that changes in the microbiome diversity are linked to diseases such as diabetes, rheumatoid arthritis, muscular dystrophy, multiple sclerosis, fibromyalgia, and possibly certain cancers.

In addition to contributions by the ENS to the presence of DA within the gut, there is evidence that endogenous gut microflora add to its levels. For example, butyrate, which is synthesized by colonic bacteria, can enhance the transcription of TH, the rate-limiting enzyme in catecholamine biosynthesis. Some bacteria can actually synthesize catecholamines that are analogs of the mammalian hormones, while other bacteria release enzymes such as β-glucuronidase, which converts glucuronide-conjugated catecholamines to free, biologically active forms [58]. Further, in germ-free mice, most DA in the gut lumen was present as a conjugated form, while free DA was predominant in control mice with gut microbiota. This study showed that gut microbiota play a critical role in the generation of free CA in the gut lumen. Whether this applies to humans is unclear, however, because most of the secreted peripheral DA in humans is in the form of DA-S, not as DA-glucuronide (see **Chapter 6**). Presently, it is unknown whether gut bacteria can produce and release an active aryl sulfatase A.

8.4 ORGANS THAT REGULATE METABOLISM: PANCREAS, ADIPOSE TISSUE, AND LIVER

Metabolic homeostasis is maintained by a network of tissues and organs that coordinately optimize the conversion of food/fuel to energy for running the cellular processes and for the generation of the building blocks of proteins, lipids, carbohydrates, and nucleic acids. The reader is referred to an excellent 2010 review on the role of peripheral DA in metabolic control and in tumor growth [59]. DA is involved to a variable extent in the regulation of the three major organs that control metabolism: the endocrine pancreas, adipose tissue, and the liver. Expression and regulation of DAR in these organs is covered in **Chapter 6**. Although skeletal muscle and the thyroid gland are also key players in metabolic homeostasis, there are no published records demonstrating a direct role of DA in their regulation, and therefore they are not covered here. The involvement of DA with the four pituitary hormones associated with metabolic regulation: growth hormone (GH), prolactin (PRL), adrenocorticotropic hormone (ACTH), and thyroid stimulating hormone (TSH) is covered in great detail in **Chapter 5**.

8.4.1 Endocrine pancreas and glucose metabolism

Figure 8.7 shows a human pancreas with a close up view of an **islet of Langerhans**. Studies with mouse pancreatic cells showed expression of TH [60]. The pancreatic cells can also produce DA by taking up L-Dopa and converting it to DA via Dopa decarboxylase (DDC), as well as by taking DA from the circulation or from neighboring exocrine cells via DAT and VMAT. In the normal human pancreas, D1R is expressed by beta cells (insulin-producing cells), D2R is expressed by alpha (glucagon), delta (somatostatin), and pancreatic polypeptide cells, D4R is expressed by beta and polypeptide cells, whereas D5R is expressed only by delta cells [61].

The **insulin-secreting beta cells**, which make up 70%–80% of the islet cells, have all the components necessary for DA synthesis, secretion, and action. Insulin is primarily released in response to elevated blood glucose levels, and its actions are fundamental to the maintenance of glucose homeostasis [60]. Binding to its glycoprotein receptor, insulin is the only hormone that lowers blood glucose levels by promoting glucose uptake by the various cells. The opposite effects (i.e., glycogenolysis and gluconeogenesis) are partly under the control of glucagon. The role of the pancreatic polypeptide in metabolism is unclear except that it serves as a satiety hormone. The main effect of somatostatin on metabolism is via its inhibition of GH, a major metabolic hormone.

There is a general agreement that DA inhibits glucose-stimulated insulin release from isolated islets, with some disagreement as to the identity of the receptors involved, from D3R in one study [62] to D2R in another [63]. DA also stimulates the proliferation of pancreatic islet cells, given that immunohistochemistry showed a reduced pancreatic beta-cell mass in *Drd2*$^{-/-}$ mice and decreased beta-cell replication in 2-month-old *Drd2*$^{-/-}$ mice [64]. In beta cells, DA and the D2R agonist quinpirole antagonized glucose-stimulated insulin release by decreasing membrane depolarization and by suppressing

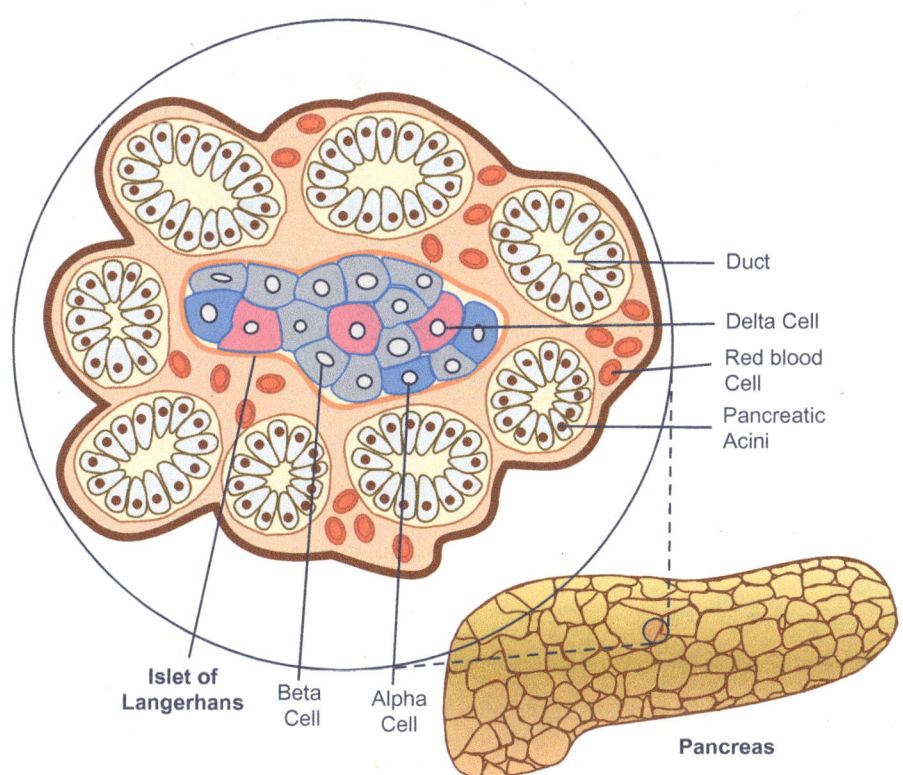

Duct

Delta Cell

Red blood Cell

Pancreatic Acini

Islet of Langerhans Beta Cell Alpha Cell

Pancreas

Figure 8.7 Structure of the human pancreas. Shown are the pancreatic acini surrounding an isle of Langerhans. Within the isle are the various hormone-producing cells. See text for other explanations.

glucose-stimulated calcium rise [63]. In addition to insulin, DA antagonized NE-induced stimulation of glucagon release [65] and inhibited the release of pancreatic polypeptide [66]. There are no published records on whether DA is involved in the direct regulation of other pancreatic hormones such as gastrin, secretin or somatostatin.

Human studies have not provided conclusive evidence with respect to the effect of DA on insulin release. In human volunteers, acute intravenous administration of metoclopramide, a peripheral antidopaminergic agent with low serotonergic activity, resulted in the suppression of basal blood insulin levels and concomitant elevation of blood glucose [67]. On the other hand, infusion of DA itself caused stimulation of insulin release in one study but its inhibition in another, reflecting either the different doses of DA employed and/or central vs. peripheral effects of DA.

8.4.2 Adipose tissue

Adipose tissue plays a pivotal role in metabolic, physiologic, and endocrine homeostasis [68,69]. This highly specialized tissue of mesenchymal origin is comprised of multiple cell types held loosely together in a collagen matrix. The predominant cell type is the terminally differentiated adipocyte, a very large cell whose size can be dramatically altered under various nutritional states. The stroma, also called the stromal vascular fraction (SVF), contains pluripotent stem cells, preadipocytes, endothelial cells, pericytes, mast cells, fibroblasts, and hematopoietic cells, mainly macrophages.

Two types of adipose tissue are recognized: **white adipose tissue (WAT)**, which stores energy in the forms of lipids, and **brown adipose tissue (BAT)**, which generates heat by uncoupling the mitochondrial aerobic respiration chain that normally results in ATP generation [70]. Human WAT is found in specific anatomical depots which differ in morphology and functions and are grossly classified as abdominal visceral (vis) and subcutaneous (sc) fat. The relative distribution of these depots determines the gender-dependent body shape and affects the development of metabolic syndrome (discussed later in this chapter). Adipose tissue also functions as an important endocrine organ that produces a multitude of adipokines that target many organs and modulate a wide range of metabolic pathways.

Non-shivering **thermogenesis** in brown adipocytes is activated by beta-AR [71]. DA also affects thermogenesis in brown adipocytes, except that both stimulatory and inhibitory actions have been reported. An early study [72] found that similar to NE, DA increased oxygen consumption in rat BAT explants. Another group, using isolated rat brown adipocytes, reported that DA antagonized the beta-AR stimulation of glycerol and nonesterified fatty acid release [73]. Both studies, however, used very high micromolar concentrations of DA. This issue was rectified in a later study employing immortalized murine brown adipocytes [74]. Within 24 h of treatment with low nanomolar doses of DA or a D1R agonist, there were increases in oxygen consumption, mitochondrial membrane potential, and uncoupling protein 1 (UCP1) levels, implicating the activation of heat production by the cells.

An early study reported that DA stimulated glucose uptake in rodent adipocytes [75], but attributed this action to the activation of beta-AR by DA. A later study compared the lipolytic action of adrenergic agonists and antagonists as well as that of DA in visceral adipocytes harvested from beta1/beta2/beta3-AR triple-knockout (beta-less) mice [76]. The authors found residual lipolytic activity of most of these agents, which they attributed to an unknown G protein-coupled receptor (GPCR) with low affinity for catecholamines. Because expression of DAR in adipocytes was not known to these authors at that time and all agents were tested at 10–100 μM, activation of lipolysis in the beta-less adipocytes was probably mediated by DAR.

In murine 3T3-L1 adipocytes, the D2R agonist bromocriptine decreased expression of both adipogenic activators and lipogenic target genes [77]. Bromocriptine also reduced intracellular nitric oxide formation and expression of pro-inflammatory genes, reflecting attenuated pro-inflammatory responses. Treatment of D2R-deficient 3T3-L1 with bromocriptine significantly decreased lipogenic activity, whereas treatment with yohimbine, an α2-AR inhibitor, showed no reduction in lipogenic activity. The authors concluded that the bromocriptine-induced attenuation of adipogenesis and lipogenesis in 3T3-L1 cells was mediated by α2-AR rather than by D2R. Yet, bromocriptine was used at 10–50 μM doses, raising the possibility of cross reactivity with ARs.

The discovery that human adipocytes express functional DAR was the consequence of our pursuit of the regulation of extrapituitary PRL [78]. Unique to humans, PRL is expressed in many extrapituitary sites, including the endometrium, myometrium, decidua, immune cells, brain, breast, prostate, skin and adipose tissue. Upon monitoring PRL release from sc and vis adipose explants from obese and lean patients, we unexpectedly noticed a gradual, time-related increase in PRL release during incubation, suggesting a removal from tonic inhibition [79]. Because a similar rise in PRL release occurs when pituitary lactotrophs are removed from hypothalamic DA inhibition, we wondered if DA also serves as an inhibitor of adipocyte PRL.

Subsequently, we used real-time polymerase chain reaction (RT-PCR) and Western blotting and detected expression of several types of DAR in human adipocytes [80]. In addition, we found that both DA and DA-S, the major circulating form of DA in humans, inhibited adipocyte PRL release (see **Figure 6.4** in **Chapter 6**). Although synthesis of DA by human adipocytes has not been determined in this study, others reported that isolated rat mesenteric adipocytes, as well SVF, expressed TH, DBH, and PNMT and were fully capable of *de novo* synthesis of catecholamines, including small amounts of DA [81]. In addition to the local production, DA can be made available to adipocytes by uptake from extracellular depots secreted by neighboring resident immune cells.

As determined by RT-PCR and Western blotting, D1R, D2R, and D4R, and to a lesser extent D5R, were expressed in human adipose tissue explants, primary adipocytes, and two adipocyte cell lines, SW872 and LS14 [80]. SW872 is a liposarcoma-derived cell line from the ATCC. LS14 is a spontaneously immortalized adipocyte cell line that we cloned from a metastatic liposarcoma [82]. The functionality of adipocyte D2R was confirmed by the inhibition of PRL release by both DA and bromocriptine, a D2R agonist, and its blockade by raclopride, a D2R antagonist [80]. The functionality of D1R was confirmed by the induction of ERK1/2 phosphorylation by DA, with a delayed effect on Akt.

Notably, as illustrated in **Figure 8.8**, Panels A and B, expression of D1R and D2R was differentially altered during early adipogenesis, with D1R increasing while D2R decreasing on day 3 of adipogenesis, before the appearance of differentiated adipocytes. This suggested a potential regulation of DAR by transcription factors such as peroxisome proliferator-activated receptor gamma (PPARγ) and CCAAT-enhancer-binding protein (cEBPα), which are critical for preadipocyte differentiation and whose expression increases during the first few days of adipogenesis [83]. Indeed, we have identified a PPARγ binding site in the promoters of both D1R and D2R and a C/EBP binding site in D1R (**Figure 8.8**C).

Expression of **serotonin receptors** 5-HT1aR, 5-HT2aR, and 5-HT2cR, considered the prime targets of **atypical antipsychotics (AAPs)** in the brain, was also examined in this study. The 5-HT1aR was low to undetectable in adipose tissue and preadipocytes but showed a strong signal in differentiated adipocytes. The 5-HT2aR was detected in adipose tissue and was lower in differentiated than in proliferating adipocytes. The 5-HT2cR was undetectable in either adipose tissue or adipocytes.

Figure 8.8 Panel A: Changes in D1R and D2R expression during differentiation of human preadipocytes. Panel B: Changes in the structure of the adipocytes in selected days of differentiation. Panel C: Identification of putative binding sites for various transcription factors in the promoters of D1R and D2R. (Redrawn and modified from Borcherding, D.C. et al. *PLoS One*, 6, e25537, 2011.)

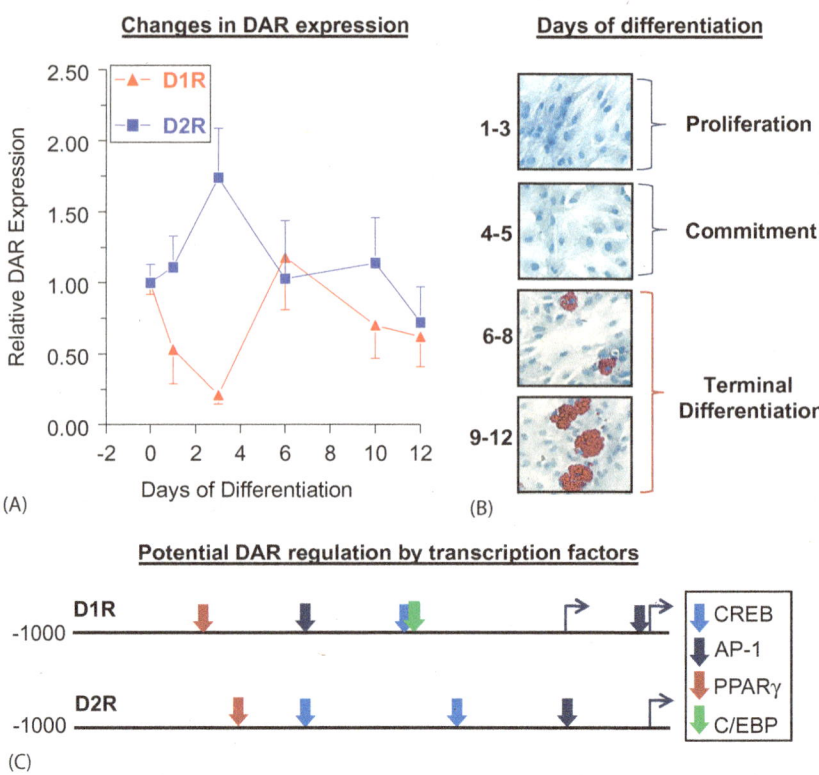

In addition to inhibiting PRL release, DA altered the release of **leptin, adiponectin, and interleukin-6** (IL-6) from human adipocytes [80]. Leptin, whose production is proportional to adipose tissue mass, plays an important role in the regulation of energy homeostasis, neuroendocrine and immune functions and in glucose, lipid and bone metabolism. Leptin administration to humans reduces appetite, while chronic leptin deficiency causes extreme obesity. Adiponectin protects against metabolic syndrome by virtue of its insulin-sensitizing, anti-inflammatory, and anti-atherogenic activities. IL-6 is a pro-inflammatory cytokine associated with the low level of inflammation that accompanies obesity. High serum IL-6 levels increase the production of C-reactive protein and can lead to coronary heart disease and atherosclerosis.

The effects of low (1 nM) and high (100 nM) doses of DA on leptin release were examined using human sc adipose explants, isolated mature adipocytes, and differentiated primary adipocytes [80]. Leptin release was inhibited 40%–80% by either dose of DA and also by the D1R/D5R agonist, SKF38393 (SKF), suggesting an action via D1R-like. Both DA and SKF caused a 60%–80% increase in adiponectin release from differentiated primary adipocytes and an increase in IL-6 release from proliferating primary preadipocytes. Others have reported that DA exerted dose-related reduction of leptin gene expression and release from adipocytes harvested from obese hypertensive patients [84]. The authors proposed that DA deficiency contributes to metabolic disorders linked to the hyperleptinemia in these patients.

In a recent study using 3T3-L1 adipocytes, quinpirole, a D2R agonist, increased both protein and mRNA expression of leptin and IL-6, but not adiponectin and visfatin, and also increased the mRNA expression of tumor necrosis factor alpha (TNF-α), monocyte chemoattractant protein-1 (MCP1), and nuclear factor kappa B p50 pathway (NF-κB-p50) [85]. Acute quinpirole treatment in C57BL/6J mice increased serum leptin concentration and leptin mRNA in visceral adipose tissue but not in subcutaneous adipose, confirming the stimulatory effect of D2R on leptin *in vivo*. The authors concluded that the stimulation of D2R increases leptin production and may have a tissue-specific pro-inflammatory effect in adipocytes.

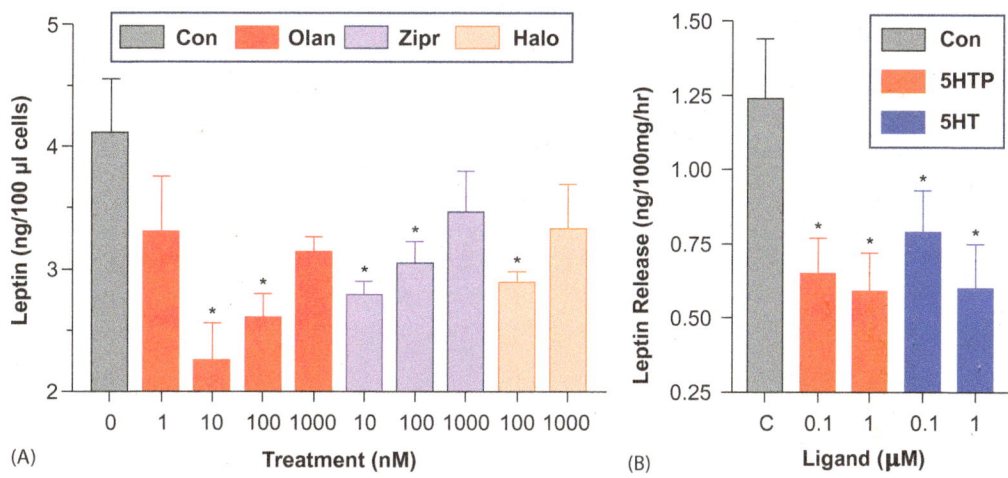

Figure 8.9 Panel A shows inhibition of leptin release by antipsychotics. Con: control; Olan: olanzapine; Zipr: ziprasidone; Halo: haloperidol. Panel B shows inhibition of leptin release by the serotonin ligands 5HTP and 5HT. Each value is a mean +/−SEM of 5 determination. * p<0.05. (Unpublished data from the author's laboratory.)

The effects of selected AAPs on leptin release from isolated primary human adipocytes were also determined in our laboratory (**Figure 8.9**A). Olanzapine, ziprasidone and haloperidol suppressed leptin release in a non-monotonic fashion. The lack of linearity and the variable responses were likely due to concomitant activation or inhibition of DAR and/or serotonergic receptor subtypes by the increasing concentrations of the AAP. **Figure 8.9**B also shows that both 5-HT and its precursor 5-HTP inhibited leptin release from sc explants, confirming not only the functionality of serotonergic receptors in human adipocytes but also demonstrating the capacity of human adipose tissue for *de novo* synthesis of serotonin from its precursor 5HTP. In addition to the regulation of PRL release from the adipocytes, our data showed that AAPs differentially affect lipolysis, lipogenesis and preadipocyte differentiation. This likely occurs by differentially activating various DAR and serotonergic receptors.

A conceptual model of the potential interactions between D1R, D2R and 5HTR in the regulation of the adipocytes is presented in **Figure 8.10**. The various receptor types may work in a cooperative or in an antagonistic manner to regulate lipogenesis, lipolysis, adipogenesis as well as adipokine release, depending on the relative expression of each receptor and availability of endogenous ligands and administered drugs.

8.4.3 The liver

The liver has multiple functions including **digestion and detoxification**, and it is a key regulator of **metabolism** via its ability to synthesize proteins and generate and store glycogen. About 70%–85% of the liver volume is occupied by parenchymal hepatocytes. Nonparenchymal cells constitute 40% of the total number of liver cells but only 6.5% of its volume. The liver has sinusoids that are lined with two types of cells—endothelial cells and phagocytic Küpffer cells—while nonparenchymal stellate cells are located in the perisinusoidal space. The liver receives a dual blood supply from the hepatic portal vein and hepatic arteries. The portal vein delivers around 75% of the liver's blood supply and carries venous blood from the spleen, the GI tract and associated organs. The hepatic arteries supply arterial blood to the liver, accounting for the remaining 25% of its blood flow.

The liver plays a major role in carbohydrate, protein, amino acid, and lipid metabolism. It synthesizes and stores glycogen from glucose by the process

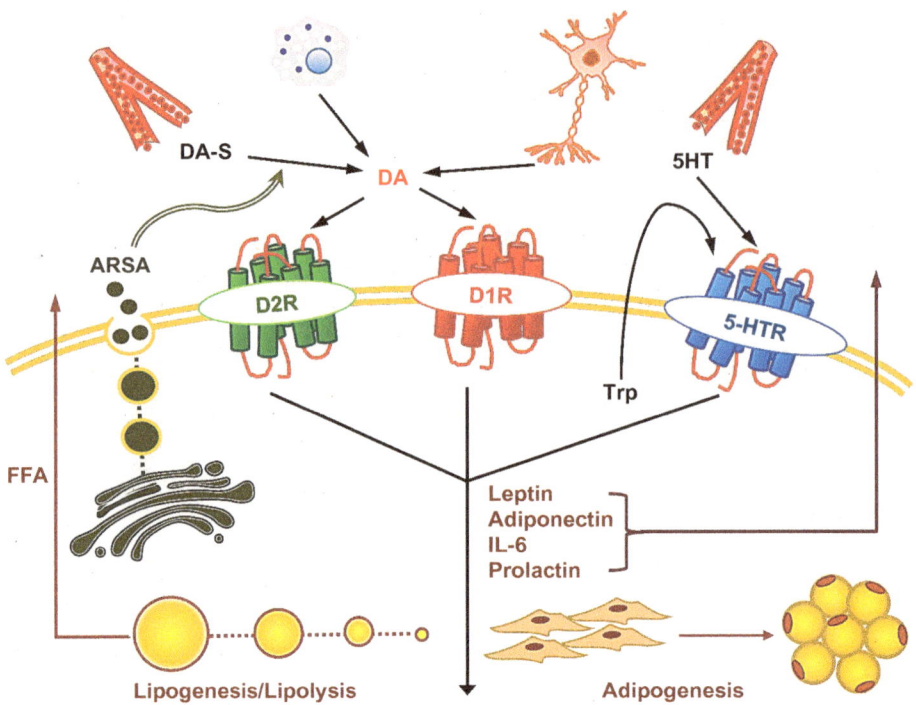

Figure 8.10 A model depicting interactions between DAR and 5HTR in the regulation of the adipocytes. 5HT: serotonin; ARSA: arylsulfatase A; DA: dopamine; DA-S: dopamine sulfate; FFA: free fatty acids; IL: interleukin; trp: tryptophan. See text for other explanations.

of glycogenesis. When the need for glucose arises, the liver releases glucose into the blood through the process of glycogenolysis, or the breakdown of glycogen into glucose. The liver is also responsible for gluconeogenesis, or the synthesis of glucose from certain amino acids, lactate or glycerol. Both adipocytes and hepatocytes can produce glycerol by the breakdown of fat.

Mouse hepatocytes, but not activated hepatic stellate cells, highly express TH [86]. In a real-time quantitative PCR (RT-qPCR) analysis, hepatocytes expressed significantly higher mRNA levels of TH than the nonparenchymal cell populations, stellate, Küpffer, or endothelial cells. However, TH mRNA levels of the normal liver were 1,800 times lower as compared with dorsal root ganglion tissue that contains sympathetic neurons. Accordingly, immunohistochemistry revealed sympathetic neurons in portal tracts as the main TH-expressing cell population in normal and fibrotic mouse or human liver. The authors concluded that sympathetic neurons in liver have the biosynthetic machinery. In addition to *de novo* synthesis from tyrosine, the liver can generate DA by decarboxylation of L-Dopa and can also take up DA from the circulation via DAT, as determined by a whole-body biodistribution of a DAT ligand by PET imaging [87].

An older study reported identification of DA binding sites in normal human liver that were reduced in diabetic patients [88]. As determined by PET imaging, D1R is present in human liver [46], but otherwise there are no published biochemical/immunochemical records on the expression of specific DARs in the normal liver except for a few studies on liver carcinomas (covered in **Chapter 13**). In cultured mouse hepatocytes, DA suppressed lipopolysaccharide-induced TNF-α production, leading the authors to suggest that DA may prevent inflammatory liver disease [89]. In another study, prolonged incubation of rat hepatocytes with DA at a high dose downregulated anti-apoptotic proteins and undermined the functional integrity of the hepatocytes [90]. On the other hand, in transplant studies, perfusion of rat livers with high DA doses increased liver viability, as judged by biochemical and histological criteria [91]. The authors concluded that postharvest DA treatment by supplementation of the preservation solution represents an easy and effective measure to reduce cold storage-associated liver injury. Evidently, there is no consensus whether DA is beneficial or harmful to liver integrity, and an extensive investigation will be required to resolve this issue.

8.5 DIABETES, OBESITY, AND ADVERSE EFFECTS OF ANTIPSYCHOTIC DRUGS

8.5.1 Dopamine, diabetes, and obesity

Two types of diabetes are recognized. **type I diabetes** is an autoimmune disease, previously called childhood diabetes, in which a damaged pancreas does not produce insulin. Treatment for this diabetes requires administration of insulin by injections or by pumps, frequent monitoring of blood glucose levels, and significant changes in eating habits and lifestyle. **Type 2 diabetes** is the most common form of diabetes, accounting for 95% of diabetes cases in adults and afflicting over 500 million people worldwide. It used to be called adult-onset diabetes, albeit with the epidemic of obesity in children, more teenagers are now diagnosed with type 2 diabetes. In this type of diabetes, the pancreas produces some insulin, but the amount is insufficient and/or there is insulin insensitivity (insulin resistance), mostly in fat, liver, and muscle cells. Although type 2 diabetes is often milder than type 1, it also causes major health complications, particularly in small blood vessels that nourish the kidneys, nerves, and eyes, and it is also associated with increased risk of heart disease and stroke. Although there is no cure for type 2 diabetes, it can be controlled with weight management, nutrition, exercise and medications that suppress gluconeogenesis, increase insulin release, or reduce insulin resistance.

Several lines of evidence implicate peripheral DA in the manifestation of diabetes. Treatment of diabetic or obese animal models with the D2R agonist bromocriptine improved glucose intolerance and reduced high serum insulin levels [92]. Although some of the bromocriptine effects were caused by its action on hypothalamic or pituitary D2R and thus affecting PRL release, direct effects of bromocriptine on insulin-producing beta cells were not ruled out. Bromocriptine is the first dopaminergic drug approved for use in type 2 diabetes [93].

Short-term treatment of Parkinson's patients with L-Dopa has reduced insulin secretion [94], except that insulin levels did not change upon co-treatment with carbidopa, a peripheral DDC inhibitor. This was interpreted as an action of peripheral DA, presumably on pancreatic beta cells that convert L-Dopa to DA. There is also evidence for reciprocal relations between insulin and brain DA [95]. Streptozotocin-induced hypoinsulinemia decreased DAT and TH transcripts in the substantia nigra, and in turn decreased insulin signaling in the basal ganglia. Collectively, PD appears to increase the risk of developing type 2 diabetes.

Obesity is defined as >20% increase over the ideal weight of an individual and is associated with higher risks of morbidity and mortality. Obesity results from enlargement of the adipocytes via enhanced lipid accumulation (hypertrophy) and/or from increased cell number (hyperplasia). The latter begins with the recruitment of stem cells to the adipocyte lineage [96], followed by adipogenesis of committed preadipocytes, which takes 7–10 days and culminates in their transformation into lipid-filled large mature adipocytes [83]. About 10% of fat cells are renewed annually, irrespective of body mass index, without an increase in their overall number in adults.

Obesity and diabetes are often associated with the **metabolic syndrome**, defined as a cluster of disorders that include insulin resistance, glucose intolerance, hypertension and dyslipidemia, and is associated with increased morbidity and high risk of mortality due to cardiovascular disease [97]. Brain DA is involved in obesity, in part, through the control of food intake. As determined by PET imaging that monitored changes in

DA or D2R, food presentation to fasting human volunteers was accompanied by increases in striatal extracellular DA [98]. In obese subjects, the reductions in striatal D2R were inversely correlated with the body weight of the subject. The association of the DA system in reward and reinforcement of food intake has led to the hypothesis that low brain DA activity in obese subjects predisposes them to excessive food intake.

As summarized in a recent review [99], mutant mice lacking D2R are glucose intolerant and have abnormal insulin secretion. In humans, administration of drugs that block DAR may cause hyperinsulinemia, increased weight gain and glucose intolerance. Conversely, treatment with L-Dopa in PD patients reduces insulin secretion upon oral glucose tolerance test, and bromocriptine improves glycemic control and glucose tolerance in obese type 2 diabetic patients as well as in nondiabetic obese animals and humans. The actions of DA on glucose homeostasis and food intake impact both the autonomic nervous system and the endocrine system.

8.5.2 Adverse effects of antipsychotic medication on body weight and metabolic syndrome

The history, development and properties of typical and **AAPs** are covered in **Chapter 2**, **Section 2.4.3**, while the receptors targeted by these drugs are listed in **Table 2.4**. To briefly summarize, all AAPs antagonize D2R (and other DARs with lesser affinity), but they also bind to a myriad of serotonergic, adrenergic, muscarinic and histaminergic receptors. In recent years, AAPs have been prescribed for a growing list of mental disorders that include schizophrenia, bipolar disorder, mania, attention deficit disorder, major depression, posttraumatic stress disorder, and autism.

Although AAPs made valuable improvement in many neuropsychiatric diseases, it was soon realized that they also caused excessive weight gain, insulin resistance, and changes in serum lipids, adipokines and inflammatory cytokines in pediatric, adult, and elderly patients [100]. Rapid weight gain in response to AAPs was especially prevalent in young patients, as was found in a survey of AAP-treated children diagnosed with a variety of neuropsychiatric disorders [101]. Although most drugs improved the behavioral problems in these children, they came at the expense of substantial weight gain. Within 12 weeks, mean weight gain was 8.5 kg for olanzapine (Zyprexa), 6.1 kg for quetiapine (Seroquel), 5.3 kg for risperidone (Risperdal), and 4.4 kg for aripiprazole (Abilify), vs. 0.2 kg in controls. Notably, old veterans treated with AAPs showed only modest weight gain but a high incidence of metabolic syndrome [102]. The effects of AAPs on body weight and metabolic syndrome are often presented as evidence for the role of DA as a contributing factor in obesity and its sequela. This may be an overextension because although all AAPs antagonize D2R, they also bind to multiple other receptors, making it difficult to assign their metabolic actions to DAR [103].

The relative metabolic disturbances caused by six of the most widely prescribed AAPs are presented in Table 8.4. Olanzapine and clozapine carry

Table 8.4 Relative metabolic disturbances associated with selected atypical antipsychotics				
Drug (Trade Name)	Weight Gain	Type 2 diabetes	Dyslipidemia	Metabolic syndrome
Olanzapine (Zyprexia)	High	High	High	High
Clozapine (Clozaril)	High	High	High	High
Risperidone (Risperdal)	Moderate	Low to moderate	Low to moderate	Medium
Quetiapine (Seroquel)	Moderate	Low to moderate	Moderate	Moderate
Ziprasidone (Geodon)	Low	Low	Low	Low
Aripiprazole (Abilify)	Low	Low	Low	Low

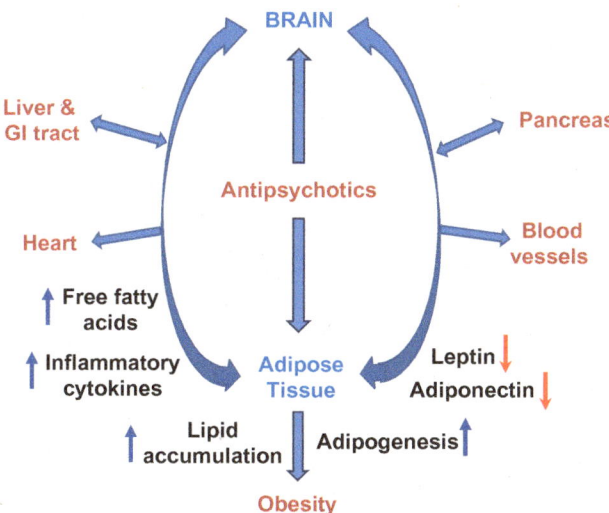

Figure 8.11 The antipsychotics affect metabolic and cardiovascular parameters and induce obesity by acting on the brain as well as on peripheral sites.

the greatest risk of metabolic syndrome, quetiapine and risperidone have an intermediate risk, while ziprasidone and aripiprazole confer lower risks [104]. Although weight gain is a known risk factor for the development of metabolic abnormalities, not all obese individuals display metabolic disturbances, and not all AAP-treated patients gain weight. Moreover, drug-induced metabolic abnormalities can occur in spite of negligible weight gain [105]. These facts underscore the importance of genetic predisposition, as well as environmental, dietary, and lifestyle factors, in determining the ultimate responsiveness of each patient to these drugs.

Oral medications can act on the brain as well as on peripheral sites, including adipose tissue. To this end, incubation of rat [106] and human [107] adipocytes with AAPs confirmed their ability to directly affect lipid metabolism and cell proliferation. Because of the multiple receptors targeted by these drugs, such actions could not be specifically ascribed to DAR, and the response of adipocytes to AAPs is, therefore, neither linear nor straightforward. Presently, it is unknown to what extent direct peripheral actions of AAPs contribute to the weight gain and metabolic disturbances seen *in vivo*. The resolution of this issue awaits the use of animals subjected to cerebro-ventricular drug delivery and, in parallel, treatment with peripheral DA-modulating drugs that do not penetrate the brain.

A better understanding of the mechanism by which AAPs act peripherally to alter adipocyte functions should facilitate the development of new and effective antipsychotic drugs that lack the disastrous metabolic consequences. Until then, excessive therapeutic use of metabolically altering AAPs, especially those prescribed for young patients who are most vulnerable to the long-term effects of childhood obesity, should be carefully considered. **Figure 8.11** is a conceptual model depicting the complex interactions between the effects of AAPs on the brain, adipose tissue and other peripheral organs that result in obesity and metabolic syndrome.

8.6 SYNOPSIS

DA plays important roles in the GI system by regulating motility, digestion and nutrient absorption. DA it synthesized by some of the GI tract components as well as by many of the accessory glands, including salivary glands, exocrine pancreas as well as the GI innate immune system, and the microbiome. DA acts via its widely expressed receptors and contributes to the regulation of several of their physiological functions and is also involved in a number of disorders such as GERD and IBS. In spite of major advances in tracing the anatomy of the ENS, there are still gaps in our knowledge of the impact of various physiological and pathophysiological conditions on innervation of the gut. DA is also involved in the regulation of the three most important organs that regulate metabolism: the liver, the endocrine pancreas and adipose tissue. The use of several of the DA-altering antipsychotic drugs is clearly associated with obesity, diabetes and metabolic syndrome. Whether this is due to central or peripheral effects of DA remains to be determined.

REFERENCES

1. Greenwood B, Davison JS. The relationship between gastrointestinal motility and secretion. *Am J Physiol.* 1987;252(1Pt 1):G1–G7.

2. Ciucci MR, et al. Early identification and treatment of communication and swallowing deficits in Parkinson disease. *Semin Speech Lang.* 2013;34(3):185–202.

3. Gould FDH, Gross A, German RZ, Richardson JR. Evidence of oropharyngeal dysfunction in feeding in the rat rotenone model of Parkinson's disease. *Parkinsons Dis.* 2018;4:22–30.

4. Jia YX, Sekizawa K, Ohrui T, Nakayama K, Sasaki H. Dopamine D1 receptor antagonist inhibits swallowing reflex in guinea pigs. *Am J Physiol.* 1998;274(1 Pt 2):R76–R80.

5. Rattan S, Goyal RK. Effect of dopamine on the esophageal smooth muscle in vivo. *Gastroenterology.* 1976;70(3):377–381.

6. Missale G, et al. Evidence for the presence of both D-1 and D-2 dopamine receptors in human esophagus. *Life Sci.* 1990;47(5):447–455.

7. Sigala S, et al. Different neurotransmitter systems are involved in the development of esophageal achalasia. *Life Sci.* 1995;56(16):1311–1320.

8. Farre R. Pathophysiology of gastro-esophageal reflux disease: A role for mucosa integrity? *Neurogastroenterol Motil.* 2013;25(10):783–799.

9. Mezey E, Eisenhofer G, Hansson S, Hunyady B, Hoffman BJ. Dopamine produced by the stomach may act as a paracrine/autocrine hormone in the rat. *Neuroendocrinology.* 1998;67(5):336–348.

10. Deng X, Zheng Z, Ye S. Localization and expression of dopamine receptors in stomach and duodenum in rats. *Zhonghua Yi Xue Za Zhi.* 1997;77(2):103–105.

11. Eisenhofer G, et al. Substantial production of dopamine in the human gastrointestinal tract. *J Clin Endocrinol Metab.* 1997;82(11):3864–3871.

12. Basu S, Dasgupta PS. Alteration of dopamine D2 receptors in human malignant stomach tissue. *Dig Dis Sci.* 1997;42(6):1260–1264.

13. Phan H, DeReese A, Day AJ, Carvalho M. The dual role of domperidone in gastroparesis and lactation. *Int J Pharm Compd.* 2014;18(3):203–207.

14. Su A, Gandhy R, Barlow C, Triadafilopoulos G. A practical review of gastrointestinal manifestations in Parkinson's disease. *Parkinsonism Relat Disord.* 2017;39:17–26.

15. Li ZS, Schmauss C, Cuenca A, Ratcliffe E, Gershon MD. Physiological modulation of intestinal motility by enteric dopaminergic neurons and the D2 receptor: Analysis of dopamine receptor expression, location, development, and function in wild-type and knock-out mice. *J Neurosci.* 2006;26(10):2798–2807.

16. Kirschstein T, et al. Dopamine induces contraction in the proximal, but relaxation in the distal rat isolated small intestine. *Neurosci Lett.* 2009;465(1):21–26.

17. Dive A, Foret F, Jamart J, Bulpa P, Installe E. Effect of dopamine on gastrointestinal motility during critical illness. *Intensive Care Med.* 2000;26(7):901–907.

18. Zhang X, et al. Dopamine receptor D1 mediates the inhibition of dopamine on the distal colonic motility. *Transl Res.* 2012;159(5):407–414.

19. Anderson G, et al. Loss of enteric dopaminergic neurons and associated changes in colon motility in an MPTP mouse model of Parkinson's disease. *Exp Neurol.* 2007;207(1):4–12.

20. Lanfranchi GA, Marzio L, Cortini C, Osset EM. Motor effect of dopamine on human sigmoid colon. Evidence for specific receptors. *Am J Dig Dis.* 1978;23(3):257–263.

21. Bokic T, Storr M, Schicho R. Potential causes and present pharmacotherapy of irritable bowel syndrome: An overview. *Pharmacology.* 2015;96(1–2):76–85.

22. Chojnacki C, et al. Evaluation of serotonin and dopamine secretion and metabolism in patients with irritable bowel syndrome. *Pol Arch Intern Med.* 2018;128(11):711–713.

23. Iovino P, Bucci C, Tremolaterra F, Santonicola A, Chiarioni G. Bloating and functional gastro-intestinal disorders: Where are we and where are we going? *World J Gastroenterol.* 2014;20(39):14407–14419.

24. Koziolek M, Garbacz G, Neumann M, Weitschies W. Simulating the postprandial stomach: Physiological considerations for dissolution and release testing. *Mol Pharm.* 2013;10(5):1610–1622.

25. Carpenter GH. The secretion, components, and properties of saliva. *Annu Rev Food Sci Technol.* 2013;4:267–276.

26. Dodds MW, Johnson DA, Yeh CK. Health benefits of saliva: A review. *J Dent.* 2005;33(3):223–233.

27. Okumura T, Nakajima Y, Matsuoka M, Takamatsu T. Study of salivary catecholamines using fully automated column-switching high-performance liquid chromatography. *J Chromatogr B Biomed Sci Appl.* 1997;694(2):305–316.

28. Sauer JR, McSwain JL, Bowman AS, Essenberg RC. Tick salivary gland physiology. *Annu Rev Entomol.* 1995;40:245–267.

29. Tomassoni D, et al. Dopamine, vesicular transporters, and dopamine receptor expression in rat major salivary glands. *Am J Physiol Regul Integr Comp Physiol.* 2015;309(5):R585–R593.

30. Sundstrom S, Carlsoo B, Danielsson A, Henriksson R. Differences in dopamine- and noradrenaline-induced amylase release from the rat parotid gland. *Eur J Pharmacol.* 1985;109(3):355–361.

31. Hatta S, Amemiya N, Takemura H, Ohshika H. Effects of dopamine on adenylyl cyclase activity and amylase secretion in rat parotid tissue. *J Dent Res.*1995;74(6):1289–1294.

32. Mezey E, et al. Non-neuronal dopamine in the gastrointestinal system. *Clin Exp Pharmacol Physiol Suppl.* 1999;26:S1–S22.

33. Graham DY. History of Helicobacter pylori, duodenal ulcer, gastric ulcer and gastric cancer. *World J Gastroenterol.* 2014;20(18):5191–5204.

34. Kusters JG, van Vliet AH, Kuipers EJ. Pathogenesis of Helicobacter pylori infection. *Clin Microbiol Rev.* 2006;19(3):449–490.

35. Hunyady B, et al. Susceptibility of dopamine D5 receptor targeted mice to cysteamine. *J Physiol Paris.* 2001;95(1–6):147–151.

36. Glavin GB, Hall AM. Central and peripheral dopamine D1/DA1 receptor modulation of gastric secretion and experimental gastric mucosal injury. *Gen Pharmacol.* 1995;26(6):1277–1279.

37. Eliassi A, Aleali F, Ghasemi T. Peripheral dopamine D2-like receptors have a regulatory effect on carbachol-, histamine- and pentagastrin-stimulated gastric acid secretion. *Clin Exp Pharmacol Physiol.* 2008;35(9):1065–1070.

38. Tsai LH, Cheng JT. Stimulatory effect of dopamine on acid secretion from the isolated rat stomach. *Neurosci Res.* 1995;21(3):235–240.

39. Goto Y, Berelowitz M, Frohman LA. Effect of catecholamines on somatostatin secretion by isolated perfused rat stomach. *Am J Physiol.* 1981;240(3):E274–E278.

40. Eldrup E, Richter EA. DOPA, dopamine, and DOPAC concentrations in the rat gastrointestinal tract decrease during fasting. *Am J Physiol Endocrinol Metab.* 2000;279(4):E815–E822.

41. Case RM. Is the rat pancreas an appropriate model of the human pancreas? *Pancreatology.* 2006;6(3):180–190.

42. Mezey E, et al. A novel nonneuronal catecholaminergic system: Exocrine pancreas synthesizes and releases dopamine. *Proc Natl Acad Sci USA.*1996;93(19):10377–10382.

43. Vayssette J, Vaysse N, Ribet A. Dopamine receptors in pancreatic acinar cells from dog. *Eur J Pharmacol.* 1986;122(3):321–328.

44. Delcenserie R, Devaux MA, Sarles H. Action of dopamine on the exocrine pancreatic secretion of the intact dog. *Br J Pharmacol.* 1986;88(1):189–195.

45. Iwatsuki K, Ren LM, Chiba S. Effects of YM435, a novel dopamine D1 receptor agonist, on pancreatic exocrine secretion in anesthetized dogs. *Eur J Pharmacol.* 1992;218(2–3):237–241.

46. Cropley VL, et al. Whole-body biodistribution and estimation of radiation-absorbed doses of the dopamine D1 receptor radioligand 11C-NNC 112 in humans. *J Nucl Med.* 2006;47(1):100–104.

47. Slifstein M, et al. Biodistribution and radiation dosimetry of the dopamine D2 ligand 11C-raclopride determined from human whole-body PET. *J Nucl Med.* 2006;47(2):313–319.

48. Demol P, Sarles H. Action of catecholamines on exocrine pancreatic secretion of conscious rats. *Arch Int Pharmacodyn Ther.* 1980;243(1):149–163.

49. Tankurt E, et al. The prokinetic effect of domperidone in gallbladder—not upon dopaminergic receptors. *Pharmacol Res.* 1996;34(3–4):153–156.

50. Zhang LN, et al. Crosstalk between dopamine receptors and the Na(+)/K(+)-ATPase (review). *Mol Med Rep.* 2013;8(5):1291–1299.

51. Feng XY, et al. Dopamine enhances duodenal epithelial permeability via the dopamine D5 receptor in rodent. *Acta Physiol (Oxf).* 2017;220(1):113–123.

52. Vieira-Coelho MA, Soares-da-Silva P. Comparative study on sodium transport and Na+,K+-ATPase activity in Caco-2 and rat jejunal epithelial cells: Effects of dopamine. *Life Sci.* 2001;69(17):1969–1981.

53. Mittal R, et al. Neurotransmitters: The critical modulators regulating gut-brain axis. *J Cell Physiol.* 2017;232(9):2359–2372.

54. Udit S, Gautron L. Molecular anatomy of the gut-brain axis revealed with transgenic technologies: Implications in metabolic research. *Front Neurosci.* 2013;7:134–149.

55. Li ZS, Pham TD, Tamir H, Chen JJ, Gershon MD. Enteric dopaminergic neurons: Definition, developmental lineage, and effects of extrinsic denervation. *J Neurosci.* 2004;24(6):1330–1339.

56. Barrenschee M, et al. Distinct pattern of enteric phospho-alpha-synuclein aggregates and gene expression profiles in patients with Parkinson's disease. *Acta Neuropathol Commun.* 2017;5(1):1–14.

57. Sandrini S, Aldriwesh M, Alruways M, Freestone P. Microbial endocrinology: Host-bacteria communication within the gut microbiome. *J Endocrinol.* 2015;225(2):R21–R34.

58. Asano Y, et al. Critical role of gut microbiota in the production of biologically active, free catecholamines in the gut lumen of mice. *Am J Physiol Gastrointest Liver Physiol.* 2012;303(11):G1288–G1295.

59. Rubi B, Maechler P. Minireview: New roles for peripheral dopamine on metabolic control and tumor growth: Let's seek the balance. *Endocrinology.* 2010;151(12):5570–5581.

60. Ustione A, Piston DW, Harris PE. Minireview: Dopaminergic regulation of insulin secretion from the pancreatic islet. *Mol Endocrinol.* 2013;27(8):1198–1207.

61. Persson-Sjogren S, Forsgren S, Taljedal IB. Tyrosine hydroxylase in mouse pancreatic islet cells, in situ and after syngeneic transplantation to kidney. *Histol Histopathol.* 2002;17(1):113–121.

62. Ustione A, Piston DW. Dopamine synthesis and D3 receptor activation in pancreatic beta-cells regulates insulin secretion and intracellular [Ca(2$^+$)] oscillations. *Mol Endocrinol.* 2012;26(11):1928–1940.

63. Rubi B, et al. Dopamine D2-like receptors are expressed in pancreatic beta cells and mediate inhibition of insulin secretion. *J Biol Chem.* 2005;280(44):36824–36832.

64. Garcia-Tornadu I, et al. Disruption of the dopamine d2 receptor impairs insulin secretion and causes glucose intolerance. *Endocrinology.* 2010;151(4):1441–1450.

65. Lechin F, Dijs B, Pardey-Maldonado B. Insulin versus glucagon crosstalk: Central plus peripheral mechanisms. *Am J Ther.* 2013;20(4):349–362.

66. Linnestad P, Guldvog I, Schrumpf E. Effects of dopamine and dopamine antagonists on postprandial release of pancreatic polypeptide in dogs and in healthy volunteers. *Scand J Gastroenterol.* 1983;18(1):81–85.

67. Morricone L, Murari M, Bombonato M, Caviezel F. Effect of acute administration of metoclopramide on insulin secretion in man. *Acta Diabetol Lat.* 1990;27(1):53–57.

68. Ahima RS. Adipose tissue as an endocrine organ. *Obesity (Silver Spring).* 2006;14(Suppl 5):242S–249S.

69. Ailhaud G. Adipose tissue as a secretory organ: From adipogenesis to the metabolic syndrome. *C R Biol.* 2006;329(8):570–577.

70. Scheja L, Heeren J. Metabolic interplay between white, beige, brown adipocytes and the liver. *J Hepatol.* 2016;64(5):1176–1186.

71. Bargut TC, Aguila MB, Mandarim-de-Lacerda CA. Brown adipose tissue: Updates in cellular and molecular biology. *Tissue Cell.* 2016;48(5):452–460.

72. Maxwell G, Crompton S, Smyth C. The effect of dopamine upon oxidative metabolism of brown fat adipocytes. *Eur J Pharmacol.* 1985;116(3):293–297.

73. Nisoli E, Tonello C, Memo M, Carruba MO. Biochemical and functional identification of a novel dopamine receptor subtype in rat brown adipose tissue. Its role in modulating sympathetic stimulation-induced thermogenesis. *J Pharmacol Exp Ther.* 1992;263(2):823–829.

74. Kohlie R, et al. Dopamine directly increases mitochondrial mass and thermogenesis in brown adipocytes. *J Mol Endocrinol.* 2017;58(2):57–66.

75. Lee TL, Hsu CT, Yen ST, Lai CW, Cheng JT. Activation of beta3-adrenoceptors by exogenous dopamine to lower glucose uptake into rat adipocytes. *J Auton Nerv Syst.* 1998;74(2–3):86–90.

76. Tavernier G, et al. Norepinephrine induces lipolysis in beta1/beta2/beta3-adrenoceptor knockout mice. *Mol Pharmacol.* 2005;68(3):793–799.

77. Mukherjee R, Yun JW. Bromocriptine inhibits adipogenesis and lipogenesis by agonistic action on alpha2-adrenergic receptor in 3T3-L1 adipocyte cells. *Mol Biol Rep.* 2013;40(5):3783–3792.

78. Ben-Jonathan N, Mershon JL, Allen DL, Steinmetz RW. Extrapituitary prolactin: Distribution, regulation, functions, and clinical aspects. *Endocr Rev.* 1996;17(6):639–669.

79. Hugo ER, Borcherding DC, Gersin KS, Loftus J, Ben-Jonathan N. Prolactin release by adipose explants, primary adipocytes, and LS14 adipocytes. *J Clin Endocrinol Metab.* 2008;93(10):4006–4012.

80. Borcherding DC, et al. Dopamine receptors in human adipocytes: Expression and functions. *PLoS One.* 2011;6(9):e25537.

81. Vargovic P, et al. Adipocytes as a new source of catecholamine production. *FEBS Lett.* 2011;585(14):2279–2284.

82. Hugo ER, et al. LS14: A novel human adipocyte cell line that produces prolactin. *Endocrinology.* 2006;147(1):306–313.

83. Rosen ED, Walkey CJ, Puigserver P, Spiegelman BM. Transcriptional regulation of adipogenesis. *Genes Dev.* 2000;14(11):1293–1307.

84. Alvarez-Aguilar C, et al. Effects of dopamine on leptin release and leptin gene (OB) expression in adipocytes from obese and hypertensive patients. *Int J Nephrol Renovasc Dis.* 2013;6:259–268.

85. Wang X, Villar VA, Tiu A, Upadhyay KK, Cuevas S. Dopamine D2 receptor upregulates leptin and IL-6 in adipocytes. *J Lipid Res*. 2018;59(4):607–614.

86. Wojtalla A, et al. The endocannabinoid N-arachidonoyl dopamine (NADA) selectively induces oxidative stress-mediated cell death in hepatic stellate cells but not in hepatocytes. *Am J Physiol Gastrointest Liver Physiol*. 2012;302(8):G873–G887.

87. Lizana H, et al. Whole-body biodistribution and dosimetry of the dopamine transporter radioligand (18)F-FE-PE2I in human subjects. *J Nucl Med*. 2018;59(8):1275–1280.

88. Nassar CF, Karkaji EG, Habbal ZM, Nasser MG. Dopamine receptors in normal and diabetic liver plasma membrane. *Gen Pharmacol*. 1986;17(3):367–370.

89. Zhou H, et al. Dopamine alleviated acute liver injury induced by lipopolysaccharide/d-galactosamine in mice. *Int Immunopharmacol*. 2018;61:249–255.

90. Sun CK, et al. Dopamine impairs functional integrity of rat hepatocytes through nuclear factor kappa B activity modulation: An in vivo, ex vivo, and in vitro study. *Liver Transpl*. 2015;21(12):1520–1532.

91. Koetting M, Stegemann J, Minor T. Dopamine as additive to cold preservation solution improves postischemic integrity of the liver. *Transpl Int*. 2010;23(9):951–958.

92. Liang Y, Jetton TL, Lubkin M, Meier AH, Cincotta AH. Bromocriptine/SKF38393 ameliorates islet dysfunction in the diabetic (db/db) mouse. *Cell Mol Life Sci*. 1998;54(7):703–711.

93. Chamarthi B, Ezrokhi M, Rutty D, Cincotta AH. Impact of bromocriptine-QR therapy on cardiovascular outcomes in type 2 diabetes mellitus subjects on metformin. *Postgrad Med*. 2016;128(8):761–769.

94. Rosati G, Maioli M, Aiello I, Farris A, Agnetti V. Effects of long-term L-dopa therapy on carbohydrate metabolism in patients with Parkinson's disease. *Eur Neurol*. 1976;14(3):229–239.

95. Lima MM, et al. Does Parkinson's disease and type-2 diabetes mellitus present common pathophysiological mechanisms and treatments? *CNS Neurol Disord Drug Targets*. 2014;13(3):418–428.

96. Tang QQ, Lane MD. Adipogenesis: From stem cell to adipocyte. *Annu Rev Biochem*. 2012;81:715–736.

97. Grundy SM. Metabolic syndrome pandemic. *Arterioscler Thromb Vasc Biol*. 2008;28(4):629–636.

98. Wang GJ, Volkow ND, Fowler JS. The role of dopamine in motivation for food in humans: Implications for obesity. *Expert Opin Ther Targets*. 2002;6(5):601–609.

99. Lopez VF, et al. Dopaminergic drugs in type 2 diabetes and glucose homeostasis. *Pharmacol Res*. 2016;109:74–80.

100. American Diabetes Association. Consensus development conference on antipsychotic drugs and obesity and diabetes. *J Clin Psychiatry*. 2004;65(2):267–272.

101. Correll CU, et al. Cardiometabolic risk of second-generation antipsychotic medications during first-time use in children and adolescents. *JAMA*. 2009;302(16):1765–1773.

102. Mittal D, et al. Monitoring veterans for metabolic side effects when prescribing antipsychotics. *Psychiatr Serv*. 2013;64(1):28–35.

103. Farah A. Atypicality of atypical antipsychotics. *Prim Care Companion J Clin Psychiatry*. 2005;7(6):268–274.

104. American Psychiatric Association. Practice guide for the treatment of patients with schizophrenia. *Am Psychiatr Assoc*. 1997;1:1–20.

105. Newcomer JW. Metabolic considerations in the use of antipsychotic medications: A review of recent evidence. *J Clin Psychiatry*. 2007;68(Suppl 1):20–27.

106. Minet-Ringuet J, et al. Alterations of lipid metabolism and gene expression in rat adipocytes during chronic olanzapine treatment. *Mol Psychiatry*. 2007;12(6):562–571.

107. Sertie AL, et al. Effects of antipsychotics with different weight gain liabilities on human in vitro models of adipose tissue differentiation and metabolism. *Prog Neuropsychopharmacol Biol Psychiatry*. 2011;35(8):1884–1890.

108. Pfeiffer RF. Gastrointestinal dysfunction in Parkinson's disease. *Lancet Neurol*. 2003;2(2):107–116.

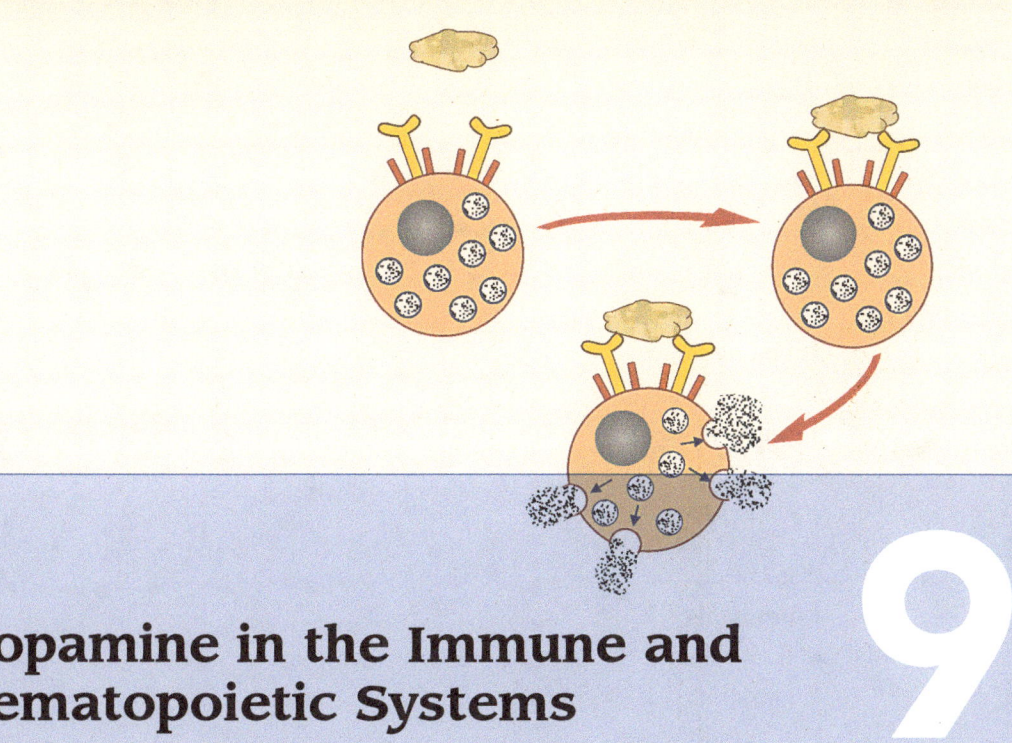

Dopamine in the Immune and Hematopoietic Systems

9

9.1 INTRODUCTION

Although often considered as separate entities, the myeloid and lymphoid families are not totally unrelated but, rather, represent two partially interwoven family trees. Immunity is operationally classified as innate and adaptive immunity. Innate immunity is the system already present in the body, whereas adaptive immunity is created in response to exposure to a foreign substance and is composed of B cells and T cells. Once activated against a specific type of antigen, immunity remains throughout life.

The first two sections of this chapter review the production and action of dopamine (DA) in cells that are derived from the lymphoid and myeloid lineages. **Table 9.1** summarizes the available information on the expression of catecholamine biosynthetic enzymes, transporters, and the five DA receptors (DARs) in various hematopoietic and immune cells. The next three sections describe the involvement of DA in autoimmune and neuropsychiatric diseases, inflammation, coagulation, and allergic reactions. Endothelial cells, angiogenesis, inflammation, and allergic reactions, under both normal and malignant conditions, are covered in the final sections of the chapter.

9.2 CELLS DERIVED FROM THE LYMPHOID LINEAGE

Information presented in this section is largely based on several reviews [1–4]. **Hematopoiesis** is the process by which the cellular constituents of blood are formed in the bone marrow, thymus, spleen or liver from multipotential hematopoietic stem cells. In a healthy adult person, 10^{11}–10^{12} new blood cells are produced every day to maintain steady-state levels in the peripheral circulation. As depicted in **Figure 9.1**, two main common progenitors are recognized: myeloid and lymphoid. These give rise to the various hematopoietic and immune cell types shown in the figure, which differ in structure, size, biomarkers, number of nuclei, chemical composition, half-life, and

Table 9.1 Dopamine biosynthetic enzymes, transporters, and receptors in hematopoietic cells

Cell type	TH	DDC	DAT	VMAT	D1R	D2R	D3R	D4R	D5R
PBL	+	+	+	+		+	+	+	+
Thymocytes	+						+		
Splenocytes	+	+							
T lymphocytes	+	+	?		+	+	+	+	+
B lymphocytes	+	+			+	+	+	+	+
Natural killer cells					+	+	+	+	+
Dendritic cells	+	+		+	+	+	+		+
Eosinophils	?				+	+	+	+	+
Macrophages	+	+	+	+	+	+	+	+	+
Mast cells	+				+		+		
Neutrophils	?		+		+	+	+	+	+
Erythrocytes			?						
Platelets			+			+	+		+

DAT: dopamine transporter; DDC: dopamine decarboxylase; PBL: peripheral blood lymphocytes; TH: tyrosine hydroxylase; VMAT: vesicular monoamine transporter.

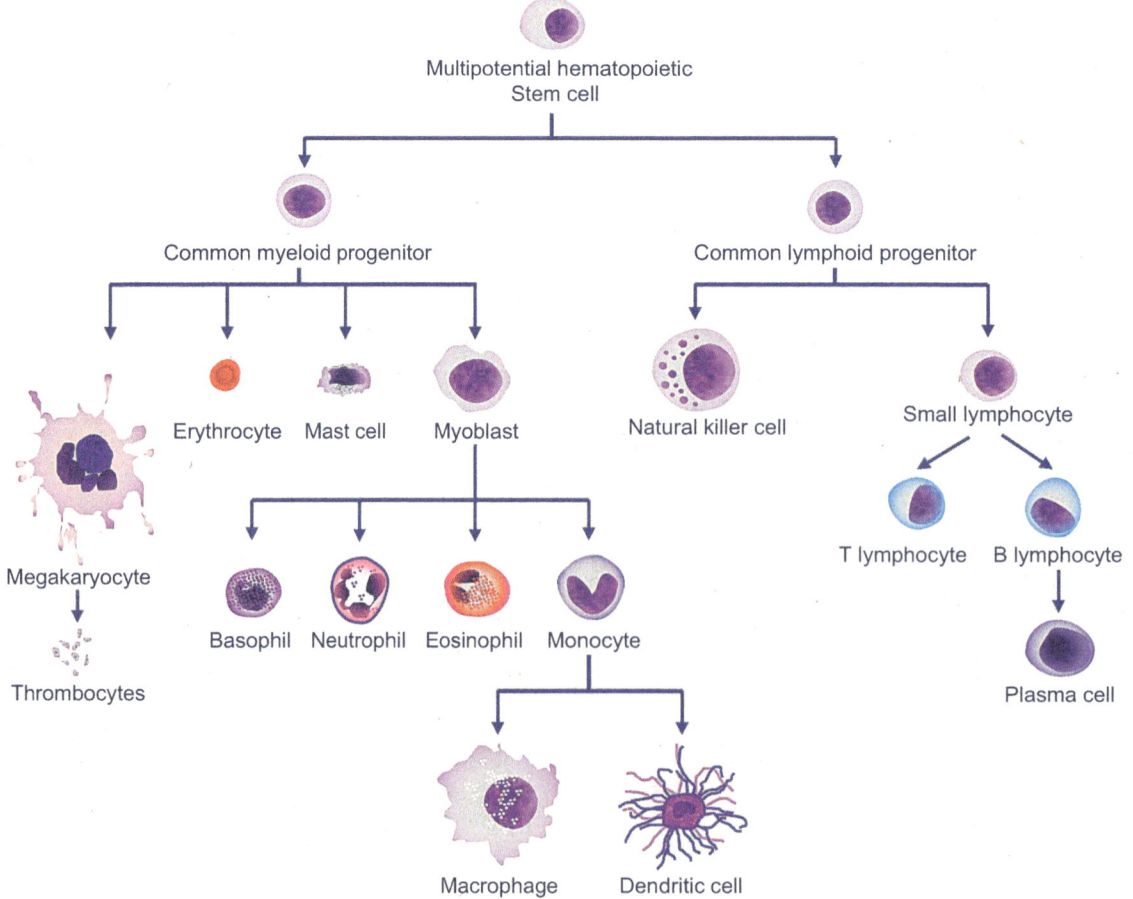

Figure 9.1 Diagram of hematopoiesis. Shown are circulating and tissue-resident cells derived from the common lymphoid and myeloid progenitors.

functions. The lymphoid lineage is composed of **T lymphocytes** (thymus cells or T cells), **B lymphocytes** (bone marrow cells or B cells), and **natural killer cells** (NK cells), all of which are mono-nucleated cells. Upon maturation, these cells enter the general circulation and are referred to as **peripheral blood lymphocytes** or PBLs.

Another common categorization of immune cells is as antigen-presenting cells (APCs), which include B cells, dendritic cells (DCs), and macrophages. These cells recognize a protein antigen and internalize it, either by phagocytosis or by a receptor-mediated endocytosis. The protein is then broken down into 10- to 30-amino acid peptides. These are loaded onto HLA class I and II molecules and are displayed on the cell surface where they can interact with appropriate **T-cell receptors (TCRs)**.

9.2.1 T cells, B cells, and NK cells

T cells are created by lymphopoiesis from a common lymphoid progenitor in the bone marrow. The newly generated cells migrate to the thymus where they undergo extensive maturation and screening. Through a combination of positive and negative selection processes, the cells are differentiated, yielding a repertoire of mature T cells that tolerate self-antigens and are capable of mounting strong responses to foreign antigens. Within the thymus, T cells undergo a V(D)J recombination, a unique mechanism of genetic recombination that occurs only in developing lymphocytes during maturation. It involves somatic recombination that results in a highly diverse assortment of the TCRs. Cell selection in the thymus is accompanied by an extensive cell death by apoptosis and phagocytosis, with only a very small percent of the cells surviving the selection process, and these are exported to extra-thymic sites.

T lymphocytes attack invaders such as bacteria and viruses, augment the antibody-producing B lymphocytes, and produce cytokines that affect the responses and/or activity of many cells [5]. They are divided into several classes: (1) CD4+ T-helper (Th17) cells, which are involved in the activation of B cells, cytotoxic T cells, and macrophages; (2) CD8+ T-cytotoxic cells, which kill virally infected cells; (3) T-memory cells, which remember previously encountered antigens; (4) regulatory T cells (Tregs), which moderate the response of other immune cells; and (5) NK cells, less common T cells, which are involved in rapid immune responses.

B lymphocytes also develop from hematopoietic progenitors in the bone marrow, where they undergo the initial process of differentiation and selection [6]. They are distinguished from T cells and NK cells by the expression of the B cell receptor (BCR) on their plasma membrane. Development of B cells occurs in several stages, marked by various patterns of gene expression, and an arrangement of the immunoglobulin H chain and L chain gene loci. They also undergo the V(D)J recombination of their immunoglobulins. To complete their development, immature B cells migrate from the bone marrow to the spleen. Once fully differentiated, B cells can secrete antibodies and comprise the humoral component of the adaptive immunity. In addition, B cells present antigen and secrete cytokines. The BCR enables each B cell clone to bind a specific antigen and initiate an antibody response. Activation of B cells occurs via antigen recognition by BCRs and requires a secondary activation signal provided either by T helper cells or by the antigen itself.

NK cells are large granular lymphocytes that constitute the third cell type derived from the common lymphoid progenitor discussed above and shown in **Figure 9.1**. NK cells differentiate and mature in the bone marrow, lymph nodes, spleen, tonsils, and thymus. Acting as a component of the adaptive immune system, NK cells provide rapid responses to virus-infected cells, and they also respond to tumor formation. Typically, immune cells detect MHC presented on the surface of infected cells, triggering the release of cytokines that induce cell lysis or apoptosis. NK cells are unique in that they can recognize stressed cells even in the absence of antibodies and MHC, allowing for a much faster immune reaction.

9.2.2 The dopaminergic system in lymphocytes and NK cells

The first indication that lymphocytes possess a dopaminergic system came in the 1980s, when pharmacological data reveled that treatment with bromocriptine prevented or reduced several humoral- or cell-mediated responses [7]. Yet, the authors of the study attributed the immune suppressive effects to the inhibition of prolactin (PRL), an immune-modulating hormone, by bromocriptine.

Following the subsequent development of molecular biological techniques such as radioligand assays and immunocytochemistry, attention has focused on the dopaminergic system in PBLs. Given the ease of blood withdrawal and the established methods for lymphocyte isolation by cytospinning and their analysis by microscopic and biochemical assays, PBLs have long served as diagnostic components for many immune disorders. PBLs were also used for the initial establishment of the biosynthesis and action of DA in immune cells. Based on the use of radiolabeled ligands as well as Western blotting, it was found that tyrosine hydroxylase (TH) was present in the cytoplasma of isolated PBLs, whereas the transporters DA transporter (DAT) and vesicular monoamine transporter (VMAT)-1 were located in the plasma membrane and the cytoplasm, respectively [8]. The content of DA differs among distinct lymphocyte subsets [9], with $CD4^+$ $CD25^-$ T lymphocytes containing an average of 0.2 pM of DA and $CD4^+$ $CD25^+$ Tregs containing 37.4 pM. The production and release of DA from lymphocytes were increased by interferon beta (IFN-β), reserpine (a VMAT inhibitor), high potassium, and protein kinase C (PKC) activators.

A search in PubMed revealed more than 50 publications, published since 2000, with the words "dopamine" and "lymphocytes" in their title. These attested to the growing interest of the research and clinical communities on the role of the dopaminergic system in lymphocyte pathobiology. As discussed in several reviews [2,3,10], lymphocytes synthesize DA through the TH/dopamine decarboxylase (DDC) pathway, and express several types of DAR as well as transporters for DA on their plasma membrane (**Table 9.1**). Most immune cells can synthesize and store DA in intracellular vesicles and release it in response to specific stimuli. The general notion is that DA operates as a bidirectional communicator and mediator between the nervous and immune systems.

In spite of good agreement among various reports with respect to the ability of lymphocytes to synthesize and/or take up DA, they did not arrive at an overall unified concept about the exact functions of DA these cells [1,11]. The lack of consensus results from several issues. One is the heterogeneity of lymphocytes and the fact that PBLs comprise a mixture of distinct immune cell types with different properties and dissimilar abundance. A second issue is the complementary as well as the opposite actions resulting from activation of the five DARs. A third problem is the dynamic nature of activation vs. inactivation of lymphocytes that can rapidly alter receptor expression and signaling pathways. Finally, as discussed in several sections of this book, the effects of DA are dose dependent, exhibiting distinct differences in the type of responses resulting from low vs. high doses.

In spite of the above limitations, several DA functions in lymphocytes have been identified [1]. As shown in **Figure 9.2**, activation and functions of the DA/DAR axis in T lymphocytes are dynamic and context-sensitive processes. DA is a potent activator of resting effector T cells (Teffs) through

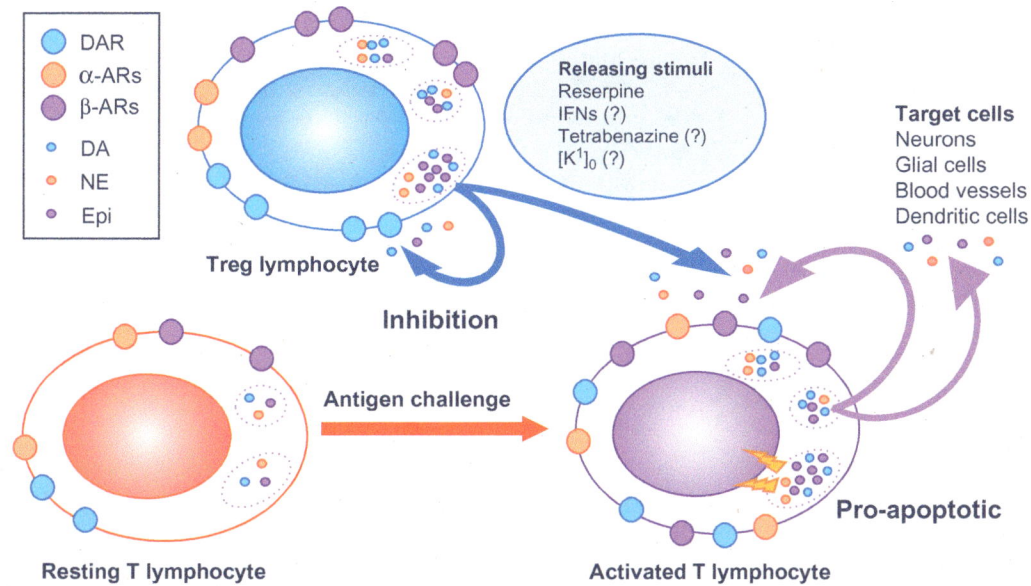

Figure 9.2 Endogenous catecholamines in human T-lymphocytes. DA, norepinephrine (NE), and epinephrine (EPI) are released in response to various stimuli, and exert autocrine/paracrine actions on lymphocytes and endocrine actions on multiple target cells. AR: adrenergic receptor; IFN: interferon. DAR, dopamine receptor; Treg, regulatory T cell. (Redrawn and modified from Cosentino, M., and Marino, F., *J. Neuroimmune. Pharmacol.*, 8, 163–179, 2013.)

two mechanisms: direct and indirect. An indirect activation occurs by the suppression of Tregs that inhibit the Teffs. DA is released from T cells following their activation by antigen, mitogen, IFN-β, anti-CD3 antibodies, or PKC activators. The released DA can then act as an autocrine/paracrine factor on the producing cells or on neighboring cells.

The DAR subtypes differentially affect the activity of lymphocytes under various conditions. For instance, D1R-like inhibits the cytotoxic functions of CD8[+] T cells, impairs the differentiation of Tregs [3], and affects the polarization of naïve CD4[+] T cells toward Th17 cells [12]. D2R enhances the production of interleukin-10 (IL-10), which negatively regulates Teffs [13]. D3R facilitated the differentiation of naïve CD8[+] T cells into a cytotoxic phenotype [13], and contributes to the development of naïve CD4[+] T cells toward the Th1 (T helper) phenotype [14]. D3R is also involved in the migration and adhesion of T cells by inducing $\beta 1$ integrin-dependent adhesion to fibronectin, potentially modulating their homing mechanism [15]. D4R affects T-cell quiescence by upregulating Krüppel-like factor-2 expression [16].

Inhibition of the proliferation of T and B lymphocyte by DA has been reported in several studies, but most have used very high concentrations of DA or its agonists, leading to potential misinterpretations. In one study [17], DA at 100 μM inhibited, while norepinephrine (NE) enhanced, lipopolysaccharaide (LPS)-stimulated proliferation of murine lymphocytes from the spleen and mesenteric lymph nodes. The effects of DA were not blocked by D1R or D2R antagonists, or by alpha- or beta-AR antagonists. The conclusion was that the two catecholamines exert opposite effects on B-cell activation via unidentified mechanisms.

An unclear mechanism of DA was observed in another study, reporting that micromolar concentrations of the neuroleptic haloperidol and spiperone potentiated anti-mu-induced murine B lymphocyte proliferation *in vitro* [18]. DA and NE inhibited, while serotonin enhanced, B-cell proliferation. Antagonists of DAR, 5HT2, 5HT1A, and alpha 1 AR did not mimic the effects of haloperidol and spiperone. The study concluded that the anti-proliferative effects of the DA drugs were due to their actions at unknown receptors. On the other hand, specific receptor-mediated effects were observed when

DA at physiological doses was incubated with human T-cell subpopulations separated by panning [19]. DA significantly inhibited both proliferation and cytotoxicity of CD8+ T cells, showing a lesser effect on CD4+ T cells. The underlying mechanism was D1R-mediated stimulation of intracellular cAMP.

Data on the *in vitro* effects of DA on NK cells are highly variable, likely due to the use of excessively high doses of DA and/or to the reliance on drugs that are not entirely specific for DAR subtypes. An early study [20] reported that cytotoxicity of murine splenic NK cells was suppressed by some, but not by all, neuroleptics, which were used at 2.5 to 15 µM doses. The conclusion that drugs such as thiothixine, fluphenazine, and trifluoperazine act by mechanisms other than DARs is consistent with the notion that antipsychotics activate multiple, often antagonistic, receptors.

A later study [21], using selective DAR agonists and antagonists at 10–100 nM, found that stimulation of D1R-like enhanced NK cell cytotoxicity via the cAMP-PKA-CREB (CAAP response element binding protein) signaling pathways, while activation of D2R-like suppressed cytotoxicity by inhibiting this pathway. Several *in vivo* studies have focused on NK cells. In one study, mice were injected with DA agonists and antagonist [22]. The results showed that haloperidol, a D2R antagonist, decreased NK cell activity in a PRL-independent manner, while bromocriptine, a D2R agonist, decreased NK cell activity via the reduction in PRL release. Another study, using normal and tumor-bearing mice, found that DA at 50 mg/kg increased the tumor-killing capacity of NK cells [23].

9.3 CELLS DERIVED FROM THE MYELOID LINEAGE

As shown in **Figure 9.1**, cells derived from common myeloid progenitors include erythrocytes, mast cells, and cells developed from myeloblasts—basophils, neutrophils, and eosinophils—as well as macrophages, which develop from monocytes. In addition, megakaryocytes give rise to thrombocytes or platelets. Another way of looking at circulating blood cells is to categorize them as erythrocytes, platelets and leukocytes. Erythrocytes are by far the most abundant circulating blood cells. Each microliter of blood contains 4–6 million erythrocytes, several hundred thousand platelets, and 4,000–6,000 leukocytes. Of the total number of leukocytes, 40%–75% are neutrophils and 1%–6% are eosinophils and basophils, while mononuclear cells, which include monocytes and lymphocytes, account for 30%–50% of the total leukocytes.

9.3.1 The dopaminergic system in erythrocytes and platelets

Human erythrocytes have a structure of a biconcave disk with a diameter of 7 μm. They lack a nucleus and most organelles, and their average life in the circulation is 120 days. The cytoplasm of erythrocytes is rich in hemoglobin, an iron-containing, 68-kDa biomolecule that binds oxygen via its heme subunit and gives rise to the red color of the cells and blood. The cell membrane of erythrocyte is composed of a mesh of fibrous proteins that confer the cells with great flexibility and the ability to deform while traversing the narrow capillary network of the circulatory system. The erythrocyte membrane contains ion pumps that maintain high levels of intracellular potassium and low levels of calcium. The main function of erythrocytes is to carry oxygen from the lungs to the tissues and carbon dioxide, as a waste product, away from the tissues and back to the lungs.

The dopaminergic system interacts with erythrocytes at several levels. One interaction is through the stimulation of erythropoiesis in bone marrow cells [4]. Another is via the ability of erythrocytes to degrade DA, based on their high expression of catechol-O-methyltransferase (COMT), the major peripheral DA metabolizing enzyme [24]. This property of erythrocytes is of particular relevance to patients with Parkinson's disease (PD) who are treated with L-Dopa in combination with inhibitors of Dopa decarboxylase and COMT in order to maximize L-Dopa's uptake by the brain. Erythrocytes also have high expression levels of phenol sulfotransferase [25], the enzyme that catalyzes sulfo-conjugation of DA and is responsible for the generation of dopamine sulfate (DA-S), the major circulating form of DA in humans. Although mature human erythrocytes express over 200 genes that are associated with signal transduction, there is no documentation that they express any of the DARs. Yet, erythrocytes are fully capable of uptake and accumulation of DA via a saturable transporter, possibly via the choline transporter whose relationship to DAT is unclear [26].

A well-studied function of DA in erythrocytes is inhibition of apoptosis, which is normally triggered by oxidative stress, hypertonic shock, removal of extracellular Cl^-, or energy depletion. To distinguish between apoptosis of nucleated cells and that in erythrocytes, the term "eryptosis" was coined [27]. In this study, treatment of erythrocytes with ionomycin, a Ca^{2+} ionophore, led to cell shrinkage, cell membrane blebbing and breakdown of membrane phosphatidylserine asymmetry, all typical features of apoptosis in nucleated cells. Several catecholamines, including DA, inhibited the activated entry of Ca^{2+} into erythrocytes by removal of Cl^-, thus preventing the increase of cytosolic Ca^{2+} activity that causes cell shrinkage. The authors concluded that the effect of catecholamines on apoptosis is due to a direct or an indirect inactivation of the calcium-permeable, nonselective cation conductance.

The above results uncovered a novel mechanism by which catecholamines can modify the half-life of erythrocytes as well as their adhesion to the vascular wall. The antagonistic action of DA may also operate during several stress situations that increase erythrocyte apoptosis, all of which occur by the activation of a Ca^{2+}-permeable nonselective cation conductance. The effects of DA on eryptosis do not appear to involve a DAR and, thus, may be receptor independent.

Platelets are small in size, 2–3 μm in the greatest diameter, and have no cell nucleus. They are derived from megakaryocytes of the bone marrow and enter the circulation as fragments of cytoplasm [28]. Activated platelets have cell membrane projections covering their surface. The average life span of circulating platelets is 8–9 days and is controlled by an internal apoptotic regulating pathway that involves $Bcl-x_L$. Old platelets are destroyed by phagocytosis in the spleen and liver. A major function of platelets is in hemostasis, or the process of stopping bleeding at the site of interrupted endothelium. Platelets accumulate at the site of vascular injury and plug the hole after undergoing several sequential events: adhesion, activation, and aggregation. Formation of the platelet plug (primary hemostasis) is accompanied by activation of the coagulation cascade, resulting in fibrin deposition and linking (secondary hemostasis).

Human platelets express D2R, D3R, and D5R [29], as well as DAT [30]. Depending upon the dose used, variable effects of DA have been reported on platelet aggregation [31]. In the micromolar range, DA induces platelet aggregation and inhibits epinephrine (Epi)-induced aggregation. An enhancing effect of DA on ADP-induced platelet aggregation is also seen in the nanomolar range. At very high concentrations (outside the physiological range), DA inhibits ADP-induced platelet aggregation. More detail on DA and wound healing is presented in **Chapter 11**.

9.3.2 The dopaminergic system in various leucocyte subtypes

Granulocytes (neutrophils, eosinophils, basophils, and mast cells), monocytes/macrophages, and DCs are collectively called myeloid cells and represent a subgroup of leukocytes. During the embryonic development of mammals, myelopoiesis occurs in a stepwise fashion that begins in the yolk sac and ends up in the bone marrow. During this process, early monocyte progenitors colonize various organs such as the brain, liver, skin, and lungs where they differentiate into resident macrophages that will self-maintain throughout life.

DCs have a similar appearance as lymphocytes but carry different markers and can originate from the lymphoid as well as the myeloid lineages. Their name comes from the protrusions, similar to dendrites in neurons, that grow on their surface during maturation. They are considered the most potent antigen-presenting cells and act by stimulating naïve T cells toward effector T cells (Th1), and cytotoxic T lymphocytes (CTLs) and initiate primary immune responses [32]. DCs play critical roles in the induction of immunological tolerance, regulate the types of responses of T cells, and act as sentinels in innate immunity against microbes. Such diverse functions depend on their heterogeneity and functional plasticity. DCs possess high phagocytic activity as immature cells, and high cytokine-producing activity as mature cells [2]. The role played by DCs in various tumors is covered in **Chapters 12** and **13**.

DCs have the complete ability to synthesize, store, and degrade DA by expressing TH, vesicular monoamine transporter (VMAT), and monoamine oxidase (MAO), respectively. Notably, DCs do not express dopamine β-hydroxylase (DBH), indicating the DA rather than NE or Epi is their final product of the catecholamine biosynthetic pathway [33]. DA can act in an autocrine manner to stimulate DARs in DCs, or it can affect other lymphocytes. D3R deficiency in DCs enhances the expansion of CTLs *in vivo* and induces a stronger antitumor immunity [34]. Coculture data has showed that D3R-inhibition in DCs potentiated antigen cross-presentation and CTLs activation. The authors suggested that D3R in DCs should become a new therapeutic target for strengthening antitumor immunity.

The role of D5R in DCs functions has also been examined [35]. Deficiency of D5R in DCs impairs LPS-induced IL-23 and IL-12 production and attenuates activation and proliferation of antigen-specific CD4+ T cells (**Figure 9.3**).

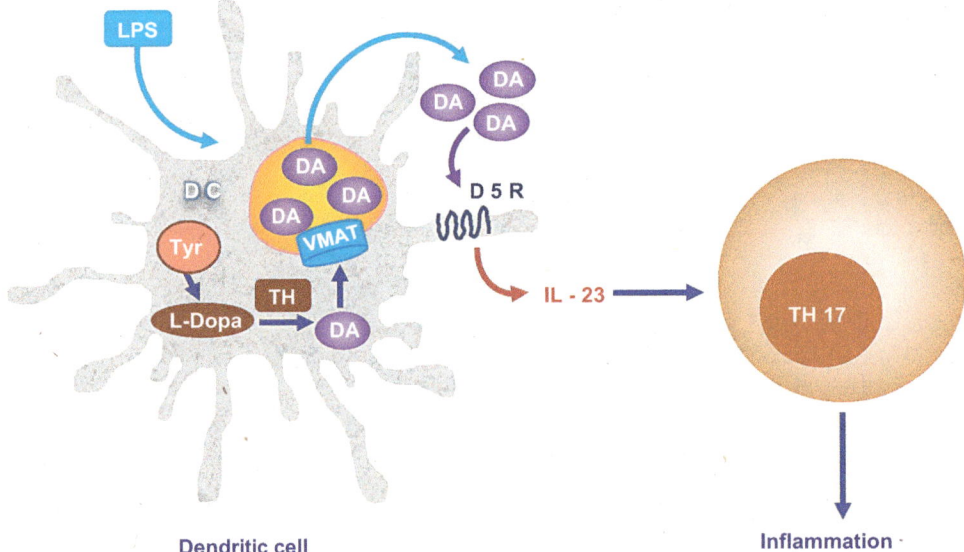

Dendritic cell   **Inflammation**

Figure 9.3 Amplification of inflammation by locally-produced DA in dendritic cells (DC). In response to antigen presentation or lipopolysaccharaide (LPS) stimulation, dendritic cells release dopamine (DA), which acts on D5Rs and promotes the production of IL-23 and enhances the responses of proinflammatory Th17 cells. TH: tyrosine hydroxylase; VMAT: vesicular monoamine transporter; DC, dendritic cell; IL, interleukin. (Redrawn and modified from Amenta. F., et al. *J. Neuroimmunol.*, 117, 133–142, 2001.)

To determine the relevance of D5R expression in DCs *in vivo*, D5R-deficient DCs were prophylactically transferred into wild-type mice recipients and were found to reduce the severity of experimental autoimmune encephalomyelitis. The mice also showed a significant reduction in the percentage of Th17 cells that infiltrate the central nervous system (CNS).

Macrophages, and their CNS counterparts the **microglia**, detect and clear microbial pathogens and injured tissue [36,37]. They are phagocytic cells that adapt their phenotype by their activity (i.e., whether they participate in acute defense against pathogenic organisms) ("M1"-phenotype) or become engaged in clearing damaged tissues and performing repair activities ("M2"-phenotype), as shown in **Figure 9.4**. M1-polarization is promoted by stimulation of pattern recognition receptors by viruses (vaccines), bacterial membrane components (LPS), or long-chain saturated fatty acids. Human macrophages express mRNA for TH, DDC, VMAT, DAT, and all DAR subtypes [38]. DA alters the production of some cytokines in both untreated and LPS-treated macrophages. In untreated cells, DA increases IL-6 and CCL2, while in LPS-treated cells, DA increases IL-6, CCL2, CXCL8, and IL-10, and decreases tumor necrosis factor alpha (TNF-α).

Fenoldopam, the peripheral D1R agonist, causes robust activation of AMP-activated protein kinase (AMPK) in a PLC-dependent mechanism in human macrophages and monocytes [39]. D1R activation also prevents Thr172-AMPK dephosphorylation and kinase inactivation in LPS-treated macrophages. Fenoldopam also reduces the severity of LPS-induced acute lung injury, including development of pulmonary edema, lung permeability, and production of inflammatory cytokines [39]. These data suggest that DA is coupled to AMPK activation, which provides a substantial anti-inflammatory advantage and reduces the severity of endotoxin-induced acute lung injury.

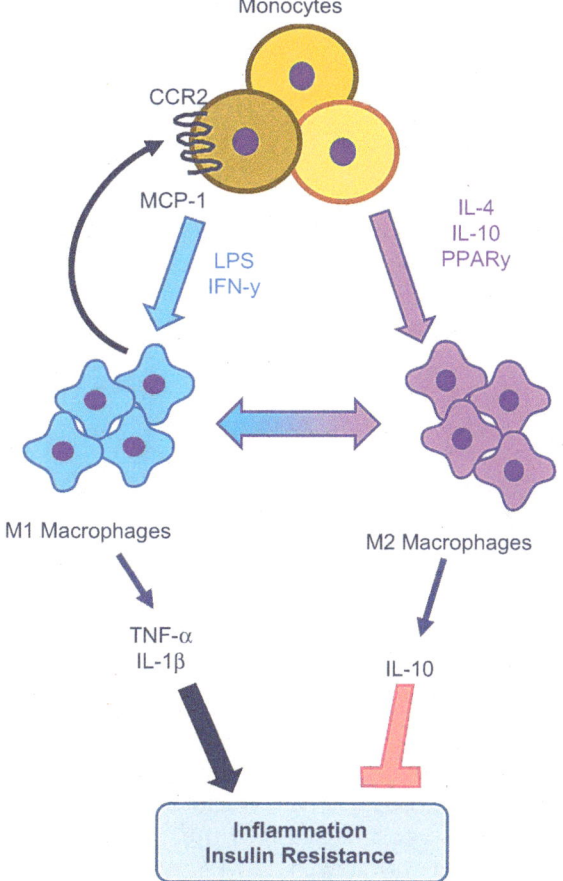

Figure 9.4 Reversible transition between the M1 and M2 phenotypes of macrophages in response to various stimuli. The chemokines and cytokines released by the macrophages affect inflammation, insulin resistance and tumor microenvironment. CCR2: chemokine receptor type 2; IFN-γ: interferon gamma; IL: interleukin; LPS: lipopolysaccharide; MCP-1: monocyte chemoattractant protein-1; PPARγ: peroxisome proliferator-activated receptor gamma; TNF-α: tumor necrosis factor alpha. See text for other explanations. (Redrawn and modified from Ohashi, W., et al., Control of macrophage dynamics as a potential therapeutic approach for clinical disorders involving chronic inflammation. *J. Pharmacol. Exp. Ther.*, 354, 240–250.)

Another study reported that D1R activation significantly inhibits low-density lipoprotein-induced proliferation of macrophages, possibly by inhibiting the PI3K/Akt and MAP kinase (MAPK)/ERK signaling pathways [40].

Neutrophils, which are the most abundant leucocytes, form an essential part of the innate immune system. They are short-lived and highly mobile and can enter parts of tissues where other cells cannot. Together with basophils and eosinophils, they belong to the family of polymorphic-nuclear cells. The names of these cells come from their staining properties of hematoxylin and eosin (H&E) cytological preparations: basophils stain dark blue, eosinophils stain bright red, and neutrophils stain pink. During the acute phase of inflammation following bacterial infection, environmental exposure, and some cancers, neutrophils are among the first-responders that migrate toward the site of inflammation [2]. They go through blood vessels and interstitial tissue by chemotaxis, following chemical signals such as IL-8 and leukotriene B4. Neutrophils are the predominant cells in pus, accounting for its whitish/yellowish appearance.

As discussed in an extensive review [2], neutrophils contain DA, albeit it is unclear whether they synthesize DA *de novo* or take it up from the circulation and/or from neighboring cells. As judged by immunohistochemistry and flow cytometry, neutrophils express DAR in the following order of magnitude: D4R>D3R>D1R>D5R.D2R. In general, *in vitro* studies have reported inhibitory effects of DA on several functions of the neutrophils, including a diminished ability of human neutrophils to adhere to the endothelium and decreases in reactive oxygen species production, cell migration, and phagocytic activity. Studies with mouse immune cells showed that DA suppressed LPS-stimulated TNF-α secretion by neutrophils and upregulated IL-10 [41]. *In vivo* studies with rat cells have found that dopaminergic agonists increase the number of neutrophils during anaphylactic shock and reduce the ovalbumin antigen-induced activation of neutrophils in a mouse model of airway inflammation [12].

Eosinophils are responsible for combating multicellular parasites and certain infections. Along with mast cells and basophils they are associated with the occurrence of allergy and asthma. Once generated during hematopoiesis in the bone marrow, eosinophils migrate into blood as terminally differentiated cells that do not multiply. There are no published records as to their ability to synthesize DA, but they express on their membrane all five DARs [42]. Older studies with rats have showed that treatment with L-Dopa or apomorphine causes a biphasic effect on eosinophil counts. At low concentration, the eosinophil count was increased, while it was decreased at high concentrations. Administration of DA in patients awaiting heart transplant caused a reduction in the explanted heart of eosinophilic myocarditis and peripheral eosinophilia [43].

Basophils and mast cells are very similar in appearance. Although mast cells were once thought to be tissue-resident basophils, it was later found that the two cell types develop from different hematopoietic lineages and cannot be the same cells. There is no information on a potential association of DA with basophils. DA synthesis and its actions in mast cells and the involvement of these cells with inflammation and allergic reactions are covered in **Section 9.6**.

9.4 AUTOIMMUNE DISEASES AND NEUROPSYCHIATRIC DISORDERS

Autoimmunity is a complex condition where the immune system does not distinguish between self and nonself antigens, as a result of a loss of immune tolerance. To date, as many as 80 autoimmune diseases have been identified, with an overall prevalence of 5%, thus constituting a serious health issue. As detailed in several reviews [1,2,44,45], dopaminergic pathways are considered as key regulators in multiple sclerosis (MS), inflammatory bowel disease (IBD), rheumatoid arthritis (RA), systemic lupus erythematosus (SLE), and basal ganglia encephalitis. In addition, a number of neurological and psychiatric disorders in which DA is involved are associated with abnormalities of specific immune cell subpopulations. The neurological disorders include PD, Alzheimer's disease (AD), HIV-related encephalitis, and migraine. Psychiatric disorders include schizophrenia, anxiety disorders, and Tourette's syndrome.

9.4.1 Dopamine and autoimmune diseases

MS is a chronic disease that leads to a progressive neurological disability and is caused by an autoimmune response against the myelin sheath of axons in different areas of the brain and spinal cord. It mainly afflicts young adults, with a prevalence of 58.3 per 100,000 individuals, affecting about 2.4 million persons worldwide. MS risk-conferring genes, together with environmental risk factors (i.e., smoking and infectious agents), account for the development and progression of the diseases. The course of MS and its symptomatology are heterogeneous and include impairments of visual sensory, motor, cognitive, and autonomic functions. The most common form is the relapsing/remitting MS, characterized by alternating phases: a neurological impairment and inflammation, followed by a remission period of clinical recovery.

The pathophysiology of MS involves myelin-reactive CD4+ T cells, imbalanced CD8+ T cells, inflammatory B cells, altered NK function, production of autoantibodies, genetic predisposition, and nongenetic factors [45]. The main etiology is the violation of immunological tolerance and an active penetration of immune cells sensitized to myelin antigens through the blood–brain barrier (BBB) into the brain tissue. There is a critical pathogenic role of the Th1- and Th17-cells that respectively produce the pro-inflammatory cytokines IFN-γ and IL-17 [46]. There are also data showing that NK and B cells serve as players in the disease. For example, immune-modulating therapies targeting NK cells in MS have suggested a regulatory role for these cells on CD4+ T cell proliferation and activity, which are dysregulated during disease progression. A single administration of rituximab (an anti-CD20 monoclonal antibody that depletes B cells) have been shown to reduce brain lesions and clinical relapses for 48 weeks. Hence, B cell subsets appear to play antibody-dependent or -independent pathogenic functions in MS.

Several lines of evidence indicate that DA influences MS pathogenesis by modulating immune cell activity and cytokine production [46]. Studies have shown that DA can either enhance or inhibit the functions of the innate and adaptive immune system, depending on activation of different DARs and, thus, influence the course of the disease. Compared with healthy controls, peripheral blood mononuclear cells (PBMCs) from MS patients had higher IL-17 production. This was mediated *in vitro* by activation of D2R-like and was reverted by stimulation of D1R-like. PBMCs from IFN-β-treated MS patients had increased production of catecholamines, higher mRNA levels of TH and D5R, and reduced levels of D2R compared with baseline levels before treatment.

As was emphasized in previous sections of this chapter, the effect of DA on T cells is complex, being receptor- and context-dependent. Consequently, the levels of DA influence the degree of immune inflammatory reaction in MS. Low DA levels are pro-inflammatory due to their effects on D3R and D5R, which have the highest affinity for DA. On the other hand, high DA levels are anti-inflammatory due to their effect on D4R, D2R, and D5R, which have relatively low affinities for DA.

More than 15 drugs are currently available for MS patients, yet treatment of the disease remains a difficult problem in clinical neurology. Although some medications are quite effective against some symptoms of the disease, most are also associated with serious side effects such as opportunistic infections, and some are very expensive. Hence, it is important to develop MS therapeutics that are clinically effective, with an acceptable tolerability and safety profile, and low cost. This problem could be solved by the development of drugs that modulate immune response in the CNS without affecting the peripheral immune system. Currently, there are no drugs for treating MS that target the immune system of the brain. It has been suggested [46] that dopaminergic therapy has the potential to become a future disease-modifying therapy of MS. Yet, to achieve this goal, more knowledge is needed, and more preclinical studies of drug efficacy and safety of such treatments are required.

IBD is a group of chronic inflammatory conditions triggered by a break of immune tolerance in the gastrointestinal tract. The most common pathologies are Crohn's disease (CD) and ulcerative colitis (UC). Inflammatory lesions in CD patients are deep and multifocal and can affect the entire GI tract from the mouth to the anus. In UC patients, inflammation is more superficial and is limited to the colon. IBD is associated with abdominal pain, weight loss, diarrhea, passage of blood or mucus or both, as well as alternating phases of relapse and remission. There is also an increased risk of developing intestinal cancer. The prevalence of IBD in Europe and North America fluctuates between 605 and 827 cases per 100,000 persons.

Development and progression of IBD result from impairment of barrier function and loss of permeability (defective tight junctions in the intestinal epithelial cell layer), altered immune response (T lymphocyte signaling), environmental factors (smoking and intestinal microbiota), and genetic predisposition (IL-23 receptor and ATG16L mutations). In biopsy specimens of inflamed gut mucosa from IBD patients, DA levels were reduced as compared with healthy controls [45]. The decrease in DA levels was attributed to IFN-γ-mediated inhibition of L-Dopa uptake by epithelial cells and decreased expression of TH and DAT levels in the colon. Moreover, a genetic polymorphism in D2R has been associated with susceptibility to develop refractory CD in patients.

In rodent models of IBD, the disease was enhanced by treatment with the peripheral DA antagonist domperidone and was ameliorated by the DA agonist bromocriptine [44]. In addition, in a rat model of the disease, UC correlated with enhanced inflammatory and damaging effects of LPS on the nigrostriatal dopaminergic neurons. Collectively, these data show that inflammatory responses that trigger IBD are associated with both local and distal dopaminergic systems, leading the authors to conclude that changes in DA may be central to disease development.

RA is a chronic autoimmune disease in which the destruction of bone tissue and the articular structures of the joints leads to a progressive disability. Macrophages infiltrating the synovial tissues play a crucial role in the progression of this disease, while DCs, neutrophils, and NK cells in the synovial fluid serve as contributing factors [44]. Although DA levels in synovial fluid are relatively low (i.e., in the picomolar to nanomolar range), the dopaminergic system strongly influences the progression of the disease [33]. DA released from DCs mediates IL-6-dependent differentiation of Th17 lymphocytes, resulting in an exacerbated cartilage destruction that can be blocked by the D1R-like antagonist. In addition, the D2R antagonist haloperidol and the agonist cabergoline can ameliorate progression of the disease, and this occurs either through their effects on synovial fibroblasts or on osteoblasts, two cell types that express TH and DAR. Expression of TH in osteoblasts indicates that DA locally synthesized in the bone could influence disease progression and that DA is involved not only in bone formation but also in bone remodeling and joint erosion in RA.

SLE is an autoimmune disease that involves multiple organs, including the kidneys and the brain [47]. The course of the disease is variable and unpredictable, and it is often difficult to treat. SLE is characterized by immune deregulation and the production of autoantibodies directed toward certain nuclear antigens. The disease is multifactorial, with genetic and environmental factors contributing to its pathogenesis. Neurologic manifestations are common and represent a more severe form of the disease. They range from acute confusion, psychosis, anxiety, and depressive disorders to clinical and subclinical illnesses such as seizures, cerebrovascular disease, chorea and myelopathy, demyelinating syndrome, and aseptic meningitis. The neurologic manifestations may also include Parkinsonian-like deficits and changes in the basal ganglia. Thus, a significant loss of central neurons seems to underlie changes in sensorimotor function and behavior in many cases.

Several observations suggest that female-related hormones are crucial regulators of SLE activity, including its preponderance in women, the fluctuations of disease activity with the menstrual cycle, the tendency of the disease to flare during pregnancy, and its remission after menopause. In addition, there are alterations in neuromediators associated with this disease. In animal models of SLE, co-treatment with estrogen and bromocriptine prevented the development of lupus-like syndrome. Bromocriptine prevented the estrogen-rescued DNA-reactive B cells from maturation and activation. Treatment with bromocriptine was beneficial to patients with mild to moderately active disease, leading to decreased serum immunoglobulin and anti-DNA antibody levels, while the discontinuation of bromocriptine treatment was followed by a flare of the disease. An open-label, double-blind study comparing bromocriptine to hydroxychloroquine (a well-accepted treatment for SLE) showed that bromocriptine therapy similarly reduced flares and suppressed disease activity in SLE [48].

Whereas the *in vivo* results with bromocriptine could be due, in part, to its indirect effects via the regulation of PRL, evidence for a potential direct involvement of DA in SLE comes from studies that have examined DAR expression in peripheral T and B lymphocytes from SLE patients [47]. The *Drd2* gene was underexpressed, while *Drd4* was overexpressed in lupus patients compared with controls. Cell sorting showed that the altered D2R and D4R expressions were by T cells. The authors concluded that the distorted expressions of these DAR could influence immune functions in lupus through several mechanisms. Because D2R can be effective in regulating the activation and differentiation of naïve CD4+ cells by promoting polarization toward Tregs, its underexpression may account, at least in part, for the reduction of Treg function and/or their numbers in lupus. In addition, because D4R is effective in triggering T-cell quiescence, its overexpression on lupus T cells suggests that induction of quiescence by D4R-specific agonists may represent a useful strategy in the treatment of lupus.

Basal ganglia encephalitis is a rare form of autoimmune encephalitis, a group of disorders characterized by symptoms of limbic and extralimbic dysfunction occurring in association with antibodies against synaptic antigens and proteins localized on the neuronal cell surface. Patients with basal ganglia encephalitis harbor serum autoantibodies against D2R, without evidence for antibodies against other DARs or DAT [49]. These patients have movement disorders characterized by parkinsonism, dystonia and chorea. They also have psychiatric disturbances with emotional lability, attention deficit disorder and psychosis. Magnetic resonance imaging has showed lesions localized to the basal ganglia in 50% of the patients. Elevated D2R immunoglobulin G was also found in 10 of 30 patients with Sydenham's chorea, 0 of 22 patients with pediatric autoimmune neuropsychiatric disorders associated with streptococcal infection, and in 4 of 44 patients with Tourette's syndrome. No D1R immunoglobulin G was detected in any disease or control groups. The authors concluded that assessment of D2R antibodies can help define autoimmune movement and psychiatric disorders

9.4.2 Dopamine, immune system, and neurological disorders

PD is characterized by the degeneration of dopaminergic neurons in the substantia nigra, pathological protein aggregates known as Lewy bodies, reduced DA concentrations in the striatum, neuroinflammation, and motor disturbances. In addition to the neurological symptoms, there are duodenal ulcers, GI absorptive motile functions, and the generation of Lewy bodies in the gut [44]. PD has also been positively correlated with IBD, leading to the hypothesis that T-cell-driven inflammation, which mediates dopaminergic neurodegeneration, is triggered in the gut. In both PD patients and Parkinson's rodent models, changes in DA concentrations have been found in the periphery as well as in the CNS.

Activation of immunologic responses appears to have an important role in PD pathology [50]. For example, microglia show greater phagocytic activity in the substantia nigra of the brain of PD patients than in controls. Microglia are bone marrow-derived macrophage-like cells that are resident within the CNS. They constitute 20% of the total glial population and represent the first line of immune defense in the brain. Microglia activation is characterized by proliferation, migration, expression of the immune-related antigens and transformation into phagocytes. In addition to the phagocytosis of dopaminergic neurons by microglia, intraneuronal accumulation of alpha-synuclein (α-Syn), as the main pathological hallmark of PD, potentially mediates initiation of the autoimmune and inflammatory events through auto-reactive T cells.

Collectively, both the innate and adaptive immune systems, via microglial activation and T-cell infiltration, are now considered pathological hallmarks of PD at sites of neuronal degeneration. However, it has been debated whether this neuroinflammation is the consequence or the cause of neuronal injury. Nonetheless, the CNS should no longer be viewed as an immune-privileged organ, as has been thought for many years, given that it contains many immune system components, including humoral and cellular mediators, that affect the homeostatic environment of the brain.

AD is a chronic neurodegenerative disease that usually starts slowly and gradually worsens over time. It is considered the cause of 60%–70% of cases of dementia. The most common early symptom is a difficulty in remembering recent events. As the disease advances, symptoms include problems with language, disorientation, mood swings, loss of motivation, and behavioral issues. As a person's condition declines, they often withdraw from family and society. Gradually, bodily functions are lost, ultimately leading to death. Although the speed of progression can vary, the typical life expectancy following diagnosis is 3–9 years. The causes for AD are mostly unknown, but genetic heritability, traumatic brain injury, hypertension, and depression are important risk factors. At the brain, formation of plaques by β-amyloid protein and neurofibrillary tangles formed by τ protein determines a loss of connection between neurons and neuron death.

As discussed in several reviews [1,38,51], peripheral monocytes of AD patients have an inflammatory phenotype and are involved in the activation of both innate and adaptive immunity. Using *in vivo* imaging techniques in a mouse model of AD, it was observed that neutrophils infiltrate the amyloid plaques, with similar observations made in transgenic models of AD. The neutrophils present in the amyloid plaques release neutrophil extracellular net (NET) formations and IL-17, thus contributing to neuronal damage and BBB disruption. Indeed, neutrophil depletion in animal models reduced the severity of AD progression. Although the frequency of NK cells of AD patients is not modified, they are less active, while peripheral DC are decreased in patients with AD, compared with controls, and this correlates with the severity and progression of pathology.

The role of the dopaminergic system in AD is still debated [2,51]. Degeneration of dopaminergic neurons is linked to the deterioration of cognitive tasks in 30%–40% of AD patients. In a mouse model of AD, a dysregulation of dopaminergic system has been implicated in the decline of memory and learning. In addition, alterations of striatal D2R may participate in the extrapyramidal manifestations that occur in AD. Emerging evidence indicates a role played by neuroinflammation in the course of AD. There are indications that chronic deposition of amyloid aggregates in the brain activates the microglia, which release cytokines and mediators, exacerbating the neurotoxicity.

HIV-encephalitis. The HIV virus attacks cells of the immune system, primarily CD4+ T cells and macrophages [52,53]. Several lines of evidences suggest the participation of the innate immunity in the early phases of HIV infections. Macrophages are the primary target of HIV in the CNS. The virus infiltrates the BBB early after infection, using HIV-infected macrophages as a "Trojan horse" for entry into the brain (**Figure 9.5**). The viral reservoir within the brain presents a substantial clinical problem because present therapies with antiretroviral drugs have variable BBB penetration capabilities. Notably, the potential to use macrophages as carriers of antiretroviral agents was tested in mice with HIV-1 encephalitis. It was found that the administration of bone marrow macrophages loaded with nanoparticles carrying an antiretroviral agent could reach certain brain areas for reducing the replication of the virus.

DA has a strong impact on macrophage functions during HIV infection. In drug abusers, increased plasma DA levels have been correlated with enhanced entry of HIV into macrophages. This effect required the activation of DAR expressed on monocyte-derived macrophages and was suppressed by the DAR antagonist flupenthixol. Similarly, CD14+ CD16+ monocytes express the mRNA of all five DARs, all of which were functional, as judged by the upregulation of Erk2. Treatment with DA and the D1R-like agonist SKF-38393 increased cell migration and transmigration across an *in vitro* BBB model. Two other cell types of the innate immunity, DCs and NK cells, also play a role in HIV progression. During HIV infections, DCs play a pivotal role as sentinels that alert other cells through the secretion of cytokines. IFN-α–producing DCs activate T cells and exert an antiviral effect. Epidemiologic data strongly support the role of NK cells in the antiviral control through the activation of killer immunoglobulin-like receptors. Once in the brain, HIV disrupts central dopaminergic pathway by acting on the DATs.

The success of antiretroviral therapy has clearly improved the life span as well as the quality of life of HIV-infected individuals by transforming the infection into a chronic condition. However, this success also comes at a cost in terms of the emergence of new neuropathological and neurocognitive

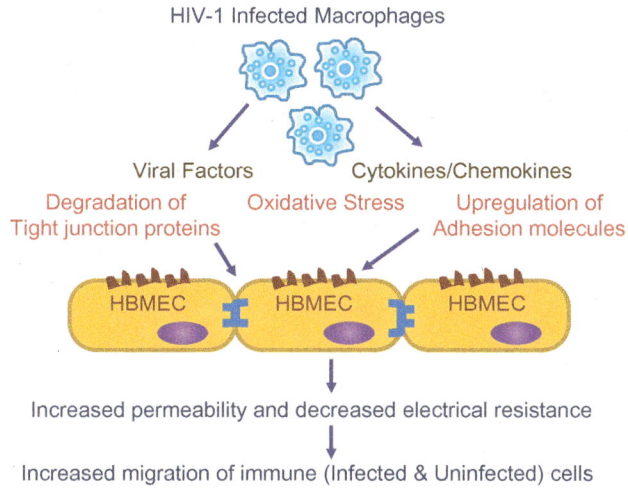

Figure 9.5 Invasion of the blood brain barrier by HIV-infected macrophages. The infected macrophages release toxic viral products and inflammatory cytokines resulting in the degradation of tight junctions and increased permeability to the brain proper by both infected and uninfected cells. HBMEC: human brain microvascular endothelial cells.

pathologies, collectively known as HIV-associated neurocognitive disorders or NeuroHIV [52]. These neuropathologies persist even in the absence of viral replication, suggesting that they result from long-term changes in the CNS induced by the HIV infection rather than by an active viral replication. Prominent among these are changes in catecholaminergic neurotransmission. HIV-infected individuals not treated with antiretroviral drugs (ARVs) show significant neuropathology, especially in DA-rich brain regions, as well as altered autonomic nervous system activity. These changes persist even in infected individuals on therapy, while the elevated stress and NE levels correlate with decreased efficacy of ARV. The authors [52] concluded that additional characterization of catecholamines in the context of HIV-associated disease should open the door to new therapeutic strategies that specifically ameliorate the effects of catecholaminergic dysregulation on NeuroHIV.

Migraine is a primary headache disorder characterized by recurrent episodes of moderate-to-severe headaches. Typically, the headaches affect one half of the head, are pulsating in nature, and last from 2 to 72 hours. Symptoms include nausea, vomiting and sensitivity to light, sound or smell. The pain is generally made worse by physical activity. About one-third of people have an aura: a short period of a visual disturbance that signals that the headache will soon occur. Globally, approximately 15% of people are affected by migraines, with a higher incidence in women than men. Migraine most often starts at puberty and is worst during middle age. In some women, migraine episodes become less common following menopause.

Migraine pain has been attributed to local sterile meningeal inflammation and subsequent activation of trigeminal primary afferent nociceptive neurons that supply the intracranial meninges and their large blood vessels [54]. Increasing evidence reveals that a specific class of inflammatory cells residing within the intracranial milieu, known as meningeal mast cells, play a critical role in migraine pathophysiology. Mast cells are tissue-resident granulocytes that originate from $CD34^+/CD117^+$ cells and circulate in the blood as immature cells. Stem cell growth factor and other cytokines (IL-3, -4, and -9] help the maturation of mast cells in the tissue. Mast cells lodge in all vascularized tissues such as intestines, respiratory tract and skin as well as in the dura mater. When activated by pathogens or allergens, mast cells degranulate and release mediators that include proteases, cytokines, and biogenic amines [55].

DA appears to be involved in several aspects of migraine. A close look at migraine episodes reveals that head pain is only one event among several experienced during a migraine attack. Almost 30% of migraineurs report having prodromal symptoms hours before the headache begins [56]. Patients typically feel tired, somnolent and drowsy, yawn excessively, and crave for foods. During the pain phase, nausea, vomiting, and sometimes hypotension may induce severe disability. In the postdromal phase, patients are tired and weak but sometimes feel euphoric. Notably, most, if not all, the symptoms preceding, accompanying, and following the headache in migraineurs resemble those that are induced by stimulation of central or peripheral DARs.

A number of studies demonstrate a dopaminergic dysfunction in migraine. Pharmacological challenges with DA agonists showed DA hypersensitivity as a specific trait in migraine patients. In addition, receptor binding studies revealed that PBLs from migraineurs contain higher D3R, D4R, and D5R densities than control lymphocytes but no differences in binding affinity. This finding indicates central or peripheral DAR hypersensitivity or both due to dopaminergic system hypofunction. In addition, a polymorphism in the *DRD4* gene is associated with predisposition to episodic migraine without aura, whereas susceptibility to chronic daily headaches seems to involve genetic variability in the *DAT* gene [56].

Migraine is one of the most common causes of headache presentations at emergency rooms. Patients with migraine attack need rapid pain relief rather than diagnostic modalities. Metoclopramide, a DA antagonist, was first used as an antiemetic agent to treat nausea in migraine attacks. It has since been widely used as a single agent for fast pain relief in migraine attacks seen in emergency rooms. Other DA antagonists such prochlorperazine and chlorpromazine have also been studied and used for migraine attacks [57]. After reviewing multiple clinical trials with metoclopramide, the authors concluded that while it seems reasonable to use metoclopramide in migraine attacks in emergency departments, further studies with high methodological quality are needed to clearly establish whether and how much metoclopramide is superior to placebo.

9.4.3 Dopamine, immune system, and psychiatric disorders

Schizophrenia is a pervasive neurodevelopmental disorder with unknown etiopathology. As indicated in several recent reviews [58–60], the DA hypothesis of schizophrenia postulates that hyperactivity of D2R neurotransmission in subcortical and limbic brain regions contributes to positive symptoms of the disease, whereas negative and cognitive symptoms of the disorder are attributed to an hypofunctionality of D1R neurotransmission in the prefrontal cortex. Although the DA hypothesis remains the driving force behind most translation research in schizophrenia, emerging evidence suggests that aberrant immune mechanisms in both peripheral and central nervous systems influence the etiology of schizophrenia and the pathophysiology of psychotic symptoms that define the illness.

The initial interest in inflammatory processes in schizophrenia came from epidemiological data and historical observations dating back several decades. Immunopathogenesis has emerged as a compelling etiological model of schizophrenia, based on recent studies on developmental exposure to infection, stress-induced inflammatory response, glial cell signaling, structural and functional brain changes, and therapeutic trials. Research using animal models of psychosis has helped to advance clinical and basic science investigations of the immune mechanisms disrupted in schizophrenia. Nonetheless, animal models are limited by the inability to recapitulate the human experience of hallucinations, delusions and thought disorder that define psychosis.

Findings of multiple studies have helped to conceptualize schizophrenia as a chronic low-grade inflammatory disorder. Although the contribution of adaptive immune responses has also been emphasized, the precise role of T cells in the underlying neurobiological pathways of schizophrenia is yet to be fully ascertained. T cells have the ability to infiltrate the brain and to mediate neuro-immune cross-talk. Conversely, the CNS and neurotransmitters can regulate the immune system. It is now well recognized that a neurotransmitter like DA, implicated widely in both the risk and progression in schizophrenia, can modulate the proliferation, trafficking, and functions of T cells. Within the brain, T cells activate microglia, induce production of pro-inflammatory cytokines as well as reactive oxygen species, and lead to neuroinflammation. Such processes contribute to neuronal injury/death and are gradually being implicated as mediators of neuro progressive changes in schizophrenia.

Antipsychotic drugs used to treat schizophrenia are also known to affect the adaptive immune system by interfering with the differentiation and functions of T cells. To date, translational studies focusing on inflammatory mechanisms in human subjects have not been executed in great detail. Understanding the neuroinflammatory mechanisms involved in schizophrenia may be essential for identifying potential therapeutic targets to minimize the morbidity and mortality of schizophrenia by interrupting disease development.

Anxiety disorders are a group of mental disorders characterized by significant feelings of anxiety and fear, which can cause physical symptoms such as fast heart rate and shakiness. Several types of anxiety disorders are recognized, including generalized anxiety disorder (GAD), specific phobia, social anxiety disorder, separation anxiety, agoraphobia, panic disorder, and childhood selective mutism. People often have more than one anxiety disorder.

T cells from healthy individuals or from those with GAD were used to evaluate the effect of stress-related dose of DA on cell proliferation and cytokine production [61]. The results showed that cell cultures from the GAD group showed lower proliferation following T cell activation than those from the control group. Addition of DA reduced the proliferative response in cell cultures from healthy but not from GAD individuals. The cytokine profile in GAD individuals revealed Th1 and Th2 deficiencies associated with a dominant Th17 phenotype, which was enhanced by DA. Unlike the control, the DA-enhanced Th17 cytokine production in GAD individuals was not affected by glucocorticoids. The authors concluded that functional dysregulation of T cells in GAD individuals is significantly amplified by DA. These immune abnormalities can have an impact by increasing the susceptibility of individuals with anxiety disorders to infectious diseases and inflammatory/autoimmune disorders.

Tourette's syndrome (TS) is a neuropsychiatric disorder characterized by chronic motor and vocal tics that usually begin in childhood. A prevalence of TS is estimated as 4–5 of 10,000 individuals. TS patients frequently show comorbidity with other psychiatric disorders such as obsessive compulsive disorder (OCD), attention deficit hyperactivity disorder (ADHD), anxiety, and affective disorders. Some forms of OCD seem to share a common genetic etiology with TS and to be a facultative part of the TS phenotypic spectrum.

Increasing evidence shows that infections and an activated immune status might be involved in the pathogenesis of TS, as signs of inflammation and immunological abnormalities have been described in this tic disorder [62]. Infections with group A streptococci, *Borrelia burgdorferi* or *Mycoplasma pneumoniae*, seem to be associated with symptoms of the disease. Studies have shown that immunologic treatment improves clinical symptoms in Tourette's syndrome and prevents their reoccurrence. Postinfectious events by cross-reactive antibodies against M-protein, altered dopaminergic neurotransmission, and inflammatory/immunological dysregulations were considered as possible mechanisms that cause symptoms. Forty years of research and clinical practice have shown that DAR antagonists are effective agents in the treatment of TS by allowing a significant tic reduction of about 70% [63]. Their main effect appears to be mediated by the blockade of striatal D2Rs. Various typical and atypical agents are available, although there is discord among experts on which agents should be considered as first choice.

9.5 ENDOTHELIAL CELLS, ANGIOGENESIS, AND COAGULATION

9.5.1 Endothelial cells and dopamine

Endothelial cells are squamous cells that line the interior surface of blood and lymphatic vessels, forming a monolayer interface between circulating blood or lymph in the lumen and the vessel wall. Cells in direct contact with blood are called vascular endothelial cells, whereas those in direct contact with lymph are known as lymphatic endothelial cells. Endothelium within the interior surfaces of the heart chambers is called endocardium. Vascular endothelial cells line the entire circulatory system, from the heart to the smallest capillaries.

Endothelial cells have multiple functions in vascular biology, including fluid filtration (e.g., in kidney glomeruli), hormone trafficking, neutrophil recruitment, control of blood pressure, hemostasis, and wound repair. They are the main cell type that controls angiogenesis (see below). They also serve as a semi-selective barrier between the vessel lumen and surrounding tissue, thereby regulating the passage of materials and the transit of leukocytes into and out of the bloodstream. Excessive or prolonged increases in permeability of the endothelial monolayer, as occurs during chronic inflammation, may lead to tissue edema and swelling. The endothelium also provides some protection against inappropriate coagulation by having a non-thrombogenic surface that contains heparan sulfate. The latter acts as a cofactor for activating antithrombin, a protease that inactivates several factors in the coagulation cascade.

Fenestrated capillaries have pores called fenestrae (windows in Latin) in endothelial cells that are 60–80 nm in diameter [64]. The pores are spanned by a diaphragm of radially oriented fibrils that allow small molecules and some proteins to pass through. Fenestrated capillaries are more permeable than continuous capillaries that are connected by tight junctions. They are found in tissues where there is extensive molecular exchange with the blood such as the small intestine, endocrine glands, kidney, and certain regions of the brain known to be outside the BBB.

Vascular endothelial growth factor (VEGF), also known as vascular permeability factor (VPF), is the main factor that regulates mitogenesis, migration and permeability of endothelial cells [65]. VEGF is produced by many cell types, including macrophages, platelets, keratinocytes, renal mesangial cells, and tumor cells. The activities of VEGF are not limited to the vascular system as it plays a role in normal physiological functions such as bone formation, hematopoiesis, wound healing, and development.

As was previously reported [66], all three catecholamines were detected by radioimmunoassay in the rat aorta and superior mesenteric arteries (SMAs). Chemical sympathectomy with 6-hydroxydopamine resulted in a significant reduction of NE and Epi, while DA levels remained unaffected, indicating a neuronal-independent local synthesis of DA. Indeed, isolated endothelial cells synthesize and release DA in response to cAMP stimulation. Consistent with these data, mRNAs coding for catecholamine biosynthetic enzymes, TH, DDC, and DBH, were detected by RT-PCR in cultured endothelial cells from SMAs. Exposure of endothelial cells to hypoxia [1% O_2] increased TH mRNA. Vascular smooth muscle cells partially expressed catecholaminergic traits. A physiological role of endogenous vascular DA was shown in the SMA, where a D1R blockade abrogated hypoxic vasodilatation. The authors concluded that endothelial cells, as well as cells of the underlying vascular wall, synthesize and release DA in an oxygen-regulated manner. This intrinsic nonneuronal DA in the splanchnic vasculature is the dominating vasodilator released upon lowering of oxygen tension.

In addition to the biosynthetic enzymes, endothelial cells from human coronary [67], aortic, and umbilical vein [68] express all five DAR subtypes. The actions of DA on endothelial cells that affect angiogenesis and coagulation are discussed in the sections below, while its actions on vascularization in tumors are covered in **Chapters 12** and **13**.

9.5.2 Angiogenesis and dopamine

The first blood vessels in the developing embryo are formed by vasculogenesis, defined as the differentiation of precursor cells (angioblasts) into endothelial cells and the *de novo* formation of a primitive vascular network. On the other hand, **angiogenesis** is the physiological process that is responsible for the formation of new blood vessels from preexisting ones. Angiogenesis plays critical roles in human physiology, ranging from reproduction and fetal growth to wound healing and tissue repair. The multistep process of angiogenesis is tightly regulated in a spatial and temporal manner by "on-off switch signals" between angiogenic factors, extracellular matrix components, and endothelial cells [69]. Uncontrolled angiogenesis may lead to several disorders, including vascular insufficiency (myocardial or critical limb ischemia) as well as vascular overgrowth (hemangiomas, vascularized tumors, and retinopathies). Thus, numerous therapeutic opportunities have been devised based on the understanding and subsequent manipulation of angiogenesis.

Angiogenesis is a highly regulated process that takes place through two nonexclusive events of microvascular growth: sprouting or splitting [70]. Sprouting differs from splitting angiogenesis by forming entirely new vessels as opposed to splitting existing vessels. As illustrated in **Figure 9.6**, sprouting proceeds in several well-characterized stages. First, angiogenic factors [VEGF and fibroblast growth factor (FGF)], released from neighboring cells, bind to their respective receptors on endothelial cells and activate signal transduction pathways. Matrix metalloproteinases (MMPs), produced by the endothelial cells are then activated and degrade the extracellular matrix, enabling an escape of endothelial cells from the parental vessel walls. This is followed by their migration and proliferation. The integrins, expressed by endothelial cells, facilitate their adhesion to the extracellular matrix and the formation of solid sprouts that connect to neighboring vessels. Angiopoietin 1 (Ang-1), binding to Tie-2 receptors, stimulates pericyte recruitment and vessel stabilization. Final vessel maturation and stabilization necessitate additional morphological changes that include lumen formation and perfusion, network establishment, remodeling, and pruning to become full-fledged functional vessels.

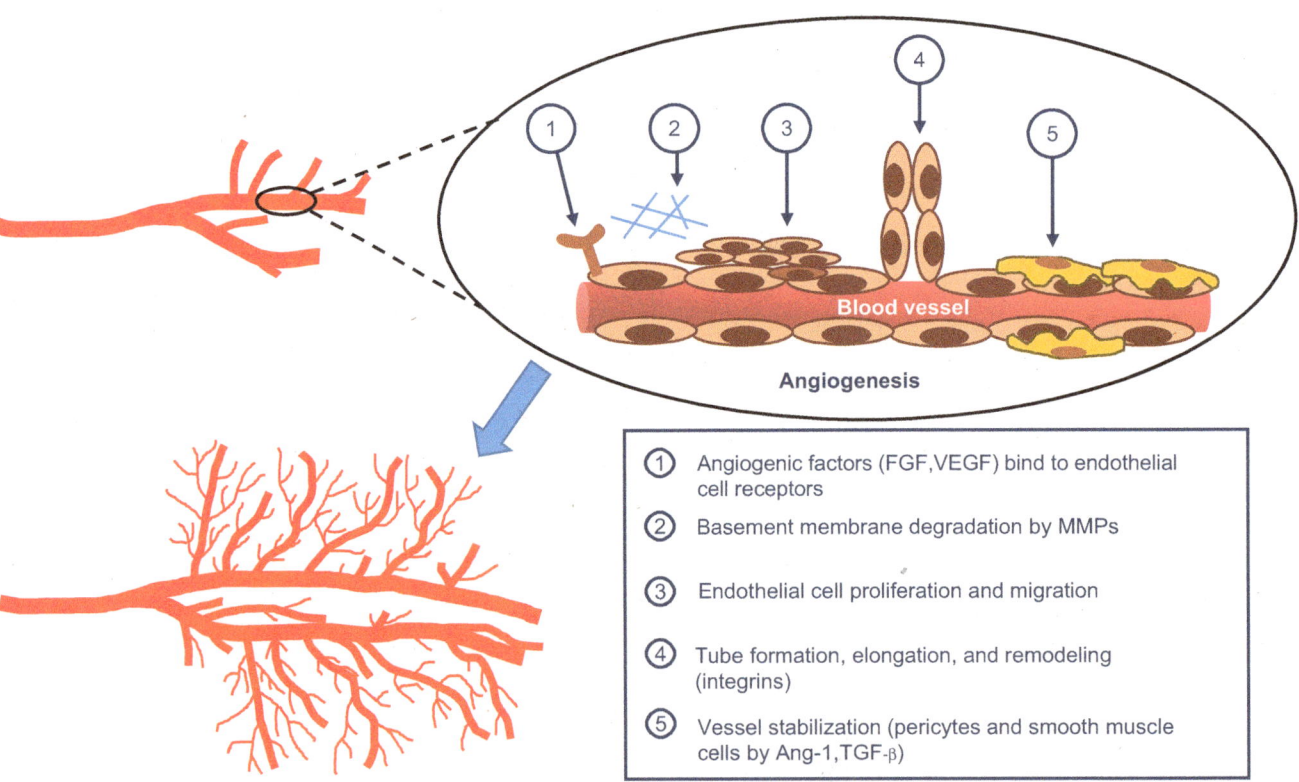

Figure 9.6 The process of sprouting angiogenesis. Binding of angiogenic factors such as FGF and VEGF to their respective preceptors in endothelial cells stimulates the release of matrix metalloproteinases (MMPs) that degrade the basement membrane. This is followed by migration and proliferation of endothelial cells, tube formation, elongation and remodeling. Pericytes and smooth muscle cells stabilize the new vessels, which become functional. Ang-1: angiopoietin 1; FGF: fibroblast growth factor; TGF-β: transforming growth factor beta; VEGF: vascular endothelial growth factor. (Redrawn and modified from Yoo, S.Y., and Kwon, S.M. Mediators. Inflamm., 127–170, 2013.)

Early studies found that DA, acting via D2R, inhibited both the angiogenic and permeability activities of VEGF in the mouse peritoneum [71]. This occurred by the induction of endocytosis of VEGF receptor 2 (VEGFR2], which is critical for promoting angiogenesis, thus preventing VEGF binding, receptor phosphorylation and signaling. Such a DA action was specific for VEGF and did not affect other mediators of microvascular permeability or endothelial-cell proliferation or migration. These findings were reinforced by another study from the same group showing that the VEGF-induced VEGFR2 phosphorylation, focal adhesion kinase, and MAPK in endothelial cells were strikingly increased in both DA-depleted and D2R knockout mice compared with normal controls [72]. Together, these data provided a mechanistic insight into the DA-mediated inhibition of VEGF and suggested that endogenous DA is an important physiological regulator of VEGF activities *in vivo*.

A widely used model for studying angiogenesis is **endometriosis**, defined as the presence of endometrial tissue (glands or stroma) at sites other than the uterus. In women, endometriosis involves the ovaries, Fallopian tubes, or the pelvic lining. The growth of ectopic endometriotic lesions is hormonally regulated, causes periodic bleeding and inflammation, and is associated with pelvic pain and infertility. Numerous peritoneal blood vessels surround active endometriotic lesions, and the implant itself is supplied with rich vascularization, demonstrating the essential role played by angiogenesis in the development and continuation of endometriotic lesions.

As an experimental animal model for endometriosis, human endometrial fragments were engrafted in the peritoneum of immunocompromised mice. This resulted in the promotion of neo-vascularization from the surrounding host vascular network into the engrafted tissue [73]. Mice were then treated with vehicle, oral cabergoline, or oral quinagolide for 14 days. At the end of treatment, implants were excised and assessed for lesion size, cell proliferation, degree of vascularization, and angiogenic gene expression. Neo-angiogenesis was inhibited and the size of active endometriotic lesions, cellular proliferation index, and angiogenic gene expression were significantly reduced by both DA agonists when compared with the placebo. The authors concluded that because the two antidopaminergic drugs were equally effective in inhibiting angiogenesis and reducing lesion size, these results provide a good rationale for pilot studies to explore the use of DA agonists for the treatment of endometriosis in humans.

Ischemia is a diminished blood supply to certain tissues, including the heart, intestines, kidneys, or limbs. It is often caused by atherosclerotic plaques and the formation of emboli that lodge in vessels of various tissues and block their blood flow. One way of treating ischemia is by the application of therapeutic angiogenesis to stimulate new blood vessel growth within the affected tissues. A recent study examined the role played by DA during a postischemic healing of unilateral hind limb ischemia (HLI) [74]. Although both TH and DA were considerably high in muscle tissues of wild-type and D2R knockout (KOD2) mice with HLI, recovery was significantly faster in KOD2 mice than in wild-type controls, indicating that peripheral DA inhibits the healing process. There were also significant differences in postischemic angiogenesis between the two groups. The elevated DA suppressed the activation of local renin–angiotensin system (RAS), which stimulates angiogenesis in ischemia. Angiotensin II (ATII) and its receptor, angiotensin receptor type 1 (AT1R), are the key players in RAS-mediated angiogenesis. DA, acting via D2R in endothelial cells, inhibited the ATII-mediated angiogenesis by suppressing AT1R expression. This study showed that DA prolongs postischemic recovery, suggesting that pharmacological intervention aimed at inhibiting DA actions holds promise as a future therapeutic strategy for the treatment of HLI and other peripheral arterial diseases.

9.5.3 Blood clotting and the coagulation cascade

Coagulation is a multistep process of blood clot formation aimed at stopping bleeding (**Figure 9.7**). When the entire coagulation cascade works properly, blood holds together firmly at an injury site and bleeding stops. People who have a bleeding disorder are unable to make strong clots quickly or at all. To minimize bleeding and prevent blood loss after tissue injury, three components are coordinately activated: blood platelets, endothelial cells, and circulating coagulation factors. The clotting process can be viewed as four interrelated events: (1) vessel compression and vasoconstriction, (2) formation of a platelet plug, (3) blood coagulation, and (4) clot retraction and thrombus dissolution [28].

When a blood vessel is injured, the surrounding tissue exerts compression on the vessel's walls, while the underlying damaged cells release potent vasoconstrictors such as serotonin, thromboxane A2, and Epi. Acting together, these processes can quickly reduce the flow of blood to the damaged area. Next, platelets are activated and undergo sequential processes of adherence, aggregation, and release. The platelets stick to one another and to the wound site and form a plug. At the same time, the coagulation cascade is initiated, with each coagulation factor activated in a specific order, leading to the formation of the blood clot. Platelet adherence is enhanced by two proteins: von Willebrand factor (vWF or factor VIII), which is synthesized by endothelial cells and megakaryocytes, and thrombin, which is generated by the plasma coagulation pathway. Many different clotting factors on the surface of the activated platelets, work together in a series of complex chemical reactions to form a fibrin clot that acts like a mesh to stop the bleeding. More detail on the overall process of wound healing is presented in **Chapter 11**.

The coagulation cascade uses as many as 10 different clotting factors (**Table 9.2**), which are normally present in the circulation in an inactive form. These proteins are synthesized in the liver and are referred to by a sequence of Roman numerals based on the order of their discovery. Two separate coagulation cascades, extrinsic and intrinsic, result in coagulation under different circumstances. In the intrinsic system, all the factors required

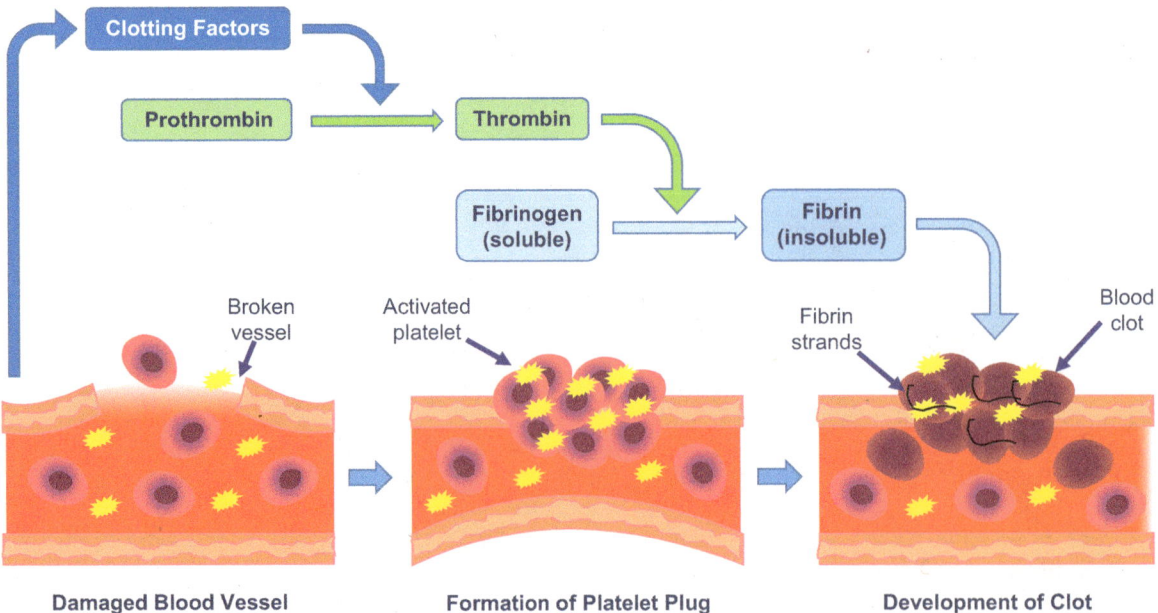

Figure 9.7 The clotting process. Injury to a vessel lining triggers the release of local and circulating clotting factors. Vasoconstriction limits blood flow, and activated platelets aggregate near the injury site and form a sticky plug. Circulating prothrombin is then converted to thrombin, which promotes the formation of insoluble fibrin from fibrinogen. The fibrin strands adhere to the plug to form an insoluble clot.

Table 9.2 Factors of the clotting cascade		
Scientific name	**Common name**	**Other names**
Factor I	Fibrinogen	
Factor II	Prothrombin	
Factor III	Tissue thromboplastin	Tissue factor
Factor IV	Calcium	
Factor V	Proaccelerin	Labile factor
Factor VII	Proconvertin	Serum prothrombin conversion accelerator
Factor VIII	Antihemophilic factor	Platelet cofactor 1
Factor IX	Christmas factor	Platelet thromboplastin component
Factor X	Stuart factor	
Factor XI	Plasma thromboplastin antecedent	
Factor XII	Hageman factor	Contact factor
Factor XIII	Fibrin stabilizing factor	

for coagulation are already present in the circulation. For initiation of the extrinsic coagulation cascade, a factor released from injured tissue, called thromboplastin or factor III, is required. The two pathways converge at a certain point, from where activation of the common pathway results in the conversion of the soluble fibrinogen to insoluble fibrin, which helps in the clotting process.

In addition to the inhibitory actions of DA on hypoxia-driven angiogenesis, DA also exerts inhibition when angiogenesis is activated by the coagulation system [75]. As indicated above, after an injury to a blood vessel, vWF promotes platelet aggregation and adhesion to the subendothelial tissue, resulting in the release of DA and VEGF that are stored in the dense granules of platelets. DA inhibits the histamine-induced vWF secretion by the endothelial cells, thereby inhibiting primary hemostasis. As has also been reported [68], DA, D2/D3, and D4-specific agonists inhibit histamine, but not thrombin-induced VWF secretion from both human aortic and umbilical vein endothelial cells. The D2–D4 agonists inhibit the histamine-induced secretion even in the absence of extracellular calcium.

Antipsychotic drugs acting as DA antagonists are associated with an increased risk of thromboembolic events, defined as the formation of a clot (thrombus) somewhere in the vasculature that breaks loose and is carried by the blood to plug another vessel [76]. The authors postulated that one plausible mechanism for the increased of thromboembolism is the association of antipsychotic drugs with enhanced platelet aggregation. An older study reported the development of a gangrene in the extremities in five patients as a complication of DA therapy, infused at dosages of 5.1–10.2 pg/kg/min [77]. A disseminated intravascular coagulation appeared to be the cause of the gangrene, as was supported by the detection of abnormalities of platelet count, prothrombin time, fibrinogen, fibrin degradation products, and fibrin monomers. The authors concluded that peripheral vasoconstriction from administered DA, even at low doses, may set the stage for the thrombotic complications of disseminated intravascular coagulation, leading to tissue damage.

Interactions between the dopaminergic and the coagulation systems were also suggested by some of the results of a study that examined serum proteins in patients with PD using quantitating proteomics [78]. The expression level of 8 proteins that included sero-transferrin and clusterin increased, while the expression level of 18 proteins, including complement component 4B, apolipoprotein A-I, alpha-2-antiplasmin, and coagulation factor V decreased. Alpha 2-antiplasmin is a serine protease inhibitor responsible for inactivating plasmin, an important enzyme that participates in fibrinolysis. Coagulation Factor V interacts with other clotting proteins such as activated factor X and prothrombin to increase the production of thrombin, the key hemostatic enzyme that converts soluble fibrinogen to a fibrin clot.

9.6 INFLAMMATION AND ALLERGIC REACTIONS

9.6.1 Inflammation and dopamine

Inflammation is a biological response of body tissues to harmful stimuli such as pathogens, damaged cells, or irritants. Inflammation is a protective physiological response that involves immune cells, blood vessels, and molecular mediators. The ultimate function of inflammation is to eliminate the initial cause of cell injury, clear out necrotic cells and tissues damaged from the original insult and the inflammatory process, and initiate tissue repair. There are five classical signs of inflammation: heat, pain, redness, swelling, and loss of function. Inflammation primarily involves the innate immunity, while the response of the adaptive immunity is specific for each pathogen. Insufficient inflammation can be harmful because of the potential for progressive tissue damage due to insults that can compromise the survival of the organism. On the other hand, chronic inflammation is associated with various diseases, such as hay fever, periodontal disease, atherosclerosis, and osteoarthritis.

As presented in **Table 9.3**, inflammation is classified as *acute* or *chronic*. Acute inflammation is the initial response to harmful stimuli achieved by an increased movement of plasma and leukocytes from the blood into the injured tissues. A series of biochemical events advance the inflammatory process, involving the local vasculature, immune system, and various cells within the injured tissue. Chronic inflammation leads to a progressive shift in the type of cells that are present at the site of inflammation, such as mononuclear cells, and is characterized by the simultaneous destruction and healing of the tissue from the inflammatory process.

Convincing evidence on the involvement of DA in inflammation comes from a thorough review of the pathology of PD. As discussed in **Section 9.4.2**,

Table 9.3 Comparison between acute and chronic inflammation		
Inflammation	**Acute**	**Chronic**
Causative agents	Pathogens, injured tissues	Nondegradable pathogens, foreign bodies, or autoimmune reactions
Major cells involved	Neutrophils, monocytes, macrophages	Monocytes, macrophages, lymphocytes, plasma cells, fibroblasts
Primary mediators	Vasoactive amines, eicosanoids	Interferon gamma (IFN-γ), cytokines, growth factors, reactive hydrogen species, hydrolytic enzymes
Onset	Immediate	Delayed
Duration	Few days	Up to many months or years
Outcomes	Resolution, abscess formation, chronic inflammation	Tissue destruction, fibrosis

PD is associated with a breach of the BBB, infiltration of T lymphocytes into the brain, and neuroinflammation. A correlation between peripheral inflammation and neurodegeneration has been reported in patients with PD [79]. Clinical evidence revealed elevated levels of TNF-α, TNF-α receptor 1, and IL-6 in blood of patients with PD compared with control subjects, supporting a general anti-inflammatory action of DA. Experimental models of PD also show an association of peripheral inflammation with PD pathogenesis. For example, a rat model of PD with long-term loss of dopaminergic neurons and low expression of TH also showed time-dependent alterations in the peripheral immune system. Collectively, both basic and clinical data have suggested that peripheral inflammation is an early event in PD pathogenesis.

Allergies are caused by hypersensitivity of the immune system to typically harmless substances in the environment such as pollen, certain food, metals, and other substances. Allergic diseases include allergic rhinitis, allergic asthma, allergic dermatitis, allergic conjunctivitis, anaphylaxis, and food or drug allergies. Together, they constitute major diseases involving as much as 22% of the world population [80]. Symptoms include red eyes, itchy rash, sneezing, runny nose, shortness of breath, or swelling. The underlying mechanism involves binding of immunoglobulin E antibodies (IgE) to an allergen and then to an IgE receptor on mast cells or basophils, where they trigger the release of inflammatory chemicals such as histamine.

Diagnosis of the specific allergy is based on a person's medical history, added by further skin and blood tests. The prick test pierces the skin surface while depositing a tiny amount of the allergen. The test is done on the back or the inside of the arms with several allergens tested at once. If positive, redness and swelling appear at the site of the prick. The intradermal test injects the allergen with a very fine needle under the first few skin layers. This type of skin test may be used when the result of a prick test is not clear. Although the skin test is very sensitive, on some occasion, blood may be tested for the presence of antibodies against suspected specific allergens.

Mast cells, which are derived from myeloid progenitors, contain many secretory granules rich in histamine, serotonin, heparin, proteolytic enzymes, and cytokines [81,82]. Best known for their role in allergy and inflammation, mast cells are also involved with wound healing, angiogenesis, and in both innate and adaptive immune mechanisms. As presented in **Figure 9.8**, mast cells and basophils are the initiating cells for IgE-mediated allergic reactions, which are triggered when allergens cross-link preformed IgE bound to the high-affinity receptor on mast cells [81]. Once activated, mast cells induce inflammatory reactions by secreting the stored

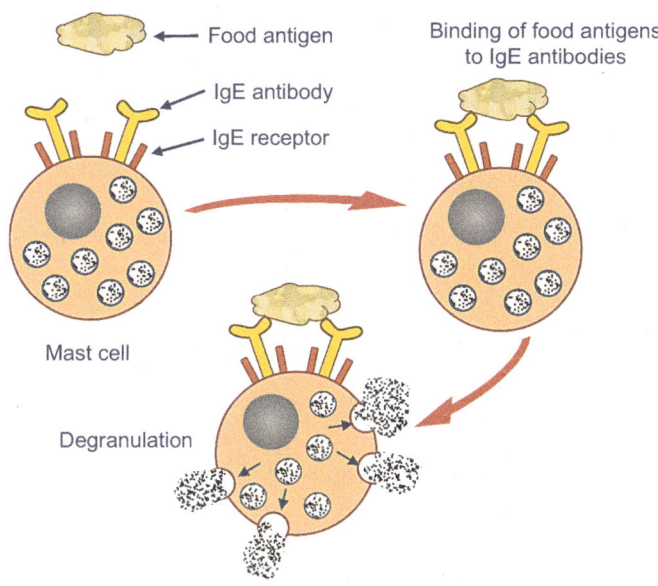

Food antigen

Binding of food antigens to IgE antibodies

IgE antibody

IgE receptor

Mast cell

Degranulation

Figure 9.8 Mast cells and the mechanism of food allergy. The combination of a food antigen with IgE antibodies on the surface of a mast cell triggers the fusion of lysosomes with the cell wall, resulting in degranulation and release of inflammatory mediators. (Redrawn and modified from Yoo, S.Y., and Kwon, S.M. *Mediators. Inflamm.*, 127–170, 2013.)

compounds, as well as by *de novo* synthesis of leukotrienes and cytokines. The consequences of mast cell activation depend on the dose of antigen and its route of entry (i.e., by inhalation, ingestion, or skin contact). Symptoms of allergic reactions range from the irritating sniffles of hay fever when pollen is inhaled, to the life-threatening circulatory collapse that occurs in systemic anaphylaxis. The immediate allergic reaction caused by mast cell degranulation is followed by a sustained inflammation, known as the late-phase response. The late response involves recruitment of other effector cells, notably Th2 lymphocytes, eosinophils, and basophils, which contribute to the immunopathology of an allergic response.

Mast cells have the capacity to store and release DA. In murine bone marrow-derived mast cells, TH expression was induced during cell maturation, resulting in increased DA content that was depleted upon mast cell activation [83]. In mast cells from human adenoidal tissues, the DA antagonist metoclopramide at high concentrations caused a biphasic effect on histamine release [84]. Both D1R [85] and D3R [86] have been implicated as mediators of the DA effects on mast cells. Cloxacepride, an amidated form of metoclopramide, significantly inhibited Con A-induced histamine release from isolated human adenoidal mast cells [84].

Treatment of ovalbumin-induced acute asthma in mice with a D1R-like antagonist, attenuated inflammatory infiltration in the airways, and repressed goblet cell hyperplasia and mucus production, while a D1R agonist had the opposite effect [87]. Furthermore, blockade of D1R-like signaling impaired Th17 function, as manifested by a significant reduction of Th17 cells in the spleen and bronchoalveolar lavage fluid. Mechanistic studies revealed that D1R-like signaling regulated Th17 differentiation to promote the development of allergic asthma. Another study reported that D3R deficiency resulted in exacerbated eosinophil infiltration into the airways of mice undergoing house dust mite–induced allergic response [14]. A D3R deficiency not only affected Th1 response, but also the frequency of Th17 cells, suggesting that D3R signaling contributed to Th17 expansion under chronic inflammatory conditions. Taken together, these data suggest that several DARs are involved in allergic asthma reaction and that drugs that alter their signaling could constitute an effective therapeutic strategy for the prevention and treatment of allergic asthma in clinical practice.

9.7 SYNOPSIS

The complete dopaminergic system, including biosynthetic and metabolizing enzymes, transporters and receptors, is present in almost all cells that are derived from lymphoid and myeloid lineages. DA is involved in a number of neurological and psychiatric disorders, as well as in several autoimmune diseases. Altered expression of DAR in peripheral lymphocytes from patients with SLE, MS, and RA supports the importance of DA regulations in autoimmunity. In some autoimmune diseases, T cells seem to have abnormal DAR expression, DA production, and/or altered responsiveness to DA. Multiple DA-altering drugs such as L-Dopa, bromocriptine, haloperidol, quinpirole, pergolide, pimozide, amantadine, tetrabenazine, and butaclamol had variable beneficial effects in these diseases. The DA-induced activation of resting Teffs and suppression of Tregs could be useful in cancer immunotherapy and infectious diseases, while the suppression of certain DAR in autoimmune diseases, pro-inflammatory, and cancerous T cells, could be advantageous. Endothelial cells, which play critical roles in vascular biology and angiogenesis, also can synthesize and release DA and variably express all DAR.

REFERENCES

1. Levite M. Dopamine and T cells: Dopamine receptors and potent effects on T cells, dopamine production in T cells, and abnormalities in the dopaminergic system in T cells in autoimmune, neurological and psychiatric diseases. *Acta Physiol (Oxf)*. 2016;216(1):42–89.

2. Pinoli M, Marino F, Cosentino M. Dopaminergic regulation of innate immunity: A review. *J Neuroimmune Pharmacol*. 2017; 12(4):602–623.

3. Buttarelli FR, Fanciulli A, Pellicano C, Pontieri FE. The dopaminergic system in peripheral blood lymphocytes: From physiology to pharmacology and potential applications to neuropsychiatric disorders. *Curr Neuropharmacol*. 2011;9(2):278–288.

4. Basu S, Dasgupta PS. Dopamine, a neurotransmitter, influences the immune system. *J Neuroimmunol*. 2000;102(2):113–124.

5. Comrie WA, Lenardo MJ. Molecular classification of primary immunodeficiencies of T lymphocytes. *Adv Immunol*. 2018;138):99–193.

6. Chen X, Jensen PE. The role of B lymphocytes as antigen-presenting cells. *Arch Immunol Ther Exp (Warsz)*. 2008;56(2):77–83.

7. Nagy E, Berczi I, Wren GE, Asa SL, Kovacs K. Immunomodulation by bromocriptine. *Immunopharmacology*. 1983;6(3):231–243.

8. Amenta F, et al. Identification of dopamine plasma membrane and vesicular transporters in human peripheral blood lymphocytes. *J Neuroimmunol*. 2001;117(1–2):133–142.

9. Cosentino M, et al. Human CD4+CD25+ regulatory T cells selectively express tyrosine hydroxylase and contain endogenous catecholamines subserving an autocrine/paracrine inhibitory functional loop. *Blood*. 2007;109(2):632–642.

10. Levite M. Dopamine in the immune system: Dopamine receptors in immune cells, potent effects, endogenous production and involvement in immune and neuropsychiatric diseases, in *Nerve-Driven Immunity*, Levine M (ed.), Springer, Vienna, 2019;1–25.

11. Cosentino M, et al. Unravelling dopamine (and catecholamine) physiopharmacology in lymphocytes: Open questions. *Trends Immunol*. 2003;24(11):581–582.

12. Nakagome K, et al. Dopamine D1-like receptor antagonist attenuates Th17-mediated immune response and ovalbumin antigen-induced neutrophilic airway inflammation. *J Immunol*. 2011;186(10):5975–5982.

13. Besser MJ, Ganor Y, Levite M. Dopamine by itself activates either D2, D3 or D1/D5 dopaminergic receptors in normal human T-cells and triggers the selective secretion of either IL-10, TNFalpha or both. *J Neuroimmunol*. 2005;169(1–2):161–171.

14. Contreras F, et al. Dopamine receptor D3 signaling on CD4+ T cells favors Th1- and Th17-mediated immunity. *J Immunol*. 2016;196(10):4143–4149.

15. Watanabe Y, et al. Dopamine selectively induces migration and homing of naive CD8+ T cells via dopamine receptor D3. *J Immunol*. 2006;176(2):848–856.

16. Sarkar C, et al. Cutting Edge: stimulation of dopamine D4 receptors induce T cell quiescence by up-regulating Kruppel-like factor-2 expression through inhibition of ERK1/ERK2 phosphorylation. *J Immunol*. 2006;177(11):7525–7529.

17. Kouassi E, Li YS, Boukhris W, Millet I, Revillard JP. Opposite effects of the catecholamines dopamine and norepinephrine on murine polyclonal B-cell activation. *Immunopharmacology*. 1988;16(3):125–137.

18. Liu Y, Wolfe SA, Jr. Haloperidol and spiperone potentiate murine splenic B cell proliferation. *Immunopharmacology*. 1996;34(2–3):147–159.

19. Saha B, Mondal AC, Majumder J, Basu S, Dasgupta PS. Physiological concentrations of dopamine inhibit the proliferation and cytotoxicity of human CD4+ and CD8+ T cells in vitro: A receptor-mediated mechanism. *Neuroimmunomodulation*. 2001;9(1):23–33.

20. Won SJ, Chuang YC, Huang WT, Liu HS, Lin MT. Suppression of natural killer cell activity in mouse spleen lymphocytes by several dopamine receptor antagonists. *Experientia*. 1995;51(4):343–348.

21. Zhao W, et al. Dopamine receptors modulate cytotoxicity of natural killer cells via cAMP-PKA-CREB signaling pathway. *PLoS One*. 2013;8(6):e65860.

22. Nozaki H, Hozumi K, Nishimura T, Habu S. Regulation of NK activity by the administration of bromocriptine in haloperidol-treated mice. *Brain Behav Immun*. 1996;10(1):17–26.

23. Basu B, Dasgupta PS, Ray MR, Lahiri S. Stimulation of NK activity in Ehrilch Ascites carcinoma-bearng mice following dopamine treatment. *Biogenic Amines*. 1992;8:191–197.

24. Soares-da-Silva P, Vieira-Coelho MA, Parada A. Catechol-O-methyltransferase inhibition in erythrocytes and liver by BIA 3–202 (1-[3,4-dibydroxy-5-nitrophenyl]-2-phenyl-ethanone). *Pharmacol Toxicol*. 2003;92(6):272–278.

25. Anderson RJ, Garcia MJ, Liebentritt DK, Kay HD. Localization of human blood phenol sulfotransferase activities: Novel detection of the thermostable enzyme in granulocytes. *J Lab Clin Med*. 1991;118(5):500–509.

26. Azoui R, Cuche JL, Renaud JF, Safar M, Dagher G. A dopamine transporter in human erythrocytes: Modulation by insulin. *Exp Physiol*. 1996;81(3):421–434.

27. Lang PA, et al. Inhibition of erythrocyte "apoptosis" by catecholamines. *Naunyn Schmiedebergs Arch Pharmacol*. 2005;372(3):228–235.

28. Periayah MH, Halim AS, Mat Saad AZ. Mechanism action of platelets and crucial blood coagulation pathways in Hemostasis. *Int J Hematol Oncol Stem Cell Res*. 2017;11(4):319–327.

29. Ricci A, et al. Dopamine receptors in human platelets. *Naunyn Schmiedebergs Arch Pharmacol.* 2001;363(4):376–382.

30. Frankhauser P, et al. Characterization of the neuronal dopamine transporter DAT in human blood platelets. *Neurosci Lett.* 2006;399(3):197–201.

31. Schedel A, Schloss P, Kluter H, Bugert P. The dopamine agonism on ADP-stimulated platelets is mediated through D2-like but not D1-like dopamine receptors. *Naunyn Schmiedebergs Arch Pharmacol.* 2008;378(4):431–439.

32. Pakalniskyte D, Schraml BU. Tissue-specific diversity and functions of conventional dendritic Cells. *Adv Immunol.* 2017;134:89–135.

33. Pacheco R, Contreras F, Zouali M. The dopaminergic system in autoimmune diseases. *Front Immunol.* 2014;5:117–135.

34. Figueroa C, et al. Inhibition of dopamine receptor D3 signaling in dendritic cells increases antigen cross-presentation to CD8+ T-cells favoring anti-tumor immunity. *J Neuroimmunol.* 2017;303:99–107.

35. Prado C, et al. Stimulation of dopamine receptor D5 expressed on dendritic cells potentiates Th17-mediated immunity. *J Immunol.* 2012;188(7):3062–3070.

36. Cao L, He C. Polarization of macrophages and microglia in inflammatory demyelination. *Neurosci Bull.* 2013;29(2):189–198.

37. Martinez FO, Sica A, Mantovani A, Locati M. Macrophage activation and polarization. *Front Biosci.* 2008;13:453–461.

38. Gaskill PJ, Carvallo L, Eugenin EA, Berman JW. Characterization and function of the human macrophage dopaminergic system: Implications for CNS disease and drug abuse. *J Neuroinflammation.* 2012;9:203–220.

39. Bone NB, Liu Z, Pittet JF, Zmijewski JW. Frontline Science: D1 dopaminergic receptor signaling activates the AMPK-bioenergetic pathway in macrophages and alveolar epithelial cells and reduces endotoxin-induced ALI. *J Leukoc Biol.* 2017;101(2):357–365.

40. Yao Y, et al. Dopamine D1-like receptors suppress the proliferation of macrophages induced by Ox-LDL. *Cell Physiol Biochem.* 2016;38(1):415–426.

41. Kawano M, Takagi R, Saika K, Matsui M, Matsushita S. Dopamine regulates cytokine secretion during innate and adaptive immune responses. *Int Immunol.* 2018;30(12):591–606.

42. McKenna F, et al. Dopamine receptor expression on human T- and B-lymphocytes, monocytes, neutrophils, eosinophils and NK cells: A flow cytometric study. *J Neuroimmunol.* 2002;132(1–2):34–40.

43. Takkenberg JJ, et al. Eosinophilic myocarditis in patients awaiting heart transplantation. *Crit Care Med.* 2004;32(3):714–721.

44. Matt SM, Gaskill PJ. Where is dopamine and how do immune cells see it? Dopamine-mediated immune cell function in health and disease. *J Neuroimmune Pharmacol.* Published online, May 2019.

45. Vidal PM, Pacheco R. Targeting the dopaminergic system in autoimmunity. *J Neuroimmune Pharmacol.* Published online, June 2019.

46. Melnikov M, Rogovskii V, Cyrillic A, Pashenkov M. Dopaminergic therapeutics in multiple sclerosis: Focus on Th17-Cell functions. *J Neuroimmune Pharmacol.* Published online, April 2019.

47. Jafari M, et al. Distorted expression of dopamine receptor genes in systemic lupus erythematosus. *Immunobiology.* 2013;218(7):979–983.

48. Walker SE. Treatment of systemic lupus erythematosus with bromocriptine. *Lupus.* 2001;10(3):197–202.

49. Dale RC, et al. Antibodies to surface dopamine-2 receptor in autoimmune movement and psychiatric disorders. *Brain.* 2012;135(Pt 11):3453–3468.

50. Panaro MA, Cianciulli A. Current opinions and perspectives on the role of immune system in the pathogenesis of Parkinson's disease. *Curr Pharm Des.* 2012;18(2):200–208.

51. Boyko AA, Troyanova NI, Kovalenko EI, Sapozhnikov AM. Similarity and differences in inflammation-related characteristics of the peripheral immune system of patients with parkinson's and Alzheimer's Diseases. *Int J Mol Sci.* 2017;18(12):2633–2649.

52. Nolan R, Gaskill PJ. The role of catecholamines in HIV neuropathogenesis. *Brain Res.* 2019;1702):54–73.

53. Rao VR, Ruiz AP, Prasad VR. Viral and cellular factors underlying neuropathogenesis in HIV associated neurocognitive disorders (HAND). *AIDS Res Ther.* 2014;11:13–29.

54. Levy D. Migraine pain, meningeal inflammation, and mast cells. *Curr Pain Headache Rep.* 2009;13(3):237–240.

55. Koyuncu ID, Kilinc E, Tore F. Shared fate of meningeal mast cells and sensory neurons in migraine. *Front Cell Neurosci.* 2019;13:136–152.

56. Barbanti P, Fofi L, Aurilia C, Egeo G. Dopaminergic symptoms in migraine. *Neurol Sci.* 2013;34(Suppl 1): S67–S70.

57. Eken C. Critical reappraisal of intravenous metoclopramide in migraine attack: A systematic review and meta-analysis. *Am J Emerg Med.* 2015;33(3):331–337.

58. Severance EG, Dickerson FB, Yolken RH. Autoimmune phenotypes in schizophrenia reveal novel treatment targets. *Pharmacol Ther.* 2018;189:184–198.

59. Debnath M. Adaptive immunity in schizophrenia: Functional implications of T cells in the etiology, course and treatment. *J Neuroimmune Pharmacol.* 2015;10(4):610–619.

60. Watkins CC, Andrews SR. Clinical studies of neuroinflammatory mechanisms in schizophrenia. *Schizophr Res.* 2016;176(1):14–22.

61. Ferreira TB, et al. Dopamine up-regulates Th17 phenotype from individuals with generalized anxiety disorder. *J Neuroimmunol.* 2011;238(1–2):58–66.

62. Krause DL, Muller N. The relationship between Tourette's syndrome and infections. *Open Neurol J.* 2012;6:124–128.

63. Mogwitz S, Buse J, Ehrlich S, Roessner V. Clinical pharmacology of dopamine-modulating agents in Tourette's syndrome. *Int Rev Neurobiol.*2013;112):281–349.

64. Stan RV, et al. The diaphragms of fenestrated endothelia: Gatekeepers of vascular permeability and blood composition. *Dev Cell.* 2012;23(6):1203–1218.

65. Apte RS, Chen DS, Ferrara N. VEGF in signaling and disease: Beyond discovery and development. *Cell.* 2019;176(6):1248–1264.

66. Pfeil U, et al. Intrinsic vascular dopamine—a key modulator of hypoxia-induced vasodilatation in splanchnic vessels. *J Physiol.* 2014;592(8):1745–1756.

67. Tonnarini G, et al. Dopamine receptor subtypes in the human coronary vessels of healthy subjects. *J Recept Signal Transduct Res.* 2011;31(1):33–38.

68. Zarei S, et al. Dopamine modulates von Willebrand factor secretion in endothelial cells via D2-D4 receptors. *J Thromb Haemost.* 2006;4(7):1588–1595.

69. Yoo SY, Kwon SM. Angiogenesis and its therapeutic opportunities. *Mediators Inflamm.* 2013:127170.

70. Ribatti D, Crivellato E. "Sprouting angiogenesis", a reappraisal. *Dev Biol.* 2012;372(2):157–165.

71. Basu S, et al. The neurotransmitter dopamine inhibits angiogenesis induced by vascular permeability factor/vascular endothelial growth factor. *Nat Med.* 2001;7(5):569–574.

72. Sarkar C, et al. Dopamine in vivo inhibits VEGF-induced phosphorylation of VEGFR-2, MAPK, and focal adhesion kinase in endothelial cells. *Am J Physiol Heart Circ Physiol.* 2004;287(4):H1554–H1560.

73. Delgado-Rosas F, et al. The effects of ergot and non-ergot-derived dopamine agonists in an experimental mouse model of endometriosis. *Reproduction.* 2011;142(5):745–755.

74. Sarkar C, Ganju RK, Pompili VJ, Chakroborty D. Enhanced peripheral dopamine impairs post-ischemic healing by suppressing angiotensin receptor type 1 expression in endothelial cells and inhibiting angiogenesis. *Angiogenesis.* 2017;20(1):97–107.

75. Osinga TE, et al. Emerging role of dopamine in neovascularization of pheochromocytoma and paraganglioma. *FASEB J.* 2017;31(6):2226–2240.

76. Zornberg GL, Jick H. Antipsychotic drug use and risk of first-time idiopathic venous thromboembolism: A case-control study. *Lancet.* 2000;356(9237):1219–1223.

77. Winkler MJ, Trunkey DD. Dopamine gangrene. Association with disseminated intravascular coagulation. *Am J Surg.* 1981;142(5):588–591.

78. Zhang X, et al. Quantitative proteomic analysis of serum proteins in patients with Parkinson's disease using an isobaric tag for relative and absolute quantification labeling, two-dimensional liquid chromatography, and tandem mass spectrometry. *Analyst.* 2012;137(2):490–495.

79. Joshi N, Singh S. Updates on immunity and inflammation in Parkinson disease pathology. *J Neurosci Res.* 2018;96(3):379–390.

80. He SH, Zhang HY, Zeng XN, Chen D, Yang PC. Mast cells and basophils are essential for allergies: Mechanisms of allergic inflammation and a proposed procedure for diagnosis. *Acta Pharmacol Sin.* 2013;34(10):1270–1283.

81. Amin K. The role of mast cells in allergic inflammation. *Respir Med.* 2012;106(1):9–14.

82. Yong LC. The mast cell: Origin, morphology, distribution, and function. *Exp Toxicol Pathol.* 1997;49(6):409–424.

83. Ronnberg E, Calounova G, Pejler G. Mast cells express tyrosine hydroxylase and store dopamine in a serglycin-dependent manner. *Biol Chem.* 2012;393(1–2):107–112.

84. Schmutzler W, Greven T, Braam U. The effects of metoclopramide and cloxacepride on human mast cells from adenoidal tissues. *Agents Actions.* 1989;27(1–2):110–112.

85. Mori T, et al. D1-like dopamine receptors antagonist inhibits cutaneous immune reactions mediated by Th2 and mast cells. *J Dermatol Sci.* 2013;71(1):37–44.

86. Xue L, et al. The dopamine D3 receptor regulates the effects of methamphetamine on LPS-induced cytokine production in murine mast cells. *Immunobiology.* 2015;220(6):744–752.

87. Nakagome K, Matsushita S, Nagata M. Neutrophilic inflammation in severe asthma. *Int Arch Allergy Immunol.* 2012;158(Suppl 1):96–102.

88. Cosentino M, Marino F. Adrenergic and dopaminergic modulation of immunity in multiple sclerosis: Teaching old drugs new tricks? *J Neuroimmune Pharmacol.* 2013;8(1):163–179.

89. Ohashi W, Hattori K, Hattori Y. Control of macrophage dynamics as a potential therapeutic approach for clinical disorders involving chronic inflammation. *J Pharmacol Exp Ther.* 2015;354(3):240–250.

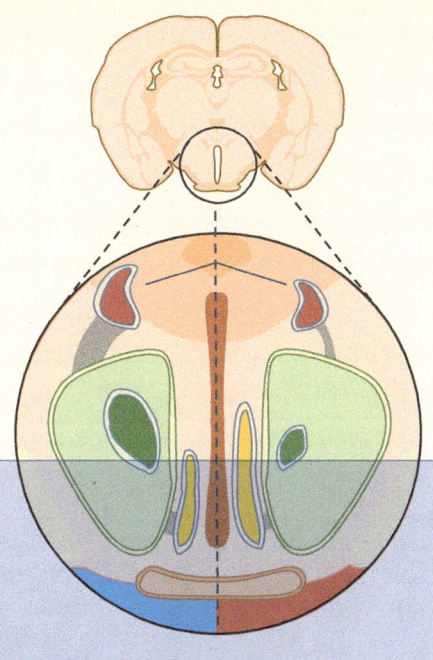

Regulation of Reproduction by Dopamine

10

10.1 INTRODUCTION

Almost all higher terrestrial organisms, with the exception of a few insects, worms, and reptiles, utilize the complicated process of sexual reproduction instead of the simpler process of offspring generation by a single parent. Sexual reproduction requires that two individuals come together at the optimal time for procreation. A compelling reason for the universal occurrence of sexual (two-parent) reproduction is the valuable reshuffling of the elements of inheritance. Thus, each zygote receives a new combination of genes from each parent, making the offspring genetically unique. In the long run, this provides a better chance for survival of the species, especially when facing gradual, periodic, and even catastrophic changes in the environment. The genetic reshuffling, coupled with the adaptive advantages of individuals with the best combination of traits, are the driving forces of evolution of the species.

The first section of this chapter covers the process of sexual differentiation of the brain and the central regulation of reproduction and behavior in both sexes. From there, the discussion moves to separate reviews of the male and female reproductive systems with an emphasis on the role of dopamine (DA) in gonadal and reproductive tract functions. The final two sections focus on the subjects of pregnancy and fetal development, followed by a discussion on parturition, lactation, neonatal development, and puberty.

10.2 SEX DETERMINATION AND SEXUAL DIFFERENTIATION OF THE BRAIN

10.2.1 Genetic and hormonal components of sexual determination and differentiation

Sexual differentiation is a complex and prolonged process. It begins at the time of fertilization by a random unification of an egg with an X- or Y-bearing spermatozoon and continues during early embryonic life with the development of male or female gonads. Whereas sexual determination is clearly under genetic control, the subsequent sexual differentiation of the reproductive organs and the brain is dictated by gonadal hormones that act on multiple sites at critical times of organogenesis. Testicular-derived transcription factors and steroid hormones impose masculinization on an inherent female state, which does not require hormonal intervention for a development toward a terminal differentiated state. In humans, sexual differentiation is incomplete at birth given that acquisition of secondary sex characteristics and attainment of functional reproductive capacity occur at puberty. **Figure 10.1** presents the genetic and hormonal components of sexual determination and differentiation during embryonic life.

Three elements—chromosomal sex, gonadal sex, and phenotypic sex—must act in coordination for a successful execution of the process of sex determination and sexual differentiation. **Chromosomal sex** refers to the sex chromosome complement (46XY in males; 46XX in females), which is determined at the time of fertilization. An individual having a normal Y chromosome is destined to develop testes, even in the presence of multiple X chromosomes. **Gonadal sex** refers to the actual commitment of gonadal tissue as testis or ovary from the bipotential embryonic gonads that can develop into either structure, depending on which genes are expressed at the critical times of organogenesis. **Phenotypic sex** refers to the type of internal and external genitalia, expression of secondary sex characteristics, and conventional gender-dependent behavior. Abnormalities in any of these elements can result in disorders of sexual development, gender ambiguities, and various degrees of reproductive dysfunctions, as detailed later in this chapter.

At about 40 days of gestation in humans, development of the testis is triggered by the expression of a specific gene on the Y chromosome that encodes **SRY**, a sex-determining region Y transcription factor [1]. SRY is transiently expressed in progenitors of Sertoli cells and serves as a pivotal switch that establishes the testis lineage. Other genes, including SRY-related gene 9 (*SOCS9*), Wilms' tumor-related gene 9 (*WTI*), and Müllerian-inhibiting substance (*MIS*), are also involved in embryonic gonadal development.

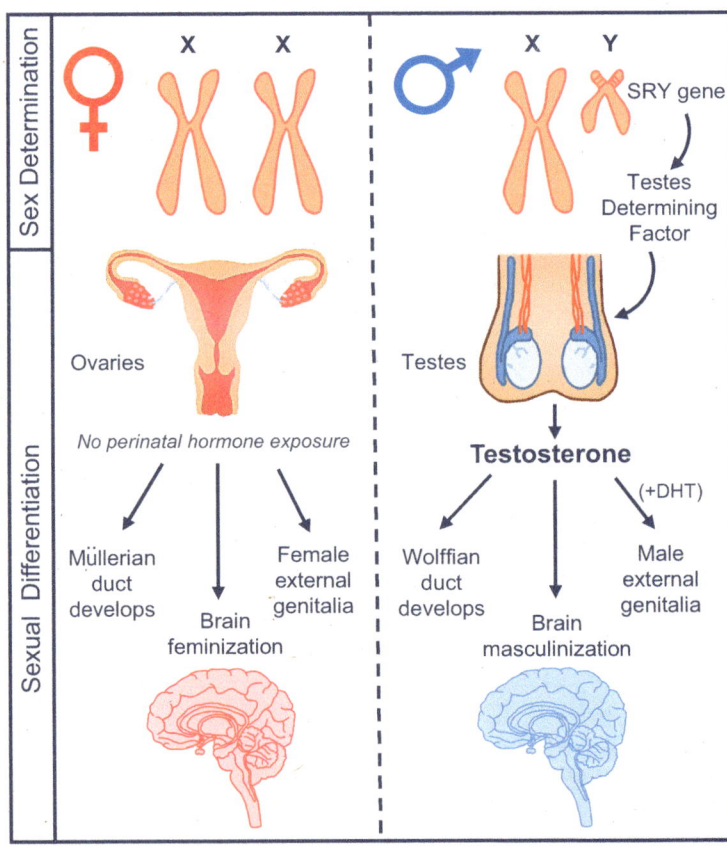

Figure 10.1 Sex determination and sexual differentiation of the brain. Sex determination occurs at the time of conception, when a fertilized ovum receives either two X chromosomes (female embryo) or an X and a Y chromosome (male embryo). The SRY gene in the Y chromosome is expressed early in development and encodes the testes-determining factor. In males, this factor induces differentiation of the bipotential gonads into testes. In females, in the absence of an SRY, the gonads differentiate into ovaries. The testes later begin to secrete testosterone and dihydrotestosterone (DHT), which induce the differentiation of the genitalia and brain into a male phenotype. Without any testosterone, the brain is feminized.

Between 6 and 8 weeks of gestation, the Leydig cells of the testis, either autonomously or under regulation by decidual human chorionic gonadotropin (hCG), begin to produce testosterone, dihydrotestosterone (DHT), and androgen binding protein (ABP). Peak production of these hormones occurs between 8 and 12 weeks of gestation, at which time differentiation of the internal genitalia along the male line takes places. The ovaries, which differentiate later, do not produce hormones and have a passive role. Within the brain, testosterone is converted to estrogen by the aromatase enzyme before it exerts its masculinizing effects. In the periphery, however, DHT is the primary form of androgen that confers the secondary sex characteristics of the male after puberty, as discussed in detail in **Section 10.6**.

Multiple physiological and behavioral responses are gender-specific. Although especially well recognized in the pattern of reproductive hormone secretion and in sexual behavior, gender-selective responses are not limited to reproduction, and several hormones, neurotransmitters and neuropeptides show clear dissimilarities between men and women. Gender differences are also seen in pathophysiological mechanisms that govern the adrenal response to stress, pain threshold, and susceptibility to certain diseases. In addition, nonsexual behaviors such as spatial orientation, verbal fluency, and aggressive behavior are also gender-biased [2].

The hypothalamus is a key player in the sex-specific control of reproductive hormone secretion and sexual behaviors. It is also a brain structure that contains nuclei with prominent gender-dependent differences in volume and morphology. As shown in **Figure 10.2**, notable examples are the sexually dimorphic nucleus of the preoptic area (SDN-POA), which is several

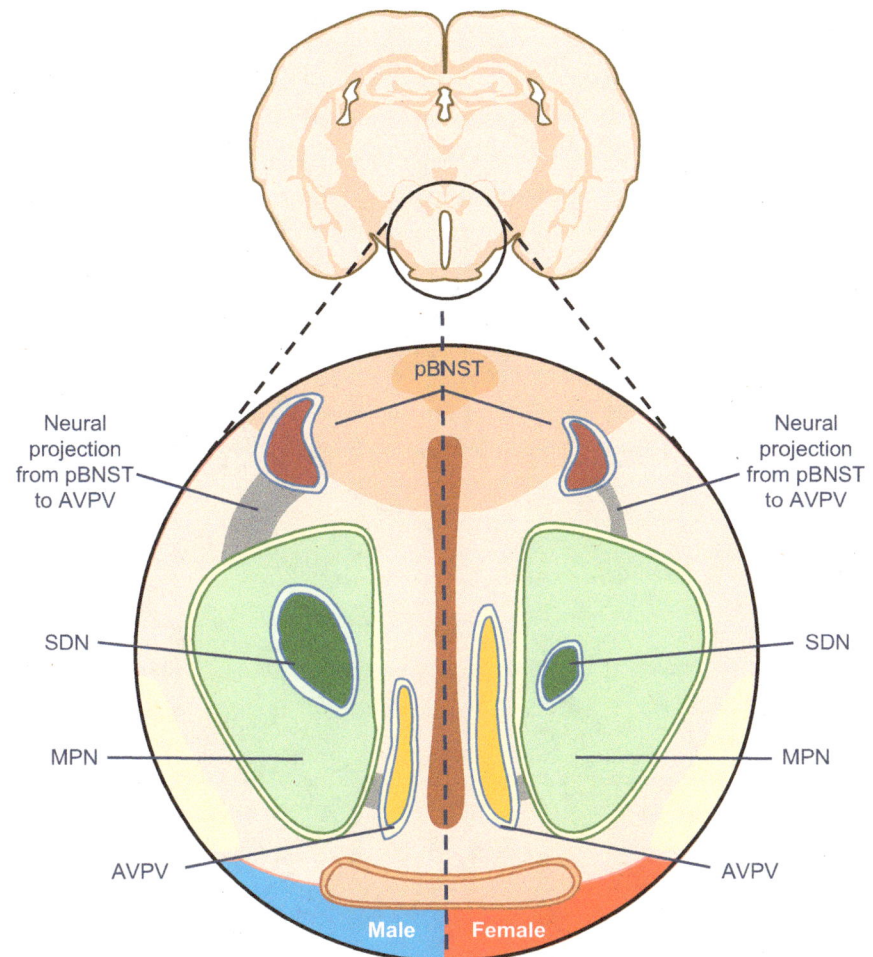

Figure 10.2 Size differences of specific nuclei in a mouse brain. The sexually dimorphic nucleus (SDN) near the medial preoptic nucleus (MPN) is bigger in males due to death of cells in the female's brain in the absence of androgens. The principal bed nucleus of the stria terminalis (pBNST) is also larger in males. In contrast, the anteroventral periventricular nucleus (AVPN) is larger in females, while males have 10-fold more neural projections from the pBNST to AVPV. (Redrawn and modified from: https://www.the-scientist.com/infographics/sex-on-the-brain-34730.)

times larger in males than in females, and the anteroventral periventricular nucleus (AVPN) which is larger in females [3]. In addition, the principle bed nucleus of the stria terminalis (PBNST), a part of the limbic system that is involved in stress-related behaviors, is denser in male brains. A list of neuroanatomical sex differences in the rat brain is presented in **Table 10.1**.

Although the sex-dependent difference in size of neuroanatomical structures is intriguing, its functional significance is unclear, and has been difficult to ascertain, even in rodents. The POA has been implicated in the control of male copulatory behavior, but the link (if any) between the size of the SDN-POA and behavior remains elusive. Experimental masculinization in female rats that altered their SDN-POA size did not result in corresponding masculinization or defeminization of their behavior. It has been speculated, however, that such structural differences account for sex differences in neurological diseases and cognitive abilities.

10.2.2 Temporal and spatial developmental stages of sexual differentiation in the brain

The two sexes have similar, but not identical, brains. During the prenatal period, a progressive sequence of "organizational" events induces

Table 10.1 Selected neuroanatomical parameters of sex differences in the rat

Structure/region	Known roles	Sex difference	Basis of difference
Sexually dimorphic nucleus of the preoptic area (SDN-POA)	The POA is implicated in the regulation of male copulatory behavior. Lesions of the SDN alone slow acquisition of this behavior.	2.6 times larger in males.	Perinatal aromatized androgen decreases neuronal apoptotic rates in males.
Anteroventral periventricular nucleus (AVPN)	Involved in the regulation of luteinizing hormone (LH) surge in females, and in male copulatory behavior.	2.2 times larger in females, with a higher cell density.	Degeneration of cells is greater in males due to prenatal action of androgen.
Bed nucleus of stria terminalis (BNST)	Plays a role in the control of male sexual behavior and gonadotropin release, and modulation of stress.	Larger volume in males.	The larger volume in males is due to different apoptotic rates caused by testosterone.
Corpus callosum	Conducts information between the two halves of the cortex.	Larger in neonatal males.	Organizational effects of testosterone lead to masculinization, while feminization depends on estrogens.
Arcuate nucleus (ARC)	Helps to regulate the estrus cycle, appetite and body weight.	Neurokin-B neurons innervate capillaries in the ventromedial ARC in postpubertal males only.	Dihydrotestosterone is responsible for the masculine projection pattern.
Amygdala	Strongly associated with emotion, decision-making and Pavlovian conditioning.	Adult males have larger medial nucleus than adult females.	Treatment of females with estradiol masculinizes this nucleus.
Cerebral cortex	Connected to many processes, from memory to language, to emotional processing.	Right posterior cortex is thicker than the left but only in males.	Gonadal hormones play a role because ovariectomy masculinizes the cortex of females.
Ventromedial hypothalamic nucleus (VMN)	Involved in control of lordosis, mounting, and norepinephrine (NE) release. High levels of steroid receptors in the ventrolateral VMN.	Females have less synapses in the ventrolateral VMN compared to males.	Organizational effects of aromatized testosterone are crucial in establishing the masculine trait.
Substantia nigra pars compacta	Made up of dopamine (DA) neurons involved in control of motor activity.	Females have 20% fewer dopaminergic neurons.	A genetic component has been demonstrated in mice.

permanent modifications of cell number and morphology of selected brain nuclei. This process eventually generates modified neuronal networks which govern the morphological and functional processes of sex-specific responses of the adult brain to hormonal and environmental inputs. The timing of the sexual development program is species-specific. For instance, the brain of a newborn rodent continues its sexual maturation for 2–3 weeks after birth, while the maturity of the primate's brain at birth is more advanced, albeit not terminal. The delayed brain maturation of the newborn rodent enables a number of experimental interventions such as administration of various gonadal steroids to neonates that can still induce reorganization in the brain; such an approach is ineffective in the primate neonate.

The embryonic brain, much like the reproductive organs, is feminine by default, and specific stimuli are needed to drive it toward a male phenotype. Two processes must be activated for the formation of a male-specific brain: (1) development of neural circuits that are permissive to the expression of male-specific responses to testicular hormones (masculinization) and (2) loss of circuits able to respond to ovarian hormones (defeminization) during puberty and adulthood. In sum, a cooperation between genetic and hormonal inputs is necessary for implementing the process of masculinization/defeminization of the developing brain.

The fetal brain is protected from the effects of circulating maternal estrogens by α-fetoprotein, which is produced by the fetal liver and binds strongly to estrogens but not to testosterone. However, estrogens can reach the brain not only via the circulation, but also from local sources, because the brain itself is capable of producing estrogens. During development, there are also sex differences in brain steroid receptor distribution. In rats, the formation of estradiol in the brain by aromatization of circulating testosterone is the critical mechanism for brain virilization, but it is not the ultimate determinant that shapes gender identity or sexual orientation in humans.

10.2.3 Interference of the sexual differentiation program by endocrine disruptors

Because of the high responsiveness of the fetal brain to gonadal hormones and its inherent plasticity, an inadvertent exposure to substances that interfere with these processes during a critical period of sexual differentiation can cause long-lasting effects on several aspects of reproduction. A noteworthy case is the misfortune of human exposure to **diethylstilbestrol (DES)**. DES was produced in 1938 as the first potent synthetic estrogen. From 1940 to 1971, DES was prescribed to several million pregnant women to prevent miscarriage and to correct some complications of pregnancy. Astute observations by several physicians in the early 1970s, who noticed increased incidence of a rare vaginal cancer in adolescent women, drew attention to deleterious effects of DES in the offspring. In retrospect, there was no good reason for treating pregnant women with DES, except that it was readily available and was considered safe. Similar to the genotoxic thalidomide that caused birth defects, DES had little impact on the treated women themselves, but had substantial, albeit delayed, effects on their children. This incidence established the notion that drug actions in the fetus cannot be predicted from its lack of effects in adults.

Prenatal exposure to DES causes multiple structural and functional abnormalities in both sons and daughters of treated mothers [4]. A large percentage of daughters had benign anomalies of the reproductive tract, with increased incidence of infertility and ectopic pregnancy. The most common cancer was vaginal clear cell adenocarcinoma, affecting 0.1%–0.15% of the exposed women, but there was also increased incidence of breast cancer. Prenatal DES exposure also increased the probability of expressing bisexuality or homosexuality in girls. Structural abnormalities in sons included cryptorchidism, malformed urethras and epididymal cysts, often accompanied by decreased sperm count, increased sperm deformities, and prostatic inflammation. There was no evidence, however, for either homosexuality or ambiguous gender identity in the exposed sons.

Recent years have witnessed a growing concern among the research community and the public at large about the adverse effects of **endocrine-disrupting chemicals (EDCs)**, defined as exogenous compounds that interfere with hormonal actions [5]. EDCs can act either as agonists or as antagonists of receptors of multiple hormones and can also alter the availability of active endogenous hormones by affecting their synthesis, metabolism, and serum half-life.

Xenoestrogens represent a subclass of EDCs that specifically mimic or interfere with the actions of estrogens. They encompass a wide array of compounds, some of which are naturally made by plants (phytoestrogens) or fungi (mycoestrogens), while others are synthetic chemicals (**Table 10.2**). The latter group includes pharmaceuticals, cosmetics, plastic additives, industrial solvents, pesticides, herbicides, and by-products of combustion and industrial manufacturing processes. The times at which exposure to xenoestrogens occurs (i.e., during fetal, neonatal, pubertal, or adult life) are of critical importance in the manifestation of their effects on expression of sexual traits.

As discussed in our review [6], one of the best studied EDCs is **bisphenol A (BPA)**, a synthetic small molecule composed of two phenol groups with a similar structure to DES (**Figure 10.3**A). BPA is produced in very large quantities (over 6 million tons worldwide per year) in the manufacture of polycarbonate plastics and epoxy resins, both of which are made of repeating BPA monomers. BPA-containing consumer products include plastic bottles, food utensils, lining of beverage and food cans, thermal receipt papers, and water pipes as well as dental cements and intravenous medical tubing. Although the carbonate linkages between the monomers are rather

Table 10.2 Different classes of xenoestrogens, their sources in the environment, and major pathways of human exposure

Class/Compounds	Sources	Exposure pathways
Phytoestrogens	Plants	Ingestion
Genestein, Resveratrol, Daidzein		
Micoestrogens	Fungi	Ingestion
Zeranol, Fusartins		
Pharmaceuticals	Drugs, cosmetics	Skin
DES, HRT, Phthalates		
Pesticides and Herbicides	Livestock, crops	Skin, inhalation
DDT, Atrazine, Dieldrin		
Plastics	Polycarbonates, epoxy resins	Skin, intravenous, oral
PCBs, bisphenols		
Food additives	Variety of foods	Ingestion
Propyl gallate, 4-hexyl resorcinol		
Detergents and Preservatives	Household goods	Skin
Parabens, Alkylphenols		
Combustion products	Car exhausts	Inhalation
Dioxin, aromatic hydrocarbons		

DES: diethylstilbestrol; DDT: Dichlorodiphenyltrichloroethane; HRT: hormone replacement therapy; PCB: polychlorinated biphenyl.

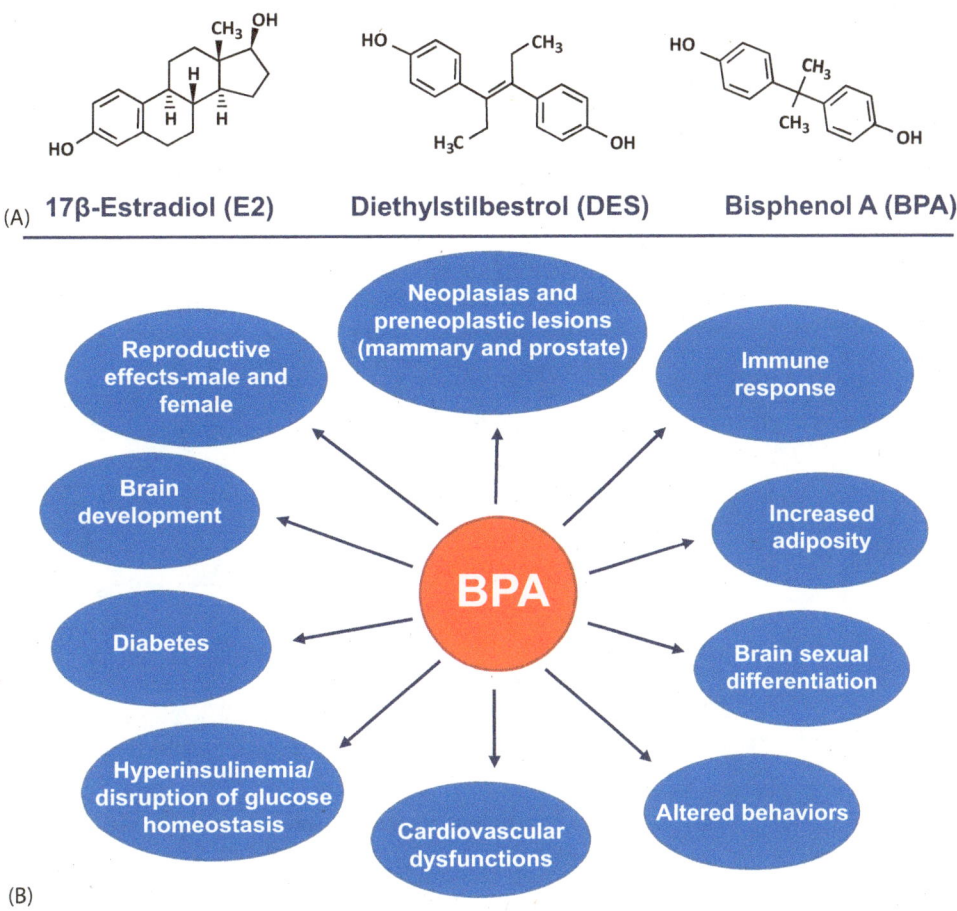

Figure 10.3 Structure and biological activities of bisphenol A (BPA). Panel A compares the structures of 17β estradiol (E2), diethylstilbestrol (DES), and BPA. Panel B shows the wide range of the effects of BPA on multiple systems in both experimental animals and humans.

stable and the polymers are chemically inert, BPA can leach out as a result of an incomplete polymerization and/or because of polymer degradation by elevated temperatures or by acidic conditions. Humans can be exposed to BPA through ingestion, skin absorption, inhalation, and through intravenous catheters. BPA at nanomolar concentrations has been detected in the blood or urine of most individuals tested. BPA has also been detected in the amniotic fluid and fetal plasma, indicating its passage across the placenta during pregnancy and underlying its potential to affect fetal development. The wide range of the documented biological actions of BPA is shown in **Figure 10.3**B.

10.2.4 Association of dopamine with sexual differentiation in the brain

Most studies on sexuality of the brain have focused on gross signs of gender differences (i.e., the size of specific regions or nuclei). However, there is a growing evidence that sex differences are also manifested at finer levels such as neurotransmitter content and synaptic connections [3]. A major caveat for understanding human sexual attributes is that most of the information is based on pharmacological and genetic manipulations conducted with rodents. Although some studies were done with other species, including zebra finches, chicken, and wallabies, only limited research has been done with nonhuman primates [7]. Hence, extrapolation of information obtained from classical rodent studies to the mechanism of sexual differentiation in humans must be done with caution.

Sex differences in dopaminergic neurons in the rodent's brain are already apparent before the time of embryonic exposure to endogenous gonadal steroid hormones. Rat embryos have a plasma surge of testicular hormones around embryonic day 17 or 18 (E17 or E18]. Yet, as early as E14, dissociated cell cultures of dopaminergic neurons are fundamentally different in morphology and function between males and females [3]. Females have a higher number of dopaminergic cells in the midbrain, and their mesencephalic and diencephalic neurons produce more DA than those of males. On the other hand, the soma of diencephalic neurons from male cultures contain larger dopaminergic neurons. The clear gender-dependent differences at such early times of development suggest an involvement of sex chromosome complement and/or nonsteroidal, sex-specific gene expression.

Indeed, as described in the above review [3], SRY has direct effects on tyrosine hydroxylase (TH) expression in the brain. In mice, *Sry* mRNA was detected in the midbrain and hypothalamus at all developmental stages. In rats, *in situ* hybridization has revealed specific *Sry* labeling in the substantia nigra, medial mammillary bodies of the hypothalamus, and cortex of males only [8]. The transcripts were co-localized with the TH protein, and all the *Sry*-positive neurons in the substantia nigra were also positive for TH. Knocking down Sry expression in the substantia nigra of male rats led to 38% fewer TH-immunoreactive neurons. This reduction was not due to neurodegeneration but most likely due to reduced TH expression. When Sry expression was knocked down in the striatum, there was a 26% decrease in TH-positive cells. In contrast, the number of TH-immunoreactive neuron was not affected in females treated with a Sry antisense. SRY expression in the adult human brain is found in the hypothalamus, frontal, and temporal cortex.

10.2.5 Gender identity and sexual preference

Resolution of the mechanisms underlying gender self-identity, transsexuality, and homosexuality remains a major challenge. A vast array of genetic and hormonal factors could lead to gender identity problems [9]. Twin and family studies have shown that genetic factors are important. For example, rare chromosomal abnormalities may lead to transsexuality, and polymorphisms of genes encoding estrogen and androgen receptors as well as aromatase also carried an increased risk. Abnormal hormone levels during early development also appear to play a role. This is suggested by the high frequency of androgen-producing polycystic ovaries, oligomenorrhea, and amenorrhea in female-to-male (FtM) transsexuals, indicating that an early intrauterine exposure of a female fetus to abnormally high levels of testosterone may be a causative factor, although this issue remains controversial. There is also a higher prevalence of hyperandrogenism in FtM transsexuals, supporting the possible involvement of high testosterone levels with transsexuality. Girls with congenital adrenal hyperplasia (CAH), who are exposed to extreme levels of testosterone *in utero*, also have an increased probability of being transsexual. Although the likelihood of developing transsexuality in such cases is 300–1,000 greater than normal, the risk for transsexuality in CAH is only 1%–3%, whereas the probability of serious gender problems in the general population is 5.2%. The consensus is that girls with CAH should be raised as girls, even when they are masculinized.

Sexual preference is a complex, sexually dimorphic trait found across animal species. Recent studies with mice revealed an association of sexual preference with DA in the nucleus accumbens [10]. The study used genetically engineered male mice that lack a functioning vomeronasal organ, a scent-sensing olfactory structure that responds to pheromones. These males showed no aggression toward other males and were equally interested in mating with males and females. When exposed to negative conditioning against female pheromones, the mutant male mice avoided mating with females while displaying sexual behavior toward other males. Importantly, female pheromones presented to intact males activated a reward mechanism linked to DA release from the nucleus accumbens, the so-called "pleasure center" in the basal forebrain. In the absence of this stimulus, males showed no preference for females. In spite of the interesting association of DA with sexual preference in rodents, there are no published records to support a dopaminergic association with homosexuality or gender identify in humans.

10.3 HYPOTHALAMO–PITUITARY REGULATION OF REPRODUCTION IN BOTH SEXES

10.3.1 Common features of the central control of reproduction

The reproductive systems of both sexes are controlled by reciprocal inter-actions between hypothalamic and anterior pituitary hormones and periph-eral hormones produced by reproductive organs and tissues (**Figure 10.4**). In both sexes, the hypothalamus monitors the levels of peripheral hormones, integrates those with internal signals and external environmental cues, and responds by regulating the release of anterior pituitary hormones. A second level of control is done by the gonadotrophs of the anterior pituitary, which also receive direct feedback information from the periphery. The primary hypothalamic hormone that regulates reproduction in both sexes is the deca-peptide gonadotropin-releasing hormone (GnRH). GnRH promotes the pro-duction and release of the gonadotropins, follicle-stimulating hormone (FSH) and luteinizing hormone (LH) from the anterior pituitary. More details on GnRH and the gonadotropins are found in **Chapters 4** and **5**, respectively.

Although FSH and LH are named after their functions in female repro-duction, they are chemically identical in males and females and are pro-duced in the pituitaries both sexes. In the testes, FSH and LH stimulate sperm production by Sertoli cells and testosterone production by the interstitial cells of Leydig, respectively. Testosterone, in turns, supports spermatogen-esis and sperm maturation, provides negative feedback information to the hypothalamo–pituitary complex, and is responsible for the expression of the secondary sex characteristics of the male and other metabolic actions.

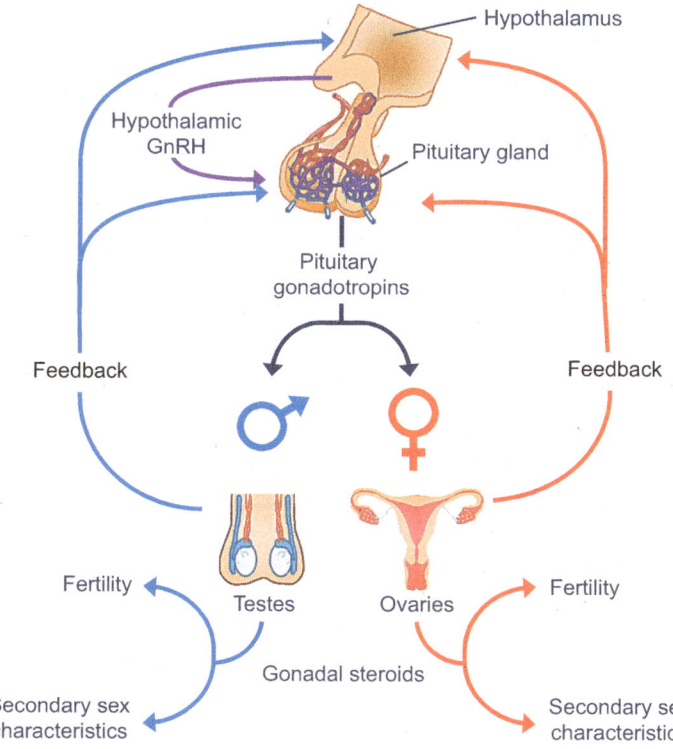

Figure 10.4 Endocrine regulation of reproduction in males and females. Hypothalamic GnRH stimulates the production and release of the pituitary gonadotropins, LH and FSH. The gonadotropins target the testes and the ovaries and promote the release of testosterone in males and estradiol and progesterone in females. The gonadal steroids regulate fertility and secondary sex characteristics and provide feedback input to the hypothalamus and pituitary. FSH: follicle-stimulating hormone; GnRH: gonadotropin-releasing hormone; LH: luteinizing hormone.

In the ovaries, FSH and LH support the development and growth of ovarian follicles, induce ovulation, and stimulate the production of estrogen and progesterone. The ovarian steroids, in turns, regulate the female reproductive system, provide both negative and positive feedback information to the hypothalamo–pituitary complex, affect the secondary sex characteristics of the female, prepare the reproductive tract for potential pregnancy, and support gestation and fetal development. Unlike the reproductive system of men, which is continuously active from puberty to old age, in women menopause occurs by the fifth decade of life. At this time, the ovaries lose their capacity to ovulate and become unresponsive to FSH and LH, resulting in the cessation of the female reproductive cyclicity.

10.3.2 Involvement of dopamine in hypothalamic control of reproduction and sex behavior

As discussed in **Chapter 4** and illustrated in **Figures 4.11** and **4.12**, kisspeptin neurons in the rostral periventricular area of the hypothalamus play an essential role in ovulation by activating the GnRH neurons. In addition, kisspeptin neurons in the arcuate nucleus affect pulsatile GnRH release in both sexes. Kisspeptin neurons also mediate the negative feedback of sex steroids on the hypothalamus because the GnRH neurons themselves do not express receptors for gonadal steroids. A subpopulation of kisspeptin neurons, located in the rostral periventricular area that synapse on the GnRH cell bodies and dendrites, express TH. The presence of TH-positive terminals that synapse on GnRH neurons is a consistent ultrastructural feature of these neurons from several species [11].

DA acts on GnRH neurons in the mediobasal hypothalamus and the preoptic area to modulate LH secretion. Application of electrophysiological approaches to brain slice preparations from male and female mice showed that half of the GnRH neurons were inhibited by DA, with a third of these neurons showing tonic inhibition by endogenous DA [12]. The effects of DA were mediated by D1- and D2-like receptors expressed by the GnRH neurons. Inhibition by DA was not altered by sex or by estrous cycle stage. The authors concluded that DA, alongside gamma-aminobutyric acid (GABA), glutamate, and kisspeptin, is used to signal to GnRH neurons.

To function properly, GnRH has to be delivered in a pulsatile manner to the gonadotrophs, where it stimulates corresponding pulses of LH release [13]. The modulation of LH pulse frequency is essential for pubertal maturation as well as for a number of reproductive functions. In infancy, LH pulsatile secretion is transiently increased, likely reflecting pulsatile GnRH secretion, but soon becomes quiescent. The prepubertal suppression of the hypothalamo–pituitary–gonadal (HPG) axis occurs in agonadal humans and primates, suggesting that hypothalamo–hypophysial factors play a role in the postnatal quiescence of the reproductive axis, until puberty sets in. In women, the pattern of GnRH secretion is essential for the regulation of the menstrual cycle. LH pulse frequency is slow in the luteal phase but speeds up during the follicular and preovulatory phases, presumably reflecting changes in GnRH pulse frequency. Abnormalities in GnRH, and hence LH, pulse frequency, are associated with a number of reproductive endocrine disorders such as hypothalamic amenorrhea, hyperprolactinemia, and polycystic ovary syndrome (PCOS).

Evidence indicates that hypothalamic DA is involved in the regulation of the pattern of pulsatile of GnRH release. Studies with ewes during the breeding season and at periods of reproductive quiescence showed that activation of hypothalamic D2R inhibited GnRH pulse frequency by suppressing kisspeptin-positive neurons in the arcuate nucleus [14]. In addition, the A15 dopaminergic neurons in the periventricular nucleus (PVN) mediated the negative feedback of estradiol on GnRH neurons [15]. The sex steroids also affected alternative splicing of the long and short D2R isoforms in rat hypothalamus [16].

DA also affects reproductive processes in females by modulating circadian rhythms. As discussed in **Chapter 4**, circadian rhythms are controlled by a master pacemaker, the suprachiasmatic nucleus of the hypothalamus, and are also affected by clock genes located in various tissues [17]. The central clock integrates photic and nonphotic signals and generates rhythmic outputs through neuroendocrine and behavioral signals. The circadian rhythm is especially important for reproductive proficiency of female rodents by entraining the timing of the preovulatory gonadotropin surge and sexual receptivity [18]. Although circadian rhythm has little consequences on ovulation or sexual receptivity in women, disruption of the circadian clock is associated with some pathological processes, including infertility [17]. A recent review [19] details the involvement of DA/DA receptor (DAR) in five brain areas associated with rhythmicity: the retina, olfactory bulb, striatum, midbrain, and hypothalamus (see also **Chapter 4**, **Section 4.2**).

Sexual receptivity in females is also affected by the hypothalamic dopaminergic neurons [20]. Ovarian steroids, estrogen (E), and progesterone (P) regulate a number of cellular functions in the brain that control sexual behavior in female rats. The progesterone receptor (PR) in the preoptic area, which is primed by pre-exposure to E, plays a critical role in the induction of sexual behavior in female rats, as was determined by the use of PR antagonists and antisense oligonucleotides. These data showed that DA, acting via D1R, activated the PRs by a ligand-independent mechanism, demonstrating cross-talk between steroid hormone- and DA-initiated pathways in sexual behavior in female rats and mice.

DA also affects sexual behavior in males. Neurochemical studies detected a significant increase in DA levels in monoamine oxidase (MAO) in response to sexual activity in male rats, while DAR antagonism impaired this response [21]. In addition, intracerebral administration of the D2R antagonists raclopride and spiperone abolished contractions of muscles that regulate erection and ejaculation, and prolonged the ejaculatory latency, while the D2R-agonist quinelorane induced rhythmic contractions of these muscles [22]. These findings implicate hypothalamic DAR as an important mediator of male sexual arousal and ejaculatory response. In addition, DA can trigger penile erection by acting on oxytocinergic neurons located in the PVN, and perhaps on the pro-erectile sacral parasympathetic nucleus within the spinal cord.

10.3.3 Involvement of dopamine in the control of reproduction by the pituitary

Unlike the wealth of information on DA role in the regulation of hypothalamic neurohormones that affect the gonadotropins, there are only few publications on the direct actions of DA on the pituitary gonadotropes. Studies with two murine gonadotroph cell lines, αT3-1 and LβT2, showed that they expressed D2R [23]. Bromocriptine, a D2R agonist, increased basal LH release but also blocked GnRH-stimulated α-subunit and LHβ-subunit expression in these cell lines, revealing an uncoupling between LH synthesis and release. Another important action of D2R in the gonadotrophs is the suppression of pituitary adenylate cyclase-activating polypeptides (PACAP), which is highly expressed in the fetal pituitary [24]. A D2R-induced decline in PACAP during the perinatal period reduced pituitary follistatin, a single-chain gonadal protein hormone that specifically inhibits FSH release, increased the expression of GnRH and FSH receptors, and facilitated activation of the neonatal gonads [24].

10.3.4 Dopamine indirectly regulates reproductive functions through the control of prolactin

The regulation of prolactin (PRL) by DA is highly relevant to reproductive competence. As discussed in **Chapter 5**, DA acting via D2R is the primary inhibitor of PRL synthesis and release. Unlike the gonadotropins, PRL is not essential for the control of reproduction. Rather, PRL has multiple modulatory effects on reproduction, which differ markedly between rodents and humans [25]. In rodents, serum PRL levels increase during the preovulatory phase of the estrous cycle, and PRL is essential for the support of progesterone production by the corpus luteum (luteotropic action) which maintains gestation. The situation in humans is quite different. Serum PRL does not change significantly during the menstrual cycle, PRL is not luteotropic, and, with the exception of lactation, PRL has no discernible effects on reproductive processes under normal physiological conditions [25]. On the other hand, hyperprolactinemia in both men or women results in many reproductive disturbances, as detailed below, suggesting that PRL contributes in a subtle manner to optimal reproduction in humans.

Hyperprolactinemia in women can lead to amenorrhea, anovulation, reduced libido, and orgasmic dysfunction. As discussed in our extensive review [25], up to 20% of secondary amenorrhea cases is due to high PRL levels. There are several mechanisms by which elevated PRL induces reproductive dysfunctions (**Figure 10.5**). One is by inhibiting GnRH production as well as pulsatility through a PRL-induced increase in hypothalamic dopaminergic tone. In fact, LH pulse frequency in hyperprolactinemic women is lower than in controls, requiring treatment with bromocriptine or cabergoline to regulate PRL secretion and restore LH pulse frequency. At the pituitary level, elevated PRL suppresses the LH secretory response to GnRH and also reduces the positive estrogen feedback on gonadotropin release. At the gonads, high PRL levels interfere with follicular development and/or with progesterone production. Chronic drug-induced hyperprolactinemia in rats, or PRL elevation in D2R-deficient mice, caused some estrous cycle irregularities, but no major deteriorating effects on fertility. These data reinforce the notion that rodents are not suitable models for hyperprolactinemia-related infertility in women.

In men, elevated serum PRL levels reduce pulsatile GnRH secretion, suppress gonadotropin release, have direct inhibitory effects on testicular steroidogenesis and spermatogenesis through local PRL receptors, and can cause gynecomastia and decrease libido (**Figure 10.5**). Collectively, the suppression of the hypothalamo–pituitary–testicular axis in men results in hypogonadism, reduced testosterone levels, and erectile dysfunction [25].

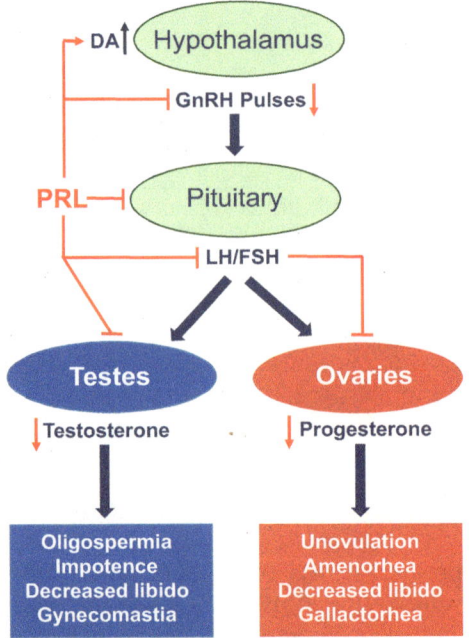

Figure 10.5 Mechanisms by which hypersecretion of prolactin (PRL) interferes with reproduction. At the hypothalamus, PRL increases the activity of dopaminergic neurons, resulting in the suppression of GnRH production and pulsatility. At the pituitary, PRL suppresses LH release in response to GnRH. At the gonads, PRL interferes with testosterone production in males and progesterone in females. Overall, hyperprolactinemia can cause low sperm count (oligospermia), impotence, decreased libido, and breast enlargement (gynecomastia) in men, and infertility, decreased libido, and inappropriate lactation (galactorrhea) in women. DA: dopamine; FSH: follicle-stimulating hormone; GnRH: gonadotropin-releasing hormone.

10.4 MALE REPRODUCTION: TESTES AND THE GENITAL TRACT

A diagram of the hormonal regulation of the hypothalamo–pituitary–testicular axis is shown in **Figure 10.6**. Both GnRH and the gonadotropins, LH and FSH, provide stimulatory inputs to the testes, while testosterone and inhibin provide negative feedback to the hypothalamus and the pituitary. The male genitalia consist of the gonads (testis), duct system (epididymis and vas deferens), an intromittent organ (penis), and accessory glands (seminal vesicles, prostate gland, and bulbourethral glands), as illustrated in **Figure 10.7**. Some of these structures are located within the pelvis and others are located outside the body cavity. The overall function of the male reproductive system is to produce sperm, synthesize and secrete androgens, and regulate the secondary sex characteristics. A comprehensive review on the male reproductive system

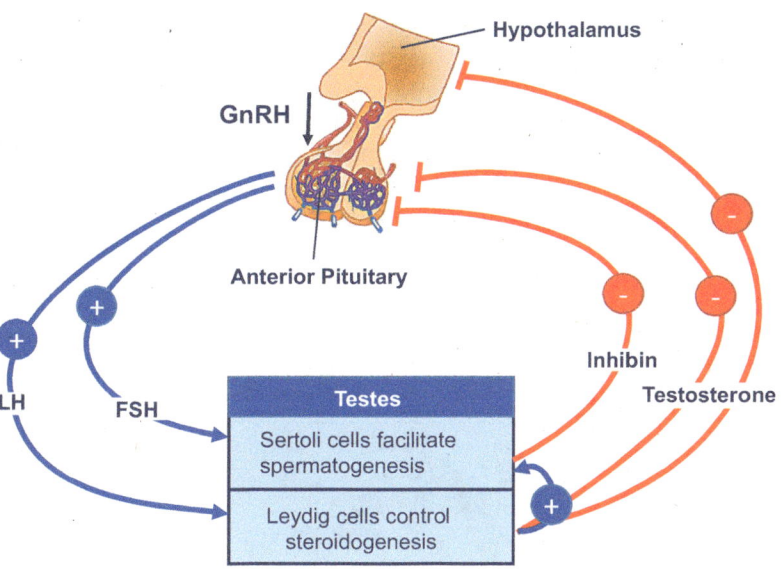

Figure 10.6 The hypothalamo–pituitary–testicular axis in males. Pulsatile GnRH release increases the production and release of the gonadotropins: LH and FSH. At the testes, FSH targets Sertoli cells and regulates spermatogenesis, while LH targets Leydig cells and regulates steroidogenesis. Testosterone provides negative feedback information to the hypothalamus and anterior pituitary and positive feedback to the Sertoli cells in the support of spermatogenesis. Inhibin provides negative feedback input to the pituitary. FSH: follicle-stimulating hormone; GnRH: gonadotropin-releasing hormone; LH: luteinizing hormone.

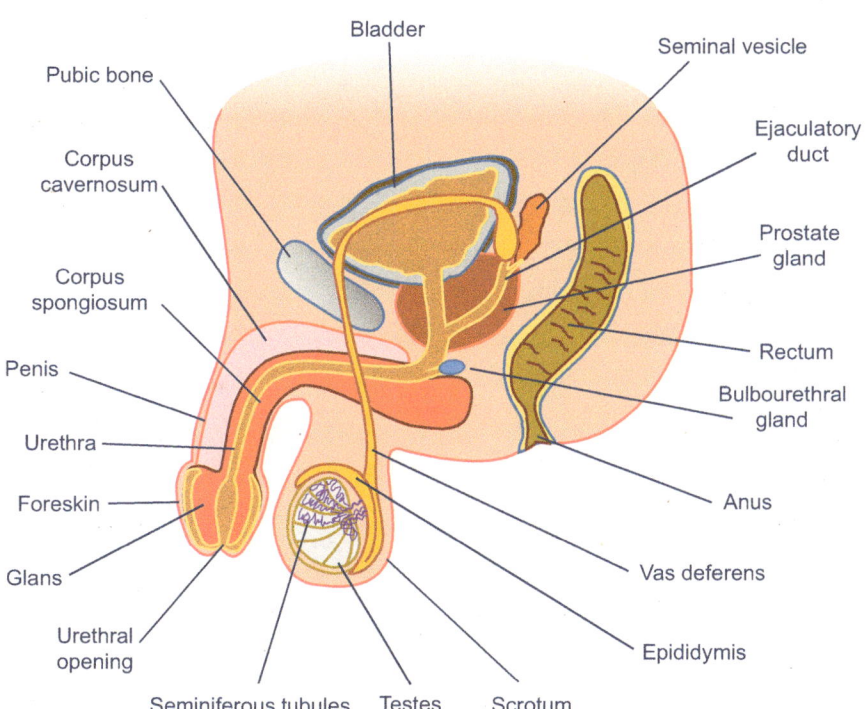

Figure 10.7 Anatomy of the male genitalia. The testes, enclosed within the scrotum, contain the seminiferous tubules. Sperm is stored and matured in the epididymis before it is transported into the vas deferens. During transport, sperm is mixed with secretions from the seminal vesicles, prostate and bulbourethral glands, forming the semen. The various tissues associated with the penis are also shown.

can be downloaded from *Endotext* at https://www.endotext.org/chapter/endocrinology-of-the-male-reproductive-system-and-spermatogenesis/.

10.4.1 The male gonads: Spermatogenesis

In most male mammals (whales and elephants being an exception) the testes are enclosed within a scrotum and are located outside the body cavity in order to maintain lower temperature. The cremaster muscle raises and lowers the testes and regulates the scrotal temperature for optimal spermatogenesis and survival of the spermatozoa. Increased testicular temperature is known to reduce both the quantity and quality of sperm produced.

Within the testes, fine coiled tubes, called the seminiferous tubules, are lined with multiple germ cells that continuously divide and differentiate into spermatozoa, from puberty through old age. Sertoli cells inside the tubules support germ cell development and are the main target of FSH. The interstitial cells of Leydig, located between the tubules, produce testosterone and are the main target of LH. Spermatogonia undergo a complex developmental program to become mature spermatozoa, a process that takes about 2 months in humans (**Figure 10.8**).

Spermatogenesis can be divided into three phases: (1) mitosis, or regular cell division; (2) meiosis, or reductional division; and (3) spermiogenesis, or cell differentiation. During the first phase, the **diploid** (2n) spermatogonia divide by mitosis. Primary spermatocytes then enter meiosis, a specialized cell division that is unique to male and female gametes. Meiosis differs from mitosis by three criteria. One is the halving of the number of chromosomes, so that each daughter cell is **haploid** (n), with only 23 chromosomes. This ensures reconstitution of the diploid status upon unification with

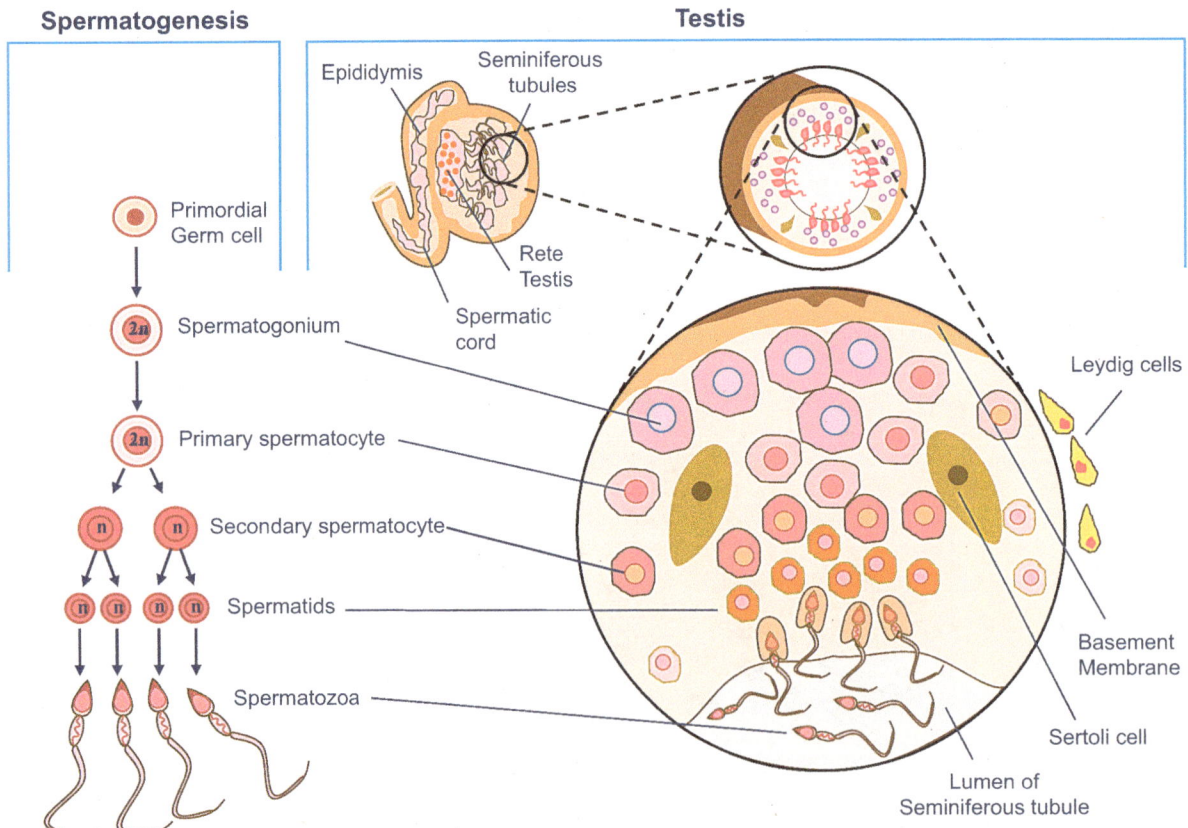

Figure 10.8 The process of spermatogenesis. The left panel shows the conversion of diploid primordial germ cells (2n) into haploid secondary (n) spermatocytes and spermatids. The latter undergo structural reorganization, becoming spermatocytes. The right panel shows the testis and epididymis and a cross section of a seminiferous tubule (insets). The various forms of germ cells divide and differentiate as they move from the periphery toward the lumen. Sertoli cells are located within the seminiferous tubules while Leydig cells are located outside the tubules.

a gamete of the opposite sex. The second involves an exchange of genetic material between homologous chromosomes during crossing over. The third is a random segregation of paternal and maternal chromosomes between the daughter cells. Hence, the number of genetically distinct gametes that can be produced by an individual is almost unlimited.

A germ cell entering meiosis undergoes two consecutive cell divisions but only one replication of chromosomes. The first cell division (**meiosis I**) is dominated by a prolonged prophase (24 days in humans), while the second (**meiosis II**) resembles regular mitosis except that it is not proceeded by DNA replication. As they undergo cell divisions and cytoplasmic reorganization and differentiation, the germ cells progressively migrate away from the walls of the seminiferous tubules toward the lumen (**Figure 10.8**). From there, fully formed spermatozoa move to the rete testes and efferent ducts and into the epididymis, where they undergo additional maturation.

10.4.2 The male gonads: Steroidogenesis and protein hormone production

Testosterone, a steroid hormone from the androgen group, is found in mammals, reptiles, birds, and other vertebrates. In mammals, testosterone is primarily secreted by the testes, but also by the ovaries, and in small amounts by the adrenal glands. Like other steroid hormones, testosterone is derived from cholesterol (**Figure 10.9**A). The first step in its biosynthesis

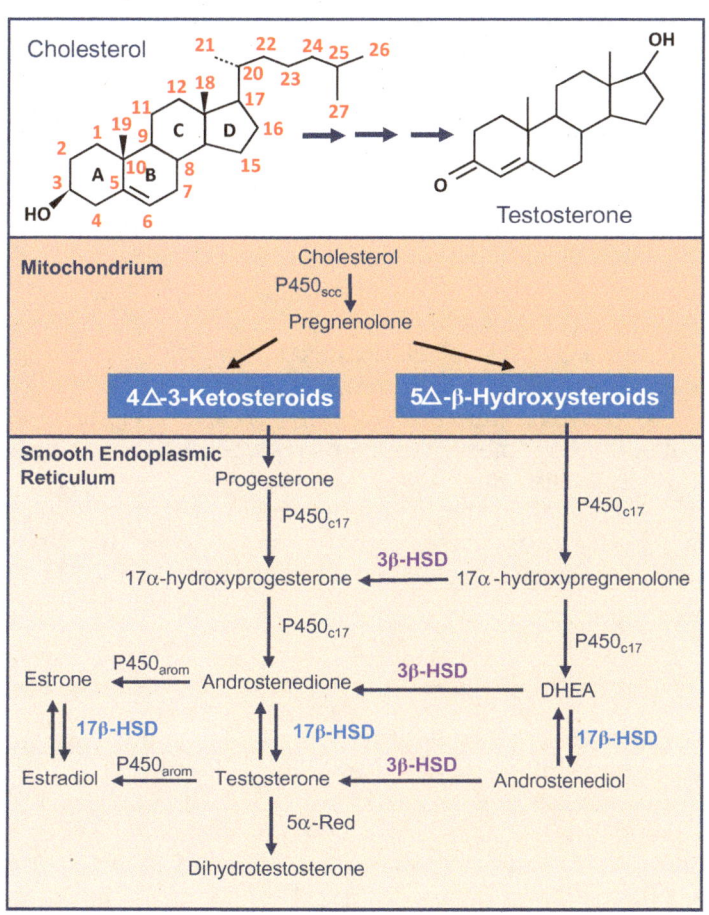

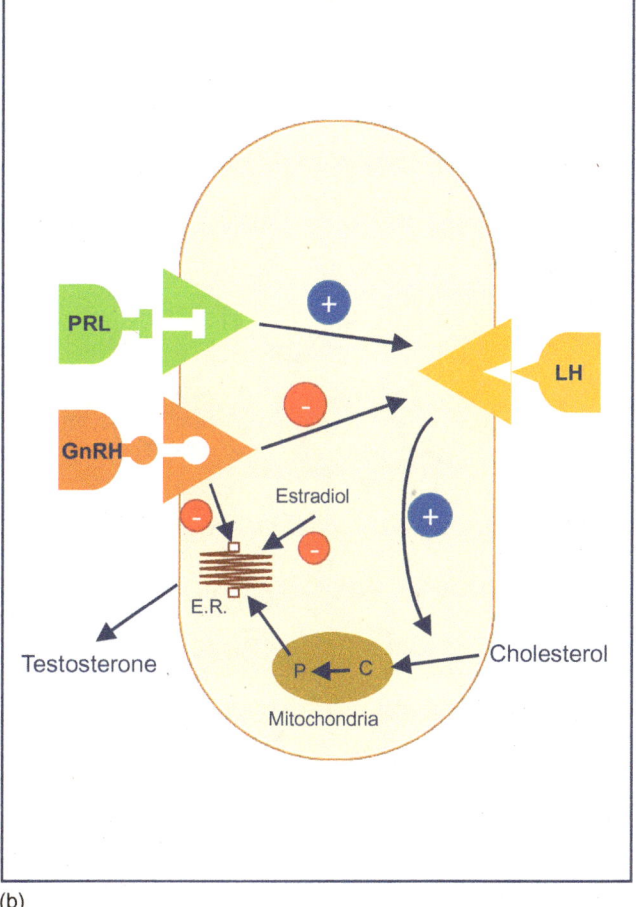

(a)   (b)

Figure 10.9 Testicular steroidogenesis and its hormonal regulation. Panel A shows the structures of cholesterol and testosterone, and steroidogenesis in Leydig cells along the delta 4 and delta 5 pathways. DHEA: dehydroepiandrosterone; HSD: hydroxysteroid dehydrogenase (3β and 17β); P450 scc, P450c17, and P450 arom: cytochrome P450 side chain cleavage, 17α-hydroxylase C17–C20 lyase, and aromatase, respectively; 5αRed: 5α reductase. Panel B shows interactions of three hormones: LH: luteinizing hormone; PRL: prolactin; and GnRH: gonadotropin-releasing hormone, in the regulation of steroidogenesis in Leydig cells.

is an oxidative cleavage of the cholesterol sidechain by mitochondrial cytochrome P450 oxidase (P450scc) with a loss of six carbon atoms to generate pregnenolone. In the next step, two additional carbon atoms are removed by the CYP17A enzyme in the endoplasmic reticulum to yield a variety of C_{19} steroids. In addition, the 3-hydroxyl group is oxidized by hydroxyl steroid dehydrogenase (3-β-HSD) to produce androstenedione. In the final step, the C-17 keto androstenedione is reduced by 17-β-HSD to yield testosterone.

Steroidogenesis involves bidirectional interactions between Sertoli and Leydig cells. The Sertoli cell cannot produce testosterone, but it has testosterone receptors and P450 aromatase and can convert testosterone, diffusing from the Leydig cells, into estrogen. In men, estradiol is produced in relatively small quantities but is important for modulating libido, erectile function, and spermatogenesis. Of the serum levels of estrogen in men, only about 20% come from the testes and the remainder is the result of aromatization of androgens in adipose tissue, skin, and bone.

LH regulates the expression of 17-β-HSD, which converts androstenedione to testosterone. The response of Leydig cells to binding of LH to its receptor is mediated by several signaling pathways, including a cAMP cascade and an inositol trisphosphate–diacylglycerol (PI3-DAG) molecule, as well as the arachidonic acid–prostaglandin pathways. LH is also involved in the control of Leydig cell development, as differentiation of Leydig cells fails to occur in the absence of LH. As illustrated in **Figure 10.9**B, modifiers of LH action in the control Leydig cell function include PRL, which at normal physiological concentrations exerts trophic effects on LH receptors and potentiates the effects of LH on steroidogenesis and spermatogenesis, and epidermal growth factor (EGF), which suppresses 17α-hydroxylase activity. In addition, GnRH, acting on local GnRH receptors, decreases the number of LH and PRL receptors

The testes also produce several protein hormones that include **activin**, **inhibin**, and **follistatin** [26]. Inhibins are released into the blood and suppress pituitary FSH secretion, whereas activins mostly exert their actions as local paracrine/autocrine growth factors. Follistatin is a potent activin binding protein, which also modulates local biological functions. Inhibin is a dimeric glycoprotein existing in two bioactive forms: A and B. Inhibin B shows temporal changes in its expression with the changing role of the Sertoli cell in immature and adult testes. In the adult, the levels of inhibin B are positively correlated with sperm number and spermatogenic status and are negatively correlated with serum FSH levels. Production of inhibin B is regulated by complex interactions between FSH, Sertoli cells, Leydig cells, and germ cells. Inhibin may also play a role at autocrine or paracrine levels in modulating the actions of activin. Other compounds such as opioids, GnRH-like peptides, vasopressin, oxytocin, and several growth factors have also been detected in the testes and appear to function as paracrine agents.

10.4.3 Involvement of dopamine with testicular functions

The testes can synthesize DA from tyrosine and also take up L-Dopa from the circulation and convert it to DA. TH and Dopa decarboxylase (DDC) were expressed in Leydig cells in adult and prenatal human [27], and rhesus monkeys [28] and throughout the testicular parenchyma and epididymis in pigs [29]. All types of sperm—spermatogonia, spermatids and spermatozoa—express D1R, D2R, and DA transporter (DAT). Confocal analysis by indirect immunofluorescence revealed that in non-capacitated spermatozoa of rat, mouse, bull, and human, D2R is mainly localized in the flagellum and is observed at lower levels in the acrosomal region of the sperm head [30]. The authors concluded that the presence of D2R in male germ cells implies new and unsuspected roles for DA signaling in testicular and sperm physiology.

There is only limited information on DA functions in spermatogenesis or steroidogenesis. In one study, chemical sympathectomy with 6-hydroxy-dopamine in rats provided some protection against the suppression of spermatogenesis in response to testicular injury [31]. In mice, psychostimulants such as cocaine, which blocks DAT, reduced the volume of the seminiferous tubules and lowered the number of spermatogonia [32]. These effects were associated with the down-regulation of D1R and D2R and the up-regulation of TH expression in Leydig cells, leading the authors to suggest that local DA plays a role in psychostimulant-induced testicular pathology.

To examine the *in vivo* effects of DA on testicular functions, 8-week-old rats were given a daily DA agonist in food for several weeks, followed by analysis of purified Leydig cells [33]. The treatment caused a decrease in hCG binding to Leydig cells due a reduction in the LH/hCG receptor numbers. Basal and LH-stimulated cAMP and testosterone production were reduced. Testosterone formation in response to dibutyryl cAMP, a cell-permeable cAMP, was also decreased, indicating additional lesions in the signal transduction pathway. The treatment also increased aromatase activity in the Leydig cells and thus the potential to produce estrogens. However, a direct addition of the same agonist to cultured Leydig cells did not inhibit cAMP production or testosterone production except at high concentrations. The authors concluded that treatment of rats with the DA agonist indirectly (i.e., via pituitary PRL) affected Leydig cell function, resulting in rapid decreases in LH receptors and cAMP and testosterone production. On the other hand, aromatase activity increased and with it the capacity to produce estrogens.

Collectively, in spite of the mounting evidence for local production of DA and the presence of its receptors in all constituents of the testis, including the sperm, it is unfortunate that so little effort has been invested to put in perspective the local actions of DA independent of PRL.

10.4.4 The male genital tract

The male genital tract transports sperm from the site of production to the exterior of the body, provides storage capacity for the sperm and assists in its maturation. The **epididymis** is a coiled segment of the spermatic ducts that stores spermatozoa while they mature and then transports them to the vas deferens, which connects to the urethra. The epididymis is composed of a head (caput), body (corpus), and tail (cauda). A significant portion of sperm maturation (i.e., the acquisition of motility and fertility) is carried out in the caput, while storage occurs in the cauda. Final sperm maturation (capacitation) is completed within the female reproductive tract, as discussed later in this chapter. The ampulla of the vas deferens serves as an accessory storage site for sperm. Heightened sexual activity and frequent ejaculation can result in the appearance of immature and immotile sperm in the ejaculate.

The **vas deferens** carries sperm from the epididymis to the ejaculatory ducts by peristalsis. During ejaculation, the sperm mixes with fluids from the **seminal vesicles**, **prostate gland**, and **bulbourethral glands**, which form the semen and provide an environment that promotes sperm survival and fertility. Semen contains only 10% sperm by volume, with the remainder consisting of the combined secretion of the accessory glands; seminal vesicles contribute about 60% of this fluid. The secretion of the seminal vesicles contains fructose (the main substrate for sperm glycolysis), ascorbic acid, and prostaglandins. Prostaglandins were first discovered in seminal fluid and were mistakenly considered the product of the prostate gland, and hence their name. The prostate produces polyamines, citric acid, cholesterol, fibrinolysis, and acid phosphatase, which are necessary for the coagulation and subsequent liquefaction of semen.

DA levels in seminal plasma from infertile men have been found to be lower than those from fertile men [34]. The decreases in DA were associated with lower serum LH and testosterone levels in all infertile groups, and with increases in serum PRL and FSH in oligozoospermic men. Whether the reduced DA levels in seminal plasma were the cause or the effects of infertility was unclear, however. In addition, in most *in vivo* studies that have manipulated the dopaminergic system, reproductive alterations were often attributed to changes in serum PRL levels by DA, underscoring the need for studying the direct effects of DA on male reproductive organs, independent of PRL.

Several studies support a role for local DA in testicular descent, in erection, and in ejaculation. Topical application of DA near the cremaster muscle in rats antagonized norepinephrine (NE)-induced muscular contraction [35]. In addition, chemical sympathectomy by 6-hydroxydopamine (6-OH-DA) during fetal life in rats resulted in inguinal testis by altering cremasteric muscle contractility [36]. This indicated that sympathetic activity, either through DA or NE, plays a role in testicular localization. DA was also effective in inducing contractions of isolated rat seminal vesicles, presumably being involved in the delivery of their secretory products to the seminal fluid [37].

Successful reproduction depends on the male's ability to achieve and maintain **penile erection**, enabling the delivery of sperm to the female reproductive tract. Penile erection begins with sensory and mental stimulation (arousal). The male sexual response reflects a dynamic balance between exciting and inhibiting forces of the autonomic nervous system within the penis and throughout the central nervous system (CNS). The sympathetic component tends to inhibit erections, whereas the parasympathetic system is one of several excitatory pathways.

Neural input to the penis, acting together with local factors, relaxes the muscles of the corpora cavernosa that surrounds the penis. This allows blood to flow into local blood vessels where it becomes entrapped, creating rigidity and an erection [38]. Vascular vasodilatation during erection is caused by nitric oxide (NO), which activates guanylate cyclase (GC), increased cGMP levels, and enhanced K^+ efflux into the vasculature. Phosphodiesterase 5 (PDE5) inhibitors such as Viagra, Cialis, and Levitra, stimulate erection because of their ability to prevent cGMP metabolism and sustain its elevated levels [39]. D1Rs and to a lesser extent D2Rs, are expected in the rat [40], rabbit [41], and human [42] corpora cavernosa. Application of the DA agonist apomorphine [42], or the D1R agonist fenoldopam [41], caused relaxation of corpus cavernosum strips from humans and rabbits, respectively. Although not specifically examined in these studies, we [43] and others [44] have shown that D1R activation increases cGMP levels in several tissues.

Erection and ejaculation during sexual intercourse are closely associated but are independently regulated [45]. Ejaculation represents the climax of the sexual cycle and comprises of emission (secretion of semen) and expulsion (propulsion of semen) phases. Various DAR are expressed in tissues that participate in ejaculation, including vas deferens [46], prostate [47], and seminal vesicles [40]. Although DA has been considered an important factor in the control of ejaculation [22], given the necessity to study ejaculation in live animals, which cannot be replicated *in vitro*, it has been difficult to specifically assign central vs. peripheral effects of DA on this process.

10.5 FEMALE REPRODUCTION: OVARIES AND GENITAL TRACT

10.5.1 Unique characteristics of female reproduction

The reproductive system of women, unlike that of men, is intermittent, and the production of mature germ cells, or ova, is not continuous. In particular, the process of meiosis in female germ cells extends over many years and is interrupted twice: first during fetal life, and then at the time of ovulation.

There is also a substantial difference among the sexes in terms of germ cell quantity. In contrast to the millions of spermatozoa produced daily by a man, only one or few ova are produced by a woman at any given time. In addition, the release of ova is not daily, but monthly, and is driven by cyclic alterations in the secretion of multiple hormones. These, in turn, cause profound cyclic changes in both structure and functions of the reproductive organs.

As shown in **Figure 10.10**, ovarian steroids exert both **negative and positive feedback** input to the hypothalamus and the pituitary, the combination of which results in the generation of the cyclic pattern of gonadotropin release that defines the female reproductive system. As discussed in more detail later in the chapter and is depicted in **Figure 10.13**B, the reproductive cycle is divided into several phases: **follicular**, **ovulatory**, and **luteal** phases that represent the combined influence of the pituitary and ovarian hormones. Given its dependency on a precise synchronization of multiple components, reproduction in the female is readily affected by stressful, environmental, psychological and social factors. The human reproductive cycle, also called the **menstrual cycle**, is characterized by menses, which is unique to humans, apes, and old-world monkeys. Menses, or menstrual bleeding, results from the withdrawal of hormonal support at the end of each cycle, leading to shedding of the superficial layer of the uterine lining. Menstrual cycles begin at puberty, are interrupted during pregnancy and lactation, and cease at menopause.

10.5.2 The female gonads: Oogenesis, folliculogenesis, and ovulation

The ovarian follicles constitute the basic female reproductive units, with each follicle containing a single ovum (egg) arrested in meiosis. **Oogenesis** is the process that produces mature ova from primordial germ cells. As illustrated in **Figure 10.11**A, the first few steps in oogenesis, up to the point of generating primary oocytes, occur prenatally. Because primary oocytes do not divide further, a woman is born with all of the oocytes that she will ever have during

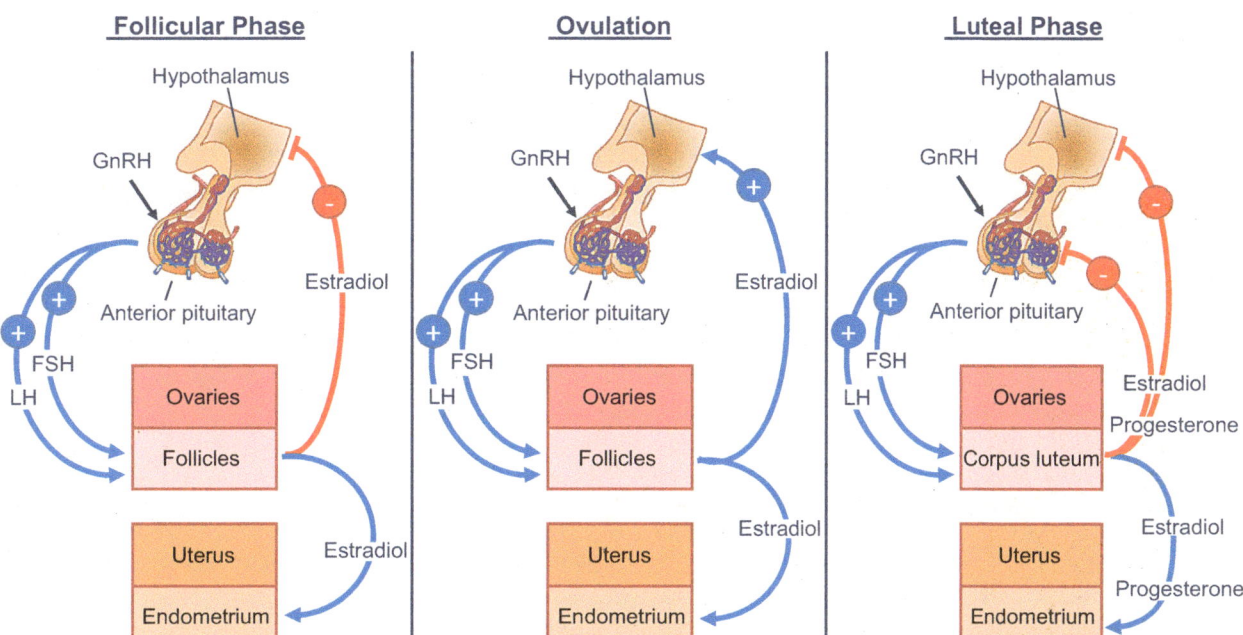

Figure 10.10 Differential regulation of reproduction during the follicular, ovulatory and luteal phases of the menstrual cycle. LH and FSH stimulate follicular growth and steroidogenesis during the follicular and ovulatory phases and support the corpus luteum during the luteal phase. Just before ovulation, the normal negative feedback by estradiol changes into a positive input on GnRH release. Estradiol promotes hyperplasia and hypertrophy of the endometrium during late follicular phase and ovulation, whereas both estradiol and progesterone prepare the endometrium for potential implantation after ovulation. See text for other explanations.

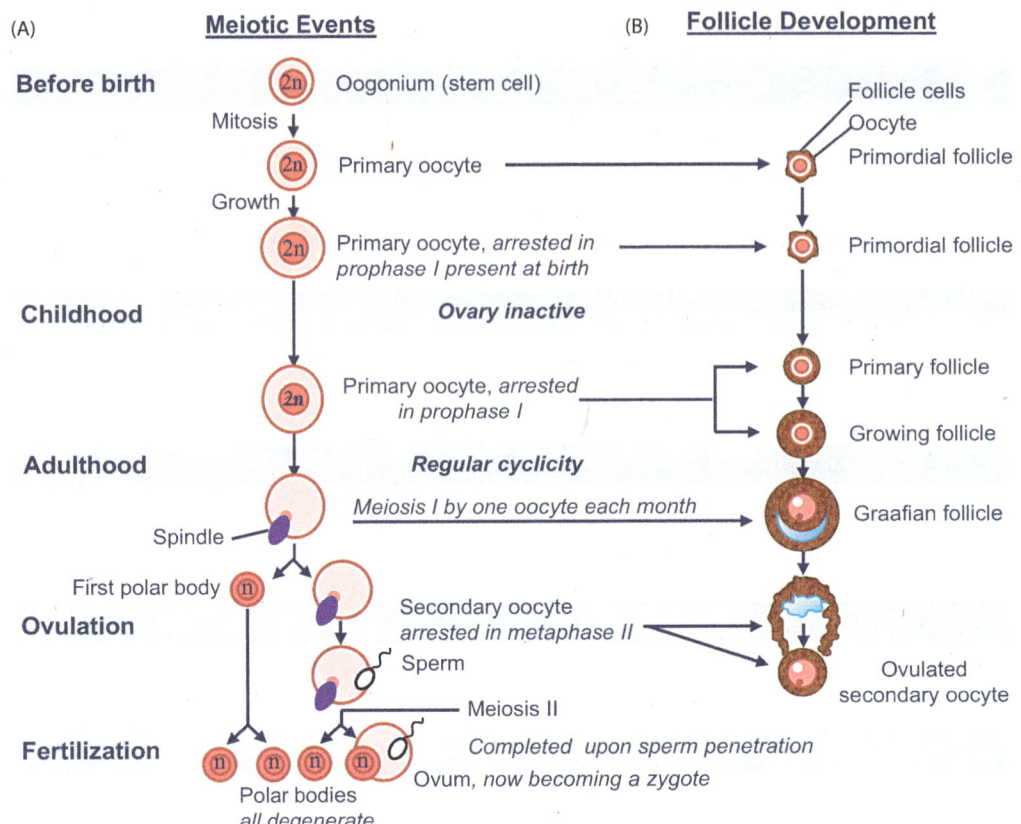

Figure 10.11 Diagrams of oogenesis (Panel A) and folliculogenesis (Panel B) throughout life. During fetal life, oogonia undergo mitosis and then begin the meiotic division, become enclosed in primordial follicles, enlarges in size, and remain arrested in prophase I. After birth and during childhood, the ovary is quiescent. Following puberty and throughout adulthood, menstrual cycles become regular and follicles grow under the effects of the gonadotropins. At ovulation, the oocyte within the Graafian follicle resumes meiotic division, becomes arrested in metaphase II, and is released. During fertilization, the meiotic division is complete, polar bodies form and degenerate, and the fertilized ovum becomes a zygote.

her life. Oocyte development occurs within a follicle, and oogenesis and folliculogenesis work in tandem (**Figure 10.11**B). Granulosa cells that surround the oocyte produce important growth factors and supporting substances that facilitate oocyte development and vice versa.

Oogenesis is a prolonged process, characterized by distinct nuclear and cytoplasmic transformations that occur at prescribed time points [48]. During the first 3–5 weeks of fetal life, about 2,000 primordial germ cells migrate from the yolk sac to the genital ridges, invade the cortex of the developing ovary, and start an active mitosis. By 8 weeks, their number reach ~600,000. Between 8 and 13 weeks, some oogonia cease mitosis and enter the first meiotic division, while others continue to proliferate. By 5–6 months, the fetal ovary contains ~6–7 million oocytes, by then all are classified as primary oocytes that are arrested in meiosis.

Survival of an oocyte in the fetal ovary depends on its encapsulation by granulosa cells. Oocytes that fail to become enclosed within primordial follicles undergo atresia and degenerate. At birth, there are about 1–2 million oocytes, and from then on, their number continues to decline with advancing age. At puberty, the ovary contains about 400,000 oocytes. During her fertile life, a woman will ovulate no more than few hundred mature ova. At the time of menopause, a woman usually has only few hundred to a few thousand oocytes left in the ovary and ovulation ceases.

Oogenesis can be divided into five stages. The first three occur prenatally, stage 4 occurs during ovulation, and stage 5 is at the time of fertilization. In **stage 1**, each primordial germ cell undergoes regular mitosis to produce two diploid oogonia. In **stage 2**, each oogonium undergoes mitosis to produce two diploid primary oocytes. In **stage 3**, each primary oocyte, enclosed in a primordial follicle consisting of a single layer of squamous follicular cells, starts to undergo the first meiotic division. This involves alignment of the homologous chromosomes, pairing and formation of chiasmata, crossing over, and exchange of genetic material. The oocyte then becomes arrested at prophase of the first meiotic division and remains in a genetically dormant state for many years to come.

Stage 4 occurs at the onset of puberty in response to the monthly LH surges. From the time of their formation to ovulation, primary oocytes grow from less than 20 μm in diameter to about 100–150 μm, becoming the largest cell in the human body. During this period, oocytes are not totally quiescent but, rather, synthesize several unique substances. These include cortical granules and components of the zona pellucida, a transparent glycoprotein layer that surrounds the plasma membrane of oocyte and is a vital constituent of its fertile capabilities.

The resumption of meiosis (oocyte maturation) is triggered by the **preovulatory LH surge**. The meiotic block is removed and the primary oocyte completes the first meiotic division. An asymmetric cytoplasmic division generates a large secondary oocyte and a small discarded cell, called the polar body. The secondary oocyte then enters meiosis II and becomes arrested at metaphase II. The resumption of meiosis at the time of ovulation is characterized by the dissolution of the nucleus, commonly referred to as germinal vesicle breakdown (GVBD). After GVBD, the chromatin is condensed into chromosomes, microtubules are organized into meiotic spindles, and the homologous chromosomes are separated to emit the first polar body. The post-GVBD maturational processes are controlled by histone modifications, centrosome proteins, various protein kinases/phosphatases, spindle checkpoints, and cytoskeletons.

A fully grown oocyte is tightly surrounded by compact layers of specialized granulosa cells (cumulus cells), forming a cumulus–oocyte complex. The process of meiosis resumption requires integration of endocrine, paracrine and autocrine signaling pathways involving interactions between the oocyte and the surrounding granulosa and cumulus cells. Ovulation occurs only once the first meiotic division has completed and a secondary oocyte has been formed within the dominant follicle. As discussed in **Section 10.6**, the completion of oogenesis (**stage 5**) occurs at the time of fertilization, when the secondary oocyte completes meiosis II and produces a very large, mature haploid ovum that unites with the sperm. The different stages of oogenesis and folliculogenesis are presented in Table 10.3 and illustrated in **Figure 10.11**.

At any given time, ovarian follicles are found at four states: resting, growing, degenerating, or ready to ovulate. The conversion from primordial

Table 10.3 Different stages in the development of ova and follicles			
Stage	Process	Ovum	Follicle
Fetal life	Migration, mitosis, first meiotic division begins	Primordial germ cells, oogonia, primary oocytes	Primordial follicle, primary follicle
Birth	Arrest in prophase, growth of oocyte and follicle	Primary oocyte	Primary follicle
Puberty	Follicular maturation	Primary oocyte	Secondary follicle
Cycle	Growth	Primary oocyte	Graafian follicle
Ovulation	Resumption of meiosis, emission of first polar body, arrest in metaphase	Secondary oocyte	Graafian follicle, corpus luteum
Fertilization	Second meiotic division complete, emission of second polar body	Zygote	Corpus luteum
Implantation	Mitotic division, blastocyst formation	Embryo	Corpus luteum

to primary follicles appears to be independent of the gonadotropins, while development beyond this step is gonadotropin-dependent, beginning at puberty and continuing in a cyclic manner throughout the reproductive years. A comprehensive review entitled "The normal menstrual cycle and the control of ovulation" is downloadable from *Endotext* at https://www.ncbi.nlm.nih.gov/books/NBK279054/.

During the early phase of the ovulatory cycle in sexually mature women, the rising FSH levels stimulate several dormant follicles to grow into the antral stage. Only the dominant, or Graafian follicle, will release an egg in response to the preovulatory LH surge, while the other developing follicles undergo atresia. Anti-Müllerian hormone, a product of granulosa cells, is believed to play a role in the selection of the dominant follicle. The atresia of nonselected follicles is caused by the suppression of pituitary FSH in response to increased estrogen release by the dominant follicle, and the accumulation of non-aromatizable androgenic compounds such as DHT.

The Graafian follicle has a cavity filled with follicular fluid (FF), which contains proteins, steroids, polysaccharides, metabolites, reactive oxygen species (ROS), and antioxidants. These molecules modulate oocyte maturation and protect the follicular cells from physical or oxidative damage. The egg is surrounded by granulosa cells and is enclosed by a basement membrane, theca interna and theca externa cells (**Figure 10.12**A). Theca cells provide androgen precursors to the granulosa cells, which produce estrogen by aromatization. During ovulation, an oocyte surrounded

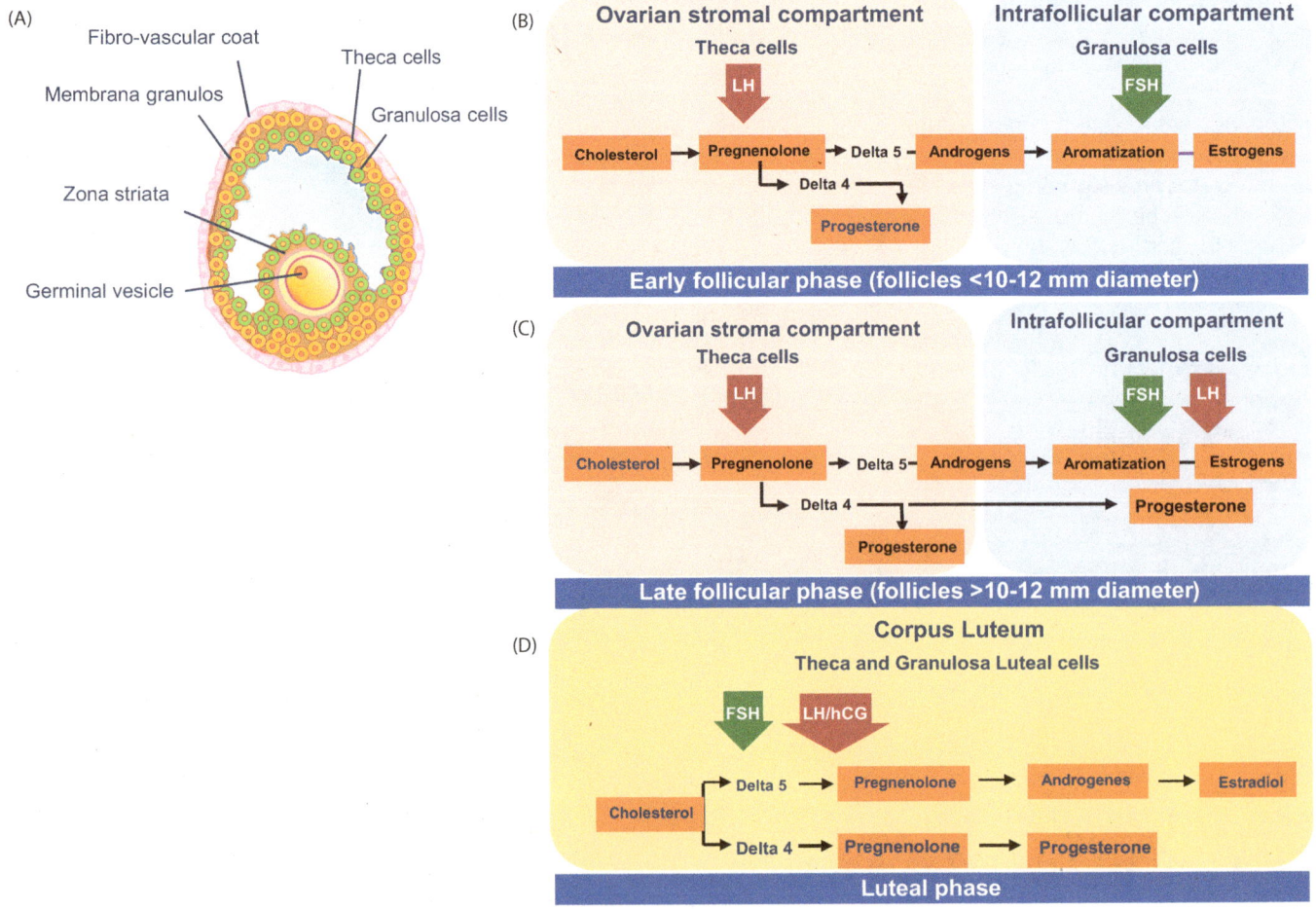

Figure 10.12 Structure of a Graafian follicle and the process of steroidogenesis. Panel A shows a Graafian follicle, where the ovum is surrounded by the various membranes, theca and granulosa cells. Panels B, C, and D depict the variable regulation of steroidogenesis in the different ovarian compartments during the early follicular, late follicular, and luteal phases, respectively. hCG: human chorionic gonadotropin. See text for other explanations.

by cumulus cells is released by the actions of proteolytic enzymes such as plasmin, collagenase, and metalloproteinases as well as by the decrease in gap junctions. The empty follicle is then transformed into an estrogen- and progesterone-producing **corpus luteum**.

Luteinization involves increased proliferation of capillaries and an irreversible differentiation of granulosa and theca cells to a luteal cell phenotype. A breakdown of the membrane that separates between theca and granulosa cells enables the mingling of the luteinized cells. The biochemical hallmark of luteinization is the capacity for progesterone biosynthesis in response to LH. If the egg is not fertilized, the corpus luteum degenerates after about 10 days and becomes scar tissue called the corpus albicans. If the egg is fertilized and uterine implantation ensues, the developing syncytiotrophoblast secretes **human chorionic gonadotropin (hCG)**, which signals the corpus luteum to survive throughout pregnancy. The corpus luteum of pregnancy continues to release progesterone, which maintains an appropriate endometrial milieu for embryonic development. More discussion on these processes is presented in the next sections.

10.5.3 The female gonads: Steroidogenesis

Ovarian steroids are critical for normal uterine function, for the establishment and maintenance of pregnancy, mammary gland development, and for the support of the female secondary sex characteristics. In most mammals (including humans) ovarian steroidogenesis occurs by a two-hormone/two-cell mechanism. Under this paradigm, the two gonadotropins, LH and FSH, and two cell types, theca and granulosa, work coordinately to produce the ovarian steroids. Theca cells, which are derived from the ovarian stroma, differentiate into endocrine cells under the influence of the growing dominant follicle (**Figure 10.12**A). The main steroidal product of the follicle is estrogen, while progesterone is the main product of the corpus luteum.

Steroidogenesis in the ovary proceeds along the same biosynthetic pathway as in the testes, but with some modifications, as shown in **Figure 10.12** panels B, C, and D. Cholesterol is either produced locally from acetate or is taken up from the circulation via low density lipoprotein (LDL) receptors. The conversion of cholesterol to pregnenolone by a side chain cleavage is a rate-limiting step that is regulated by the gonadotropins. LH acts on both cell types while the action of FSH is limited to the granulosa cells. Granulosa cells are deficient in the C-21 side chain cleavage enzyme and therefore cannot generate C-19 androgenic compounds but, rather, express the aromatase enzyme. Under LH regulation, theca cells produce androgenic substrates, primarily androstenedione, which diffuses into the granulosa cells and is converted to estradiol by aromatization. The Delta 4 pathway is used for the production of progesterone. Small amounts of androgens are also secreted by the ovary.

10.5.4 Involvement of dopamine with ovarian functions

DA affects both folliculogenesis and steroidogenesis. DA is detected in ovaries and FF from several species, including humans [49,50], monkeys [51], pigs [52], and rats [53]. In mouse ovaries, immunoreactive TH is detected around follicles, in theca cells, and in corpora lutea, but not in granulosa cells or oocytes [54]. In rhesus monkeys, the oocytes contain neither TH mRNA nor protein, indicating that they are unable to synthesize DA [51]. The oocytes, however, express a *DAT* gene identical to that found in the human brain. The relevance of DAT and dopamine β-hydroxylase (DBH) in the oocytes is indicated by the ability of isolated oocytes and granulosa cells to metabolize exogenous DA into NE. Isolated follicles containing oocytes, but not those from which the oocytes had been removed, respond to DA with increased cAMP levels.

Follicular fluid from human preovulatory follicles contain up to 200 ng/mL DA [50], about 400-fold higher than the basal serum free DA levels. In patients undergoing *in vitro* fertilization (IVF), DA levels are significantly higher in FF from cycles that result in pregnancy than in FF from samples not resulting in pregnancy [55]. Ovarian DA can act via D1R, D2R, D4R, or D5R, which are co-expressed in the follicle and corpus luteum, and are linked to different signaling pathways, suggesting a complex role of DA in the control of ovarian functions [56].

Polycystic ovary syndrome (PCOS) is the most common disorder of the ovaries, affecting 5%–10% of women of reproductive age. In PCOS, the follicles mature up to a certain stage, but then stop growing, fail to ovulate, and appear as cysts in the ovaries on an ultrasound scan. DA levels and DAT transcript levels were higher in granulosa cells derived from PCOS than in non-PCOS cells [57]. Moreover, DA significantly induces reactive oxygen species (ROS) in PCOS-derived granulosa cells and induces apoptosis. ROS play an important role in the process of ovarian senescence and in the pathogenesis of endometriosis and PCOS. Collectively, these results have identified DA as a factor that plays a role in human ovarian physiology and pathology. The full functional importance of DA-induced ROS in small follicles and other ovarian compartments, especially in PCOS samples, remains to be determined.

The regulation of **corpora lutea (CL)** functions by DA was examined using CL isolated from pseudopregnant rabbits [58]. From early to late pseudopregnancy, D1R was decreased while D3R was increased in both luteal cells and blood vessels; DA immunoreactivity was seen only in luteal cells. A D1R agonist increased the release of progesterone and prostaglandin E2 (PGE2] from early CL, whereas a D3R agonist decreased progesterone and increased PGF2α by mid and late CL. The authors concluded that the DA/DAR system has a dual modulatory function in the life span of CL: the DA/D1R is luteotropic, while the DA/D3R is luteolytic. These results shed new light on the mechanisms regulating luteal activity that might improve the ability to optimize reproductive efficiency in mammal species, including humans.

DA stimulated estradiol release from human granulosa cells [59] and raised intracellular calcium levels in the absence of extracellular calcium, an effect which was abolished by a D2R antagonist [56]. In addition, a 24-h treatment of human granulosa cells with DA caused a small, but a significant, enlargement of the viable cells. In pregnant mare's serum gonadotropin (PMSG)-treated rats, DA increased progesterone release via D1R in mixed ovarian cells [60]. The PMSG preparation had been extensively used in the past as an effective approach for stimulating follicular growth and estrogen production in immature rats. Cross-talk between D1R and the progesterone receptor (PR) has also been suggested, based on the report that DA increased the transcriptional activity of PR by promoting its translocation from the cytoplasm to the nucleus [61].

Direct effects of DA on aromatase expression was observed in neuronal progenitor cells from goldfish [62]. In this study, activation of D1R by SKF 38393 up-regulated aromatase B gene expression through phosphorylation of cAMP response element binding protein (CREB). This up-regulation was enhanced by low concentration of estradiol through increased expression of D1R and p-CREB protein. There are no similar reports on the action of DA on specific steroidal biosynthetic enzyme using mammalian ovarian constituents.

10.5.5 The female genital tract

As shown in **Figure 10.13**, the female genital tract is composed of the oviducts (Fallopian tubes), uterus, cervix, and vagina. As discussed in earlier sections, under the cyclic release of the gonadotropins and the ovarian steroids, the different components of the reproductive tract undergo profound structural and functional changes. The temporal relationships between plasma

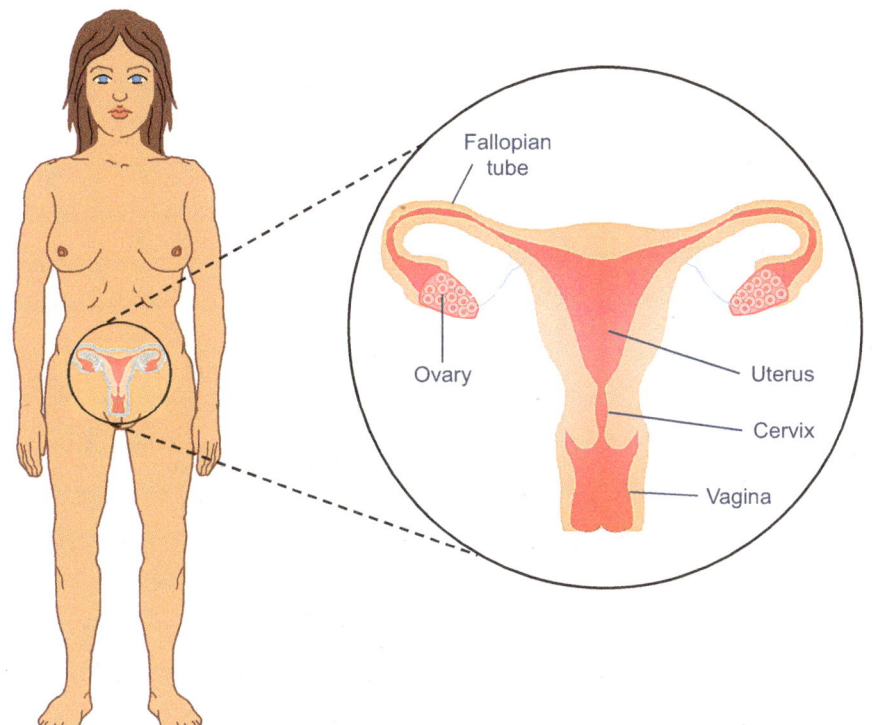

Figure 10.13 Anatomy of the woman's genitalia. Shown are the ovary, situated near the fimbria of the oviduct, which is connected to the uterus. The narrow cervix and wider vagina are located below the uterus.

hormone levels, uterine and ovarian cycles during the follicular, ovulatory, and luteal phases of the menstrual cycle are shown in **Figure 10.14**.

Human **oviducts** are long narrow tubes whose main role is to transport the egg from the ovary to the uterus, transport sperm from the vagina and cervix in the opposite direction, and provide a supporting environment for fertilization and early zygote development. The oviduct is not attached to the ovary. Rather, its funnel-shaped structure, the infundibulum, is equipped with finger-like projections, called fimbria, which collect the ovulated egg from the vicinity of the ovary. The ampulla, where fertilization normally occurs, is the widest part of the oviducts and is connected through a narrower isthmus to the uterine cavity. In cross section, the oviduct consists of an innermost ciliated and secretory epithelium, surrounded by outer circular and longitudinal muscles. Both ciliary movement and muscular peristalsis aid in egg and sperm transport.

Ectopic pregnancy is a complication of pregnancy where the zygote attaches to a site outside the uterus. Most ectopic pregnancies occur in the oviducts, but they can also attach to various abdominal structures. Signs and symptoms of ectopic pregnancy include abdominal pain and vaginal bleeding and can be diagnosed by ultrasound. Ectopic pregnancies are rare but serious because they can cause rupture of the oviduct and internal bleeding. They are treated by surgery or by cytotoxic drugs such as methotrexate.

DA biosynthetic enzymes have been measured in oviducts from pigs [63] and rabbits [64]. Catecholamines were also detected in the human oviduct at different stages of the menstrual cycle [65]. NE was the predominant catecholamine, and epinephrine (Epi) was barely detectable. DA levels were highest in the isthmus and were similar in the ampulla and fimbria. In all oviductal regions, DA level was lowest during the follicular phase, increased during ovulation, and increased further during the luteal phase. The cycle-associated changes raised the prospect that DA is involved in zygote transport to the site of implantation. In sow oviducts, which express D2R, DA depressed the contractile activity of ampule segments while causing relaxation in the isthmic region [66]. This suggested differential DAR expression along the oviduct and

Figure 10.14 Ovarian, uterine, and hormonal changes that occur during the menstrual cycle. During the follicular phase, the ovary is dominated by the growing follicles, the uterus begins proliferation, pituitary hormones start rising, and estrogen levels are elevated. During ovulation, an ovum is released from the Graafian follicle, the uterus undergoes significant cellular and metabolic changes, the gonadotropins surge, and estrogen levels decline. The luteal phase is characterized by corpus luteum formation, the secretory phase of the endometrium in preparation for implantation, low gonadotropin levels, and elevated serum estrogen and progesterone levels. FSH: follicle-stimulating hormone; LH: luteinizing hormone.

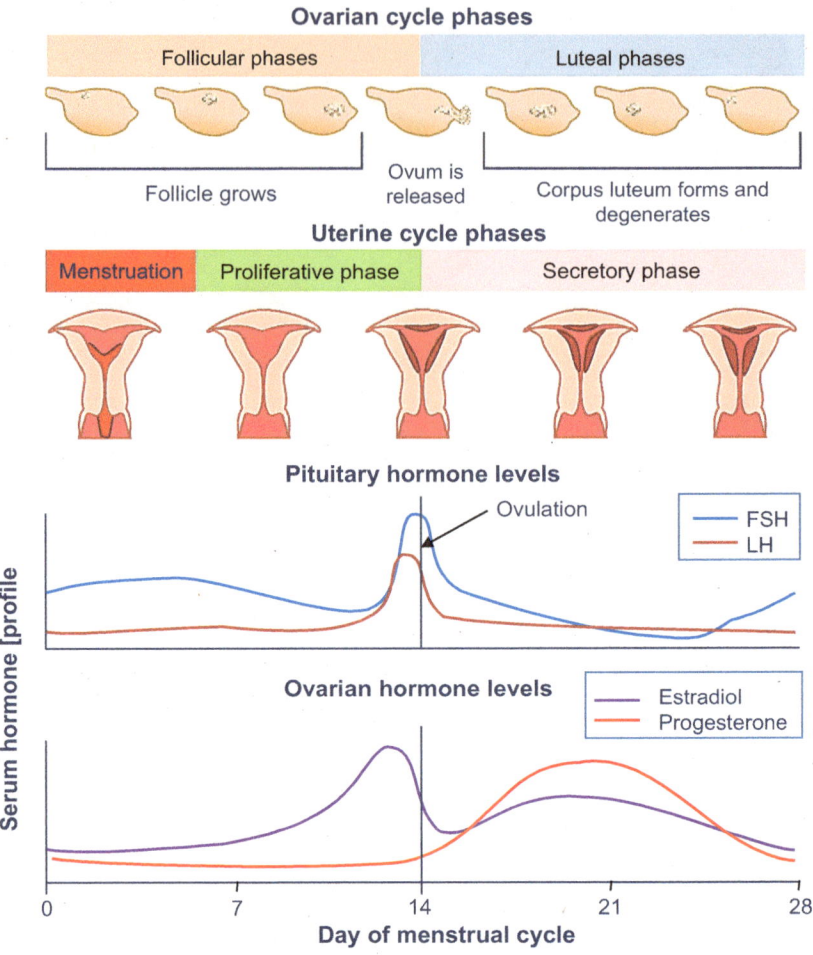

a role for DA in peristaltic activity, which coordinates the transport of both egg and sperm in the female reproductive tract.

The human **uterus** is a hollow muscular organ with two special properties—(1) great distensibility, enabling it to markedly expand to accommodate a full-term baby, and (2) a strong contractile capability for a forceful expulsion of the baby during delivery. The wall of the uterus is made of three layers: the outermost perimetrium, which protects the uterus from friction; next to it is the myometrium, which contains several muscle layers; and the innermost layer is the endometrium. The latter is made of a columnar epithelial tissue embedded with exocrine glands, and a highly vascularized connective tissue. Both endometrium and myometrium are sensitive to sex steroids and their structure and functions depend greatly on the hormonal milieu.

During the woman's ovulatory cycle, the **endometrium** undergoes four sequential phases: proliferative, secretory, ischemic, and menstrual. The rising levels of estrogen before ovulation stimulate hyperplasia and hypertrophy of the endometrium, cause elongation of the endometrial glands, and increase PR expression. The secretory phase begins on the day of ovulation. Under the combined actions of estradiol and progesterone, the endometrial glands store glycogen and secrete carbohydrate-rich mucus. The stroma increases in vascularity and becomes edematous, and the spiral arteries become tortuous. Peak secretory activity, edema formation and overall thickness of the endometrium are reached 6–8 days after ovulation in preparation for blastocyst implantation.

The ischemic phase is initiated by the declining levels of the steroids because of regression of the corpus luteum. There are necrotic changes in the mucosal layer, reduced blood supply resulting from constriction of the arteries, and infiltration by leukocytes and macrophages that phagocytize the ischemic tissue. Desquamation and sloughing of the entire functional layer of the endometrium occur during the menstrual phase. This is caused by the liberation of proteolytic enzymes and increased production of prostaglandins, which induce vasospasm and breakdown of the spiral arteries. Menstrual blood flow lasts 4–5 days, averages 30–50 mL, and does not clot because of the presence of fibrinolysin.

TH and DBH were found to be expressed in the guinea pig myometrium and endometrium, indicating local synthesis of DA and NE [67]. Studies with sheep uteri found large quantities of DA in the myometrium of virgin, pregnant and postpartum ewes [68], with DA levels decreasing during pregnancy. The DA/NE ratio in the sheep uterus was greater than that in the cervix and was not affected by the stage of pregnancy. The authors suggested that uterine innervation is supplied by "short" nerve fibers whose activity is regulated by the steroids of pregnancy.

Incubation of isolated uteri from DES-treated rats with DA produced dose-dependent relaxation in the K^+-depolarized uterus [69]. On a molar basis, DA was less potent than Epi in relaxing the uterus, but both compounds had a similar maximal relaxation. A study that focused on the dopaminergic system in the human uterus found expression of VMAT2, NET, but not DAT or serotonin receptor (SERT), in the endometrium throughout the menstrual cycle [70]. In another study [71], DA and morphine were potent stimuli for a transient surge of NO release from isolated human endometrial glandular cells; NO is an important intracellular and intercellular signal transduction pathway in the reproductive cycle.

Endometriosis is a growth of endometrial tissue outside the uterine cavity, affecting 6%–10% of women [72]. It is most common in women in their 30s and 40s and it is located primarily in the ovaries, oviducts, and their vicinity, and in rare cases in other parts of the body. Endometriosis is an estrogen-dependent disease, associated with chronic pelvic pain, dysmenorrhea, dyspareunia, and infertility. Treatment of endometriosis includes surgery, hormonal therapies, or their combination. Adverse effects of hormonal treatments limit their long-term use, and recurrence rates are high after cessation of therapy.

Because D2R was expressed in an ectopic human endometrium [73], the effects of DA agonists on animal models of endometriosis were examined [74]. Human uterine fragments were implanted in the peritoneal cavity of immune-deficient female mice and the mice were treated with the D2R agonist cabergoline. D2R activation decreased the number of active endometriotic lesions and suppressed cell proliferation. The effects were attributed to the suppression of angiogenesis via the reduced expression and phosphorylation of vascular endothelial growth factor receptor 2 (VEGFR-2). The authors concluded that these data support the testing of peripheral DA agonists as novel therapeutic approaches for peritoneal endometriosis in humans. A recent review on effective treatments of endometriosis listed D2R agonists among the most promising drugs [75].

The **cervix** in nonpregnant women is rather narrow, allowing passage of sperm in and menstruation out. Cervical mucus undergoes cyclic changes in composition and volume. During the follicular phase, estrogen increases the quantity, alkalinity, viscosity, and elasticity of the cervical mucus. Mucosal elasticity is greatest at ovulation, providing a suitable environment for sperm survival and transport. With the rising progesterone levels at the luteal phase, the quantity and elasticity of the mucus decline and it becomes thicker. At this time and throughout pregnancy, the cervical mucus provides protection against infection.

The **vagina** is an elastic, muscular tube that connects the cervix to the body exterior. The inner surface of the vagina is folded, providing greater elasticity and increased friction during sexual intercourse. Under the influence of estrogen, the vaginal epithelium proliferates. Basophilic cells predominate in the early follicular phase and then become cornified. During the postovulatory period, progesterone makes the vaginal mucus thicker, and the epithelium becomes infiltrated with leukocytes. Nerve fibers containing TH and DBH immunoreactivity were abundant in bovine [76] and equine [77] cervix and vagina. Yet, there are no reports on DA involvement in the functions of these structures during the reproductive cycle or throughout pregnancy.

10.6 FERTILIZATION, PREGNANCY, AND FETAL DEVELOPMENT

10.6.1 Overview of conception, pregnancy, and embryonic development

Figure 10.15 shows the oviduct, where fertilization and zygote formation take place, and the transport of the blastocyst toward the uterus, the site of implantation. Pregnancy begins with the union of a sperm and an egg. The life span of germ cells is rather short, necessitating their rapid transport to the ampulla of the oviduct. Immediately after fertilization, the zygote begins to divide, becoming a morula and then a blastocyst while still surrounded by the zona pellucida. Because the blastocyst contains only a limited energy supply, it must reach the uterus within a week. Implantation can occur only in a steroidal-primed, receptive uterus. Upon implantation, the embryo immediately produces and secretes hCG, which informs the woman that she is pregnant. The hormone extends the life of the corpus luteum and

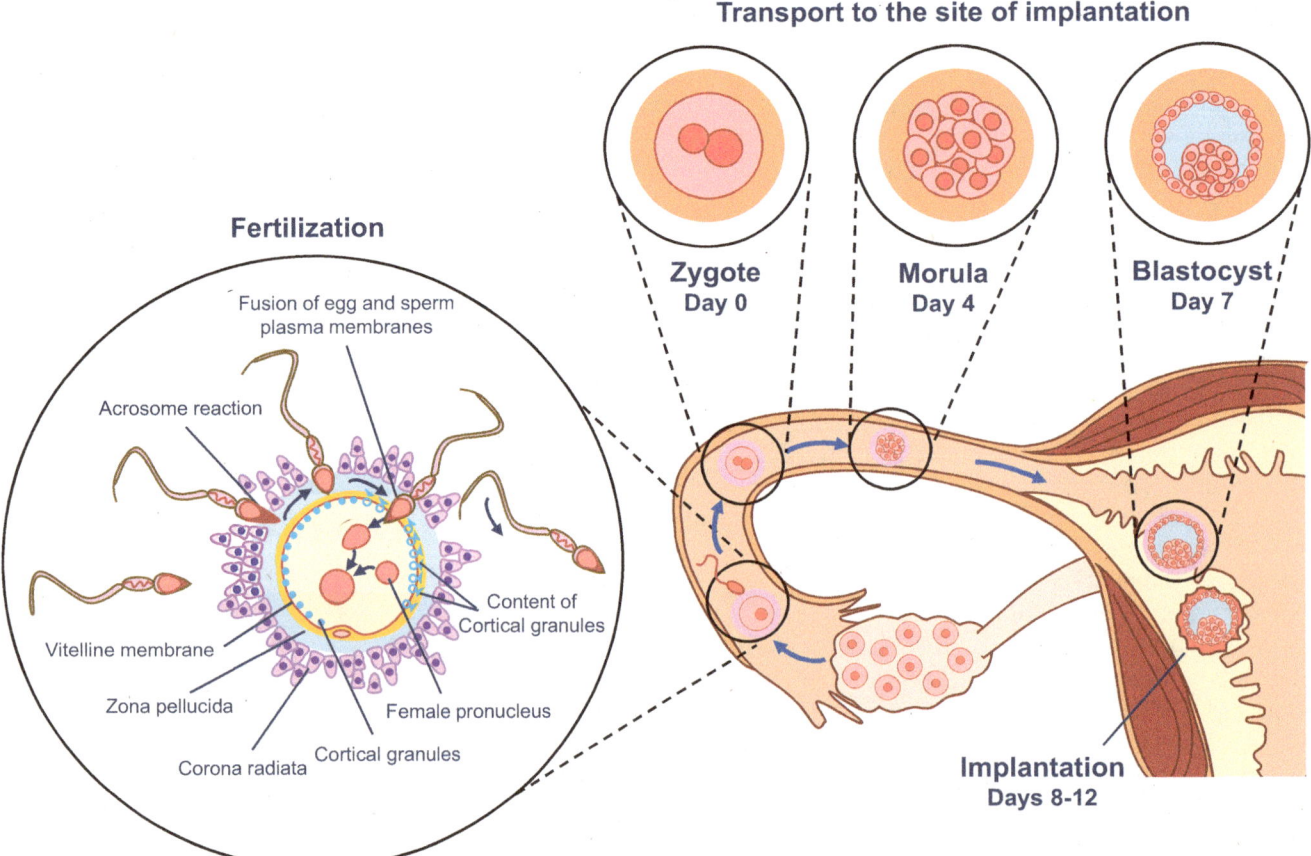

Figure 10.15 Timing of the cellular changes that occur in the oviduct following ovulation and fertilization. After fertilization (large inset) and zygote formation, the developing embryo undergoes an extensive mitosis that generates a morula within 4 days. This is followed by cellular differentiation and cavitation, which generate the blastocyst. Implantation in a primed endometrium occurs within 8–12 days after fertilization.

prevents onset of the next ovulatory cycle and menstruation. As detailed below, there is some involvement of DA in all these functions.

The placenta is a unique organ that exists only during pregnancy, where it serves as the lung, kidneys, and food supply of the fetus. The placenta produces protein and steroid hormones, thus duplicating the functions of both the pituitary gland and the gonads. Several of the fetal endocrine glands are active during embryonic development and participate in unique functions such as sexual differentiation. By bringing the maternal and fetal blood into contact, the placenta delivers maternal hormones and antibodies to the developing baby. Placental communication is bidirectional, and the fetus also sends messages back to its mother via this organ.

10.6.2 Fertilization

The natural course of **fertilization** in humans begins with the deposition of sperm-containing ejaculate in the woman's vagina. The volume of ejaculate in fertile men is 2–6 mL, containing 50–250 million spermatozoa/mL. Of these, only a few hundreds, usually spaced in time, will reach the ampulla of the oviduct as rapidly as 20–30 min after coitus. The innate motility of sperm alone cannot account for its fast transport within the woman's genital tract. Rather, sperm is propelled by muscular contractions of the vagina, cervix and uterus and by ciliary movement, peristaltic activity and fluid flow in the oviduct. Major losses of sperm occur in the vagina (due to acidity) and the uterine lumen (due to phagocytosis by leukocytes). Sperm can be stored for several hours in the cervical crypts, where it forms a pool for slow release. Although sperm may remain motile for as long as 4 days in the woman's genital tract, their fertilizing capacity is limited to 1–2 days. The sperm of most mammals can be cryopreserved for years, provided that agents such as glycerol are included in the medium to prevent crystal formation during freezing.

The intricate process of fertilization can be divided into four stages: (1) sperm preparation, (2) sperm-egg recognition, (3) sperm–egg fusion, and (4) zygote formation. During their passage within the female genital tract, spermatozoa must undergo a series of changes before they can fertilize an egg. The first step is capacitation, which is a prerequisite for the next step of acrosome reaction. Capacitation includes destabilization of the acrosomal membrane and changes in the tail that increase sperm mobility [78]. Capacitation normally occurs inside the female genital tract but can be achieved *in vitro* by sperm culturing in a defined medium.

Incubation of boar sperm, which expresses D2R, with DA or bromocriptine increased sperm viability and motility via Akt phosphorylation [79]. Immunofluorescence analysis showed a dynamic D2R expression in sperm, which was related to the stage of capacitation, and was accompanied by expression of many phosphorylated proteins in the acrosome and mid-piece regions. These data demonstrated that DA within a physiological range supports sperm capacitation and motility. The authors speculated that during transport in the female genital tract, DA is available to the sperm from both semen and oviductal fluid.

ZP3, a glycoprotein in the zona pellucida (ZP), serves as a sperm recognition site. Species-selective sperm "receptors" account, in part, for the failure of fertilization between germ cells from unrelated species. The acrosome reaction is triggered by contact of the sperm with the egg, enabling sperm penetration into the ZP and its fusion with the oocyte's membrane [80]. The acrosome reaction involves calcium-dependent release of hyaluronidase and exposure of masked membrane domains in the sperm's head. In one study, heat-solubilized human ZP was used to study the induction of the acrosome reaction in capacitated human spermatozoa [81]. Pimozide, a D2R antagonist, prevented the activation of voltage-operated calcium channels (VOCCs), which mediate acrosomal exocytosis in response to contact with the ZP. In contrast, L-type VOCCs inhibitors such as nifedipine and verapamil failed to inhibit the ZP-mediated acrosome reaction.

After digesting a hole in the ZP, the sperm head enters the egg's parenchyma and the membranes of the two cells fuse. The next critical step is prevention of **polyspermy**, given that polyploidy is not compatible with life. Within seconds of contact with a sperm, Na+ channels in the egg's membrane are activated, and a rushed entry of sodium depolarizes the egg's resting potential of –70 mV. The fast, although transient, electrical block to polyspermy is followed within 10–15 min by the cortical reaction. The lysosome-like cortical granules are released from the egg's membrane, destroying the oocyte's sperm receptors, and modifying the extracellular matrix around the egg, which becomes impenetrable to additional sperm entry. Although there are several publications on DAR expression and putative functions of DA in *Drosophila*, *Xenopus*, and Zebrafish oocytes, there are no comparable records on mammalian oocytes.

The oocyte, which because of ovulation has been arrested in metaphase of meiosis II, responds to sperm entry by completing meiosis. This leads to an unequal division of the cytoplasm and the generation of a large ovum, and a very small second cell, the polar body. The egg now includes two haploid nuclei, referred to as pronuclei, that are derived from the sperm and oocyte (**Figure 10.15**). The pronuclei decondense, expand, and replicate their DNA. Upon migration toward each other, the nuclear envelopes of the pronuclei disintegrate, and male- and female-derived genetic material intermingles. This step completes the process of fertilization, yielding a single-celled, diploid zygote endowed with all the genetic instructions needed for developing into a human being.

***In vitro* fertilization (IVF)** is the process of fertilizing an egg by sperm outside the body. In 1978, Louise Brown was the first child born after her mother received a natural cycle IVF treatment. In recognition of this accomplishment, Dr. Robert Edwards, the developer of this technique, was awarded the Nobel Prize in Physiology and Medicine in 2010. Since the 1980s, IVF has been practiced worldwide. In 2018 alone, over 8 million children were born using IVF and other assisted reproduction techniques. Patients as old as 65 have benefited from IVF.

In preparation for IVF, ovulation in the donor woman is stimulated or hyperstimulated with gonadotropins, and the ovulatory cycle is monitored. Using ultrasound-guided transvaginal approach, 5–20 eggs can be retrieved from antral follicles. Eggs are stripped of cumulus cells and may be subjected to chromosomal evaluation. Concurrently, sperm washing is used to remove inactive cells and seminal fluid from a sample of semen. Eggs and sperm cells are then co-incubated for 4–8 h. In cases of low sperm count or motility, a single sperm can be injected directly into an egg, using an intracytoplasmic sperm injection method.

Fertilized eggs showing two pronuclei are incubated in a special medium for about 48 h when the zygotes consist of six to eight cells. After additional cell divisions, embryos are transferred to the patient's uterus by a thin transvaginal catheter. The number of embryos to be transferred depends on the number available, the age of the woman, and other health and diagnostic factors. Either freshly prepared or thawed frozen embryos can be transferred. Luteal support for the mother is then provided by treatment with progestins, hCG, or GnRH agonists, and may be accompanied by estradiol. The success rate of IVF (i.e., the percentage of IVF cycles resulting in a live birth) is from 5% to 40%, depending on the maternal age, cause of infertility, embryo status, reproductive history, lifestyle factors, and proficiency of the IVF clinic.

A major complication of IVF is a risk of multiple births, which is directly related to the practice of implanting multiple embryos. Another risk factor is the development of ovarian hyperstimulation syndrome (OHSS), especially if hCG is used to induce final follicular maturation. In spite of some early reports that IVF was also associated with increased birth defects in infants, these claims were not confirmed in subsequent large epidemiological surveys. See **Chapter 7**, **Section 7.5.2**, for additional discussion on OHSS.

It has been reported that DA agonists reduce the extravasation of fluids caused by OHSS by minimizing the phosphorylation of VEGF-2 [82]. The DA agonists cabergoline or bromocriptine have been used by some clinics as an adjuvant to IVF treatment primarily because of their effectiveness to reduce OHSS but also because of their positive effects on the microenvironment of the follicular fluid. A recent survey found that on univariate analysis, the DA agonists showed beneficial effects of increased pregnancy rates, live births and reduced pregnancy losses, but these became nonsignificant after logistic regression and inclusion of additional patients [83]. The authors concluded that any association between DA agonists and the success of embryo transfers merits further exploration of the potential benefits.

Birth control methods to prevent pregnancy include physical barriers, hormone contraceptives, and surgical ablation methods, which vary in efficacy, side effects, and permanence. Physical barrier methods include intrauterine devices (IUDs) that can be impregnated with copper or progestin, sponges, cervical caps, vaginal rings, and diaphragms and also spermicides for women and condoms for men. Hormonal contraceptives include pills that contain various combinations of estrogen and progesterone, intradermal implantable progesterone rods, injectable Depo-Provera (medroxy-progesterone acetate), and estrogen/progestin-containing skin patches. Emergency contraceptives (also known as the "morning-after pill") refer to the use of drugs or devices as an emergency measure to prevent unwanted pregnancy. These include levonorgestrel (Plan B One-Step), a progestin only pill that prevents ovulation. Surgical ablation methods include tubal ligation or sterilization in women and vasectomy for men. Vasectomy, or the disconnection of the vas deferens in the scrotal area, is an effective method of male contraception. Because of sperm storage in the ampulla, men remain fertile for 4–5 weeks after vasectomy.

10.6.3 Blastocyst transport and implantation

After fertilization, the zygote undergoes several mitotic divisions and forms a solid ball, the morula, which travels down the oviduct toward the uterus (**Figure 10.15**). Upon reaching the 20–30 cell stage, the morula starts a process of differentiation and cavitation to form a blastocyst. The blastocyst, still surrounded by the ZP, is composed of three regions: (1) an inner cell mass, which will develop into the embryo; (2) a surrounding outer layer called the trophoblast, which will develop into the placenta; and (3) a fluid-filled cavity called the blastocoele. DA-producing pluripotent cells isolated from the inner cell mass of preimplantation monkey blastocysts provided evidence for a very early expression of DA, even before implantation [84]. Yet, what functions are fulfilled by this DA in implantation and/or in growth of the embryo, remain to be determined.

Within 8–12 days after fertilization, the blastocyst implants unto the endometrial lining of the uterus (**Figure 10.15**). To enable implantation, the endometrium must undergo decidualization (i.e., increased thickness, vascularization, and secretory activity), driven by the rising estrogen and progesterone levels. The blastocyst first breaches the surrounding coat that prevented its attachment to the oviduct during transport. It then releases digestive enzymes that degrade the endometrial lining and it also releases hormones (gonadotropins and steroids), which together with support by maternal hormones, facilitate its embedding within the uterine wall.

The human **decidua** represents the maternal part of the placenta. It is composed of the decidua basalis, which interacts with the trophoblast; the decidua capsularis, which grows over the embryo on the luminal side;

and the decidua parietalis, which fuses with the decidua capsularis by the fourth month of gestation. The placenta connects the developing fetus to the uterus to allow nutrient uptake, thermoregulation, waste elimination, and gas exchange via the maternal blood supply. It also protects the fetus from infection and immune rejection, and produces hormones that support the pregnancy. The human placenta is classified as hemochorial because the chorionic epithelium is in direct contact with the maternal blood. The placenta is one of the organs with the highest evolutionary diversity among animal species, and no animal models exactly represents human placentation [85].

10.6.4 Endocrine functions of the placenta

Progesterone production for the first 7–8 weeks of gestation is carried out by the corpus luteum. From then on, steroidogenesis is taken over by the placenta, an active, though an incomplete, steroid-producing organ. Production of estrogens (estradiol, estrone, and estriol) during gestation requires a close cooperation between the maternal, placental, and fetal compartments, which together have been termed the **fetoplacental unit**. Because the placenta lacks the p450c17 enzyme for converting pregnenolone to androgens, it utilizes androgenic precursors such as dehydroepiandrosterone sulfate (DHEAS), which is made by the very active fetal zone of the adrenal in the fetus. Conjugation of androgenic precursors to sulfate ensures greater water solubility and also reduces their bioactivity while within the fetal circulation. The placenta has an active sulfatase for deconjugation of sulfated steroids and an aromatase for producing estrogenic compounds. Pregnenolone sulfate is provided to the placenta from the maternal compartment, and after deconjugation, it is converted to progesterone.

The maternal endocrine system undergoes significant changes during pregnancy. The profile of several important hormones in the maternal plasma during gestation is shown in **Figure 10.16**A. Of total plasma estrogens, the most abundant is estriol, and the least abundant is estradiol. Human placental lactogen (hPL), also called human chorionic somatomammotropin (hCS), is produced by the syncytiotrophoblast and has both growth hormone (GH)-like and PRL-like activities [86]. hCG, made by the trophoblast, maintains progesterone production by the corpus luteum during the first trimester, until the placenta takes over progesterone production by 10 weeks of gestation and its levels are then reduced. The detection of hCG in the woman's urine is used as an early test for pregnancy.

As discussed in our previous review [25], the pattern of PRL secretion during human pregnancy involves three independently regulated compartments: **maternal**, **fetal**, **and decidual**. Maternal serum PRL levels start rising at 6–8 weeks of gestation and progressively increase to reach 200–300 ng/mL at term. Concurrently, the pituitary gland enlarges due to increases in lactotroph size and number. Increased PRL release and lactotroph hyperplasia are driven by estrogens, which suppress the hypothalamic DA and stimulate lactotroph proliferation. PRL begins to rise in the fetal circulation at 20–24 weeks and increases steeply from week 30 to term, when it reaches levels similar to those in the maternal serum. The fetal PRL rise is autonomous given that there is no evidence for PRL transfer from mother to fetus and vice versa. Despite the profound changes in PRL in fetal serum, there is little knowledge of its importance in fetal physiology. Unique to humans, the decidua produces very large amounts of PRL, which accumulates in the amniotic fluid (AF) and attains peak levels as high as 4,000–5,000 ng/mL between 16 and 22 weeks of gestation, reducing to 400–500 ng/mL at term. Whereas much is known about PRL production by the human decidua, little is understood about the exact functions of decidual or AF PRL.

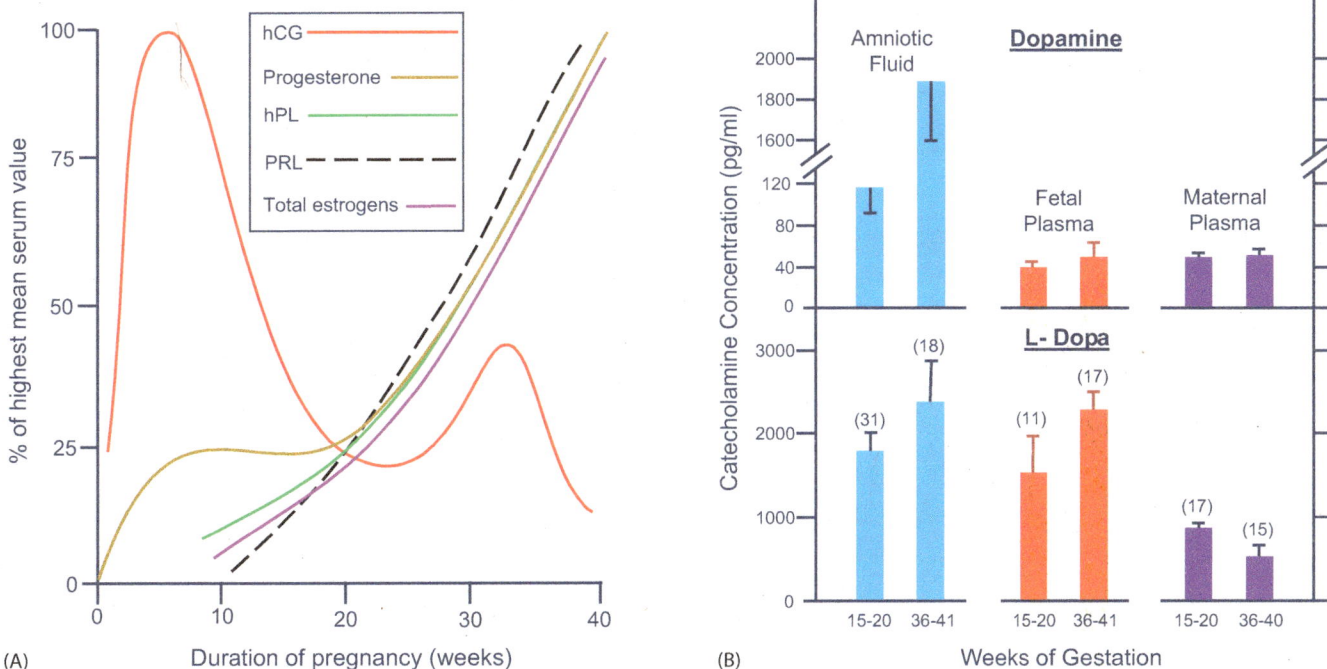

Figure 10.16 Hormonal changes during human pregnancy. Panel A shows the profile of protein and steroid hormones in the maternal plasma throughout human pregnancy. Panel B depicts changes in dopamine and L-Dopa levels in amniotic fluid, fetal and maternal plasma that occur from early (weeks 15–20) to late (weeks 36–41) pregnancy. hCG: human chorionic gonadotropin; hPL: human placental lactogen; PRL: prolactin.

D2R expression in the human placenta show a decrease at 9–16 weeks of gestation, a return to baseline levels at 17–18 weeks, a secondary increase at week 19 of gestation, and maximal values at term [87]. *In situ* hybridization has revealed expression of D2R in cytotrophoblasts, syncytial trophoblasts, vascular endothelial cells, Hafbauer cells, and fibroblasts in the chorionic villi [88]. There are no published records on expression of DAR other than D2R in the human placenta.

Catecholamine levels have been analyzed in rat, rabbit, sheep, and guinea pig uteri before and during pregnancy [89]. NE was the major uterine catecholamine and its levels were reduced during pregnancy in all species examined. The DA/dihydroxyphenylacetic acid (DOPAC) ratio in the guinea pig uterus fell dramatically with pregnancy, indicating increased DA release and/or catabolism. The different DA/NE ratios suggested that DA is stored with NE in adrenergic neurons in the guinea pig myometrium, but within other neuronal or cellular stores in the sheep uterus. Whether the precipitous fall in uterine DA before the end of gestation designates its involvement in keeping a quiescent uterus prior to parturition remains to be determined.

10.6.5 L-Dopa and dopamine levels in different compartments of pregnancy

The growing fetus is surrounded by **amniotic fluid (AF)**, a dynamic milieu that changes in volume and constituents as pregnancy progresses. The AF is generated from the maternal plasma and passes through the fetal membranes by osmotic and hydrostatic forces. When the fetal kidneys begin to function at week 16, fetal urine also contributes to this fluid. The AF is absorbed through fetal tissues and skin. After the 18th–25th week of pregnancy, when keratinization of the embryo's skin takes place, the fluid is primarily absorbed by the fetal gut after swallowing. AF contains nutrients and growth factors that facilitate fetal growth, provide mechanical cushioning, antimicrobial effectors that protect the fetus, and shed fetal cells [90]. **Amniocentesis**, or sampling of AF, is a widely practiced method which allows assessments of fetal gender, maturity, genetic background, and disease conditions.

Catecholamine levels during gestation in fetal and maternal plasma and the AF have been measured in rats [91,92], sheep [93], and humans [94,95]. In our human study [95], AF, fetal, and maternal blood were obtained from two groups of pregnant women with uncomplicated pregnancies: one at 15–20 weeks of gestation, and another during labor at 36–41 weeks of gestation. As shown in **Figure 10.16**B, L-Dopa constituted 80% of the total circulating fetal catecholamines, and its levels were two- to three-fold higher in fetal than in maternal plasma. No changes in L-Dopa or DA concentrations occurred in either fetal or maternal plasma from mid- to late gestation. On the other hand, DA in the AF was quite high and increased 10- to 15-fold toward the end of gestation. DA in the AF was found to be biologically active, as was judged by its ability to inhibit pituitary PRL secretion. However, unlike its action on pituitary PRL, DA did not inhibit decidual PRL release *in vitro*, suggesting that the decidua is not one of its target tissues, explaining the co-existence of elevated levels of both PRL and DA in this compartment. Marked increases in both NE and Epi were also seen in human fetal plasma, but no changes in maternal plasma. NE in the AF increases three- to four-fold while Epi remains undetectable.

Another group, using AF obtained by vaginal amniocentesis from primiparous women, reported that DA was the main catecholamine in the AF, where its concentrations increased markedly before parturition [96]. L-Dopa was not measured in this study. The authors concluded that DA represents a fetal signal of facilitated intrauterine prostaglandin production and the eventual onset of labor.

In addition to suppressing maternal pituitary PRL, DA inhibits hPL release from trophoblastic cells, an action that is mediated by inhibiting calcium influx through D2R [97]. hPL is secreted into both the maternal and fetal circulations after the sixth week of pregnancy. In the mother, hPL stimulates insulin-like growth factor (IGF) production and modulates intermediary metabolism, increasing glucose and amino acid availability to the fetus. hPL also affects changes in the breast in preparation for lactation. In the fetus, hPL modulates embryonic development, regulates intermediary metabolism, and stimulates the production of IGFs, insulin, adrenocortical hormones, and pulmonary surfactant. Although a unique receptor for hPL has been sought, it has never been identified and the hormone appears to act by binding to the prolactin receptor (PRLR).

During late gestation in the rat, L-Dopa has been found to constitute 50% of total fetal plasma catecholamines and be significantly higher in fetal than in maternal plasma [92]. Although DA in fetal plasma was 10-fold lower than L-dopa, it was significantly higher in fetal than maternal plasma; NE levels were similar in both. Maternal plasma Epi levels remained relatively constant, whereas fetal plasma Epi increased 50-fold from day 17 to day 22. L-Dopa concentrations in the rat AF fluid were 10-fold higher than those of DA, and both increased markedly during the last 2 days of gestation. We concluded that similar to humans, L-Dopa is the predominant catecholamine in fetal plasma and the AF during gestation in the rat.

We also used chronically cannulated sheep fetuses to examine for catecholamine levels during gestation [93]. L-Dopa concentrations in the fetal circulation were 10–25 times higher than DA, 5–10 times higher than NE, and 100 times higher than Epi. Dopa was the only catecholamine that was significantly higher in fetal than in maternal plasma. These data indicated that Dopa is the predominant circulating catecholamine in the sheep fetus and may represent a delayed maturation of DDC. Although the physiological importance of this observation is unknown, fetal L-Dopa likely serves as the source of DA to fetal tissues that express DDC.

Collectively, both the origin and functions of elevated L-Dopa in different compartments of pregnancy in several species remain enigmatic and raise the following questions: (1) Is L-Dopa a critical factor for the maintenance and/or progression of successful pregnancy? (2) Does L-Dopa play an important role in fetal development or in parturition? (3) Must L-Dopa be converted to DA to exert its actions, or does it have a unique function on its own during pregnancy? Given the complexity of maternal–fetal relationship, the inability to manipulate individual pregnancy compartments, or to reduce only L-Dopa without altering the synthesis of other catecholamines, the above questions may remain unresolved for a long time.

10.6.6 Development of the fetal reproductive organs

The specific roles of genetics vs. hormones in sexual differentiation of the brain is covered in **Section 10.2** and **Figure 10.1**. Here the focus is on the development and differentiation of the male and female internal and external genitalia. Whether possessing an XX or and XY karyotype, every human embryo goes initially through an ambisexual stage with the potential to acquire either masculine or feminine characteristics. As illustrated in **Figure 10.17**A, a 4- to 6-week-old human embryo has indifferent gonads, a pair of Wolffian ducts capable of forming male internal genitalia, a pair of Müllerian ducts serving as the anlage of the internal female genitalia, and bipotential structures of the external genitalia (**Figure 10.17**B). Depending on the genetic program, the inner medullary tissue of the indifferent gonad will become the testicular components (in response to SRY), while the outer cortical tissue will develop into an ovary in the absence of SRY. Accordingly, the primordial germ cells will become spermatogonia or oogonia.

In an XY fetus, the testes differentiate first under the influence of SRY. During the 6–8 weeks of gestation, the seminiferous tubules become distinguishable and the Leydig cells start producing testosterone. The Sertoli cells produce two nonsteroidal compounds: **Müllerian inhibiting factor (MIF)**, a large glycoprotein which inhibits cell division of the Müllerian ducts, and **androgen binding protein (ABP)**, which binds testosterone. Under the influence of testosterone, the Wolffian ducts give rise to the epididymis, vas deferens, seminal vesicles and ejaculatory duct while the Müllerian duct becomes vestigial (**Figure 10.17**A). Development of the ovary, which has a passive role, begins only in weeks 9–10 of gestation. Without androgen in the normal XX female fetus, the Müllerian ducts fuse in the midline and develop into the oviducts, uterus, cervix and the upper portion of the vagina, while the Wolffian ducts regress.

The primordial external genitalia include the genital tubercle, genital swelling, urethral folds, and urogenital sinus (**Figure 10.17**B). Differentiation of the external genitalia also occurs between 8 and 12 weeks of gestation and is determined by the presence or absence of male sex hormones. Notably, differentiation along the male line requires an active 5α-reductase that converts testosterone to DHT. Without DHT, regardless of the genetic, gonadal or hormonal sex, the external genitalia develop along the female pattern. The penis, penile urethra and scrotum develop from the genital tubercle, urogenital folds and genital swelling, respectively, while the clitoris, labia majora and labia minora develop from these primordial structures in the female. The lower vagina develop from the vaginal cord. Androgen-dependent differentiation occurs only during fetal life and is thereafter irreversible. However, exposure of females to high androgens either before or after birth can cause clitoral hypertrophy. Testicular descent into the scrotum, which occurs during the third trimester, is also controlled by androgens. There are many disorders of sexual development that can be manifested either before or after birth. Although of great interest, these are beyond the scope of this book. The interested reader is referred to a concise review [98].

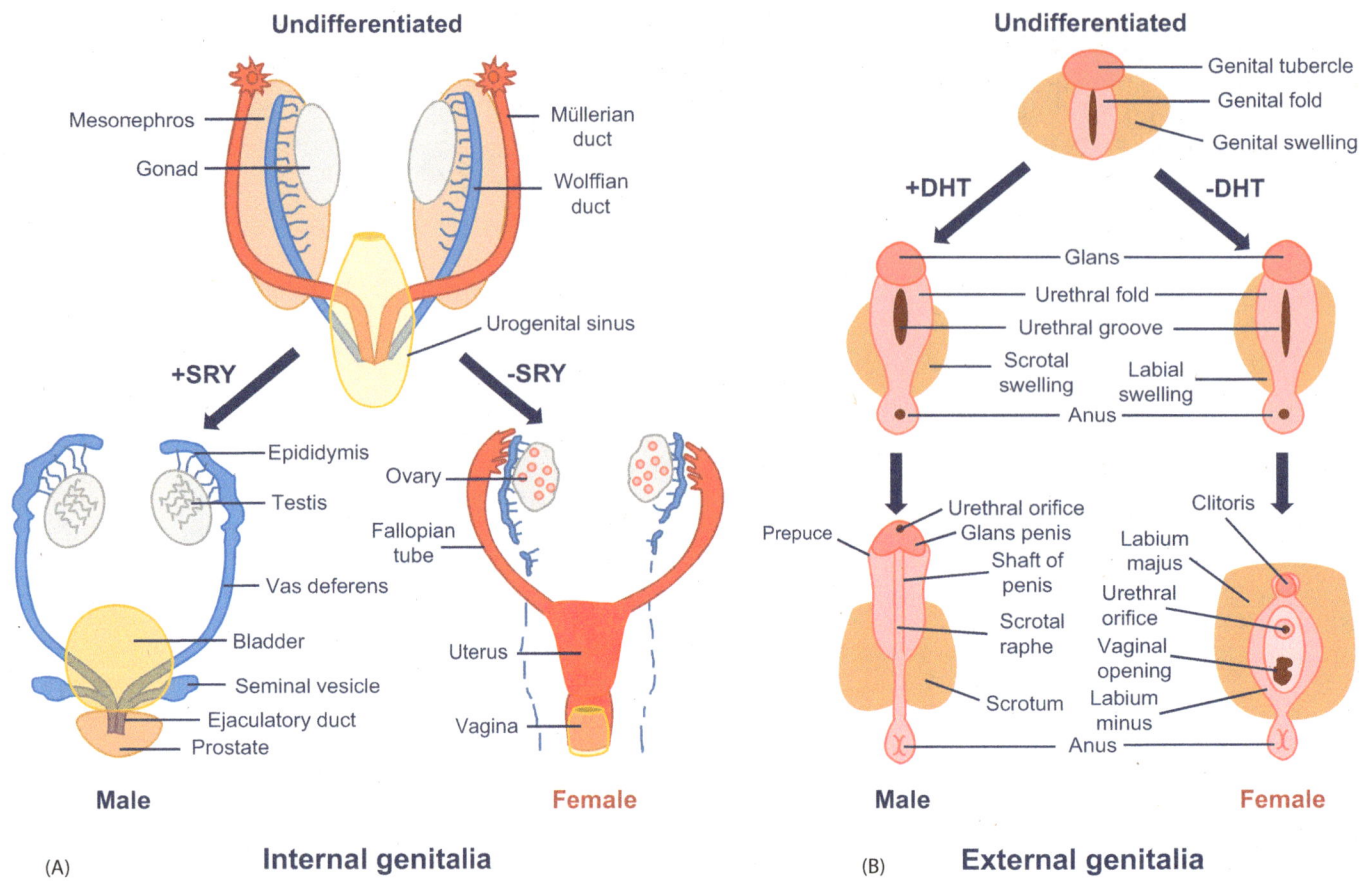

Figure 10.17 Development of genitalia from the bipotential primordia in the human fetus. Panels A and B show the development of male and female phenotypes of the internal and external genitalia, respectively, during early fetal life. See text for explanations.

10.7 PARTURITION AND LACTATION

10.7.1 The process of parturition

The myometrium must remain quiescent during pregnancy for the uninterrupted growth and development of the fetus. At the time of labor, the myometrium is transformed into a highly coordinated, strongly contracting organ, suitable for a successful expulsion of the newborn. The control of the onset of parturition is complex, involving interactions between mother, fetus and the placenta, and differs among species. Two comprehensive reviews cover the multitude of paracrine/autocrine events, fetal hormonal changes, and overlapping maternal/fetal control mechanisms that trigger parturition in women [99,100].

Briefly, the "final common pathway" that results in delivery is composed of interactive inflammatory and endocrine paths that alter the balance in favor of coordinated uterine contractility and cervical dilation. These mechanisms involve a shift from progesterone to estrogen dominance, corticotropin releasing hormone and cortisol actions, increased sensitivity to oxytocin, formation of gap junctions, and increased activity of prostaglandins. In parallel, changes in the cervix include a decrease in progesterone dominance and the actions of prostaglandins and relaxin, via connective tissue alterations, leading to cervical softening and dilation.

Two studies have reported expression of D1R [101] and D2R [102] in the human term placenta. Although both Dopa and DA concentrations increase in several compartments of pregnancy just before labor, only a few and somewhat inconsistent studies have examined the role of DA in parturition. In samples of decidua obtained during an elective C-section, DA stimulated the synthesis of prostaglandin F, which is involved in the initiation of labor [103]. The use of myometrial strips from women during parturition showed that metoclopramide, the peripheral D2R antagonist, relaxed myometrial contractions, and exhibited different responses to subsequent oxytocin treatment [104]. However, opposite results were found in a study that used perifused myometrial specimens from term pregnant human uteri [105]. In this case, DA increased myometrial contractility, which was reduced by pretreatment with prostaglandin synthase inhibitor. The discrepancy among these studies is likely due to different experimental conditions or dissimilar drug dosages.

The influence of DA on uterine activity was studied by external tocography in women at the end of a normal pregnancy [100]. In women not in labor, DA infusion dose-dependently induced regular uterine contractions. For women in spontaneous labor, DA at a low dose also caused a significant increase in the frequency of contractions, but in those receiving an oxytocin infusion, no further stimulation was seen. DA did not have any effect on fetal heart rate, maternal pulse rate, or blood pressure.

10.7.2 Hormonal regulation of lactation

Lactation is the final critical step of mammalian reproduction, as it provides nutrition and initial immune protection to the newborn [106]. Several processes—mammogenesis, lactogenesis, galactopoiesis, and galactokinesis (milk ejection)—assure proper lactation. Mammogenesis refers to the differentiation, growth and maturation of the mammary glands. Lactogenesis designates milk production, while galactopoiesis is the maintenance of established lactation. Milk ejection is the process by which the stored milk is released. PRL is the single most important hormone that promotes secretion and continued milk production by the mammary glands, while oxytocin is the most powerful hormone that affects the discharge of milk from the mammary gland. As discussed in **Chapter 5**, hypothalamic DA is a major inhibitor of the production and release of both PRL and oxytocin. Hence, DA is indirectly involved in multiple aspects of lactation.

Production of breast milk is controlled by an interplay of various hormones, with PRL being the predominant hormone. However, the remodeling and maturation of breast tissue, resulting in milk production, is controlled by many other hormones besides PRL, including hPL, estrogen, progesterone, insulin, GH, cortisol, and thyroxine [107]. During pregnancy, high levels of estrogen and progesterone inhibit the effects of PRL on breast milk production, preventing lactation during pregnancy. A dramatic reduction in progesterone following delivery removes this blockade and triggers lactation.

After delivery, suckling or manual nipple stimulation induces PRL release from the anterior pituitary and oxytocin release from the posterior pituitary. Other sensory pathways (e.g., baby crying) also affect oxytocin release. PRL increases the production and secretion of breast milk, while oxytocin promotes the contraction of breast myoepithelial cells, resulting in milk letdown. PRL regulates the volume of milk produced. However, once lactation is established, infant demand drives the process. Notably, GH, not PRL, is the primary regulator of lactation in the cow, whereas PRL, not GH, serves this function in humans and laboratory species.

In the absence of suckling, lactation ceases in 2 to 3 weeks after delivery. Lactation is also associated with the suppression of cyclicity and unovulation. The contraceptive effect of lactation is strong in rodents but is only moderate in humans, where it is primarily due to the actions of elevated PRL on various sites along the HPG axis (see **Section 10.4.3** and **Figure 10.5**). In non-breast-feeding women after parturition, cyclicity is resumed within

1 month after delivery, whereas fully lactating women have a period of several months of lactational amenorrhea, with the first few menstrual cycles being unovulatory.

The mammary gland is composed of stromal and epithelial components that communicate with each other through the extracellular matrix. Crosstalk between the mammary epithelium and stroma is crucial for the proper patterning and functions of the normal mammary gland during lactation as well as during tumorigenesis [108]. The distribution of TH-positive nerve fibers have been studied in normal mammary glands from rats and humans [109]. TH was detected around blood vessels and near lactiferous ducts and alveoli. The mammary nerve, artery, and vein from rats also contained TH-positive cells, and they were also found at the base of the nipple. Similar findings were reported in porcine [110] and canine [111] mammary glands. These studies showed that the mammary gland is capable of catecholamine synthesis, but whether the final product was DA or NE, as well as whether the TH immunoreactivity also occurs in the lactating mammary gland, have not been determined. Although we [112] and others [113,114] found expression of various DAR in breast cancer cells, DAR were not detected in normal breast epithelial cells. There are also no published records of detection of DAR in normal mammary epithelial cells from other species.

The absence of DAR in milk-producing alveoli or in lactiferous ducts that deliver milk to the nipple does not preclude the possibility that local or circulating DA affect lactation through DAR-expressing stromal adipocytes [43]. Two comprehensive reviews cover the adaptation of maternal adipose tissue to lactation [115], emphasizing the diverse and active roles of adipocytes during mammary gland growth and function [116]. Shortly after parturition in mice, adipocytes located in close proximity to epithelial cells undergo rapid depletion of lipid stores, presumably providing lipid precursors for milk synthesis. The mammary adipose tissue is also the source of paracrine and endocrine molecules, including estrogen synthesis from androgen precursors by aromatization. An indirect regulation of lactation by DA also includes its actions on the mammary vasculature [117], and lymphatic vessels [118]. Unfortunately, very little is known about the lactating human breast, which has a different cytoarchitecture than the mammary glands in rodents.

10.8 NEONATAL DEVELOPMENT, PUBERTY, AND AGING

10.8.1 The reproductive axis during the neonatal and prepubertal periods

After birth, the newborn is cut off from the maternal and placental steroids. The removal of steroidal negative feedback stimulates the secretion of gonadotropins, the level of which exhibit pulsatility with wide fluctuations during the first few months of life [119]. These stimulate the gonads, resulting in a transient increase in serum testosterone in male infants and estradiol in females. FSH levels are usually higher in females than males. By 3 months of age, the levels of both the gonadotropins and gonadal steroids are at the low normal adult range. Circulating gonadotropins then decline to very low levels by 6–7 months of life in males and 1–2 years in females and remained suppressed until the onset of puberty.

Throughout childhood, the gonads are quiescent and plasma steroid levels are low. Yet, gonadotropin release is also suppressed. The prepubertal restraint of gonadotropin secretion is explained by two mechanisms that affect the hypothalamic GnRH pulse generator. One is a sex-steroid-dependent mechanism that renders the pulse generator extremely sensitive to negative feedback by steroids. The other is an intrinsic CNS inhibition of the GnRH pulse generator that involves kisspeptin neurons, DA, opioids, GABA, and melatonin [120]. Together, they suppress the amplitude and frequency of GnRH pulses, resulting in a diminished secretion of the gonadotropins and gonadal steroids. Throughout this period of quiescence, the pituitary and gonads can respond to exogenous GnRH and gonadotropins but at a relatively low sensitivity.

10.8.2 Hormonal control of puberty and disorders of puberty

The onset of puberty depends on the reactivation and maturation of the hypothalamic pulse generator. As a result of disinhibition by the hypothalamic factors listed above, the frequency and amplitude of GnRH pulses increase. Initially, pulsatility is most prominent at night, entrained by deep sleep. Later, it becomes established throughout the 24-h period. GnRH increases the number of its own receptors on the gonadotrophs and augments the synthesis, storage and release of the gonadotropins. This initiates a cascade of events, including rising circulating levels of gonadal steroids, a progressive development of secondary sex characteristics, and establishment of an adult pattern of negative feedback of the hypothalamo–pituitary unit.

As illustrated in **Figure 10.18**, puberty in humans begins at 10–12 years of age. It is a slow process, lasting 4–6 years, and involves the development of secondary sex characteristics, growth spurt, and acquisition of fertility. Over the last 150 years, the age of onset of puberty has declined by 2–3 months per decade, reflecting improvements in nutrition and general health. The first signs of puberty in girls is breast budding (thelarche) and appearance of pubic hair. Menarche, or establishment of menstruation, is a

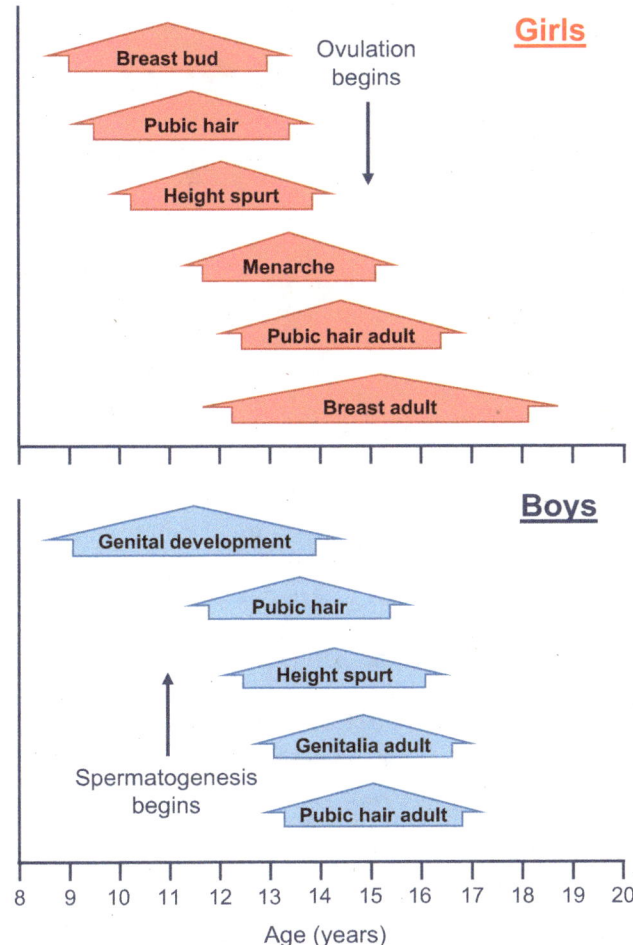

Figure 10.18 The process of puberty in girls and boys. Shown is the timeline of pubertal development from childhood to sexual maturity.

later event. The peak growth spurt and appearance of axillary hair in boys usually occur 2 years later than in girls. Growth of facial hair, deepening of the voice and broadening of the shoulders are late events in male puberty maturation.

In addition to the reproductive hormones, puberty is regulated by other hormones. The adrenal androgens dehydroepiandrosterone (DHEA) and its sulfated form, DHEAS, are primarily responsible for the development of pubic and axillary hair. Adrenal maturation precedes gonadal maturation by about 2 years. The pubertal growth spurt requires a concerted action by sex steroids and GH. The principle regulators of GH are hypothalamic GHRH and IGF-1. The gonadal steroids act by augmenting GH, which is turn stimulates the production of IGF-1 in liver and other tissues. Plasma concentrations of IGF-1 increases markedly during puberty, with peak levels observed earlier in girls than in boys.

Disorders of puberty are classified into **precocious puberty**, defined as sexual maturation before the age of 8 years, and **delayed puberty**, when menses does not start by age 17, or testicular development is delayed beyond age 20. True precocious puberty results from premature activation of the HPG axis, leading to the development of secondary sex characteristics as well as gametogenesis. The most frequent causes are CNS lesions or infections, hypothalamic disease or hypothyroidism. The youngest confirmed mother in history was a Peruvian girl, who in the 1930s gave birth to a healthy baby boy when she was only 5.5 years old. The causes of her precocious puberty were never determined. Pseudoprecocious puberty is defined as an early development of secondary sexual characteristics without gametogenesis. It can result from abnormal exposure of immature boys to androgens and of immature girls to estrogens. Augmented steroid production can be of gonadal or adrenal origin.

10.8.3 Aging of the reproductive system

Aging encompasses irreversible changes that occur over time at the level of molecules, cells, tissues, organs, and systems and is observed in all eukaryotic organisms. Some long-lived female mammals can continue to breed until the end of life (e.g., elephants breed into their 60s and baleen whales into their 90s). On the other hand, the mean age of last birth in natural-fertility in humans cluster around 38 years of age. After this, the reproductive system of women declines and culminates in menopause about 10 years later [121,122]. Yet, even in hunter–gatherer societies without access to modern medicine or technology, women who reach menopausal age can expect to live well into their 60s. The disparity between reproductive and total life span is puzzling because there is no selection advantage for nonreproductive survival.

In women, the potential for reproduction ceases as the ovaries are depleted of follicles. A transition period in mid-life, when menstruation becomes less regular, is called perimenopause. During this time, women often experience hot flushes, which may be associated with shivering, sweating, and reddening of the skin. Other symptoms include vaginal dryness, trouble sleeping, and mood changes. **Menopause** is associated with a significant decline in plasma concentrations of sex hormones, an increase in the concentrations of the gonadotrophins and changes in other hormones such as the inhibins. These changes are superimposed with effects of aging, social and metabolic factors, daily activity and well-being. Although menopause is entirely natural, in some cases, ovarian failure can occur earlier than usual. Elderly females are also affected by a range of clinical disorders including endocrine, cardiovascular, skeletal, urogenital tract and immunological systems, body mass, vasomotor tone, mood and sleep pattern disruption.

Compared with women, the decline in male reproductive capacity with age is less pronounced and men can usually father a child throughout their life span. The decline of the seminiferous tubules with age is characterized by a decrease in the number of Sertoli cells and germ cells. Vascular changes play a role in testicular fibrosis, and its progression with age causes separation of the germinal epithelium from the blood supply. The development of tubular involution with age is thought to account for testicular atrophy [123]. Aging men have a mild decline of total testosterone (T), an increase of sex hormone binding globulin (SHBG), a more pronounced decline of free T, and a moderate increase of LH. Comorbidities, in particular the increased prevalence of obesity, play an important contributory role in the hormonal changes in aging men.

10.9 SYNOPSIS

Sexual reproduction necessitates not only the successful production of viable gametes but also the bringing together of the two sexes. In subhuman species, sexual behavior is the key for enabling copulation at the optimal time for fertilization. In humans, sexual attractions also serve to cement a bond between two parents for raising the helpless babies. Beyond that, hormone-regulated sexuality has enriched the life of humans in many ways, socially and culturally. The hormonal theory of sexuality holds that, just as exposure to certain hormones plays a role in fetal sex differentiation, such exposure also influences the sexual orientation that emerges later in the adult. Three classes of hormones—hypothalamic (GnRH), pituitary (LH, FSH, and PRL), and gonadal (sex steroids and protein hormones)—exert both positive and negative influences on the HPG axis and regulate sexual behavior, secondary sex characteristics, fertility, pregnancy, and lactation. The DARs are expressed in all components of the reproductive systems in both males and females, and DA appears to act as a modulator in many of the reproductive processes. Nonetheless, many gaps in our understanding of the full involvement of DA in pathophysiology of reproduction remains, which could provide rewarding areas of future basic and clinical research.

REFERENCES

1. Larney C, Bailey TL, Koopman P. Switching on sex: Transcriptional regulation of the testis-determining gene Sry. *Development*. 2014;141(11):2195–2205.

2. Negri-Cesi P, et al. Sexual differentiation of the rodent hypothalamus: Hormonal and environmental influences. *J Steroid Biochem Mol Biol*. 2008;109(3–5):294–299.

3. Ngun TC, Ghahramani N, Sanchez FJ, Bocklandt S, Vilain E. The genetics of sex differences in brain and behavior. *Front Neuroendocrinol*. 2011;32(2):227–246.

4. Swan SH. Intrauterine exposure to diethylstilbestrol: Long-term effects in humans. *APMIS*. 2000;108(12):793–804.

5. Gore AC, Krishnan K, Reilly MP. Endocrine-disrupting chemicals: Effects on neuroendocrine systems and the neurobiology of social behavior. *Horm Behav*. 2019;111:7–22.

6. Ben-Jonathan N. Endocrine disrupting chemicals and breast cancer: The saga of bisphenol A, in *Estrogen and Breast Cancer*, Zhang X (ed.), Springer, New York, 2018;343–377.

7. Wallen K. Hormonal influences on sexually differentiated behavior in nonhuman primates. *Front Neuroendocrinol*. 2005;26(1):7–26.

8. Dewing P, et al. Direct regulation of adult brain function by the male-specific factor SRY. *Curr Biol*. 2006;16(4):415–420.

9. Fisher AD, Ristori J, Morelli G, Maggi M. The molecular mechanisms of sexual orientation and gender identity. *Mol Cell Endocrinol*. 2018;467:3–13.

10. Beny-Shefer Y, et al. Nucleus accumbens dopamine signaling regulates sexual preference for females in male mice. *Cell Rep*. 2017;21(11):3079–3088.

11. Chen WP, Witkin JW, Silverman AJ. Gonadotropin releasing hormone (GnRH) neurons are directly innervated by catecholamine terminals. *Synapse*. 1989;3(3):288–290.

12. Liu X, Herbison AE. Dopamine regulation of gonadotropin-releasing hormone neuron excitability in male and female mice. *Endocrinology*. 2013;154(1):340–350.

13. Tsutsumi R, Webster NJ. GnRH pulsatility, the pituitary response and reproductive dysfunction. *Endocr J*. 2009;56(6):729–737.

14. Goodman RL, Jansen HT, Billings HJ, Coolen LM, Lehman MN. Neural systems mediating seasonal breeding in the ewe. *J Neuroendocrinol*. 2010;22(7):674–681.

15. Hardy SL, Anderson GM, Valent M, Connors JM, Goodman RL. Evidence that estrogen receptor alpha, but not beta, mediates seasonal changes in the response of the ovine retrochiasmatic area to estradiol. *Biol Reprod*. 2003;68(3):846–852.

16. Guivarc'h D, Vernier P, Vincent JD. Sex steroid hormones change the differential distribution of the isoforms of the D2 dopamine receptor messenger RNA in the rat brain. *Neuroscience*. 1995;69(1):159–166.

17. Boden MJ, Varcoe TJ, Kennaway DJ. Circadian regulation of reproduction: From gamete to offspring. *Prog Biophys Mol Biol*. 2013;113(3):387–397.

18. Christian CA, Moenter SM. The neurobiology of preovulatory and estradiol-induced gonadotropin-releasing hormone surges. *Endocr Rev*. 2010;31(4):544–577.

19. Korshunov KS, Blakemore LJ, Trombley PQ. Dopamine: A modulator of circadian rhythms in the central nervous system. *Front Cell Neurosci*. 2017;11:91–107.

20. Mani SK, Mitchell A, O'Malley BW. Progesterone receptor and dopamine receptors are required in Delta 9-tetrahydrocannabinol modulation of sexual receptivity in female rats. *Proc Natl Acad Sci USA*. 2001;98(3):1249–1254.

21. Will RG, Hull EM, Dominguez JM. Influences of dopamine and glutamate in the medial preoptic area on male sexual behavior. *Pharmacol Biochem Behav*. 2014;121:115–123.

22. Wolters JP, Hellstrom WJ. Current concepts in ejaculatory dysfunction. *Rev Urol*. 2006;8(Suppl 4):S18–S25.

23. Henderson HL, Townsend J, Tortonese DJ. Direct effects of prolactin and dopamine on the gonadotroph response to GnRH. *J Endocrinol*. 2008;197(2):343–350.

24. Winters SJ, et al. Dopamine-2 receptor activation suppresses PACAP expression in gonadotrophs. *Endocrinology*. 2014;155(7):2647–2657.

25. Ben-Jonathan N, Lapensee CR, Lapensee EW. What can we learn from rodents about prolactin in humans? *Endocr Rev*. 2008;29(1):1–41.

26. De Kretser DM, et al. The role of activin, follistatin and inhibin in testicular physiology. *Mol Cell Endocrinol*. 2004;225(1–2):57–64.

27. Davidoff MS, et al. Catecholamine-synthesizing enzymes in the adult and prenatal human testis. *Histochem Cell Biol*. 2005;124(3–4):313–323.

28. Mayerhofer A, et al. Testis of prepubertal rhesus monkeys receives a dual catecholaminergic input provided by the extrinsic innervation and an intragonadal source of catecholamines. *Biol Reprod*. 1996;55(3):509–518.

29. Lakomy M, Kaleczyc J, Majewski M. Noradrenergic and peptidergic innervation of the testis and epididymis in the male pig. *Folia Histochem Cytobiol*. 1997;35(1):19–27.

30. Otth C, et al. Novel identification of peripheral dopaminergic D2 receptor in male germ cells. *J Cell Biochem*. 2007;100(1):141–150.

31. Oguzkurt P, Okur DH, Tanyel FC, Buyukpamukcu N, Hicsonmez A. The effects of vasodilatation and chemical sympathectomy on spermatogenesis after unilateral testicular torsion: A flow cytometric DNA analysis. *Br J Urol*. 1998;82(1):104–108.

32. Gonzalez CR, et al. Psychostimulant-induced testicular toxicity in mice: Evidence of cocaine and caffeine effects on the local dopaminergic system. *PLoS One*. 2015;10(11):e0142713.

33. Dirami G, Cooke BA. Effect of a dopamine agonist on luteinizing hormone receptors, cyclic AMP production and steroidogenesis in rat Leydig cells. *Toxicol Appl Pharmacol*. 1998;150(2):393–401.

34. Shukla KK, et al. Mucuna pruriens improves male fertility by its action on the hypothalamus-pituitary-gonadal axis. *Fertil Steril*. 2009;92(6):1934–1940.

35. Matsubayashi H. Inhibitory effects of dopamine on noradrenaline-induced constriction of arterioles in vivo in the striated cremaster muscle. *Hiroshima J Med Sci*. 1993;42(1):1–7.

36. Tanyel FC, et al. Chemical sympathectomy by 6-OH dopamine during fetal life results in inguinal testis through altering cremasteric contractility in rats. *J Pediatr Surg*. 2003;38(11):1628–1632.

37. Castelli M, Rossi T, Baggio G, Bertolini A, Ferrari W. Characterization of the contractile activity of dopamine on the rat isolated seminal vesicle. *Pharmacol Res Commun*. 1985;17(4):351–359.

38. Courtois F, Carrier S, Charvier K, Guertin PA, Journel NM. The control of male sexual responses. *Curr Pharm Des*. 2013;19(24):4341–4356.

39. Hakky TS, Jain L. Current use of phosphodiesterase inhibitors in urology. *Turk J Urol*. 2015;41(2):88–92.

40. Hyun JS, et al. Localization of peripheral dopamine D1 and D2 receptors in rat corpus cavernosum. *BJU Int*. 2002;90(1):105–112.

41. Senbel AM. Interaction between nitric oxide and dopaminergic transmission in the peripheral control of penile erection. *Fundam Clin Pharmacol*. 2011;25(1):63–71.

42. d'Emmanuele di Villa Bianca R, et al. Peripheral relaxant activity of apomorphine and of a D1 selective receptor agonist on human corpus cavernosum strips. *Int J Impot Res*. 2005;17(2):127–133.

43. Borcherding DC, et al. Dopamine receptors in human adipocytes: Expression and functions. *PLoS One*. 2011;6(9):e25537.

44. Ramirez AD, Smith SM. Regulation of dopamine signaling in the striatum by phosphodiesterase inhibitors: Novel therapeutics to treat neurological and psychiatric disorders. *Cent Nerv Syst Agents Med Chem*. 2014;14(2):72–82.

45. Giuliano F. Neurophysiology of erection and ejaculation. *J Sex Med*. 2011;8 Suppl 4):310–315.

46. Dixon JS, Jen PY, Gosling JA. Structure and autonomic innervation of the human vas deferens: A review. *Microsc Res Tech*. 1998;42(6):423–432.

47. Amenta D, Cavallotti C, Amenta F. Dopamine receptors mediating the stimulation and the inhibition of adenylate cyclase in rat prostate gland. *Neurosci Lett*. 1987;77(1):66–70.

48. Hillier SG, Smitz J, Eichenlaub-Ritter U. Folliculogenesis and oogenesis: From basic science to the clinic. *Mol Hum Reprod*. 2010;16(9):617–620.

49. Saller S, et al. Dopamine in human follicular fluid is associated with cellular uptake and metabolism-dependent generation of reactive oxygen species in granulosa cells: Implications for physiology and pathology. *Hum Reprod*. 2014;29(3):555–567.

50. Bodis J, Bognar Z, Hartmann G, Torok A, Csaba IF. Measurement of noradrenaline, dopamine and serotonin contents in follicular fluid of human Graafian follicles after superovulation treatment. *Gynecol Obstet Invest*. 1992;33(3):165–167.

51. Mayerhofer A, et al. Oocytes are a source of catecholamines in the primate ovary: Evidence for a cell-cell regulatory loop. *Proc Natl Acad Sci USA*. 1998;95(18):10990–10995.

52. Fernandez-Pardal J, Gimeno MF, Gimeno AL. Catecholamines in sow Graafian follicles at proestrus and at diestrus. *Biol Reprod*. 1986;34(3):439–445.

53. Bahr JM, Ben-Jonathan N. Preovulatory depletion of ovarian catecholamines in the rat. *Endocrinology*. 1981;108(5):1815–1820.

54. Feng Y, et al. CLARITY reveals dynamics of ovarian follicular architecture and vasculature in three-dimensions. *Sci Rep*. 2017;7:44810.

55. Bodis J, et al. Relationship between the monoamine, progesterone and estradiol content in follicular fluid of preovulatory Graafian follicles after superovulation treatment. *Gynecol Obstet Invest*. 1993;35(4):232–235.

56. Rey-Ares V, et al. Dopamine receptor repertoire of human granulosa cells. *Reprod Biol Endocrinol*. 2007;5:40–50.

57. Musali N, et al. Follicular fluid norepinephrine and dopamine concentrations are higher in polycystic ovary syndrome. *Gynecol Endocrinol*. 2016;32(6):460–463.

58. Parillo F, et al. Evidence for a dopamine intrinsic direct role in the regulation of the ovary reproductive function: In vitro study on rabbit corpora lutea. *PLoS One*. 2014;9(8):e104797.

59. Bodis J, et al. Effect of noradrenaline and dopamine on progesterone and estradiol secretion of human granulosa cells. *Acta Endocrinol (Copenh)*. 1993;129(2):165–168.

60. Mori H, et al. The involvement of dopamine in the regulation of steroidogenesis in rat ovarian cells. *Horm Res*. 1994;41(Suppl 1):36–40.

61. Power RF, Mani SK, Codina J, Conneely OM, O'Malley BW. Dopaminergic and ligand-independent activation of steroid hormone receptors. *Science*. 1991;254(5038):1636–1639.

62. Xing L, McDonald H, Da Fonte DF, Gutierrez-Villagomez JM, Trudeau VL. Dopamine D1 receptor activation regulates the expression of the estrogen synthesis gene aromatase B in radial glial cells. *Front Neurosci*. 2015;9:310–320.

63. Czaja K, Wasowicz K, Klimczuk M, Podlasz P, Lakomy M. Distribution and immunohistochemical characterisation of paracervical neurons innervating the oviduct in the pig. *Folia Morphol (Warsz)*. 2001;60(3):205–211.

64. Khatchadourian C, Menezo Y, Gerard M, Thibault C. Catecholamines within the rabbit oviduct at fertilization time. *Hum Reprod*. 1987;2(1):1–5.

65. Helm G, Owman C, Rosengren E, Sjoberg NO. Regional and cyclic variations in catecholamine concentration of the human fallopian tube. *Biol Reprod*. 1982;26(4):553–558.

66. Chaud M, Fernandez PJ, Viggiano M, Gimeno MF, Gimeno AL. Is there a role for dopamine in the regulation of motility of sow oviducts? *Pharmacol Res Commun*. 1983;15(10):923–936.

67. Mitchell BS, Ahmed E. An immunohistochemical study of the catecholamine synthesizing enzymes and neuropeptides in the female guinea-pig uterus and vagina. *Histochem J*. 1992;24(6):361–367.

68. Renegar RH, Rexroad CE, Jr. Uterine adrenergic and cholinesterase-positive nerves and myometrial catecholamine concentrations during pregnancy in sheep. *Acta Anat (Basel)*. 1990;137(4):373–381.

69. Estan L, Martinez-Mir I, Rubio E, Morales-Olivas FJ. Relaxant effect of dopamine on the isolated rat uterus. *Naunyn Schmiedebergs Arch Pharmacol*. 1988;338(5):484–488.

70. Bottalico B, et al. Plasma membrane and vesicular monoamine transporters in normal endometrium and early pregnancy decidua. *Mol Hum Reprod*. 2003;9(7):389–394.

71. Tseng L, Mazella J, Goligorsky MS, Rialas CM, Stefano GB. Dopamine and morphine stimulate nitric oxide release in human endometrial glandular epithelial cells. *J Soc Gynecol Investig*. 2000;7(6):343–347.

72. Attar R, Attar E. Experimental treatments of endometriosis. *Womens Health (Lond).* 2015;11(5):653–664.

73. Novella-Maestre E, et al. Identification and quantification of dopamine receptor 2 in human eutopic and ectopic endometrium: A novel molecular target for endometriosis therapy. *Biol Reprod.* 2010;83(5):866–873.

74. Novella-Maestre E, et al. Dopamine agonist administration causes a reduction in endometrial implants through modulation of angiogenesis in experimentally induced endometriosis. *Hum Reprod.* 2009;24(5):1025–1035.

75. Goenka L, George M, Sen M. A peek into the drug development scenario of endometriosis—A systematic review. *Biomed Pharmacother.* 2017;90:575–585.

76. Lakomy M, Kaleczyc J, Majewski M, Sienkiewicz W. Peptidergic innervation of the bovine vagina and uterus. *Acta Histochem.* 1995;97(1):53–66.

77. Bae SE, Corcoran BM, Watson ED. Immunohistochemical study of the distribution of adrenergic and peptidergic innervation in the equine uterus and the cervix. *Reproduction.* 2001;122(2):275–282.

78. Stival C, et al. Sperm capacitation and acrosome reaction in mammalian sperm. *Adv Anat Embryol Cell Biol.* 2016;220:93–106.

79. Ramirez AR, et al. The presence and function of dopamine type 2 receptors in boar sperm: A possible role for dopamine in viability, capacitation, and modulation of sperm motility. *Biol Reprod.* 2009;80(4):753–761.

80. Brucker C, Lipford GB. The human sperm acrosome reaction: Physiology and regulatory mechanisms. An update. *Hum Reprod Update.* 1995;1(1):51–62.

81. Bhandari B, Bansal P, Talwar P, Gupta SK. Delineation of downstream signalling components during acrosome reaction mediated by heat solubilized human zona pellucida. *Reprod Biol Endocrinol.* 2010;8:7–13.

82. Ferrero H, et al. Dopamine receptor 2 activation inhibits ovarian vascular endothelial growth factor secretion in an ovarian hyperstimulation syndrome (OHSS) animal model: Implications for treatment of OHSS with dopamine receptor 2 agonists. *Fertil Steril.* 2014;102(5):1468–1476.

83. Shirlow R, Healey M, Volovsky M, MacLachlan V, Vollenhoven B. The effects of adjuvant therapies on embryo transfer success. *J Reprod Infertil.* 2017;18(4):368–378.

84. Takagi Y, Takahashi J. Primate embryonic stem cells as a source of dopaminergic neurons: A novel transplantation for Parkinson's disease. *Discov Med.* 2005;5(26):219–223.

85. Schmidt A, Morales-Prieto DM, Pastuschek J, Frohlich K, Markert UR. Only humans have human placentas: Molecular differences between mice and humans. *J Reprod Immunol.* 2015;108:65–71.

86. Handwerger S, Freemark M. The roles of placental growth hormone and placental lactogen in the regulation of human fetal growth and development. *J Pediatr Endocrinol Metab.* 2000;13(4):343–356.

87. Vaillancourt C, Petit A, Belisle S. Expression of human placental D2-dopamine receptor during normal and abnormal pregnancies. *Placenta.* 1998;19(1):73–80.

88. Kim HJ, Koh PO, Kang SS, Paik WY, Choi WS. The localization of dopamine D2 receptor mRNA in the human placenta and the anti-angiogenic effect of apomorphine in the chorioallantoic membrane. *Life Sci.* 2001;68(9):1031–1040.

89. Arkinstall SJ, Jones CT. Regional changes in catecholamine content of the pregnant uterus. *J Reprod Fertil.* 1985;73(2):547–557.

90. Underwood MA, Gilbert WM, Sherman MP. Amniotic fluid: Not just fetal urine anymore. *J Perinatol.* 2005;25(5):341–348.

91. Ben-Jonathan N, Maxson RE. Elevation of dopamine in fetal plasma and the amniotic fluid during gestation. *Endocrinology.* 1978;102(2):649–652.

92. Peleg D, Arbogast LA, Peleg E, Ben-Jonathan N. Predominance of L-dopa in fetal plasma and the amniotic fluid during late gestation in the rat. *Am J Obstet Gynecol.* 1984;149(8):880–883.

93. Ben-Jonathan N, et al. Plasma catecholamines in the chronically cannulated sheep fetus: Predominance of L-dihydroxyphenylalanine. *Endocrinology.* 1983;113(1):216–221.

94. Ben-Jonathan N, Munsick RA. Dopamine and prolactin in human pregnancy. *J Clin Endocrinol Metab.* 1980;51(5):1019–1025.

95. Peleg D, Munsick RA, Diker D, Goldman JA, Ben-Jonathan N. Distribution of catecholamines between fetal and maternal compartments during human pregnancy with emphasis on L-dopa and dopamine. *J Clin Endocrinol Metab.* 1986;62(5):911–914.

96. Godziejewski J, Maruchin JE. Catecholamine concentration in amniotic fluid: Possible role of dopamine in parturition. *Am J Obstet Gynecol.* 1988;159(6):1600.

97. Vaillancourt C, Petit A, Belisle S. D2-dopamine agonists inhibit adenosine 3′:5′-cyclic monophosphate (cAMP) production in human term trophoblastic cells. *Life Sci.* 1994;55(20):1545–1552.

98. Barthold JS. Disorders of sex differentiation: A pediatric urologist's perspective of new terminology and recommendations. *J Urol.* 2011;185(2):393–400.

99. Menon R, Bonney EA, Condon J, Mesiano S, Taylor RN. Novel concepts on pregnancy clocks and alarms: Redundancy and synergy in human parturition. *Hum Reprod Update.* 2016;22(5):535–560.

100. Urban J, Radwan J, Laudanski T, Akerlund M. Dopamine influence on human uterine activity at term pregnancy. *Br J Obstet Gynaecol.* 1982;89(6):451–455.

101. Yanagawa T, et al. Presence of dopamine DA-1 receptors in human decidua. *Placenta.* 1997;18(2–3):169–172.

102. Arai F, Kishimoto Y, Tada K, Kondo Y, Kudo T. The presence and role of the dopamine DA-2 receptor in the human decidua. *J Obstet Gynaecol Res*. 2000;26(6):449–454.

103. Tada K, Kudo T, Kishimoto Y. Effects of L-dopa or dopamine on human decidual prostaglandin synthesis. *Acta Med Okayama*. 1991;45(5):333–338.

104. Tang YY, et al. Relaxant effects of metoclopramide and magnesium sulfate on isolated pregnant myometrium: An in vitro study. *Int J Obstet Anesth*. 2014;23(2):131–137.

105. Wikland M, Lindblom B, Wiqvist N. Catecholamines and contractility of the human myometrium at term: A possible role for prostaglandins. *Acta Physiol Hung*. 1985;65(3):331–334.

106. Buhimschi CS. Endocrinology of lactation. *Obstet Gynecol Clin North Am*. 2004;31(4):963–979.

107. Gabay MP. Galactogogues: Medications that induce lactation. *J Hum Lact*. 2002;18(3):274–279.

108. Smith BA, Welm AL, Welm BE. On the shoulders of giants: A historical perspective of unique experimental methods in mammary gland research. *Semin Cell Dev Biol*. 2012;23(5):583–590.

109. Eriksson M, Lindh B, Uvnas-Moberg K, Hokfelt T. Distribution and origin of peptide-containing nerve fibres in the rat and human mammary gland. *Neuroscience*. 1996;70(1):227–245.

110. Franke-Radowiecka A, Kaleczyc J, Klimczuk M, Lakomy M. Noradrenergic and peptidergic innervation of the mammary gland in the immature pig. *Folia Histochem Cytobiol*. 2002;40(1):17–25.

111. Pinho MS, Gulbenkian S. Innervation of the canine mammary gland: An immunohistochemical study. *Histol Histopathol*. 2007;22(11):1175–1184.

112. Borcherding DC, et al. Expression and therapeutic targeting of dopamine receptor-1 (D1R) in breast cancer. *Oncogene*. 2016;35(24):3103–3113.

113. Minami K, et al. Inhibitory effects of dopamine receptor D1 agonist on mammary tumor and bone metastasis. *Sci Rep*. 2017;7:45686.

114. Sachlos E, et al. Identification of drugs including a dopamine receptor antagonist that selectively target cancer stem cells. *Cell*. 2012;149(6):1284–1297.

115. Vernon RG, Pond CM. Adaptations of maternal adipose tissue to lactation. *J Mammary Gland Biol Neoplasia*. 1997;2(3):231–241.

116. Hovey RC, Aimo L. Diverse and active roles for adipocytes during mammary gland growth and function. *J Mammary Gland Biol Neoplasia*. 2010;15(3):279–290.

117. Katai R, et al. The variable effects of dopamine among human isolated arteries commonly used for coronary bypass grafts. *Anesth Analg*. 2004;98(4):915–920.

118. Dobbins DE. Receptor mechanisms of prenodal lymphatic constriction by dopamine. *Regulatory Peptides*. 2003;114(1):7–13.

119. Choi JH, Yoo HW. Control of puberty: Genetics, endocrinology, and environment. *Curr Opin Endocrinol Diabetes Obes*. 2013;20(1):62–68.

120. Sizonenko PC, Aubert ML. Neuroendocrine changes characteristic of sexual maturation. *J Neural Transm Suppl*. 1986;21:159–181.

121. Cant MA, Johnstone RA. Reproductive conflict and the separation of reproductive generations in humans. *Proc Natl Acad Sci USA*. 2008;105(14):5332–5336.

122. Honour JW. Biochemistry of the menopause. *Ann Clin Biochem*. 2018;55(1):18–33.

123. Kaufman JM, Lapauw B, Mahmoud A, T'Sjoen G, Huhtaniemi IT. Aging and the male reproductive system. *Endocr Rev*. 2019;40(4):906–972.

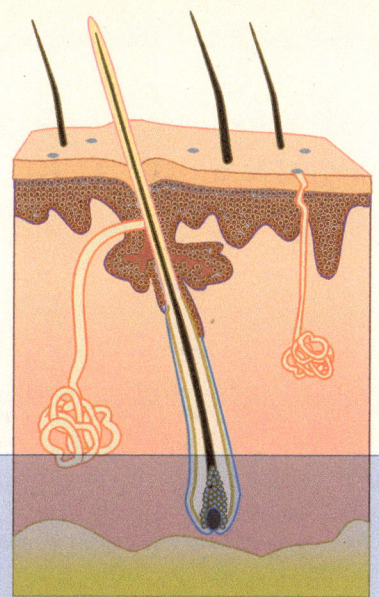

Actions of Dopamine on the Skin and the Skeleton

11

11.1 INTRODUCTION

The skin is the outer boundary of the body, the part which is in direct contact with the external environment. It is subjected to continuous wear and tear, to drying and to temperature changes and is often exposed to cuts and wounds. The skin helps to preserve the homeostasis of the internal environment and presents a barrier to foreign substances that might interfere with bodily functions. It also prevents internal molecules and fluids from leaking out. To fulfill its multiple functions, the skin is composed of many cell types, some of which are capable of continuous renewal, while others are terminally differentiated. The first four sections of the chapter cover the basic properties of the skin and then review the involvement of dopamine (DA) in several important functions of the skin, including wound healing, pigmentation, sweating, and hair growth.

The musculoskeletal system provides form, support, stability, and movement to the body. It is made up of the bones of the skeleton, muscles, and joints. Bones provide shape, hold the body upright, and protect organs from injury. They also store minerals and contain the bone marrow where new blood cells are made. The three types of muscle—skeletal, cardiac and smooth—differ in cellular structure, location, and mode of action. Joints are the physical points of connection between two bones. Joints contain a variety of fibrous connective tissue, ligaments that connect bones to each other, tendons that connect muscle to bone, and cartilage that covers the ends of bone and provides cushioning. After briefly reviewing the basic properties of the musculoskeletal system and its regulation by the nervous system, the last two sections of the chapter focus on the involvement of dopamine in pathophysiology of this system.

11.2 SKIN STRUCTURE AND WOUND HEALING

11.2.1 Overview of skin structure

The skin is the largest organ of the human body, occupying a total area of ~20 square feet [1]. It provides protection from microorganisms and the elements, is essential for thermoregulation, provides insulation, and permits the sensations of touch, heat, and cold. Despite a constant exposure to physical, biochemical, and radiation insults, this functional integumentary system is capable of counteracting these forces and maintaining a relative state of homeostasis. A dynamic balance underlies the remarkable plasticity of the skin and has been effectively exploited in reconstructive medicine, including tissue expansion, scar revision surgery, and skin grafting.

Multiple cell populations and matrix components form distinct, yet interdependent, compartments that regulate skin behavior during development and throughout life [2]. As illustrated in **Figure 11.1**, the skin is composed of three anatomically discrete layers: (1) the epidermis, or the outermost layer of the skin, which provides a waterproof barrier and creates the skin tone; (2) the dermis, which contains connective tissue, hair follicles, sweat glands and sebaceous glands; and (3) the hypodermis (subcutis), which is composed of adipose and connective tissues.

The **epidermis** has a 5- to100-μm thickness, depending upon its location, and is composed of stratified squamous epithelium devoid of blood or nerve supplies. Keratinocytes are the major cell type, constituting 95% of the epidermis, which also contains melanocytes. The deepest section of the

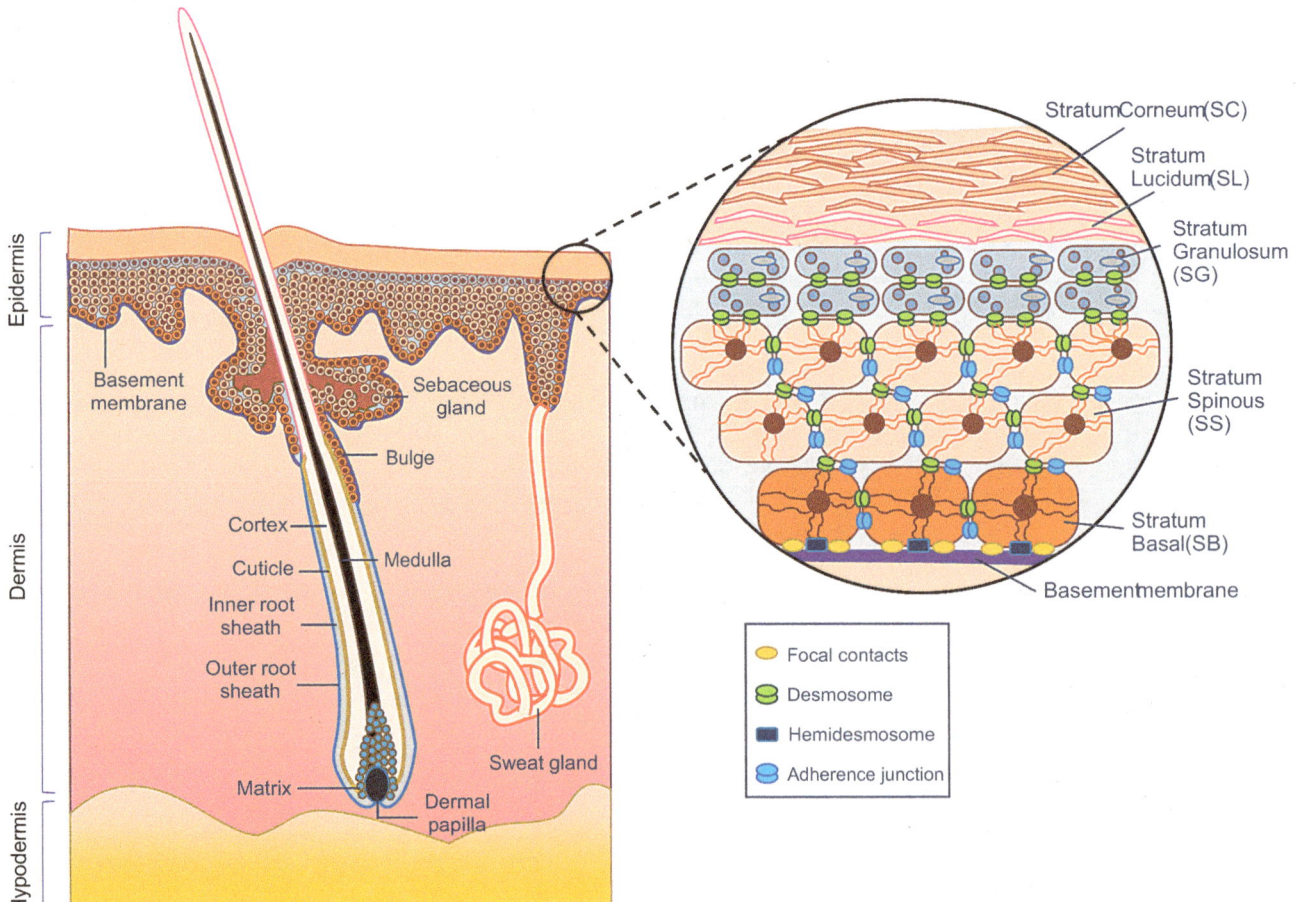

Figure 11.1 Structure of the human skin. Shown are the three layers of the skin: epidermis, dermis, and hypodermis. The inset shows details of the epidermis, highlighting the different shape of the keratinocytes that populate its substructures, and the various intercellular contacts between them. (Redrawn from Fuchs, E. and Raghavan, S. *Nat. Rev. Genet.*, 3, 199–209, 2003.)

epidermis, the stratum basale (basal layer or BL), is the reproductive layer of the epidermis. Its cells constantly divide and provide a continuous supply of new cells to the upper strata. This layer also contains neuroendocrine mechanoreceptors (Merkel cells). The proliferating keratinocytes are pushed upward to form the stratum spinosum (spinous layer or SL). The SL consists of about 10 rows of cells that fit closely together and are connected by desmosomes, or specialized structures for cell-to-cell adhesion. Also found in the SL are bone marrow-derived sentinel cells of the immune system called Langerhans' cells, which are the antigen-presenting cells of the skin and play a role in immunological reactions such as allergic contact dermatitis. As they move upward toward the skin surface, keratinocytes gradually flatten and become part of the stratum granulosum, where the nondividing keratinocytes produce of a protein called keratinohyalin. The next layer, stratum lucidum, is present only in the thick skin of palms and soles.

The outer surface of the skin, the stratum corneum (SC), consists of about 20 layers of closely packed cells in various stages of disintegration. The cell nuclei and other organelles of the keratinocytes are destroyed by lysosomal enzymes and the cells die. Keratin, a fibrous protein, fills the interior of each cell in the SC. It takes about 2 weeks for a basal cell to reach the stratum corneum, and another 2 weeks to slough off. Thousands of cells slough off the surface of the human skin each day and are precisely replaced by new ones from the deepest layers of the epidermis. Skin color is determined by the melanocytes, which produce melanin.

The **dermis**, a 2- to 4-mm thick tissue layer, contains irregular dense connective tissue, nerves, and blood vessels. The main cell type, the fibroblast, is responsible for the synthesis and degradation of the extracellular matrix (ECM). This matrix is composed of highly organized collagen, elastic, and reticular fibers that give the skin mechanical and tensile strength and elasticity. The dermis also hosts multifunctional cells of the immune system such as macrophages and mast cells, which trigger allergic reactions by secreting bioactive mediators such as histamine.

Structures within the dermis include

1. Excretory and secretory glands (sebaceous, eccrine, and apocrine). Sebaceous glands secrete triglyceride and cholesterol-rich sebum that lubricate the skin and keep it supple and waterproof. They are often associated with hair shafts.
2. Hair follicles and nails. The hair follicle provides a protective niche to several stem cell populations in the skin, including keratinocyte stem cells, melanocyte stem cells, a population of epidermal neural crest stem cells, and the dermal stem cell compartment, the dermal papilla. These stem cells are most active during wound healing.
3. Sensory nerve receptors of Merkel and Meissner's corpuscles (touch receptors), Pacinian corpuscles (pressure receptors), and Ruffini corpuscles (mechanoreceptors). Fingerlike projections within the papillary region of the dermis extend toward the epidermis. The thick reticular region has collagenous, elastic, and reticular fibers.

The **hypodermis** lies below the dermis and attaches the skin to the underlying bone and muscle. It supplies the skin with blood vessels and nerves and serves as padding and insulation. The hypodermis consists of connective tissue and elastin, and its main cell types are fibroblasts, macrophages and adipocytes. Notably, subcutaneous tissue contains about 50% of total body fat.

Multiple **stem cell** populations exist throughout the skin and are surrounded by a unique milieu that enables each stem cell type to function [3]. Epithelial stem cells are derived from the interfollicular epithelium, sebaceous glands, and the bulge area of hair follicles. Dermal stem cells originate from the dermal papilla of hair follicles or from perivascular regions.

Adipose tissue also harbors multipotent cell populations that originate from the perivascular space. Acting together, the various stem cell populations facilitate skin repair and renewal throughout life. An important concept in stem cell biology is the niche, which describes the dynamic cellular and noncellular microenvironment of stem cells that regulates their "stemness." This includes neighboring cells, soluble signaling molecules, ECM, mechanical forces, oxygen tension, and other factors that enable a stem cell to maintain its regenerative potential.

The current understanding of skin biology and its response to injury provides an insight into intrinsic restorative pathways in complex organs. For example, the epithelial layer of skin is continuously renewed throughout life, and autologous skin grafts can be transplanted and survive long-term without major adverse effects. Success in hair follicle transfer supports the concept that the skin appendages can also promote regenerative pathways. Thus, human skin represents a unique paradigm for organ homeostasis that enables researchers to study putative repair mechanisms for regenerative medicine.

11.2.2 The complexity of wound healing

The skin is a remarkably plastic organ that sustains insults and injury throughout life. Its ability to expeditiously repair wounds is paramount to the survival of the organism. Wound healing is regulated by differentiated cells, stem cells, cytokine networks, ECM, and mechanical forces [4,5]. When the skin barrier is broken, a regulated sequence of cellular and biochemical events is set into motion to repair the damage. Cells that are actively involved in wound healing include those residing within the skin, i.e., keratinocytes, macrophages, myofibroblasts, endothelial cells, adipocytes, and mast cells, as well as those that are recruited to the injury site, i.e., platelets, neutrophils, and mesenchymal stem cells (MSCs).

As illustrated in **Figure 11.2**A, the process of wound healing is divided into several interacting stages: (1) hemostasis and coagulation, (2) inflammation, (3) proliferative and migration, and (4) tissue remodeling. These stages are not mutually excluding but, rather, overlap in time and space. The various phases of wound healing can take from few minutes to several months (**Figure 11.2**B). The process is not only complex but also fragile, being

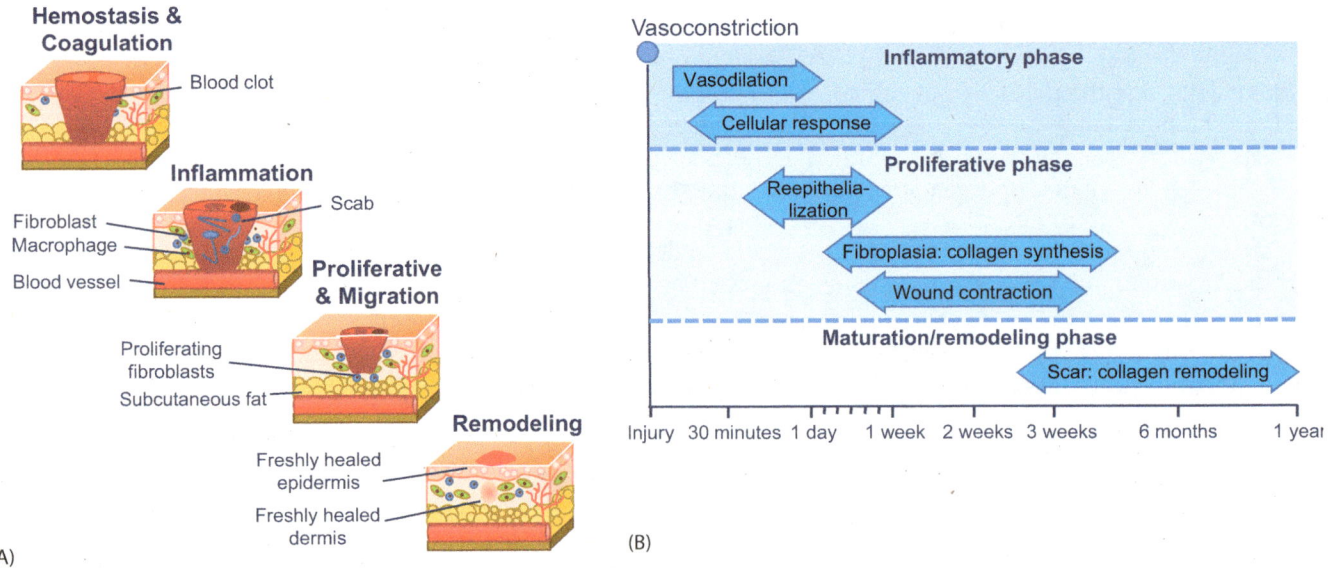

Figure 11.2 The process of wound healing. Panel A shows the various structural stages of wound healing. Panel B depicts the initial, followed by a prolonged, time course of healing, until full recovery.

susceptible to interruption or to failure, which may lead to the formation of nonhealing chronic wounds. Factors that contribute to nonhealing chronic wounds are diabetes, venous or arterial disease, infections, and some metabolic deficiencies associated with old age.

Hemostasis and coagulation: Within the first few seconds of injury, local blood vessels undergo an intense vasoconstriction aimed at minimizing blood loss. Platelets, which normally circulate in blood in an inactive (quiescent) form, are recruited to the wound site and become activated (**Figure 11.3**). Activated platelets release chemical signals that trigger the coagulation cascade. Subsequently, the platelets induce clot formation by converting on their surface prothrombin into thrombin, a serine protease enzyme. Thrombin converts fibrinogen, a soluble plasma protein, into the insoluble fibrin. At the injury site, fibrin forms a mesh of fibers and acts as an adhesive that binds the platelets to each other. The resulting clot plugs the break in the blood vessel, slowing/preventing further bleeding. Additional detail on the coagulation cascade is presented in **Chapter 9**, **Section 9.5.3**, **Table 9.2**, and **Figure 9.7**.

Inflammation: This is a critical phase of healing that protects wounds from invading bacteria and assists in the tissue repair process. Damaged and dead cells are cleared out along with bacteria and other pathogens or debris. This occurs through phagocytosis by neutrophils which engulf cell debris. Platelet-derived growth factors are then released into the wound, inducing migration and division of cells during the proliferative phase.

Proliferation and migration: This phase includes angiogenesis, collagen deposition, granulation tissue formation, epithelialization, and wound contraction. For angiogenesis, actively dividing vascular endothelial cells form new blood vessels. In addition, fibroblasts grow and form new ECM by releasing collagen and fibronectin. In parallel, there is reepithelialization of the epidermis by the proliferation of epithelial cells that cover the wound bed.

Wound contraction and remodeling: myofibroblasts decrease the size of the wound by gripping the wound edges and contracting, acting like smooth muscle cells. Collagen is then realigned along tension lines, and cells that are no longer needed are removed by apoptosis. Freshly healed epidermis and dermis are formed by the appropriate cells.

11.2.3 Role of dopamine in wound healing

Studies with animal models have demonstrated multiple actions of DA in wound healing, albeit the reported results have not always been in a complete agreement. In one study with mice, immunohistochemistry and real-time polymerase chain reaction analysis revealed expression of D2R in the basal epidermis, and D4R in the uppermost layer of the epidermis [6].

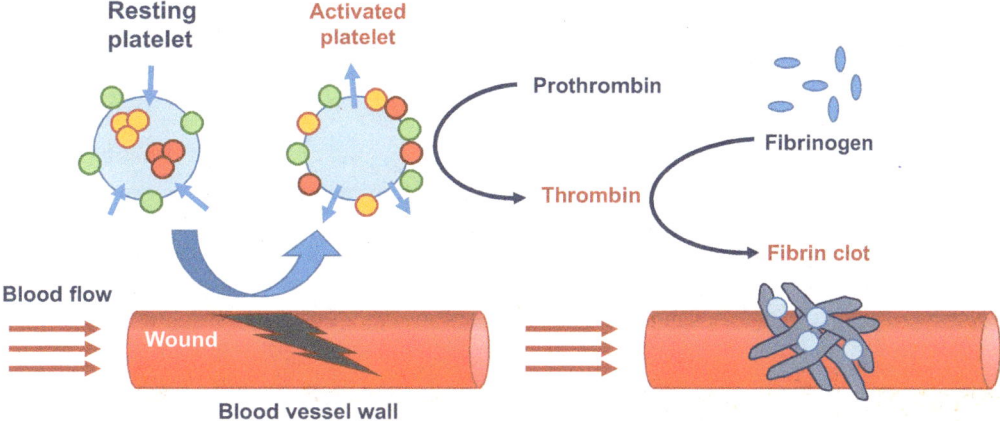

Figure 11.3 Hemostasis and clot formation. Activated platelets release various chemicals which convert prothrombin to thrombin. Thrombin then triggers the transformation of a circulating soluble fibrinogen into fibrin fibers which form a clot that binds the platelets and plug the break in the blood vessel.

Topical application of D2R agonists to skin wounds accelerated wound healing, whereas the D2R antagonists delayed it. The D2R agonists also reduced epidermal hyperplasia induced by barrier disruption. The authors concluded that DA and its receptors are involved in epidermal barrier homeostasis.

Using D2R knockout mice (KOD2), another study reported the opposite effects of DA on wound healing [7]. KOD2 mice were subjected to hind limb wounding, followed by an induction of ischemia by blood vessel ligation. The recovery from ischemia was much faster in KOD2 mice than in wild-type controls, indicating that DA suppressed the healing process. There was also a significant difference in the postischemic angiogenesis between the two groups. Collectively, the two studies suggested that DA receptors (DARs) are involved in epidermal barrier homeostasis, but the exact mechanism by which DA acts, as well as the potential involvement of DARs other than D2R, remain to be determined.

Several studies have established direct actions of DA on the most critical cells involved in wound healing: platelets, keratinocytes, endothelial cells, and MSCs. Upon activation, such cells either physically produce a blood clot that plugs the wound, or secrete growth factors and cytokines that contribute to the different phases of wound healing [8]. Human platelets express D2R, D3R, and D5R [9], as well as DA transporter (DAT) [10], and DA was reported to promote platelet aggregation [11]. Human keratinocytes express TH and are capable of *de novo* synthesis of DA [12]. DA added to cultured keratinocytes increased the production of interleukin-6 (IL-6) and IL-8 in a dose-dependent manner [13]. The DA- and cabergoline-induced IL-6 release was partially reduced by sulpiride, the D2R antagonist, and was abrogated by propranolol, a beta-adrenergic receptor (β-AR) antagonist. Cell viability was not affected by any of these drugs. The authors concluded that DA agonists induce the keratinocytes to produce cytokines that are associated with the inflammatory cutaneous processes. These effects were mediated by both DAR and β-AR, as well as by receptor-independent oxidative mechanisms.

Angiogenesis is a critical step in the complex sequence of events leading to dermal wound healing. Interactions between DA and vascular endothelial growth factor (VEGF) in the control of angiogenesis were studied using human umbilical vein endothelial cells (HUVECs), an established endothelial cell line [14]. DA inhibited VEGF-induced phosphorylation of the VEGF receptor 2 (VEGFR-2), and suppressed endothelial cell proliferation, migration, and microvascular permeability, resulting in the overall inhibition of angiogenesis. The postulated mechanism by which DA affects VEGF actions is illustrated in **Figure 11.4**. Under normal conditions, D2R co-localizes with VEGFR-2 at the cell surface. DA pretreatment increased the translocation of Src-homology-2-domain-protein tyrosine phosphatase (SHP-2) to the cell surface, where it co-localizes with D2R. The subsequent VEGF treatment stimulates SHP-2 phosphorylation, and with it, its phosphatase activity which dephosphorylates VEGFR-2. The involvement of SHP-2 in this process was confirmed by showing that SHP-2 knockdown impaired the DA-regulated inhibition of VEGF-induced phosphorylation of VEGFR-2 and cell migration. The authors concluded that this study established a novel role for SHP-2 phosphatase in the DA-mediated inhibition of angiogenesis.

The role of MSCs in wound healing is of particular interest, given their capacity for transdifferentiation and their pro-angiogenic potential [3]. Adult MSCs are multipotent stem cells found in bone marrow and adipose tissue. Chemokines released by the wound increase the mobility of MSCs from their tissues of origin, promote their migration into blood, and from there into the wound, where they can differentiate into various cell types and accelerate the process of wound repair. MSCs also release proangiogenic factors like VEGF, which support the growth, survival and differentiation of endothelial cells that form new vasculature.

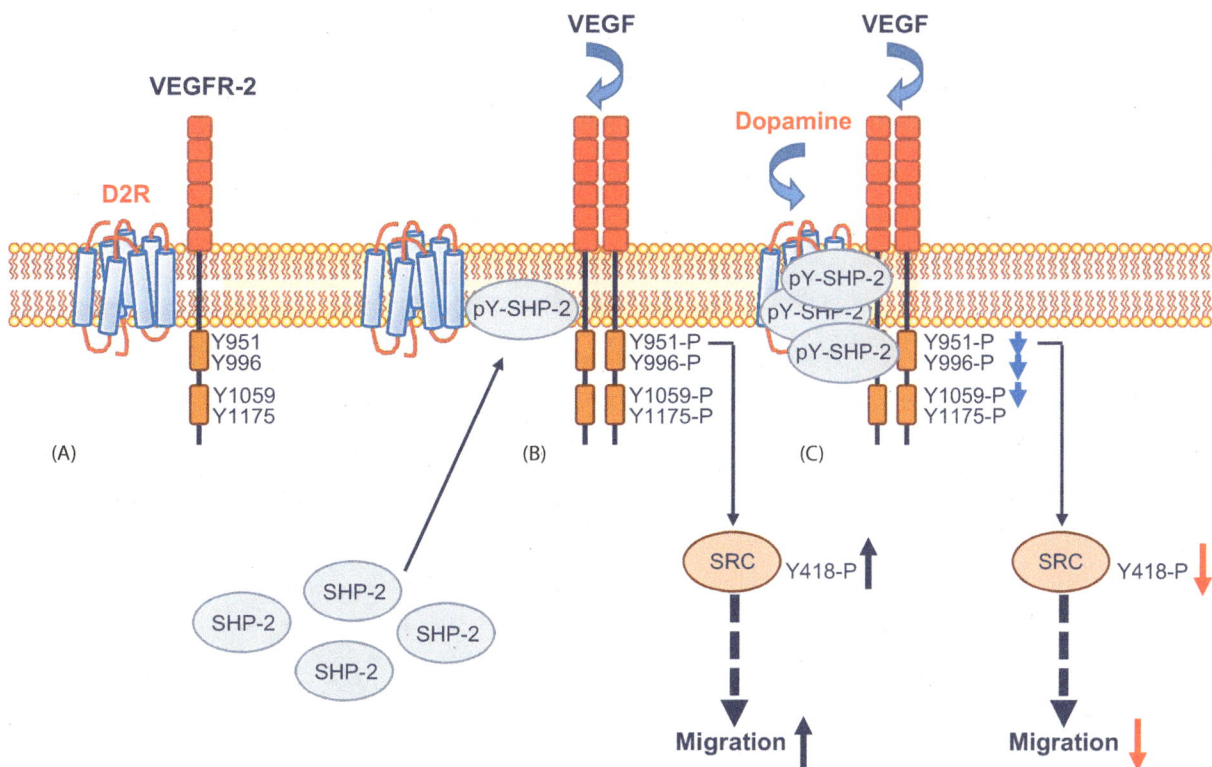

Figure 11.4 Interactions between the receptors for dopamine and VEGF. Panel A shows that in untreated human vascular endothelial cells (HUVECs), D2R and VEGFR-2 are associated with each other. Panel B shows that VEGF induces translocation of SHP-2 from the cytosol to the cell surface, where it becomes associated with both VEGFR-2 and D2R. VEGF also promotes the dissociation of VEGFR-2 from D2R, inducing VEGFR-2 phosphorylation and downstream signaling. Panel C shows that DA pretreatment, followed by VEGF stimulation, leads to increased association of D2R with VEGFR-2 and stimulation of SHP-2 phosphorylation which increases its phosphatase activity. Activated SHP-2 inhibits the phosphorylation of VEGFR-2 at several residues. Decreased phosphorylation of VEGFR-2 at Y951 leads to loss of Src phosphorylation at Y418 and lowering of its kinase activity, effectively blocking VEGF-induced endothelial cell migration. SHP-2: Src-homology-2-domain-protein tyrosine phosphatase; VEGF: vascular endothelial growth factor; VEGFR: vascular endothelial growth factor receptor. (Redrawn and modified from Strobl, J.S. et al., *Breast Cancer Res. Treat.*, 51, 83–95, 1998.)

Using *in vivo* and *in vitro* approaches, two laboratories have examined the effects of DA on MSCs. In one study, DA inhibited the mobilization of D2R-expressing murine MSCs into wound beds and their incorporation into new vessels by suppressing Akt phosphorylation and actin polymerization [15]. The inhibitory effect of DA was reversed by eticlopride, a specific D2R antagonist. The accelerated mobilization of MSCs to the wound site was accompanied by increased angiogenesis. The second study [16] used similar approaches but arrived at an entirely different conclusion, i.e., that DA induced MSCs migration via D2R. Administration of catecholamines *in vivo* also induced the mobilization of colony-forming fibroblasts. Both *in vitro* and *in vivo* migration of MSCs were suppressed by D2R antagonists and blocking antibodies. The latter study also found a significant increase in the MSC-like population (CD45⁻CD31⁻CD34⁻CD105⁺) in blood obtained from patients treated with L-Dopa or with catecholaminergic agonists. The reasons for the marked disagreement between the two studies are not clear, but could be due to site-specific involvement of DARs other than D2R.

11.2.4 Polydopamine nanopolymers used for drug delivery in wound healing

Hydrogels loaded with antibiotics or with various drugs have been used for several years to treat infections, or tissue damage caused by surgery, burns or ulcers. The catechol motif in Dopa-containing mussel adhesive proteins, which is responsible for the enormous ability of these shells to adhere to various surfaces, inspired the development of catechol-containing molecules and polymers for coating various materials for a wide range of applications. Among the catechol-containing coatings, **polydopamine (PDA)** has attracted the most attention [17]. PDA, synthesized by DA oxidation, has been used to coat metals, oxides, ceramics, polymers, carbon nanotubes, and magnetite nanoparticles. Although the exact molecular linkages within the polymers of PDA are still under debate, the consensus is that they are composed of indole units of different states of hydrogenation and are connected by CC bonds between the benzene rings. In addition to the indole units, there is evidence that PDA also contains DA units in the polymer backbone that are not cyclized, but have aminoethyl side chains [18].

A validation study [19] used gold nanoparticles (GNPs) with PDA nanolayers that formed uniform core/shell nanostructures (GNP/PDA). *In vitro* experiments showed that the PDA-coated GNPs had low cytotoxicity and were taken up by cancer cells. Analysis by transmission electron microscopy showed that the PDA nanoshells remained intact within the cells after 24 h incubation and were stable in liver and spleen cells for at least 6 weeks. When delivered to mice, PDA nanoshells showed no long-term toxicity in the main organs. These results established that PDA surface modification can serve as an effective strategy to form ultrastable coatings on nanoparticles (NPs) *in vivo*, which improve the intracellular delivery capacity and biocompatibility of NPs for many biomedical applications.

Another study developed photothermal therapy-assisted nanocatalytic antibacterial system, utilizing PDA coating on hydroxyapatite (HAp) incorporated with gold nanoparticles (Au-HAp). The PDA/Au-HAp nanoparticles produced hydroxyl radicals by catalyzing H_2O_2, rendering the bacteria vulnerable to killing by increased temperature [20]. At a controlled photo-induced temperature of 45°C, which caused no damage to normal tissues, the antibacterial efficacy against *Escherichia coli* and *Staphylococcus aureus* was better than 95%. Surgical wounds, made on the flanks of rats, were infected with bacterial suspension of *S. aureus*. A solution of the above nanoparticles as well as H_2O_2 were applied onto the wounds and were irradiated with a laser. Compared with controls, the synergistic antibacterial and PDA system not only killed bacteria effectively, but also promoted scar generation that protected the wounded tissues. The authors concluded that this system accelerated wound healing by stimulating tissue repairing-related gene expression to facilitate formation of granulation tissues and collagen synthesis.

A three-dimensional (3D) printer was used to generate functionalized 3D-printed bioceramic (BC) scaffolds coated with PDA [21]. Adipose-derived mesenchymal stem cells (Ad-MSCs) were cultured on the polydopamine-modified BC scaffolds (Dopa-BC), and conditioned media were collected and analyzed (**Figure 11.5**). The cells produced more immunomodulatory and pro-angiogenic factors than those cultured on regular BC scaffolds or on microplates. Functional assays such as endothelial cell migration, tube formation, and macrophage polarization, confirmed the enhanced paracrine functions of the secreted trophic factors from Ad-MSCs cultured on Dopa-BC scaffolds. Additional studies found that focal adhesion kinase- and extracellular signal-related kinase signaling were the transduction pathways through which the mussel-inspired surface stimulated the paracrine effect of Ad-MSCs.

The same investigators [21] also used diabetic rats with defective skin healing properties (**Figure 11.5**). Application of conditioned media harvested from the Ad-MSCs cultured on Dopa-BC, accelerated wound closure in the diabetic rats, enhanced vascularization, and promoted macrophage switching from a proinflammatory M1 to a pro-healing and anti-inflammatory M2 phenotype in the wound bed. The authors concluded that coating with PDA represents an effective method for enhancing the beneficial paracrine function of MSCs. These findings offer a novel strategy for accelerating tissue regeneration by guiding the paracrine-signaling network.

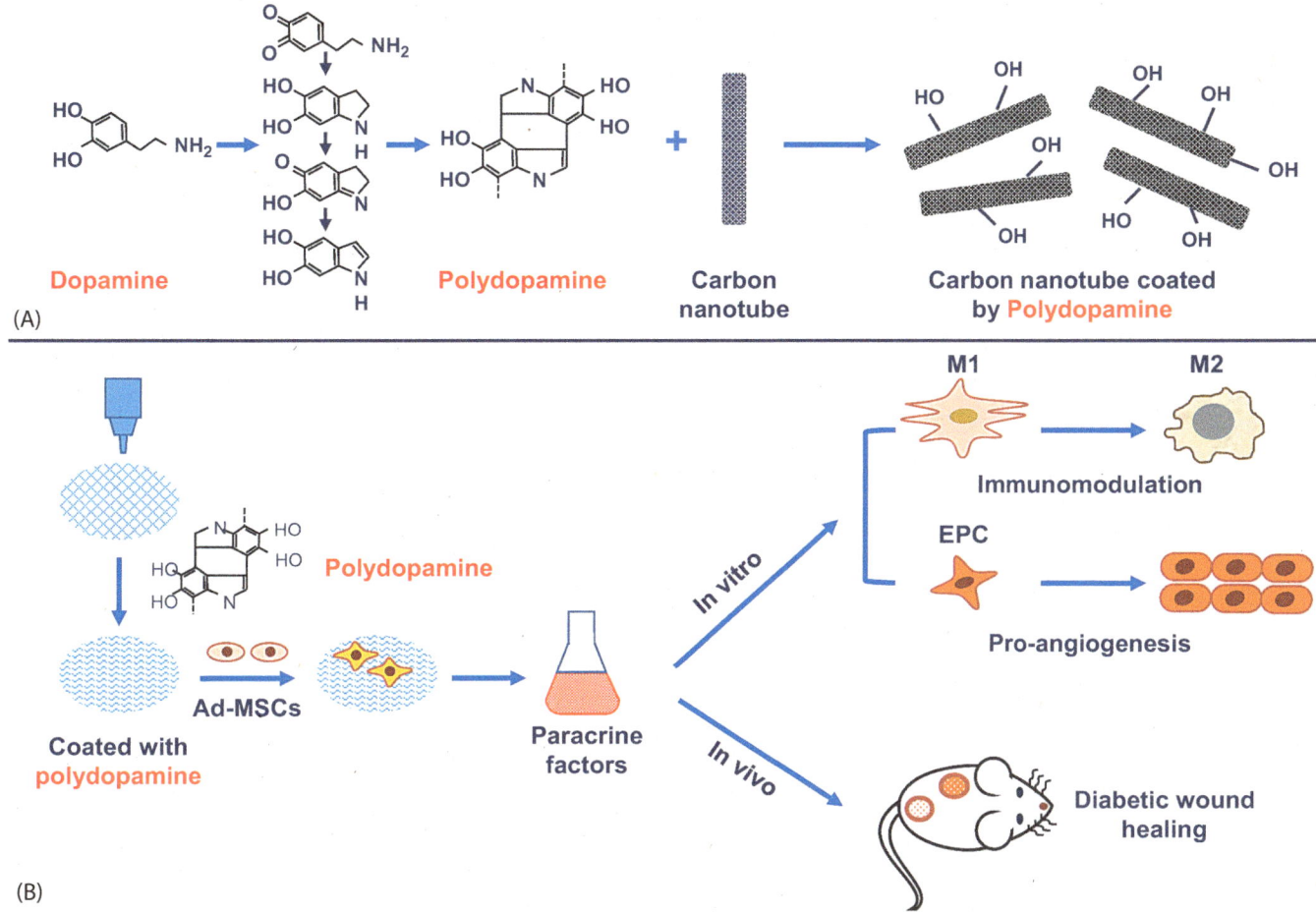

Figure 11.5 Use of polydopamine nanoparticles in wound healing. Panel A shows that oxidation of dopamine (DA) results in the formation of polydopamine, which is used for coating substrates such as carbon nanotubes. Panel B shows that adipose-derived mesenchymal stem cells (Ad-MSCs) cultured on polydopamine modified bioceramic scaffolds produced more immunomodulatory and pro-angiogenic factors than those cultured on control scaffolds or on microplates. Functional assays such as migration of endothelial cells, macrophage polarization, and wound healing *in vivo* revealed enhanced paracrine functions of secreted factors from these cells. EPC: endothelial progenitor cells. (Redrawn and modified from Li, T. et al., *ACS Appl. Mater. Interfaces*, 11, 17134–17146, 2019.)

In conclusion, the main appeal of PDA in biomedical applications is that it is a robust, relatively inert, nonpoisonous, and biocompatible coating material. The developers of this technology also emphasize that the structure of PDA allows an easy functionalization through reactions with amino- or mercapto-nucleophiles. Thus, PDA coatings can serve as a medium for incorporating various antibiotics, biomolecules or enzymes. It is surprising, however, that no attempts have been made to determine whether bioactive DA is released from the polymers, similar to the situation with bisphenol A being released from polycarbonate plastics, as discussed in **Chapter 10**. This possibility can be easily determined by analyzing the potential activation of DARs as well as testing whether the various PDA nanoparticles are as effective in DAR-deficient cells or mutant mice.

11.3 PIGMENTATION AND MELANOGENESIS

11.3.1 Overview of skin pigmentation and melanogenesis

As discussed in an extensive review [22], the human skin comes in a wide range of colors and gradations, from white to brown to black; these skin shades are commonly used to define ethnicity. Skin pigmentation is due to the presence of a chemically inert and stable pigment known as melanin, which is produced deep inside the skin but is displayed as a mosaic at its surface. Minor changes in the physiological status of the human body or exposure to harmful external factors can affect the pattern of pigmentation either in a transitory (e.g., in pregnancy), or in a permanent (e.g., age spots) manner. Skin pigmentation is determined by intricate cellular and molecular interactions between melanocytes and keratinocytes.

During embryogenesis, progenitor melanoblasts originating from the neural crest migrate to their final location in the interfollicular epidermis and epidermal hair follicles [23,24]. In the adult skin, functional melanocytes are continuously repopulated by the differentiation of melanocyte stem cells (McSCs) residing in the epidermis. To date, knowledge of McSCs largely comes from studying the stem cell niche of mouse hair follicles. Because of anatomical differences between the mouse and human skin, there are distinct features associated with mouse and human McSCs, as well as their niches in the skin. Given the neuroectodermal origin of melanocytes, they share structural, biochemical, and cellular features with neurons.

Three types of melanin are recognized: eumelanin (brown and black color), pheumelanin (red color), and neuromelanin (found in DA neurons in the substantia nigra). Without melanin, the human skin is pale white with shades of pink caused by blood flow through the skin. Fair-skinned people produce little melanin, darker-skinned people produce moderate amounts, and very dark-skinned people produce the most [22]. Notably, the difference in skin color between lightly and darkly pigmented individuals is not due to the number of melanocytes in their skin but rather to the quantity and relative amounts of eumelanin and pheomelanin.

Melanocytes and keratinocytes form a functional epidermal pigment unit. As illustrated in **Figure 11.6**, melanin-containing organelles named melanosomes are transferred from the melanocytes to the keratinocytes through dendrite-like projections [25]. As the keratinocytes move from the lower part of the epidermis up to the surface, they carry the ingested melanosomes with them, resulting in a deposition of the melanin-filled melanosomes near the skin's surface. Melanin dissipates most of the absorbed ultraviolet (UV) radiation and protects the skin from damage by UVB radiation.

Melanocytes interact with endocrine, immune, and inflammatory systems as well as the central nervous system, and their activity can be regulated by drugs [26]. Melanogenesis is under hormonal control, primarily by α-melanocyte stimulating hormone (α-MSH), and adrenocorticotropic

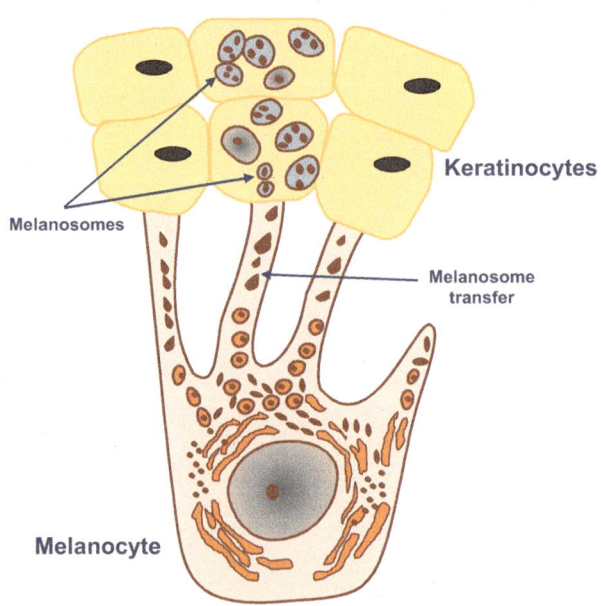

Figure 11.6 Epidermal pigment unit composed of melanocytes and keratinocytes. (Redrawn and modified from http://www.midwestscc.org/blog2/wp-content/uploads/presentations/Teamworks2012HypopigmentingAgents.pdf.)

hormone (ACTH). As discussed in **Chapter 5**, both hormones are cleavage products of the proopiomelanocortin (POMC) precursor and are produced in the pituitary gland as well as in skin cells, primarily keratinocytes. Within their molecular structure, α-MSH and ACTH share the tetrapeptide His-Phe-Arg-Trp, which is essential for the melanotropic activity.

The melanocortin 1 receptor (MC1R) is a G_s protein-coupled receptor that regulates skin pigmentation, UV responses, and also the risk of melanoma. It is a highly polymorphic gene whose loss of function correlates with a fair, UV-sensitive, and melanoma-prone phenotype due to defective epidermal melanization and suboptimal DNA repair. MC1R signaling, executed via adenylyl cyclase activation and cAMP generation, is hormonally controlled by positive agonists such as α-MSH and ACTH, a negative agonist agouti signaling protein, and the neutral antagonist β-defensin 3. Activation of cAMP signaling up-regulates melanin production and deposition in the epidermis.

Skin exposure to UV radiation stimulates melanogenesis by activating tyrosinase, a key enzyme of the biosynthetic pathway (**Figure 11.7**). Tyrosinase is a copper-containing glycoprotein enzyme located in the membrane of the melanosome [27]. It has an inner melanosomal domain that contains the catalytic region, followed by a short transmembrane domain and a cytoplasmic domain composed of about 30 amino acids. Tyrosinase catalyzes the first two steps of melanin production: hydroxylation of L-tyrosine to L-Dopa and the subsequent oxidation of L-Dopa to the corresponding quinone, L-dopaquinone. The concentration of L-tyrosine for melanogenesis depends on the conversion of L-phenylalanine by intracellular phenylalanine hydroxylase (PAH) activity. In contrast to L-tyrosine, which can only be transported into the melanosome by facilitated diffusion, L-phenylalanine is actively transported through the melanosomal membrane to ensure a high content of L-tyrosine inside this organelle.

Following the formation of dopaquinone, the melanin pathway divides into production of the black-brownish eumelanin and red-yellow pheomelanin. In the eumelanin pathway, dopachrome is either spontaneously converted to 5,6-dihydroxyindole or is enzymatically converted to 5,6-dihydroxyindole-2-carboxylic acid via enzymatic conversion by dopachrome tautomerase (DCT), also referred to as tyrosine-related protein-2 (TRP-2). The two TRP enzymes, TRP-1 and TRP-2, share ~40% amino acid homology with tyrosinase. They reside within the melanosomes and, like tyrosinase, span the melanosomal membrane. It has been suggested that TRP-1 increases the

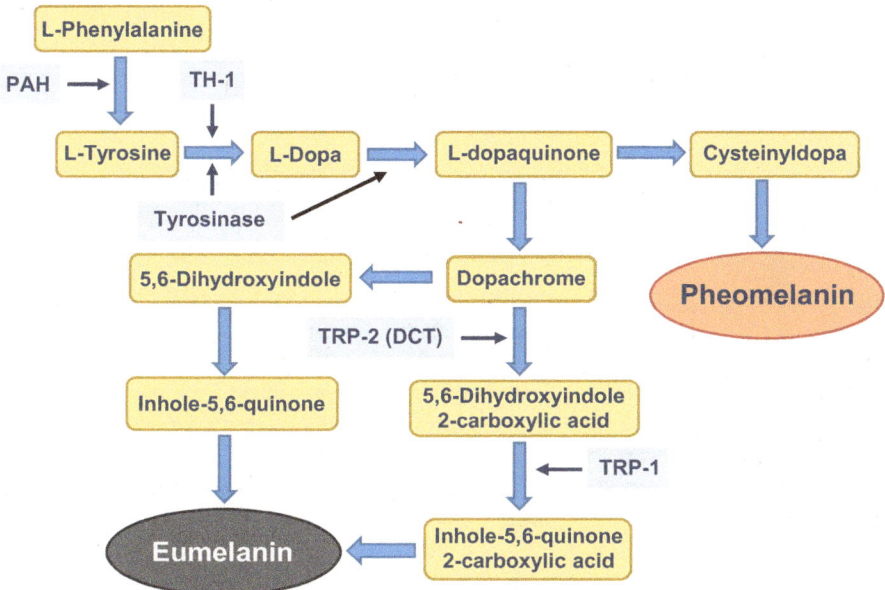

Figure 11.7 Biosynthesis of pheomelanin and eumelanin. DCT: dopachrome tautomerase; PAH: phenylalanine hydroxylase; TH-1: tyrosinase hydroxylase-1; TRP-1 and TRP-2: tyrosine-related protein-1 and -2. (Redrawn and modified from Gillbro, J.M. and Olsson, M.J. *Int. J. Cosmet. Sci.*, 33, 210–221, 2011.)

ratio of eumelanin to pheomelanin and also increases tyrosinase stability [27]. Finally, polymerization of indoles and quinones leads to eumelanin formation. The pheomelanin pathway branches from the eumelanin pathway at the L-dopaquinone step and depends on the presence of cysteine, which is actively transported through the melanosomal membrane. Cysteine reacts with L-dopaquinone to form cysteinyl-dopa. The latter is converted to quinoleimine, alanine-hydroxyl dihydrobenzothazine, and polymerizes to pheomelanin.

11.3.2 Selected disorders of pigmentation

Disorders of pigmentation are diagnosed as skin that is discolored, blotchy, darker, or lighter than normal. These changes occur when the body either produces too little (hypopigmentation) or too much (hyperpigmentation) melanin. The disorders can be localized in restricted sites or can diffusely spread throughout the body. Some pigmentation disorders such as liver spots, are common, whereas others, such as albinism, are rare. In some disorders, the cause of dyspigmentation is readily identified as sun exposure, drug reactions, or inflammation. In other cases, the etiology is not as clear. Evidently, most disorders can be diagnosed by appearance.

Albinism is a group of rare congenital diseases of pigmentation characterized by a complete or a partial absence of pigment in the skin, hair and eyes. It affects people of all ethnic backgrounds and its worldwide frequency is about 1 in 17,000. The prevalence of the different forms of albinism varies by population and is highest overall in people of sub-Saharan African descent. Albinism results from an absence or a defect of tyrosinase or other enzymes involved in melanin synthesis [28]. The lack of melanin or its reduction can affect the eyes, skin, and hair (oculocutaneous albinism, OCA), or the eyes only (ocular albinism, OA). Notably, OA is characterized by inadequate L-Dopa and DA pathways in the eyes, as was determined by studying DA-related chiasm projection patterns in the albino mouse and the role of ocular albinism type 1 (OA1) and dopamine 1A (D1A) receptors in the optic pathway [29].

Albinism is associated with several vision defects such as photophobia and nystagmus. The retina can be damaged by unfiltered light entering the eye through the iris because of the lack of normal pigmentation of the iris. The genetic mutation associated with albinism may also render the optic nerve within the eye dysfunctional due to improper development. Sunlight can cause damage to the skin more easily, and people with albinism are at a much greater risk for developing skin cancer.

Although there is no cure for albinism, it can be managed depending on the severity of its symptoms. Because the skin is more prone to sunburn, wearing long sleeves and full-length garments is mandatory to protect the skin against UV exposure. These individuals should also use a sunscreen with a high sun protection factor. They are also advised to check for changes on the skin such as a new or a changing mole, an abnormal growth or a lump. Glasses or contact lenses are often used to correct eyesight issues such as near-sightedness, long-sightedness and astigmatism. Sun glasses that filter out UV rays help patients with photophobia or sensitivity to light. Eye exercises may be recommended to deal with squinting and lazy eye.

Melasma is an acquired hypermelanosis of sun-exposed areas of the skin. It is a patchy brown discoloration of facial skin and is more common in women than men [30]. The exact cause is elusive, but the two most important factors implicated in its pathogenesis are sunlight and genetic predisposition. Melasma can also be caused by various endocrine disorders and is especially prevalent in patients with primary adrenal cortical insufficiency or Addison's disease. The negative feedback by corticosteroids on the hypothalamo–pituitary axis in these patients is reduced or absent, leading to increased production of ACTH and α-MSH, whose melanocyte-stimulating effect is responsible for the brown color of the skin. Hyperpigmentation usually precedes other symptoms of Addison's disease and is intensified at sites that are exposed to light and pressure: skin folds, lines of the hands, nipples, and areas of scarring. Other factors that induce melasma include pregnancy and exogenous hormones such as oral contraceptives, hormone replacement therapy, and thyroid dysfunction. Treatments include sun protection, topical formulations (hydroquinone, tretinoin), chemical peels, and fractional laser therapy.

Vitiligo is characterized by an uneven distribution of melanin in spots or patches. It is a rather common disease, affecting 0.5%–2% of the general population, with no clear sex predilection. Vitiligo is an autoimmune disorder that causes destruction of the melanocytes. Patients with vitiligo are more likely than average to have other autoimmune disorders. It generally begins on the hands and feet and then spreads to the extensor aspects of the joints and to the face. Segmental vitiligo has a more favorable prognosis than the symmetrical, diffuse variant. The absence of functional melanocytes in a vitiligo skin is probably the result of their destruction. The course of the disease is unpredictable, with periods of stabilization followed by periods of exacerbation. Treatment modalities include topical corticosteroids, topical calcineurin inhibitors and phototherapy.

11.3.3 Role of dopamine in skin pigmentation and its disorders

In addition to having a distinct tyrosinase-to-melanin biosynthetic pathway, melanocytes express tyrosine hydroxylase (TH), Dopa decarboxylase (DDC), monoamine oxidase (MAO), D1R and D2R [31,32], thus having a full capacity for both production and responsiveness to DA. There are uncertain clinical associations between DA, vitiligo, and skin pigmentation, although some studies reported increased urinary levels of homovanylic acid (HVA) and vesicular monoamine transporter (VMAT), which are derived respectively from DA and norepinephrine/epinephrine (NE/Epi) in vitiligo patients [33,34]. As discussed in **Chapter 13**, an uncontrolled proliferation of the melanocytes results in melanoma, the most aggressive type of skin cancer.

The capacity of DA to alter melanogenesis and skin pigmentation had been demonstrated in three *in vitro* studies. In an early study [35], frozen sections of vitiliginous skin were treated with various substrates presumed to be involved in melanogenesis, followed by analysis of their effects on dendritic melanocytes. Tyrosine, when used alone, had a weak melanogenic reaction, while Dopa and DA enhanced melanogenesis; such changes were not seen in non-dendritic melanocytes. A second study [36] reported that a DA conjugate [niacinamide, N-nicotinoyl dopamine (NND)] caused pigment lightening in a reconstructed skin model by decreasing melanin production without affecting the viability and morphology of the melanocytes or the overall tissue histology. Depigmentation and whitening of skin by the conjugate were also seen by applying nonirritating 0.1% NND on the forearm skin in human volunteers. A third study used cultured human melanocytes [37]. Among the four monoamines tested, only DA caused a dose-dependent decrease in melanocyte viability at doses from 0.01 to 100 μM. Treatment with DA led to the generation of reactive oxygen species (ROS) and resulted in apoptosis. The authors concluded that DA-induced apoptosis may play a role in development or progression of vitiligo, which is often viewed as a disease closely related to apoptosis of melanocytes.

A potential role of the cGMP/PKG pathway in melanin synthesis was also explored [38]. Incubation of B16 melanoma cells with the phosphodiesterase 5 (PDE5) inhibitors sildenafil and vardenafil, or with the cGMP analog 8-CPT-cGMP, stimulated cAMP response element binding protein

(CREB) phosphorylation, increased tyrosinase expression, and enhanced melanin synthesis. These changes were counteracted by KT5823, a selective PKG inhibitor. Because KT5823 did not affect melanin synthesis induced by cAMP-elevating agents, it indicated a selective inhibition of a cGMP-induced melanin synthesis. The authors concluded that PKG-dependent CREB phosphorylation stimulates tyrosinase expression, and that melanin synthesis is promoted when cGMP degradation is blocked. They also proposed that PDE5 inhibitors could be beneficial for the treatment of hypopigmentation diseases.

11.4 SWEATING

11.4.1 Overview of sweating, apocrine, and eccrine glands

Secretion of sweat is an important mechanism for temperature regulation. When body temperature rises, the sympathetic nervous system stimulates the sweat glands to secrete water to the skin surface, where it cools the body by evaporation. Sweating works in concert with cutaneous vasodilation to optimize heat dissipation during hyperthermia. Sweating also provides a skin protective mechanism because antibodies and immunoglobulins in sweat can prevent bacterial colonization on the surface of skin.

Although sweating is found in many mammals, only relatively few, e.g., humans and horses, produce large amounts of sweat in order to cool down. Animals with few sweat glands such as dogs, accomplish temperature regulation by panting, which evaporates water from the moist lining of the oral cavity and pharynx. Rodents, on the other hand, do not pant and have only a few sweat glands in their foot pads. Instead of sweating, they deal with higher temperatures by decreasing their metabolism.

As shown in **Figure 11.8** and is detailed in **Table 11.1**, the skin has two types of sweat glands. One type are the **apocrine sweat glands**, which secrete fluid into the sac of the hair follicle through which it comes out on the skin. The other type are the **eccrine sweat glands**, which secrete sweat directly onto the surface of the skin. Accordingly, apocrine glands are located deep in the layers of the skin while eccrine glands are more superficially placed. Sweat glands are composed of an intraepidermal spiral duct, a dermal duct made of a straight and a coiled portion, and a secretory tubule, coiled deep in the dermis or hypodermis. The coiled portion is formed by two concentric layer of columnar or cuboidal epithelial cells. The epithelial cells are interposed by myoepithelial cells that support the secretory epithelial cells. The duct of the gland is formed by two layers of cuboidal epithelium and open through the sweat pore. The secretory part of the apocrine glands is larger than that of eccrine glands.

Apocrine glands are found in the armpits, areola of the breast, perineum (area between anus and genitals), in the ear, and in the eyelids. Eccrine glands are present all over the body, except for the abovementioned parts. Apocrine sweat glands secrete a fluid containing pheromone-like compounds that attract the opposite sex. This phenomenon occurs in all mammals, including humans. A distinguishing feature of apocrine glands is that they are inactive before puberty. The hormonal surge during puberty brings about a change in the size of the apocrine glands and starts their functioning. In sum, apocrine glands are considered as modified glands that secrete wax in the ears, milk in the breast and secretions of the ciliary glands in the eyelids, while eccrine glands are the major sweat glands of the body and are widely distributed.

11.4.2 Dopamine and sweating

Sweat glands in most mammalian species are innervated by sympathetic cholinergic neurons, but both the distribution and type of innervation of the sweat glands differ markedly among species. During early development

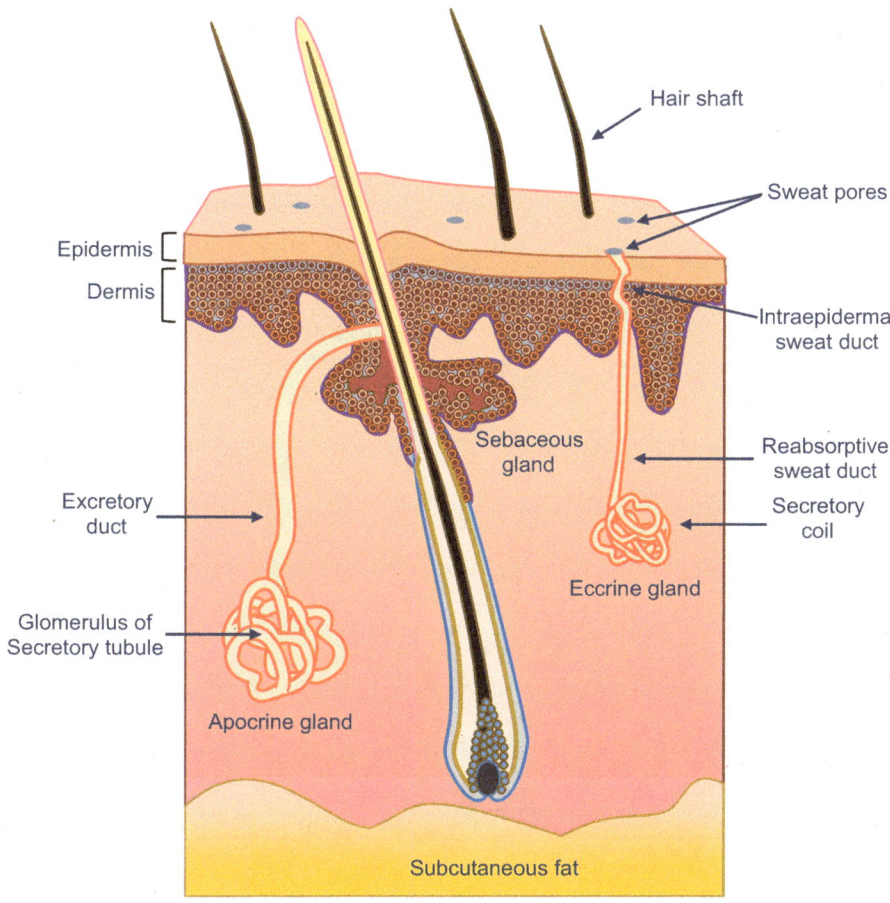

Figure 11.8 Diagram of the structure and location of sweat glands in human skin. Shown are an apocrine gland with an opening into the sac of an hair follicle, and an eccrine gland with an opening on the surface of the skin.

Table 11.1 Characteristics of apocrine and eccrine glands in humans	
Apocrine glands	**Eccrine glands**
Occur in the armpits and perianal areas	Occur all over the skin
Diameter is 80–100 μm	Diameter is 30–40 μm
Secretory unit is made up of simple columnar epithelium	Secretory unit is made up of mixed clear and dark cells
Secretes viscous fluid into hair follicle	Secretion is watery
Duct opens to a hair follicle	Duct opens to the skin surface
Secretion consists of proteins and fatty acids	Secretion consists of water and sodium chloride
Secretion is responsible for body odor	Secretion is responsible for body cooling

in rodents, the sweat glands transiently expresses noradrenergic innervation, which is lost in adult animals [39]. On the other hand, sweat glands in adult humans maintain the expression of all catecholamine biosynthetic enzymes, including TH, DDC, dopamine β-hydroxylase (DBH), as well as vesicular monoamine transporter 2 (VMAT2) [40]. The authors stated that cholinergic and noradrenergic co-transmission is a unique feature of sweat gland innervation in primates, although they did not consider the possibility that in addition to NE, DA also functions as a local neurotransmitter.

A number of non-motor dysfunctions occur early in the development of Parkinson's disease (PD) that exert profound effects on the patient's quality of life. Among these are excessive salivation, profuse sweating, sensory denervation, and altered pain sensation [41–43]. PD is characterized by depletion of DA within selective areas of the brain and presumably also in some peripheral sites. Therapeutic targeting of PD with DA agents, as well as the effects of DA deficit itself, impact upon a number of skin functions. Evidence indicates that PD patients have alterations in sweat gland functions and decreased innervation to blood vessels and erector pilli muscles in the skin [44]. There was a marked reduction in sweat gland innervation in the majority of skin biopsy samples from PD patients, as manifested by the presence of nerve-devoid (naked) sweat glands or sparse fiber density around the glands. PD patients also had increased thresholds for cold and heat pain, mechanical pain, and touch, and increased ulceration. However, the study did not determine whether the observed innervation deficit in the sweat glands was specifically due to loss of cholinergic and/or dopaminergic/noradrenergic fibers.

11.5 HAIR GROWTH

11.5.1 Overview of hair growth and its hormonal regulation

Hair serves some protective functions and plays a role in sexual attraction [45,46]. It is found everywhere on the external body except for mucus membranes, palms and soles. Hair can be categorized as vellus (fine, soft and nonpigmented) and terminal (long, coarse and pigmented). The part of hair seen on the surface of the skin is the shaft, while that below the skin surface is the root. The root, together with its epithelial and connective tissue covering, is called the hair follicle. Hair is made of a tough protein called keratin and is anchored into the skin by the hair follicle. The hair follicle is a dynamic organ which resides in the dermal layer of the skin and is made of multiple cell types, each with distinct functions. For example, terminal hair grow on the scalp and lanugo (fine, soft hair) cover the body of fetuses and some newborn babies. The hair follicle regulates hair growth via a complex interaction between hormones, neuropeptides and immune cells.

The base of the hair follicle, the hair bulb, contains cells that divide and grow and build the hair shaft (**Figure 11.9**A). Blood vessels nourish the cells in the hair bulb, and deliver hormones that modify hair growth and structure at different times of life. The bulge is located in the outer root sheath at the insertion point of the arrector pili muscle. The arrector muscles contract in response to cold or fear, making the hair stand up straight. The bulge houses several types of stem cells, which supply the entire hair follicle with new cells and take part in healing the epidermis after a wound. Hair grows at different rates in different people; the average rate is around one-half inch per month. Hair color is created by pigment cells producing melanin in the hair follicle. With aging, pigment cells die, and hair turns gray.

The process of hair growth occurs in three sequential stages (**Figure 11.9**B). The first stage is the active growth phase called *anagen*. The second stage is *catagen*, representing a transition phase, when the hair follicle detaches from the blood supply. *Telogen*, or the resting phase, is the phase when the hair dies. After that, new hair begins to grow in the same follicle, pushing the old hair out. On average, the scalp has more than 100,000 hairs, 90% of which are in the growth phase, and about 90–100 hairs are shed every day. All stages occur simultaneously—while one strand of hair may be in the anagen phase, another may be in the telogen phase. Each has specific characteristics that determine the length of the hair.

The number of hair follicles does not change over the individual's lifetime, but the follicle size and type of hair can change in response to numerous factors and hormones, particularly to androgens [47]. For example, eyebrow, eyelashes and vellus hair are androgen-insensitive, whereas the axillary and

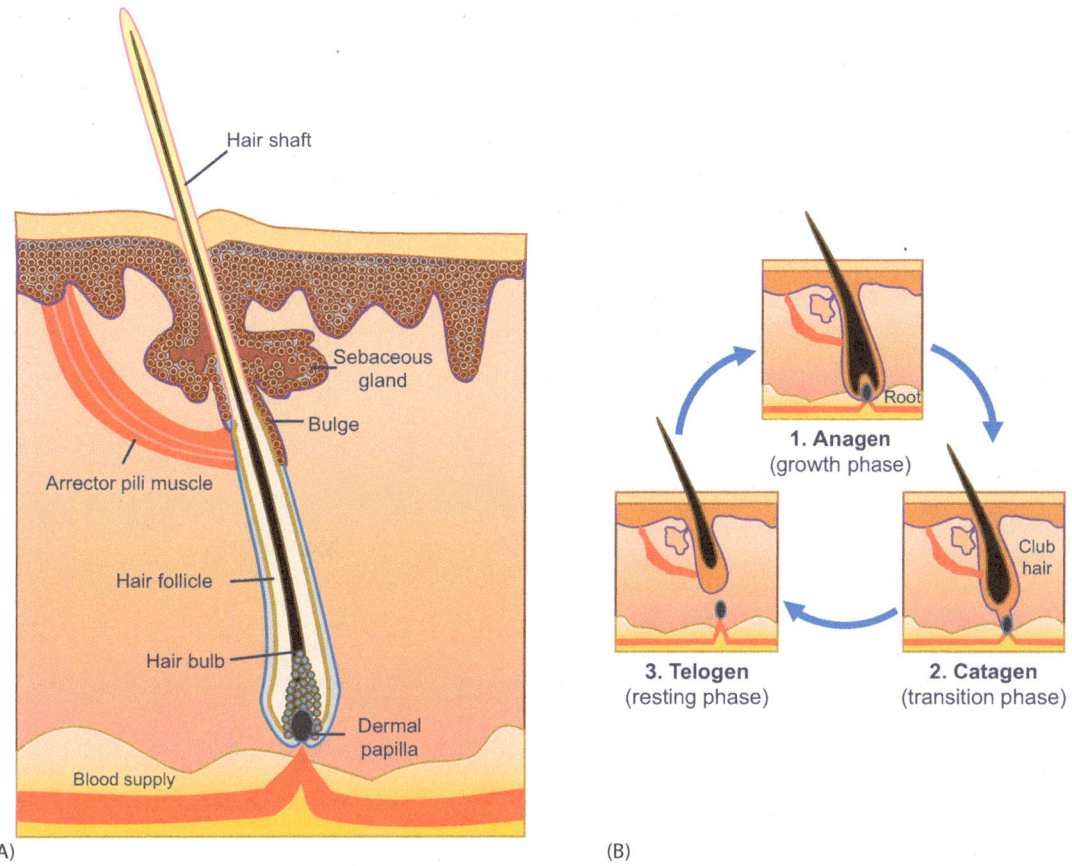

Figure 11.9 Diagram of an hair follicle. Panel A shows the dermal papilla, blood supply, hair shaft, hair bulge arrector muscle, and adjacent sebaceous gland. Panel B depicts the three sequential stages of hair growth: anagen, catagen, and telogen. See text for other explanations.

pubic areas are sensitive to low levels of androgens. Hair growth on the face, chest, upper abdomen and back requires greater levels of androgens and is, therefore, more characteristic of the hair distribution in men. Androgen excess in women (called hirsutism), which occurs in polycystic ovary syndrome (PCOS) and idiopathic hyperandrogenism, leads to increased hair growth in most androgen-sensitive sites. On the scalp, however, there is hair loss because androgens cause scalp hair to spend less time in the analgen phase.

11.5.2 Role of dopamine in hair growth

Alopecia refers to loss of hair from part of the head or body. The severity of hair loss can vary from a small area to the entire body [45,48]. Alopecia is an autoimmune skin disease that affects as many as 6.8 million people in the United States, with a lifetime risk of 2.1%. People of all ages, both sexes and all ethnic groups can develop the disease. Alopecia often first appears during childhood and can be different for various individuals. The causes that trigger the immune system to attack healthy hair follicles are not clear but could be internal (virus or bacteria), due to stress, or environmental. Some medications, including chemotherapy, HIV/AIDS, hypothyroidism, and malnutrition (e.g., iron deficiency), can also cause hair loss. The causes of hair loss accompanied by scarring or inflammation include fungal infection, lupus erythematosus, radiation therapy, and sarcoidosis. Diagnosis of hair loss is partly based on the areas affected. Common interventions include medications such as minoxidil and hair transplant surgery.

Treatment with L-Dopa or DA agonists is associated with development of alopecia, mainly in women [49,50]. In most cases, hair loss was reversible after cessation of medication, but persisted in some patients. Microdissected scalp hair follicles (HFs) from women were used to determine whether DA has a direct impact on hair growth [51]. Tissues were incubated with increasing doses of DA (10–1,000 nmoles), followed by analysis of hair shaft production, HF cycling, keratinocyte proliferation, and apoptosis. Whereas DA had no consistent effects on hair shaft production, it promoted HF regression, which was associated with reduced proliferation of HF keratinocytes, and lower intrafollicular melanin production. The study also found expression of D1R transcripts in HFs and skin. The authors concluded that DA inhibits hair growth by promoting catagen, which lasts between 10 and 20 days in humans. During catagen, the lower portion of hair follicle regresses, and hair growth ceases. The authors stated that these data offer a plausible explanation for the induction of hair loss in some women by bromocriptine. They proposed that DA agonists should be explored as inhibitors of unwanted hair growth such as hirsutism and hypertrichosis (an abnormal, werewolf-like hair growth).

An indirect role of DA in the control of hair growth occurs via its inhibitory effects on prolactin (PRL) release [52]. Hyperprolactinemia in women is associated with diffuse hair loss in the scalp in 82% of patients with high PRL levels. This hair loss was characterized by uniform shedding of the hair in central, parietal and occipital region of the head. During the hair follicle transformation from growth (anagen) to regression (catagen), PRL and prolactin receptor (PRLR) immunoreactivity were up-regulated. Treatment of the hair follicles with high doses of PRL caused a significant inhibition of hair shaft elongation and premature catagen development, along with reduced proliferation and increased apoptosis of hair bulb keratinocytes. From these data it was suggested that PRL acts as an autocrine hair growth modulator with hair growth inhibitory effects.

11.6 THE SKELETON AND BONE REMODELING

11.6.1 Overview of the skeleton and bone remodeling

The human skeleton provides the internal framework of the human body and is divided into the axial skeleton and appendicular skeleton. The axial skeleton is composed of the vertebral column, rib cage, skull, and other associated bones. The appendicular skeleton is attached to the axial skeleton and is composed of the shoulder and pelvic girdles, and upper and lower limbs. The bone mass in the skeleton reaches its maximal density around age 21, followed by progressive loss of bone tissue with advanced age. The aging bone has reduced mineral content, and is prone to osteoporosis—a condition in which bones are less dense, more fragile, and predisposed to fractures [53].

The human skeleton performs six major functions: support, movement, protection, production of blood cells, storage of minerals, and endocrine regulation. The skeleton protects many vital organs from being damaged. For instance, the skull protects the brain, the vertebrae protect the spinal cord, and the rib cage, spine and sternum protect the lungs, heart and major blood vessels. The skeleton is also the site of hematopoiesis, which takes place in the bone marrow. The bone matrix stores calcium, iron and ferritin and is involved in calcium and iron metabolism. Bone is not only a target of several endocrine signals, but also acts as an endocrine tissue by secreting hormones such as fibroblast growth factor 23 (FGF23) and osteocalcin, which are implicated in the regulation of phosphate homeostasis and of energy metabolism, respectively [54].

Bone is a metabolically active tissue composed of several cell types, including osteoblasts, osteocytes, and osteoclasts. Osteoblasts are involved in the creation and mineralization of bone tissue while osteocytes are mostly inactive, and are in contact with other cells in the bone through gap junctions. Osteoclasts are responsible for the breakdown of bone by the process of bone resorption. Osteoblasts and osteocytes are derived from osteoprogenitor cells. They are connective tissue cells found at the surface of bone, which can be stimulated to proliferate and differentiate. Osteoclasts are large, multinucleate cells formed through the fusion of precursor cells. They are derived from a monocyte stem-cell lineage and similar to macrophages have phagocytic properties. As discussed in **Chapter 9**, the bone marrow contains hematopoietic stem cells which give rise to white blood cells, red blood cells and platelets.

Bone remodeling occurs throughout a person's life, and involves bone resorption followed by replacement, with little change in shape [55]. Repeated stress, such as weight-bearing exercise or bone healing, results in bone thickening at the points of maximum stress. Bone remodeling involves removal of mineralized bone by osteoclasts followed by formation of matrix that becomes mineralized by osteoblasts. As shown in **Figure 11.10**, the remodeling cycle consists of several phases: resorption, during which osteoclasts digest old bone; reversal, when mononuclear cells appear on the bone surface; and formation, when osteoblasts lay down new bone. Remodeling adjusts the bone architecture to meet changing mechanical needs and helps to repair micro-damages in the bone matrix. It also plays a role in maintaining plasma calcium homeostasis.

The regulation of bone remodeling is both systemic and local. The major systemic regulators are parathyroid hormone (PTH), calcitriol, and hormones such as growth hormone, glucocorticoids, thyroid hormones, and sex hormones. Insulin-like growth factors (IGFs), prostaglandins, tumor growth factor-beta (TGF-β), and bone morphogenetic proteins (BMPs) are involved as well. Local regulation of bone remodeling is carried out by a large number of cytokines and growth factors. Through the RANK/receptor activator of nuclear factor kappa B (NF-κB) ligand (RANKL)/osteoprotegerin system, the processes of bone resorption and formation are tightly coupled, allowing a wave of bone formation to follow each cycle of bone resorption, thereby maintaining skeletal integrity.

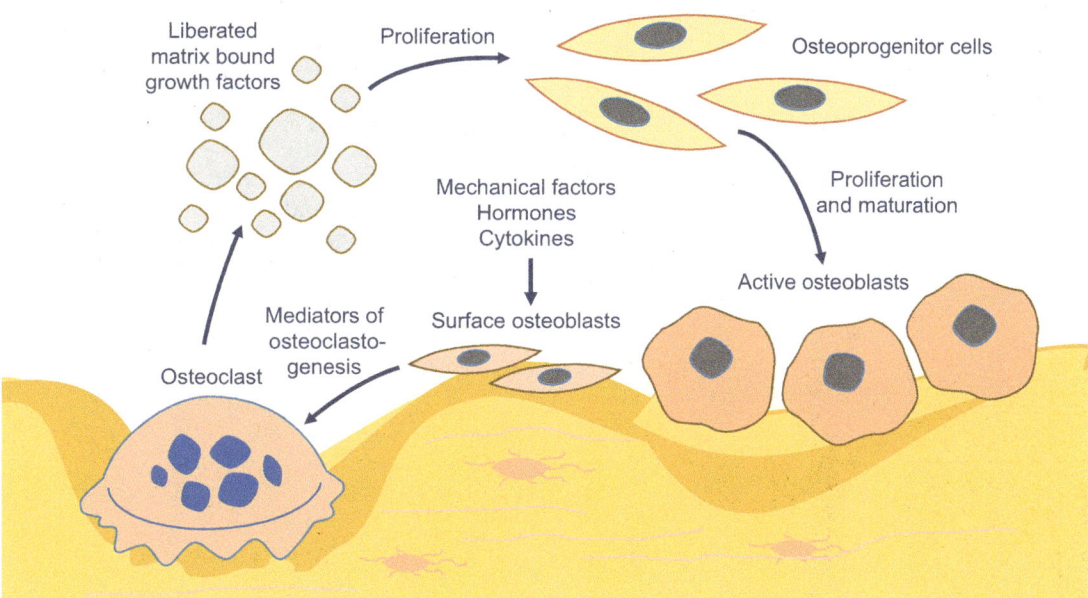

Figure 11.10 The cycle of bone remodeling. Shown are bone digestion by osteoclasts and bone rebuilding by osteoblasts. (Redrawn and modified from https://afrahalmutairi.wordpress.com/2015/12/12/bone-remodeling/.)

11.6.2 Involvement of dopamine in bone homeostasis

Atypical antipsychotics are widely used to treat many neuropsychological disorders that include schizophrenia, bipolar disorder, and autism. In addition to inducing obesity and causing adverse metabolic side effects, drugs with a strong antidopaminergic activity such as risperidone (RIS) also cause bone loss and increased fracture risk in both men and women. One potential mechanism of bone loss by RIS is hypogonadism due to hyperprolactinemia that results from D2R antagonism. However, many patients have normal PRL levels, suggesting a direct action of the drug on bone constituents. To examine for this possibility, mice were treated daily for 8 weeks with RIS or vehicle [56]. RIS caused a significant trabecular bone loss, which was due to increased bone turnover. All five DARs were expressed in primary mouse osteoclasts, and all but D3R were expressed in primary osteoblasts. Treatment of these cells with DA during differentiation caused the suppression of osteoblast mineralization and marker gene expression. The authors concluded that the reduced bone mineral density (BMD) and increased fracture risk caused by antidopaminergic drugs are due, in part, to their direct activation of DARs on bone cells.

In another study [57], human osteoclast precursor cells were found to express all DAR subtypes. DA inhibited cAMP formation and subsequent osteoclast differentiation and these inhibitory effects of DA were mimicked by D2-like receptor agonists. In addition, a D2R agonist suppressed lipopolysaccharide (LPS)-induced osteoclast formation in murine bone marrow culture *ex vivo*. These results indicated that D2-like receptors, potentially including D3R and D4R, are functionally dominant for the inhibition of osteoclastogenesis. On the other hand, stimulatory, rather than inhibitory effects by DA on MC3T3-E1 cells (a mouse pre-osteoblast cell line) were reported by another study, which confirmed expression of all five DARs in these cells [58]. In this case, DA treatment increased both cell proliferation and osteogenic mineralization. The discrepancy between the two studies can be explained by differences in abundance of D1R-like vs. D2R-like in their cell cultures.

Indirect effects of DA on bone density via its inhibition of PRL release were also examined [59]. Given that schizophrenia is associated with high rates of osteoporosis, the investigators set out to determine whether this was due to the PRL-raising properties of some antipsychotic medications. To this end, they compared the effects of PRL-raising vs. PRL-sparing antipsychotic medication in premenopausal women. The subjects had lumbar spine and hip BMD evaluation by dual-energy X-ray absorptiometry (DEXA), and blood samples were taken for analysis of PRL and sex hormones. The group taking PRL-raising drugs had significantly higher rates of bone pathology, as compared with those taking olanzapine. High PRL levels were related to measures of hypogonadism and low BMD values. Within the PRL-raising group, those taking newer atypical compounds had higher levels of PRL, lower levels of sex hormones, and lower BMD values than the group taking conventional antipsychotics. The authors concluded that the high rates of osteoporosis associated with schizophrenia may result from hypogonadism, secondary to antipsychotic-induced hyperprolactinemia. They also proposed that the PRL-raising profile of antipsychotic drugs should be considered when choosing an antipsychotic.

PET imaging of rhesus monkeys, using a DAT-specific ligand (^{18}F-LBT-999), showed high uptake of the ligand in the vertebral bone and the skull, with a relatively large dose accumulating in osteogenic cells [60]. Because DAT controls the temporal and spatial activity of released DA by its rapid uptake into the producing cells, its genomic deletion results in elevated levels of DA at multiple sites. The effects of DAT deletion on bone metabolism was studied using DAT-deficient mice [61]. The DAT$^{-/-}$ mice had reduced bone mass and strength, shorter femur length and dry weight, and lower ash calcium content. The volume of cancellous bones (bone found at the end of long bones) in the proximal tibial metaphysis of the knockout mice was significantly decreased with reduced trabecular thickness. Femoral strength in the DAT$^{-/-}$ mice was 30% lower than that of wild-type mice. The authors concluded that deletion of the DAT gene results in deficiencies in skeletal structure and integrity. They attributed these effects to activation of the serotonergic system, but without considering direct effects of peripheral DA on bone metabolism.

11.7 JOINTS, BONES, AND MUSCLES

11.7.1 Overview of joints and muscles

The adult human body contains 206 bones and approximately 300 joints, defined as points where two bones meet. Several types of joints are recognized. Synovial joints are found in arms and legs and enable bones to move over each other. Cartilaginous joints, such as those found in the spine and pelvis, provide more stability and less movement. Fibrous joints, as found in the skull, offer stability but do not allow movement at all in the adult. The joints between bones allow movement, with some allowing a wider range of movement than others, e.g., the ball and socket joint of the shoulder allows a greater range of movement than the pivot joint at the neck. Movement is powered by skeletal muscles, which are attached to the skeleton at various sites on bones.

A joint is constructed to allow for different degrees and types of movement. Some joints, such as those in the knee, elbow, and shoulder, are self-lubricating and almost frictionless. They can withstand compression and maintain heavy loads, while still executing smooth and precise movements. Other joints, such as the sutures between bones of the skull, permit very little movement (only during birth) in order to protect the brain and the sense organs.

The majority of joints are classified as **synovial joints**, as exemplified by those in knees and knuckles. As shown in **Figure 11.11**, a synovial joint links bones with a fibrous capsule that is continuous with the periosteum of the bones, constitutes the outer boundary of a synovial cavity, and surrounds the bones' articulating surfaces. The capsule is typically surrounded by soft tissue structures that support the joint and help facilitate movement. These structures include: tendons that attach muscles to bone, ligaments that attach bone to bone, bursae, and small sacs of synovial fluid that provide additional cushioning and lubrication. The health of a synovial joint is intertwined with the health of the supporting soft tissue structures. For example, damage to a ligament can skew joint alignment and eventually lead to joint degeneration (osteoarthritis) and vice versa.

The synovial cavity/joint is filled with **synovial fluid**, a dialysate of blood plasma which gives the fluid viscosity and elasticity, and contains large amounts of hyaluronic acid, released by the synovial lining cells. Synovial fluid has two main functions: lubrication for the joint, and nutrition for the cartilage. **Cartilage** is a resilient and smooth elastic tissue composed of specialized cells called chondrocytes, which produce a large amount of collagenous ECM rich is proteoglycan and elastin fibers. Because cartilage does not contain blood vessels or nerves, nutrition is supplied to the chondrocytes by diffusion. Compared with other connective tissues, cartilage has a very slow turnover of its ECM and does not repair.

Muscles are composed of cells specialized for contraction or shortening. They are broadly divided into three types—skeletal, cardiac, and smooth—which differ in structure, location and control of functions. Skeletal muscles are normally under conscious control. When the brain sends a signal for a

Figure 11.11 Diagram of a synovial joint. Shown are the different components surrounding a synovial joint. (Redrawn and modified from https://en.wikipedia.org/wiki/Joint.)

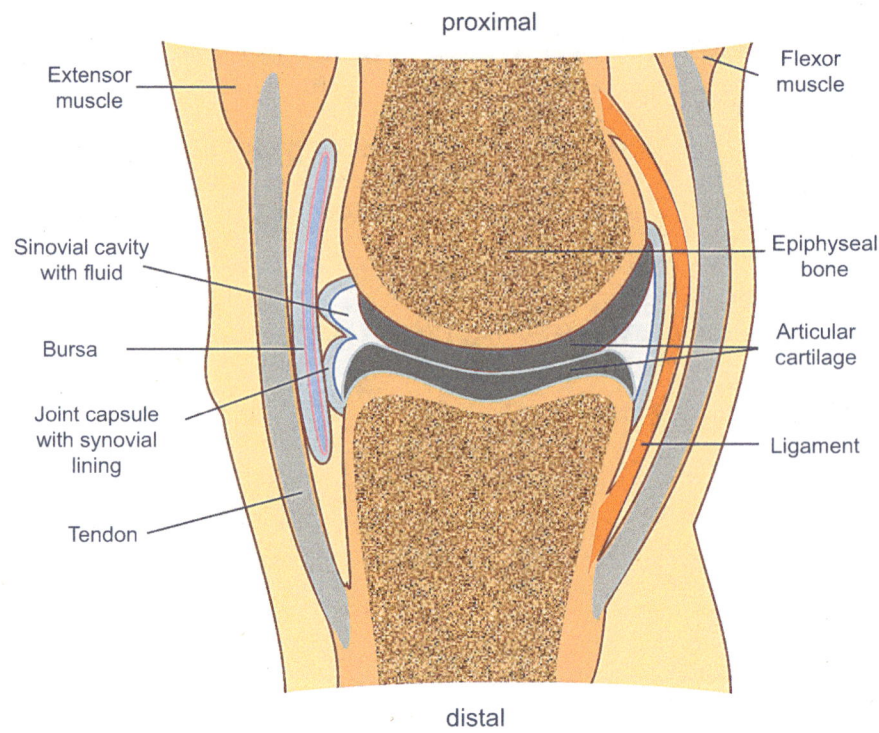

muscle to contract, it shortens and pulls one bone toward another across a joint. Muscles work in pairs, i.e., when one shortens, a corresponding muscle lengthens. Skeletal muscles run from one bone to another, usually passing at least one joint. They are connected to bones by tendons, which are the long thin ends of the muscles. Physical activity maintains or increases the strength of skeletal muscles. Smooth muscles are under autonomic rather than conscious control. They usually sit in or around blood vessels and organs, and regulate blood pressure, airways, and digestion. Cardiac muscle is made of striated fibers, similar in appearance to skeletal muscle fibers, but instead of connecting to bones they are attached to each other and are specialized for rhythmic contraction not under conscious control.

11.7.2 Involvement of dopamine in muscle disorders

As illustrated in **Figure 11.12**, disorders that affect muscular functions can be due to a pathology that directly impacts on a muscle, or indirectly, by affecting the motor neurons that innervate the muscle or the neuromuscular junction. Whereas the role of DA in modulating neuronal functions is well established, much less is known about its actions on the neuromuscular junctions or muscles themselves. A recent study reported that DA can potentiate neuromuscular transmission by activating presynaptic D1-like receptors, rather than adrenoceptors, and that this effect involves enhanced calcium release via the ryanodine calcium release [62]. As described in **Chapters 6** and **8**, direct actions of DA on cardiac muscle or smooth muscle (e.g., intestinal, airway, mesenteric, esophageal) are well established, but there are very few reports on comparable actions of DA on skeletal muscles. In one study, activation of the D1R and, to a lesser extent, D5R resulted in increased cAMP and modulation of muscle mass and force of isolated mouse lower leg muscles under both atrophying and non-atrophying conditions [63].

Neuromuscular disorders can be caused by autoimmune diseases, genetic/hereditary disorders, some forms of collagen disorder, failure of the electrical insulation surrounding nerves (the myelin), or resulting from

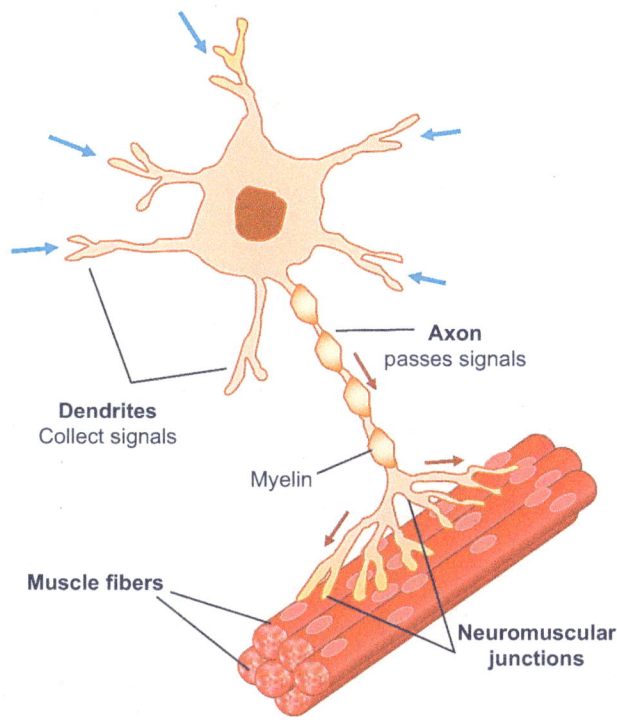

Axon
passes signals

Dendrites
Collect signals

Myelin

Muscle fibers

**Neuromuscular
junctions**

Figure 11.12 Diagram of a motor neuron, neuromuscular junction and muscle. Shown are the dendrites, which collect signals, an axon, which passes the signal and is covered by a myelin sheath, and the neuromuscular junction. A chemical synapse passes the signals from the neuron to the muscle fibers.

poisoning with heavy metals. As summarized below, there is evidence for the involvement of central DA and, less well documented, peripheral DA in the pathogenesis of some of these diseases.

Amyotrophic Lateral Sclerosis (ALS) is a chronic progressive neuromuscular disorder of unknown etiology. It is characterized by muscular weakness, muscle wasting, and increased reflexes, with conserved intellect and higher functions. The neuropathology of ALS is attributed to damage of the motor neurons in the cerebral cortex, some motor nuclei of the brainstem, and anterior horns of the spinal cord. However, there is also evidence for involvement of other neuronal systems. In particular, neurochemical and imaging studies have shown damage of DA neurons in the brain and spinal cord of ALS patients [64]. There are only limited number of therapeutic strategies that effectively relieve symptoms and improve the quality of life for ALS patients. Among these is bromocriptine, a D2R agonist that has been shown to confer neuroprotection, sustained motor function and slowed disease progression in mouse models of ALS, and exhibiting promising benefits in sustaining motoneural functions in phase II clinical trial in ALS patients [65].

Multiple Sclerosis (MS) is a chronic disease that leads to a progressive neurological disability and is caused by autoimmune responses against the myelin sheath of axons in different areas of the brain and spinal cord. The significant decline in the muscle strength in MS subjects is due in part, to changes in skeletal muscle fiber type and myosin heavy chain (MHC) expression. One of the prevailing theories on MS pathogenesis is a violation of immunological tolerance and an active penetration of immune cells, sensitized to myelin antigens, through the blood–brain barrier into brain tissue [66]. Evidence shows that DA influences MS pathogenesis by modulating immune cell activity and cytokine production [66,67]. However, depending on the activation of different DARs, DA can either enhance or inhibit immune functions. Compared with healthy controls, peripheral blood mononuclear cells (PBMCs) from MS patients had higher IL-17 production. This was mediated *in vitro* by activation of D2R-like and was reverted by stimulation of D1R-like. PBMCs from IFN-β–treated MS patients had increased production of catecholamines, higher mRNA levels of TH and D5R, and reduced D2R compared with baseline levels before treatment.

Fibromyalgia, characterized by chronic widespread pain and bodily tenderness, is not strictly considered a neuromuscular disease. The disorder occurs mainly in middle-aged women and is often associated with other symptoms including chronic fatigue, morning stiffness, and affective disturbances. In one study, fibromyalgia patients and healthy controls were subjected to deep muscle pain produced by injection of hypertonic saline into the anterior tibialis muscle [68]. PET imaging with [^{11}C]-raclopride (a D2/D3 ligand) was used to determine release of endogenous DA in response to the painful stimulation. The fibromyalgia patients experienced the hypertonic saline as more painful than healthy controls. The healthy patients released DA in the basal ganglia during the painful stimulation, whereas the fibromyalgia patients did not. In the control subjects, the amount of DA release correlated with the amount of perceived pain, while this was not the case with the fibromyalgia patients. The authors concluded that fibromyalgia is associated with an abnormal DA response to pain, suggesting that the therapeutic effects of dopaminergic treatments for this intractable disorder should be explored. Because an older study reported a direct facilitatory action of DA at the neuromuscular junction of a cat tibia [69], the possibility of abnormality of peripheral DA in fibromyalgia should also be examined.

11.7.3 Involvement of dopamine in bone and synovial joint disorders

Rheumatoid arthritis (RA) is a chronic, systemic inflammatory disorder that affects many tissues and organs, but principally attacks synovial joints. The process involves an inflammatory response of the capsule around the joints (synovium), secondary to hyperplasia of synovial cells, excess synovial fluid, and development of fibrous tissue in the synovium. As a chronic inflammation, RA can lead to articular bone erosion and consequently to joint destruction [70], which can cause severe disability that affects all aspects of motor function, from walking to fine movements of the hand. Another disease with a similar presentation is **Osteoarthritis (OA)**, the most common form of arthritis, which affects millions of people worldwide. It occurs when the protective cartilage that cushions the ends of bones wears down over time. Although osteoarthritis can damage any joint, it mostly affects joints in hands, knees, hips, and spine.

The major complaint by individuals with arthritis is joint pain, which can be constant and localized to the joint affected. Arthritis is the most common cause of disability in the United States. More than 20 million individuals with arthritis have severe limitations in function on a daily basis. Treatment options vary depending on the type of arthritis and include physical therapy, lifestyle changes, orthopedic bracing, and medications. The medications can help reduce inflammation in the joint which decreases pain and slow the joint damage. Joint replacement surgery may be required in eroding forms of arthritis.

There is increasing evidence on the involvement of DA in the pathogenesis of RA [71–73]. A recent extensive review [70] covered multiple studies showing direct effects of DA on the systemic immune response as well as on bone remodeling and joint inflammation, both in humans and in animal models of arthritis. These data are briefly summarized below, and are presented as a diagram in **Figure 11.13** on human (Panel A) and animal (Panel B) studies. The author concluded that while more research is necessary to accurately determine the effect of DA in RA, these results support a possible use of dopaminergic drugs for the future treatment of arthritis. In addition, they point out that dopaminergic agents use to treat comorbidities, might influence the immune response and the disease progression in RA patients.

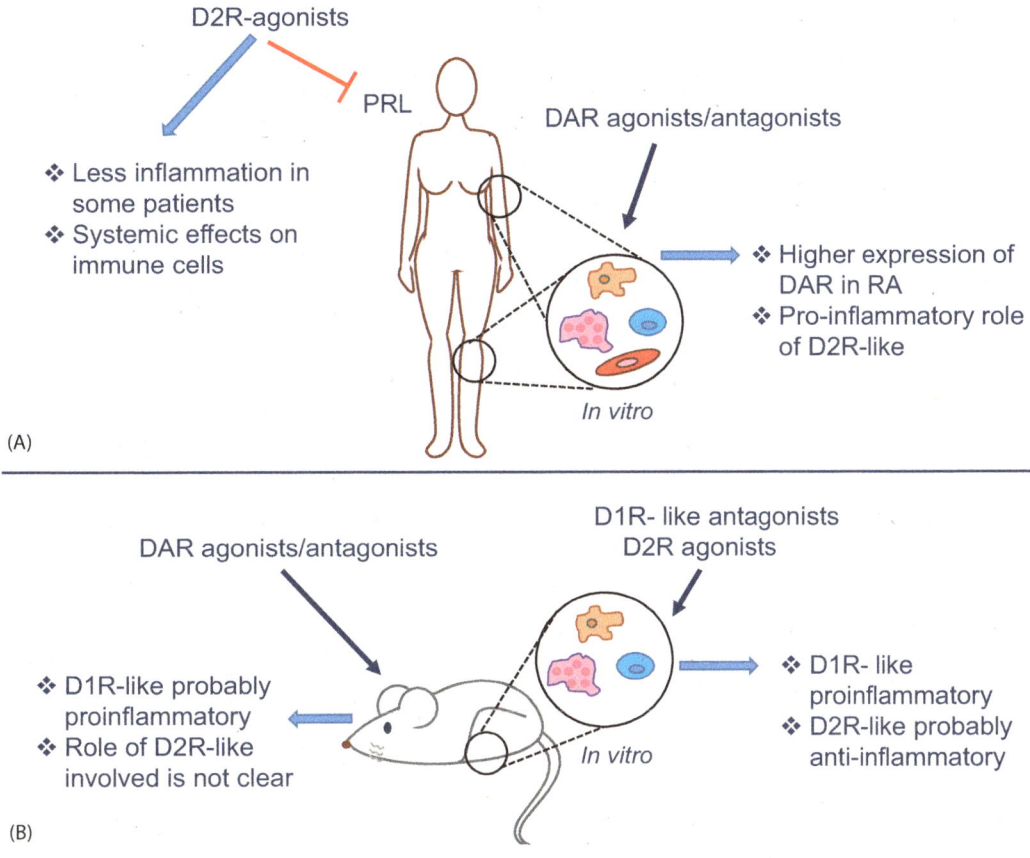

Figure 11.13 Role of dopaminergic agents in rheumatoid arthritis. Panel A shows both direct and indirect (via inhibition of PRL) of DA in rheumatoid arthritis in humans. Panel B depicts evidence derived from animal studies. DA: dopamine; DAR: DA receptor; PRL: prolactin. (Redrawn from Capellino, S., *J. Neuroimmune Pharmacol.*, 1–9, 2019.)

Given that synovial fibroblasts (SFs) contribute to inflammation and joint destruction in RA, one study examined the dopaminergic system in SFs obtained from patients with RA or OA undergoing knee joint replacement surgery [72]. The SFs were found to possess a complete intrinsic dopaminergic system, including DAR, DAT, and TH, and also containing DA, its precursor (L-Dopa), and its main metabolites [3-MT, dihydroxyphenylacetic acid (DOPAC), and HVA]. SFs from patients with RA showed higher expression of D1R and D5R than those from patients with OA. Moreover, exogenous DA strongly inhibited the production of IL-8 in patients with RA. The authors concluded that increased expression of DARs results in anti-inflammatory effects, as shown *in vitro* after treatment with DA. They also suggested that DA and its receptors should be considered as future therapeutic targets in RA and other autoimmune diseases.

A more recent study was focused on mast cells, which have been identified as important effector cells in the synovial inflammation of RA [73]. D3R-positive mast cells in the synovial fluid from RA patients were gradually reduced along with the increase in disease severity, DAS28 score of RA patients. D3R-positive mast cells in the synovial fluid were strongly correlated with disease severity.

As was also summarized in the above review [70], a potential direct role of DA was investigated in several *in vivo* and *in vitro* studies with animal models of arthritis. D2-like receptor activation led to reduced cartilage destruction and synovial hyperplasia in severe combined immunodeficient (SCID) mice engrafted with human synovium, as well as in the collagen-induced arthritis (CIA) mouse model. In addition, Drd2$^{-/-}$ mice manifested a more severe CIA compared with wild-type mice. *In vitro*, the stimulation of D2-like receptors had anti-inflammatory effects on lymphocytes from CIA mice. In rats, the role of D2-like receptors is controversial because their blockade in the CIA model reduced the amount of proinflammatory biomarkers, suggesting a proinflammatory role of D2R. In contrast, treatment with pergolide, a DAR agonist with higher affinity to D2-like than to D1-like receptors led to anti-inflammatory effects in the carrageenan-induced arthritis model.

11.8 SYNOPSIS

DARs are detected in all layers of the skin, and several lines of evidence indicate that DA is involved in major cutaneous functions, including wound healing, pigmentation, sweating and hair growth. In wound healing, DA has direct interactions with platelets, keratinocytes and endothelial cells. In pigmentation, the biosynthetic pathways for the production of DA and melanin share the same precursors (tyrosine and L-Dopa). Innervation deficit of sweat glands and their dysfunction in patients with PD support the notion that DA contributes to the function of sweat glands. Both direct and indirect (via the regulation of PRL) role for DA on hair growth have been demonstrated in several studies. Studies with DAT-deficient mice and other evidence indicate that DA affects bone metabolism and is also involved in synovial joint disorders and several muscular dysfunctions.

REFERENCES

1. Ojeh N, Pastar I, Tomic-Canic M, Stojadinovic O. Stem cells in skin regeneration, wound healing, and their clinical applications. *Int J Mol Sci.* 2015;16(10):25476–25501.

2. Fuchs E, Raghavan S. Getting under the skin of epidermal morphogenesis. *Nat Rev Genet.* 2002;3(3):199–209.

3. Strong AL, Neumeister MW, Levi B. Stem cells and tissue engineering: Regeneration of the skin and its contents. *Clin Plast Surg.* 2017;44(3):635–650.

4. Gonzalez AC, Costa TF, Andrade ZA, Medrado AR. Wound healing—A literature review. *An Bras Dermatol.* 2016;91(5):614–620.

5. Velnar T, Bailey T, Smrkolj V. The wound healing process: An overview of the cellular and molecular mechanisms. *J Int Med Res.* 2009;37(5):1528–1542.

6. Fuziwara S, Suzuki A, Inoue K, Denda M. Dopamine D2-like receptor agonists accelerate barrier repair and inhibit the epidermal hyperplasia induced by barrier disruption. *J Invest Dermatol.* 2005;125(4):783–789.

7. Sarkar C, Ganju RK, Pompili VJ, Chakroborty D. Enhanced peripheral dopamine impairs post-ischemic healing by suppressing angiotensin receptor type 1 expression in endothelial cells and inhibiting angiogenesis. *Angiogenesis.* 2017;20(1):97–107.

8. Holman DM, Kalaaji AN. Cytokines in dermatology. *J Drugs Dermatol.* 2006;5(6):520–524.

9. Ricci A, et al. Dopamine receptors in human platelets. *Naunyn Schmiedebergs Arch Pharmacol.* 2001;363(4):376–382.

10. Frankhauser P, et al. Characterization of the neuronal dopamine transporter DAT in human blood platelets. *Neurosci Lett.* 2006;399(3):197–201.

11. Schedel A, Schloss P, Kluter H, Bugert P. The dopamine agonism on ADP-stimulated platelets is mediated through D2-like but not D1-like dopamine receptors. *Naunyn Schmiedebergs Arch Pharmacol.* 2008;378(4):431–439.

12. Ramchand CN, Clark AE, Ramchand R, Hemmings GP. Cultured human keratinocytes as a model for studying the dopamine metabolism in schizophrenia. *Med Hypotheses.* 1995;44(1):53–57.

13. Parrado AC, Canellada A, Gentile T, Rey-Roldan EB. Dopamine agonists upregulate IL-6 and IL-8 production in human keratinocytes. *Neuroimmunomodulation.* 2012;19(6):359–366.

14. Sinha S, et al. Dopamine regulates phosphorylation of VEGF receptor 2 by engaging Src-homology-2-domain-containing protein tyrosine phosphatase 2. *J Cell Sci.* 2009;122(Pt 18):3385–3392.

15. Shome S, Dasgupta PS, Basu S. Dopamine regulates mobilization of mesenchymal stem cells during wound angiogenesis. *PLoS One.* 2012;7(2):e31682.

16. Mirones I, et al. Dopamine mobilizes mesenchymal progenitor cells through D2-class receptors and their PI3K/AKT pathway. *Stem Cells.* 2014;32(9):2529–2538.

17. Liebscher J, et al. Structure of polydopamine: A never-ending story? *Langmuir.* 2013;29(33):10539–10548.

18. Mrowczynski R, Bunge A, Liebscher J. Polydopamine—An organocatalyst rather than an innocent polymer. *Chemistry.* 2014;20(28):8647–8653.

19. Liu X, et al. Mussel-inspired polydopamine: A biocompatible and ultrastable coating for nanoparticles in vivo. *ACS Nano*. 2013;7(10):9384–9395.

20. Xu X, et al. Controlled-temperature photothermal and oxidative bacteria killing and acceleration of wound healing by polydopamine-assisted Au-hydroxyapatite nanorods. *Acta Biomater*. 2018;77:352–364.

21. Li T, et al. Mussel-inspired nanostructures potentiate the immunomodulatory properties and angiogenesis of mesenchymal stem cells. *ACS Appl Mater Interfaces*. 2019;11(19):17134–17146.

22. Costin GE, Hearing VJ. Human skin pigmentation: Melanocytes modulate skin color in response to stress. *Faseb J*. 2007;21(4):976–994.

23. Mull AN, Zolekar A, Wang YC. Understanding melanocyte stem cells for disease modeling and regenerative medicine applications. *Int J Mol Sci*. 2015;16(12):30458–30469.

24. Bonaventure J, Domingues MJ, Larue L. Cellular and molecular mechanisms controlling the migration of melanocytes and melanoma cells. *Pigment Cell Melanoma Res*. 2013;26(3):316–325.

25. Wu X, Hammer JA. Melanosome transfer: It is best to give and receive. *Curr Opin Cell Biol*. 2014;29:1–7.

26. Videira IF, Moura DF, Magina S. Mechanisms regulating melanogenesis. *An Bras Dermatol*. 2013;88(1):76–83.

27. Gillbro JM, Olsson MJ. The melanogenesis and mechanisms of skin-lightening agents—Existing and new approaches. *Int J Cosmet Sci*. 2011;33(3):210–221.

28. Montoliu L, et al. Increasing the complexity: New genes and new types of albinism. *Pigment Cell Melanoma Res*. 2014;27(1):11–18.

29. Chen T, et al. Dopamine signaling regulates the projection patterns in the mouse chiasm. *Brain Res*. 2015;1625:324–336.

30. Nicolaidou E, Katsambas AD. Pigmentation disorders: Hyperpigmentation and hypopigmentation. *Clin Dermatol*. 2014;32(1):66–72.

31. Marles LK, Peters EM, Tobin DJ, Hibberts NA, Schallreuter KU. Tyrosine hydroxylase isoenzyme I is present in human melanosomes: A possible novel function in pigmentation. *Exp Dermatol*. 2003;12(1):61–70.

32. Reimann E, et al. Expression profile of genes associated with the dopamine pathway in vitiligo skin biopsies and blood sera. *Dermatology*. 2012;224(2):168–176.

33. Cucchi ML, Frattini P, Santagostino G, Preda S, Orecchia G. Catecholamines increase in the urine of non-segmental vitiligo especially during its active phase. *Pigment Cell Res*. 2003;16(2):111–116.

34. Morrone A, et al. Catecholamines and vitiligo. *Pigment Cell Res*. 1992;5(2):65–69.

35. Iyengar B, Misra RS. Reaction of dendritic melanocytes in vitiligo to the substrates of tyrosine metabolism. *Acta Anat (Basel)*. 1987;129(3):203–205.

36. Kim B, et al. N-Nicotinoyl dopamine, a novel niacinamide derivative, retains high antioxidant activity and inhibits skin pigmentation. *Exp Dermatol*. 2011;20(11):950–952.

37. Chu CY, Liu YL, Chiu HC, Jee SH. Dopamine-induced apoptosis in human melanocytes involves generation of reactive oxygen species. *Br J Dermatol*. 2006;154(6):1071–1079.

38. Zhang X, et al. PDE5 inhibitor promotes melanin synthesis through the PKG pathway in B16 melanoma cells. *J Cell Biochem*. 2012;113(8):2738–2743.

39. Habecker BA, Malec NM, Landis SC. Differential regulation of adrenergic receptor development by sympathetic innervation. *J Neurosci*. 1996;16(1):229–237.

40. Weihe E, et al. Coexpression of cholinergic and noradrenergic phenotypes in human and nonhuman autonomic nervous system. *J Comp Neurol*. 2005;492(3):370–379.

41. Beitz JM. Skin and wound issues in patients with Parkinson's disease: An overview of common disorders. *Ostomy Wound Manage*. 2013;59(6):26–36.

42. Modugno N, et al. A clinical overview of non-motor symptoms in Parkinson's Disease. *Arch Ital Biol*. 2013;151(4):148–168.

43. Kass-Iliyya L, et al. Small fiber neuropathy in Parkinson's disease: A clinical, pathological and corneal confocal microscopy study. *Parkinsonism Relat Disord*. 2015;21(12):1454–1460.

44. Dabby R, et al. Skin biopsy for assessment of autonomic denervation in Parkinson's disease. *J Neural Transm (Vienna)*. 2006;113(9):1169–1176.

45. Sehgal VN, Srivastava G, Aggarwal AK, Midha R. Hair biology and its comprehensive sequence in female pattern baldness: Diagnosis and treatment modalities—Part I. *Skinmed*. 2013;11(1):39–45.

46. Bernard BA. Hair biology: An update. *Int J Cosmet Sci*. 2002;24(1):13–16.

47. Brodell LA, Mercurio MG. Hirsutism: Diagnosis and management. *Gend Med*. 2010;7(2):79–87.

48. Sasaki GH. Review of human hair follicle biology: Dynamics of niches and stem cell regulation for possible therapeutic hair stimulation for plastic surgeons. *Aesthetic Plast Surg*. 2019;43(1):253–266.

49. Grauer MT, Sieb JP. Alopecia induced by dopamine agonists. *Neurology*. 2002;59(12):2012–2025.

50. Tabamo RE, Di RA. Alopecia induced by dopamine agonists. *Neurology*. 2002;58(5):829–830.

51. Langan EA, et al. Dopamine is a novel, direct inducer of catagen in human scalp hair follicles in vitro. *Br J Dermatol*. 2013;168(3):520–525.

52. Lutz G. Hair loss and hyperprolactinemia in women. *Dermatoendocrinol*. 2012;4(1):65–71.

53. Demontiero O, Vidal C, Duque G. Aging and bone loss: New insights for the clinician. *Ther Adv Musculoskelet Dis*. 2012;4(2):61–76.

54. Han Y, You X, Xing W, Zhang Z, Zou W. Paracrine and endocrine actions of bone-the functions of secretory proteins from osteoblasts, osteocytes, and osteoclasts. *Bone Res.* 2018;6:16–27.

55. Hadjidakis DJ, Androulakis II. Bone remodeling. *Ann N Y Acad Sci.* 2006;1092):385–396.

56. Motyl KJ, et al. A novel role for dopamine signaling in the pathogenesis of bone loss from the atypical antipsychotic drug risperidone in female mice. *Bone.* 2017;103:168–176.

57. Hanami K, et al. Dopamine D2-like receptor signaling suppresses human osteoclastogenesis. *Bone.* 2013;56(1):1–8.

58. Lee DJ, et al. Dopaminergic effects on in vitro osteogenesis. *Bone Res.* 2015;3:1–10.

59. O'Keane V, Meaney AM. Antipsychotic drugs: A new risk factor for osteoporosis in young women with schizophrenia? *J Clin Psychopharmacol.* 2005;25(1):26–31.

60. Varrone A, et al. Imaging of the striatal and extrastriatal dopamine transporter with (18)F-LBT-999: Quantification, biodistribution, and radiation dosimetry in nonhuman primates. *J Nucl Med.* 2011;52(8):1313–1321.

61. Bliziotes M, Gunness M, Eshleman A, Wiren K. The role of dopamine and serotonin in regulating bone mass and strength: Studies on dopamine and serotonin transporter null mice. *J Musculoskelet Neuronal Interact.* 2002;2(3):291–295.

62. Elnozahi NA, AlQot HE, Mohy El-Din MM, Bistawroos AE, Abou Zeit-Har MS. Modulation of dopamine-mediated facilitation at the neuromuscular junction of Wistar rats: A role for adenosine A1/A2A receptors and P2 purinoceptors. *Neuroscience.* 2016;326:45–55.

63. Reichart DL, et al. Activation of the dopamine 1 and dopamine 5 receptors increase skeletal muscle mass and force production under non-atrophying and atrophying conditions. *BMC Musculoskelet Disord.* 2011;12:27.

64. Takahashi H, et al. Evidence for a dopaminergic deficit in sporadic amyotrophic lateral sclerosis on positron emission scanning. *Lancet.* 1993;342(8878):1016–1018.

65. Nagata E, et al. Bromocriptine mesylate attenuates amyotrophic lateral sclerosis: A phase 2a, randomized, double-blind, placebo-controlled research in Japanese patients. *PLoS One.* 2016;11(2):e0149509.

66. Melnikov M, Rogovskii V, Cyrillic A, Pashenkov M. Dopaminergic Therapeutics in multiple sclerosis: Focus on Th17-Cell functions. *J Neuroimmune Pharmacol.* 2019. doi:10.1007/s11481-019-09852-3.

67. Levite M, Marino F, Cosentino M. Dopamine, T cells and multiple sclerosis (MS). *J Neural Transm (Vienna).* 2017;124(5):525–542.

68. Wood PB, et al. Fibromyalgia patients show an abnormal dopamine response to pain. *Eur J Neurosci.* 2007;25(12):3576–3582.

69. Gallagher JP, Karczmar AG. A direct facilitatory action for dopamine at the neuromuscular junction. *Neuropharmacology.* 1973;12(8):783–791.

70. Capellino S. Dopaminergic agents in rheumatoid arthritis. *J Neuroimmune Pharmacol.* 2019:1–9. doi:10.1007/S11481-019-09850-5.

71. Wei L, et al. The effects of dopamine receptor 2 expression on B cells on bone metabolism and TNF-alpha levels in rheumatoid arthritis. *BMC Musculoskelet Disord.* 2016. doi:10.1186/s12891-016-1220-7.

72. Capellino S, et al. Increased expression of dopamine receptors in synovial fibroblasts from patients with rheumatoid arthritis: Inhibitory effects of dopamine on interleukin-8 and interleukin-6. *Arthritis Rheumatol.* 2014;66(10):2685–2693.

73. Xue L, et al. Associations between D3R expression in synovial mast cells and disease activity and oxidant status in patients with rheumatoid arthritis. *Clin Rheumatol.* 2018;37(10):2621–2632.

74. Strobl JS, Melkoumian Z, Peterson VA, Hylton H. The cell death response to gamma-radiation in MCF-7 cells is enhanced by a neuroleptic drug, pimozide. *Breast Cancer Res Treat.* 1998;51(1):83–95.

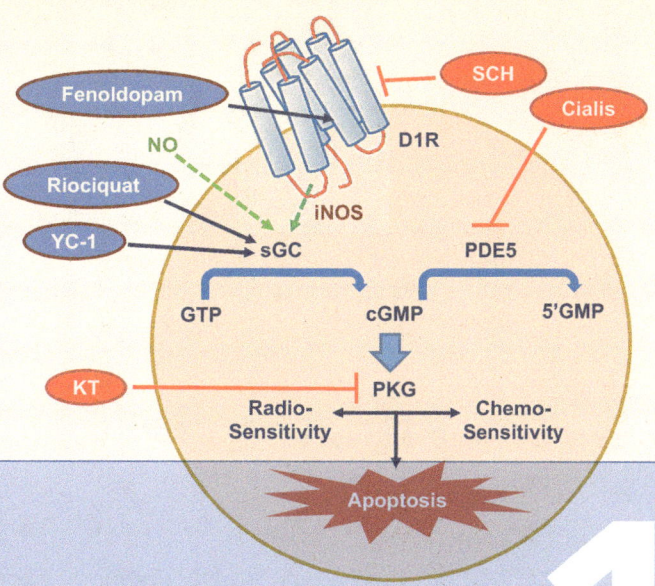

Dopamine and Tumorigenesis in Reproductive Tissues

12

12.1 INTRODUCTION

Tumors are masses of mutated and dysfunctional cells that may cause pain and disfigurement, invade organs, and potentially spread throughout the body. Tumors grow because of DNA malfunctions, mainly in genes that regulate cell growth or cell death. Some of these mutations can lead to rapid, unchecked growth by producing tumors that expand quickly and damage nearby organs and tissue. Transformed cells can produce enzymes that dissolve the surrounding tissue and grow beyond their normal boundaries, a process that is defined as invasiveness. Moreover, many tumor cells can be released into the blood stream, invade remote organs and grow at distant locations, a process that is defined as metastasis.

From a histological standpoint, there are hundreds of different cancers, which can be generally categorized by the type of tissue of origin into (1) carcinomas (epithelial tissue), (2) sarcomas (soft tissue and bone), (3) gliomas (brain), and (4) hematopoietic (bone marrow and lymph nodes). Some cellular mutations are less aggressive, forming slow-growing tumors that are not cancerous. These benign tumors do not metastasize, but can cause harm by pushing and compressing adjacent normal tissues. Overall information on cancer occurrence, risks, diagnosis, and treatments can be found in an online review on human cancers, grouped by different organs at https://www.iarc.fr/en/publications/pdfs-online/wcr/2003/wcr-5.pdf, which is also listed in the bibliography as a reference [1].

This chapter covers both benign and malignant tumors that occur along the pituitary–gonadal–reproductive tract axis in both females and males. The involvement of dopamine (DA) in cancers of reproductive organs was revealed in a large retrospective survey conducted in Sweden in 2013 [2]. The study examined for associations between schizophrenia (representing overactivity of the dopaminergic systems) and increased risk of cancer in various organs. Of the 59,233 schizophrenic patients surveyed, 6,137 developed cancer during the 40 years of the survey. The overall cancer incidence among patients with schizophrenia and their first-degree relatives was *significantly lower* than that in the general population. This suggested that some genetic factors, which presumably contribute to the development of schizophrenia, may protect against certain cancers. On the other hand, female patients had *increased* incidence of breast, cervical, and endometrial cancers, but only after the first diagnosis of schizophrenia. Although not directly examined in this study, the authors speculated that when patients are diagnosed with schizophrenia, most, if not all, would be treated with antipsychotics. Thus, the anti-DA receptor (anti-DAR) activity of such medications could have been a major reason for the increased cancer incidence in reproductive tissues.

12.2 PITUITARY TUMORS

12.2.1 Prevalence, classification, and molecular characterization of pituitary tumors

Pituitary tumors, classified as adenomas, represent 10%–15% of primary intracranial neoplasms. Unlike carcinomas, which are invasive malignant epithelial cells with a capacity for metastasis, adenomas are benign epithelial tumors of glandular origin that do not metastasize. Although both malignant and benign tumors exhibit dysregulated cellular proliferation, they differ in growth rates, degree of differentiation, invasiveness, and metastasis. Senescence, a cellular defense mechanism against malignant transformation, may explain the benign nature of many pituitary tumors [3]. The occurrence, diagnosis, cellular composition, and treatment of pituitary tumors are covered in several extensive reviews [4–6], and are briefly summarized below.

Pituitary adenomas are detectable in about 20% of random autopsies, but only a small fraction of these were symptomatic when the individuals were alive, indicating that many of these tumors can remain silent for many years without noticeable symptoms. Pituitary adenomas can arise from any pituitary cell type, grow rather slowly, and are classified by size either as microadenomas (<10 mm) or as macroadenomas (>10 mm). Macroadenomas are characterized by expansion of the tumor into surrounding structures and a rapid growth. They are difficult to manage and are associated with poor prognosis, limited therapeutic options, and tumor unresponsiveness to therapy. Pituitary tumors arise sporadically and are rarely hereditary [4]. Pituitary carcinomas are extremely rare, with only about 100 cases reported in the literature.

Morbidity associated with pituitary adenomas can be due either to the sequela of hormone hypersecretion or as a consequence of intracranial invasion of large tumors. In spite of their relative prevalence, the pathophysiology of pituitary tumors has not been well characterized [7]. Many, but not all, of the pituitary tumors are composed of monoclonal cell populations with a disrupted control of replication pathways. Oncogenes and tumor suppressor genes that are rather common in other malignancies (e.g., *jun, fos, myc, p53*) are seldom associated with pituitary tumors.

Table 12.1 presents the classification of pituitary tumor subtypes in large surveys in four European countries [5]. The prevalence of pituitary adenomas ranges from 1 in 865 adults to 1 in 2,688 adults. Approximately 50% are macroadenomas (>10 mm), causing mass effects such as headache, hypopituitarism, and visual field defects resulting from a compression of the optic chiasm. Prolactinomas account for 32%–66% of the total pituitary adenomas, while growth hormone (GH)-secreting tumors account for 8%–16% of the tumors. Adrenocorticotropic hormone (ACTH)-secreting tumors are less common, accounting for 2%–6% of the adenomas, while thyroid stimulating hormone (TSH)- or gonadotropin-secreting tumors are rather rare, accounting for no more than for 1%–2% of the tumors. About 40% of pituitary tumors are classified as clinically nonfunctional adenomas (CNFAs). Although such adenomas may show excess expression of certain hormones, these hormones are not secreted in larger than usual amounts into the circulation.

Table 12.1 Classification of pituitary tumors by prevalence, hormone production, and size in four studies in different European countries

Patients (n)	Country	Prevalence	PRL (%)	CNFA (%)	GH (%)	ACTH (%)	Macro (%)
164	Finland	1/1,471	51	37	8	3	46
316	Malta	1/1,321	46	34	16	2	43
592	Sweden	1/2,688	32	54	9	4	65
471	Iceland	1/865	40	43	11	6	55

Modified from Molitch, M.E., *JAMA*, 317, 516–524, 2017.
ACTH: adrenocorticotropin hormone; CNFA: clinically nonfunctional adenomas; GH: growth hormone; Macro: macroadenomas.

In 1997, a pituitary tumor transforming gene (PTTG) was identified [8]. PTTG was found to be overexpressed in 23 of 30 nonfunctioning pituitary tumors, in all 13 GH-producing tumors, in 9 of 10 prolactinomas, and in 1 ACTH-secreting tumor [9]. More than 10-fold increases in PTTG expression were seen in some tumors. Moreover, PTTG expression was significantly higher in hormone-secreting tumors that had invaded the sphenoid bone than in tumors confined to the pituitary fossa. The authors concluded that PTTG is a molecular marker for invasiveness in hormone-secreting pituitary tumors.

Subsequent studies revealed that PTTG is overexpressed in other tumors, including thyroid, breast and ovarian carcinomas. The PTTG protein, also known as human securin, induces cellular transformation, is involved in cell cycle regulation, and controls the segregation of sister chromatids during mitosis [10]. PTTG is an anaphase inhibitor that prevents premature chromosome separation through inhibition of separase activity; hence, its degradation is required to start anaphase. Through this important function, PTTG participates in several key cellular events such as mitosis, cell cycle progression, DNA repair, and apoptosis. Another putative oncogene, *gsp* (Gαs subunit) was overexpressed in ~40% of hormonally active pituitary adenomas [7]. Both cAMP [11] and cGMP [12] signaling cascades also appear to be involved in pituitary tumors.

12.2.2 Prolactinomas and dopamine: Pathogenesis and treatments

Prolactin (PRL)-secreting tumors are the most common pituitary tumor subtype, constituting 40%–45% of the total pituitary adenomas and 60% of the hormone-secreting adenomas. Because there are causes of hyperprolactinemia other than prolactinomas, a history of reproductive issues, physical examination, record of treatment with certain drugs (i.e., neuroleptics), and magnetic resonance imaging (MRI) are used to determine whether a tumor is present and its size. If no tumor is found, the hyperprolactinemia is considered idiopathic. The logistics of the diagnosis of prolactinomas in women and the selection of treatment modalities are presented in **Figure 12.1**. Depending upon the extent of PRL hypersecretion, a prolactinoma can cause amenorrhea, loss of libido, galactorrhea, and infertility in women and loss of libido, erectile dysfunction, and infertility in men. Patients with prolactinoma can also have increased levels of fasting plasma glucose, low-density lipoprotein cholesterol, and triglycerides [13], underscoring some of the actions of PRL as a metabolic hormone.

With respect to the nature of PRL-secreting pituitary tumors, a distinction should be made between prolactinomas that release excess amounts of PRL because of increased number and/or activity of the lactotrophs and large macroadenomas that compress the pituitary stalk or invade into the

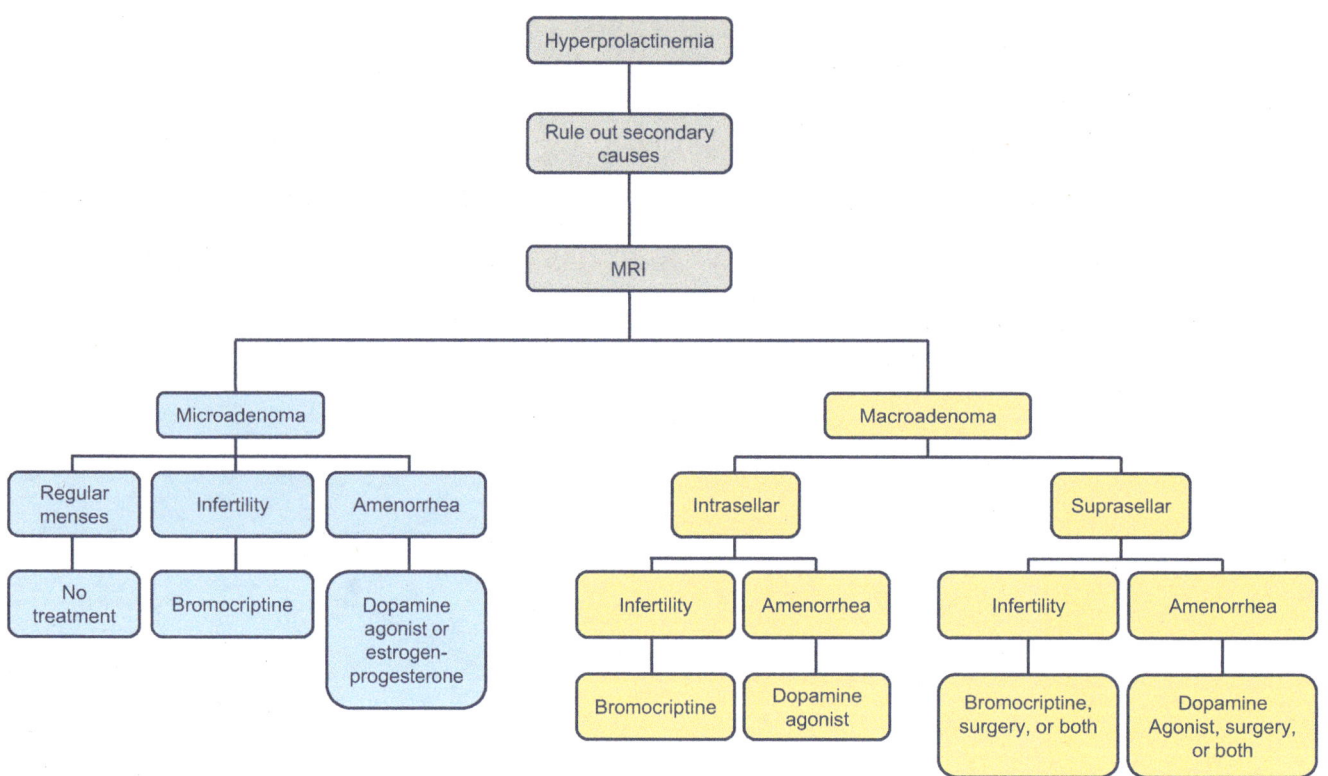

Figure 12.1 The diagnostic process for identifying the causes of hyperprolactinemia. Treatment modalities are determined by the tumor size and whether it interferes with neurological and/or reproductive functions. MRI: magnetic resonance imaging.

Figure 12.2 The wide spectrum of circulating prolactin (PRL) values. Shown are physiological, pharmacological, or pathological conditions.

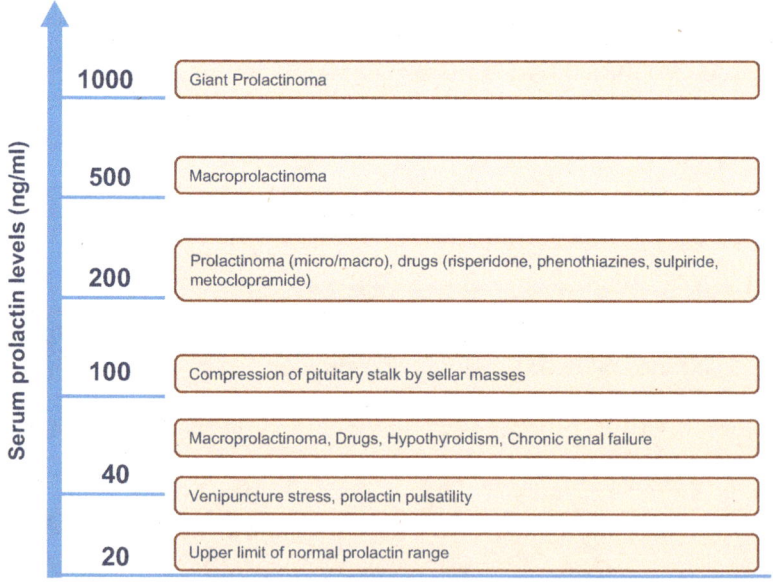

hypothalamus, resulting in increased PRL release because of DA deprivation. Radiologically diagnosed large pituitary adenomas with serum PRL levels below 100–150 ng/mL are suggestive of non-PRL secreting tumors rather than prolactinomas (**Figure 12.2**). On the other hand, the levels of PRL that are secreted by prolactinomas are usually proportional to tumor size and can reach circulating levels as high as 5,000–10,000 ng/mL.

More women are diagnosed with prolactinomas than men. However, as discussed in our previous review [6], it is uncertain whether women have a *higher incidence* of prolactinomas than men or women are diagnosed more frequently at a premenopausal age because of reproductive disturbances such as amenorrhea (an abnormal absence of menstruation). There is also no conclusive evidence for a direct correlation between exposure to excess estrogens and development of prolactinomas in women. As was concluded in a survey that included 72 women with prolactinomas and 303 controls [14], the risk of prolactinoma for women who had used oral contraceptives (OCs) for birth control was very low. The authors indicated that previous findings of an association between OC use and occurrence of prolactinomas may have resulted from OC treatment of menstrual irregularity in women with an undiagnosed prolactinoma. In addition, lactotroph hyperplasia in estrogen-treated male-to-female transgender patients is a rare event.

Statements in the literature on the pathogenesis of human prolactinomas that are based on rodent models are overreached. Spontaneous prolactinomas, which occur in aged rats of some strains (e.g., Fischer 344), invariably express p450 aromatase, indicating abnormally high conversion of testosterone to estradiol. Moreover, pituitary tumorigenesis, induced in mice by overexpression of oncogenes or knocking down of tumor suppressor genes, occurs almost exclusively in females, and this is preceded by a long phase of hyperplasia. Thus, the strong estrogenic component in the induction of prolactinomas in rodents is not evident in humans, although both estrogen receptor alpha and beta (ERα and ERβ) and their various isoforms are expressed in both normal and tumorous PRL-secreting cells [15].

The mechanisms underlying prolactinoma formation are enigmatic. More than any other pituitary cell type, lactotrophs have considerable plasticity, and increase both in number and size under various physiological conditions, e.g., during human pregnancy, and in response to inappropriate estrogen exposure in certain strains of rats. The increase in lactotroph number is attributed to the combined effects of increased cell division, reduced apoptosis, and trans-differentiation of lactotrophs from other pituitary cell types, primarily bifunctional somatolactotrophs. Yet, unlike the typical malignant transformation of epithelial cells elsewhere, lactotrophs undergo only the initial stage of tumorigenesis, i.e., uncontrolled cell growth, but do not progress into *bona fide* carcinomas. Consequently, prolactinomas lack most markers of malignancy such as high mitotic index, dedifferentiation, and metastasis.

The most highly expressed DAR in prolactinomas are D2R. As detailed in **Chapter 2**, D2Rs are composed of two isoforms, D2 long (D2L or $D2_{444}$) and D2 short (D2S or $D2_{415}$), which are generated by an alternative splicing of the *Drd2* gene. D2L differs from D2S by an additional 29 amino acids encoded by exon 6. This insertion is localized within the third cytoplasmic loop of the receptor, the same region that interacts with G proteins. This insertion accounts for the differential interactions of D2L and D2S with G proteins, the activation of distinct downstream signaling pathways, and the diversity of functions. Studies with mice showed that D2S overexpression, compared with wild-type or D2L-overexpression, reduced PRL levels and pituitary size. It was then suggested that a decrease in D2S expression may play a role in D2R agonist resistant prolactinomas [16].

A major hindrance in understanding the involvement of DA in prolactinomas has been the lack of stable human lactotroph cell lines. Most studies on the regulation of the lactotrophs have been carried out with the widely popular rat GH3 lactotroph cell line, which does not express DAR. Only a few studies have used the mouse lactotroph cell line MMQ, which expresses D2R and responds to DA. One study found that GH3 and MMQ cell lines express D5R and respond to its activation by reduced colony formation and autophagic cell death. In addition, DA increased PRL release in GH4C1 cells stably transfected with human D1R or D5R, whereas in cells transfected with either the long or the short D2R isoforms, DA, as expected, caused PRL inhibition.

Two hypotheses have been put forward to explain the pathogenesis of prolactinomas: DA dysregulation, and local somatic mutation [17]. Formation of prolactinomas due to loss of DA inhibition is supported by several observations. Among these are the development of large prolactinomas in transgenic mice with D2R deletion and estrogen-induced prolactinomas in Fischer 344 rats that involve pituitary neovascularization and an escape from dopaminergic inhibition. In humans, the resistance of some prolactinomas to DA has been taken as evidence for altered dopaminergic receptor or post-receptor regulatory controls. On the other hand, prolonged DA deficits in humans, caused by neuroleptics or by pituitary stalk dysfunction, do not induce prolactinomas. In addition, most tumors are confined to only a part of the pituitary rather than presenting as a widespread hyperplasia. Further, there is a relatively low frequency of recurrence after successful tumor resection. Collectively, these observations argue against loss of DA inhibition as the primary cause of prolactinoma formation in humans, despite evidence to the contrary in some animal studies.

The local somatic mutation hypothesis is based on X-chromosomal inactivation analysis showing that almost all human pituitary adenomas are monoclonal. This is consistent with tumor origin from a single cell that has undergone genomic mutation(s), followed by clonal expansion. The clonality of adenomas has prompted a search for genetic abnormalities such as loss of tumor suppressor genes or activation of protooncogenes that may explain the abnormal cell proliferation. However, alterations in p53 or retinoblastoma genes are very uncommon in prolactinomas, as are rearrangement, amplification, or mutations of oncogenes such as myc, ras, c-erbB2 (neu), and Gα proteins. Despite evidence for loss of heterozygocity on the long arm of chromosome 11, which is the location of the multiple endocrine neoplasia type 1 (MEN1) tumor suppressor gene, inactivation of this gene in sporadic pituitary adenomas is rare. Moreover, although PTTG is overexpressed in most PRL- and GH-secreting human pituitary adenomas, its potential role as a causative factor in the pathogenesis of prolactinomas is unclear.

In view of these conflicting findings, an integrated hypothesis that reconciles the two theories has emerged. It stipulates an occurrence of a certain genomic alteration (the identity of which is unknown) that transforms a single lactotroph as the initial step. The transformed cell escapes normal cell cycle regulation and starts proliferating in response to local growth factors that are abundant in the pituitary, e.g., fibroblast growth factor (FGF), epidermal growth factor (EGF), transforming growth factor alpha (TGFα), or TGFβ. Because transformed cells proliferate faster than normal cells, they are at an increased risk of additional genetic alterations, further promoting tumor progression. Among these could be loss of responsiveness to DA due to mutations in the D2R or its signaling pathways. Although some reduction in D2R expression and receptor density has been reported in prolactinomas, this applies only to the 10%–15% of those tumors that are resistant to bromocriptine therapy and does not offer a universal mechanism that explains the pathogenesis of prolactinomas. It is also important to emphasize that micro-prolactinomas grow very slowly in most patients and do not normally progress to become macroprolactinomas. In fact, the risk of progressing from a microadenoma to a macroadenoma is estimated to be no more than 4%–7% [18]. This suggests that aggressive macroadenoma represent a different and distinct entity, indicating a different etiology of the two types of tumors.

Treatments of prolactinomas vary according to the size of the tumor and the patient's desire to restore fertility (**Figure 12.1**). The main goals are to suppress tumor growth, normalize serum PRL levels, correct visual abnormalities, and preserve pituitary function. D2R agonists, including bromocriptine, pergolide, and cabergoline, are the mainstay treatment for prolactinomas [16,19]. These, however, can cause some side effects such as moderate gastrointestinal dysfunctions, irregular heartbeats, and lightheadedness, which usually subside after long treatment and dose adjustments.

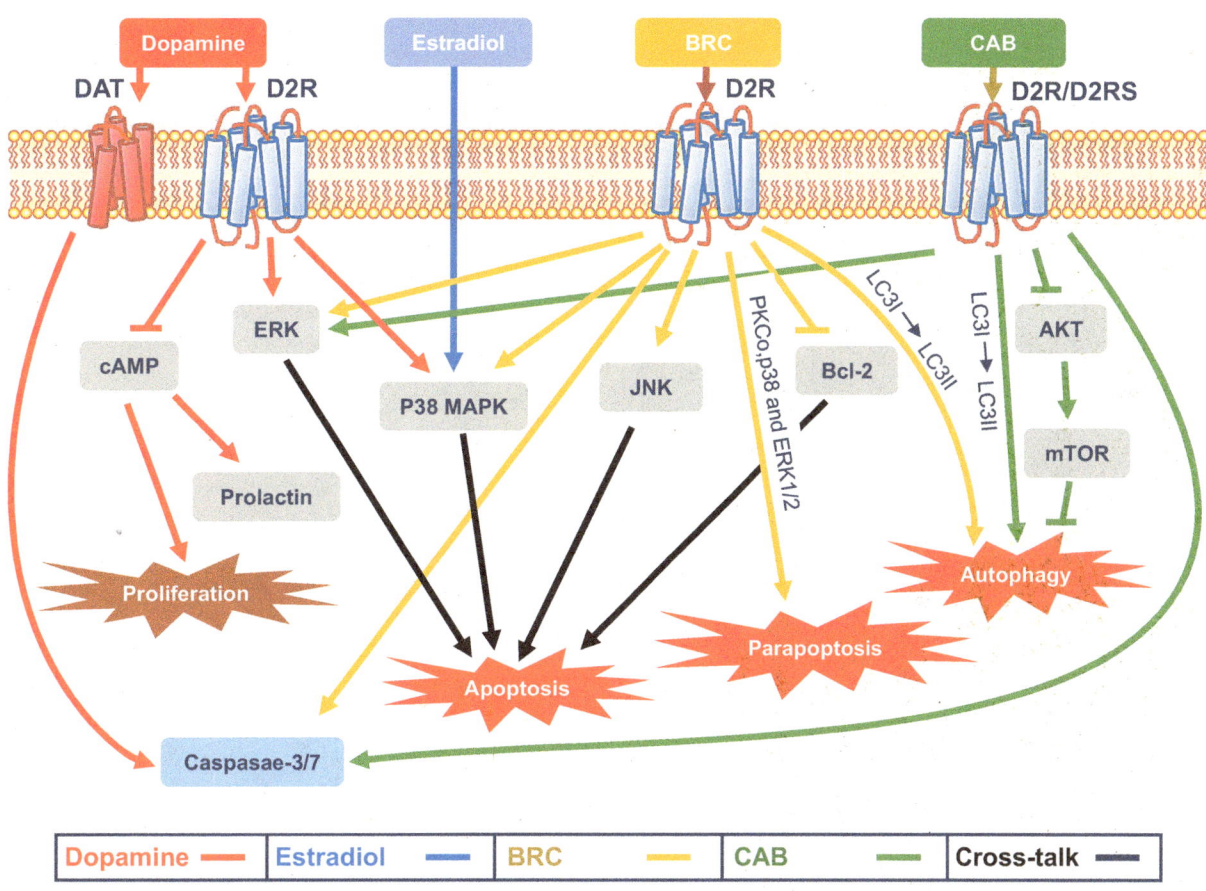

Figure 12.3 Proposed mechanisms by which dopamine (DA), DA agonists and estradiol interact to affect prolactinomas. Some aspects, i.e., role of estradiol, are more relevant to rodents than humans. BRC: bromocriptine; CAB: cabergoline; DAT: dopamine transporter; LC3: microtubule-associated protein 1A/1B-light chain 3; MAPK: MAP kinase; mTOR: target of rapamycin. (Redrawn from Liu, X. et al., *Front Endocrinol. (Lausanne)*, 9, 1–6, 2019.)

Cabergoline has been more effective than bromocriptine in normalizing PRL levels and in reducing tumor size, because of its longer half-life and lesser adverse effects. Overall, most small- or moderate-sized prolactinomas are treatable and the majority of patients achieve a positive response. The cure rate for patients with invasive macro-prolactinomas is less satisfactory and presents a challenge to the physician.

As illustrated in **Figure 12.3**, DA and its agonists cause tumor shrinkage by different mechanisms that include autophagy, apoptosis and parapoptosis, representing different modes of cell death that share some regulators. Notably, this model is based on data obtained with rodent pituitary cells, and the conclusions, including a putative cross-talk between DA and estrogen, are better established in rodent than in human prolactinomas.

About 15%–20% of patients with prolactinomas are resistant to DA-targeted therapy due to decreased expression and/or dysfunctional signaling of D2R. DA-resistant prolactinomas are more common in men than women, are more likely to be invasive macroadenomas than DA agonist-responsive prolactinomas, and are also more angiogenic, more proliferative, and more likely to exhibit cellular atypia [20]. Treatment of patients with resistance includes alternative DA agonists, transsphenoidal surgery, or radiotherapy.

12.2.3 Dopamine and other pituitary adenomas subtypes

GH-secreting tumors account for 15%–20% of total pituitary tumors and about two-thirds are macroadenomas. They are diagnosed by the elevated serum GH and insulin-like growth factor-1 (IGF-1) levels, and present with an acromegalic phenotype, i.e., enlargement of hands, feet, and face [5]. If the GH excess begins prior to the closure of the epiphyses during puberty, it can result in gigantism (excessive height). Comorbidities associated with GH hypersecretion include diabetes mellitus, hypertension, arthritis, and sleep apnea. GH-secreting tumors commonly express D2R, albeit at lower levels than those in the normal pituitary [21]. DA treatment of cultured somatotropinomas suppresses GH secretion by 20%–25% or more, while tumors lacking D2R expression show resistance. Because D2R expression was positively correlated with *in vitro* but not with *in vivo* suppression of GH by quinagolide in somatotroph adenomas, the link between D2R expression and treatment response is unclear. A combination therapy of cabergoline with somatostatin analogs (octreotide, lanreotide, or pasireotide) has been successful in select patients with GH-secreting tumors [22], presumably because of heterodimerization of somatostatin and DAR [23]. Interestingly, somatic mutations in G proteins (the so-called *gsp* oncogene) have been identified in 30%–40% of GH-secreting adenomas.

Tumors that secrete excessive amounts of **ACTH** account for 15% of total pituitary tumors. They typically result in a Cushing's syndrome due to excess cortisol release from the adrenals. See **Chapter 5**, **Section 5.7.1** and **Figure 5.14**, for more detail on Cushing's Syndrome. Up to 75% of human corticotroph adenomas express D2R [21]. In 15 of 20 ACTH-secreting adenomas, D2R immunostaining was moderate or strong, and D2R mRNA was detected in 10 of 12 cases. DR2-expressing primary tumor cell cultures showed dose-dependent inhibition of ACTH with bromocriptine and cabergoline, whereas tumors not expressing D2R did not respond. Clinically, DA therapy in Cushing's disease results in normalization of urinary cortisol levels in ~25% of patients, but the reported rates of tumor shrinkage are very low. TSH and luteinizing hormone/follicle-stimulating hormone (LH/FSH)-secreting tumors account for only 1%–2% of total pituitary adenomas; such tumors have little, if any, association with DA.

The occurrence of **nonfunctional pituitary adenomas** varies from 20% to 30% of total pituitary tumors. These tumors are often composed of gonadotrophs, although they are devoid of humoral hypersecretory syndromes [24]. They are usually large at the time of diagnosis, presenting with headaches, visual field defects, and hypopituitarism. Transsphenoidal surgery remains the treatment of choice for a rapid decompression of neighboring structures, often improving visual and pituitary function. Patients need long-term follow-up for the detection and treatment of hypopituitarism, visual dysfunction and tumor growth that may develop over time. In a small study, D2R was expressed in 67% of nonfunctional pituitary adenomas, among which the long isoform was found in 50%, the short isoform in 17%, and both isoforms in 33% of cases; D4R was expressed in 17% of the cases [25]. After 1 year of cabergoline treatment, tumor shrinkage was evident in 56% of the patients and was associated with D2R expression. The role of DA in the promotion of tumorigenesis was reinforced by a recent report that D2R agonists reduced both migration and invasion of cultured pituitary cells obtained from patients with nonfunctional pituitary adenomas [26].

12.3 BREAST CANCER

12.3.1 Prevalence, classification, and molecular characterization of breast cancer

Breast cancer is the most common malignancy among women, with more than 1 million cases occurring worldwide annually, presenting a critical global health challenge. Breast cancer primarily arises in the terminal

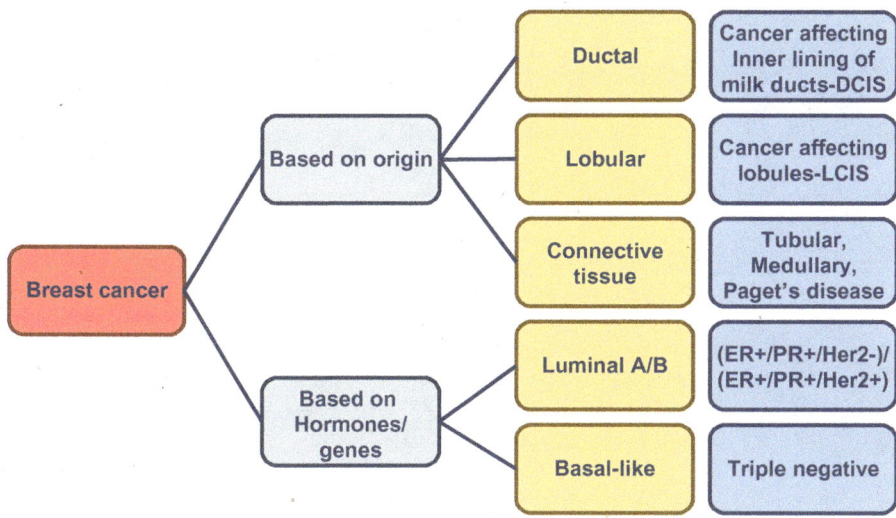

Figure 12.4 Classification of breast cancer. This classification is based on the origin of tumor cells and expression of cellular markers. DCIS: ductal carcinoma *in situ*; LCIS: lobular carcinoma *in situ*.

ductal-lobular units, which represent only 10% of the total volume of the mature human breast. Increased risk of developing breast cancer is correlated with early menarche, nulliparity, late age at first childbirth, and late menopause, as well as with obesity and exposure to exogenous hormones (e.g., OCs and xenoestrogens). Mutations in *BRCA1* and *BRCA2* genes are associated with increased risk of both breast and ovarian cancers, but heritable breast cancer accounts for no more than 10%–15% of all cases [27]. Mutations in ataxia telangiectasia (*ATM*) are also indicative of increased risk of breast cancer.

Breast cancer is classified by different criteria, with the major purpose of classification being to select the best treatment. Historically, patient management decisions have been based on histologic features, i.e., tumor size, histologic grade, lymph node status, proliferation indices, and hormone (ER and PR) and epidermal growth factor receptor 2 (HERs) status in conjunction with patient characteristics such as age and reproductive history. **Figure 12.4** illustrates the general categorization of breast cancer by site of origin and receptor status.

In recent years, various molecular techniques, particularly gene expression profiling, have been used increasingly to refine the classification of breast cancer and to assess prognosis and response to therapy. As shown in **Figure 12.5**, based on the genes expressed by the cancer cells, four molecular subtypes of breast cancer are recognized: (1) **Luminal A** is hormone-receptor positive [ER and progesterone receptor (PR)], HER2 negative, with low levels of the proliferation index. This is the most common breast cancer subtype, occurring in half of the patients. These low-grade cancers tend to grow slowly and have the best prognosis. (2) **Luminal B** is hormone-receptor positive and is either HER2 positive or HER2 negative, with high levels of proliferation index. These cancers, which occur in about 25% of the cases, generally grow slightly faster than luminal A cancers, and their prognosis is slightly worse. (3) **HER2-enriched** is hormone-receptor negative and HER2 enriched. These cancers tend to grow faster than luminal cancers, and can have a worse prognosis. However, they are often successfully treated with targeted therapies aimed at the HER2 protein. (4) **Basal-like triple-negative** is hormone-receptor negative and HER2 negative. This cancer subtype is more common in women with *BRCA1* gene mutations, as well as in younger and African-American women. Most triple-negative breast cancers are more aggressive, with poor prognosis. As covered in [28], there are additional refinements of breast cancer classification, including the

Figure 12.5 Classification, prognosis and treatment of breast cancer based on hormone receptors. BC: breast cancer; GEP: gene expression profiling; TIL: tumor infiltrating lymphocytes. (Redrawn from http://www.pathophys.org/breast-cancer/.)

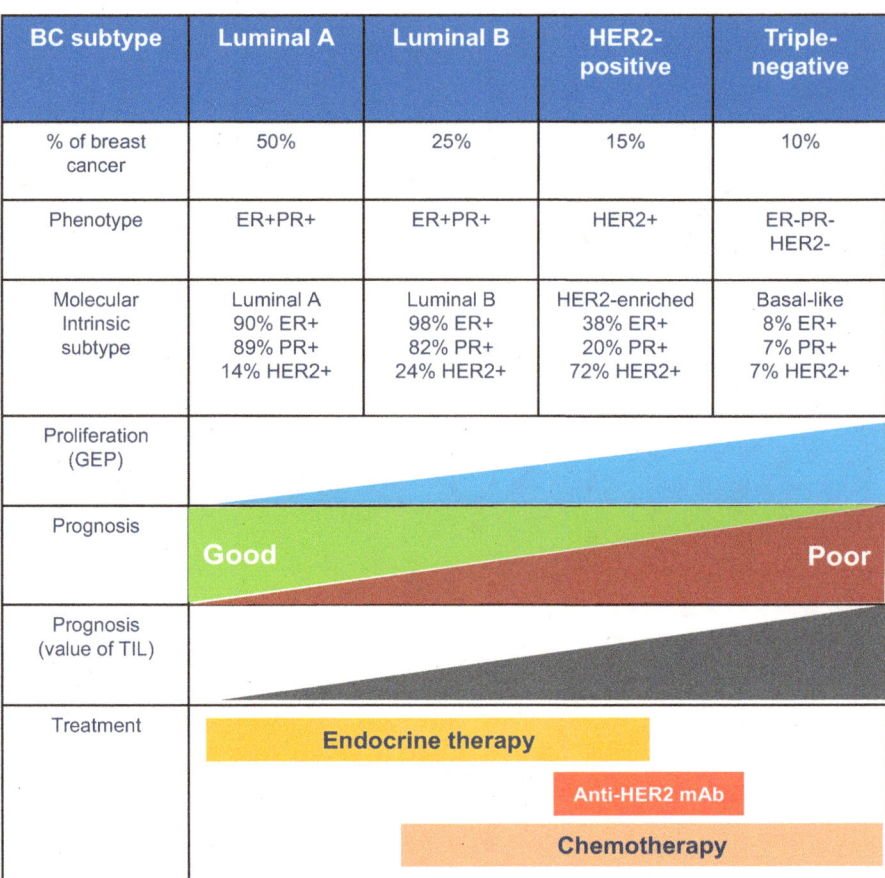

BC subtype	Luminal A	Luminal B	HER2-positive	Triple-negative
% of breast cancer	50%	25%	15%	10%
Phenotype	ER+PR+	ER+PR+	HER2+	ER-PR-HER2-
Molecular Intrinsic subtype	Luminal A 90% ER+ 89% PR+ 14% HER2+	Luminal B 98% ER+ 82% PR+ 24% HER2+	HER2-enriched 38% ER+ 20% PR+ 72% HER2+	Basal-like 8% ER+ 7% PR+ 7% HER2+

status of claudins, transmembrane proteins that are enriched in tight junctions involved in the migration and epithelial mesenchymal transition (EMT), which have been considered as tumor suppressors.

12.3.2 Chemotherapy and immunotherapy in breast cancer

The biomedical community is confronted by breast cancer on two fronts: One is how to implement an early accurate diagnosis, and another is how to provide an effective clinical management [29]. Some breast cancers are aggressive and life-threatening and must be managed robustly, i.e., by surgery, radiotherapy, and chemotherapy. Neoadjuvant chemotherapy is used to reduce tumor size before surgery, while adjuvant chemotherapy is used after tumor excision. Chemotherapy is the mainstay treatment for patients with triple-negative tumors, which are resistant to hormone or targeted therapy, and for those with advanced metastatic disease [30]. Over the years, dozens of anticancer drugs have been developed, with treatment options taking into account tumor grade and histology and whether the desired outcome is curative or palliative. Most regimens combine drugs acting by different mechanisms so as to improve the odds of suppressing tumor growth.

The most common classes of drugs used for chemotherapy include (1) anthracyclines, such as **doxorubicin** (Adriamycin), which stop DNA replication and induce apoptosis; (2) antimitotics such as **paclitaxel** (Taxol), which stabilize the microtubules and prevent cell division; (3) antimetabolites such as **5-fluorouracil** (5-FU) and its prodrug **capecitabine**, which block the action of thymidylate synthase and stop DNA production; (4) alkylating agents such as **cyclophosphamide** (Cytoxan), which form interstrand and intrastrand DNA crosslinking, leading to apoptosis; and (5) platinum agents such as **carboplatin** (Paraplatin) and **cisplatin**, which form DNA adducts that induce cell cycle arrest.

Chemotherapeutic drugs can cause side effects, depending upon the type and dose of drugs given, and the length of treatment. Among the most common side effects are hair loss, mouth sores, loss of appetite, nausea, vomiting, and diarrhea. Chemotherapy can also affect blood-forming cells of the bone marrow, leading to increased risk of infections (low white blood cell counts), easy bruising or bleeding (low blood platelet counts), and fatigue (low red blood cell counts and other reasons). For younger women, changes in menstrual periods are a common side effect of chemotherapy. Premature menopause and infertility can occur, and may be permanent. Some drugs are more likely to cause this than others. The older a woman is when she gets chemotherapy, the more likely it is that she will go through menopause or become infertile as a result. Consequently, there also is increased risk of bone loss and osteoporosis.

Immunotherapy. Evading antitumor immunity is a hallmark of the emergence and progression of cancer. Tumors employ multiple mechanisms to avoid recognition by the host immune system, including expression of the negative T cell regulatory molecule programmed death ligand-1 (PD-L1). The PD-1 receptor and ligands PD-L1 and PD-L2, which are members of the CD28 and B7 families, play critical roles in T cell co-inhibition and exhaustion [31]. Overexpression of PD-L1 and PD-1 on tumor cells and tumor-infiltrating lymphocytes, respectively, correlates with poor disease outcome in several human cancers. Monoclonal antibodies (mAbs) that block the PD-1/PD-L1 pathway have been developed for cancer immunotherapy via an enhancement of T cell functions. Clinical trials with mAbs to PD-1 and PD-L1 have shown impressive response rates, particularly for melanoma, non-small-cell lung cancer, renal cell carcinoma, and bladder cancer.

As indicated above, about 15%–20% of breast cancer patients have triple-negative tumors, an aggressive disease associated with shorter time to recurrences and death and propensity for metastasis to the brain and lungs [32,33]. Recently, several preclinical and clinical studies demonstrated that immunotherapy has the potential to improve outcomes for some triple-negative breast cancer patients. In March 2019, the FDA approved the first checkpoint inhibitor immunotherapy drug, an anti-PD-L1 antibody called atezolizumab, in combination with chemotherapy, for the treatment of about 20% of triple-negative, metastatic breast tumors that express the PD-L1 protein. When functioning properly, activated T lymphocytes can attack tumor cells and curb their growth, but tumors that express PD-L1 can evade the immune system through inhibition of the antitumor activity of the effector T cells. The immunotherapy works by blocking PD-L1, enabling the immune system to attack the tumors.

12.3.3 Hormone and targeted therapies in breast cancer

The identification of molecular markers in breast cancer such as overexpression of receptors for the gonadal steroids and HER2 presented an important opportunity for defining druggable targets. ERs are members of the large nuclear steroid receptor family, which include PR, androgen (AR), and glucocorticoid (GR)/mineralocorticoid (MR) receptors, as well as less related molecules such as thyroid receptors, retinoic acid, and orphan receptors. Humans have two ERs: ERα and ERβ, which are encoded by *ESR1* and *ESR2* genes, located on chromosomes 6q25.1 and 14q23-24.1, respectively. *ESR1* and *ESR2* comprise eight exons separated by seven intronic regions and span more than 140 kb and approximately 40 kb, respectively.

As illustrated in **Figure 12.6**A, the classical ER is an intracellular ligand-dependent transcription factor consisting of an N-terminal-activating

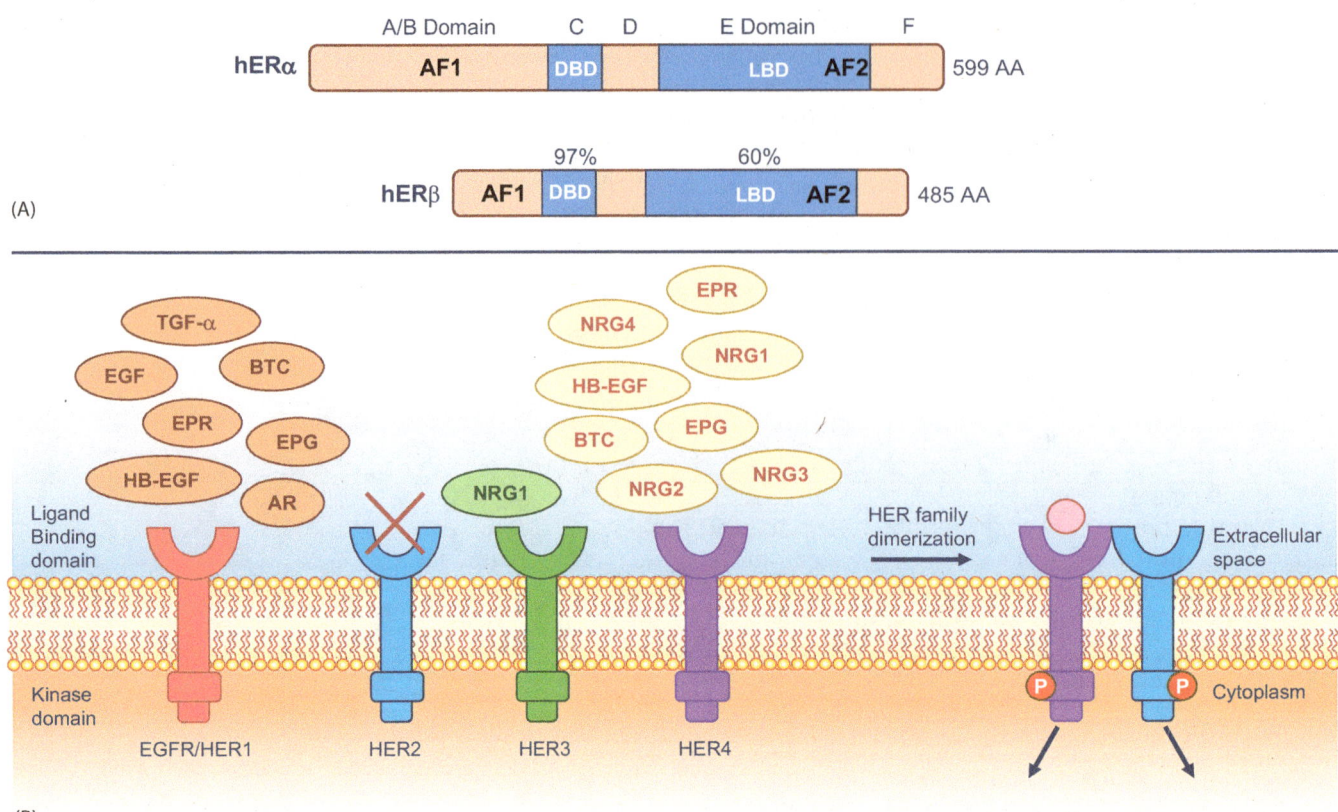

Figure 12.6 Receptors for estrogen (ER) and epidermal growth factor (EGFR) family members. Panel A is a schematic representation of the protein structures of human ERα and ERβ. Functional domains of the receptors include: DNA binding domain (DBD), ligand binding domain (LBD), and two transcriptional activation functions, AF-1 and AF-2. The D domain is the hinge region. The percent homology shared between ERα and ERβ in the C and E domains is also shown. Panel B is a schematic presentation of the epidermal growth factor (EGF) receptor family. HER2 has no known ligand and is the preferred partner for heterodimerization. AR: amphiregulin; BTC: betacellulin; EPG: epigen; EPR: epiregulin; HB-EGF: heparin-binding EGF-like growth factor; NRG1, 2, 3 or 4: neuregulin 1, 2, 3, or 4; TGF-α: transforming growth factor-α. (Redrawn from multiple sources.)

function (AF1) domain, which is involved in protein–protein interactions important for transcriptional activity of ER, a DNA-binding domain, and a ligand-binding domain (AF2). ERα and ERβ have high homology in the DNA binding domain and lower homology in the ligand binding domain, underlying the dissimilar binding affinities of the various estrogenic ligands to the two receptors. Following ligand binding, the ERs dimerize, enter the nucleus, and bind to estrogen response elements (EREs) in the promoters of target genes. Recruitment of co-regulators results in the formation of complexes that mediate transcription. The plethora of cell-specific co-activators and co-repressors account, in part, for the partial agonist versus antagonist activities of ligands such as tamoxifen in the uterus, breast, bone, and cardiovascular system.

Endocrine therapy is aimed at suppressing the tumorigenic actions of estrogens. It has become an established adjuvant therapy, which is given after surgery or radiotherapy but also prior or subsequent to chemotherapy. Effective drugs for hormone-sensitive cancers belong to three main classes: (1) selective estrogen receptor modulators (SERMs) such as **tamoxifen**, (2) aromatase inhibitors such as **anastrozol**, and (3) selective estrogen receptor down-regulators such as **fulverstrant** [34]. ERα is expressed at low levels in the normal breast epithelium, but its expression increases in carcinomas [35]. The expression pattern of ERβ is opposite that of ERα, suggesting that loss of ERβ expression indicates breast cancer development and/or progression.

In addition to the intracellular ERs, estrogenic ligands can bind to other types of receptors. A subpopulation of ERs in breast cancer cells (BCCs) is localized to the cell membrane and mediate rapid, non-genomic actions of estrogens. Because steroid receptors do not have transmembrane or kinase domains, they are unlikely to be incorporated into the cell membrane as integral proteins. Instead, they interact via palmitoylation of membrane-associated proteins such as caveolin, striatin, and Sch [36]. Both IGF1 and EGF receptors are also involved in tethering ERa to the membrane and in initiating MAP kinase (MAPK), phosphoinositide 3-kinase (PI3K) activation.

GPR30 (G protein-coupled receptor 30) is a seven-transmembrane domain receptor that directly binds estrogens and signals through trimeric G proteins [37]. Estrogen binding to GPR30 stimulates the cAMP pathways through Gαs, and Src through Gβγ. Subsequently, heparan-bound EGF is released, and activates the epidermal growth factor receptor (EGFR) and its downstream signaling that include MAPK, PI3K, and phospholipase C (PLC). Both tamoxifen and ICI 182,780 act as agonists, rather than antagonists, of GPR30. Expression of GPR30 is higher in invasive breast carcinoma than in normal breast tissue and is associated with larger tumor size, suggesting that it may be a predictor of aggressive disease. The relatively high binding affinity of GPR30 to estradiol (Kd of 3 nM) makes this receptor a likely mediator of estrogen actions in ER-negative BCCs, but its relative role in cells that also express ERα and ERβ is unclear.

The role of the **PR**, which exists in two isoforms, PRA and PRB, in breast cancer has been controversial. Whereas there is no doubt about the clinical benefit of measuring ERα, the value of measuring PR expression has been questioned [38]. PR expression is considered a surrogate marker for ERα integrity and endocrine response given that high total PR levels correlate with an improved tamoxifen response, longer disease-free and overall survival. Moreover, the ERα$^+$PR$^-$ group has a worse prognosis, and this phenotype has been associated with impaired ERα function or aberrant growth factor signaling that could contribute to Tamoxifen resistance. On the other hand, the existence of the ERα$^-$PR$^+$ group of breast cancer patients remains controversial. Although there is a consensus that PRA is the prevailing isoform in breast cancer, the utility of assessing the PRA/PRB ratio as a prognostic marker needs to be validated in larger cohorts of breast cancer patients. In addition, clinical trials using antiprogestins have yielded inconsistent results.

HER2, also known as erbB-2, is a member of the human EGFR family, consisting of four plasma membrane bound receptor tyrosine kinases (**Figure 12.6**B). HER1 is the EGF receptor, HER2 has no known ligand, HER3, which lacks the kinase domain, binds to neuroregulin 1, and HER4 binds to multiple ligands [39]. All four receptors contain an extracellular ligand binding domain, a transmembrane domain, and an intracellular domain that can interact with a multitude of signaling molecules and exhibit both ligand-dependent and ligand-independent activity. HER2 can heterodimerize with any of the three receptors and is considered a preferred dimerization partner of the other receptors.

HER2-positive tumors are more likely to have a higher tumor grade, tend to grow faster, and are more likely to spread to the lymph nodes. HER2-positive early breast cancers (stage I and stage II) are two to five times more likely to recur than HER2-negative tumors. Drugs used to treat HER2-positive tumors belong to two classes: monoclonal antibodies against the receptor, e.g., trastuzumab (**herceptin**), and small molecule inhibitors of receptor signaling, e.g., **lapatinib**. Over time, however, many treated tumors develop resistance to these therapies, curbing the success of treatment.

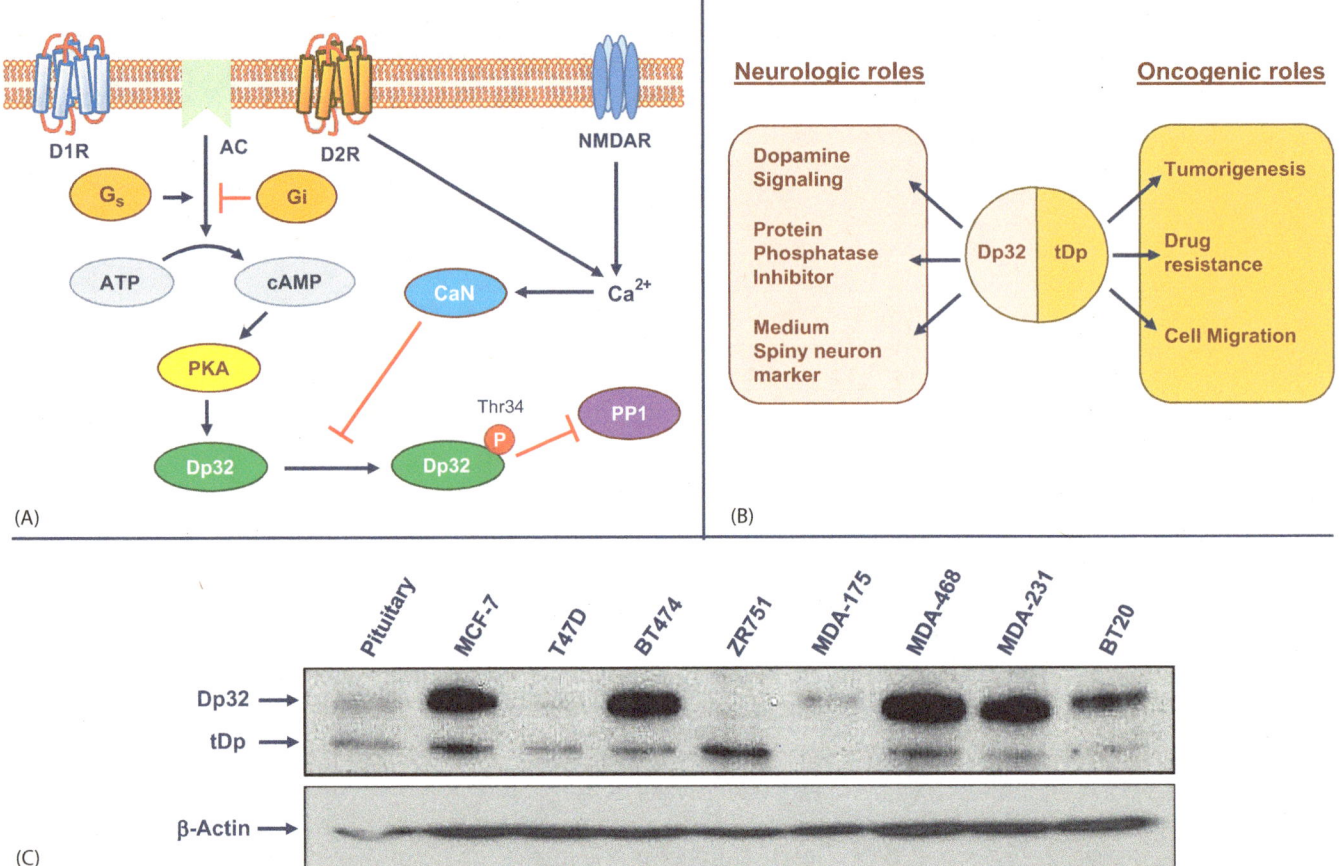

Figure 12.7 Regulation, actions and expression of DARPP-32 (Dp32). Panel A shows dopamine and glutamate signaling pathways that modulate the phosphatase 1 inhibitory activity (PPI) of Dp32 in neurons. AC: adenylyl cyclase; CaN: calcineurin; CtDP aN: calcineurin; Dp32: DARPP-32; NMDAR: N-methyl-D-aspartate receptor; PP1: protein phosphatase 1; tDP: truncated DP32. (Redrawn from Avanes, A. et al., *Biochem. Pharmacol.*, 160, 71–79, 2019.) Panel B shows the multiple neurologic and oncogenic actions of Dp32 and tDp. Panel C shows the expression of Dp32 and tDp in multiple breast cancer cell lines. (Unpublished data from the author's laboratory.)

12.3.4 DARPP: A dopamine-regulated phosphatase involved in tumorigenesis

Dopamine- and cAMP-regulated phosphoprotein-32, (DARPP-32 or Dp32), also known as phosphoprotein phosphatase-1 regulatory subunit 1B (PPP1R1B), was discovered in the 1980s as a 32-kDa protein substrate of DA-activated PKA in the brain [40,41]. Dp32 is expressed in DA-responsive brain regions such as the caudate nucleus, putamen, nucleus accumbens and cerebral cortex. As shown in **Figure 12.7**A, Dp32 mediates downstream signaling of DA via D1R, and is negatively regulated by DA via D2R, as well as by glutamate signaling through the N-methyl-D-aspartate receptor (NMDAR). Dp32 is expressed from the *PPP1R1B* gene on chromosome 17q12.1. Phosphorylation at Thr-34 by PKA converts Dp32 into a potent inhibitor of protein phosphatase 1 (PP1), while phosphorylation at Thr-75 by other mediators transforms it into a PKA inhibitor. The general consensus is that Dp32 is a tightly regulated hub molecule that mediates many actions of DA.

The action of Dp32 was thought to be limited to neurons until 2002, when working with gastric cancer cells, El-Rifai's group discovered another PPP1R1B transcript with a start site within intron 1 that codes for tDp, an N-terminal-truncated Dp32 isoform that lacks the first 36 residues [40]. The tDp and/or Dp32 transcripts and proteins have since been found to be overexpressed in breast, colon, esophageal, gastric, lung and prostate cancers where they exert variable oncogenic actions (**Figure 12.7**B). In breast cancer, overexpressed tDp was linked to acquired resistance to trastuzumab [42] and lapatinib [43] in HER2-positive BCC lines and to increased cellular proliferation and inhibition of apoptosis [44]. These data suggest that the Dp32 proteins can serve as biomarkers for drug resistance, as well as therapeutic targets. The authors concluded that future efforts should focus on the development of inhibitors that target Dp32 and its signaling pathways. Studies in our laboratory, using Western blotting, have shown high expression of both Dp32 and tDP in most, but not all, of eight BCCs examined (**Figure 12.7**C).

12.3.5 D1R expression and actions in breast cancer

Evidence from our laboratory [45] demonstrated that D1R is overexpressed in breast cancer and could serve as a novel biomarker for advanced disease. Moreover, activation of D1R or antagonism of D2R could be used for suppressing tumor growth and invasiveness, for enhancing chemosensitivity, and for inhibiting tumor angiogenesis. A brief description of these data follows below.

Using both real-time polymerase chain reaction (RT-PCR) and immunocytochemistry, we discovered overexpression of D1R in a small sample of primary breast tumors, but not in their adjacent normal breast tissue [45]. Subsequently, we used tissue microarrays containing 751 breast tumors and 30 normal breast tissue samples to score D1R expression by immunocytochemistry. As evident in **Figure 12.8**A, strong to intermediate D1R staining was seen in ~30% of the tumors, 15% had a weak signal, and the remainder, as well as all normal breast tissue samples, were D1R-negative. Further analysis revealed that D1R staining was significantly linked to premenopausal age, ER-negative, PR-negative, but HER2-positive tumors, indicating that D1R-overexpressing tumors do not fit within the conventional "triple negative" category and constitute a high risk category. Indeed, D1R-positive tumors were significantly associated with hallmarks of advanced disease: higher tumor stage, higher tumor grade, and node metastases. Moreover, Kaplan-Meyer analysis of 508 tumors revealed that patients with D1R-positive tumors had a significantly shorter median survival time: 6.5 years versus 12.5 years for those with D1R-negative tumors (Figure 12.8B).

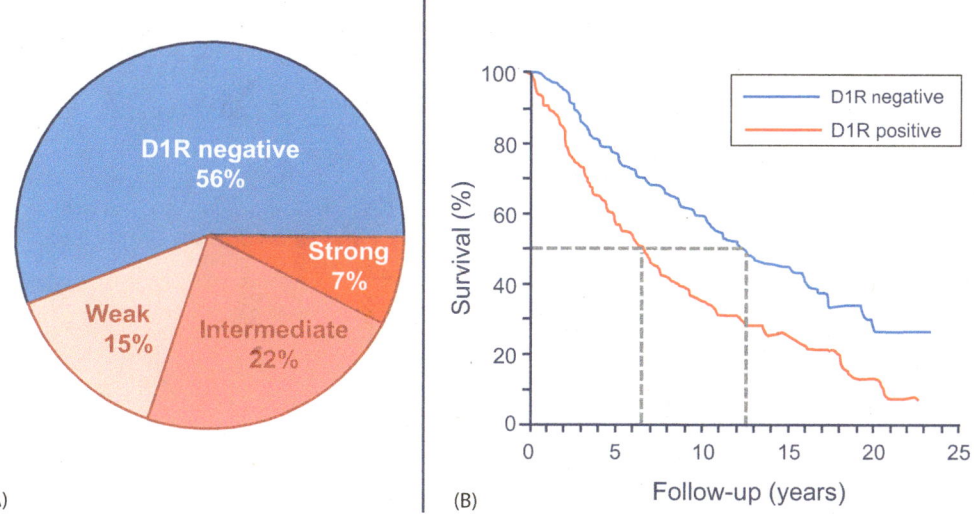

(A)   (B)

Figure 12.8 D1R in breast cancer. Panel A shows the distribution of immunoreactive D1R in tissue microarrays containing 751 breast carcinomas and 30 normal breast samples. Data shown are percent of total tumor number. All 30 normal tissue samples were also D1R-negative. Panel B shows that positive D1R expression is associated with shorter patient survival, as determined by Kaplan–Meier analysis of 508 tumors. (Redrawn from Borcherding, D.C. et al., *Oncogene*, 35, 3103–3113, 2016.)

One question was whether metastases retain expression of D1R. In preliminary studies, we used tissue microarrays with 120 matched breast carcinomas and lymph node metastases and found an almost identical D1R expression in both samples. These are interesting data because estrogen or androgen receptors are often not expressed in metastases [46,47]. However, it is imperative to verify that systemic metastases in liver, lung, or bone also maintain functional D1R.

We also compared D1R and D2R expression in eight BCC lines. D1R was more abundant in aggressive, triple-negative cells than in ER-positive cells. All cells also expressed variable amounts of D2R. Cloning of the *DRD1* transcript from MDA-MB-231 cells confirmed its identity with the published sequence. The putative roles of D1R in breast cancer were then examined [45]. Based on data obtained with other overexpressed receptors in breast cancer (ER, EGFR), we fully expected that D1R-agonists would stimulate cell growth. Surprisingly, low nM doses of DA or three D1R-selective agonists, but not cabergoline, a D2R agonist, suppressed cell viability, which was due to increased apoptosis. D1R activation also inhibited cell invasion, and increased cell sensitivity to cytotoxicity by doxorubicin.

Because D1R agonists are classified as cAMP activators, the effect of D1R activation on cAMP accumulation was next examined. Unexpectedly, short-term cell incubation with DA or fenoldopam, a peripheral D1R agonist, increased cGMP rather than cAMP. The role of the cGMP/protein kinase G (PKG) axis was verified by finding that a direct stimulation of soluble guanylate cyclase (sGC) by YC-1, as well as a blockade of phosphodiesterase 5 (PDE5; which breaks down cGMP) by Cialis (tadalafil), augmented cGMP levels, increased PKG activity, and induced apoptosis. Our observations were indeed supported by the report that cGMP, via PKG activation, suppressed both ER-positive and ER-negative BCC [48]. In addition, sildenafil was reported to increase cytotoxicity by doxorubicin, cisplatin and paclitaxel, and its co-administration with doxorubicin enhanced the suppression of xenograft growth [49].

Athymic nude mice were orthotopically implanted with two highly tumorigenic BCC—MDA-MB-231 and SUM159—followed by treatment with fenoldopam, delivered by subcutaneously implanted Alzet osmotic mini-pumps (**Figure 12.9**A). The D1R agonist dramatically suppressed tumor growth by increasing both apoptosis and necrosis [45]. The combination of apoptosis and necrosis likely explains the more robust suppressive effects of fenoldopam *in vivo* than *in vitro*. Increased necrosis could have been due to inhibition of angiogenesis given that DA was reported to reduce tumor angiogenesis in rats [50] and mice [51] by inhibiting vascular endothelial growth factor (VEGF) [52] and its receptor [53] via endothelial DAR [54]. Notably, the suppression of tumor growth by fenoldopam was long lasting: Treated tumors remained quiescent for at least 2 additional weeks after removal of the Alzet pumps (**Figure 12.9**A). A fluorescent imaging method for visualizing D1R-expressing tumors and metastases was also developed [45]. **Figure 12.9**B shows an intense fluorescence of the primary tumors and axillary metastases in mice with MDA-MB-231-derived tumors that were intravenously injected with human anti-D1R antibody conjugated to Alexa-Fluor 647.

These studies have established D1R as a contributing factor to breast tumorigenesis. However, it is *counter-intuitive* that increased D1R expression correlates with advanced disease and shorter patient survival, while D1R activation, rather than its suppression, causes apoptosis and tumor shrinkage. This enigma raises the following questions: (1) Does the *DRD1* gene become rearranged, mutated, or is fused to an oncogene during tumorigenesis, enabling it to activate an oncogenic pathway(s)? The *DRD1* gene is located on chromosome 5q35.2, a region at the end of a chromosome known to have several breakpoints and somatic rearrangements. Deletions in this region are associated with myelodysplastic syndrome and acute myeloid leukemia [55]. (2) Do some D1R-expessing tumors in patients fail to respond to induction

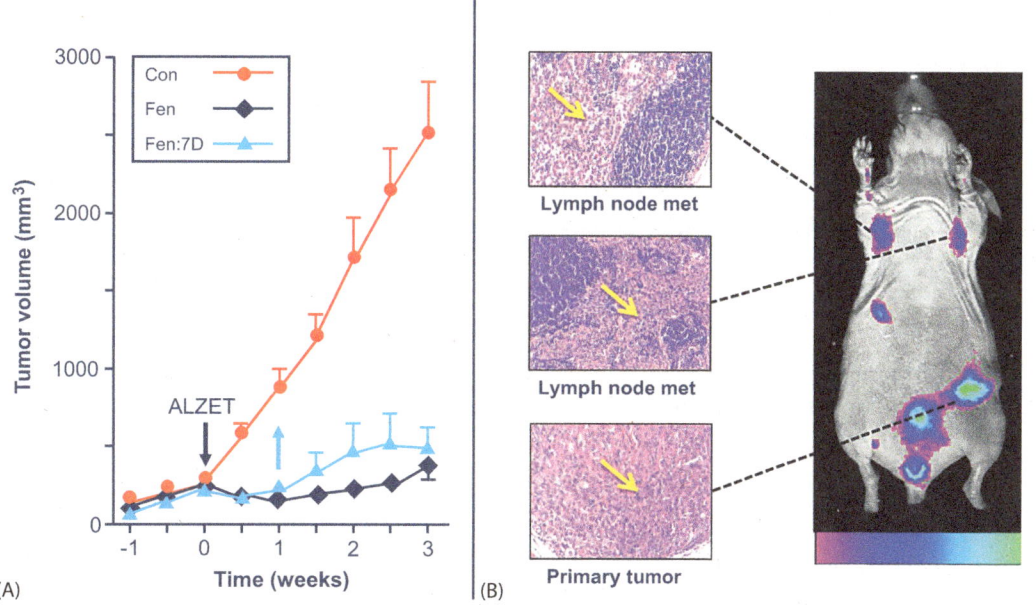

Figure 12.9 Effects of Fenoldopam, a D1R agonist, on breast cancer in mice and tumor imaging. Panel A shows that treatment with Fenoldopam (Fen), delivered by Alzet osmotic mini-pumps, markedly reduced growth of SUM159-derived xenografts in athymic nude mice. One group (Fen) had the pumps for 3 weeks, while another (Fen:7D), had the pumps removed after 7 days. Panel B Fluorescence imaging of D1R-expressing xenografts. Mice with MDA-MB-231-derived tumors were injected intravenously with human anti-D1R antibody conjugated to Alexa-Fluor 647. *In vivo* fluorescence imaging after 24 h shows intense fluorescence of the primary tumor and metastases. Insets show histological preparations of primary tumor and axillary lymph node metastases. (Redrawn from Borcherding, D.C. et al., *Oncogene*, 35, 3103–3113, 2016.)

of apoptosis by circulating DA because they do not express arylsulfatase A (ARSA) (for converting DA-S to DA) and/or because they overexpress D2R, which antagonizes D1R-induced apoptosis? (3) What is the molecular linkage between D1R and the cGMP/PKG/apoptotic pathway? D1R is likely associated with the heterotrimeric Gα12 or Gα13 subunits, known to be coupled to inducible nitric oxide synthase (iNOS), which increases cGMP [56,57]. (4) Does D1R activation affect the functions of Dp32 or tDp in BCCs?

Several other reports support the involvement of both D1R-like and D2R-like in breast cancer. An early small study found binding of ^{3}H-spiperone, a D2R-like antagonist, in breast tumors [58], while a more recent report identified expression of D3R and D5R in breast cancer stem cells [59]. D1R activators also inhibited BCC migration and bone metastases in xenograft-bearing nude mice [60]. Focusing on D2R, another study showed that bromocriptine, a D2R agonist, induced apoptosis in MCF-7 cells [61]. In addition, the combination of sulpiride, an atypical antipsychotic, and dexamethasone was more effective than each alone in inhibiting colony formation, cell migration, and invasion of murine and human BCC *in vitro*, and suppressed the growth of xenografts in mice [62]. Notably, tamoxifen, the most widely used anti-estrogen in the treatment of ER-positive breast cancer, inhibited DAT activity [63], raising the possibility that tamoxifen increases DA availability to tumors, with the potential to alter their growth, depending on the ratio of D1R-like/D2R-like in any given tumor.

In a recent study, aripiprazole, a D2R partial agonist, enhanced the radiosensitivity of MCF-7 cells and induced their apoptosis via AMPK activation [64]. Another recent study found that fenoldopam decreased lung metastasis in a mouse model of mammary cancer by elevating cGMP [65]. Fenoldopam also impacted on the cancer-induced increase of white blood cell increase, resulting in decreased neutrophils but increased lymphocytes in primary tumors. The authors concluded that D1R is a potential target for metastatic breast cancer treatment and even other cancers at a late stage.

Figure 12.10 A model of the interactions of D1R with the cGMP/PKG system. The association of D1R with inducible nitric oxide synthase (iNOS) is assumed but not proven. KT: inhibitor of protein kinase G (PKG); PDE5: phosphodiesterase 5; PKC: protein kinase C; SCH: D1R inhibitor; sCG: soluble guanylate cyclase; YC-1: stimulator of sCG. See text for other explanations.

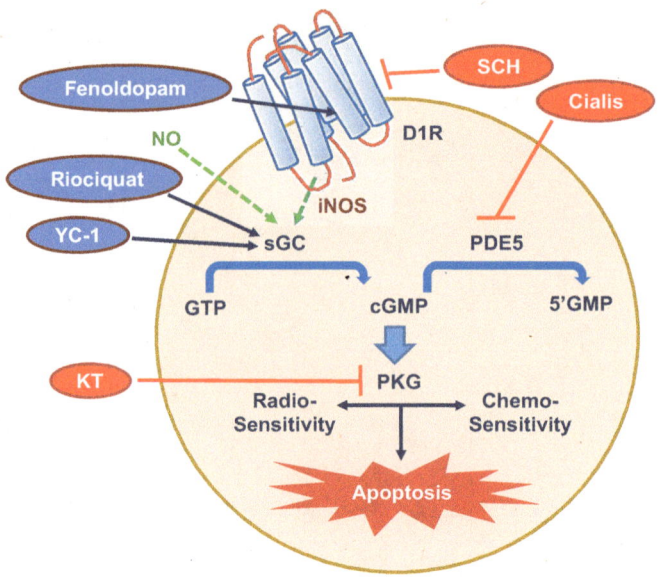

In sum, FDA-approved drugs, especially those with high selectivity for DAR subtypes that do not cross the blood–brain barrier, could be repurposed to treat breast cancer patients. Drugs that target downstream signaling of D1R, e.g., selective PDE5 inhibitors, could also be used as targeted therapy in breast cancer. Cialis, a selective PDE5 inhibitor, which is widely used to treat erectile dysfunction, is currently in phase I/II clinical trials in patients with head and neck cancer [66]. In addition to suppressing tumor growth, Cialis can boost the capacity of the immune system to eliminate cancer cells [67].

Identification of breast cancer patients as candidates for DAR-targeted therapy could be done by two approaches: (1) analysis of DAR immunostaining in tumor biopsies, as is routinely done for ER and HER2; and (2) positron emission tomography (PET imaging), which is effectively used to diagnose neuropsychiatric disorders [68]. The advantages of PET imaging include the ability to assess receptor expression for the entire tumor burden, thus avoiding some sampling errors that occur with heterogeneous receptor expression in primary tumors. PET imaging can also be used for a noninvasive assessment of distal metastases that are not accessible to sampling by biopsy, and for serial monitoring of drug effects on designated targets.

Putative mechanisms by which D1R activation suppresses breast cancer are conceptualized in **Figure 12.10**. According to this model, the binding of agonists such as fenoldopam to D1R activates sGC, resulting in increased cGMP levels. Nitric oxide (NO), as well as a number of drugs (i.e., Riociguat and YC-1, which activate sGC), and Cialis, an inhibitor of PDE5, can also increase cGMP levels. Elevated cGMP activates PKG, which, depending upon the cell context, can lead to apoptosis, as well as to increased radio- and chemosensitivity. The role of the cGMP system in breast cancer is supported by others [69], reporting that inhibition of PDE5 and activation of PKG by exisulind-induced apoptosis in a number of BCCs, via a reduction of Wnt/β transcriptional activity and subsequent down-regulation of target genes such as cyclin D and survivin.

12.4 OVARIAN CANCER

12.4.1 Characteristics of ovarian cancer

Cancer of the ovary is the fourth leading cause of cancer mortality among women, causing more deaths than other female reproductive cancers [70,71]. The typical age of diagnosis is 63. The confounding problem is that ovarian cancer is difficult to detect early, because women with this cancer often have no symptoms, or just mild symptoms until the disease is in an advanced stage. Ovarian cancer is highly heterogeneous, and tumors can occur in the stroma, follicles, and germ cells. There are no reliable biomarkers for ovarian cancer, although CA-125 has been used with a mixed success. In addition, relatively little is known about risk factors or the tumorigenic mechanisms that underlie this disease [70]. Diagnosis can be done by imaging, followed by laparoscopic evaluation or laparotomy to determine tumor staging. Standard treatments include aggressive debulking surgery, radiotherapy, chemotherapy, and their combinations.

The clinical and biological behavior of epithelial ovarian cancers differs from most other types of cancers [70]. Although the dissemination of cancerous cells in other carcinomas utilizes vascular routes of intra- and extravasation for cell migration from the primary site to distant organs, ovarian carcinoma metastasis is a more passive event. Hence, ovarian tumor cells often migrate to peritoneal organs with the peritoneal fluid and rarely generate metastases outside the peritoneum. This passive mechanism is evident by some molecular changes in cell elements for its anchoring to remote metastatic sites. The surrounding cellular and tissue environment also promotes hosting of the cells outside the ovarian tumor.

12.4.2 Involvement of dopamine in ovarian cancer

Two studies have used thioridazine, a D2R blocker, to examine the role of DA in ovarian cancer. In one study [72], injection of thioridazine into nude mice engrafted with xenografts derived from 2,774 human ovarian cancer cells, significantly inhibited tumor growth by approximately five-fold and also decreased tumor vascularity. Thioridazine inhibited the phosphorylation of signaling molecules downstream of PI3K, including Akt, phosphoinositide-dependent protein kinase 1 (PDK1), and mammalian target of rapamycin (mTOR). These data also provide evidence that thioridazine regulates endothelial cell function and subsequent angiogenesis by inhibiting VEGF receptor 2(VEGFR-2)/PI3K/mTOR signal transduction.

A more recent study reported that two human ovarian cancer cell lines, A2780 and SKOV3, express D2R, as determined by Western blotting analysis [73]. Cell incubation with thioridazine induced apoptosis and autophagy, which were attributed to increased reactive oxygen species (ROS) and the associated DNA damage. In addition, heme oxygenase 1, NAPDH quinone dehydrogenase 1, hypoxia inducible factor-1α, and phosphorylated (p)-protein kinase B expression all were significantly decreased, while the expression of MAP kinase increased. Moreover, treatment with thioridazine for 3 weeks in mice engrafted with SKOV3 xenografts caused a 50% reduction in tumor volume. The authors concluded that thioridazine may be used as a potential drug in ovarian cancer therapy.

A preclinical model of stress-induced ovarian cancer growth was used to examine the effects of DA [54]. In this model, periodic immobilization strongly activates the hypothalamo–pituitary adrenal axis and the sympathetic nervous system, characteristic of chronic stress. Both endothelial and ovarian cancer cells expressed all DAR subtypes, with the exception of D3R, which was not detected in the ovarian cancer cells. DA significantly inhibited cell viability and stimulated apoptosis by reducing cAMP and by inhibiting NE- and VEGF-induced Src kinase activation. The inhibitory effects of DA on xenograft growth and microvessel density, as well as its stimulatory effect on apoptosis, were prevented by the D2R antagonist eticlopride. The authors concluded that depletion of DA under chronic stress creates a permissive microenvironment for tumor growth that can be reversed by DA replacement.

Another study from the same group found that whereas the antiangiogenic effect of DA was mediated by D2R, DA acted via D1R to increase vessel stabilization by recruiting pericytes to tumor endothelial cells [74]. DA stimulated the migration of mouse pericyte-like cells *in vitro* and increased cAMP levels in these cells. In addition, both DA or the D1R agonist SKF 82958 increased cisplatin uptake in an ovarian tumor xenograft model. Tumor vascular stabilization by DA was observed by another group, using a glioma tumor model [75].

The involvement of the NO/cGMP/PKG axis in ovarian cancer was examined in two studies from the same laboratory. The first study explored for associations between NO and cisplatin resistance in several human ovarian cancer cell lines [76]. An NO donor caused apoptosis and sensitized the cisplatin-resistant C13 ovarian cancer cells to cisplatin by up-regulating p53. Basal iNOS levels in the cisplatin-sensitive OV2008 cells were higher than those in the resistant C13 cells, and cisplatin up-regulated iNOS in OV2008 cells only. The second study examined the role of PKG-Iα in the promotion of ovarian cancer [77]. A combined use of signaling pathway inhibitors and knockdown approaches revealed interactions between Src kinase and PKG-Iα in the promotion of DNA synthesis and cell proliferation in human ovarian cancer cells. The authors concluded that activation of the NO/PKG signaling pathway provides a novel therapeutic target for disrupting ovarian cancer cell proliferation.

Teratomas are germ cell tumors that often contain tissues such as hair, muscle, and bone [78,79]. Mature teratomas are typically benign and found mostly in the ovaries, while immature teratomas are usually malignant and are more often found in the testes. Malignant teratomas represent 2%–3% of all ovarian cancers and mostly occur in younger patients. An interesting association between DA and teratomas was revealed in a study that used human embryonic stems cells (hESCs) to generate DA neurons intended for transplantation in Parkinson's disease (PD) [80]. Co-culturing of the hESCs cells with mouse stromal cells lead to their *in vitro* differentiation into DA neurons. When these neurons were engrafted into the brain of 6OH-DA-lesion rat model of PD, they developed severe teratomas. As illustrated in **Figure 12.11**, one approach to partially overcome this occurrence was to treat the hESCs with a chemotherapeutic agent such as vinblastine, which minimizes the proliferative, but not the quiescent, teratoma. This report raised safety concerns for future applications of pleuripotent stem cells in regenerative medicine [81]. Nonetheless, whether DA is the driver for primary teratoma formation in humans remains to be determined.

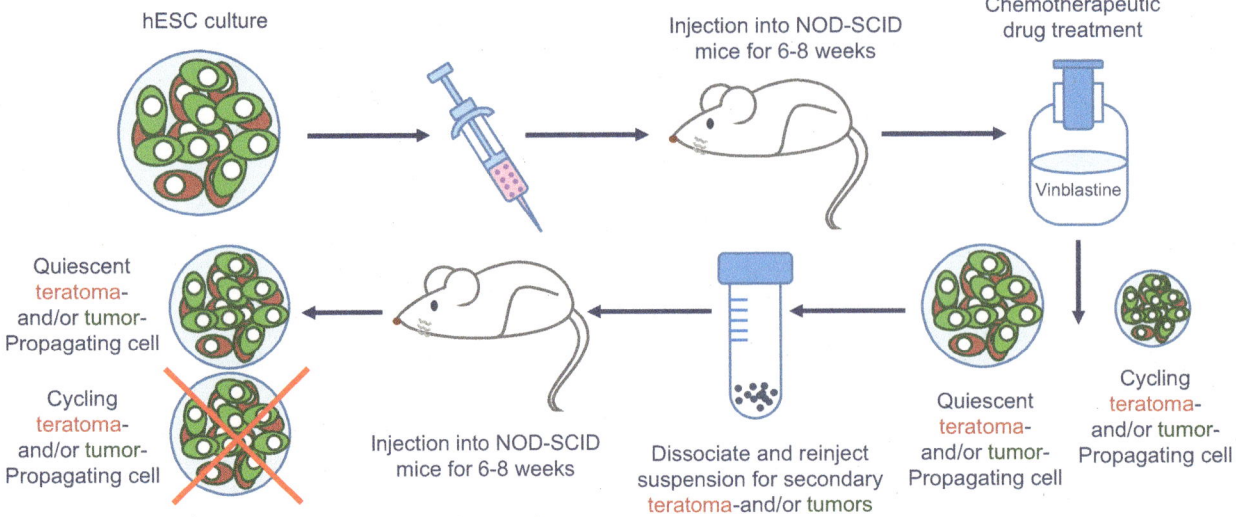

Figure 12.11 Experimental manipulations of dopamine-associated teratomas. Shown is an approach for reducing the occurrence of propagating teratomas in human embryonic stem cells (hESCs) treated to generate dopaminergic neurons. Cultures are injected to non-obese diabetic/severe combined immunodeficient (NOD-SCID) mice and then treated with vinblastine 1 week before extraction. A quiescent teratoma- and/or tumor-propagating cell will not respond to the drug, whereas a cycling propagating cell would respond to vinblastine and the teratoma size should decrease. (Redrawn from Werbowetski-Ogilvie, T.E. and Bhatia, M., *Trends Mol. Med.*, 14, 323–332, 2008.)

12.5 ENDOMETRIAL AND CERVICAL CANCER

Endometrial cancer is the seventh most common cancer of women, with ~190,000 new cases and 45,000 death occurring worldwide each year. Endometrial cancer is a heterogeneous disease [82]. Type I cancers are hormonally driven, and typically present with a low grade at an early stage. These cancers can often be cured by surgery and have a low rate of recurrence. Type II cancers are less differentiated, often appear at a later stage, and are of serous, clear cell, or high grade histology, with a high risk of recurrence. Most of the risk factors for endometrial cancer involve high levels of estrogens. About 40% of the cases are thought to be related to obesity. Excess adipose tissue increases the conversion of androstendione into estrone, with the high circulating levels of estrone exposing the endometrium to continuously high estrogenic activity. Obesity also causes less estrogen to be removed from the blood. Polycystic ovary syndrome is also associated with higher rates of endometrial cancer for the same reasons as obesity.

The most common management of endometrial cancer is total abdominal hysterectomy and bilateral salpingo-oophrectomy (removal of ovaries and oviducts). In selected patients, hormonal therapy can be as effective as chemotherapy without the toxicity and at a much lower cost. Currently, progesterone is the hormonal treatment of choice, but other drugs include SERMs, aromatase inhibitors, and GnRH antagonists. Because tumors that express these receptors are most sensitive to targeted therapy, patient selection is vitally important for a successful treatment.

A recent study examined the effects of thioridazine, alone or in combination with medroxy-progesterone acetate (MPA), on endometrial cancer [83]. The combined treatment increased apoptosis in two endometrial cancer cell lines, ISK and KLE. Expression of both progesterone receptor B (PRB) and D2R increased, while that of EGFR and p-AKT decreased. These data suggested that thioridazine plus MPA suppress tumor growth by inhibiting the PI3K/AKT signal transduction pathway, which is mediated by PRB, D2R and EGFR. A functional association between DA and NO was seen in endometrial glandular epithelial cells isolated from human endometrial specimens [84]. DA caused a dose-dependent transient surge of NO, whereas estrogen, progesterone, or relaxin did not. Cells treated with relaxin or with DA for 4 days enhanced the DA-induced NO release four- to six-fold. It remains to be determined whether DA-induced NO release activates the cGMP/PKG apoptotic pathway also in endometrial cancer cells.

Cervical cancer is the second most common cancer in women worldwide and also the fourth most common cause of death [85]. More than 90% of the cases are caused by human papillomavirus (HPV) infections. The discovery of HPV as the causal agent of cervical cancer has led to the development of HPV-based diagnostics. Prophylactic vaccines, based on the oncogenic HPV type virus-like particles, have also been introduced as a preventive approach. With the introduction of extensive screening programs (Pap smears), based on exfoliative cervical cytology, the incidence and mortality has declined significantly in the last 40 years. Squamous cell carcinomas account for 75% of the cases, while adenocarcinomas account for 25%. Treatments include radical hysterectomy, neoadjuvant chemotherapy, and combined chemotherapy and radiation. Because cervical adenocarcinoma is sufficiently distinct from squamous cell carcinoma, they warrant specific treatment modalities.

Using primary cervical tumors, a correlation between D2R expression and development of cervical cancer was established [86]. Immunostaining for D2R showed a progressively increased signal from normal tissue, to cervical intraepithelial neoplasia (CIN), to full carcinomas. Treatment of SiHa cells, a human cervical squamous carcinoma cell line, with thioridazine induced apoptosis and necrosis and suppressed D2R expression. The authors concluded that increased D2R expression is associated with cervical cancer progression and that D2R may serve as a novel tumor marker and a potential therapeutic target for cervical cancer. In another study [87], treatment of Hela, human cervical cancer cells, with thioridazine induced Bax-Bak-dependent and -independent apoptosis by enhancing ROS production followed by endoplasmic reticulum stress.

12.6 PROSTATE AND TESTICULAR CANCER

12.6.1 Characteristics of prostate cancer

Prostate cancer is among the most frequently diagnosed solid tumors in men, and is one of the leading causes of cancer-related deaths in Western countries [88–90]. Age is the strongest risk factor for prostate cancer. Diagnosis for prostate cancer include digital rectal exam and blood **PSA (prostate specific antigen)** analysis. If these suggest an abnormality, ultrasound, prostate biopsy, or MRI are employed for confirmation. Given that testosterone has a strong mitogenic activity in prostate cells, **androgen deprivation therapy (ADT)** is the standard therapy for men with *de novo* or recurrent metastatic prostate cancer. ADT can be achieved by the use of antiandrogens (e.g., flutamide, bicalutamide, and enzalutamide), androgen synthesis inhibitors (e.g., ketoconazole, aminoglutethamide, and abiraterone), or immunotherapy.

ADT usually leads to good initial biochemical and clinical responses. Over time, however, patients inevitably develop resistance to treatment, and their disease continues to progress, resulting in the so-called castration-resistant prostate cancer. At this stage, tumors are no longer curable, and treatment options are only palliative, as bone metastases cause pain, spinal cord compression, numbness, and increased risk for fractures. Taxanes, such as docetaxel and cabazitaxel, and novel androgen receptor-targeting agents such as abiraterone acetate and enzalutamide, have variable beneficial effects on castration-resistant prostate cancer. The concept and principle of primary hormonal therapy for metastatic hormone-sensitive prostate cancer remained unchanged for decades. Clearly, identification of novel targets and development of effective drugs would be highly desirable for prostate cancer patients.

12.6.2 Role of dopamine in prostate cancer

Emerging evidence shows the potential roles of DA/DAR and their downstream signaling pathways in prostate cancer. Using the androgen-dependent human prostate cancer cell line LNCap, direct effects of DA on cell proliferation were examined [91]. The cells express significant amounts of D2R protein, but DA at 1 nM only moderately inhibited cell proliferation. However, the inhibitory activity of DA was greatly enhanced when chimera heterodimers, consisting of D2R and somatostatin receptors (SSTRs), were transfected into the cells. Using DU145 and LNCap prostate cancer cells, another study found that the D2R antagonist pimozide at 10–20 μM inhibited cell proliferation, migration and colony formation [92]. Pimozide also suppressed STAT3, which is not considered a typical signaling for D2R, indicating an unclear mechanism. Risperidone, an antipsychotics which binds to multiple receptors in addition to DAR, suppressed the proliferation of PC3 cells *in vitro*, and retarded their growth as xenografts in nude mice [93].

Although there is only limited documentation on a direct effect of DAR activation on prostate cancer, similar to breast cancer, DA may activate the NO/cGMP/PKG pathway in prostate cancer. The involvement of this pathway in chemosensitivity of DU145 and PC-3 prostate cancer cells was confirmed by showing that inhibitors of sGC and PKG increased drug resistance [94]. Moreover, direct activation of PKG with 8-Br-cGMP attenuated the acquisition of hypoxia-induced resistance to doxorubicin in these cells. Another study found that exisulind, which acts by activating PKG, had antineoplastic activity in rodent models of colon, prostate, bladder, mammary and lung cancer [95]. The authors pointed out that in a randomized, placebo-controlled trial of prostate cancer patients, exisulind inhibited the rise of PSA in men with PSA progression after radical prostatectomy. Both preclinical and early clinical data suggested that exisulind and similar drugs in this class could be useful in treating various cancers both as monotherapy and in combination with chemotherapy and other targeted agents.

The blockade of cGMP breakdown by PDE inhibitors also has beneficial effects on prostate cancer [96]. PDE5 and PDE11 were the most prominent PDEs in DU145 and PC-3 cells, representing 90% of the total cGMP-specific PDE activity. Treatment of DU145 cells with a PDE inhibitor reduced hypoxia-associated acquisition of resistance to doxorubicin, and attenuated the growth of human prostate cancer xenografts in mice. The authors concluded that these data offer a rationale for future therapeutic applications of PDE inhibitors in men with prostate cancer.

Another study examined the effects of sildenafil on prostate cancer [97]. Co-treatment with sildenafil increased doxorubicin-induced apoptosis in PC-3 and DU145 cells. This action was mediated by enhanced generation of ROS, increased caspase-3 and caspase-9 activities, as well as reduced expression of Bcl-xL, and Bad phosphorylation. In addition, co-treatment with sildenafil and doxorubicin inhibited growth of prostate tumor xenografts via caspase 3 activation. Doppler echocardiography showed that sildenafil treatment ameliorated doxorubicin-induced left ventricular dysfunction. The potential role of Dp32 in prostate cancer was also evaluated, using tissue microarrays [98]. After screening 43 normal human epithelium tissue samples and 187 breast, prostate, colon, and stomach carcinomas, both Dp32 and tDp32 were found to be overexpressed in 60% of primary prostate tumors, suggesting their association with tumorigenesis.

12.6.3 Testicular cancer

Unlike prostate cancer, which is more prevalent in older men, **testicular cancer** is the most common cancer in men 15–34 years of age, although it accounts for no more than 1.5% of all male-related cancers [99]. The two main types of testicular cancer are seminoma and non-seminoma. Non-seminomas grow and spread more quickly than seminomas [99,100]. Less common are Leydig and Sertoli cell tumors, and testicular non-Hodgkin lymphomas in the elderly. Most testicular cancers can be cured even when diagnosed at an advanced stage. Treatment includes surgery, radiotherapy, chemotherapy or their combination. Teratomas are the most common testicular tumors in prepubertal children [101]. The potential association between DA and teratomas was covered earlier in this chapter under ovarian cancer.

There are only a few reports on DA involvement in testicular cancer. The incidence of cancer was studied in a cohort of 9,156 patients admitted with diagnosis of schizophrenia in Denmark in 1970–1987 [102]. The overall cancer incidence was reduced, being most pronounced in testicular cancer. Another study performed a whole genome screening of seminoma specimens [100], with the most promising gene candidates confirmed by real-time quantitative PCR (RT-qPCR). A two-fold decrease in *DRD1* gene expression was associated with a two-fold higher probability of having a metastasized seminoma, leading the authors to suggest that D1R expression in seminomas provides some protection from metastasis.

12.7 SYNOPSIS

There is increasing evidence that DA is involved in both benign and malignant tumorigenesis along the pituitary–gonadal axis and accessory reproductive organs in both males and females. Within the pituitary, DA and D2R are intimately associated with the development and progression of prolactinomas. Several D2R antagonists are currently widely utilized in the treatment of prolactinomas as well as several other types of pituitary adenomas. DAR are expressed in almost all female reproductive organs. A special case is made for the overexpression and action of D1R in a significant number of aggressive breast cancer and the potential for using this receptor and its agonists for the diagnosis and treatment of this disease. The role of D1R in this disease has also been emphasized. There is also good evidence for expression of various DARs in ovarian, endometrial, and cervical cancers and the involvement of DA, primarily via regulation of angiogenesis, in tumorigenesis in these organs. In males, some antidopaminergic drugs have shown promise in the treatment of prostate cancer, while there is only scant information on the involvement of DA in testicular cancer.

REFERENCES

1. Human cancers by organs. https://www.iarc.fr/en/publications/pdfs-online/wcr/2003/wcr-5.pdf. 2003:181–269.

2. Ji J, et al. Incidence of cancer in patients with schizophrenia and their first-degree relatives: A population-based study in Sweden. *Schizophr Bull.* 2013;39(3):527–536.

3. Chesnokova V, Melmed S. Pituitary senescence: The evolving role of Pttg. *Mol Cell Endocrinol.* 2010;326(1–2):55–59.

4. Dworakowska D, Grossman AB. The molecular pathogenesis of pituitary tumors: Implications for clinical management. *Minerva Endocrinol.* 2012;37(2):157–172.

5. Molitch ME. Diagnosis and treatment of pituitary adenomas: A review. *JAMA.* 2017;317(5):516–524.

6. Ben-Jonathan N, Lapensee CR, Lapensee EW. What can we learn from rodents about prolactin in humans? *Endocr Rev.* 2008;29(1):1–41.

7. Sonabend AM, Musleh W, Lesniak MS. Oncogenesis and mutagenesis of pituitary tumors. *Expert Rev Anticancer Ther.* 2006;6(Suppl 9):S3–S14.

8. Pei L, Melmed S. Isolation and characterization of a pituitary tumor-transforming gene (PTTG). *Mol Endocrinol.* 1997;11(4):433–441.

9. Zhang X, et al. Pituitary tumor transforming gene (PTTG) expression in pituitary adenomas. *J Clin Endocrinol Metab.* 1999;84(2):761–767.

10. Smith VE, Franklyn JA, McCabe CJ. Pituitary tumor-transforming gene and its binding factor in endocrine cancer. *Expert Rev Mol Med.* 2010;12:e38.

11. Mantovani G, et al. Effect of cyclic adenosine 3',5'-monophosphate/protein kinase a pathway on markers of cell proliferation in nonfunctioning pituitary adenomas. *J Clin Endocrinol Metab.* 2005;90(12):6721–6724.

12. Kukstas LA, et al. Different expression of the two dopaminergic D2 receptors, D2415 and D2444, in two types of lactotroph each characterised by their response to dopamine, and modification of expression by sex steroids. *Endocrinology.* 1991;129(2):1101–1103.

13. Pala NA, Laway BA, Misgar RA, Dar RA. Metabolic abnormalities in patients with prolactinoma: Response to treatment with cabergoline. *Diabetol Metab Syndr.* 2015;7:99–106.

14. Shy KK, McTiernan AM, Daling JR, Weiss NS. Oral contraceptive use and the occurrence of pituitary prolactinoma. *JAMA.* 1983;249(16):2204–2207.

15. Shupnik MA, et al. Selective expression of estrogen receptor alpha and beta isoforms in human pituitary tumors. *J Clin Endocrinol Metab.* 1998;83(11):3965–3972.

16. Liu X, et al. The mechanism and pathways of dopamine and dopamine agonists in prolactinomas. *Front Endocrinol (Lausanne).* 2019;9:1–6.

17. Ben-Jonathan N, Hnasko R. Dopamine as a prolactin (PRL) inhibitor. *Endocr Rev.* 2001;22(6):724–763.

18. Gurlek A, Karavitaki N, Ansorge O, Wass JA. What are the markers of aggressiveness in prolactinomas? Changes in cell biology, extracellular matrix components, angiogenesis and genetics. *Eur J Endocrinol.* 2007;156(2):143–153.

19. Molitch ME. Medical management of prolactin-secreting pituitary adenomas. *Pituitary.* 2002;5(2):55–65.

20. Oh MC, Aghi MK. Dopamine agonist-resistant prolactinomas. *J Neurosurg.* 2011;114(5):1369–1379.

21. Cooper O, Greenman Y. Dopamine agonists for pituitary adenomas. *Front Endocrinol (Lausanne).* 2018;9:469–482.

22. Lim DS, Fleseriu M. The role of combination medical therapy in the treatment of acromegaly. *Pituitary.* 2017;20(1):136–148.

23. Saveanu A, Jaquet P. Somatostatin-dopamine ligands in the treatment of pituitary adenomas. *Rev Endocr Metab Disord.* 2009;10(2):83–90.

24. Greenman Y, Stern N. Non-functioning pituitary adenomas. *Best Pract Res Clin Endocrinol Metab.* 2009;23(5):625–638.

25. Pivonello R, et al. Dopamine receptor expression and function in clinically nonfunctioning pituitary tumors: Comparison with the effectiveness of cabergoline treatment. *J Clin Endocrinol Metab.* 2004;89(4):1674–1683.

26. Peverelli E, et al. Dopamine receptor type 2 (DRD2) inhibits migration and invasion of human tumorous pituitary cells through ROCK-mediated cofilin inactivation. *Cancer Lett.* 2016;381(2):279–286.

27. Streff H, et al. Cancer incidence in first- and second-degree relatives of BRCA1 and BRCA2 mutation carriers. *Oncologist.* 2016;21(7):869–874.

28. Schnitt SJ. Classification and prognosis of invasive breast cancer: From morphology to molecular taxonomy. *Mod Pathol.* 2010;23(Suppl 2):S60–S64.

29. Ligresti G, et al. Breast cancer: Molecular basis and therapeutic strategies (Review). *Mol Med Rep.* 2008;1(4):451–458.

30. Coley HM. Mechanisms and strategies to overcome chemotherapy resistance in metastatic breast cancer. *Cancer Treat Rev.* 2008;34(4):378–390.

31. Ohaegbulam KC, Assal A, Lazar-Molnar E, Yao Y, Zang X. Human cancer immunotherapy with antibodies to the PD-1 and PD-L1 pathway. *Trends Mol Med.* 2015;21(1):24–33.

32. Amos KD, Adamo B, Anders CK. Triple-negative breast cancer: An update on neoadjuvant clinical trials. *Int J Breast Cancer.* 2012;38:1–7.

33. Teng YH, Thike AA, Wong NS, Tan PH. Therapeutic targets in triple negative breast cancer—Where are we now? *Recent Pat Anticancer Drug Discov.* 2011;6(2):196–209.

34. Fan W, Chang J, Fu P. Endocrine therapy resistance in breast cancer: Current status, possible mechanisms and overcoming strategies. *Future Med Chem.* 2015;7(12):1511–1519.

35. Huang B, Warner M, Gustafsson JA. Estrogen receptors in breast carcinogenesis and endocrine therapy. *Mol Cell Endocrinol.* 2015;418(Pt 3):240–244.

36. Song RX, Fan P, Yue W, Chen Y, Santen RJ. Role of receptor complexes in the extranuclear actions of estrogen receptor alpha in breast cancer. *Endocr Relat Cancer.* 2006;13(Suppl 1):S3–S13.

37. Barton M, et al. Twenty years of the G protein-coupled estrogen receptor GPER: Historical and personal perspectives. *J Steroid Biochem Mol Biol.* 2018;176:4–15.

38. Lamb CA, Fabris VT, Jacobsen B, Molinolo AA, Lanari C. Biological and clinical impact of imbalanced progesterone receptor isoform ratios in breast cancer. *Endocr Relat Cancer.* 2018;25:605–624.

39. Wieduwilt MJ, Moasser MM. The epidermal growth factor receptor family: Biology driving targeted therapeutics. *Cell Mol Life Sci.* 2008;65(10):1566–1584.

40. Belkhiri A, Zhu S, El-Rifai W. DARPP-32: From neurotransmission to cancer. *Oncotarget.* 2016;7(14):17631–17640.

41. Avanes A, Lenz G, Momand J. Darpp-32 and t-Darpp protein products of PPP1R1B: Old dogs with new tricks. *Biochem Pharmacol.* 2019;160:71–79.

42. Belkhiri A, et al. Expression of t-DARPP mediates trastuzumab resistance in breast cancer cells. *Clin Cancer Res.* 2008;14(14):4564–4571.

43. Christenson JL, Denny EC, Kane SE. t-Darpp overexpression in HER2-positive breast cancer confers a survival advantage in lapatinib. *Oncotarget.* 2015;6(32):33134–33145.

44. Denny EC, Kane SE. t-Darpp promotes enhanced EGFR activation and new drug synergies in Her2-positive breast cancer cells. *PLoS One.* 2015;10(6):e0132267.

45. Borcherding DC, et al. Expression and therapeutic targeting of dopamine receptor-1 (D1R) in breast cancer. *Oncogene.* 2016;35(24):3103–3113.

46. Cimino-Mathews A, et al. Androgen receptor expression is usually maintained in initial surgically resected breast cancer metastases but is often lost in end-stage metastases found at autopsy. *Hum Pathol.* 2012;43(7):1003–1011.

47. Idirisinghe PK, et al. Hormone receptor and c-ERBB2 status in distant metastatic and locally recurrent breast cancer. Pathologic correlations and clinical significance. *Am J Clin Pathol.* 2010;133(3):416–429.

48. Fallahian F, Karami-Tehrani F, Salami S, Aghaei M. Cyclic GMP induced apoptosis via protein kinase G in oestrogen receptor-positive and -negative breast cancer cell lines. *FEBS J.* 2011;278(18):3360–3369.

49. Di X, et al. Influence of the phosphodiesterase-5 inhibitor, sildenafil, on sensitivity to chemotherapy in breast tumor cells. *Breast Cancer Res Treat.* 2010;124(2):349–360.

50. Teunis MA, et al. Reduced tumor growth, experimental metastasis formation, and angiogenesis in rats with a hyperreactive dopaminergic system. *FASEB J.* 2002;16(11):1465–1467.

51. Basu S, et al. Ablation of peripheral dopaminergic nerves stimulates malignant tumor growth by inducing vascular permeability factor/vascular endothelial growth factor-mediated angiogenesis. *Cancer Res.* 2004;64(16):5551–5555.

52. Chakroborty D, Sarkar C, Basu B, Dasgupta PS, Basu S. Catecholamines regulate tumor angiogenesis. *Cancer Res.* 2009;69(9):3727–3730.

53. Sarkar C, et al. Dopamine in vivo inhibits VEGF-induced phosphorylation of VEGFR-2, MAPK, and focal adhesion kinase in endothelial cells. *Am J Physiol Heart Circ Physiol.* 2004;287(4):H1554–H1560.

54. Moreno-Smith M, et al. Dopamine blocks stress-mediated ovarian carcinoma growth. *Clin Cancer Res.* 2011;17(11):3649–3659.

55. Douet-Guilbert N, et al. Molecular characterization of deletions of the long arm of chromosome 5 (del(5q)) in 94 MDS/AML patients. *Leukemia.* 2012;26(7):1695–1697.

56. Kitamura K, Singer WD, Star RA, Muallem S, Miller RT. Induction of inducible nitric-oxide synthase by the heterotrimeric G protein Galpha13. *J Biol Chem.* 1996;271(13):7412–7415.

57. Kim SG, Lee CH. G-protein signaling in iNOS gene expression. *Methods Enzymol.* 2005;396:377–387.

58. Carlo RD, et al. Steroid, prolactin, and dopamine receptors in normal and pathologic breast tissue. *Ann N Y Acad Sci.* 1986;464:559–562.

59. Sachlos E, et al. Identification of drugs including a dopamine receptor antagonist that selectively target cancer stem cells. *Cell.* 2012;149(6):1284–1297.

60. Minami K, et al. Inhibitory effects of dopamine receptor D1 agonist on mammary tumor and bone metastasis. *Sci Rep.* 2017;7:1–12.

61. Pornour M, Ahangari G, Hejazi SH, Deezagi A. New perspective therapy of breast cancer based on selective dopamine receptor D2 agonist and antagonist effects on MCF-7 cell line. *Recent Pat Anticancer Drug Discov.* 2015;10(2):214–223.

62. Li J, et al. Dopamine D2 receptor antagonist sulpiride enhances dexamethasone responses in the treatment of drug-resistant and metastatic breast cancer. *Acta Pharmacol Sin.* 2017;38:1282–1296.

63. Mikelman SR, Guptaroy B, Gnegy ME. Tamoxifen and its active metabolites inhibit dopamine transporter function independently of the estrogen receptors. *J Neurochem.* 2017;141(1):31–36.

64. Lee H, Kang S, Sonn JK, Lim YB. Dopamine receptor D2 activation suppresses the radiosensitizing effect of aripiprazole via activation of AMPK. *FEBS Open Bio.* 2019: 1580–1588. doi:10.1002/2211-5463.12699.

65. Yang L, et al. Dopamine D1 receptor agonists inhibit lung metastasis of breast cancer reducing cancer stemness. *Eur J Pharmacol.* 2019;859:172499.

66. Califano JA, et al. Tadalafil augments tumor specific immunity in patients with head and neck squamous cell carcinoma. *Clin Cancer Res.* 2015;21(1):30–38.

67. Serafini P, et al. Phosphodiesterase-5 inhibition augments endogenous antitumor immunity by reducing myeloid-derived suppressor cell function. *J Exp Med.* 2006;203(12):2691–2702.

68. Okubo Y, et al. Decreased prefrontal dopamine D1 receptors in schizophrenia revealed by PET. *Nature.* 1997;385(6617):634–636.

69. Tinsley HN, et al. Inhibition of PDE5 by sulindac sulfide selectively induces apoptosis and attenuates oncogenic Wnt/beta-catenin-mediated transcription in human breast tumor cells. *Cancer Prev Res (Phila).* 2011;4(8):1275–1284.

70. Longuespee R, et al. Ovarian cancer molecular pathology. *Cancer Metastasis Rev.* 2012;31(3–4):713–732.

71. Poole EM, Konstantinopoulos PA, Terry KL. Prognostic implications of reproductive and lifestyle factors in ovarian cancer. *Gynecol Oncol.* 2016;142(3):574–587.

72. Park MS, et al. Thioridazine inhibits angiogenesis and tumor growth by targeting the VEGFR-2/PI3K/mTOR pathway in ovarian cancer xenografts. *Oncotarget.* 2014;5(13):4929–4934.

73. Yong M, et al. DR2 blocker thioridazine: A promising drug for ovarian cancer therapy. *Oncol Lett.* 2017;14(6):8171–8177.

74. Moreno-Smith M, et al. Biologic effects of dopamine on tumor vasculature in ovarian carcinoma. *Neoplasia.* 2013;15(5):502–510.

75. Qin T, et al. Dopamine induces growth inhibition and vascular normalization through reprograming M2-polarized macrophages in rat C6 glioma. *Toxicol Appl Pharmacol.* 2015;286(2):112–123.

76. Leung EL, Fraser M, Fiscus RR, Tsang BK. Cisplatin alters nitric oxide synthase levels in human ovarian cancer cells: Involvement in p53 regulation and cisplatin resistance. *Br J Cancer.* 2008;98(11):1803–1809.

77. Leung EL, Wong JC, Johlfs MG, Tsang BK, Fiscus RR. Protein kinase G type Ialpha activity in human ovarian cancer cells significantly contributes to enhanced Src activation and DNA synthesis/cell proliferation. *Mol Cancer Res.* 2010;8(4):578–591.

78. Gobel U, et al. Germ-cell tumors in childhood and adolescence. GPOH MAKEI and the MAHO study groups. *Ann Oncol.* 2000;11(3):263–271.

79. Kraggerud SM, et al. Molecular characteristics of malignant ovarian germ cell tumors and comparison with testicular counterparts: Implications for pathogenesis. *Endocr Rev.* 2013;34(3):339–376.

80. Brederlau A, et al. Transplantation of human embryonic stem cell-derived cells to a rat model of Parkinson's disease: Effect of in vitro differentiation on graft survival and teratoma formation. *Stem Cells.* 2006;24(6):1433–1440.

81. Kriks S, et al. Dopamine neurons derived from human ES cells efficiently engraft in animal models of Parkinson's disease. *Nature.* 2011;480(7378):547–551.

82. Carlson MJ, Thiel KW, Leslie KK. Past, present, and future of hormonal therapy in recurrent endometrial cancer. *Int J Womens Health.* 2014;6:429–435.

83. Meng Q, Sun X, Wang J, Wang Y, Wang L. The important application of thioridazine in the endometrial cancer. *Am J Transl Res.* 2016;8(6):2767–2775.

84. Tseng L, Mazella J, Goligorsky MS, Rialas CM, Stefano GB. Dopamine and morphine stimulate nitric oxide release in human endometrial glandular epithelial cells. *J Soc Gynecol Investig.* 2000;7(6):343–347.

85. Bava SV, Thulasidasan AK, Sreekanth CN, Anto RJ. Cervical cancer: A comprehensive approach towards extermination. *Ann Med.* 2016;48(3):149–161.

86. Mao M, Yu T, Hu J, Hu L. Dopamine D2 receptor blocker thioridazine induces cell death in human uterine cervical carcinoma cell line SiHa. *J Obstet Gynaecol Res.* 2015;41(8):1240–1245.

87. Seervi M, Rani A, Sharma AK, Santhosh Kumar TR. ROS mediated ER stress induces Bax-Bak dependent and independent apoptosis in response to Thioridazine. *Biomed Pharmacother.* 2018;106:200–209.

88. Katzenwadel A, Wolf P. Androgen deprivation of prostate cancer: Leading to a therapeutic dead end. *Cancer Lett.* 2015;367(1):12–17.

89. Shiota M, Eto M. Current status of primary pharmacotherapy and future perspectives toward upfront therapy for metastatic hormone-sensitive prostate cancer. *Int J Urol.* 2016;23(5):360–369.

90. Armstrong CM, Gao AC. Drug resistance in castration resistant prostate cancer: Resistance mechanisms and emerging treatment strategies. *Am J Clin Exp Urol.* 2015;3(2):64–76.

91. Arvigo M, et al. Somatostatin and dopamine receptor interaction in prostate and lung cancer cell lines. *J Endocrinol.* 2010;207(3):309–317.

92. Zhou W, et al. The antipsychotic drug pimozide inhibits cell growth in prostate cancer through suppression of STAT3 activation. *Int J Oncol.* 2016;48(1):322–328.

93. Dilly SJ, et al. A chemical genomics approach to drug reprofiling in oncology: Antipsychotic drug risperidone as a potential adenocarcinoma treatment. *Cancer Lett.* 2017;393:16–21.

94. Frederiksen LJ, et al. Chemosensitization of cancer in vitro and in vivo by nitric oxide signaling. *Clin Cancer Res.* 2007;13(7):2199–2206.

95. Goluboff ET. Exisulind, a selective apoptotic antineoplastic drug. *Expert Opin Investig Drugs.* 2001;10(10):1875–1882.

96. Hamilton TK, et al. Potential therapeutic applications of phosphodiesterase inhibition in prostate cancer. *World J Urol.* 2013;31(2):325–330.

97. Das A, et al. Sildenafil increases chemotherapeutic efficacy of doxorubicin in prostate cancer and ameliorates cardiac dysfunction. *Proc Natl Acad Sci USA.* 2010;107(42):18202–18207.

98. Beckler A, et al. Overexpression of the 32-kilodalton dopamine and cyclic adenosine 3',5'-monophosphate-regulated phosphoprotein in common adenocarcinomas. *Cancer.* 2003;98(7):1547–1551.

99. Viatori M. Testicular cancer. *Semin Oncol Nurs.* 2012;28(3):180–189.

100. Ruf CG, et al. Predicting metastasized seminoma using gene expression. *BJU Int.* 2012;110(2Pt 2):E14–E20.

101. Akiyama S, Ito K, Kim WJ, Tanaka Y, Yamazaki Y. Prepubertal testicular tumors: A single-center experience of 44 years. *J Pediatr Surg.* 2016;51(8):1351–1354.

102. Mortensen PB. The occurrence of cancer in first admitted schizophrenic patients. *Schizophr Res.* 1994;12(3):185–194.

103. Werbowetski-Ogilvie TE, Bhatia M. Pluripotent human stem cell lines: What we can learn about cancer initiation. *Trends Mol Med.* 2008;14(8):323–332.

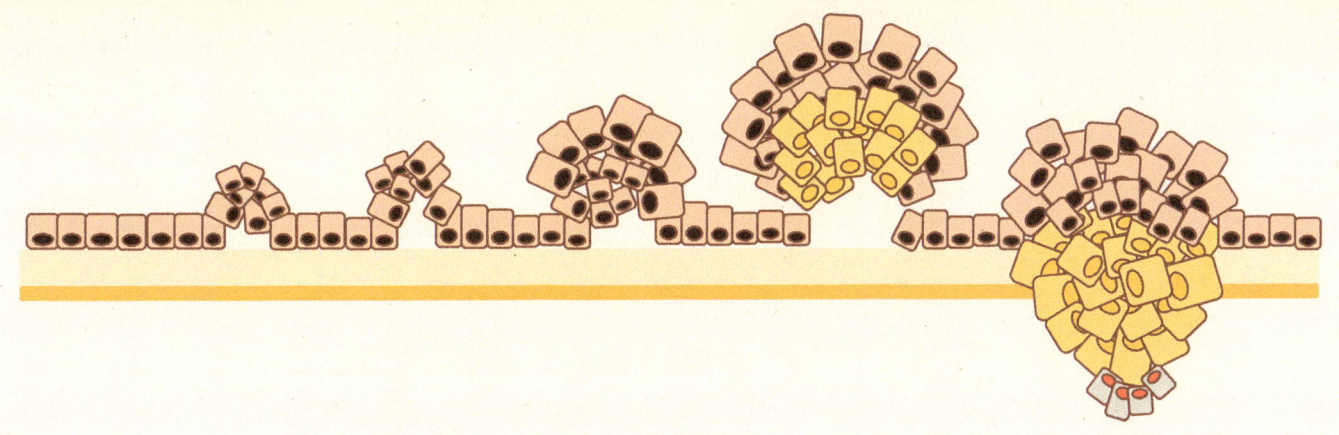

Involvement of Dopamine with Various Cancers

13

13.1 INTRODUCTION

Dopamine (DA) is involved to a variable degree with malignancies in most, though not all, body systems. The selection of various cancers for discussion was dictated by the availability of information on the role of DA in such tumors. Each section is preceded by a brief review of the attributes of the relevant cancers and is then followed by summary of available information on the role of DA in these malignancies. The first section covers hematological malignancies that account for about 10% of new cancers diagnosed annually and are the most common malignancies of childhood. The next section briefly reviews cancers that occur in the gastrointestinal (GI) tract and associated structures such as the liver and pancreas. This is followed by lung, kidney and bladder cancers, and then by tumors that afflict the skin, especially melanomas. The final sections cover what is known on the involvement of DA in head and neck cancer and in primary brain tumors as well as peripheral neurological and neuroendocrine tumors.

13.2 HEMATOLOGICAL MALIGNANCIES

13.2.1 Prevalence and classification of hematological malignancies

As illustrated in **Figure 13.1**, hematologic malignancies are derived from the two major blood cell lineages: myeloid and lymphoid [1]. The classification of leukemias and lymphomas is complex and can be confusing because of the variety of cell types involved, the different sites of origin of the neoplastic process (bone marrow, lymph node, GI tract, etc.), and the relative frequency or infrequency of circulating tumor cells. Leukemia and lymphoma are terms that reflect the primary behavior and often the primary site of a neoplasm.

    Table 13.1 presents the designation of hematopoietic malignancies by cells of origin and duration of the disease. Leukemias have cells that circulate in peripheral blood and can originate in lymph nodes or bone marrow. Lymphomas generally form solid masses in lymph nodes or organs containing lymphoid tissue but may occasionally have circulating tumor cells as

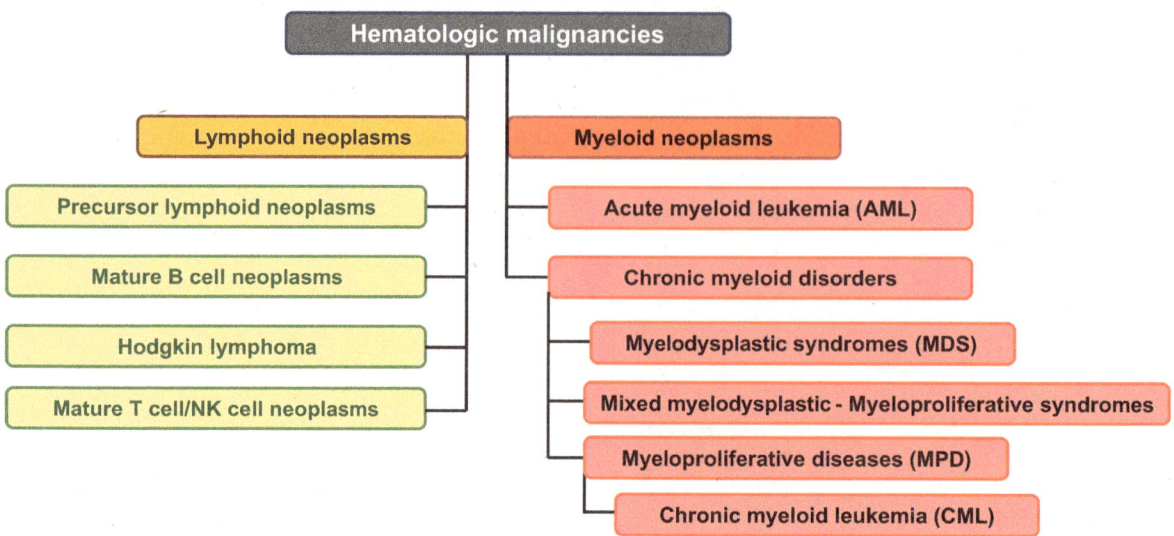

Figure 13.1 Classification of hematological malignancies. The various hematological cancers are grouped by their origins from lymphoid or myeloid progenitors. See text for other explanations. NK: natural killer.

Table 13.1 Different types of leukemia and their attributes

Types	Cells involved	Cytology	Symptoms	Statistics	Prevalence
Acute myeloid leukemia (AML)	Immature myeloid white blood cells	Oncogene mutations, single myeloid mutation, cytogenetic abnormalities	Anemia, spontaneous bleeding	Adults and children	80%
Chronic myeloid leukemia (CML)	Myeloid stem cells	Chromosomal translocation, granulocytes	Anemia, low platelet counts, enlarged spleen	Rare in children	90%
Acute lymphocytic leukemia (ALL)	Immature B or T cells and macrophages	Chromosomal aberrations	Disturbed marrow functions	Common in children	33%
Chronic lymphocytic leukemia (CLL)	Lymphoid B or T cells	Chromosomal abnormalities	Swelling of lymph nodes, spleen increase	Commonly affects people over age 55	30%

well. Leukemias and lymphomas can also be defined as being chronic or acute. Chronic neoplasms are of longer duration and are slowly progressive, while acute neoplasms are of shorter duration and are rapidly progressing.

13.2.2 Attributes and classification of leukemias

Leukemia is the eleventh most common cancer worldwide, with more than 250,000 new cases diagnosed annually. Leukemia results from a malignant transformation of white blood cells (WBCs) or their precursors. Typically it entails clonal neoplastic proliferation of immature cells or blasts of the hematopoietic system, characterized by aberrant or arrested differentiation. The cause of most leukemias is unknown, but a combination of genetic factors and non-inherited environmental factors appear to play a role. Risk factors include smoking, ionizing radiation, certain chemicals (e.g., benzene), some chemotherapeutic agents, and infection with the human T-lymphotropic virus 1 (HTLV-1) virus. Given the lack or deficiency of normal blood cells, symptoms may include bleeding and bruising, chronic fatigue, fever, and increased risk of infections (**Figure 13.2**). Diagnosis is made by blood tests and bone marrow biopsy. To identify the specific hematopoietic or lymphoid neoplasm, more testing is usually required, including immunophenotyping or genetic information to identify the specific histology. Treatments include a combination of chemotherapy, radiation therapy, targeted therapy, and bone marrow transplants.

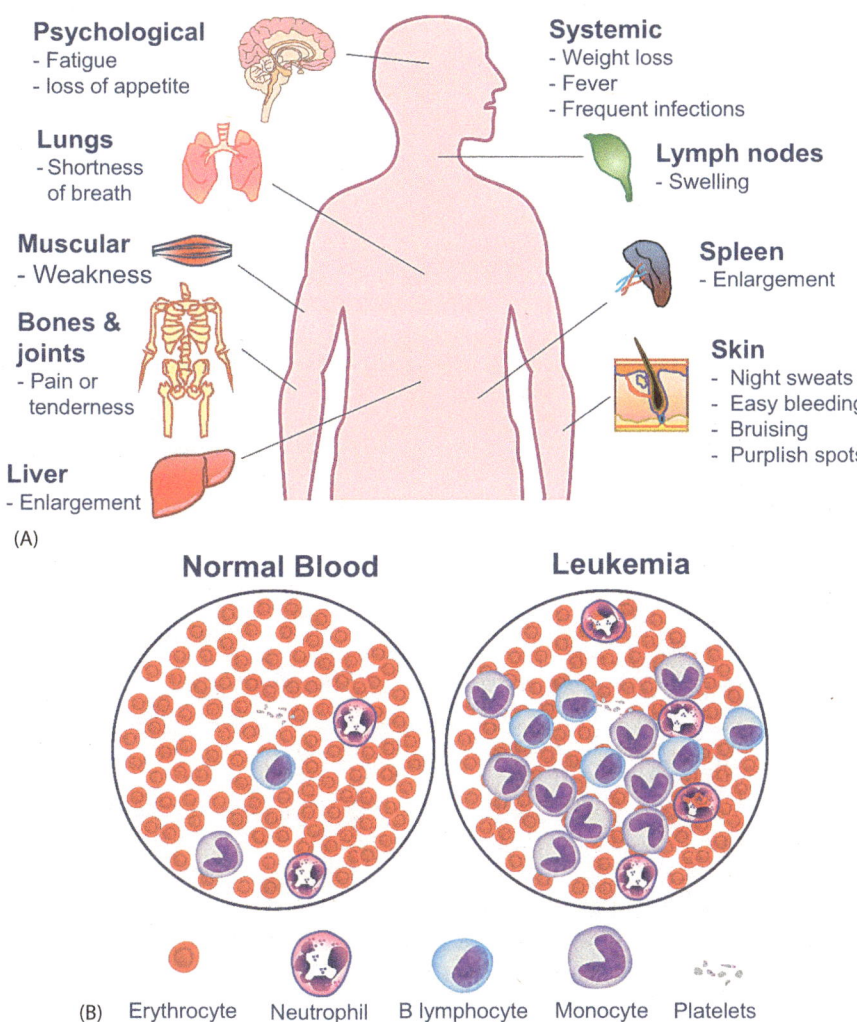

Psychological
- Fatigue
- loss of appetite

Lungs
- Shortness
  of breath

Muscular
- Weakness

**Bones &
joints**
- Pain or
  tenderness

Liver
- Enlargement

(A)

Systemic
- Weight loss
- Fever
- Frequent infections

Lymph nodes
- Swelling

Spleen
- Enlargement

Skin
- Night sweats
- Easy bleeding
- Bruising
- Purplish spots

Normal Blood   **Leukemia**

(B)   Erythrocyte   Neutrophil   B lymphocyte   Monocyte   Platelets

Figure 13.2 Symptoms and diagnosis of leukemia. Panel A shows the various organs and systems that are affected by leukemia. The prevalence of symptoms depends upon the type of leukemia. Panel B is a microscopic display of normal vs. leukemic blood samples. The leukemic sample has abnormal levels of monocytes, and very low levels of erythrocytes and platelets.

Based on whether they are acute or chronic, myeloid or lymphocytic, leukemias are divided into four main types, as listed in **Table 13.1**: (1) acute myeloid leukemia (AML), (2) chronic myeloid leukemia (CML), (3) acute lymphocytic leukemia (ALL), and (4) chronic lymphocytic leukemia (CLL). Acute leukemia is the most common cancer in children, representing half of all cancers in persons younger than 15 years of age. Three-quarters of the leukemia cases in children are of the acute lymphoblastic type 2. About 90% of all leukemias are diagnosed in adults, with AML and CLL being most common in adults. AML begins in bone marrow cells and spreads into the blood system, most often occurring in immature white blood cells. AML is a fast growing form of leukemia that if left untreated can be fatal. More information on the criteria for leukemia classification, molecular signature, and management can be obtained in several reviews [2–4].

13.2.3 Involvement of dopamine in leukemias

As described in **Chapter 9**, most immune cells express tyrosine hydroxylase (TH) and/or Dopa decarboxylase (DDC), and can produce dopamine (DA) either from tyrosine or from Dopa, respectively, while cells that express the DA transporter (DAT) can take up extracellular DA [5,6]. The effects of phenothiazines such as trifluoperazine, thioridazine, and chlorpromazine were examined using multiple leukemia and lymphoma cell lines: Daudi and Raji (derived from Burkitt's lymphoma), K-562 (derived from myelogenous leukemia), and BALL-1, MOLT-4, HPB-ALL, and CCRF-HSB-2 (derived from acute lymphoblastic leukemia) [7]. The drugs showed strong cytotoxicity and anti-proliferative activity in the leukemic cells while having no effects on normal lymphocytes. The most sensitive cell lines were MOLT-4 and Raji, and the most resistant were HPB-ALL and CCRF-HSB-2. The drugs induced DNA fragmentation by inhibiting mitochondrial DNA polymerase and by decreasing ATP production, both of which are crucial events for the viability of cancer cells.

Using HL-60, a human promyelocytic leukemia cell line, one study reported that DA induced apoptosis, as judged by DNA fragmentation, that was coupled to a transient activation of MAP kinase [8]. However, others found that the DA-induced DNA fragmentation in HL-60 cells was caused by tetrahydro-papaveroline (THP) a metabolite of DA, rather than by DA itself [9]. The authors suggested that some of the cytotoxicity ascribed to high concentrations of L-Dopa or DA could be due to the effects of their metabolites. Treatment of THP1, an acute myeloid leukemia cell line, with SCH39166, a D1R/D5R antagonist, caused significant reductions in cell proliferation, colony formation, and migration, and slowdown of cell cycle progression, but no increase in apoptosis [10].

An older study reported that blood levels of both DA and cortisol were elevated in patients with CML, suggesting that these increases were caused by the stress induced by the disease [11]. A large-scale chemical screen for drugs that alter the behavior of cancer stem cells (CSCs) has identified thioridazine, an antipsychotic drug with D2R antagonist properties, as having potent antileukemic properties [12]. Thioridazine selectively targeted the pleuripotent CSCs that express DA receptors (DARs) and impaired their ability to initiate *in vivo* leukemic disease while having no effects on normal blood stem cells. In addition, a combination of thioridazine and the chemotherapy agent cytarabine resulted in a more dramatic suppression of leukemic clonogenicity than that caused by thioridazine alone. The authors concluded that DAR may serve as a biomarker for leukemia and other malignancies and that human pleural stem cells can be used for identifying cancer-targeting drugs.

The above results have inspired a recent phase I clinical trial to evaluate the safety and initial therapeutic efficacy of thioridazine in combination with cytarabine in a small number of patients with AML [13]. Although there were positive results in the majority of patients (i.e., reduced levels of peripheral blasts), there were also adverse cardiac side effects (i.e., QTc prolongation). The authors concluded that D2R antagonism holds promise as a novel therapeutic target to combat AML. They suggested that larger patient cohorts and modified thioridazine formulation with reduced neurological and cardiac liability should be considered for future trials.

13.2.4 Association of dopamine with lymphomas

Lymphomas are the most common blood cancer. Two main forms of lymphoma are recognized: Hodgkin lymphoma and non-Hodgkin lymphoma (NHL), which differ in etiology and response to treatment [14]. In addition, two other subtypes of lymphoma are multiple myeloma and immunoproliferative diseases. A lymphoma occurs when lymphocytes multiply uncontrollably, acquire a malignant phenotype, and can travel to lymph nodes, spleen, bone marrow, and other organs, where they localize and form tumors. Both B-cells and T-cells can develop into lymphomas. The primary presentation of lymphoma is swelling of the lymph nodes. Systemic symptoms include fever, night sweats, loss of appetite and weight loss, fatigue, respiratory distress and itching. The incidence of NHL is higher in males than in females and increases with age. Burkitt lymphoma is an aggressive subtype of B-cell NHL. It is particularly prevalent in sub-Saharan Africa, middle eastern countries, and in patients with HIV/AIDS. The endemic form of Burkitt lymphoma is linked to malaria and to infection with the Epstein-Barr virus (EBV), a common virus that also causes glandular fever.

T-cell lymphomas account for ~15% of all NHLs. There are many forms of T-cell lymphomas, which can be aggressive or indolent [14]. Jurkat cells are an immortalized human T-lymphocyte cell line that was originally obtained from the peripheral blood of a boy with T cell leukemia. They are commonly used to study T-cell signaling, apoptosis, and expression of chemokine receptors susceptible to vial entry, particularly HIV. Jurkat lymphocytes express many components of the DA system: DAT and vesicular monoamine transporter (VMAT), D1-like and D2-like receptors, confirming their ability to internalize, store and respond to DA. Yet, they lack TH expression and do not synthesize DA *de novo* [15]. To examine the effects of DA on HIV-1 gene transcription, a reporter containing the full length long terminal repeat of HIV-1 was transfected into Jurkat cells [16]. Large doses of DA (50–200 µM) stimulated HIV-1 transcription via nuclear factor kappa B (NF-κB) activation. The combination of DA and tumor necrosis factor alpha (TNF-α) increased HIV-1 transcription as well as replication. Others reported that D1R activation inhibited cell proliferation in T-cells from normal volunteers but failed to do so in Jurkat cells because of overexpression of phosphodiesterase (PDE) activity [17].

An older study found that nitric oxide (NO) was involved in apoptosis and functional impairment of lymphocytes from patients with AIDS [18]. A short report, using CD8+ cytotoxic T lymphocytes, showed that hypoxia-induced acquisition of malignant phenotypes was in part due to impaired NO-mediated activation of cGMP signaling [19]. Restoration of the cGMP signaling prevented the hypoxic response by interfering with hypoxia inhibiting factor-1α (HIF-1α). The authors concluded that activation of NO/cGMP signaling could be useful in cancer therapy. Collectively, as seen in other malignancies, lymphocyte tumorigenesis may be associated with a switch from the canonical signaling pathways (i.e., adenylate cyclase/cAMP) that mediate DAR action in non-transformed cells to alternative pathways such as guanylate cyclase/cGMP.

B-cell lymphomas have several subtypes, including diffuse large B cell lymphoma, follicular lymphoma, Burkitt lymphoma, and mantle cell lymphoma [20,21]. Normal and malignant human B lymphocytes were examined for expression of DA pathway components and for cytostatic responses to DA and related compounds [22]. DA in the micromolar range rapidly arrested the proliferation of both normal and malignant lymphocytes. L-Dopa and apomorphine also showed antiproliferative activity. With the exception of D4R, all other DAR subtypes were variably expressed in normal and neoplastic B cells, as was DAT. The authors concluded that the impact of DA on lymphocytes provides an opportunity for therapeutic intervention in B cell neoplasia and other lymphoproliferative disorders. Another study reported that addition of DA to cultured Burkitt lymphoma cells (Raji, Namalwa, Daudi, and Jijoye) caused a rapid and an extensive cell death, while a myeloma cell line, SKO, was unresponsive [23]. Additions of fetal calf serum or supernatant from cultures of Raji or T24 bladder carcinoma cells partially counteracted the effect of DA, suggesting that DA acted by inhibiting the production of unidentified endocrine/autocrine factors.

Multiple myeloma is a progressive plasma cell tumor leading to overproduction of monoclonal immunoglobulins, osteolytic bone lesions, renal disease, and immunodeficiency. It is the second most prevalent hematologic malignancy in the United States. The dysregulated plasma cells can accumulate in the bone marrow, where they crowd out healthy blood cells. Within bones, myeloma cells increase the activity of osteoclasts (which break down bone) and decrease the activity of osteoblasts (which form new bone), causing bones to dissolve at a faster rate than they are formed. This damages and weakens the bones, causing pain and lesions. Multiple myeloma can also form tumors in the skin and other soft tissue.

Figure 13.3 Suppression by dopamine of osteolytic lesions induced by multiple myeloma. Dopamine acts by inhibiting VEGF release from myeloma cells and IL-6 release from stromal cells. This results in a reduction of bone destruction by the myeloma cells. IL-6: interleukin 6; MMP9: matrix metallopeptidase 9; VEGF: vascular endothelial growth factor.

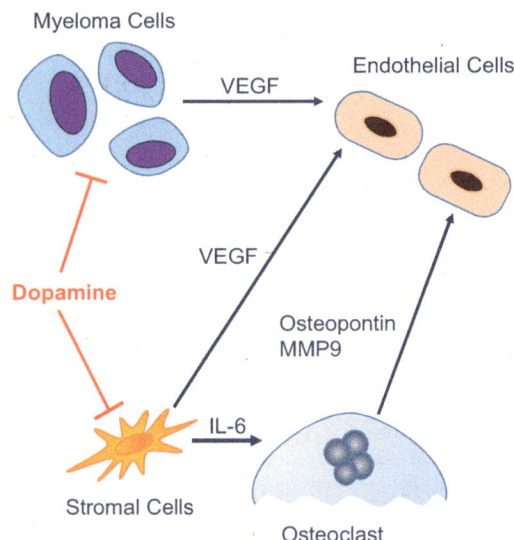

Vascular endothelial growth factor (VEGF) is a critical cytokine responsible for an induction of angiogenesis in several solid tumors and in hematological malignancies like multiple myeloma. A recent study examined the role of VEGF in the induction of bone marrow angiogenesis, which promotes the survival, growth and proliferation of multiple myeloma, and the role of DA on this process [24]. The results showed that VEGF significantly increased the secretion of interleukin-6 (IL-6) from isolated primary bone marrow (BM) stromal cells as well as from HS-5 human BM cells. Addition of DA or quinpirole, the D2R agonist, to VEGF-treated BM cells resulted in the suppression of IL-6 production, and this effect was abrogated by pretreatment with the D2R antagonist, eticlopride. DA also inhibited the proliferation of MM1.S human multiple myeloma cells. The authors concluded that DA may be an effective approach for retarding the progression of multiple myeloma. Because anti-angiogenic monotherapy cannot completely eradicate the disease, strategies using DA with more traditional modalities at the time of initial treatment might be beneficial. Interactions between DA and VEGF in myeloma-induced bone lesions are depicted in **Figure 13.3**.

13.3 GASTROINTESTINAL-RELATED CANCERS

The GI tract is a succession of structures composed of concentric epithelial, mesenchymal and myometrial cell types that undergo continuous cell divisions driven by "wear and tear" and exposure to diet and environmental factors. These features make the system susceptible to tumorigenesis, which can occur throughout its structures, from mouth to anus. This section covers the involvement of DA in gastric, intestinal, hepatic, and pancreatic cancers. Cancers in other parts of the GI system, i.e., mouth and esophagus, are covered under head and neck cancer in **Section 13.6**, while some forms of pancreatic cancer are covered in **Section 13.7** under neurological and neuroendocrine tumors.

13.3.1 Dopamine and overall incidence of GI malignancies

A case–control survey in Taiwan examined the prevalence of cancer among patients with Parkinson's disease (PD) [25]. The aim was to evaluate the effects of medication on cancer incidence, focusing on the use of ergot and non-ergot DA agonists. Of the 6,211 patients with PD, 329 had cancer. Among all cancer types, there was a particularly high occurrence of liver cancer in users of ergot-derived DA agonist. It was concluded that DA replacement in PD is especially associated with liver cancer, but a better understanding of the cause-and-effect was needed. Another epidemiological study in Taiwan, which included a 10-fold larger population base, looked at risks of having 19 different cancers in patients with PD [26]. Among the 62,023 patients with Parkinson disease, increased risk was found primarily in digestion-associated cancers (esophageal, stomach, colorectal, gallbladder, pancreas, and liver). The risk of developing liver cancer was among the highest, and it was primarily seen in the 50- to 59-year-old group. There was no significant association of PD and breast, ovarian, or thyroid cancers. The study concluded that the difference between these data and most previous cohorts underlined the importance of ethnicity and environmental exposures in pathogenesis of the disease.

13.3.2 Gastric cancer

Stomach cancer is among the most common malignancies worldwide, with 870,000 new cases diagnosed annually [27]. Mortality from gastric cancer is second only to lung cancer. The incidence of gastric cancer has been declining worldwide, likely due to changes in dietary practices such as increased intake of fresh fruits and vegetable and reduced consumption of smoked and cured meat and fish. Infection with *Helicobacter pylori*, which causes chronic gastritis, is a risk factor in the development of this cancer. Most patients diagnosed with stomach cancer have an advanced disease and poor prognosis, with survival rates not exceeding 15%. Management depends on tumor staging. Small intramucosal tumors can be resected endoscopically. Standard treatment for invasive gastric cancer is gastrectomy with regional lymph node dissection. For patients with advanced disease, neo-adjuvant or adjuvant chemotherapy is employed, using 5-FU, doxorubicin, or cisplatin, and their combination.

An early study, using a radioreceptor binding assay, found high-affinity D2R in normal, benign, and malignant human stomach tissues [28]. Both the concentration (Bmax) and affinity (Kd) of the DA binding sites were similar in the normal and benign tumor tissues, while the Bmax was lower in malignant tissue, without a change in Kd. Another study found loss of TH and DA in gastric tumors, and examined the effects of DA on oncogene-induced gastric cancer in rats, and on human gastric cancer xenografts in nude mice [29]. DA inhibited tumor growth in both animal models by suppressing angiogenesis. A second study from the same group found that human gastric tumors had higher expression of phosphorylated IGF-IR and its downstream signaling than normal tissues [30]. Acting via D2R, DA inhibited IGF-I-induced proliferation of AGS gastric cancer cells by up-regulating Krüppel-like factor 4 (KLF4), a cell cycle suppressor.

Analysis of immunoreactive D2R expression in paired gastric carcinomas and adjacent normal tissue showed that 50% of the cases had high D2R expression, as compared with 40% with low expression [31]. Patients with high D2R expression had shorter survival. Thioridazine inhibited the growth of AGS gastric cancer cells. The disagreement between the above results could be due to the size of the patient population among the two studies, the heterogeneity of gastric tumors, and/or because of variable effects of different doses of the D2R ligands.

The potential role of DAR other than D2R in gastric cancer was examined in a recent study after finding that SCG7910, a human gastric cancer cell line, expressed D5R [32]. Receptor activation by the D1R/D5R agonist SKF83959 suppressed cell proliferation by inhibiting mammalian target of rapamycin (mTOR) activity and by inducing autophagy. SKF83959 also inhibited the growth of human gastric cancer xenografts in mice. The authors concluded that this study revealed novel mechanisms for the tumor suppressive effects of D5R agonists and suggested their potential use as a therapeutic approach in the treatment of gastric cancer.

As detailed in **Chapter 12**, DARPP-32 (DA- and cAMP-regulated phosphoprotein) or Dp32, and its truncated isoform, tDP, are overexpressed in a variety of cancers, including human gastric tumors [33,34]. In one study, Dp32 expression in gastric tumors increased during the transition from normal tissue to metaplasia, to neoplasia [35]. However, analysis of Dp32 functions in gastric cancer generated controversial data. Induced t-Dp overexpression in AGS cells protected against ceramide-induced apoptosis by up-regulating Bcl2 and by blocking caspase activation via an Akt-dependent mechanism [36]; overexpression of Dp32 acted in the opposite manner. Transfection of Dp32 into vincristine-resistant human gastric cancer cells enhanced cell sensitivity to vincristine, adriamycin, 5-FU, and cisplatin and decreased efflux of adriamycin [37]. A follow-up study found that Dp32 mediated drug resistance by down-regulating the P-glycoprotein membrane transporter [38].

In disagreement with the above data, other studies showed promotion of chemoresistance by Dp32. For instance, overexpression of Dp32 increased gastric cancer cell resistance to gefitinib, an inhibitor of EGFR, by promoting interaction between EGFR and ERBB3 via the PI3K-AKT signaling [39]. Another study [40] reported that Dp32 overexpression increased resistance to apoptosis by TNF-related apoptosis-inducing ligand (TRAIL) by increasing BCL-xL via Src/signal transducer and activator of transcription 3 (Src/STAT3) signaling. Dp32 overexpression in AGS cells also stimulated invasion by up-regulating matrix metalloproteinase and CXCR4 [41]. In a recent study, analysis of human gastric cancer tissue microarrays showed high levels of Dp32 and positive immunostaining for nuclear STAT3 in cancer tissues as compared with non-cancer histologically normal tissues [42]. Collectively, the above studies indicate that Dp32 signaling plays a key role in regulating the STAT3 signaling, a critical step in gastric tumorigenesis. However, it remains to be determined whether Dp32 overexpression is beneficial or detrimental for the progression of gastric cancer and to what extent it is associated with the DA/DAR axis.

13.3.3 Colorectal cancer

Cancers of the colon and rectum are rare in developing countries, but they are the second most frequent malignancy in affluent societies. Over 940,000 cases occur annually worldwide [27]. A major etiological factor is diet rich in fat, refined carbohydrates and animal protein combined with low physical activity. This risk can be reduced by decreasing meat consumption and by increasing the intake of vegetables and fruit. Inflammatory bowel disease as well as Crohn's disease predispose the individuals to development of colorectal cancer. As illustrated in **Figure 13.4**, colon cancer progresses through

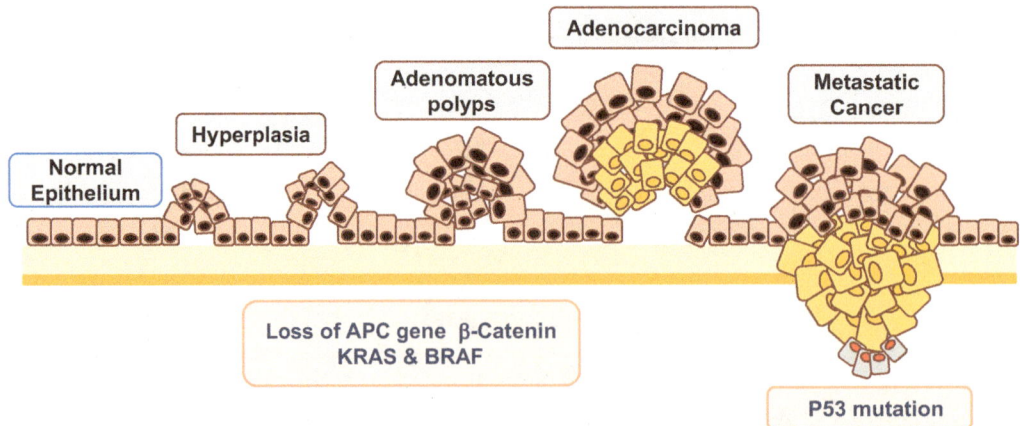

Figure 13.4 Development of sporadic colon cancer. Cumulative mutations due to chromosomal instability, microsatellite instability, and epigenetic changes result in DNA damage and loss of tumor suppressor genes including *APC*. These changes lead to clonal expansion of mutated intestinal epithelial cells. Sporadic colon cancer development initiates from hyperplasia to adenoma and eventually into adenocarcinoma. (Redrawn and modified From Subramaniam, R. et al., *Cancer. Res. Front.*, 2, 1–21, 2016.)

sequential genetic alterations, with the earliest change being mutation of the *APC* gene, followed by mutations in some oncogenes and loss of tumor suppressor genes [43]. Colonoscopy is the most reliable means for an early detection. The progressively improved treatment that may include resection, radiotherapy and chemotherapy, has resulted in a 5-year survival rate of about 50%.

Two studies have examined for genetic alterations in the *DRD2* gene in colorectal cancer (CRC). An early study analyzed 101 CRC specimens for allelic loss at the *DRD2* gene, located at chromosome 11q22-23 [44]. Loss of heterozygosity (LOH) was seen in 16.7% of adenomas and 33.8% of carcinomas. The authors concluded, however, that inactivation of any putative tumor-suppressor gene at 11q22-23 by LOH is not a common event in the development of CRC, but may be significant if accompanied by chromosome 14 deletions. A second study examined whether polymorphisms within the *DRD2* gene are associated with risk of sporadic CRC [45]. Following genotyping of 370 cancer cases and 327 controls for seven single-nucleotide polymorphisms (SNPs) of *DRD2*, three SNPs were found within *DRD2* that were associated with CRC. Because these SNPs are known to be associated with reduced levels of D2R, these data supported a role for D2R in modulating the risk of CRC. In addition to D2R, others reported that the human colon cancer cell line SW480 expressed D5R and that the D1R/D5R agonist SKF83959 inhibited cell proliferation [32].

Given some reports on the presence of pro-inflammatory gut microbiota in both CRC and PD, a large epidemiological study with 22,093 CRC cases and 85,833 matched controls sought to evaluate associations between PD and CRC [46]. The emphasis was on diagnosis of PD prior to CRC, disease duration, and Parkinson's-specific therapies. Data analysis showed a lower risk for CRC in patients with PD. This association was independent of age of PD diagnosis, disease duration, or therapy used. Of note, the effect was more prominent in females than in males. However, the number of patients receiving DA therapy was not large enough to generate statistically significant results. The authors concluded that PD is inversely associated with CRC and that more data are needed on specific effects of DA treatment on CRC.

The prognostic significance of Dp32 protein expression in primary CRC was also examined, using histopathological analysis of 100 patients operated on for CRC between 1994 and 1997 [47]. Dp32 expression in primary tumors was significantly greater in patients with distant metastases than in those without distant metastases. The data indicated that Dp32 is a potential marker of worse prognosis and a valuable tool for managing adjuvant treatment in patients with advanced colorectal cancer.

Studies from several laboratories found anti-tumorigenic effects of activated cGMP/protein kinase G (PKG) axis in colon cancer cells, including inhibition of tumor growth and anti-angiogenesis [48]. The use of a cDNA array revealed lower PKG type-1 expression in colon tumors than in normal tissue [49]. In SW620 colon carcinoma cells engineered for inducible expression of PKG1β, enzyme activation decreased tumor growth and invasiveness in nude mouse xenografts. Sulindac sulfide is a nonsteroidal anti-inflammatory drug that inhibits PDE, resulting in increased cGMP levels and PKG activation [50]. Sulindac sulfide inhibited the proliferation of colon tumor cells and induced apoptosis without affecting normal colon cells. PKG knockdown and use of tadalafil and sildenafil also inhibited growth of colon tumor cells that expressed higher PDE5 levels than controls. Sulindac exerted its effects by inhibiting the Wnt/beta-catenin T-cell factor transcriptional activity, leading to down-regulation of cyclin D1 and survivin.

The targeting of VEGF or its receptors has been used in the clinics to treat several cancers [51]. Three reports from the same laboratory focused on anti-angiogenic actions of DA in colon cancer. The first study showed that DA not only inhibited tumor angiogenesis and growth of colon cancer xenografts but also did not cause hypertension, hematological, renal, and hepatic toxicities as did sunitinib, a commonly used anti-angiogenic agent [52]. DA also prevented 5-FU-induced neutropenia in the HT29 colon cancer-bearing mice. The second study showed that DA in combination with 5-FU inhibited tumor growth and increased the life span of tumor bearing mice, as compared with each alone [53]. DA had no direct effects on growth and survival of tumor cells *in vitro*, whereby there is no angiogenic component. The anti-angiogenic actions of DA were mediated by inhibiting both the proliferation and the migration of tumor endothelial cells by suppressing vascular endothelial growth factor receptor 2 (VEGFR-2), MAP kinase and focal adhesion kinase phosphorylation. The authors concluded that DA enhanced the efficacy of a commonly used anticancer drug and indicated that DA might have a role as an anti-angiogenic agent for the treatment of colon cancer.

Impaired tumor blood flow, caused by structurally and functionally abnormal blood vessels, aggravates tumor hypoxia, hinders the delivery of chemotherapeutic agents, and increases resistance of tumor cells to anti-neoplastic drugs [54]. The third study from the above laboratory examined whether DA affected the stabilization of tumor vasculature [55]. The results showed loss of sympathetic innervation and DA in abnormal tumor blood vessels in malignant colon xenograft. DA administration normalized the morphology and improved the functions of tumor vessels by acting upon pericytes and endothelial cells. D2R activation up-regulated the expression of angiopoietin 1 in pericytes and expression of KLF2 in tumor endothelial cells. Vessel stabilization by DA also increased the concentration of 5-FU in tumor tissues. These results established a functional relationship between vascular stabilization and DA and indicated that D2R-selective agonists should be considered for treating colon cancer and other disorders in which normalization of blood vessels have therapeutic benefits.

13.3.4 Liver cancer

Hepatocellular carcinoma (HCC) is the sixth most common malignancy in the world, and the third leading cause of cancer deaths [27]. Over 80% of the cases occur in Asia and Africa, where HCC is most frequently caused by hepatitis B virus infection, whereas in Japan, this cancer is mainly caused by hepatitis C virus infection. In Western countries, liver cirrhosis resulting from chronic alcohol abuse, is a major etiological factor for this cancer. HCC is almost always lethal, with survival from the time of diagnosis being less than 6 months, and only 10% of the patients survive 5 years or more. Chemotherapy combining cisplatin, doxorubicin, interferon, and 5-FU elicits a response, but no agent, alone or in combination, significantly improves survival. Hormone therapy is also disappointing, although the results with octreotide (a somatostatin analog) are promising. Metastatic HCC commonly spreads to the lungs and bones. The liver is also a frequent site of metastases from other cancers, e.g., colorectal cancer. The poor prognosis and lack of effective therapies for HCC indicates that the development of prevention programs, such as hepatitis B vaccination, is of critical importance [56].

Analysis of DAR expression in paired HCC specimens and adjacent normal tissue showed up-regulation of D1R and down-regulation of D5R in the HCC samples; other DAR were low to undetectable [57]. Treatment with thioridazine caused dose-dependent suppression of cell viability and sphere formation in SNU449, LM3, and Huh7 HCC cell lines. The effects of thioridazine were mediated by induction of G0/G1 cell cycle arrest and suppression of stemness genes *CD133, OCT4*, and *EpCam*. Thioridazine also inhibited cell migration by suppressing epithelial–mesenchymal transition-related genes such as *twist2* and *E-cadherin*. Injection of thioridazine to mice bearing LM3 cell-derived xenografts suppressed tumor growth. The authors concluded that thioridazine has a potential role in the treatment of HCC.

All five DAR subtypes were expressed in HepG2 cells, an immortalized human liver carcinoma cell line [58]. Cell treatment with the D1R agonists SKF82958 or SKF38393 increased D1R expression and stimulated SULT2A1 sulfotransferase activity. These effects were partially blocked by co-treatment with the specific D1R antagonist SCH23390. These data suggest that D1R is involved in metabolism of drugs and xenobiotics through up-regulation of SULT2A1. 3,3',4',7-Tetrahydroxyflavone (fisetin) is an abundant flavonoid that is produced in some vegetables and fruits. It has a similar structure to DA and appears to mimic DA action [59]. Both fisetin and bromocriptine, albeit at high micromolar doses, suppressed the proliferation, migration, and invasion of liver cancer cells. Growth of orthotopically implanted liver cancer xenografts was inhibited by fisetin administration, which was accompanied by higher levels of serum and tumor DA and increased survival rate of the mice [59]. The authors concluded that DA represents a novel therapeutic strategy for suppressing liver cancer progression.

As discussed in a recent review [60], NO has multiple effects on liver cancer. Administration of an NO donor or overexpression of nitric oxide synthase 3 (NOS-3) increased oxidative stress and cell death in anti-Fas-stimulated HepG2 cells. Activation of this pathway also increased the expression of cell death receptors (TNF-R1, Fas, and Trail-R1), and reduced the growth of HepG2 xenografts in mice. NO donors sensitized liver cancer cells to chemotherapeutic agents such as cisplatin and melphalan by nitrosylation of critical thiols in DNA repair enzymes. It was concluded that administration of NO donors increases the therapeutic outcome in a combined treatment with chemotherapeutic agents.

13.3.5 Pancreatic cancer

Pancreatic cancer is the 14th most common cancer worldwide, with 215,000 new cases occurring every year [27,61]. The highest incidence occurs in more-developed countries, where about 40% of the cases are attributed to tobacco. Over 90% of pancreatic tumors are adenocarcinomas, arising from the ductal epithelium of the exocrine pancreas. Mutations in *kRAS* and *p53* genes are implicated in the development of ductal adenocarcinoma. Endocrine tumors of the pancreas, which are rare, arise from the islets of Langerhans. Pancreatic cancer is one of the most lethal malignancies, with most patients dying within 1 year of clinical diagnosis. The poor prognosis results from lack of apparent symptoms early during the disease, detection at late stages, aggressive tumor biology, early metastasis, and lack of effective options for chemotherapy. Gemcitabine confers a median survival advantage of only 6 months because most patients develop resistance to therapy.

The human exocrine pancreas has a complete DA system, including biosynthetic capability and DAR expression [62,63], but only two studies have examined the relevance of DA to pancreatic cancer. The first study [64] focused on potential interactions between D2R and somatostatin receptors [sst (1–5)]. The study examined specimens from 35 patients with gastro-entero–pancreatic (GEP) tumors (19 pancreatic and 16 intestinal), as well as 13 somatotroph adenomas. Whereas sst(2) levels were similar between GEP and somatotroph tumors, sst(5) and D2R were higher in GEP. The authors indicated that pancreatic tumors co-expressed sst(2) and D2R in 100% of cases, and sst(5) in 89%, supporting the testing of bi-specific agonists (sst(2)/sst(5) or sst(2)/D2R) in these tumors. Although D2R was found in all tumors, 80% express only low levels of D2R, indicating that D2R targeting alone may be beneficial to only a subset of patients with GEP tumors.

As shown in **Figure 13.5**, chimeric molecules, directed against both DA and somatostatin receptors, have been designed [65]. Such molecules combine two pharmacological moieties, a somatostatin analog and a DA agonist or antagonist. A molecule named BIM-23A760 was tested in

Figure 13.5 Induction of apoptosis in malignant cells by a somatostatin/dopamine chimera. The engineered molecule induces hetero-dimerization of somatostatin and dopamine receptors. (Redrawn and modified from Baldelli, R. et al., *Front. Endocrinol.*, (*Lausanne*), 5, 1–10, 2014.)

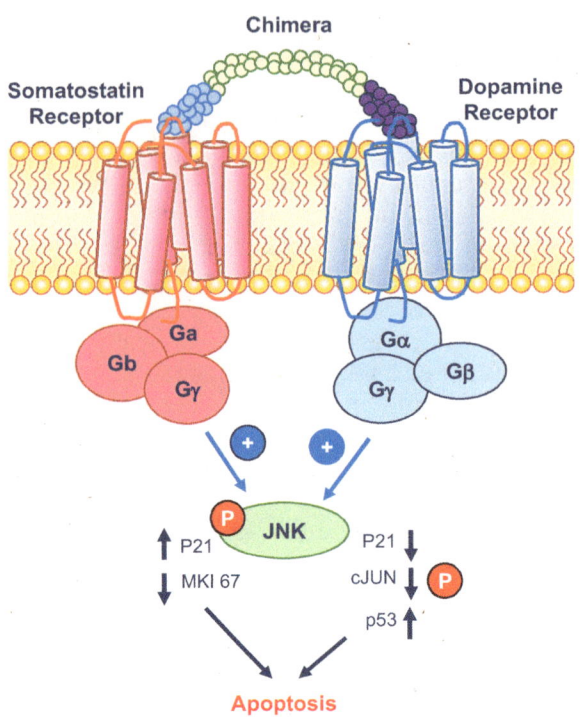

nonfunctioning pituitary adenomas and neuroendocrine tumors. The chimera induced apoptosis and inhibited cellular proliferation by activating the ERK1/2 and p38 MAPK pathways. The chimeric approach has a great potential for curbing the growth of a number of tumors that express both DA and somatostatin receptors.

Microarrays with 195 pancreatic ductal adenocarcinomas (PDACs) and 41 non-tumor pancreatic tissues were used for gene expression profile analysis [66]. Another set of 152 samples (40 non-tumor pancreatic tissues, 63 PDAC sections, and 49 chronic pancreatitis samples) was used for tissue microarray analysis. The D2R protein was much higher in PDACs than in non-tumor tissues. Five pancreatic cancer cell lines at various degrees of differentiation were then used: moderately differentiated BxPC-3 cells, poorly differentiated Panc-1 and MiaPaCa-2, and well-differentiated Capan-1 and CFPAC-1 cells; human dermal fibroblasts served as controls. Knockdown of *DRD2* by RNAi or its inhibition with pimozide and haloperidol reduced cell proliferation and migration and induced apoptosis. In orthotopically transplanted pancreatic tumor cells, *DRD2* knockdown or haloperidol administration reduced tumor growth and metastases. The authors concluded that D2R antagonists, routinely used for management of schizophrenia, should be tested in patients with pancreatic cancer.

13.4 LUNG, KIDNEY, AND BLADDER CANCERS

13.4.1 Lung cancer and dopamine

Lung cancer is the most common tumor type worldwide and is the leading cause of cancer death, with 900,000 new cases diagnosed each year [27]. Most cases are caused by smoking, but also by certain occupational exposure. No effective treatment is currently available, and the 5-year survival rate is less than 15%. The main cancer types are squamous cell carcinoma, adenocarcinoma, large cell carcinoma, and small cell carcinoma. The first three are also referred to as "non-small cell" lung carcinomas (NSCLCs).

Squamous cell carcinomas arise primarily in the proximal segmental bronchi and are more strongly associated with smoking than are the adenocarcinomas. The mainstay treatment for small cell lung cancer is chemotherapy, with concomitant radiotherapy for patients with limited disease. Surgery is considered for small isolated lesions.

Both D1R and D2R are involved in lung cancer by targeting the tumor cells themselves, the endothelia, and the infiltrating lymphocytes. An older study [67] reported that bromocriptine inhibited the growth of human small cell lung cancer (SCLC) xenografts in mice, causing marked degenerative changes in the tumors such as densely aggregated chromatin masses and cytoplasmic vacuolization. Incubation of the SCLC cell line NCI-H69 with bromocriptine inhibited clonal growth, which was blocked by the D2R antagonists, metoclopramide, and domperidone. A more recent study sought to determine whether D2R agonists inhibit progression of lung tumors and also to identify subpopulations of patients that could benefit from D2R-targeted therapy [68]. Examination of lung cancer specimens showed positive correlations between endothelial D2R expression and tumor stage; patients with smoking history had greater levels of D2R in the lung endothelium. Moreover, D2R agonists abrogated tumor progression in syngeneic (LLC1) and orthotopic human xenografts (A549) in mice by inhibiting tumor angiogenesis and reducing tumor infiltrating myeloid-derived suppressor cells.

Focusing on NSCLC tissues rather than on endothelial cells, a recent study reported that D2R expression was lower in tumors than in adjacent normal lung tissues [69]. Moreover, *DRD2* mRNA and protein levels in NSCLC were negatively correlated with the tumor size, tumor, node, metastasis (TNM) status, and patient overall survival. *In vitro* experiments showed that disruption of *DRD2* promoted the proliferation of NSCLC cell lines A549 and SK-MES-1 by inhibiting the NF-κB signaling pathway. In addition, *DRD2* overexpression not only blocked lipopolysaccharide-induced growth of A549 and SK-MES-1 cells but also inhibited tumorigenesis in murine xenograft models. The authors concluded that *DRD2* may be a potential therapeutic target for lung cancer patients with high *DRD2* expression by ablating the NF-κB signaling pathway.

Two studies reported increased circulating DA levels in lung cancer patients. One study found significantly higher plasma DA levels in patients with lung cancer than in controls [70]. A further refinement revealed that plasma DA levels were more elevated in patients with SCLC than in those with NSCLC or in healthy controls [71]. This study also found that SCLC cells express D1-like and D2-like receptors, as well as membrane and vesicular transporters. The D1R agonist SKF38393 increased cAMP levels and Dp32 protein expression is several SCLC cell lines without affecting cell growth [71]. Treatment with the D1R antagonist SCH23390 inhibited the SKF38393 effects. In contrast, the D2R agonist quinpirole counteracted SCLC cell proliferation without affecting cAMP levels but by decreasing sulpiride-induced AKT phosphorylation. When tested *in vitro*, DA significantly inhibited the proliferation and cytotoxicity of T cells obtained from lung cancer patients and normal volunteers by a D1R-mediated cAMP elevation [70].

The effects of cGMP/PKG-Iα pathway activation on two NSCLC cells—NCI-H460 and A549—were also examined [72]. ODQ, which blocks NO-induced cGMP/PKG activation, caused apoptosis, suppressed colony formation, decreased cAMP response element binding protein (CREB) phosphorylation, and inhibited expression of cellular inhibitor of apoptosis 1 (c-IAP1), and survivin, while DT-2, an inhibitor of PKG-Iα kinase, or a PKG-Iα knocked down, caused the opposite effects. Inhibition of PKG kinase activity dramatically enhanced the pro-apoptotic effects of cisplatin. The authors concluded that PKG-Iα kinase activity is necessary for maintaining higher levels of CREB phosphorylation and expression of anti-apoptotic proteins. These prevented spontaneous apoptosis and promoted colony formation in NSCLC cells, which may limit the efficacy of chemotherapeutic agents like cisplatin.

13.4.2 Kidney and bladder cancer

Kidney cancer is the 15th most common cancer worldwide, with men generally affected more frequently than women [27]. Both tobacco smoking and obesity constitute risk factors. About 85%–90% of the cases are renal cell carcinoma (RCC), a heterogeneous group of tumors that arise from the proximal convoluted tubules. Given that RCC patients frequently present with arterial hypertension due to various causes, including DA deficiency, potential associations between DA and RCC were examined. One study evaluated the clinical significance of expression of DDC (which catalyzes DA biosynthesis) in surgically resected RCC tumors and adjacent normal renal tissue [73]. DDC mRNA levels were dramatically down-regulated in RCC tumors, with a dissimilar expression in clear cell carcinomas and other tumor subtypes such as papillary and chromophobe tumors. A significant inverse correlation was also seen between DDC expression and tumor grade.

Another study focused on the role of DAT in RCC [74]. DAT expression was much higher in RCC than in normal kidney tissue and other tumor types. The high DAT expression in clear cell renal carcinoma cell lines was associated with increased DA uptake. In addition, DAT expression was influenced by HIF-2α, and hypoxia induced DAT expression in normal renal cells. Thus, both studies showed that components of the DA system, including DDC and DAT, constitute novel biomarkers for RCC, having a diagnostic accuracy and discrimination between clear cell and non-clear cell tumor subtypes. These proteins can be exploited for diagnostic or therapeutic purposes for detection or treatment of RCC.

Bladder cancer affects about 3.5 million people globally, with 430,000 new cases diagnosed every year, resulting in close to 200,000 deaths. Age of onset is most often between 65 and 85 years of age. Blood in the urine is the most common symptom in bladder cancer. Other symptoms include pain during urination (dysuria), frequent urination, and lower back pain. However, these signs and symptoms are not specific to bladder cancer and may also be caused by noncancerous conditions, including prostate infections, overactive bladder or cystitis. Risk factors include smoking, family history, prior radiation therapy, frequent bladder infections, and exposure to certain chemicals. The most common type is transitional cell carcinoma, while other types include squamous cell carcinoma and adenocarcinoma. Diagnosis is typically made by cystoscopy and tissue biopsy, while staging is determined by computed tomography (CT) scan and bone scan.

A recent study used liquid chromatography–high resolution mass spectrometry–based metabolomics and discovered high levels of DA sulfate (DA-S) in urine from bladder cancer patients [75]. Although of interest, neither the source nor the significance of elevated DA-S to bladder cancer were determined. Another study reported significantly higher specific polymorphism of the *DrD2* gene in patients with bladder cancer who were also heavy smokers [76]. The authors suggested that the specific D2R alleles confer greater vulnerability to tobacco use and are indirectly associated with bladder cancer.

13.5 CANCERS OF THE SKIN

13.5.1 Prevalence and attributes of skin cancer

Skin cancer is the most common form of cancer, accounting for about 40% of global cancer cases. There are three main types of skin cancers: basal cell skin cancer (BCC), squamous-cell skin cancer (SCC), and melanoma. The first two (along with a few less common skin cancers) are known as non-melanoma skin cancer (NMSC), which afflicts 2–3 million people worldwide. About 80% of NMSCs are BCCs and 20% are SCCs, both of which rarely result in death.

BCC grows slowly and can damage the tissue around it but is unlikely to spread to distant areas or result in death. SCC is more likely to spread, and melanomas are the most aggressive. Globally, melanoma occurs in more than 200,000 people, resulting in 50,000 deaths. White people in Australia, New Zealand and South Africa have the highest rates of melanoma.

Greater than 90% of skin cancers are caused by exposure to ultraviolet (UV) radiation from the sun, which increases the risk of all three main types of skin cancer [77]. Exposure during childhood is particularly harmful for melanomas and BCC, while total exposure, regardless of when it occurs, is more important for SCC. Decreasing the exposure to UV radiation and the use of sunscreens appear to be effective methods of preventing melanoma and SCC but not BCC. People with light skin and those with poor immune function such as from medications or HIV/AIDS are at higher risk for melanoma.

Non-melanoma skin cancer is usually curable. Treatment is typically done by surgical removal and less commonly by radiation therapy or topical medications such as fluorouracil. Treatment of melanoma involves combination of surgery, chemotherapy, radiation therapy and targeted therapy. In people whose disease has spread, palliative care may be used to improve quality of life. Melanoma has one of the higher survival rates among cancers, with over 90% of cases in the United States surviving more than 5 years. There are little, if any, indications for an association between DA and non-melanoma skin cancer, while substantial evidence shows a strong association of melanoma with DA, as detailed below.

13.5.2 Dopamine and melanoma

Melanomas result from excessive proliferation of melanocytes, the pigment-forming cells of the skin, which is the site of >95% of primary melanomas [78,79]. About 133,000 new cases of malignant melanoma are diagnosed worldwide each year. The 5-year survival for an early-stage melanoma, located at the site where it had started, is over 98%. Survival for melanoma that has spread to nearby lymph nodes is 62%, while survival for melanoma that has spread to other parts of the body is only 18%. The risk of developing this highly malignant skin cancer varies according to racial background (dark skin pigmentation is protective) and geography (excessive sunlight-derived UV irradiation is promoting), with the highest incidence observed in white populations in Australia.

A melanoma often arises from a preexisting pigmented skin lesion (a mole or "naevus"). The major characteristics of melanoma have been designated as ABCDE: Asymmetry, Border irregularity, Color variation, Diameter >6 mm, and Elevation. The highest risk is for dysplastic (atypical) naevi that are larger than 6 mm in diameter, have irregular pigmentation and ill-defined margins, and exist in multiples. Formation of a melanoma that penetrates the dermis is shown in **Figure 13.6**A. In addition to the skin, melanomas can be found in the eye and in the mucous membranes of mouth, nose, anus and vagina and to a lesser extent in the intestine. About 20% of familial melanomas have germline mutations in the *CDKN2A* gene, which encodes p16INK4A. The mainstay treatment of systemic metastases is chemotherapy. However, no highly effective single agent or combination has yet been developed, and metastatic melanoma is characterized by drug resistance.

Several epidemiological studies reported significant co-occurrence of PD and melanoma [80,81], but a well-defined biological explanation for this association is lacking. Case reports have suggested that L-Dopa has a causal relationship with malignant melanoma because of a shared DA biochemical pathway [82]. **Figure 13.6**B shows the involvement of L-Dopa in the synthesis of both DA and melanin. Increased neuromelanin, which is synthesized in striatal DA neurons, may enhance the individual's susceptibility to oxidative stress-induced neuronal injury, which is relevant to PD.

Figure 13.6 Human skin with a melanoma and the biosynthetic pathways for dopamine and melanin. Panel A shows a cross section of skin with a penetrating melanoma. Panel B shows that conversion of L-Dopa by Dopa decarboxylase in the striatum generates dopamine, while conversion of L-dopa by tyrosinase in the skin, generates dopaquinone that further develops into melanin.

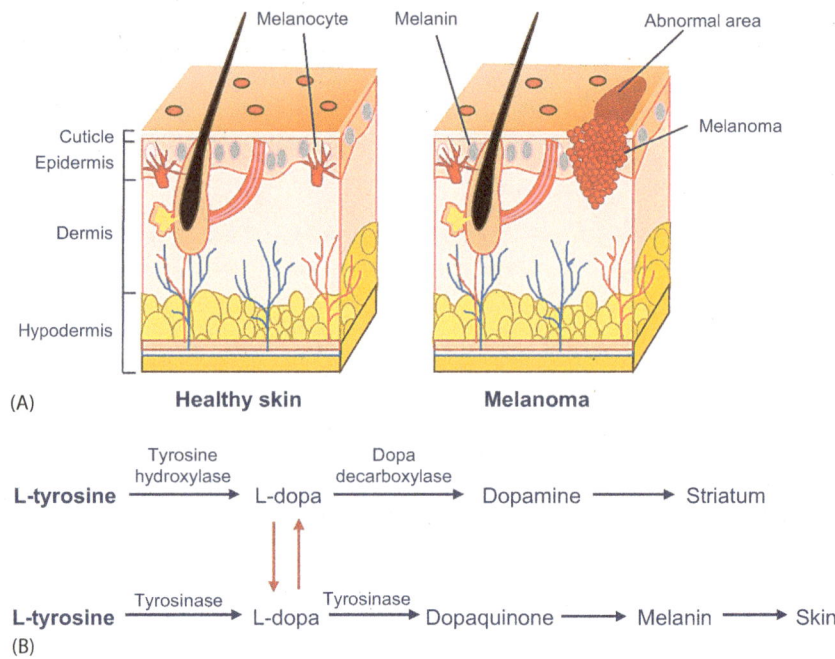

It has also been reported that α-synuclein, which is implicated in the pathogenesis of PD, is also expressed in the skin and in melanoma [83]. Given its involvement in the biosynthesis of both melanin and DA, α-synuclein may be one of the factors responsible for the positive association between PD and melanoma.

From the late 1970s to the late 1980s, eight publications by Wick et al. [84–91] focused on the effects of DA and related compounds on melanomas, using cultured cells, tumor-bearing mice, and small clinical trials. Treatment of exponentially growing human and murine melanoma cells with L-Dopa, L-Dopa methyl ester, 3,4-dihydroxy-benzylamine (DHBA), DA, and norepinephrine (NE; at 1–10 µM each) caused variable cytotoxicity. The effect was due to the rapid inhibition of thymidine incorporation into DNA, without affecting uridine or leucine incorporation. The authors postulated that DA and related compounds inhibit DNA polymerase and also act as potent reducing agents that interfere with redox reactions crucial to DNA synthesis. The conversion of these compounds into quinones further disrupted cellular metabolism and accounted for the increased cytotoxicity in melanin-producing cells.

Because melanomas can convert L-Dopa into melanin, the first *in vivo* experiment by the Wick's group examined whether L-Dopa incorporation into melanin was enhanced by pretreating mice bearing melanoma xenografts with Ro4-4602, a potent DDC inhibitor. The results showed enhanced incorporation of the label into the tumor, with only limited entry into adrenal tissue, indicating a diversion of metabolism to melanin, away from catecholamine formation. When the anti-tumorigenic effects of L-dopa methyl ester, DA, and DHBA on growth of xenografts were compared, DHBA was more effective than DA in prolonging the life span of tumor-bearing mice, likely due to its lower systemic toxicity.

The effects of DA were tested in four patients with advanced melanoma [88]. DA was infused at 20 µg/kg/min for 48–120 h. Tumor and bone marrow samples were obtained before and after treatment. Each of the patients showed a significant reduction in the labeling index of tumor cells following the treatment. In spite of the positive responses, adverse cardiovascular effects of DA precluded a repetitive use. Instead, a combination of levodopa/carbidopa (a DDC inhibitor) was used with another cohort of patients in an attempt to circumvent the DA-induced toxicity and deliver the drug to the central nervous system (CNS) where metastases occurred [90]. The rationale was that L-Dopa is oxidized by tyrosinase to melanin only in melanocytes. Because melanoma cells contain more tyrosinase than normal melanocytes, the authors hoped to take advantage of this fact and identify drugs that are specifically toxic to malignant melanomas. Of the 12 patients treated, 4 showed significant clinical responses. One patient had a complete resolution of a CNS lesion, as measured by CT scan and a corresponding improvement in symptoms.

The above studies were published more than 30 years ago, with promising results in two small clinical trials. To-date, however, DA-altering drugs have not been clinically used against a disease that has few effective treatment options. One wonders whether the biomedical community has inadvertently overlooked these data or perhaps some follow-up clinical trials with a larger number of patients yielded negative outcome and/or unacceptable adverse effects.

13.6 HEAD AND NECK CANCER

13.6.1 Prevalence and attributes of head and neck cancer

Head and neck cancer is the sixth most common cancer, with 600,000 new cases diagnosed annually worldwide [92,93]. Tumors can arise in the oral cavity (tongue and salivary glands), oropharynx, hypopharynx, and larynx. Esophageal cancer is usually considered as a distinct category. Over 90% of the tumors are classified as squamous cell carcinomas (HNSCCs). Risk factors include tobacco use and heavy alcohol consumption, and tumors are significantly more common in men than women. Human papillomavirus (HPV)-positive tumors are increasing in prevalence and, in general, have improved prognosis. The overall 5-year survival rate for HNSCC is less than 50%. Treatments include surgery, radiotherapy, chemotherapy, or their combinations. Disease progression as well as surgical resection can result in multiple adverse side effects such as facial disfigurement, difficulty swallowing, and speech impediment, which severely reduce the quality of life. Hence, more effective therapeutic approaches for HNSCC are urgently needed.

13.6.2 Dopamine and DARPP-32 in head and neck and esophageal cancers

Head and neck cancer exhibits several components of the DA/DAR signaling pathways. Analysis of surgical specimens from HNSCC patients showed lower DDC expression in squamous cell carcinomas of the larynx and tongue than in adjacent normal tissues [94]. The low DDC expression was seen in tumors of advanced TNM stage or bigger size, as compared with early-stage or smaller tumors. No differences were observed between tumors from pharynx, buccal mucosa, or nasal cavity and their normal counterparts. These data suggested that DDC expression constitutes a potential prognostic biomarker in tongue and/or larynx SCCs, which represent the majority of HNSCC cases. We have detected D1R expression in primary tumors and HNSCC cell lines (unpublished observations). There are, however, no published records on DAR expression in primary head and neck tumors.

Expression of DARPP-32 (Dp32) and tDp was examined in specimens from patients with esophageal squamous cell carcinomas [95]. Positive Dp32 immunostaining was detected in 30% of the tumors and was inversely correlated with advanced pathologic stage and tumor size; all adjacent normal tissue samples were Dp32-negative. The overall survival rate was worse in patients with Dp32-negative tumors than those with Dp32-positive tumors. Dp32 expression was also observed in four of seven HNSCC cell lines examined, whereas t-Dp expression was not seen in any. These findings suggest a slower progression of Dp32-expressing tumors than Dp32-32-negative tumors. Whether Dp32 and tDp are regulated by local DA in head and neck tumors had not been determined.

13.6.3 Role of the cGMP/PDE/PKG pathway in head and neck cancer

Our laboratory as well as others have examined the relevance of the cGMP/PKG/apoptotic pathway in head and neck cancer. See **Chapter 12** and **Figure 12.10** for more detail on the cGMP/PDE5/PKG system in breast cancer. Several compounds, including sodium nitroprusside, an NO donor, BAY 41-2272, a soluble guanylate cyclase (sGC) activator, and tadalafil, a PDE5 inhibitor, all increase intracellular cGMP levels in several HNSCC cell lines (**Figure 13.7**). The combination of sGC activators and PDE5 inhibitor was especially powerful, increasing cGMP levels 20- to 150-fold [96]. The functionality of the cGMP/apoptotic axis was verified by the ability of tadalafil and sildenafil alone and in combination with BAY 42-2272 to suppress cell viability in four cell lines representing the diversity of the disease in terms

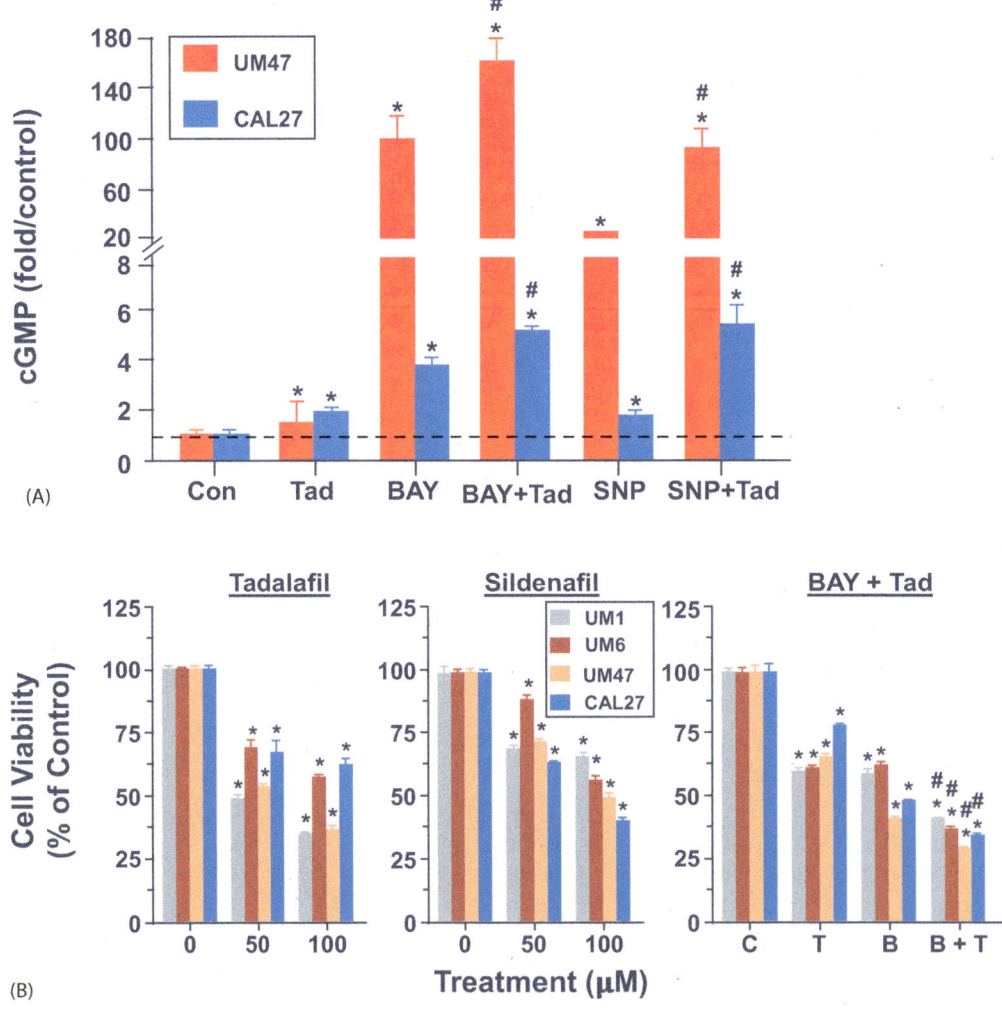

Figure 13.7 Effects of activators of the cGMP/PKG pathway on head and neck squamous cell carcinoma (HNSCC) cell lines. Panel A: Stimulation of cGMP accumulation in UM47 and CAL27 HNSCC cell lines. Con: control; Tad: tadalafil; Bay: Bay 41-2272; PKG: protein kinase G; SNP: sodium nitroprusside; * significant vs. control; # significant vs. cells treated with tadalafil. Panel B: Suppression of cell viability of four HNSCC cell line by the phosphodiesterase 5 (PDE-5) inhibitors tadalafil and sildenafil, and effects of co-treatment with Bay 41-2272, a stimulator of soluble guanylate cyclase. * significant vs. control; # significant vs. cells treated with tadalafil. See text for other explanations. (Redrawn and modified from Tuttle, T.R., et al., *Cancer. Lett.*, 370, 279–285, 2016.)

of anatomical origin and HPV expression: UM1 (oral cavity, HPV$^-$), UM6 (oro-pharynx, HPV$^-$), UM47 (tongue, HPV$^+$), and CAL27 (tongue, HPV$^-$). In addition, Tadalafil was very effective in suppressing CAL27-derived xenograft in nude mice [96]. We also found that selective sGC stimulators reduced HNSCC cell viability synergistically with cisplatin and enhanced apoptosis by cisplatin. BAY 41-2272, an sGC stimulator, reduced expression of the survival proteins EGFR and β-catenin and increased pro-apoptotic Bax, suggesting a potential mechanism for the anti-tumorigenic effects of these drugs.

In a follow up study [97], we found differential expression of critical components of the cGMP/PKG pathway in four HNSCC cell lines. The sGC stimulator BAY 41-2272 reduced expression of the survival proteins EGFR and beta-catenin and increased pro-apoptotic Bax, suggesting a potential mechanism for the anti-tumorigenic effects of these drugs. Moreover, the sGC stimulators effectively reduced viability in cells with acquired cisplatin resistance and were synergistic with cisplatin in susceptible cells.

Three reports from the same group focused on the efficacy of PDE5 inhibition in head and neck cancer. The first study [98] examined potential mechanisms by which tadalafil suppressed tumor growth. In several mouse tumor models, the PDE5 inhibitors delayed tumor progression by reversing tumor-induced immunosuppressive mechanisms. Sildenafil down-regulated arginase 1 and inducible NOS (iNOS) expression and reduced the suppressive effects of CD11b$^+$/Gr-1$^+$ myeloid-derived suppressor cells (MDSCs), which were recruited by the growing tumors. By removing these tumor escape mechanisms, silde-nafil enhanced intra-tumoral T cell infiltration and activation, reduced tumor outgrowth, and improved the antitumor efficacy of adoptive T cell therapy. Sildenafil also restored T cell proliferation of peripheral blood mononuclear cells, obtained from patients with multiple myeloma and head and neck cancer.

The second study [99] explored whether PDE5 inhibitors augment immune functions in patients with head and neck cancer by inhibiting MDSC. To that end, a prospective phase II clinical trial was conducted to determine the effects of systemic PDE5 inhibition on immune function in patients with HNSCC. Tadalafil increased T-cell expansion, reduced peripheral MDSC numbers, and increased general immunity as assessed by a delayed hypersensitivity response. Tumor-specific immunity in response to HNSCC tumor lysate from tadalafil-treated patients was augmented. The authors concluded that tadalafil increases general and tumor-specific immunity in patients with HNSCC. They emphasized that evasion of immune surveillance and suppression of systemic and tumor-specific immunity are among the major features of head and neck cancer.

The third study further refined the previous findings [100]. MDSCs in HNSCCs were characterized retrospectively to determine whether their presence at the tumor site correlated with recurrence. A prospective, three-arm study was conducted in which patients undergoing surgical resection of oral and oropharyngeal tumors were treated preoperatively with two doses of tadalafil or placebo for 20 days. Tadalafil reduced both MDSCs and Treg concentrations in blood and in tumors. The concentration of blood CD8$^+$ T cells reactive to autologous tumor antigens was increased after treatment. It was concluded that tadalafil beneficially modulate the tumor micro- and macro-environment in patients with HNSCC by lowering MDSCs and Tregs and by increasing tumor-specific CD8$^+$ T cells.

13.7 NEUROLOGICAL AND NEUROENDOCRINE TUMORS

A neurological tumor occurs when abnormal cells form a mass within the brain or in association with the peripheral nervous system. There are two main types of tumors: cancerous and benign. Cancerous brain tumors can be divided into primary tumors, which start inside the brain, and secondary tumors, known as brain metastases, which have spread from tumors localized elsewhere. The category of primary brain cancers is second only to leukemia in people in the United States under 20 years of age. Brain tumors may produce symptoms that vary depending on the part of the brain involved. These include headaches, seizures, problems with vision, vomiting and mental changes. Of more than 120 types of brain and peripheral nervous system tumors, DA is involved in the following: glioblastomas, neuroblastomas, meningiomas, and neuroendocrine tumors.

13.7.1 Dopamine in brain tumors: Glioblastoma and meningioma

Glioblastoma (GBM) is the most prevalent CNS tumor in adults, with 10,000 new diagnoses annually [101]. It is also one of the most aggressive cancers, with a median survival of only 15 months, despite the use of robust combination therapy, including surgery, radiation treatment, and chemotherapy. Although this treatment regimen is successful in the short term, recurrent tumors with resistance to anti-glioma modalities are nearly universal and fatal. Glioblastoma occurs more commonly in males than females, and it most often begins around 64 years of age. Initially, the signs and symptoms of a glioblastoma are nonspecific and may include headaches, personality changes, nausea and stroke-like symptoms. The worsening of symptoms is often rapid and may progress to unconsciousness.

The cellular origin of glioblastoma is unknown. Given the similarities in immunostaining of glial cells and glioblastoma, the latter have long been assumed to originate from glial cells. More recent studies suggest that astrocytes, oligodendrocytes, progenitor cells, and neural stem cells could all serve as the cell of origin [102]. An aggressive and therapy-resistant subpopulation of glioblastoma cells—glioma-initiating cells (GICs)—resemble in their protein expression profile and self-renewal capacities healthy neural stem cells and progenitor cells. Progenitor cells in the developing brain express various DARs, and activation of these receptors influences cell proliferation as well as differentiation [101]. Specifically, ablation of dopaminergic neurons reduces the proliferation of progenitors in the sub-ventricular zone (SVZ), a site of neurogenesis in the adult brain; this reduction can be counteracted by DA agonists. In some studies, treatment with DA agonists increased proliferation of progenitors in the SVZ, while other studies found that blocking DAR can also increase proliferation. Although these results may appear diametrically opposed, they actually exemplify the diversity of the DAR, because D1-like and D2-like receptors often act in opposite directions. A model showing stimulatory effects of DA on both glioblastoma and endothelial cells is presented in **Figure 13.8**.

Bioinformatics has provided evidence that monoamines are involved in the initiation and growth of glioblastomas in patients. Several DA and serotonin receptors, as well as enzymes responsible for the synthesis of these neurotransmitters, predicted survival outcomes in GBM patients. For example, *DRD2* mRNA and protein expression were significantly increased in GBM samples from patients' biopsies as compared with controls. One factor that complicates understanding of the role of these receptors is the great heterogeneity of GBM tumors. In addition, cell states are not static but, rather, fluctuate and shift in response to the dynamics of the microenvironment. The variable expression levels may represent differences in the differentiation status of individual tumors. DA, either from surrounding neurons or self-generated, is available in the tumor microenvironment and participates in tumor growth.

Meningiomas, also known as **meningeal tumors**, are typically slow-growing tumors that form from the meninges or the membranous layers

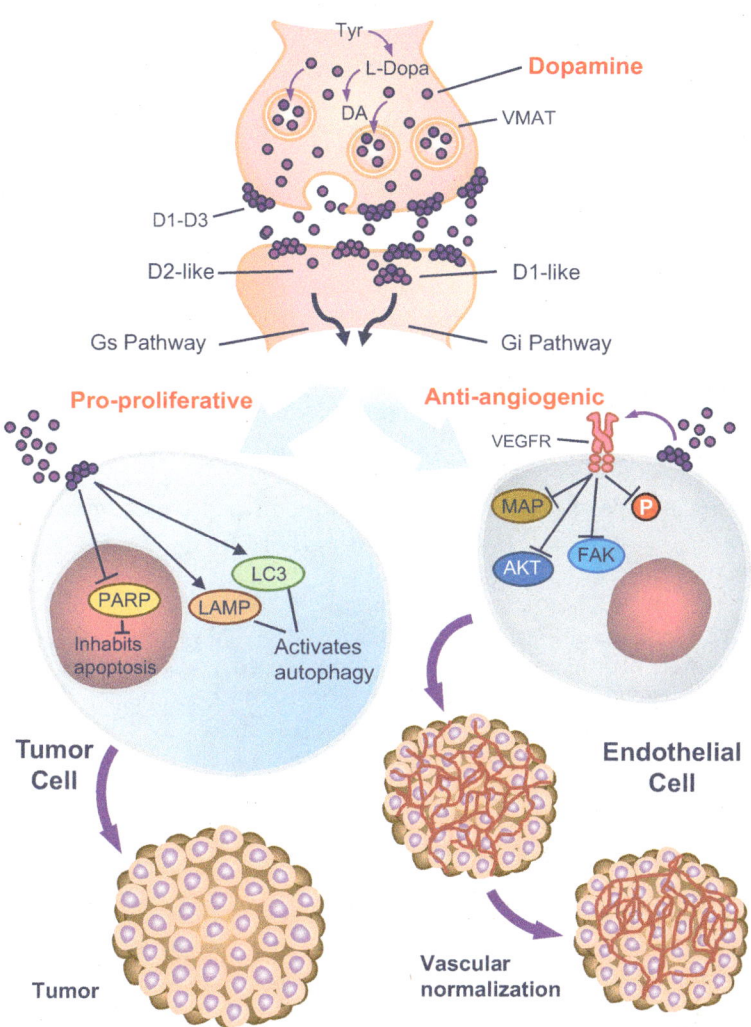

Figure 13.8 Pro-proliferative and anti-angiogenic activities of DA in glioblastoma. Dopamine has both a direct effect on the tumor cells and an indirect effect via the tumor-associated endothelial cells. DA: dopamine; tyr: tyrosine; VEGFR: vascular endothelial growth factor receptor; VMAT: vesicular monoamine transporter. (Redrawn and modified from Caragher, S.P. et al., *Neuro. Oncol.*, 20, 1014–1025, 2018.)

surrounding the brain and spinal cord. They account for about 35% of all intracranial neoplasms, with an annual incidence of ~7.22 cases per 100,000 individuals. The peak incidence is between the sixth and seventh decades of life, afflicting females more often than males [103]. About 90%–95% of meningiomas are benign in nature, 5%–7% are atypical, and only 2% are malignant. These tumors have a high recurrence rate, reaching up to 50% after incomplete surgical resection. Symptoms depend on the location and usually result from the tumor pressing on nearby tissue. Many cases never produce symptoms. Occasionally seizures, dementia, trouble talking, vision problems, one-sided weakness, or loss of bladder control may occur. Human meningioma cells express high affinity D1R, and DA agonists have antiproliferative effects on tumors in culture [104]. A recent study used 157 specimens of meningioma and found the presence of D2R in over 90% of the samples, including those with a higher histological malignancy [103]. The authors concluded that this finding warrants further investigation of the therapeutic potential of dopaminergic agents to treat these tumors.

13.7.2 Dopamine and neuroblastoma

Neuroblastoma (NBL) is defined as a solid cancerous tumor that begins in nerve cells outside the brain of infants and young children. NBL is the most common extra-cranial solid tumor in childhood, accounting for 10% of the pediatric malignancies and is responsible for 15% of the pediatric cancer-related deaths. About 90% of the cases occur in children less than 5 years of age, and NBL is rare in adults [105]. High-risk NBL is considered challenging and has one of the least favorable outcomes among pediatric cancers. NBL most frequently starts from one of the adrenal glands, but can also develop in the neck, chest, abdomen, or spine. Symptoms may include bone pain, a lump in the abdomen, neck, or chest, or a painless bluish lump under the skin. Advanced disease at presentation is common because 50%–60% of all neuroblastoma cases present with metastases. NBL is a complex disease that has high heterogeneity and is therefore difficult to target for successful therapy.

Diagnosis is established by tumor biopsy and by urinary catecholamines, which are elevated in about 90% of cases. Urinary catecholamines and their metabolites, i.e., DA, homovanillic acid (HVA), and 4-hydroxy-3-methoxy-mandelic acid (HMMA) are established tests that are used for two purposes: (1) to support the diagnosis of NBL disease, and (2) to monitor disease activity following treatment [106]. Both free and total 3-methoxytyramine (MTY), the O-methylated product of DA, were elevated in all patients with untreated NBL disease, leading the authors to suggest that MTY is an especially good biomarker for differentiating treated patients with advanced or residual disease from controls. The main drivers of NBL formation are neural crest cell-derived sympathoadrenal cells that undergo abnormal genetic arrangements. In the research arena, neuroblastoma-derived cell lines such as SH-SY5Y, SK-N-MC, SK-N-SH, and N2A have been extensively used as cellular models to investigate multiple aspects of DA homeostasis, including biosynthesis, reuptake, receptor expression, mechanism of action, and drug effects.

13.7.3 Prevalence and attributes of neuroendocrine tumors

Neuroendocrine tumors **(NETs)** and neuroendocrine carcinomas **(NECs)** are a heterogeneous group of tumors. They are found in a wide variety of organs, including the lung, thymus, thyroid, stomach, duodenum, small bowel, large bowel, appendix, pancreas, adrenal, and skin [107,108]. Tumors usually occur sporadically but are also associated with familial syndromes such as multiple endocrine neoplasia type 1 and 2, Von Hippel-Lindau disease, neurofibromatosis type 1, Carney complex, pheochromocytoma–paraganglioma syndrome, and familial non-medullary thyroid carcinoma. Tumors that occur as part of hereditary syndromes are characterized by specific genetic abnormalities and frequently appear at a young age.

There is insufficient information on the global incidence of NET, as exemplified by the grouping of NETs and CNS tumors in the online summary of cancer by organs [27]. A retrospective study that included data from 10 hospitals identified 149 patients with NETs/NECs [109]. As illustrated in **Figure 13.9**, about 70% of the tumors were localized in different portions of the gut, with smaller numbers found in the lung, bladder, prostate and liver.

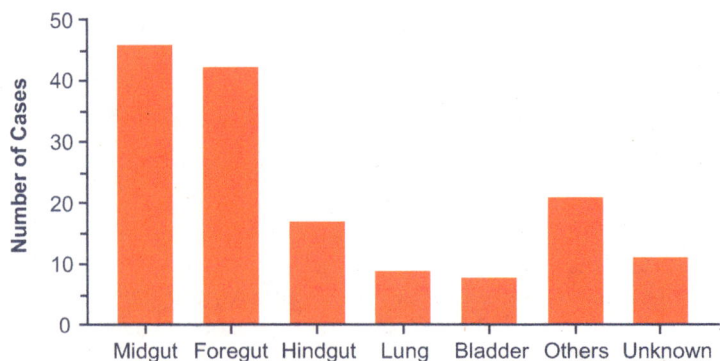

Figure 13.9 Occurrence of neuroendocrine tumors in various organs. (Redrawn and modified from Maschmeyer, G. et al., *J. Cancer. Res. Clin. Oncol.*, 141, 1639–1644, 2015.)

Histological grading were G1 ($n = 71$), G2 ($n = 27$), G3 ($n = 34$), undifferentiated "G4" ($n = 4$), and not specified ($n = 13$). Treatment modalities were surgical resection ($n = 102$), chemotherapy ($n = 49$), somatostatin analogs ($n = 39$), radiotherapy ($n = 22$), receptor-directed radionuclide therapy ($n = 12$), and systemic tyrosine kinase inhibition ($n = 5$). The authors concluded that NETs and NECs are not very rare tumors but, rather, are diagnosed with uneven frequencies. They also stated that chemotherapy, receptor-directed radionuclide application, and somatostatin analog therapy were often applied without a clear correlation with histologic grading.

Well-differentiated neuroendocrine tumors (WD-NETs) and poorly differentiated neuroendocrine carcinomas (PD-NECs) represent different families of neoplasms [110]. WD-NETs are relatively indolent, can evolve over many years, resemble non-neoplastic neuroendocrine cells, and produce neurosecretory proteins such as chromogranin A and synaptophysin. They primarily arise in the lungs and throughout the GI tract and pancreas, and are uncommon in the prostate gland. Surgical resection is the mainstay of therapy, and treatment of unresectable tumors depends on their site of origin. In contrast, PD-NECs often have alterations in P53 and Rb, exhibit an aggressive clinical course, and are treated with platinum-based chemotherapy. Only WD-NETs arise in patients with inherited neuroendocrine neoplasia syndromes, and some genetic alterations are site specific. The authors concluded that advances in understanding the molecular basis of NETs should lead to new diagnostic and therapeutic strategies.

Carcinoid tumors, a subset of NET, usually begin in the digestive tract (stomach, appendix, small intestine, colon, rectum) or in the lungs. They often do not cause signs and symptoms until late in the disease. Advanced carcinoid tumors can produce and release catecholamines or serotonin that cause signs and symptoms, collectively called a **carcinoid syndrome**, that include diarrhea, skin flushing, abdominal cramping, difficulty breathing, rapid heartbeat, and peripheral edema.

13.7.4 Dopamine in neuroendocrine tumors

Pheochromocytomas (PCCs) and **paragangliomas** (PGLs) constitute a special category within NETs [111–113]. PCCs are tumors of the adrenal medullary chromaffin cells, as well as of extra-adrenal chromaffin tissue that failed to involute after birth. PGLs originate in ganglia of the sympathetic nervous system. Both types of tumors are rather rare, and because they are characterized by catecholamine production, they have been diagnosed by measuring free or unconjugated catecholamine metabolites in plasma or urine. Until recently, attention has focused on overproduction of NE and epinephrine (Epi), with little interest in DA. However, increased DA production appears to be an independent predictor of malignancy in these tumors, and it may be associated with germline mutations of the succinate dehydrogenase subunit B (SDHB) gene [114]. DA was reported to inhibit angiogenesis and neovascularization in PGLs [111]. Notably, PC12 cells—which have been used in thousands of studies on catecholamine biosynthesis, metabolism, storage, and release—were derived from pheochromocytoma of rat adrenal medulla [115].

Immunoreactive D2R was evaluated in NET samples from 44 patients [116]. D2R was positive in 85% of the samples (100% of bronchial carcinoids and 93% of islet cell tumors), and its intensity in NETs was similar to that of pituitary samples. No differences in D2R expression were seen with respect to tumor grading, size, proliferative activity, presence of metastases, endocrine activity, or gender. However, D2R expression was significantly higher in the more aggressive tumors than in those without recurrence/progression of disease. The authors concluded that the high expression of D2R in NETs is clinically significant and that the ability of DA drugs to suppress NETs should be exploited. The unique secretory characteristics of NETs lend themselves to imaging by positron emission tomography (PET), which can target specific metabolic pathways or receptors [117,118]. Among these, C-11- or F-18-Dopa (Dopa PET), and radiolabeled DA analogs such as fluorine 18 (F18-DA), have been used to identify neuroendocrine GI tumors [119].

The human neuroendocrine pancreatic cell line BON and the neuroendocrine gut cell line STC-1 were used to study expression and functions the DA system in NET [117]. Both cell types expressed D2R and D5R as well as DAT. For functional receptor activation, intracellular cAMP levels were analyzed. DA and the D1/D5 receptor agonist SKF38393 increased cAMP in STC-1 cells, whereas DA or the D2-like agonist quinpirole decreased cAMP in BON cells. Functional DAT activity was, however, not detected in either cell line. The authors concluded that the presence of both DAR suggests autocrine and/or paracrine functions of DA in neuroendocrine GI tumor cells.

Interactions between somatostatin and DA receptors are of particular relevance to the diagnosis and treatment of NETs/NECs [107,120]. Both receptors belong to the G-protein-coupled receptor (GPCR) superfamily and share some structural and functional characteristics. Given the high expression of somatostatin receptors (SSTRs) in many NETs, somatostatin analogs have been incorporated into medical therapy for these tumors. Co-expression of SSTR and D2R was examined by immunohistochemistry in GEP neuroendocrine tumors (GEP-NETs) from 46 patients [121]. Overall, 76% of the tumors were positive for different types of SSTRs, whereas 37% were positive for D2R alone. Co-expression of SSTRs and D2R was seen in 88% of tumors. The authors concluded that the high rates of SSTR2A and SSTR2B and lower expression of SSTR5 in these tumors are of great importance for more accurate imaging, staging and targeted therapy of the disease. Moreover, the co-expression of SSTRs and D2R in a significant number of the cases offers a potential therapeutic alternative for GEP-NETs. A similar profile of DA and somatostatin receptor co-expression was reported in another study using quantitative polymerase chain reaction [118].

As summarized in a 2011 review [122], efforts have been undertaken to generate chimeric molecules that combine the structural features of both DA and somatostatin. The chimeras retain the ability to interact with receptors of both families and display enhanced potency and efficacy, as compared with those of individual SST or DAR agonists. The chimeric molecules showed excellent ability to suppress GH and PRL secretion from pituitary adenoma cells from acromegalic patients. They also suppressed adrenocorticotropic hormone (ACTH) secretion from Cushing's-causing corticotroph tumors and inhibited the proliferation of nonfunctioning pituitary adenoma cells. The chimeric SST/DA compounds were efficacious in suppressing both GH and IGF1 when tested in nonhuman primates, with no effect on either insulin secretion or glycemic control. Initial clinical studies employing subcutaneous administration of the chimeric BIM-23A760, revealed both prolonged circulating half-life and extended duration of biological effect. With chronic administration, however, BIM-23A760 produced a metabolite with DA activity that gradually accumulated and interfered with the activity of the parent compound. The authors concluded that efforts were underway to produce second-generation chimeras for treatment of neuroendocrine disease.

Several studies, using second-generation chimeras, produced mixed results. One study used a collection of sst/D2 chimeras: BIM 23A387, BIM 23A760, BIM 23A761, and BIM 23A78 [123]. When tested on primary GEP- NET cells, the chimeras did not provide an advantage over either lanreotide (an sst2 analog) or bromocriptine in their ability to decrease chromogranin A (CgA), gastrin, or serotonin release. The authors concluded that factors such as the respective levels of expression of SSTR and D2R, their organization in hetero-oligomers, or the functional correlates of these receptor aggregates underlie these tissue-specific differences, which remain to be elucidated.

Another study tested BIM-23A760 and BIM-23A758 in comparison with the selective SSTR2 agonist BIM-23023 and the selective D2R agonist BIM-53097 on human NET cell lines of heterogeneous origin [124]. Although having only minor effects on human pancreatic (BON1) and bronchus carcinoid cells (NCI-H727), BIM-23A758 induced significant antitumor effects in human midgut carcinoid cells (GOT1). These effects involved apoptosis induction as well as inhibition of MAPK and Akt signaling. Consistent with their antitumor response to BIM-23A758, GOT1 cells showed relatively high expression levels of SSTR2 and D2R mRNA. In particular, GOT1 cells highly expressed the short isoforms of D2R. In contrast to BIM-23A758, the SSTR2/D2R chimeric compound BIM-23A760 as well as the individual SSTR2 and D2R agonistic BIM-23023 and BIM-53097 had no or only minor antitumor activities in the NET cell lines. Taken together, the authors concluded that the SSTR2/D2R chimeric compound BIM-23A758 is a promising candidate for treating NETs that highly express SSTR2 and D2R. In particular, sufficient expression of the short D2R isoform might play a pivotal role in an effective treatment.

13.8 SYNOPSIS

DA is associated with a variety of tumors in different systems throughout the body. In the hematologic system, there is evidence for a role of DA in leukemia, lymphoma and multiple myeloma. Both DA and the DA and cyclic AMP-regulated phosphoprotein (DARPP or Dp32) and its truncated isoform, tDP, are expressed in GI-related tumors, including gastric, colorectal and hepatic. In pancreatic cancer, chimeric molecules, directed against both somatostatin and DARs, showed a therapeutic promise. Evidence derived primarily from studies with cultured cells, showed the involvement of DA and its receptors in lung, kidney and bladder cancers. Melanoma, the most aggressive skin cancer, shares a biosynthetic pathway in the production of melanin and DA. Older studies in the 1970s, using cultured cells, tumor-bearing mice, and small clinical trials, reported the efficacy of L-Dopa, in the suppression of melanoma. Activation of the cGMP/PKG axis was very effective in the inhibition of head and neck cancer, although a direct connection to the dopaminergic system is still missing. Within more than 120 types of brain and nervous system tumors, DA is involved in glioblastoma, neuroblastoma, meningiomas, and neuroendocrine tumors.

REFERENCES

1. Zhao XF, Reitz M, Chen QC, Stass S. Pathogenesis of early leukemia and lymphoma. *Cancer Biomark.* 2010;9(1–6):341–374.

2. Basso G, Case C, Dell'Orto MC. Diagnosis and genetic subtypes of leukemia combining gene expression and flow cytometry. *Blood Cells Mol Dis.* 2007;39(2):164–168.

3. Yu MG, Zheng HY. Acute myeloid leukemia: Advancements in diagnosis and treatment. *Chin Med J (Engl).* 2017;130(2):211–218.

4. Torkaman A, Charkari NM, Aghaeipour M. An approach for leukemia classification based on cooperative game theory. *Anal Cell Pathol (Amst).* 2011;34(5):235–246.

5. Pinoli M, Marino F, Cosentino M. Dopaminergic regulation of innate immunity: A review. *J Neuroimmune Pharmacol.* 2017;12(4):602–623.

6. Levite M. Dopamine and T cells: Dopamine receptors and potent effects on T cells, dopamine production in T cells, and abnormalities in the dopaminergic system in T cells in autoimmune, neurological and psychiatric diseases. *Acta Physiol (Oxf).* 2016;216(1):42–89.

7. Zhelev Z, et al. Phenothiazines suppress proliferation and induce apoptosis in cultured leukemic cells without any influence on the viability of normal lymphocytes. Phenothiazines and leukemia. *Cancer Chemother Pharmacol.* 2004;53(3):267–275.

8. Terasaka H, et al. Induction of apoptosis by dopamine in human oral tumor cell lines. *Anticancer Res.* 2000;20(1A):243–250.

9. Kobayashi H, Oikawa S, Kawanishi S. Mechanism of DNA damage and apoptosis induced by tetrahydropapaveroline, a metabolite of dopamine. *Neurochem Res.* 2006;31(4):523–532.

10. Fleischmann KK, Pagel P, Schmid I, Roscher AA. RNAi-mediated silencing of MLL-AF9 reveals leukemia-associated downstream targets and processes. *Mol Cancer*. 2014;13:27.

11. Singh JN, Chansouria JP, Singh VP, Udupa KN. Blood bioamines, cortisol and aminoacid levels in leukemic patients. *Indian J Cancer*. 1989;26(4):222–226.

12. Sachlos E, et al. Identification of drugs including a dopamine receptor antagonist that selectively target cancer stem cells. *Cell*. 2012;149(6):1284–1297.

13. Aslostovar L, et al. A phase 1 trial evaluating thioridazine in combination with cytarabine in patients with acute myeloid leukemia. *Blood Adv*. 2018;2(15):1935–1945.

14. Phillips AA, Owens C, Lee S, Bhagat G. An update on the management of peripheral T-cell lymphoma and emerging treatment options. *J Blood Med*. 2011;2:119–129.

15. Alberio T, et al. Proteomic characterization of Jurkat T leukemic cells after dopamine stimulation: A model of circulating dopamine-sensitive cells. *Biochimie*. 2011;93(5):892–898.

16. Rohr O, Sawaya BE, Lecestre D, Aunis D, Schaeffer E. Dopamine stimulates expression of the human immunodeficiency virus type 1 via NF-kappaB in cells of the immune system. *Nucleic Acids Res*. 1999;27(16):3291–3299.

17. Basu B, et al. D1 and D2 dopamine receptor-mediated inhibition of activated normal T cell proliferation is lost in jurkat T leukemic cells. *J Biol Chem*. 2010;285(35):27026–27032.

18. Mossalayi MD, Becherel PA, Debre P. Critical role of nitric oxide during the apoptosis of peripheral blood leukocytes from patients with AIDS. *Mol Med*. 1999;5(12):812–819.

19. Graham C, Barsoum I, Kim J, Black M, Siemens RD. Mechanisms of hypoxia-induced immune escape in cancer and their regulation by nitric oxide. *Redox Biol*. 2015;5:417–429.

20. Iqbal J, et al. Genomic signatures in B-cell lymphoma: How can these improve precision in diagnosis and inform prognosis? *Blood Rev*. 2016;30(2):73–88.

21. Gifford GK, Gill AJ, Stevenson WS. Molecular subtyping of diffuse large B-cell lymphoma: Update on biology, diagnosis and emerging platforms for practising pathologists. *Pathology*. 2016;48(1):5–16.

22. Meredith EJ, et al. Dopamine targets cycling B cells independent of receptors/transporter for oxidative attack: Implications for non-Hodgkin's lymphoma. *Proc Natl Acad Sci USA*. 2006;103(36):13485–13490.

23. Braesch-Andersen S, Paulie S, Stamenkovic I. Dopamine-induced lymphoma cell death by inhibition of hormone release. *Scand J Immunol*. 1992;36(4):547–553.

24. Sarkar C, Chakroborty D, Basu S. Dopamine inhibits growth of multiple myeloma lymphoma. *Cancer Res* 2016;76(Suppl 14):3278–3278.

25. Wang V, Chao TH, Hsieh CC, Lin CC, Kao CH. Cancer risks among the users of ergot-derived dopamine agonists for Parkinson's disease, a nationwide population-based survey. *Parkinsonism Relat Disord*. 2015;21(1):18–22.

26. Lin PY, et al. Association between Parkinson disease and risk of cancer in Taiwan. *JAMA Oncol*. 2015;1(5):633–640.

27. Human cancers by organs. https://publications.iarc.fr/_publications/media/download/4100/67cd70452750498a27d50c2f1fa6e047121a020d.pdf 2003:181–269.

28. Basu S, Dasgupta PS. Alteration of dopamine D2 receptors in human malignant stomach tissue. *Dig Dis Sci*. 1997;42(6):1260–1264.

29. Chakroborty D, et al. Depleted dopamine in gastric cancer tissues: Dopamine treatment retards growth of gastric cancer by inhibiting angiogenesis. *Clin Cancer Res*. 2004;10(13):4349–4356.

30. Ganguly S, et al. Dopamine, by acting through its D2 receptor, inhibits insulin-like growth factor-I (IGF-I)-induced gastric cancer cell proliferation via up-regulation of Kruppel-like factor 4 through down-regulation of IGF-IR and AKT phosphorylation. *Am J Pathol*. 2010;177(6):2701–2707.

31. Mu J, et al. Dopamine receptor D2 is correlated with gastric cancer prognosis. *Oncol Lett*. 2017;13(3):1223–1227.

32. Leng ZG, et al. Activation of DRD5 (dopamine receptor D5) inhibits tumor growth by autophagic cell death. *Autophagy*. 2017;13(8):1–16.

33. El-Rifai W, et al. Gastric cancers overexpress DARPP-32 and a novel isoform, t-DARPP. *Cancer Res*. 2002;62(14):4061–4064.

34. Beckler A, et al. Overexpression of the 32-kilodalton dopamine and cyclic adenosine 3′,5′-monophosphate-regulated phosphoprotein in common adenocarcinomas. *Cancer*. 2003;98(7):1547–1551.

35. Mukherjee K, et al. Dopamine and cAMP regulated phosphoprotein MW 32 kDa is overexpressed in early stages of gastric tumorigenesis. *Surgery*. 2010;148(2):354–363.

36. Belkhiri A, Dar AA, Zaika A, Kelley M, El-Rifai W. t-Darpp promotes cancer cell survival by up-regulation of Bcl2 through Akt-dependent mechanism. *Cancer Res*. 2008;68(2):395–403.

37. Hong L, et al. Reversal of multidrug resistance of vincristine-resistant gastric adenocarcinoma cells through up-regulation of DARPP-32. *Cell Biol Int*. 2007;31(9):1010–1015.

38. Hong L, et al. DARPP-32 mediates multidrug resistance of gastric cancer through regulation of P-gp and ZNRD1. *Cancer Invest*. 2007;25(8):699–705.

39. Zhu S, Belkhiri A, El-Rifai W. DARPP-32 increases interactions between epidermal growth factor receptor and ERBB3 to promote tumor resistance to gefitinib. *Gastroenterology*. 2011;141(5):1738–1748.

40. Belkhiri A, Zhu S, Chen Z, Soutto M, El-Rifai W. Resistance to TRAIL is mediated by DARPP-32 in gastric cancer. *Clin Cancer Res.* 2012;18(14):3889–3900.

41. Zhu S, et al. Regulation of CXCR4-mediated invasion by DARPP-32 in gastric cancer cells. *Mol Cancer Res.* 2013;11(1):86–94.

42. Zhu S, et al. Activation of IGF1R by DARPP-32 promotes STAT3 signaling in gastric cancer cells. *Oncogene.* 2019;38(29):5805–5816.

43. Subramaniam R, Mizoguchi A, Mizoguchi E. Mechanistic roles of epithelial and immune cell signaling during the development of colitis-associated cancer. *Cancer Res Front.* 2016;2(1):1–21.

44. Gustafson CE, Young J, Leggett B, Searle J, Chenevix-Trench G. Loss of heterozygosity on the long arm of chromosome 11 in colorectal tumours. *Br J Cancer.* 1994;70(3):395–397.

45. Gemignani F, et al. Polymorphisms of the dopamine receptor gene DRD2 and colorectal cancer risk. *Cancer Epidemiol Biomarkers Prev.* 2005;14(7):1633–1638.

46. Boursi B, Mamtani R, Haynes K, Yang YX. Parkinson's disease and colorectal cancer risk-A nested case control study. *Cancer Epidemiol.* 2016;43:9–14.

47. Kopljar M, et al. High expression of DARPP-32 in colorectal cancer is associated with liver metastases and predicts survival for dukes A and B patients: Results of a pilot study. *Int Surg.* 2015;100(2):213–220.

48. Browning DD. Protein kinase G as a therapeutic target for the treatment of metastatic colorectal cancer. *Expert Opin Ther Targets.* 2008;12(3):367–376.

49. Hou Y, et al. An anti-tumor role for cGMP-dependent protein kinase. *Cancer Lett.* 2006;240(1):60–68.

50. Li N, et al. Sulindac selectively inhibits colon tumor cell growth by activating the cGMP/PKG pathway to suppress Wnt/beta-catenin signaling. *Mol Cancer Ther.* 2013;12(9):1848–1859.

51. Amini A, Masoumi MS, Morris DL, Pourgholami MH. The critical role of vascular endothelial growth factor in tumor angiogenesis. *Curr Cancer Drug Targets.* 2012;12(1):23–43.

52. Sarkar C, Chakroborty D, Dasgupta PS, Basu S. Dopamine is a safe antiangiogenic drug which can also prevent 5-fluorouracil induced neutropenia. *Int J Cancer.* 2015;137(3):744–749.

53. Sarkar C, Chakroborty D, Chowdhury UR, Dasgupta PS, Basu S. Dopamine increases the efficacy of anticancer drugs in breast and colon cancer preclinical models. *Clin Cancer Res.* 2008;14(8):2502–2510.

54. Jain RK. Normalization of tumor vasculature: An emerging concept in antiangiogenic therapy. *Science.* 2005;307(5706):58–62.

55. Chakroborty D, et al. Dopamine stabilizes tumor blood vessels by up-regulating angiopoietin 1 expression in pericytes and Kruppel-like factor-2 expression in tumor endothelial cells. *Proc Natl Acad Sci USA.* 2011;108(51):20730–20735.

56. Tu T, Buhler S, Bartenschlager R. Chronic viral hepatitis and its association with liver cancer. *Biol Chem.* 2017;398(8):817–837.

57. Lu M, et al. Roles of dopamine receptors and their antagonist thioridazine in hepatoma metastasis. *Onco Targets Ther.* 2015;8:1543–1552.

58. Xu JJ, et al. Dopamine D1 receptor activation induces dehydroepiandrosterone sulfotransferase (SULT2A1) in HepG2 cells. *Acta Pharmacol Sin.* 2014;35(7):889–898.

59. Liu XF, Long HJ, Miao XY, Liu GL, Yao HL. Fisetin inhibits liver cancer growth in a mouse model: Relation to dopamine receptor. *Oncol Rep.* 2017;38(1):53–62.

60. Muntane J, De la Rosa AJ, Marin LM, Padillo FJ. Nitric oxide and cell death in liver cancer cells. *Mitochondrion.* 2013;13(3):257–262.

61. Karanikas M, et al. Pancreatic cancer from molecular pathways to treatment opinion. *J Cancer.* 2016;7(10):1328–1339.

62. Eisenhofer G, et al. Substantial production of dopamine in the human gastrointestinal tract. *J Clin Endocrinol Metab.* 1997;82(11):3864–3871.

63. Mezey E, et al. A novel nonneuronal catecholaminergic system: Exocrine pancreas synthesizes and releases dopamine. *Proc Natl Acad Sci USA.* 1996;93(19):10377–10382.

64. O'Toole D, et al. The analysis of quantitative expression of somatostatin and dopamine receptors in gastro-entero-pancreatic tumours opens new therapeutic strategies. *Eur J Endocrinol.* 2006;155(6):849–857.

65. Baldelli R, et al. Somatostatin analogs therapy in gastroenteropancreatic neuroendocrine tumors: Current aspects and new perspectives. *Front Endocrinol (Lausanne).* 2014;5:1–10.

66. Jandaghi P, et al. Expression of DRD2 is increased in human pancreatic ductal adenocarcinoma and inhibitors slow tumor growth in mice. *Gastroenterology.* 2016;151(6):1218–1231.

67. Ishibashi M, et al. Inhibition of growth of human small cell lung cancer by bromocriptine. *Cancer Res.* 1994;54(13):3442–3446.

68. Hoeppner LH, et al. Dopamine D2 receptor agonists inhibit lung cancer progression by reducing angiogenesis and tumor infiltrating myeloid derived suppressor cells. *Mol Oncol.* 2015;9(1):270–281.

69. Wu XY, et al. Overexpressed D2 dopamine receptor inhibits non-small cell lung cancer progression through inhibiting NF-kappaB signaling pathway. *Cell Physiol Biochem.* 2018;48(6):2258–2272.

70. Saha B, Mondal AC, Basu S, Dasgupta PS. Circulating dopamine level, in lung carcinoma patients, inhibits proliferation and cytotoxicity of CD4+ and CD8+ T cells by D1 dopamine receptors: An in vitro analysis. *Int Immunopharmacol.* 2001;1(7):1363–1374.

71. Cherubini E, et al. Genetic and functional analysis of polymorphisms in the human dopamine receptor and transporter genes in small cell lung cancer. *J Cell Physiol.* 2016;231(2):345–356.

72. Wong JC, Bathina M, Fiscus RR. Cyclic GMP/protein kinase G type-Ialpha (PKG-Ialpha) signaling pathway promotes CREB phosphorylation and maintains higher c-IAP1, livin, survivin, and Mcl-1 expression and the inhibition of PKG-Ialpha kinase activity synergizes with cisplatin in non-small cell lung cancer cells. *J Cell Biochem.* 2012;113(11):3587–3598.

73. Papadopoulos EI, et al. L-DOPA decarboxylase mRNA levels provide high diagnostic accuracy and discrimination between clear cell and non-clear cell subtypes in renal cell carcinoma. *Clin Biochem.* 2015;48(9):590–595.

74. Hansson J, et al. Overexpression of functional SLC6A3 in clear cell renal cell carcinoma. *Clin Cancer Res.* 2017;23(8):2105–2115.

75. Cheng X, et al. Metabolomics of non-muscle invasive bladder cancer: Biomarkers for early detection of bladder cancer. *Front Oncol.* 2018;8:494–505.

76. Clague J, Cinciripini P, Blalock J, Wu X, Hudmon KS. The D2 dopamine receptor gene and nicotine dependence among bladder cancer patients and controls. *Behav Genet.* 2010;40(1):49–58.

77. Saladi RN, Persaud AN. The causes of skin cancer: A comprehensive review. *Drugs Today (Barc).* 2005;41(1):37–53.

78. Volkovova K, Bilanicova D, Bartonova A, Letasiova S, Dusinska M. Associations between environmental factors and incidence of cutaneous melanoma: Review. *Environ Health.* 2012;11(Suppl 1):S12.

79. Garbe C, Eigentler TK, Keilholz U, Hauschild A, Kirkwood JM. Systematic review of medical treatment in melanoma: Current status and future prospects. *Oncologist.* 2011;16(1):5–24.

80. Huang P, Yang XD, Chen SD, Xiao Q. The association between Parkinson's disease and melanoma: A systematic review and meta-analysis. *Transl Neurodegener.* 2015;4:21–31.

81. Pan T, Li X, Jankovic J. The association between Parkinson's disease and melanoma. *Int J Cancer.* 2011;128(10):2251–2260.

82. Herrero HE. Pigmentation genes link Parkinson's disease to melanoma, opening a window on both etiologies. *Med Hypotheses.* 2009;72(3):280–284.

83. Pan T, Zhu J, Hwu WJ, Jankovic J. The role of alpha-synuclein in melanin synthesis in melanoma and dopaminergic neuronal cells. *PLoS One.* 2012;7(9):e45183.

84. Wick MM, Kramer RA, Gorman M. Enhancement of L-dopa incorporation into melanoma by dopa decarboxylase inhibition. *J Invest Dermatol.* 1978;70(6):358–360.

85. Wick MM. Dopamine: A novel antitumor agent active against B-16 melanoma in vivo. *J Invest Dermatol.* 1978;71(2):163–164.

86. Wick MM. 3,4-Dihydroxybenzylamine: A dopamine analog with enhanced antitumor activity against B16 melanoma. *J Natl Cancer Inst.* 1979;63(6):1465–1467.

87. Wick MM. Levodopa and dopamine analogs as DNA polymerase inhibitors and antitumor agents in human melanoma. *Cancer Res.* 1980;40(5):1414–1418.

88. Wick MM. Therapeutic effect of dopamine infusion on human malignant melanoma. *Cancer Treat Rep.* 1982;66(8):1657–1659.

89. Fitzgerald GB, Wick MM. 3,4-Dihydroxybenzylamine: An improved dopamine analog cytotoxic for melanoma cells in part through oxidation products inhibitory to DNA polymerase. *J Invest Dermatol.* 1983;80(2):119–123.

90. Wick MM. The chemotherapy of malignant melanoma. *J Invest Dermatol.* 1983;80(1 Suppl):61s–62s.

91. Wick MM. Levodopa/dopamine analogs as inhibitors of DNA synthesis in human melanoma cells. *J Invest Dermatol.* 1989;92(5 Suppl):329S–331S.

92. Leemans CR, Braakhuis BJ, Brakenhoff RH. The molecular biology of head and neck cancer. *Nat Rev Cancer.* 2011;11(1):9–22.

93. Denaro N, Russi EG, Adamo V, Merlano MC. State-of-the-art and emerging treatment options in the management of head and neck cancer: News from 2013. *Oncology.* 2014;86(4):212–229.

94. Geomela PA, Kontos CK, Yiotakis I, Fragoulis EG, Scorilas A. L-DOPA decarboxylase mRNA expression is associated with tumor stage and size in head and neck squamous cell carcinoma: A retrospective cohort study. *BMC Cancer.* 2012;12:484–493.

95. Ebihara Y, et al. DARPP-32 expression arises after a phase of dysplasia in oesophageal squamous cell carcinoma. *Br J Cancer.* 2004;91(1):119–123.

96. Tuttle TR, Mierzwa ML, Wells SI, Fox SR, Ben-Jonathan N. The cyclic GMP/protein kinase G pathway as a therapeutic target in head and neck squamous cell carcinoma. *Cancer Lett.* 2016;370(2):279–285.

97. Tuttle TR, Takiar V, Kumar B, Kumar P, Ben-Jonathan N. Soluble guanylate cyclase stimulators increase sensitivity to cisplatin in head and neck squamous cell carcinoma cells. *Cancer Lett.* 2017;389:33–40.

98. Serafini P, et al. Phosphodiesterase-5 inhibition augments endogenous antitumor immunity by reducing myeloid-derived suppressor cell function. *J Exp Med.* 2006;203(12):2691–2702.

99. Califano JA, et al. Tadalafil augments tumor specific immunity in patients with head and neck squamous cell carcinoma. *Clin Cancer Res.* 2015;21(1):30–38.

100. Weed DT, et al. Tadalafil reduces myeloid-derived suppressor cells and regulatory T cells and promotes tumor immunity in patients with head and neck squamous cell carcinoma. *Clin Cancer Res.* 2015;21(1):39–48.

101. Caragher SP, Hall RR, Ahsan R, Ahmed AU. Monoamines in glioblastoma: Complex biology with therapeutic potential. *Neuro Oncol.* 2018;20(8):1014–1025.

102. Zong H, Verhaak RG, Canoll P. The cellular origin for malignant glioma and prospects for clinical advancements. *Expert Rev Mol Diagn.* 2012;12(4):383–394.

103. Trott G, et al. Abundant immunohistochemical expression of dopamine D2 receptor and p53 protein in meningiomas: Follow-up, relation to gender, age, tumor grade, and recurrence. *Braz J Med Biol Res.* 2015;48(5):415–419.

104. Schrell UM, Fahlbusch R, Adams EF, Nomikos P, Reif M. Growth of cultured human cerebral meningiomas is inhibited by dopaminergic agents. Presence of high affinity dopamine-D1 receptors. *J Clin Endocrinol Metab.* 1990;71(6):1669–1671.

105. Triche TJ. Neuroblastoma and other childhood neural tumors: A review. *Pediatr Pathol.* 1990;10(1–2):175–193.

106. Lam L, Woollard GA, Teague L, Davidson JS. Clinical validation of urine 3-methoxytyramine as a biomarker of neuroblastoma and comparison with other catecholamine-related biomarkers. *Ann Clin Biochem.* 2017;54(2):264–272.

107. Hofland LJ, Vandamme T, Albertelli M, Ferone D. Hormone and receptor candidates for target and biotherapy of neuroendocrine tumors. *Front Horm Res.* 2015;44:216–238.

108. Gaal J, de Krijger RR. Neuroendocrine tumors and tumor syndromes in childhood. *Pediatr Dev Pathol.* 2010;13(6):427–441.

109. Maschmeyer G, et al. A retrospective review of diagnosis and treatment modalities of neuroendocrine tumors (excluding primary lung cancer) in 10 oncological institutions of the East German Study Group of Hematology and Oncology (OSHO), 2010–2012. *J Cancer Res Clin Oncol.* 2015;141(9):1639–1644.

110. Klimstra DS, Beltran H, Lilenbaum R, Bergsland E. The spectrum of neuroendocrine tumors: Histologic classification, unique features and areas of overlap. *Am Soc Clin Oncol Educ Book.* 2015:92–103.

111. Osinga TE, et al. Emerging role of dopamine in neovascularization of pheochromocytoma and paraganglioma. *FASEB J.* 2017;31(6):2226–2240.

112. Jimenez C, et al. Current and future treatments for malignant pheochromocytoma and sympathetic paraganglioma. *Curr Oncol Rep.* 2013;15(4):356–371.

113. Darr R, et al. Accuracy of recommended sampling and assay methods for the determination of plasma-free and urinary fractionated metanephrines in the diagnosis of pheochromocytoma and paraganglioma: A systematic review. *Endocrine.* 2017;56(3):495–503.

114. Osinga TE, et al. Catecholamine-synthesizing enzymes are expressed in parasympathetic head and neck paraganglioma tissue. *Neuroendocrinology.* 2015;101(4):289–295.

115. Greene LA, Tischler AS. Establishment of a noradrenergic clonal line of rat adrenal pheochromocytoma cells which respond to nerve growth factor. *Proc Natl Acad Sci USA.* 1976;73(7):2424–2428.

116. Grossrubatscher E, et al. High expression of dopamine receptor subtype 2 in a large series of neuroendocrine tumors. *Cancer Biol Ther.* 2008;7(12):1970–1978.

117. Lemmer K, et al. Expression of dopamine receptors and transporter in neuroendocrine gastrointestinal tumor cells. *Life Sci.* 2002;71(6):667–678.

118. Saveanu A, et al. Expression of somatostatin receptors, dopamine D(2) receptors, noradrenaline transporters, and vesicular monoamine transporters in 52 pheochromocytomas and paragangliomas. *Endocr Relat Cancer.* 2011;18(2):287–300.

119. Martiniova L, et al. Usefulness of [18F]-DA and [18F]-DOPA for PET imaging in a mouse model of pheochromocytoma. *Nucl Med Biol.* 2012;39(2):215–226.

120. Gatto F, Hofland LJ. The role of somatostatin and dopamine D2 receptors in endocrine tumors. *Endocr Relat Cancer.* 2011;18(6):R233–R251.

121. Diakatou E, et al. Somatostatin and dopamine receptor profile of gastroenteropancreatic neuroendocrine tumors: An immunohistochemical study. *Endocr Pathol.* 2011;22(1):24–30.

122. Culler MD. Somatostatin-dopamine chimeras: A novel approach to treatment of neuroendocrine tumors. *Horm Metab Res.* 2011;43(12):854–857.

123. Couvelard A, et al. Antisecretory effects of chimeric somatostatin/dopamine receptor ligands on gastroenteropancreatic neuroendocrine tumors. *Pancreas.* 2017;46(5):631–638.

124. Zitzmann K, et al. The novel somatostatin receptor 2/dopamine type 2 receptor chimeric compound BIM-23A758 decreases the viability of human GOT1 midgut carcinoid cells. *Neuroendocrinology.* 2013;98(2):128–136.

Glossary

AAP (atypical antipsychotics) Also known as second generation antipsychotics, are a group of drugs used to treat primarily psychiatric conditions such as schizophrenia and bipolar disorder, but also autism, anxiety disorder, obsessive-compulsive disorder and others. The AAPs interact with the serotonin, norepinephrine, dopamine and acetylcholine receptors. The best-known AAP include aripiprazole, olanzapine, ziprasidone, clozapine, quetiapine, paliperidone, and risperidone.

acrosome reaction The reaction that occurs in the acrosome of the sperm as it approaches the egg. The acrosome is a cap-like structure over the anterior half of the sperm's head. As the sperm approaches the zona pellucida of the egg, the membrane surrounding the acrosome fuses with the plasma membrane of the sperm's head, exposing the contents of the acrosome. The contents include surface antigens necessary for binding to the egg's cell membrane, and numerous enzymes which are responsible for breaking through the egg's tough coating and allowing fertilization to proceed.

ACTH (adrenocorticotropic hormone) A polypeptide anterior pituitary hormone. It is an important component of the hypothalamo-pituitary-adrenal (HPA) axis and is often produced in response to biological stress. ACTH is positively regulated by hypothalamic CRH (corticotropin releasing hormone) and negatively by cortisol. Its principal effects are increased production and release of cortisol by the adrenal cortex. ACTH is also related to the circadian rhythm in many organisms. Diseases associated with elevated ACTH include adrenal insufficiency (Addison's disease) and Cushing's syndrome.

adrenal gland Adrenal glands, located above the kidneys, are composed of an outer cortex which produces steroid hormones such as aldosterone and cortisol, and an inner medulla which produces catecholamines (dopamine, norepinephrine, and epinephrine). The cortex is divided into three zones: zona glomerulosa, zona fasciculate, and zona reticularis.

amphetamines CNS stimulants used to treat attention deficit hyperactivity disorder, narcolepsy and obesity. At therapeutic doses, they cause euphoria, desire for sex, increased wakefulness, and improved cognitive control. Larger doses may impair cognitive functions, and carry a great risk of psychosis and other serious side effects. Addiction to amphetamines, resulting from heavy recreational use, is a severe health risk which does not occur from long-term medical use at therapeutic doses. AMPH is a DA releaser that elevates DA by three mechanisms: (a) as a competitive inhibitor of DA uptake by the DA transporter (DAT); (b) by facilitating DA move out of vesicles, and (c) by promoting DAT-mediated reverse-transport of DA into the synaptic cleft independently of action-potential-induced vesicular release.

AMPT (α-Methyl-p-tyrosine) A non-endogenous drug involved in the catecholamine biosynthetic pathway. AMPT inhibits tyrosine hydroxylase by competing with tyrosine at the tyrosine-binding site. AMPT has been used in the treatment of pheochromocytoma, and it also inhibits the production of melanin.

angiogenesis Formation of new blood vessels from preexisting ones. Angiogenesis continues the growth of the vasculature by sprouting and splitting. A large number of angiogenic and antiangiogenic factors work in a highly coordinated manner to induce endothelial cell outgrowth and the formation of functional vessels.

arcuate nucleus An aggregation of neurons in the mediobasal hypothalamus, adjacent to the third ventricle and median eminence. The central role of the arcuate nucleus is with homeostasis (i.e., regulation of feeding, metabolism, fertility, and cardiovascular functions). It contains diverse populations of neurons that mediate different physiological functions, including neuroendocrine neurons, dopaminergic neurons, centrally projecting neurons, and astrocytes. The arcuate nucleus is also responsible for integrating information and providing inputs to other hypothalamic nuclei or inputs to areas outside this region of the brain.

β-Arrestins A small family of proteins, important for the regulation of signal transduction at G protein-coupled receptors (GPCR). The arrestins block GPCR coupling to G proteins in two ways. One, by occluding the binding site of the receptor for heterotrimeric G-protein, preventing its activation (desensitization). Two, by linking the receptor to the internalization machinery: clathrin and clathrin adaptor, which promotes receptor internalization via coated pits and subsequent transport to internal compartments, called endosomes.

ARSA (arylsulfatase A) A sulfatase enzyme that breaks down sulfatides such as dopamine sulfate and cerebroside-sulfate. ARSA is expressed in several peripheral organs and can be released from the lysosomes and is responsible for converting extracellular DA-sulfate to bioactive dopamine. A deficiency in ARSA is associated with metachromatic leukodystrophy, an autosomal recessive disease.

autoreceptors Receptors located in the membrane of presynaptic nerve cells that serve as part of a negative feedback loop in signal transduction. They can be found in any part of the cell membrane: dendrites, cell body, axon, or axon terminals.

AVP (arginine vasopressin) Also called antidiuretic hormone (ADH), is a nanopeptide. It is synthesized as a prohormone in hypothalamic magnocellular neurons, and is processed as it travels down the neurohypophysial tract to the posterior pituitary. AVP, released from storage vesicles in response to hyperosmolarity, has two main functions: (a) to increase the reabsorption of solute-free water into the circulation from the filtrate in the kidney nephron tubules, and (b) to constrict blood vessels, which increases peripheral vascular resistance and raises arterial blood pressure. Some AVP, released directly into the brain from the hypothalamus, plays a role in social behavior, sexual motivation, and pair bonding.

blood–brain barrier (BBB) A highly selective semipermeable border that separates the circulating blood from the brain and extracellular fluid in the CNS. It is formed by the endothelial cells of the capillary wall, astrocytes, and pericytes. The BBB allows the passage of some molecules by passive diffusion, as well as by a selective transport of substances such as glucose, water, and amino acids, which are critical to neural function.

carotid body A small cluster of chemoreceptors, located in the bifurcation of the common carotid artery. The carotid body is made up of two types of cells: glomus cells and sustentacular supporting cells. Glomus type I cells are derived from the neural crest. They release several neurotransmitters, including dopamine and acetylcholine, that trigger excitatory postsynaptic potentials in synapsed neurons leading to respiratory center in the brain. They are innervated by the glossopharyngeal nerve. Glomus type II cells resemble glial cells, express the glial marker S100, and act as supporting cells. The carotid body detects changes in the composition of arterial blood flowing through it, mainly the partial pressure of arterial oxygen, but also of carbon dioxide. It is also sensitive to changes in blood pH and temperature. Feedback from the carotid body is sent to cardiorespiratory centers in the medulla oblongata via the glossopharengeal nerve.

catecholestrogens Estrogens that contain a catechol (1,2-dihydroxybenzene) group within their structure. They are endogenous metabolites of estradiol and estrone, and include 2 hydroxylated (e.g., 2-hydroxyestradiol and 2-hydroxyestrone) and 4-hydroxylated (e.g., 4-hydroxestradiol and 4-hydroxestrone) compounds.

CCK (cholecystokinin) Also called pancreozymin, is a peptide hormone of the gastrointestinal (GI) system, responsible for stimulating the digestion of fat and proteins. CCK is synthesized and secreted by enteroendocrine cells in the duodenum, the first segment of the small intestine. It induces the release of digestive enzymes and bile from the pancreas and gallbladder, respectively. In the brain, CCK acts as a hunger suppressant.

chromaffin cells Neuroendocrine cells, found primarily in the adrenal medulla, which produce catecholamines and other neuroactive substances. They are located in close proximity to presynaptic sympathetic ganglia, with which they communicate, and are structurally similar to postsynaptic sympathetic neurons. They serve a variety of functions such as the response to stress, monitoring of carbon dioxide and oxygen concentrations, maintenance of respiration, and blood pressure regulation. In mammals, the largest extra-adrenal cluster of chromaffin cells is the organ of Zuckerkandl.

cocaine A strong stimulant, frequently used as a recreational drug taken by snorting, inhalation or intravenous injections. It can easily cross the blood–brain barrier and may lead to breakdown of the barrier. Mental effects include loss of contact with reality, intense feeling of happiness, or agitation. Physical symptoms include fast heart rate, sweating, and enlarged pupils. High doses can raise the blood pressure or body temperature. Cocaine acts by inhibiting the reuptake of dopamine, norepinephrine and serotonin, resulting in increased concentrations of these neurotransmitters in the brain.

COMT (catechol-O-methyltransferase) Enzymes that degrade substances having a catechol structure, that include catecholamines, catecholestrogens and various drugs. The enzyme acts by introducing a methyl group, donated by S-adenosyl-methionine (SAM), to the catecholamine. Two isoforms of COMT are produced: a soluble short form (S-COMT) and a membrane-bound long form (MB-COMT). Several pharmaceutical drugs target COMT to alter its activity and therefore the availability of catecholamines.

CRH (corticotrophin releasing hormone) A peptide hormone composed of 41 residues, which is secreted by the paraventricular nucleus (PVN) of the hypothalamus in response to stress. Its main function is the stimulation of the pituitary synthesis of ACTH, as part of the hypothalamo-pituitary-adrenal (HPA) axis. CRH is also synthesized in peripheral tissues, such as T lymphocytes, and is highly expressed in the placenta where CRH is a marker that determines the length of gestation and the timing of parturition and delivery.

CSF (cerebrospinal fluid) A clear colorless fluid within the subarachnoid space and the ventricular system in the brain and spinal cord. It is produced by specialized ependymal cells in the choroid plexuses of the ventricles. At any given time, there is about 125 mL of CSF, with 500 mL generated per day. The CSF acts as a cushion or buffer, provides mechanical and immunological protection to the brain, and participates in the autoregulation of the cerebral blood flow and the distribution of some neuroactive substances.

CTX (cholera toxin) A multimeric protein complex secreted by the bacterium *Vibrio cholerae*. CTX is responsible for the massive, watery diarrhea of cholera infection. It acts by keeping Gαs in an activated state, leading to increased adenylate cyclase activity and elevating intracellular cAMP to more than 100-fold over normal. Activated PKA phosphorylates the cystic fibrosis transmembrane conductance regulator (CFTR) chloride channel. This leads to ATP-mediated efflux of chloride ions, and secretion of H_2O, Na^+, K^+ and HCO_3^- into the intestinal lumen. The rapid fluid loss from the intestine, up to 2 liters per hour, results in the severe dehydration and other factors associated with cholera infection.

DAR (dopamine receptors) A class of G protein-coupled receptors (GPCR) that activate various effectors through G-protein coupling. There are five DAR subtypes: D1, D2, D3, D4, and D5. The D1 and D5 (D1-like) receptors are coupled to $Gs\alpha$ and $Golf$, which activate adenylate cyclase and increase intracellular cAMP concentration. The D2-like receptors are coupled to $Gi\alpha$, which suppresses cAMP synthesis by inhibiting adenylate cyclase. The DAR are involved in many neurological processes, including motivation, pleasure, cognition, memory, learning, and fine motor control, as well as in the modulation of neuroendocrine signaling. Abnormal DAR signaling is implicated in several neuropsychiatric disorders, and serve as common neurologic drug targets. Antipsychotics are often DAR antagonists while psychostimulants are typically indirect agonists.

DAR KO mice Several laboratories have used gene targeting via homologous recombination to generate mice deficient in each of the five DAR subtypes. Phenotypic analysis of these mice has been instrumental in identifying the role of dopamine receptor subtypes in mediating dopamine's effects on motor function, cognition, reward, and emotional behaviors.

DARPP-32 (32-kDa dopamine and cAMP-regulated phosphoprotein) A multifunctional phosphoprotein with a protein phosphatase 1 (PP1) inhibitory function. DARPP-32 is activated by PKA- and cyclin-dependent kinase 5-induced phosphorylation. It plays a critical role in tumorigenesis and is overexpressed in breast, prostate, colon, and stomach cancers.

DAT (dopamine transporter) A membrane-spanning transporter that pumps dopamine out of the synaptic cleft back into the cytosol. It is composed of twelve transmembrane domains (TMD) with a large extracellular loop between the third and fourth TMDs. Dopamine reuptake via DAT is the main mechanism through which it is cleared from synapses. DAT acts as a symporter, requiring the sequential binding and cotransport of two Na^+ ions and one Cl^- with the dopamine substrate. The driving force for dopamine reuptake is the concentration gradient generated by the plasma membrane NA^+/K^+ ATPase.

DAT-KO Mice deficient in the dopamine transporter (DAT) have a pronounced behavioral phenotype resulting from the permanent elevation of extracellular DA. They have hyper-locomotion, disrupted sensorimotor gating, increased impulsivity, and impaired ability to perform learning tasks. The hyperdopaminergic phenotype of these mice has been used to study the effects of pharmacological agents on DA-related functions and behaviors. They also serve as a model for understanding the pathology and pharmacology of dopamine-related disorders such as schizophrenia, bipolar disorder, Parkinson's disease, and addiction.

DBH (dopamine beta-hydroxylase) The third enzyme in the biosynthetic pathway of catecholamines, which converts dopamine to norepinephrine. DBH is expressed in noradrenergic nerve terminals of the central and peripheral nervous systems, as well as in chromaffin cells of the adrenal medulla and in few other peripheral cells. During exocytosis, DBH is released together with catecholamines from adrenal chromaffin cells. In addition to DA, DBH hydroxylates trace amines (e.g., p-tyramine to p-octopamine) and also participates in the metabolism of xenobiotics such as amphetamine.

DDC (dopa decarboxylase) Also known as aromatic L-amino acid decarboxylase, the enzyme generates dopamine from Dopa by removing the carboxyl moiety from the catechol side chain. In addition to catecholamines, DDC catalyzes the biosynthesis of serotonin (5HT) from 5-hydroxytryptophan (5-HTP), histamine from L-histidine, as well as the synthesis of some trace amines.

domperidone A selective peripheral D2R antagonist used as an anti-mimetic, gastric prokinetic agent, and galactagogue. The drug relieves nausea, enhances transit of food through the stomach by increasing peristalsis in the GI tract, and promotes lactation by increasing the release of prolactin.

DOPAC (dihydroxyphenylacetic acid) DOPAC is a metabolite of dopamine resulting from catalysis by monoamine oxidase (MAO). It can be oxidized by hydrogen peroxide to form toxic metabolites which destroy dopamine storage vesicles in the substantia nigra, which contributes to a failure of levodopa treatment of Parkinson's disease. This can be prevented by a concomitant application of MAO inhibitors such as selegiline or rasagiline.

endocrine disrupting chemicals (EDC) Found in many household and industrial products, EDCs are substances that interfere with the synthesis, secretion, transport, binding, action, or elimination of natural hormones that are responsible for development, behavior, fertility, and maintenance of homeostasis. Any system in the body controlled by hormones can be derailed by hormone disruptors. Specifically, EDCs may be associated with development of learning disabilities, cognitive and brain development problems, body deformation, breast, prostate, thyroid and other cancers, and sexual development problems such as feminization of males or masculinization of females.

endocytosis An active transport which brings substances into the cell. The substance is surrounded by an area of the cell membrane which buds off inside the cell, forming a vesicle that contains the ingested material. Endocytosis include pinocytosis (cell drinking) and phagocytosis (cell eating).

ENS (enteric nervous system) One of the main divisions of the autonomic nervous system, consisting of a mesh-like system of neurons that governs the function of the GI tract. The ENS, derived from neural crest cells, is also called the second brain. The ENS can operate independently of the brain and spinal cord, but does rely on innervation from the autonomic nervous system via the vagus nerve and prevertebral ganglia. The neurons of the ENT control the motor functions of the system, in addition to the secretion of GI enzymes. They communicate through many neurotransmitters, including dopamine, serotonin and acetylcholine. The large presence of serotonin and dopamine in the gut are key areas of research for neurogastroenterologists.

entrainment A term referring to the interaction between circadian rhythms and the environment. A central example is the entrainment of circadian rhythms to the daily light–dark cycle. Circadian clocks generate self-sustaining, cell-autonomous oscillations with a time period of approximately 24 hours. Features of a circadian clock include its persistence under constant conditions, and an entrainment to external input. The neural circuitry that regulates the circadian rhythms includes the retina, suprachiasmatic nucleus (SCN), and the pineal gland. These structures operate under bidirectional neural and hormonal to and from various organs and affect multiple physiological and endocrine functions.

ERK 1/2 (extracellular regulated kinase 1/2) Widely expressed protein kinase involved in the regulation cell division and postmitotic functions in differentiated cells. The term ERK 1/2 is sometimes used as a synonym for mitogen-activated protein kinase (MAPK), but it is actually a specific subset of the mammalian MAPK family. The ERK pathway is activated by growth factors, cytokines, virus infections, ligands for heterotrimeric G protein-coupled receptors, transforming agents, and carcinogens.

exocytosis An active transport for releasing proteins and neurotransmitters from the cell via an energy-dependent process. This process is needed since most large or polar chemicals cannot pass through the hydrophobic portion of the cell membrane by passive means. Exocytosis is also the mechanism for inserting ion channels, surface receptors, lipids, and other components into the cell membrane.

fenoldopam A synthetic benzazepine derivative which acts as a selective peripheral D1R agonist. It is used as an antihypertensive agent postoperatively to treat a hypertensive crisis. In animal studies, fenoldopam administration rapidly and significantly suppressed breast tumor xenografts.

FS-cells (folliculo-stellate) Non-endocrine cells in the anterior pituitary. They play a role in three areas of pituitary functions: autocrine/paracrine control of endocrine cells though the production of cytokines and growth factors, intrapituitary communication, and modulation of inflammatory response feedback. As judged by the expression of cell markers such S-100 and glial fibrillary acidic protein (GFAP), they are derived from glial neuroectodermic cells. Although FS cells do not secrete hormones, they affect the functions of hormone-secreting endocrine cells via gap junctions, which propagate calcium-mediated signals to coordinate the function of excitable endocrine cells throughout the gland.

FSH (follicle-stimulating hormone) A glycoprotein hormone, composed of alpha and beta subunits which is produced and secreted by the gonadotropic cells of the anterior pituitary. It regulates development, growth, pubertal maturation and reproductive processes in both sexes. Within the reproductive system, FSH often works together with luteinizing hormone (LH).

gastrointestinal (GI) tract Includes all the structures that form a continuous passageway for food: mouth, esophagus, stomach, small intestine and large intestine. The complete human digestive system is composed of the GI tract plus the accessory organs of digestion: tongue, salivary glands, pancreas, liver, and gall bladder.

GIRK (G-protein-gated inwardly rectifying potassium channels) A family of lipid-gated channels which are activated (opened) by the signaling lipid PIP2 and a signal transduction cascade, initiated by a ligand-stimulated GPCR. An activated Gβγ dimeric protein then interacts with the GIRK channels, which become permeable to potassium ions, resulting in hyperpolarization of the cell membrane. Of the four types of GIRK channels, GIRK1 to GIRK3 are distributed broadly in the CNS, where their distributions overlap, whereas GIRK4 is found primarily in the heart.

GH (growth hormone) Also known as somatotropin, is a 191-amino acid, single-chain polypeptide that is produced and secreted by somatotropic cells of the anterior pituitary gland. GH stimulates cellular growth, proliferation and regeneration. It also increases the production of IGF-1 and increases the concentration of glucose and free fatty acids. Prolonged excess of GH causes gigantism or acromegaly, while its deficiency in children results in growth failure, development of a short stature and delayed sexual maturity. hGH has been used for performance enhancement in athletes, and bovine GH (bGH) has been used for increasing milk production in dairy cows.

ghrelin A circulating hormone produced by enteroendocrine cells of the GI tract, especially the stomach, and is often called a "hunger hormone" as it increases food intake. Blood ghrelin levels are highest before meals when hungry, returning to lower levels after mealtimes. Ghrelin helps to prepare for food intake by increasing gastric motility and secretion. Acting via a specific receptor, the growth hormone secretagogue receptor 1A (GHSR-1A), Ghrelin activates NPY neurons in the arcuate nucleus, which initiate appetite. In addition to its function in energy homeostasis, ghrelin activates the cholinergic–dopaminergic reward link in inputs to the ventral tegmental area in the mesolimbic pathway, a circuit that communicates the hedonic and reinforcing aspects of rewards, such as food and addictive drugs such as ethanol.

GHRH (growth hormone releasing hormone) A 44-amino acid peptide produced in the arcuate nucleus of the hypothalamus. After its release from neurosecretory nerve terminals, GHRH is carried by the hypothalamo-hypophysial portal system to the anterior pituitary where it binds to its receptor (GHRHR) and stimulates the release of GH. GHRH is released in a pulsatile manner, stimulating a similar pulsatility of GH.

GnRH (gonadotropin-releasing hormone) A decapeptide produced in the preoptic area of the hypothalamus and is released into the hypophysial portal vasculature. After binding to its receptor (GnRHR) in the anterior pituitary, GnRH stimulates the release of both LH and FSH. The pulsatile release of GnRH is necessary for correct reproductive functions such as follicular growth, ovulation, and corpus luteum actions in females, and spermatogenesis in males. GnRH activity influences a variety of sexual and social behavior in both sexes. Synthetic analogues of GnRH have been used to treat precocious puberty, as well as breast cancer, endometriosis and prostate cancer.

GPCR (G protein-coupled receptors) A superfamily of membrane receptors that are grouped into five main subfamilies: (1) rhodopsin, (2) adhesion, (3) glutamate, (4) secretin, and (5) frizzled/smoothened. They are called seven-transmembrane receptors (7TM) because they pass through the cell membrane seven times. The ligands that bind and activate these receptors include light-sensitive compounds, odors, pheromones, neurotransmitters and hormones, varying in size from small molecules to peptides to large proteins. Two principal signal transduction pathways are involved in GPCR: the cAMP pathway and the phosphatidyl inositol pathway. GPCR are involved in many diseases and are an important drug target, with about 35% of all FDA-approved drugs target 108 members of this family.

GRKs (G-protein-coupled receptor kinases) A family of protein kinases which phosphorylate the intracellular domain of GPCR. They function in tandem with the arrestins to regulate the sensitivity of GPCRs for stimulating downstream heterotrimeric G proteins and G-protein-independent signaling pathways.

hypophysial portal system A system of blood vessels which connects the hypothalamus with the anterior pituitary, whose main function is to transport and exchange hormones between the two structures. The primary capillary plexus, located in the median eminence, coalesces into long veins that run along the pituitary stalk and form a second capillary plexus within the anterior pituitary. Short portal vessels connect the posterior and anterior lobes of the pituitary. The portal capillaries are fenestrated (have small channels with high vascular permeability), allowing for a rapid exchange between the hypothalamus and the pituitary. The main hypothalamic hormones transported by the system include GnRH, GHRH, TRH, CRH, and dopamine.

IHDA (incertohypothalamic dopaminergic pathway) A short pathway from the zona incerta in the subthalamus, which contains the A13 dopaminergic neurons, with projections to dorsal and lateral hypothalamic areas, septum, bed nucleus of the stria terminalis, and central nucleus of the amygdala. The IHDA participates in the regulation of gonadotropin secretion and also has a role in sexual behavior.

kisspeptin A neuropeptide composed of 54 amino acids that acts upstream of gonadotropin-releasing hormone (GnRH) neurons, and is critical for maturation and function of the reproductive axis. Kisspeptin-expressing neurons from the periventricular nucleus and arcuate nucleus send projections to the medial preoptic area, where there is an abundance of GnRH cell bodies. It is a ligand for G-protein-coupled receptor GPR54. Kisspeptin-GPR54 signaling has an important role in initiating the release of GnRH at puberty. Kisspeptin interacts with other neuropeptides such as neurokinin B and dynorphin to regulate GnRH pulse generation and its own signaling is regulated by nutritional status and stress. GnRH acts on the anterior pituitary and stimulates the release of the gonadotropins, LH and FSH, which control sexual maturation and gametogenesis. Disruption of GPR54 signaling can cause hypogonadotropic hypogonadism in rodents and humans. In addition, kisspeptin may also represent a novel potential therapeutic target in the treatment of fertility disorders.

leptin A 16-kDa protein of 167 amino acids, predominantly made by adipose tissue and enterocytes in the small intestine. Leptin participates in the regulation of energy balance by inhibiting hunger, which in turn diminishes fat storage in adipocytes. Leptin acts on its receptors, a single-transmembrane-domain type I cytokine receptors (LEP-R or OB-R), in the arcuate nucleus of the hypothalamus where it regulates metabolic homeostasis through the control of food intake. In obesity, there is a decreased sensitivity to leptin, resulting in an inability to detect satiety despite high energy stores and high levels of leptin. Although regulation of fat stores is considered the primary function of leptin, it also plays a role in other physiological processes, as judged by its many sites of synthesis other than fat cells, and the many cell types beyond hypothalamic cells that have leptin receptors.

LH (luteinizing hormone) Also known as lutropin is a glycoprotein hormone, composed of alpha and beta subunits, which is produced by gonadotropic cells of the anterior pituitary gland. In females, an acute rise of LH (midcycle LH surge) triggers ovulation and development of the corpus luteum. In males, LH had also been called interstitial cell–stimulating hormone (ICSH). It stimulates production of testosterone in Leydig cells of the testes. In both sexes, LH acts synergistically with FSH. LH shares structural and functional features with a placental hormone, chorionic gonadotropin (hCG).

locus ceruleus (LC) Located in the posterior area of the pons in the lateral floor of the fourth ventricle, and is mostly composed of medium-size neurons. Melanin granules within the LC generate its blue color. The LC is a part of the reticular activating system which is involves with the physiological responses to stress and panic. It is the main site for brain synthesis of norepinephrine, with long projections to the spinal cord, the brain stem, cerebellum, hypothalamus, thalamic relay nuclei, amygdala, basal telencephalon, and cortex.

magnocellular neurons Large neuroendocrine cells within the supraoptic (SON) and paraventricular (PVN) nuclei of the hypothalamus. These cells are electrically excitable, and generate action potentials in response to afferent stimuli. They produce oxytocin (OT) and vasopressin (AVP or ADH), and are distinguished by having a single long varicose axon, which projects to the posterior pituitary.

MCH (melanin concentrating hormone) A cyclic 19-amino acid orexigenic hypothalamic peptide. In mammals, MCH is involved in the regulation of feeding behavior, mood, sleep-wake cycle, and energy balance, and also has some effects on pigmentation in the periphery.

median eminence A small swelling on the tuber cinereum, posterior to and atop the pituitary stalk. The par nervosa (part of the posterior pituitary gland) is continuous with the median eminence via the infundibular stalk. As one of the seven areas of the brain devoid of a blood–brain barrier, the median eminence is a circumventricular organ with permeable capillaries. It functions as a gateway for the release of hypothalamic releasing and inhibiting hormones. It is integral to the hypophysial portal system which connects the hypothalamus with the pituitary gland. Parvocellular neurons from the hypothalamus terminate in the median eminence.

melatonin A hormone that regulates the sleep-wake cycle and is primarily released by the pineal gland. Melatonin biosynthesis starts with L-tryptophan, which is hydroxylated by tryptophan hydroxylase to produce 5-hydroxytryptophan (5-HTP), which is decarboxylated by pyridoxal phosphate and 5-hydroxytryptophan decarboxylase to produce serotonin. Serotonin is converted into N-acetylserotonin by serotonin N-acetyltransferase. Hydroxyindole O-methyltransferase and S-adenosyl methionine then convert N-acetylserotonin into melatonin through methylation of the hydroxyl group. Melatonin is often used as a supplement for the short-term treatment of trouble sleeping such as from jet lag and shift work, but evidence of benefit is unclear. In animals (including humans), melatonin is involved in synchronizing the circadian rhythm, including sleep-wake timing, blood pressure regulation, and seasonal reproduction. Many of its effects are through activation of the melatonin receptors, while others are due to its role as an antioxidant.

mesocortical dopaminergic pathway A dopaminergic pathway connecting the ventral tegmentum to the prefrontal cortex. It is essential to the normal cognitive function of the dorsomedial prefrontal cortex, and is involved in cognitive control, motivation and emotional response. Abnormality of this system is associated with psychoses such as schizophrenia, especially with the negative symptoms of the disease.

mesolimbic dopaminergic pathway Often referred to as the reward pathway, it is a dopaminergic pathway connecting the ventral tegmental area in the midbrain to the ventral striatum of the basal ganglia in the forebrain.

The ventral striatum includes the nucleus accumbens and the olfactory tubercle. Dopamine released from this pathway regulates motivation and desire for rewarding stimuli and facilitates reinforcement and reward-related motor function learning, and also play a role in the perception of pleasure. Dysregulation of this pathway plays a role in the development and maintenance of an addiction.

metoclopramide A peripheral D2R antagonist, which also acts as a mixed 5-HT3 antagonist and a 5-HT4 agonist. This prokinetic drug works by increasing movements or contractions of the stomach and intestines. It relieves symptoms such as nausea, vomiting, heartburn, a feeling of fullness after meals, and loss of appetite. Metoclopramide is also used to treat heartburn for patients with gastroesophageal reflux disease (GERD).

monoamine oxidase (MAO) Enzymes bound to the outer membrane of mitochondria that catalyze oxidation of monoamines by removing their amine group. They are important in the breakdown of ingested monoamines, and inactivate monoamine neurotransmitters. Humans have two types of MAOs: A and B. Both are differentially found in neurons, astroglia, and in several peripheral organs. MAOs are involved in several psychiatric and neurological diseases, some of which can be treated with monoamine oxidase inhibitors (MAOIs). These drugs are used as effective therapeutic agents for panic disorder, social phobia, resistant depression, atypical depression, and Parkinson's disease.

MPTP (1-Methyl-4-phenyl-1,2,3,6-tetrahydropyridine) A prodrug of the neurotoxin MPP+, which causes permanent symptoms of Parkinson's disease by destroying dopaminergic neurons in the substantia nigra. It has been used to study several disease models in various animal studies.

MSH (melanocyte stimulating hormones) Also known as melanotropins, are a family of peptide hormones and neuropeptides that are cleavage products of a large precursor protein called proopiomelanocortin (POMC). These peptides, including α-MSH, β-MSH, and γ-MSH, are produced by cells in the pars intermedia of the pituitary gland. α-MSH is the most important melanocortin for pigmentation. Acting through the melanocortin 1 receptor, α-MSH stimulates the production and release of melanin by melanocytes in skin and hair, by a process known as melanogenesis. Within the hypothalamus, α-MSH suppresses appetite and is also involved in sexual arousal.

nigrostriatal dopaminergic pathway This pathway connects the substantia nigra pars compacta in the midbrain to the caudate nucleus and putamen located in the dorsal striatum. It is associated with motor function, reward-related cognition and associated learning, and its dysregulation is involved in Parkinson's disease, addiction and chorea.

norepinephrine transporter (NET) The NET is also known as solute carrier family 6 member 2 (SLC6A2). It is composed of 617 amino acid and configures as 12 transmembrane domains (TMDs) with intracellular N and

C terminals. NET is responsible for the sodium-chloride (Na^+/Cl^-)-dependent reuptake of extracellular norepinephrine, but can also reuptake dopamine. NETs, along with other monoamine transporters, are the targets of many antidepressants and recreational drugs. An overabundance of NET is associated with attention deficit hyperactivity disorder.

NPY (neuropeptide Y) A 36 amino-acid neuropeptide involved in various physiological and homeostatic processes in both the central and peripheral nervous systems. It is the most abundant peptide in the mammalian CNS and is secreted alongside other neurotransmitters such as GABA and glutamate. In the autonomic system, NPY is produced in neurons of the sympathetic nervous system, serving as a strong vasoconstrictor and stimulating the growth of fat tissue. In the brain, NPY is produced in various locations including the hypothalamus, where it has several functions: increasing food intake and storage of energy as fat, reducing anxiety and stress, reducing pain perception, affecting the circadian rhythm, reducing voluntary alcohol intake, lowering blood pressure, and controlling epileptic seizures.

OCT (organic cation transporters) OCTs translocate endogenous (e.g., dopamine) and exogenous (e.g., drugs) substances of cationic nature. They are involved in intestinal absorption, hepatic uptake, renal excretion of hydrophilic drugs, and play an important role in the detoxification of exogenous compounds. OCT transporters belong to the SLC22 family and include three subtypes: OCT1, OCT2, and OCT3. OCT knockout mice are viable and fertile, have reduced uptake of MPP^+ into the heart, but otherwise show no significant phenotypic differences from wild type mice.

6-OHDA (6-Hydroxydopamine) Also known as oxidopamine, is a neurotoxic synthetic compound, used experimentally to destroy catecholaminergic neurons in the brain. It enters the neurons via the dopamine and noradrenaline reuptake transporters. When used together with a selective noradrenaline reuptake inhibitor such as desipramine, it selectively destroys the dopaminergic neurons.

oligomerization The formation of a macromolecular complex by non-covalent bonding of proteins or nucleic acids. A homo-oligomer is formed by few identical molecules, while a hetero-oligomer is made of more than one, different, macromolecules.

palmitoylation Covalent attachment of fatty acids such as palmitic acid to cysteine (S-palmitoylation), and less frequently to serine and threonine (O-palmitoylation) residues of certain proteins. The precise function of palmitoylation depends on the particular protein involved. Palmitoylation enhances the hydrophobicity of proteins and contributes to their membrane association. It also plays a role in subcellular trafficking of proteins between membrane compartments, as well as in modulating protein-protein interactions.

PC12 cells A rat cell line that is derived from adrenal medullary pheochromocytoma, and represents non-differentiated neuroblastic cells. PC12 cells have been used extensively to examine different aspects of catecholamine homeostasis.

PDE5 (phosphodiesterase 5) inhibitors PDE5 is a key enzyme involved in the degradation of cGMP and the suppression of its specific signaling pathways in physiological processes such as smooth muscle contraction and relaxation. The PDE5 inhibitors function by preventing the degradation of cyclic GMP and subsequently prolonging its physiological effects. PDE5 inhibitors are used primarily as remedies for erectile dysfunction, and also have some other medical applications such as treatment of pulmonary hypertension. Sildenafil (Viagra), Tadalafil, (Cialis) and Vardenafil (Levitra) selectively inhibit PDE5, which is responsible for the degradation of cGMP in the corpus cavernosum of the penis, enabling erection.

pineal gland A small endocrine gland located in the epithalamus, near the center of the brain, between the two hemispheres. It mainly consists of pinealocytes, but also contains interstitial cells, perivascular phagocytes, and neurons. The pineal is one of the neurosecretory circumventricular organs outside the blood–brain barrier. The pineal produces melatonin, a serotonin-derived hormone which modulates sleep patterns in circadian and seasonal cycles. Melatonin production is stimulated by darkness and is inhibited by light. Light-sensitive neurons in the retina detect light and send a signal to the SCN through the retinohypothalamic tract, synchronizing the SCN to the day-night cycle. Information from the SCN is relayed to the PVN, then to the spinal cord, and via the sympathetic system to the superior cervical ganglia (SCG), and from there to the pineal.

pheochromocytoma Tumors arising from chromaffin cells of the adrenal gland which make, store, metabolize, and release catecholamines. Symptoms of a pheochromocytoma are those of sympathetic nervous system hyperactivity, including skin sensations, elevated heart rate, and high blood pressure, often as paroxysmal (sporadic) episodes. There are also palpitations, anxiety, excessive sweating, headaches, pallor, and weight loss. Treatment primarily entails surgical resection. Patients with pheochromocytoma have increased lifetime risk of secondary cancers, with a slightly increased mortality risk.

PKA (protein kinase A) Also called cAMP-dependent protein kinases, are enzymes whose activity depends on the cellular levels of cAMP. PKA functions include the regulation of glycogen, glucose, and lipid metabolism, and the mediation of catecholamine receptors. PKA exists as a tetramer, made of two regulatory subunits and two catalytic subunits. Steps involved in PKA activation include: (a) increases in cytosolic cAMP, (b) binding of two cAMP molecules to each regulatory subunit, (c) dissociation of the complex, and (d) phosphorylation of Ser or Thr residues in target proteins by the free catalytic subunits.

PKC (protein kinase C) A family of enzymes that phosphorylate Ser and Thr residues in target proteins, and play important roles in several signal transduction cascades. The PKC enzymes are activated by signals such as increases in the concentrations of diacylglycerol (DAG) or calcium ions (Ca^{2+}). In humans the PKC family consists of fifteen isoenzymes which are divided into three subfamilies: conventional (or classical), novel, and atypical. Conventional (c) PKCs require Ca^{2+}, DAG, and a phospholipid such as phosphatidylserine for activation. Novel (n) PKCs require DAG, but not Ca^{2+}, for activation, while atypical (a) PKCs require neither Ca^{2+} nor DAG for activation.

PKG (cGMP-dependent protein kinase) A Ser/Thr-dependent protein kinase that is activated by cGMP. It phosphorylates a number of biologically important targets and is implicated in the regulation of smooth muscle relaxation, platelet function, sperm metabolism, cell division and nucleic acid synthesis. It has some overlapping functions with PKA. PKG mediates the actions of dopamine in some malignant and nonmalignant tissues. There are two types of PKG: I and II. PKG-I gene is expressed as cytosolic PKG-Iα or PKG-Iβ isoform, while the PKG-II gene is expressed as a membrane associated PKG-II protein.

PMAT (plasma membrane monoamine transporter) A low-affinity monoamine transporter, which is an integral membrane protein. PMAT consists of 530 amino acid residues with 10–12 transmembrane segments, and is not homologous to other known monoamine transporters. It transports serotonin, dopamine, norepinephrine, and adenosine from synaptic spaces into presynaptic neurons or neighboring cells. It is highly expressed in the human brain, heart tissue, skeletal muscle, and kidneys. PMAT is relatively insensitive to the high affinity inhibitors of the monoamine transporters such as SERT, DAT, or NET.

PNMT (phenylethanolamine-N-methyltransferase) The forth enzyme in the catecholamine biosynthetic pathway, which converts norepinephrine to epinephrine by transferring a methyl group from S-adenosine methionine (SAM) to the amine group of norepinephrine. PNMT also methylates catechol-O-methyltransferase (COMT) by adding a methyl group in another location. PNMT is found primarily in the adrenal medulla, but is also expressed in small groups of neurons in the human brain.

POMC (pro-opiomelanocortin) A precursor polypeptide with 241 amino acid residues, and is part of the central melanocortin system. POMC is synthesized from a 285-amino-acid-long polypeptide precursor (pre-pro POMC), by the removal of a 44-amino-acid-long signal peptide sequence during translation. In different tissues, POMC is cleaved to give rise to multiple peptide hormones: MSHs, ACTH, endorphins, and enkephalins, each of which is packaged in large dense-core vesicles to be released from the cells by exocytosis in response to appropriate stimulation.

PP2A (protein phosphatase 2A) A heterotrimeric protein phosphatase that is ubiquitously expressed and accounts for a large fraction of phosphatase activity in most cells. It has a broad substrate specificity and diverse cellular functions. Among its targets are proteins of oncogenic signaling cascades, such as Raf, MEK and AKT, where PP2A may act as a tumor suppressor.

PRL (prolactin) A protein hormone, best known for its role in milk production. PRL is a multifunctional hormone which affects 300 separate processes in various vertebrates, including humans. PRL is primarily secreted from the pituitary lactotrophs, but is also produced, and locally acts as a cytokine, in multiple peripheral cells in humans. The hormone acts in endocrine, autocrine or paracrine manner through the PRL receptor and plays modulating roles in metabolism, regulation of the immune system, reproduction, hematopoiesis, angiogenesis and pancreatic development, among others.

PVN (paraventricular nucleus) The PVN of the hypothalamus lies adjacent to the third ventricle within the periventricular zone. It is highly vascularized and is protected by the blood–brain barrier. The PVN contains magnocellular cells whose axons extend to the posterior pituitary, parvocellular cells that project to the median eminence, and several populations of peptide-containing cells that project to many brain regions, including the brainstem and spinal cord. The main hormone produced by the PVN is oxytocin, with additional hormones including CRH, vasopressin, and TRH. Input from many brain areas to the PVN carries information important for coordinating the regulation of electrolyte composition, stress, metabolism, energy intake, and circadian rhythm.

reserpine An indole alkaloid isolated from the plant Rauvolfia serpentine. It has been used for the treatment of high blood pressure, usually in combination with a thiazide diuretic or a vasodilator. Its antihypertensive actions are mostly due to its anti-noradrenergic effects, resulting from its ability to deplete catecholamines from peripheral sympathetic nerve endings. Reserpine irreversibly blocks the H^+-coupled VMAT1 and VMAT2. Blockade of these transporters by reserpine inhibits uptake and reduces dopamine, norepinephrine, serotonin, and histamine in synaptic vesicles of neurons.

reuptake Reabsorption of a neurotransmitter by a transporter, after it has performed its function of transmitting a neural impulse. Reuptake can take place along the plasma membrane of an axon terminal (i.e., the presynaptic neuron) or in glial cells. The process is necessary for normal synaptic actions by allowing the recycling of neurotransmitters, and by regulating their level in the synapse, thus controlling how long a signal will last.

RGS (regulators of G protein signaling) Multifunctional GTPase-accelerating proteins that promote GTP hydrolysis by the α-subunit of heterotrimeric G proteins, thus inactivating the G protein and rapidly switching off the signaling pathways of GPCR. Upon activation by receptors, G proteins exchange GDP for GTP, are released from the receptor, and dissociate into a free, active GTP-bound α-subunit and a βγ-dimer, both of which activate downstream effectors. The response is terminated upon GTP hydrolysis by the α-subunit which can rebind the βγ-dimer and the receptor. RGS proteins markedly reduce the lifespan of GTP-bound α-subunits by stabilizing the G protein transition state.

rhodopsin Found in the rods of the retina, rhodopsin is a light-sensitive GPCR involved in visual phototransduction. It is extremely sensitive to light, and thus enables vision in low-light conditions. When rhodopsin is exposed to light, it immediately photo-bleaches. In humans, it is regenerated fully in about 30 minutes, after which rods are more sensitive.

ROS (reactive oxygen species) Reactive chemical species containing oxygen, including peroxides, superoxide, hydroxyl radical and alpha oxygen. ROS are formed as a natural byproduct of oxygen metabolism and have important roles in cell signaling and homeostasis. During times of environmental stress (e.g., UV or heat exposure), ROS levels can dramatically increase, causing damage, known as oxidative stress, to cell structures.

secretory vs synaptic vesicles Secretory vesicles of neuroendocrine cells differ from synaptic vesicles of the CNS in size, morphology, content, biogenesis, and speed of release. Secretory vesicles are larger with a diameter of 270 nm, contain biogenic amines, ions, peptides, and proteins, are derived from the Golgi apparatus, and are retrieved after exocytosis, but do not directly recycle. Microscopically, they are identified as dense core secretory granules or vesicles. In contrast, synaptic vesicles are much smaller with a diameter of 40–60 nm, contain only neurotransmitters, and appear as clear vesicles in electron micrographs. Synaptic vesicles are derived from the endosome, the major sorting compartment in the cell.

serotonin (Ser) or 5-hydroxytryptamine (5-HT) is a monoamine neurotransmitter with multiple functions including cognition, reward, learning, memory, and other physiological processes. Serotonin is an indoleamine derived from tryptophan via the rate-limiting hydroxylation of the 5 position on the ring. This forms the intermediate 5-hydroxytriptophan or 5HTP, followed by decarboxylation to produce serotonin. Serotonin is primarily found in the enteric nervous system in the GI tract, and is also produced in the CNS, specifically in the Raphe nucleus located in the brain stem. In addition, serotonin is stored in blood platelets and is released during agitation and vasoconstriction.

SERT (serotonin transporter) SERT, also known as solute carrier family 6 member 4 (SLC6A4), transports serotonin from the synaptic cleft back to the presynaptic neuron. SERT terminates the action of serotonin and recycles it in a sodium-dependent manner. It is a member of the sodium:neurotransmitter symporter family. SERT is the target of many antidepressant medications of the SSRI (selective serotonin reuptake inhibitors) and tricyclic antidepressant classes. A repeat length polymorphism in the promoter of the *SERT* gene affects the rate of serotonin uptake and may play a role in sudden infant death syndrome, aggressive behavior in Alzheimer

disease patients, posttraumatic stress disorder and depression-susceptibility in people with emotional trauma.

SDN (sexually dimorphic nucleus) An ovoid, densely packed cluster of large cells located in the medial preoptic area (POA) of the hypothalamus, which is associated with sexual behavior in animals. In all species of mammals studied, the SDN is considerably larger in males than in females. In humans, the SDN volume is 2.2 times as large in males as in females and contain 2.1 times as many cells. The volume of SDN is modified by testosterone, with larger volume of male SDN correlating with the higher concentration of fetal testosterone level in males than in females. In addition, testosterone acts during specific prenatal period to organize the development of aromatase-expressing neurons into the male-typical SDN (testosterone is transformed to estrogen by aromatase).

SLC (solute carrier transporters) The SLC group of membrane transport proteins include over 400 members and is organized into 65 families. Most members of the group are located in the cell membrane. Solutes that are transported by various members are highly diverse and include both charged and uncharged organic molecules, as well as inorganic ions. Depending on the SLC, the transporters function as either monomers or obligate homo- or hetero-oligomers.

SNARE (soluble *N*-ethylmalemide-sensitive factor attachment protein receptor) A large protein complex consisting of more than 60 members in mammalian cells. The primary role of SNARE proteins is to mediate vesicle fusion with their target membrane-bound compartments. The best studied SNAREs are those that mediate docking of synaptic vesicles with the presynaptic membrane in neurons during exocytosis. These SNAREs are the targets of the bacterial neurotoxins responsible for botulism and tetanus.

SRY (sex-determining region Y protein) Also known as Testis-determining factor (TDF), is a DNA-binding protein encoded by the *SRY* gene that is responsible for the initiation of male sex determination in humans. SRY is a sex-determining gene on the Y chromosome. The function of SRY is to regulate (i.e., turn on and off) genes in order to make sure that they are expressed in the right cell at the right time and in the right amount throughout the life of the cell and the organism. Mutations in this gene lead to a range of disorders of sex development with varying effects on an individual's phenotype and genotype.

striatum A cluster of neurons in the subcortical basal ganglia of the forebrain. It has a striped (striated) appearance of grey-and-white matter. The striatum is a critical component of the motor and reward systems, receiving both glutaminergic and dopaminergic inputs from different sources, and serving as a primary input to the rest of the basal ganglia. In humans, the striatum is divided into ventral and dorsal areas, based upon function and connections. The ventral striatum consists of the nucleus accumbens and olfactory tubercle. The dorsal striatum consists of the caudate nucleus and putamen, separated by the internal capsule.

substantia nigra A basal ganglia structure located in the midbrain that plays an important role in reward and movement. Parts of the substantia nigra is dark due to high levels of neuromelanin in the dopaminergic neurons. The substantia nigra consists of two parts, pars compacta and pars reticulate, having different connections and functions. The pars compacta serves as an output to the basal ganglia circuit, supplying the striatum with dopamine, while the pars reticulate conveys signals from the basal ganglia to other brain areas. Parkinson's disease is characterized by loss of dopaminergic neurons in the pars compacta.

SULT1A3 (sulfotransferase A3) Sulfotransferases are cytosolic enzymes that differ in their tissue distributions and substrate specificities. SULT1A3 specifically catalyzes the sulfate conjugation of catecholamines and structurally related drugs.

suprachiasmatic nucleus (SCN) A small hypothalamic nucleus, situated above the optic chiasm and bilateral to the third ventricle, which is responsible for controlling circadian rhythm. The neuronal and hormonal activities it generates regulate different body functions in a 24-hour cycle. The SCN contains several cell types and different peptides, including vasopressin, VIP and neurotransmitters. It receives innervation from three main pathways: the retinohypothalamic tract, geniculo-hypothalamic tract, and projections from the raphe nuclei. Its main output is to the subparaventricular zone and dorsomedial hypothalamic nucleus.

TH (tyrosine hydroxylase) The first enzyme, and the rate-limiting step, in the biosynthetic pathway of catecholamines which converts tyrosine into Dopa. For its hydroxylation reaction the enzyme uses molecular oxygen (O_2) as well as iron (Fe^{2+}) and tetrahydrobiopterin as cofactors. TH is a tetramer of four identical subunits. Each subunit consists of three domains: a carboxy terminal domain that allows tetramerization, a catalytic core, and an amino terminal regulatory domain which controls access of substrates to the active site.

TIDA (tuberoinfundibular dopaminergic pathway) A population of dopaminergic neurons that project from the arcuate nucleus, periventricular, and paraventricular nuclei, in the tuberal region of the hypothalamus to the median eminence. It is one of the four major dopaminergic pathways in the brain. Dopamine released at this site primarily inhibits the secretion of prolactin from anterior pituitary lactotrophs by binding to D2 receptors.

THDA (tuberohypophysial dopaminergic pathway) Dopamine neuronal cell bodies in the arcuate and periventricular nuclei send projections to the neuro-intermediate lobe of the pituitary, where they regulate the production and release of POMC-derived hormones that include β-endorphin and αMSH.

TSH (thyroid stimulating hormone) Also known as thyrotropin, is a glycoprotein consisting of two subunits, alpha and beta. It is produced by thyrotrophs cells in the anterior pituitary and its main function is the stimulation of the thyroid gland to produce thyroxine (T_4), and triiodothyronine (T_3) which stimulate metabolism of almost every tissue in the body.

TRH (thyrotropin releasing hormone) Is a tripeptide with blocked termini that is produced by parvocellular neurons of the PVN of the hypothalamus. It is translated as a 242-amino acid precursor polypeptide that contains 6 copies of the sequence -Gln-His-Pro-Gly-. Its main function is to stimulate the release of TSH and prolactin from the anterior pituitary. TRH has also antidepressant and anti-suicidal properties.

UGT (UDP-Glucuronosyltransferases) A microsomal glycosyltransferase that catalyzes the transfer of the glucuronic acid component of UDP-glucuronic acid to a small hydrophobic molecule. Glucuronidation is one of the major phase II drug-metabolizing reactions that contributes to drug biotransformation.

VEGF (vascular endothelium growth factor) Originally known as vascular permeability factor (VPF), is a signal protein produced by cells that stimulates the formation of blood vessels. It is part of the system that restores the oxygen supply to tissues when blood circulation is inadequate such as in hypoxic conditions. The VEGF family comprises five members: VEGF-A, placental growth factor (PGF), VEGF-B, VEGF-C, and VEGF-D. The main function of the VEGF's is to create new blood vessels during embryonic development, after injury, in muscle following exercise, and collateral circulation that bypass blocked vessels. Overexpression of VEGF contributes to several diseases. For example, solid cancers cannot grow beyond a limited size without an adequate blood supply. Those cancers that express VEGF are able to grow and metastasize. Overexpression of VEGF can also cause vascular disease in the retina of the eye and other parts of the body. Drugs such as afilibercept, bevacizumab, pegaptinab, and ranibizumasb can inhibit VEGF and control or slow those diseases.

VMAT (vesicular monoamine transporters) Transport proteins that are integrated into the membrane of synaptic vesicles of presynaptic neurons. They transport monoamine neurotransmitters such as dopamine, norepinephrine, epinephrine, serotonin, and histamine into the vesicles. To power monoamine import, VMATs utilize a proton gradient generated by vesicular ATPase. Humans express two types of VMATs: VAMT1 and VAMT2. VMAT1 is mostly found in large dense core vesicles of the peripheral nervous system: neuroendocrine cells and adrenal medullary chromaffin cells. VMAT2 is mostly expressed in monoaminergic cells of the brain, sympathetic nervous system, mast cells, histamine-containing cells in the gut, β-cells of the pancreas and blood platelets.

VMH (ventromedial nucleus) A pear-shaped structure in the tuberal region of the hypothalamus that involved in terminating hunger. It has been designated as a satiety center given its association with the recognition of the feeling of fullness. It is also involved in fear, thermoregulation, and sexual activity.

Index

Note: Page numbers in italic and bold refer to figures and tables, respectively.

3 20